Advances in Neuroscience

Volume 1

Volume Transmission in the Brain

Novel Mechanisms for Neural Transmission

Advances in Neuroscience

ADVANCES IN NEUROSCIENCE
VOLUME 1

Volume Transmission in the Brain

Novel Mechanisms for Neural Transmission

EDITORS

Kjell Fuxe, M.D.
Department of Histology and Neurobiology
Karolinska Institute
Stockholm, Sweden

Luigi F. Agnati, M.D.
Department of Human Physiology
University of Modena
Modena, Italy

RAVEN PRESS NEW YORK

Raven Press, 1185 Avenue of the Americas, New York, New York 10036

Made in the United States of America

Library of Congress Cataloging-in-Publication Data

Volume transmission in the brain: novel mechanisms for neural transmission/editors, Kjell Fuxe, Luigi F. Agnati.
p. cm. -- (Advances in neuroscience; v. 1)
Based on the conference held in Stockholm, Sept. 28-30, 1989.
Includes bibliographical referecnes and index.
ISBN 0-88167-699-3
1. Neural transmission--Congresses. I. Fuxe, Kjell. II. Agnati, Luigi Francesco. III. Series.
[DNLM: 1. Brain--physiology--congresses. 2. Neural Transmission--physiology--congresses. WL 300 V943 1989]
QP364.5.V65 1991
612.8'2--dc20
DNLM/DLC
for Library of Congress 90-9171
CIP

The material contained in this volume was submitted as previously unpublished material, except in the instances in which some of the illustrative material was derived.

Great care has been taken to maintain the accuracy of the information contained in the volume. However, neither Raven Press nor the editors can be held responsible for errors or for any consequences arising from the use of the information contained herein.

9 8 7 6 5 4 3 2 1

The Wenner-Gren Center International Symposium, *Volume Transmission in the Brain,* held in Stockholm, September 28–September 30, 1989.

Participants in alphabetical order (see number key below): Luigi F. Agnati[1], José Aguirre[2], Christiane Ayer-LeLiévre[3], Michel Baudry[4], Alain Beaudet[5], Jean-Marie Besson[6], Garth Bissette[7], Börje Bjelke[8], Floyd Bloom[9], Béla Bohus[10], Bernd Bunnemann[11], Jean-Pierre Changeux[12], Antonio Cintra[13], Roser Cortés[14], Annica Dahlström[15], Ron DeKloet[16], Laurent Descarries[17], Gerald Edelman[18], George Fink[19], Kjell Fuxe[20], F. Georges Gonon[21], Sten Grillner[22], Menek Goldstein[23], Elisabeth Hansson[24], Miles Herkenham[25], Madhu Kalia[26], Lawrence Kruger[27], Olle Lindvall[28], Jan Lundberg[29], Michela Matteoli[30], Jean Merrill[31], Dieter Meyer[32], Viktor Mutt[33], Charles Nicholson[34], Christer Nilsson[35], David Ottoson[36], Sanford Palay[37], Andras Pellionisz[38], Alain Privat[39], Thomas Reese[40], Donald Reis[41], Jean-Charles Schwartz[42], Bo Siesjö[43], Stephen Skaper[44], Derek Smyth[45], Eva Sykova[46], Lars Terenius[47], Urban Ungerstedt[48], Sylvester Vizi[49], and Walter Zieglgänsberger[50]. *Not in photograph:* Bengt Andersson, Paul Bach-y-Rita, Anders Hamberger, Peter Kalén, Pierre Mailleux, Lars Olson, Guido Rindi, Aryeh Routtenberg, and Olli Ventä.

Preface

Since the "querelle" between the neuroscientists supporting either the "reticular theory" or the "neuron theory" almost a century ago,[1] the view of a possible diffuse mode of communication in the central nervous system has been suggested.[2]

In the 1950s and 1960s the ultimate morphological and functional evidence was obtained in favor of the neuron doctrine of Cajal.[3] Thus, the synaptic junction as "the site" for interneuronal communication was established.

However, already in the 1960s and later on in the 1970s, histochemical evidence showed that monoamines could be released outside the synaptic specialization, and in the past decade pharmacological and receptor autoradiographic studies demonstrated functional receptors outside the pre- and postsynaptic structures. All this evidence indicates that the neurons can communicate both in a focal mode (synaptic contacts) and in a widespread but still organized mode. However, the latter mode does not rest on the syncytial organization of the central nervous system (as suggested by the reticular theory) but on the possibility that electrical and, above all, chemical signals can diffuse in the extracellular fluid of the brain.[4]

Thus, the hypothesis of a widespread but still organized mode of communication among neurons has a long history, and therefore we felt it was timely to summarize the available evidence and the new ideas on this phenomenon. In fact, there is no doubt that a basic step in the understanding of physiological as well as pathological aspects of brain function is the characterization of how these two modes of interneuronal communication in the central nervous system work and interact.

The Editors

[1]See, e.g., Schäfer EA, et al., eds. *Quain's elements of anatomy*. London: Longmans, Green and Co., 1909; Warwick R, Williams PL, eds. *Gray's anatomy*. London: Longman, 1973.

[2]Golgi C. La rete nervosa diffusa degli organi centrali del sistema nervoso. Suo significato fisiologico. In: Golgi C., ed. *Opera ommia,* vol. 11. Milano: Ulrico Hoepli, 1903;579–605.

[3]See, e.g., Reinoso-Suarez F. Cajal: A modern insight in neuroscience. In: Grisolia S., et al., eds. *Ramon Cajal's contribution to the neurosciences*. Amsterdam: Elsevier, 1982;3–22; see also, e.g., Robertson JD. The ultrastructure of synapses. In: Schmitt FO, ed. *The neurosciences II. Study program*. New York: The Rockefeller University Press, 1970;715–728.

[4]Actually Golgi had a broader view than the one strictly derived from the reticular theory. In this regard, it may be interesting to translate an entire sentence of Golgi's paper quoted above: "Studies on electricity show that electrical currents can take place without direct contact between conductors and this phenomenon may well occur in the central nervous system."

Acknowledgments

We are very much in debt to the Wenner-Gren Center Foundation, and to the Fidia Research Foundation and Wenner-Gren Center Foundation secretary Prof. David Ottoson, Mrs. Gun Lennerstrand, and to Mrs. Anne Edgren, for making this book possible. This work is based on a Wenner-Gren Center International Symposium entitled *Volume transmission in the brain: Extracellular fluid as pathways for electrical and chemical communication,* a conference organized by K. Fuxe, L. F. Agnati, and B. Bjelke, September 28–September 30, 1989.

Contents

I. Introduction

II. Relationship of Pre- and Postsynaptic Features of Neurons: Indications for Volume Transmission

VII. Transplants and Volume Transmission

VIII. Integrative Aspects of Wiring and Volume Transmission Under Normal and Pathological Conditions

IX. Theoretical Models of Information Handling in the Brain

X. Conclusion

Contributors

Luigi F. Agnati
Department of Human Physiology
University of Modena
Via Campi 287
41100 Modena, Italy

J. A. Aguirre
Department of Histology and Neurobiology
Karolinska Institute
Box 60400
104 01 Stockholm, Sweden

Bengt Andersson
Department of Physiology
Faculty of Veterinary Medicine
Swedish University of Agricultural Sciences
S-75007 Uppsala, Sweden

Barbro Asplund
Laboratory for Experimental Brain Research
Lund Hospital
S-221 85 Lund, Sweden

Christiane Ayer-LeLiévre
Department of Histology and Neurobiology
Karolinska Institute
Box 60400
103 01 Stockholm, Sweden

Paul Bach-y-Rita
Departments of Neurology and Physical Medicine and Rehabilitation
Karolinska Institute
Huddinge Hospital
141 86 Huddinge, Sweden

Michel Baudry
Department of Biology
University of Southern California
Los Angeles, California 90089

Alain Beaudet
Department of Neurology and Neurosurgery
McGill University
3801 University Street
Montreal, Quebec, Canada H3A 2B4

J. Bengzon
Department of Medical Cell Research
University of Lund
Biskopsgatan 5
S-223 62 Lund, Sweden

Jean-Marie Besson
Unité de Recherches de Pharmacologie due Système Nerveux
INSERM U. 161
Ecole Practique des Hautes Etudes
2 rue d'Alesia
75014 Paris, France

Garth Bissette
Departments of Psychiatry and Pharmacology
Duke University Medical Center
Box 3859
Durham, North Carolina 27710

Börje Bjelke
Department of Histology and Neurobiology
Karolinska Institute
Box 60400
104 01 Stockholm, Sweden

Anders Björklund
Department of Medical Cell Research
Section of Neurobiology
University of Lund
Biskopsgatan 5
S-223 62 Lund, Sweden

Floyd E. Bloom
Department of Neuropharmacology
Research Institute of Scripps Clinic
10666 North Torrey Pines Road
La Jolla, California 92037

Béla Bohus
Department of Animal Physiology
University of Groningen
P.O. Box 14
9750 AA Haren, The Netherlands

Marie-Louise Bouthemet
Unité de Neurobiologie et Pharmacologie (U. 109)
Centre Paul Broca de l'INSERM
75014 Paris, France

Lennart Brodin
Nobelinstitute of Neurophysiology
Karolinska Institute
Box 60400
104 01 Stockholm, Sweden

P. Brundin
Department of Medical Cell Research
University of Lund
Biskopsgatan 5
S-223 62 Lund, Sweden

M. Buda
INSERM U171-CNRS URA 1195
Centre Hospitalier Lyon-Sud
69310 Pierre-Bénite, France

B. Bunnemann
Department of Histology and Neurobiology
Karolinska Institute
Stockholm, Sweden

J.-P. Changeux
Neurobiologie Moleculair
Institute Pasteur
28, Rue du Dr. Roux
754724 Paris Cedex 15, France

Johan Christenson
Nobelinstitute of Neurophysiology
Karolinska Institute
Box 60400
104 01 Stockholm, Sweden

Antonio Cintra
Department of Histology and Neurobiology
Karolinska Institute
Box 60400
104 01 Stockholm, Sweden

Roser Cortés
Department of Histology and Neurobiology
Karolinska Institute
Box 60400
104 01 Stockholm, Sweden

Kathryn L. Crossin
The Rockefeller University
1230 York Avenue
New York, New York 10021-6399

Annica Dahlström
Department of Histology
University of Goteborg
Box 33031
400 33 Göteborg, Sweden

P. Davalli
Institute of Chemical Biology
University of Modena
Via Campi 287
41100 Modena, Italy

P. DeCamilli
Department of Cell Biology
Yale University School of Medicine
333 Cedar Street
New Haven, Connecticut 06510

R. Deisz
Clinical Neuropharmacology
Max Planck Institute of Psychiatry
P.O. Box 401240
D-8000 Munich 40, Federal Republic of Germany

E. Ronald de Kloet
Rudolf Magnus Institute for Pharmacology
University of Utrecht
Vondellaan 6
3521 GD Utrecht, The Netherlands

Laurent Descarries
Department of Physiology
University of Montreal
CP 6128 Suc A
Montreal, Quebec, Canada H3C 3J7

H.-U. Dodt
Clinical Neuropharmacology
Max Planck Institute of Psychiatry
P.O. Box 401240
D-8000 Munich 40, Federal Republic of Germany

Ted Ebendal
Department of Developmental Biology
Uppsala University
S-751 23 Uppsala, Sweden

Gerald M. Edelman
The Rockefeller University
1230 York Avenue
New York, New York 10021-6399

Anders Ekholm
Laboratory for Experimental Brain Research
Lund Hospital
S-221 85 Lund, Sweden

P. Eneroth
Department of Applied Chemistry
Huddinge Hospital
Huddinge, Sweden

F. Ferraguti
Institute of Human Physiology
University of Modena
Via Campi 287
41100 Modena, Italy

George Fink
MRC Brain Metabolism Unit
University Department of Pharmacology
1 George Square
Edinburgh EH8 9JZ, Scotland

Kjell Fuxe
Department of Histology and Neurobiology
Karolinska Institute
Box 60400
104 01 Stockholm, Sweden

D. Ganten
Department of Pharmacology
German Institute for High Blood Pressure Research
University of Heidelberg
Im Neuenheimer Feld 366
6900 Heidelberg, Federal Republic of Germany

Bruno Giros
Unité de Neurobiologie et Pharmacologie (U. 109)
Centre Paul Broca de l'INSERM
75014 Paris, France

Menek Goldstein
Neurochemistry Research Laboratories
New York University Medical Center
School of Medicine
550 First Avenue
New York, New York 10016

F. G. Gonon
INSERM U 171-CNRS URA 1195
Groupe de Recherche de Neurochimie Fonctionnel
Centre Hospitalier Lyon-Sud Pav 4H
Chemin du Grand Revoyet
69310 Pierre-Bénite, France

Sten Grillner
Nobelinstitute of Neurophysiology
Karolinska Institute
Box 60400
104 01 Stockholm, Sweden

R. Grimaldi
Institute of Human Physiology
University of Modena
Via Campi 287
41100 Modena, Italy

Claude Gros
Unité de Neurobiologie et Pharmacologie (U. 109)
Centre Paul Broca de l'INSERM
75014 Paris, France

Anders Hamberger
Institute of Neurobiology
University of Göteborg
Box 33031
S-400 33 Göteborg, Sweden

Elisabeth Hansson
Department of Histology
Institute of Neurobiology
University of Göteborg
Box 33031
400 33 Göteborg, Sweden

Kazuho Harada
Neurochemistry Research Laboratories
New York University Medical Center
School of Medicine
550 First Avenue
New York, New York 10016

Miles Herkenham
Section on Functional Neuroanatomy
Clinical Neuroendocrinology Branch
National Institute of Mental Health
National Institutes of Health
Bethesda, Maryland 20892

Costantino Iadecola
Department of Neurology and Neurosciences
Division of Neurobiology
Cornell University Medical College
1300 York Avenue
New York, New York 10021

Marian Joëls
Institute of Molecular Biology and Medical Biotechnology
Padualaan 8
3584 CH Utrecht, The Netherlands

Jörgen Junkergård
Department of Physiology
Faculty of Veterinary Medicine
Swedish University of Agricultural Sciences
S-750 07 Uppsala, Sweden

Peter Kalén
Department of Medical Cell Research
Section of Neurobiology
University of Lund
Biskopsgatan 5
S-223 62 Lund, Sweden

Madhu Kalia
Department of Pharmacology
Thomas Jefferson University
1020 Locust Street
Philadelphia, Pennsylvania 19107

M. Kokaia
Department of Medical Cell Research
University of Lund
Biskopsgatan 5
S-223 62 Lund, Sweden

Lawrence Kruger
Department of Anatomy
Center of Health Sciences
University of California at Los Angeles
Los Angeles, California 90024

Alberta Leon
Fidia Research Laboratories
Department of CNS Research
Via Ponte della Fabbrica 3/A
35031 Abano Terme, Italy

Beth Levant
Departments of Psychiatry and Pharmacology
Duke University Medical Center
Durham, North Carolina 22710

Jow Y. Lew
Neurochemistry Research Laboratories
New York University Medical Center
School of Medicine
550 First Avenue
New York, New York 10016

Olle Lindvall
Department of Medical Cell Research
Section of Neurobiology
University of Lund
Biskopsgatan 5
S-223 62 Lund, Sweden

Maria Lindvall-Axelsson
Department of Medical Cell Research
Section of Neurobiology
University of Lund
Biskopsgatan 5
S-223 62 Lund, Sweden

Catherine Llorens-Cortes
Unité de Neurobiologie et Pharmacologie (U. 109)
Centre Paul Broca de l'INSERM
75014 Paris, France

Marie-Christine Lombard
Unité de Recherches de Pharmacologie du Système Nerveux de l'INSERM
Ecole Practique des Hautes Etudes
2 rue d'Alesia
75014 Paris, France

J. M. Lundberg
Department of Pharmacology
Karolinska Institute
Box 60400
S-104 01 Stockholm, Sweden

Pierre Mailleux
Laboratory of Neuropathology and Neuropeptide Research
Erasmus and Brugmann Hospitals
Free University of Brussels
Campus of Anderlecht
808 Route de Lennik
1070 Brussels, Belgium

M. Matteoli
Department of Cell Biology
Yale University School of Medicine
333 Cedar Street
New Haven, Connecticut 06510

Emanuel Meller
Millhauser Laboratories
New York University Medical Center
School of Medicine
550 First Avenue
New York, New York 10016

C. C. Mermet
INSERM U 171-CNRS URA 1195
Centre Hospitalier Lyon-Sud
69310 Pierre-Bénite, France

Jean E. Merrill
Department of Neurology
Reed Neurological Research Center
University of California, Los Angeles School of Medicine
710 Westwood Plaza
Los Angeles, California 90024

D. K. Meyer
Department of Pharmacology
University of Freiburg
Hermann Herderstrasse 5
D-7800 Freiburg, Federal Republic of Germany

Viktor Mutt
Department of Biochemistry II
Karolinska Institute
Box 60400
S-104 01 Stockholm, Sweden

Charles B. Nemeroff
Departments of Psychiatry and Pharmacology
Duke University Medical Center
Durham, North Carolina 27710

Charles Nicholson
Department of Physiology and Biophysics
New York University Medical Center
550 First Avenue
New York, New York 10016

Christer Nilsson
Department of Medical Cell Research
Section of Neurobiology
University of Lund
Biskopsgatan 5
S-223 62 Lund, Sweden

Ola G. Nilsson
Department of Medical Cell Research
Section of Neurobiology
University of Lund
Biskopsgatan 5
S-223 62 Lund, Sweden

Fred Nyberg
Department of Pharmacology
Uppsala University
751 24 Uppsala, Sweden

Britta Nyström
Institute of Neurobiology
P.O. Box 33031
S-400 33 Göteborg, Sweden

Lars Olson
Department of Histology and Neurobiology
Karolinska Institute
Box 60400
S-104 01 Stockholm, Sweden

David Ottoson
Wenner-Gren Foundation
Sveavagen 166
113 46 Stockholm, Sweden

Christer Owman
Department of Medical Cell Research
Section of Neurobiology
University of Lund
Biskopsgatan 5
S-223 62 Lund, Sweden

Sanford L. Palay
Department of Anatomy
Harvard Medical School
25 Shattuck Street
Boston, Massachusetts 02115

H. Pawelzik
Clinical Neuropharmacology
Max Planck Institute for Psychiatry
P.O. Box 401240
D-8000 Munich 40, Federal Republic of Germany

A. J. Pellionisz
Department of Physiology and Biophysics
New York University Medical Center
Medical Science Building
550 First Avenue
New York, New York 10016

E. Merlo Pich
Institute of Human Physiology
University of Modena
Via Campi 287
41100 Modena, Italy

Julia M. Polak
Department of Histochemistry
Royal Postgraduate Medical School
Hammersmith Hospital
DuCane Road
London, W12 ONN, England

Hélène Pollard
Unité du Neurobiologie et Pharmacologie (U. 109)
Centre Paul Broca de l'INSERM
75014 Paris, France

Alain Privat
INSERM U 249
Department of Biology
Boulevard Henri IV
34060 Montpellier Cedex, France

Thomas Reese
NINCDS
National Institutes of Health
Bethesda, Maryland 20892

A. Thomas Reetz
Department of Cell Biology
Yale University School of Medicine
333 Cedar Street
New Haven, Connecticut 06510

Donald J. Reis
Department of Neurology and Neurosciences
Division of Neurobiology
The New York Hospital
Cornell Medical Center
1300 York Avenue
New York, New York 10021

Margaret E. Rice
Department of Physiology and Biophysics
New York University Medical Center
550 First Avenue
New York, New York 10016

Guido Rindi
Department of Human Pathology
University of Pavia
Anatomic Pathology Unit
Via Forlanini 14
27100 Pavia, Italy

J. del Rio-Garcia
National Institute for Medical Research
The Ridgeway
Mill Hill
London NW7 1AA, England

Roberta Rosie
MRC Brain Metabolism Unit
University Department of Pharmacology
1 George Square
Edinburgh EH8 9JZ, Scotland

Aryeh Routtenberg
Department of Psychology
Northwestern University
Cresap Neuroscience Laboratory
2021 Sheridan Road
Evanston, Illinois 60208

Jean-Charles Schwartz
Unité de Neurobiologie et Pharmacologie (U. 109)
Centre Paul Broca de l'INSERM
2 ter rue d'Alesia
75104 Paris, France

Philippe Séquéla
Department of Physiology
University of Montreal
CP 6218 Suc A
Montreal, Quebec, Canada H3C 3J7

Bo K. Siesjö
Laboratory for Experimental Brain Research
EA-5
Lund Hospital
S-221 85 Lund, Sweden

Herbert Silfvenius
Department of Neurosurgery
University of Göteborg
Box 33031
S-400 33 Göteborg, Sweden

Stephan D. Skaper
Fidia Research Laboratories
Via Ponte della Fabbrica 3/A
35031 Abano Terme, Italy

D. G. Smyth
National Institute for Medical Research
The Ridgeway, Mill Hill
London NW7 1AA, England

Enrico Solcia
Department of Human Pathology
University of Pavia
Anatomic Pathology Unit
via Forlanini 16
27100 Pavia, Italy

Ingrid Strömberg
Department of Histology and Neurobiology
Karolinska Institute
Box 60400
S-104 01 Stockholm, Sweden

M. F. Suaud-Chagny
INSERM U171-CNRS URA 1195
Centre Hospitalier Lyon-Sud
69310 Pierre-Bénite, France

E. Sykova
Laboratory of Neurohumoral Regulation
Institute of Physiological Regulations
Czechoslovak Academy of Sciences
Bulovka, Pav. 11
180 85 Prague 8, Czechoslovakia

Lars Terenius
Department of Drug Dependence Research
Karolinska Institute
S-10401 Stockholm, Sweden

Elspeth Thomson
MRC Brain Metabolism Unit
University Department of Pharmacology
1 George Square
Edinburgh EH8 9JZ, Scotland

B. Tinner
Department of Histology and Neurobiology
Karolinska Institute
Box 60400
104 01 Stockholm, Sweden

A. Toffano
Fidia Research Laboratories
Via Ponte della Fabbrica 3/A
35031 Abano Terme, Italy

Gino Toffano
Fidia Research Laboratories
Via Ponte della Fabbrica 3/A
35031 Abano Terme, Italy

Urban Ungerstedt
Department of Pharmacology
Karolinska Institute
Box 60400
S-104 01 Stockholm, Sweden

Jean-Jacques Vanderhaeghen
Laboratory of Neuropathology and Neuropeptide Research
Erasmus and Brugmann Hospitals
Free University of Brussels
Campus of Anderlecht
808 Route de Lennik
1070 Brussels, Belgium

Guido Vantini
Fidia Research Laboratories
Via Ponte della Fabbrica 3/A
35031 Abano Terme, Italy

Olli Ventä
Helsinki University of Technology
Rakentajanaukio 2 C
SF-02150 Espoo, Finland

Monique Verslijpe
Laboratory of Neuropathology and Neuropeptide Research
Erasmus and Brugmann Hospitals
Free University of Brussels
Campus of Anderlecht
808 Route de Lennik
1070 Brussels, Belgium

E. Sylvester Vizi
Institute of Experimental Medicine
Hungarian Academy of Sciences
P.O.B. 67
H-1450 Budapest, Hungary

Peter Wallén
Nobelinstitute of Neurophysiology
Karolinska Institute
Box 60400
104 01 Stockholm, Sweden

Kenneth C. Watkins
Department of Physiology
University of Montreal
CP 6128 Suc A
Montreal, Quebec, Canada H3C 3J7

Cynthia Wetmore
Department of Histology and Neurobiology
Karolinska Institute
Box 60400
S-104 01 Stockholm, Sweden

C. Wikkelsø
Department of Neurology
University of Göteborg
Box 33031
S-400 33 Göteborg, Sweden

W. Zieglgänsberger
Clinical Neuropharmacology
Max Planck Institute of Psychiatry
P.O. Box 401240
D-8000 Munich 40, Federal Republic of Germany

I. Zini
Institute of Human Physiology
University of Modena
Via Campi 287
41100 Modena, Italy

M. Zoli
Institute of Human Physiology
University of Modena
Via Campi 287
41100 Modena, Italy

Volume Transmission in the Brain: Novel Mechanisms for Neural Transmission,
edited by K. Fuxe and L.F. Agnati.
Raven Press, Ltd., New York © 1991.

1

Two Principal Modes of Electrochemical Communication in the Brain: Volume versus Wiring Transmission

*Kjell Fuxe and **Luigi F. Agnati

**Department of Histology and Neurobiology, Karolinska Institute, Stockholm, Sweden and*
***Department of Human Physiology, University of Modena, Modena, Italy.*

This book represents a collection of papers dealing with the potential role of the brain extracellular fluid as a pathway for electrical and chemical communication (volume transmission, VT). According to classical work, chemical communication between cells has been characterized on the basis of the distance between release sites and target cells. Other relevant features are concentration of the signals in the medium and the affinity of the receptors for the signal.

Endocrine communication takes place via a hormone, which "is a substance that is secreted by one cell and travels through the circulation where it exerts actions on other cells" (1). Usually hormones travel for a long distance, have low plasma concentrations (lower than 10–7 M), and high-affinity receptors. However, "hormones may also diffuse to neighbouring target cells, bypassing the blood stream in what is called paracrine communication. Or, more simply, a given cell may respond to its own hormone output, often as a form of feedback inhibition, in autocrine communication" (2). In these latter cases signal concentration is usually higher and receptor affinity may be somewhat lower than the ones observed for endocrine communication.

The present hypothesis proposes that all these types of communication are present in the central nervous system (CNS) besides the synaptic one and have been called VT in contrast to the known wiring transmission (WT) operating via synapses (Fig. 1). This hypothesis is based in part on our early observations in 1969 to 1972 using monoamine fluorescence histochemistry, showing that monoaminergic drugs can lead to the appearance of extraneuronal monoamine fluorescence surrounding the monoamine nerve cells (3–6), probably reflecting diffusion of monoamines in extracellular fluid (ECF). Injected dopamine (DA) was also shown to diffuse over large distances, especially along the fiber tracts (7; Routtenberg A, *this volume*). Other related concepts have been put forward in the 1980s (see Table 1). Our hypothesis provides original developments and an unitary frame for the electrochemical communication in the brain. For example, no author has previously suggested the possibility of a long to very long (>1 mm) distance action of chemical signals released in the brain to act on other brain cells (either neurons or glial cells) (Table 2). Particularly in this case, ECF characteristics are the key to understanding communication in brain outside the synapse.

The most striking feature of brain cell ag-

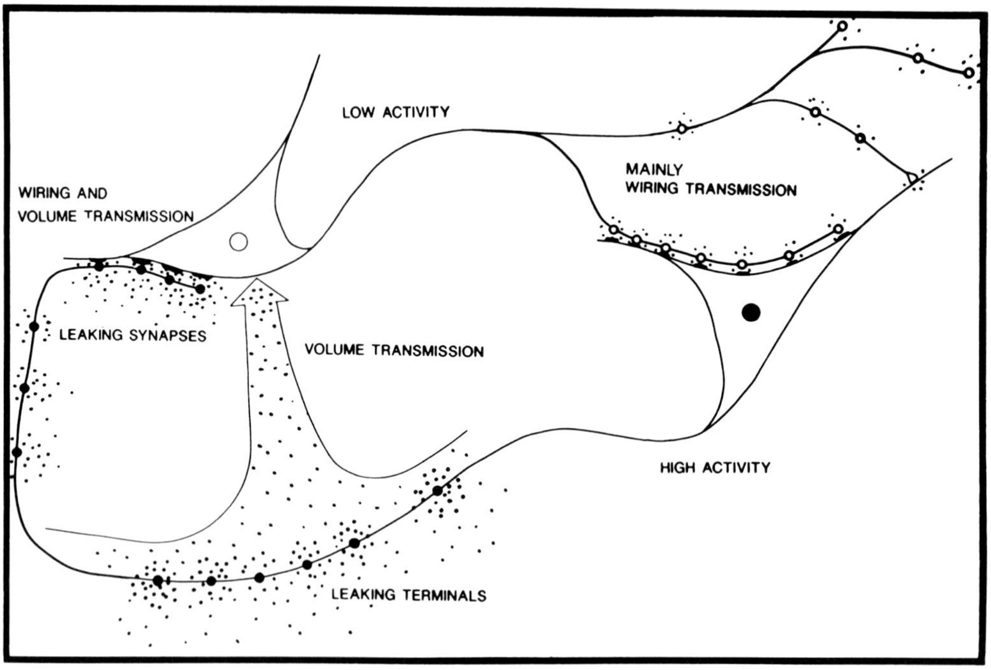

FIG. 1. Schematic illustration of WT and VT in chemical communication in the CNS. It is illustrated that chemical signals in VT originate both from leaking nerve terminals and from leaking synapses. It seems likely that the vast majority of synapses operate not only via WT, but also via VT. When the nerve cell is in low activity, mainly WT mode is in operation, whereas in high activity states, VT becomes more important in view of available data in the literature (see Lundberg, *this volume*). Thus, results suggest that neuropeptides are mainly released when a high frequency of nerve impulses exists. The nucleus in the nerve cell with high activity is shown in *black,* while that in the nerve cell with low activity is shown as a *large open circle. Small black spots* along the axons represent varicosities operating via VT or VT/WT (when synapses are formed) mode. *Open circles* along the axons of the neuron with low activity represent varicosities sometimes forming synapses mainly operating via WT mode. *Very small black dots* surrounding the varicosities and synapses illustrate variations in the concentrations of transmitters, especially neuropeptides, diffusing in ECF. Synapses are indicated by postsynaptic thickenings.

gregates is the vast number of interfaces provided by the plasma membrane area of the cells. "Across these interfaces streams a flux of chemical messengers, ranging from ions to macromolecules" (8). Thus, a fundamental role in the information handling is played by the plasma membrane of neurons and glia cells. However, of basic importance is also ECF, whose function should not be "relegated to that of one constituent of the electrochemical battery that drives neural activity" (8). Furthermore, in the CNS it is possible to distinguish the spreading of electrical signals along the cell processes from the diffusion of electrical signals from one cell to another via electrotonic currents. A particular case is the so-called ephaptic transmission, which consists of the leakage of electrical signals between neighboring processes. The sites at which such interactions occur are called "ephapses" (9).

On this basis we have suggested that two main modes of electrochemical transmission can be distinguished in the brain: VT and WT (10,11). VT is characterized by the conduction of chemical or electrical signals via a leaking pathway (Table 2). Thus, the

TABLE 1. *Concepts related to the existence of chemical and electrical communication in the extracellular fluid (ECF) of the brain*

Authors	Year	Hypothesis
Branton, Mayeri, Brownell, and Simon (34)	1978	Local hormonal communication between neurons in aplysia
Beaudet and Descarries (35)	1978	Sites of transmitter release not matched by synaptic specialization
Guillemin (36)	1978	Paracrine secretion in brain
Nicholson (37)	1979	Diffusion of electrical signals within the interconnected microenvironments of the ECF
Cuello (38)	1983	Nonclassical neuronal communication
Vizi (39)	1984	Nonsynaptic interactions between neurons
Schmitt (40)	1984	Parasynaptic system in brain
Agnati, Fuxe, and Zoli (41)	1985	Volume vs. wiring transmission
Agnati, Fuxe, Zoli, Zini, Toffano, and Ferraguti (6)	1986	
Herkenham and McLean (42)	1986	Parasynaptic intercellular communication
Herkenham (43)	1987	
Fuxe et al. (13)	1980	Cascade of peptide fragments form chemical pathways to produce syndromic responses
Agnati et al. (10,11)	1980 1983	
DeWied and Jolles (9)	1982	
Fuxe et al. (15)	1986	
Agnati, Zoli, Merlo Pich, Ruggeri, and Fuxe (12)	1988	
Fuxe, Agnati, Zoli, Bjelke, and Zini (16)	1989	

physical substrate for the conduction involves the ECF outside the specialized sites represented by the synaptic cleft. A clear example of such a phenomenon is the electrocorticogram recording, which can be understood on the basis of a theory called "volume conduction" (see, e.g., ref. 12). This theory describes the flow of current through ECF under various conditions of brain activity. The theory of volume conduction was one of the reasons that inspired us to propose the term "VT" for a broader phenomenon, that is, the spreading of chemical and electrical signals in ECF. This new view emphasizes the importance of understanding the structure of the extracellular space with its intracellular matrix of polyionic molecules, which may highly contribute to the tortuosity of ECF pathways (13). Thus, the structure of brain cell microenvironment may depend on physical and chemical constraints represented by not only cellular elements such as astroglial cells, but also by the extracellular matrix. In addition, the functional characteristics of ECF are probably dependent on the functional polarity of the glial cells (see ref. 8) and the chemical characteristics of the extracellular matrix (e.g., capability of buffering ions or of controlling molecular mobility). New efforts should therefore be devoted to the study of the morphofunctional characteristics of ECF in terms of diffusion of electrochemical signals. This

TABLE 2. *The main concept of volume transmission*

is the conduction of the message along a leaking pathway. This conduction is characterized by the absence of the physical constraints linked to the wiring of neurons.

For this reason this concept includes

1. The so-called nonsynaptic transmission, which is a very short- to short-distance chemical phenomenon
2. The ephaptic transmission, which is a very short-distance electrical phenomenon
3. The paracrine transmission, which is a long-distance phenomenon
4. The neuroendocrine transmission, which is a very long-distance chemical phenomenon

will allow a better understanding of the communicational features of the brain.

Wiring transmission is characterized by the conduction of the signal via a nonleaking pathway. Wiring transmission is based on the evidence that the communication in brain rests on networks, and therefore the classical wiring diagram of axon, synapses, and dendrites can be recognized as the physical substrate for the flow of information. The best example is the conduction of action potentials along the axon. Wiring transmission imposes much stricter physical constraints to the communication than VT (see Fig. 1). However, by comparing the respective advantages and disadvantages of WT and VT, it is possible to surmise that each of them is particularly suited for specific tasks (see Table 3).

The present dichotomic subdivision allows a precise classification of all the intercellular communication modalities that have been so far described in the CNS (see Tables 2 and 4). Furthermore, the concept of VT is the basis to appreciate fully the proposal of the existence of "chemical networks" (14–21) in the brain (Figs. 2 and 3). In fact, it has been demonstrated that a parent peptide can be split in the ECF by proteolytic enzymes into active fragments with their own spectrum of activity. These chemical signals obviously diffuse to the target structures via the ECF.

Evidence for the existence of the VT mode in the CNS is severalfold (see Table 5) and even more is provided by the chapters of this volume. Already in the 1960s and early 1970s (3–6), studies on monoamine neurons indicated that their large plexa of varicose terminals and dendrites could release monoamines to be accumulated in the extracellular space (see Fig. 1). The maintenance of striatal DA neurotransmission in Parkinson's disease despite a massive disappearance of DA nerve terminals has also early been explained on the basis of a concept related to VT (22,23). Thus, increased DA release from surviving DA nerve terminals with a high firing rate and an up-regulated DA synthesis could maintain DA transmission because of diffusion of increased amounts of DA in ECF, reaching supersensitive high-affinity DA receptors present all over the neostriatum because of the massive degeneration of DA nerve terminals. Thus, the function of diffuse neuronal systems may be maintained despite massive degeneration by increasing the proportion of volume transmission, so that individual DA terminals can influence larger spheres of striatal tissue. These studies promoted new ideas for chemical communication, which are in line with the present proposal claiming that electrical and chemical intercellular communications may also involve ECF. Furthermore, the fact

TABLE 3. *Main features of morphofunctional roles of glial cells and neurons in wiring transmission and volume transmission*[a]

		Glial functions		Neuron functions	
		Physical	Biochemical	Physical	Biochemical
Wiring transmission	Electrical synapse	Synapse segregation	Modulation of coupling via ECF	Location of synapse	Modulation of coupling via ICF
	Chemical synapse	Synapse segregation	Modulation of synaptic cleft microenvironment	Location of synapse	Control of the efficacy of transmission lines and their interplay
Volume transmission	Electrotonic signals	Shaping of ECF pathways for signal diffusion	Control of ECF ion composition	Location of sources and sinks for signals	Control of sources and sinks for signals
	Humoral signals	Shaping of ECF pathways for signal diffusion	Release, uptake, and metabolism of signals	Location of release and recognition sites for signals	Release, uptake, and metabolism of signals

[a]ECF/ICF, extracellular/intracellular fluid.

TABLE 4. *Overview of the organization of communication in the CNS*[a]

	Coding	Channel	Class	Mode
Communication in the CNS	Chemical	Cell structure	Axonal transport	WT
			Dendritic transport	WT
			Intramembrane diffusion	WT
			Gap junction diffusion	WT
		ECF pathway	Autocrine	WT/VT
			Synaptic	WT/VT
			Paracrine	VT
			Endocrine	VT
	Electrical	Cell structure	Postsynaptic potential	WT
			Action potential	WT
			Gap junction potential	WT
		ECT pathway	Ephaptic interactions	VT

[a]WT, wiring transmission; VT, volume transmission; ECF, extracellular fluid.

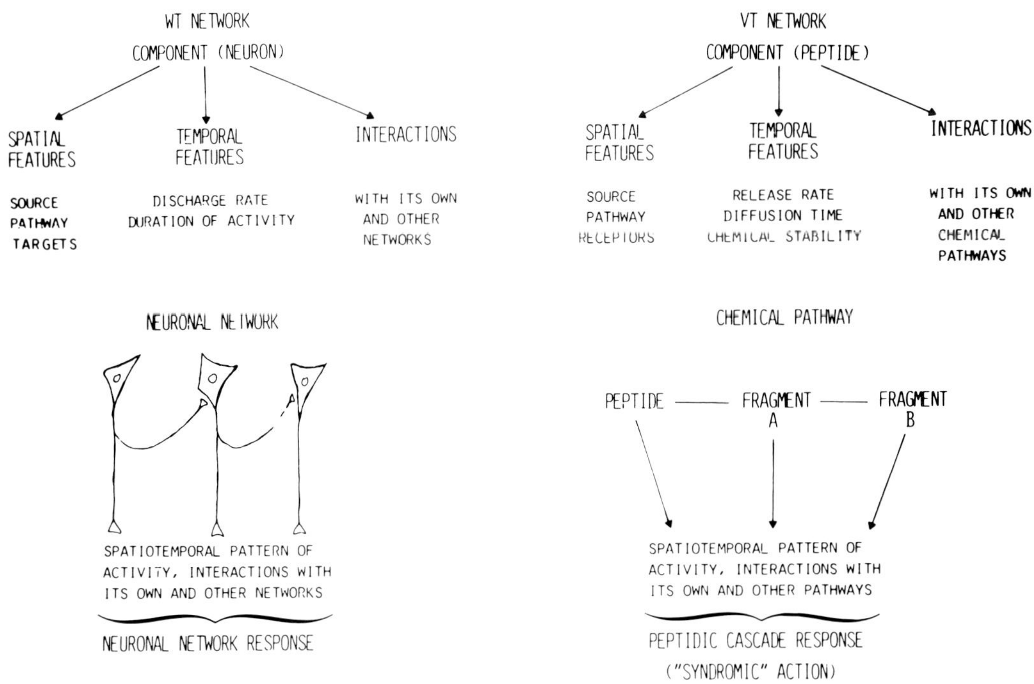

FIG. 2. Schematic representation of WT and volume transmission network. Although WT networks rely on chemicophysical as well as structural features of the neurons, VT networks rely on chemical characteristics of the message and chemicophysical as well as structural features of the extracellular fluid.

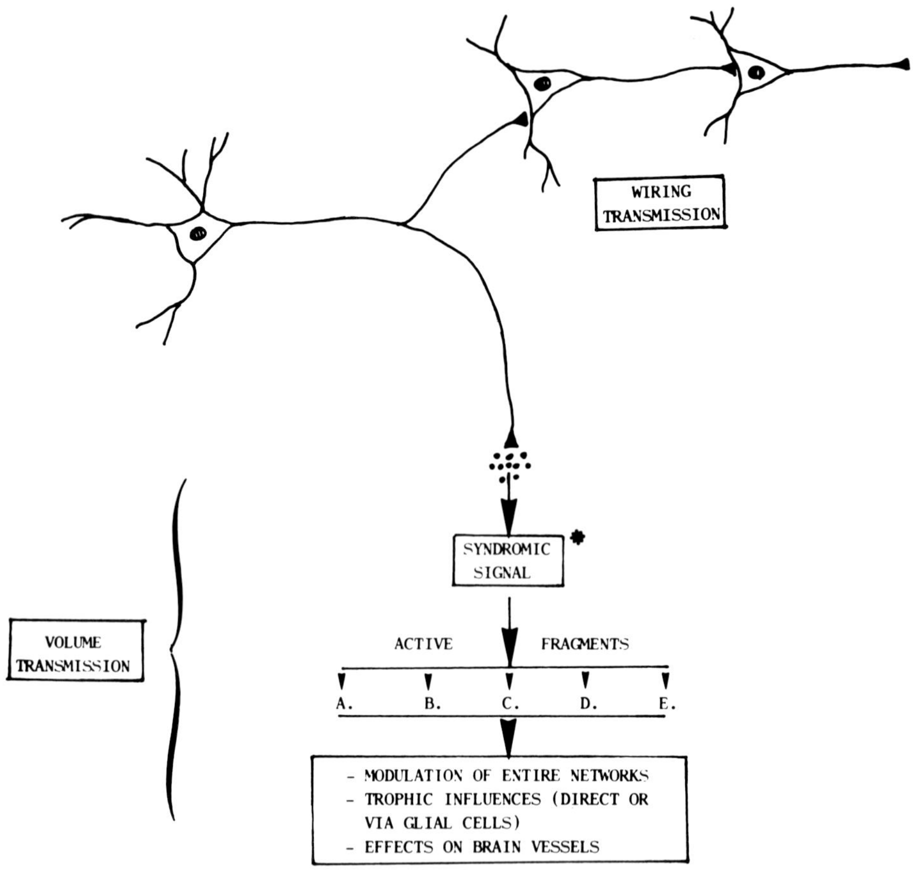

FIG. 3. Schematic illustration of WT and VT processes within the CNS. It is illustrated that neuropeptides importantly may participate in VT and represent syndromic signals in ECF pathways by being cleaved into active fragments by enzymes in the nerve and glial cell membranes facing ECF. The parent peptide and active fragments may modulate entire networks and exert trophic influences on nerve and glial cells and also control cerebral blood flow. In the aged brain, VT may be preferentially reduced in view of degeneration of peptide neurons as well as reduction in their biosynthetic capability of neuropeptides, leading then also to reduction of syndromic signals in VT (15,32,33). *, in an aged brain the syndromic signals decrease: A. reduction of biosynthetic capability of peptide neurons, B. reduction of surviving peptide neurons.

that adeno- and neurohypophyseal hormones exist in central neurons and exert biological effects (24–29) support this idea.

Volume transmission is of particular importance for the possibility of a long span (even millimeters) diffusion of chemical signals in the CNS, leading to long-term actions (see chapters by Olson L, et al., Grimaldi R, et al., Bjelke B, et al., *this volume*). A preliminary characterization of such a long distance diffusion for neuropeptide Y (NPY) has recently been performed (see Table 6). It has been shown that NPY (10 fmol/0.5 μl) may diffuse more than 1 mm when injected into the striatum and hypothalamus (see Table 5). Finally, this diffusion seems to be anisotropic, showing that the ECF matrix with its own chemical and physical characteristics may control the spreading of the signal.

TABLE 5. *Experimental support for the existence of volume transmission*

Type of signal for volume transmission	Experimental evidence or indications	Author/year
Electrical signals	Intercellular microenvironment acts as a communication channel for electrical signals	Nicholson 1979, 1980 (3,37)
Chemical signals	Morphological evidence for nonsynaptic paracrine transfer of information in invertebrates	Branton, Mayeri, Brownell, and Simon 1978 (34)
	Ultrastructural evidence of mismatch of release sites, synaptic specialization, and binding sites in monoamine and neuropeptide terminals	Beaudet and Descarries 1978 (35) Hamel and Beaudet 1987 (44)
	Ultrastructural evidence of nonsynaptic axoaxonic interactions and dendroaxonic interactions	Cuello 1983 (38)
	Transmitter-receptor mismatches at light microscopical level	Kuhar 1985 (45) Agnati et al. 1985, 1986 (6,41) Herkenham and McLean 1986 (42) Herkenham 1987 (43)
	5-HT and CA in the extracellular space	Carlsson, Jonason, Lindqvist, and Fuxe 1969 (19) Butcher, Engel, and Fuxe 1970, 1972 (17,18) Fuxe and Ungerstedt 1970 (20)
	Peptide fragments produce differential responses compared with the parent peptide	Fuxe et al. 1980 (13) Agnati et al. 1980, 1983 (10,11) deWied and Jolles 1982 (9)

Volume transmission opens up the possibility for a better understanding of the mechanisms underlying long-term adjustments in brain function and may be one of the reasons why many different chemical substances are used as signals in the CNS. Furthermore, coexistence of neuropeptides and classical transmitters may reflect the possibility for a neuronal system, e.g., NPY/A, costoring neurons to operate via

TABLE 6. *Diffusion of ^{125}I-neuropeptide Y (NPY) injected into some brain regions of the rat assessed by quantitative autoradiography*[a,b]

	Stereotactic coordinates of the cannula tip (mm from Bregma)			Diffusion distance from the cannula tip (mm)			
Brain area	Anteroposterior	Lateral	Dorsoventral	RIGHT	LEFT	Anterior	Posterior
Striatum	0.9	2.6	6.0	1.07 ± 0.07	0.98 ± 0.05	1.4 ± 0.1	1.5 ± 0.1
Hypothalamus	−3.1	1.0	9.0	0.75 ± 0.14	0.70 ± 0.01	1.7 ± 0.1	2.5 ± 0.1

[a]From Merlo Pich et al. (*in preparation*), with permission.

[b]Three to four CD Sprague-Dawley rats (250 g) were used for each measurement. Microinjections of 10 fmol/0.5 μl were performed in anesthetized rats placed on a Kopf stereotactic frame using a 28-gauge cannula connected to a microinfusion pump (Carnegie Medicine). The cannula was left *in situ* for 1 hr; then the rat was killed by decapitation, the brain removed, frozen at −80°C, and cut by a cryostat. One 20-μm coronal section of five was kept, for a total rostrocaudal length of about 4 mm. Sections were exposed to a iodine-sensitive film (Amersham) for 5 days. Gray Values were assessed by using an IBAS II image analyzer (Zeiss Kontron). As an index of the diffusion along major axes has been used, the distance from the cannula tip of the section where 50% of the maximal Gray Value (i.e., the Gray Value at the cannula tip) was observed. This Gray Value corresponds approximately to 0.5 fmol/g tissue of ^{125}I-NPY. Stereotactic coordinates are given according to Paxinos & Watson (1982).

both WT and VT mode, because neuropeptides are highly suitable for VT [30,31, Fuxe K, et al., *this volume;* Lundberg J, *this volume*].

We are convinced that these old and new concepts will be further clarified from the chapters in this volume.

REFERENCES

1. Baxter JD, Frohman LA, Broadus AE, Felig P. Introduction to the endocrine system. In: Felig P, Baxter JD, Broadus AE, Frohman LA, eds. *Endocrinology and metabolism.* New York: McGraw–Hill, 1987;3–23.
2. Cook DL. The cellular biology of the endocrine system: an overview. In: Patton HD, Fuchs RF, Hille B, Scher AM, Steiner R, eds. *Textbook of physiology.* Philadelphia: Saunders Co, 1989; 1139–1159.
3. Butcher L, Engel J, Fuxe K. L-Dopa induced changes in central monoamine neurons after peripheral decarboxylase inhibition. *J Pharm Pharmacol* 1970;22:313–316.
4. Butcher LL, Engel J, Fuxe K. Behavioral, biochemical and histochemical analyses of the central effects of monoamine precursors after peripheral decarboxylase inhibition. *Brain Res* 1972;41:387–411.
5. Carlsson A, Jonason J, Lindqvist M, Fuxe K. Demonstration of extraneuronal 5-hydroxytryptamine accumulation in brain following membrane-pump blockade by chlorimipramine. *Brain Res* 1969;12:456–460.
6. Fuxe K, Ungerstedt U. Histochemical, biochemical and functional studies on central monoamine neurons after acute and chronic amphetamine administration. In: Costa E, Garattini S, eds. *Amphetamines and related compounds.* Milano: Proc Mario Negri Inst Pharmacol Res, 1970:257–288.
7. Ungerstedt U, Butcher LL, Butcher SG, Andén N-E, Fuxe K. Direct chemical stimulation of dopaminergic mechanisms in the neostriatum of the rat. *Brain Res* 1969;14:461–471.
8. Nicholson C. Measurement of extracellular ions in the brain. *Trends Neurosci* 1980;3:216–218.
9. Shepherd GM. *Neurobiology.* New York: Oxford University Press, 1988.
10. Agnati LF, Fuxe K, Zoli M, et al. Aspects on the information handling by the central nervous system: focus on cotransmission in the aged rat brain. *Prog Brain Res* 1986;68:291–301.
11. Agnati LF, Fuxe K, Zoli M, Zini I, Toffano G, Ferraguti F. A correlation analysis of the regional distribution of central enkephalin and b-endorphin immunoreactive terminals and of opiate receptor in adult and old male rats. Evidence of the existence of two main types of communication in the central nervous system: the volume transmission and the wiring transmission. *Acta Physiol Scand* 1986;128:201–207.
12. Martin JH. Cortical neurons, the EEG, and the mechanisms of epilepsy. In: Kandel ER, Schwartz JH, eds. *Principles of neural science.* Amsterdam: Elsevier, 1985:636–647.
13. Nicholson C, Rice ME. Calcium diffusion in the cerebellum of the skate (*Raja erinacea*). *Mount Desert Island Biol Lab Bull* 1986;26:149–150.
14. DeWied D, Jolles J. Neuropeptides derived from proopiocortin: behavioural, physiological and neurochemical aspects. *Physiol Rev* 1982;62: 976–1059.
15. Agnati LF, Andersson K, Fuxe K, Locatelli V, Mutt V, Hökfelt T. Differential actions of CCK-39, CCK-8 and CCK-7 on pre- and postsynaptic dopamine mechanisms in various parts of the forebrain of the rat. *Neurosci Lett* 1980;(Suppl 5):S204.
16. Agnati LF, Fuxe K, Blake C, et al. Morphometrical characterization of the ACTH-endorphin and dopamine neuron system of the mediobasal hypothalamus of the rat. Effects of neonatal monosodium glutamate treatment. *INSERM* 1983; 110:87–102.
17. Agnati LF, Zoli M, Merlo Pich E, Ruggeri M, Fuxe K. The emerging complexity of the brain. Limits on brain–computer analogy. In: Bonatti M, Decina M, eds. *Traffic engineering for ISDN design and planning.* Amsterdam: Elsevier, 1988;209–230.
18. Fuxe K, Andersson K, Locatelli V, et al. Cholecystokinin peptides produce marked reduction of dopamine turnover in discrete areas in the rat brain following intraventricular injection. *Eur J Pharmacol* 1980;67:329–331.
19. Fuxe K, Andersson K, Locatelli V, et al. Neuropeptides and central catecholamine systems: interactions in neuroendocrine and central cardiovascular regulation. In: Costa E, Trabucchi M, eds. *Neural peptides and neuronal communication.* New York: Raven Press, 1980;37–50.
20. Fuxe K, Agnati LF, Härfstrand A, et al. Studies on peptide comodulator transmission. New perspective on the treatment of disorders of the central nervous system. *Prog Brain Res* 1986; 66:341–368.
21. Fuxe K, Agnati LF, Zoli M, Bjelke B, Zini I. Some aspects of the communicational and computational organization of the brain. *Acta Physiol Scand* 1989;135:203–216.
22. Fuxe K, Hökfelt T, Ungerstedt U. Morphological and functional aspects of central monoamine neurons. *Int Rev Neurobiol* 1970;13:93–126.
23. Zigmond MJ, Stricker EM, Berger TW. Parkinsonism: insights from animal models utilizing neurotoxic agents. In: *Animal models of dementia.* New York: Alan R. Liss, 1987;1–38.
24. Fuxe K, Hökfelt T, Eneroth P, Gustafsson J-Å, Skett P. Prolactin-like immunoreactivity: localization in nerve terminals of rat hypothalamus. *Science* 1977;196:899–900.
25. Fuxe K, Andersson K, Hökfelt T, et al. Prolactin–monoamine interactions in rat brain and

their importance in regulation of LH and prolactin secretion. In: Robyn C, Harter M, eds. *Progress in prolactin physiology and pathology*. Amsterdam: Elsevier/North-Holland Biomedical Press, 1978;95–109.
26. Watson SJ, Akil H, Richard CW, Barchas JD. Evidence for two separate opiate peptide neuronal systems and the coexistence of beta-lipotropin, beta-endorphin and ACTH immunoreactivities in the same hypothalamic neurons. *Nature (Lond)* 1978;275:226–228.
27. Watson SJ, Richard CW, Barchas JD. Adrenocorticotropin in rat brain: immunocytochemical localization in the cells and axons. *Science* 1978;200:1180–1182.
28. Vandesande F, Dierickx K, Demey J. Identification of the vasopressin-neurophysin producing neurons of the rat separate opiate peptide neuronal systems and the coexistence of beta-lipotropin, beta-endorphin and ACTH immunoreactivities in the same hypothalamic neurons. *Nature* (Lond) 1978;275:226–228.
29. Swanson LW, Sawchenko PE. Paraventricular nucleus: a site for the integration of neuroendocrine and autonomic mechanisms. *Neuroendocrinology* 1980;31:410–417.
30. Härfstrand A. *Brain neuropeptide Y mechanisms. Basic aspects and involvement in cardiovascular and neuroendocrine regulation*. Thesis, 1987; ISBN: 91:7900-306-0.
31. Fuxe K, Härfstrand A, Agnati LF, von Euler G, Svensson T, Fredholm B. On the role of NPY in central cardiovascular regulation. In: Mutt V, Fuxe K, Hökfelt T, Lundberg J, eds. *Neuropeptide Y*. New York: Raven Press, 1989;201–214.
32. Agnati LF, Zoli M, Grimaldi R, Fuxe K, Toffano G, Zini I. Cellular and synaptic alterations in the aging brain. *Aging* (*in press*).
33. Agnati LF, Zoli M, Merlo Pich E, Benfenati F, Fuxe K. Experimental and theoretical aspects on neural plasticity in the central nervous system. VIII. Theoretical aspects on communication and computation in the central nervous system. *Neurochem Int* (*in press*).
34. Branton WD, Mayeri E, Brownell P, Simon SB. Evidence for local hormonal communication between neurones in aplysia. *Nature* 1978;274:70–72.
35. Beaudet A, Descarries L. The monoamine innervation of the rat cerebral cortex: synaptic and non-synaptic axon terminals. *Neuroscience* 1978;3:851–860.
36. Guillemin R. Peptides in the brain: the new endocrinology of the neuron. *Science (NY)* 1978; 202:390–402.
37. Nicholson C. Brain cell microenvironment as a communication channel. In: Schmitt FO, Worden FG, eds. *The neurosciences: fourth study program*. Cambridge, MA: The MIT Press, 1979;457–476.
38. Cuello AC. *Co-transmission*. London: MacMillan, 1982.
39. Vizi ES. *Non-synaptic interactions between neurons: modulation of neurochemical transmission*. New York: John Wiley, 1984.
40. Schmitt FO. Molecular regulators of brain function. A new view. *Neuroscience* 1984;13:991–1001.
41. Agnati LF, Fuxe K, Zoli M. Considerazioni sulle relazioni tra sistema nervoso centrale e sistema endocrino. *Crescita* 1985;14:34–38.
42. Herkenham M, McLean S. Mismatches between receptors and transmitter localizations in the brain. In: Boast CA, Snowhill EW, Altar CA, eds. *Quantitative receptor autoradiography*. New York: Alan R. Liss, 1986;137–171.
43. Herkenham M. Mismatches between neurotransmitter and receptor localization in brain: observations and implications. *Neuroscience* 1987;23:1–38.
44. Hamel E, Beaudet A. Opioid receptors in rat neostriatum: radioautographic distribution at the electron microscopy level. *Brain Res* 1987; 401:239–257.
45. Kuhar MJ. The mismatch problem in receptor mapping studies. *Trends Neurosci* 1985;8:190–191.

*Volume Transmission in the Brain:
Novel Mechanisms for Neural Transmission,*
edited by K. Fuxe and L.F. Agnati.
Raven Press, Ltd., New York © 1991.

2

An Integrative View of Information Handling in the CNS

Floyd E. Bloom

Department of Neuropharmacology, Research Institute of Scripps Clinic, La Jolla, California 92037

In part, the pleasure of participating in another of the Wenner-Gren International Symposia stems from the pleasure of renewing friendships with our Swedish colleagues and in part from the recollection that the past meeting topics have had heuristic impact on our research directions and concepts. Previous recent deliberations illuminated the facts of coexisting peptides, amines, and amino acids within specific subsets of neurons and forced us to revise our concepts of neurotransmission to include the possibility of multi-messenger cotransmission. Similarly, the observations that neurons express a mosaic of cell surface receptors for a variety of transmitters have been heuristic in leading to examination of the general class of receptor–receptor interactions.

Thus, when we consider another set of basic factual observations that were unexpected and not easily explainable, we are again forced to consider whether additional revisions to our concepts of the brain's operations are not once more needed. In the mid-1970s, maps of chemical systems in the brain proliferated as new peptide-containing systems were added to the maps of monoamines laid out largely by our Swedish hosts. At the same period, a second strategy of mapping was also being refined that disclosed the sites at which pharmacologic probes, generally receptor subtype-specific antagonists, bound with high affinity to brain sections. Detailed analysis of these two data sets disclosed that in many cases the ligand binding sites and endogenous ligand-releasing fiber systems showed consistently heterogeneous distributions at the light microscopic level—the so-called "ligand-receptor" mismatch. These data sets were extensively documented by Drs. Fuxe and Agnati by quantitative receptor autoradiography and semi-quantitative immunocytochemistry (1). Similar conclusions were comprehensively considered by Herkenham (2), based on anatomic comparison of ligand binding autoradiography and the immunocytochemistry of the relevant endogenous ligands.

As I review Drs. Fuxe and Agnati's most recent collaborations, it was these disparities in ligand content and ligand probe binding that initiated the dilemmas that spawned the present symposium. They raise the new concept of "volume transmission," defined by them as "information handling . . . based on the electrotonic currents and the spread of humoral signals," and propose that this format of regulation be added to the more classical concepts of "wiring transmission," defined as "information handling . . . based on the topological organization of the (nervous system) elements and the patterns of impulse flow along the neural networks" (3,4). In their view, the possibilities of volume transmission would offer an additional means to regulate the operations of the neural networks, and "liberate . . . (them) . . . from the constraints of neuroanatomical wiring" (1).

More recently, the volume transmission concept has been appropriately enlarged to include an additional and richly documented emphasis on intercellular communications between the adrenal cortex and gonads and the brain mediated by bloodborne steroid hormones. In these introductory comments, I wish to offer some of my initial reflections on volume transmission. I approach these concepts from the base of fundamental observations centered on the norepinephrine (NE)-containing projections from the locus ceruleus (LC) and several of the peptide systems in which my colleagues and I have worked (5–8).

HOW PRECISE IS THE INNERVATION OF TARGET FIELDS BY MONOAMINERGIC NEURONS?

A previous era in central transmission centered on the nature of the ultrastructural identification of synaptic transmission sites and the features of the specialized junctional apparatus that underly this form of communication, which Dr. Palay considers here (see *this volume*). Transmitter-specific cytochemical methods identified which neurons and pathways contained specific transmitters by exploiting their distinctive cytochemical reactivity or using autoradiography to exploit unique biological properties such as active re-uptake. Application of these approaches at the ultrastructural level allowed one to analyze the degree to which the transmitter property of a neuron specified the features of its interneuronal junctions. In these studies the presence and the absence of specialized junctions for nerve terminals identified as monoamine-containing became a contentious interlude. Many concluded that aminergic systems were not only diffuse in their terminal field innervation but also in the micro-domain of their cellular targets in these fields. These sets of data are also re-examined as well [see chapter by Descarries et al., *this volume*] and contrast with the rather abundant degree to which we (9–12) and others (see ref. 6 for references) have found these neurons to have conventional specialized junctions, the frequency of which is as much affected by methods as it is by species and region (13,14).

We based our identification of NE-containing terminals on a convergence of data derived from accumulating exogenous radioactive transmitter, selective neurotoxins, and direct cytochemistry (9–12). From these data, we became convinced that specialized junctions were a common feature of aminergic neurons, probably as common although perhaps not as extensive a proportion of the surface area per nerve terminal as with any other specifiable class identified chemically. Subsequently, ultrastructural immunoperoxidase studies have yielded similar conclusions (13–15).

We have recently re-examined this question in the primate visual cortex, a terminal field in which monoaminergic afferents form rather specifically distributed laminar profiles (16,17). We exploited the availability of an antibody to serotonin (5-HT) that achieves optimal results in well-fixed material (18) to consider the ultrastructure of these 5-HT-containing nerve terminals. Of more than passing interest, we recall that the existence of the raphe innervation of the cortex by 5-HT-containing neurons was one of the founding observations that inaugurated the monoamine era more than a quarter century ago (19).

As was clear from earlier studies at the light microscopic level, the macaque neocortex is densely innervated by 5-HT-containing fibers, reaching its highest density in the primary visual cortex. 5-HT-immunoreactive fibers are found in every layer of the visual cortex, but three well-defined bands of increased density of 5-HT innervation were aligned with layers IIIB to IVCα, with layer VA, and with layer VIB, of which layer IVCα exhibited the most dense immunoreactive elements. When these zones were explored in detail at the ultrastructural level, light microscopically defined "varicosities" were distributed identically to ultrastructurally defined 5-HT-containing boutons and fibers, with diameters of 0.5 to 0.9 μm, and intervaricose diameters of 0.14 to 0.29 μm.

Relevant to the present discussions and to the analysis of presumptive 5-HT-containing profiles identified by autoradiography after accumulation of exogenous 5-HT (20), random single-section samples of the primate visual cortex showed specialized postsynaptic junctions in only 1% to 1.5% of the 5-HT-immunoreactive profiles. This frequency of specialized junctions rose to 6% when only boutons (profiles with enlarged diameters and accumulations of synaptic vesicles) were tabulated. However, it was clear that the perimeters of the boutons (diameter range, 0.2–1.0 μm) were considerably greater than the length of the specialized synaptic junction (range, 0.85–0.5 μm).

Therefore, extensive quantitative analysis of these junctions was performed in serial sections of 27 boutons. The serial section analysis revealed that while many varicosities did not exhibit a synaptic complex in a single section, they nevertheless all consistently showed such junctions in other sections in the series. Furthermore, as suggested by the single random section samples, the synaptic complexes are often small (i.e., often contained within 3 to 400 mμ). In a separate sample of 475 5-HT-immunoreactive boutons in layer IVC_{α}, stereological methods were used to compute the total surface area of the boutons; the fraction represented by the synaptic complex represented only 0.12%, thus accounting for the apparent low distribution of contacts in random samples (18). Similar findings in the rodent neocortex (13) suggest that here as well, 5-HT-immunoreactive terminals form specialized contacts in 90% or more of the detected boutons. Because our analyses never detected more than one junctional complex for any extensively sampled 5-HT-immunoreactive bouton, we concluded that each 5-HT varicosity within the cortex is likely to form one, but only one, synapse.

Thus our data support the conclusion that in the neocortex, and probably elsewhere, 5-HT-containing terminals do show the ultrastructural features of orthodox "wiring transmission" synaptic systems. From this, one could conclude either that 5-HT and other monoamines do not require a novel set of structural principles to distinguish them from other "classical" transmitter systems or that these ultrastructural features have nothing to do with the spatial, temporal, and functional precision of the transmitter being released from these defined sites and the sites at which they can produce effects. As noted by Herkenham's analysis (1987; also see *this volume*), the 5-HT-receptor subtype most pronounced in neocortex is the low-affinity 5-HT_2 type, and further, "the 5-HT_2 receptor lamination in (rodent) cortex is more pronounced than the . . . (5-HT) fiber plexus lamination." Taken at face value, this particular discrepancy between the endogenous ligand and the ligand probe–binding maps could suggest that the binding sites *per se* extend well beyond the spatial domain in which the ligand "operates." Conversely, within the 5-HT system of cortex, the synaptically relevant receptor may well match the ultrastructural junctions. Although 5-HT cortical transmission has not been well-studied physiologically *in vivo*, 5-HT responsivity in cerebellar cortex was found to be restricted to granule cells, known also to receive 5-HT accumulating mossy fibers (20). Furthermore, as the pharmacological subtypology of the receptors evolves, the larger domain of sites may be found to have other features that unlink them from considerations of synaptic relevance. Although there is no reason to believe that the lower-affinity binding sites will have more macroscopically distinctive cellular distributions than their high-affinity forms, the degree to which the binding site mapping depends on the affinity limits easy answers to this issue.

However, these data showing that specific sites do have synaptic specializations (when extensively sampled through serial sections or through multiple samples in a specific terminal field) do not nullify previous reports that many monoamine-containing nerve terminals do not exhibit synaptic specializations (14). Specifically, it is not clear whether these systems have boutons, which are additional nonsynaptic sites, perhaps representing either vesicle clusters in sites of active re-uptake, or other intraneuronal sites of transmitter storage after re-uptake that do or do not also release transmitter.

Furthermore, the functional equation of ligand binding sites as exclusively "synaptic" receptors (in the functional, not structural, sense) may require more discriminative methods of re-examination. The cloning of the major known species of 5-HT receptor molecules (21) may provide additional tools for the ultrastructural comparison of boutons containing specific transmitters. For example, when applied to the central beta-adrenergic sites (22), the receptor protein was found to reside in postsynaptic junctional specializations opposite tyrosine hydroxylase–positive terminals. However, the quantitative proportions of such junctions in brain have not been reported.

SYNAPTIC PRECISION OF CENTRAL PATHWAYS ACTIVATED AND EVALUATED ELECTROPHYSIOLOGICALLY

Our acceptance or rejection of concepts often hinges on the data that we find most likely to reflect the properties with which we imagine the brain to run. In my case, the concepts of how specific aminergic systems operate was honed in a decade-long struggle to determine the "main" central actions of NE. To me, this meant comparing the effects of iontophoretic NE with electrical activation of the pathways that contained NE to prove that NE was in fact a central transmitter. In those days, with techniques that were crude and primitive by today's standards, we were pleased simply to be able to document that the responsive cells in the cerebellum (23,24) and in the hippocampus (25,26) were those that appeared to be the main contact target of the LC axons (also see refs. 5 and 27 for more extensive review).

In the LC–cerebellar system, extensive controls were run to verify that the activation of the noradrenergic neurons and fibers were necessary and sufficient to produce actions identical to iontophoretic NE administration to Purkinje cells, with little or no consistent effects on other cerebellar cortical neurons. The poststimulus time histograms of this pathway showed the onset of the effect to begin after approximately 150 msec and to endure for approximately 300 msec. Although these synaptic latencies are long, much of this delay is directly attributable to the slow conduction velocity of the fine unmyelinated fiber systems, with some additional tens of milliseconds perhaps consumed by the second-messenger mediation (8,27). In any case, even delays of 125 msec would seem to be too short to provide for any substantive degree of volume transmission. Stimulation of the pathway with rates approaching those found subsequently when LC neurons are activated in behaviorally responsive animals (28,29) produced even longer-lasting effects. Furthermore, the effects of NE and the LC were identical biophysically: Hyperpolarizations were accompanied by modest increases or no changes in membrane resistance (23,24).

In the LC-hippocampal projection studies (25,26), single stimulus activations of the LC gave rise to reduced spontaneous firing with latencies of 200 to 250 msec, again lasting for periods of approximately two–three times that duration. Multiple stimulations (10 Hz for 10 sec) produced 30 to 40 sec of depressed firing. However, these effects were restricted to neurons with the spontaneous firing patterns indicative of pyramidal neurons and were not seen with nonpyramidal neurons even in the same immediate vicinity (26). Similarly, in studies of LC-pathway projections to neurons of the rodent neocortex, there are similar but less well-defined properties, suggesting that the effects are somewhat tightly time-linked and target cell–selective (30,31). For this amine system as well, therefore, the basic parameters observed for its structural and functional properties suggest that it most closely follows the principles defined by our organizers as those of wiring transmission.

Despite their distinctive anatomy and physiology, the availability of these data plus a broad pharmacological armamentarium combined to allow monoamines to be among the first chemically defined systems to satisfy rigorous criteria as central transmitters (6,27). The noradrenergic and other monoaminergic systems provide excellent illustration of the time and space features by which chemically labeled neuronal systems may be distinguished and compared (6,32). The highly divergent efferent morphology of the aminergic systems has characterized these systems as having a far broader spatial and temporal domains of potential influence than other systems. It is only through evaluation of synaptic actions documented to be transmitted by specific agents that one can assess the minimum delay for onset of effects and the possibility of longer-distance volume transmission. In the intact CNS, data exist for only a few monoaminergic systems (see ref. 5 for review). These *in vivo* data suggest that endogenous noradrenergic systems themselves may have a synaptic latency that approaches 20 to 30 msec, after a conduction transit time of another 100 msec, and that sensory activation of the LC itself would require yet another 35 to 75 msec depending on the modality (4,27–29). At the upper limit, therefore, of 200 msec, the opportunities for volume transmission would seem limited.

The aminergic neurons are thought to be extensively interconnected through recurrent collateral innervation within their source nuclei. Because most monoamine neurons tend to respond to their own amine by depression of discharge, such recurrent interconnections tend in general to yield similar, succinct excitatory responses to sensory activation, followed by prolonged inhibition and refractoriness to sensory activation (4,32,33).

The monoaminergic systems also offer insight into a third comparative axis, that may be termed "mechanistic" (27) or energy (32) domain. For the LC–NE system, virtually all of its synaptic actions *in vivo* and *in vitro* can be attributed to the ability of NE to activate within target neurons the rapid synthesis of cyclic adenosine monophosphate (AMP) and thereby to modify the postsynaptic cell's properties during longer periods. For example, postulated changes in a cyclic AMP–mediated activation of a protein kinase could lead to phosphorylation of membrane proteins that would account for the observed increase in membrane resistance. This would alter the postsynaptic membrane in a manner that "enables" it to respond more effectively and in a more enduring manner to its other afferent inputs (27,35). While past data have emphasized the beta-adrenergic receptor–mediated responses of NE target neurons, more recent data suggest that the alpha 1–adrenergic receptor may well interact on the same target cells (see below). Within the LC, alpha 2–adrenergic responses are prominent; such receptors also occur throughout the NE terminal fields, but their functional roles and effects remain uncertain (36).

INTERACTIONS BETWEEN VASOACTIVE INTESTINAL POLYPEPTIDE AND NOREPINEPHRINE IN RAT CEREBRAL CORTEX

The aminergic activating systems thus possess temporal, spatial, and mechanistic interactions that can extend and lengthen the effects of this system on cortical targets. A specific example of this interaction (see ref. 37 for references) is the neurochemical and physiological interaction of the NE afferent cortical system with intrinsic cortical neurons that contain the peptide transmitter vasoactive intestinal polypeptide (VIP).

Several lines of evidence support a role for VIP as a neuronal messenger in cerebral cortex, particularly the biochemical data on its presence, release, binding, and the ability to stimulate cyclic AMP formation in cortical slices somewhat more potently than NE. Cytochemically, VIP- and NE-containing circuits show a contrasting but complementary cortical anatomy: VIP neurons are intrinsic, bipolar, radially oriented, intracortical neurons, whereas the NE innervation arises only from LC and innervates a broad expanse of cortex in a horizontal plane. The two fiber systems may have the same targets (i.e., the pyramidal cells) as do many other cortical afferents (Bloom, Morrison, and Battenberg, *unpublished data*).

Identified cortical pyramidal neurons, like most identified potential target cells, are depressed in spontaneous firing by iontophoresis of either NE or cyclic AMP (4,27). Biochemically, (37) VIP and NE can act synergistically to increase cyclic AMP in cerebral cortex. Furthermore, in single-unit electrophysiologic tests (38), application of VIP during subthreshold NE administration causes pronounced inhibitions of cellular discharge regardless of the effect of VIP (none, speeding, or slowing) before NE.

Magistretti and Schorderet (37) showed that the synergism of VIP by NE was blocked by phentolamine, an alpha-adrenergic receptor antagonist, and mimicked by phenylephrine, the alpha-receptor agonist. This result was supported by the electrophysiologic tests, showing the NE–VIP synergism could be replicated by phenylephrine, an alpha-adrenergic agonist. An apparently similar cyclic AMP–mediated enhancement by beta receptors of noradenergic target cell responsiveness to alpha-adrenergic agonists has been reported for other cellular systems (38). If NE- and VIP-containing fibers do indeed converge on the same cortical target cell, it is feasible that cyclic AMP is the intracellular mediator of their synergistic interaction, underlying the "enabling"-type action (also see ref. 8, 35).

LC–NE EFFECTS ON CORTICAL NEURONAL ACTIVITY

There have been numerous physiologic studies of the possible impact of LC–NE projection on putative target cells in various neocortical regions, based on earlier studies of LC–NE synaptic actions on identified target cells in cerebellum and hippocampus (see ref. 4 for review). In the neocortex of unanesthetized monkeys, target cells in auditory cortex with vigorous reproducible responsiveness to behaviorally relevant functional stimuli were tested before, during, and after iontophoresis of NE; these tests showed not only dose-dependent inhibition of spontaneous and vocalization-evoked activity, but also that a given dose of NE reduced spontaneous activity more than the activity evoked by acoustic stimuli (39). Those data, combined with earlier observations on cerebellar and hippocampal neurons (27), fit the hypothetical view that the LC–NE enhances elicited activity relative to spontaneous activity, thereby "enabling" (7,8,32) or increasing "signal-to-noise" characteristics of the target cells (39) to afferent stimuli active during periods of enhanced LC activity. Analogous experiments in anesthetized rats (40) showed a similar reduction of background activity relative to stimulus-elicited activity in the somatosensory cortex. As in cerebellum and and hippocampus (27), both excitatory responses and inhibitory responses were enhanced or enabled by NE application or LC activity.

Taken together, the anatomic data concerning LC projections, the functional data on activity of LC source neurons, and the synaptic effects of the transmitter on specific target neurons indicate that the LC is activated during alerting or arousal and releases NE onto target neurons in many brain regions, including specific targets in neocortex. This transmitter then acts to enhance the selectivity and vigor of responses of target neurons to contemporaneous afferent stimuli.

Levin et al. (41) recently extended these physiological implications of the integrative actions of the central NE system in the rodent brain with the use of the 2-deoxyglucose uptake autoradiographic method. They first confirmed that stimulation of a single rat facial vibrissae increased glucose use in the corresponding barrels (lamina IV) and associated columns in laminae I to VIa of the contralateral first somatosensory cortex. They then observed that 6-hydroxydopamine lesions of the LC (sufficient to have depleted ipsilateral cortical NE by more than 90%) gave modest (10–20%) but statistically significant increases in barrel and column metabolic representations, compared with the contralateral cortex. When bilateral removal of all but one vibrissa on each side was combined with a unilateral LC lesion, the barrel and column metabolic representation on the LC-intact side enlarged only on the NE-intact side. Therefore, these authors concluded that the effect of NE on the cortex depends on the status of its afferent input. When NE cortical innervation is intact, metabolic activity is increased only within the activated barrel and column in the intact cortex. The authors also suggest that this same system may independently modulate plastic enlargement of columnar representations in the partially vibrissae deafferented cortex. These data suggest that even at the macroscopic level of barrel and column organization, the impact of the effects of the LC and presumably other monoaminergic cortical projections are rather closely tied to their wiring transmission properties.

PEPTIDE-CONTAINING SYSTEMS MAY HAVE RELEASE AND RESPONSE PROPERTIES COMPATIBLE WITH LOCAL VOLUME TRANSMISSION

It remains an enigma that despite extensive efforts to study peptides as interneuronal signals in the CNS, almost all the existing data are pharmacological in nature. Specifically, it has proven to be very difficult to define CNS actions of peptide-containing pathways that can be shown conclusively to be mediated by the peptides released from these pathways despite the precedents in the peripheral nervous system (42,43). It remains at the moment unclear as to whether this shortcoming represents difficulties in selectively eliciting the release of the peptides in systems in which their effects can be well-studied or the lack of appropriate antagonists or other undefined experimental reasons.

Even more attention has been accorded to the apparent "mismatch" between the location of molecules that can specifically bind to presumptive peptidergic "receptors" and the locations of the peptides themselves (2). Some interpret these discrepant light microscopic localizations to indicate that peptides in the brain (may) act via paracrine routes of delivery over undefined distances significantly longer than the conventional synaptic cleft. Others (44) interpret data on different peptides to speak in favor of more conventional synaptic actions of peptides in the CNS.

In what may be the first study to combine pharmacological stimulation of a peptide-containing pathway with ligand displacement autoradiography, Neumaier and Chavkin (45) attempted to correlate the displacement of ^{3}H diprenorphine with the location of the hippocampal opioid peptide pathways, by releasing the opioid peptide globally with high potassium from endogenous pathways in an *in vitro* slice preparation of rat hippocampus. They observed "incomplete overlap" between the endogenous ligand-containing fiber maps and the sites from which diprenorphine was displaced. Displacement was maximal in the stratum radiatum and stratum moleculare, whereas the density of opioid immunoreactive axons was maximal in the stratum lucidum. This result suggested to them that significant diffusion of released peptides can occur under these special conditions. However, by electrophysiological detection, the cells that seem to mediate the effects of exogenous opioid peptides in the hippocampus were not the pyramidal cells, but rather the local inhibitory interneurons (27; see also Zieglgänsberger et al., *this volume*). In any case, given this likely functionally responsive element, the locations of whose dendritic trees were difficult to discern from the macroscopic resolution of ligand binding studies, we neither support nor attack the volume transmission hypothesis as applied to the peptides. Further extensions of this paradigm with more precise activation of pathways over more discrete stimulus intervals are needed to determine whether such potential diffusion suggested by the Neumaier and Chavkin (45) strategy represents an excess unphysiological release overwhelming local peptidases or an early glimpse of an actual physiological diffusional spread to distal receptors. It is also unclear how this *in vitro* system compares with the operations of the intact nervous system.

Gerfen et al. (46) demonstrated the heterogeneous transmitter patterns of afferent systems that innervate specific cellular populations of the basal ganglia. Gerfen suggested that this nonlaminar structure could also be a useful system in which to probe for diffusional effects of released amines and peptides (*personal communication*). For example, the dopaminergic system closely follows the dendritic pattern of neurons in the caudate, which have been found to express one or more neuropeptides (47). Evaluation of the changes in messenger ribonucleic acid (mRNA) expression for these peptides, mapped onto the domains of the afferent dopaminergic system, would allow for precise evaluation of the degree to which diffusion or volume conduction does occur and, if so, how limited its dimensions may be within physiological time intervals.

For example, studies on caudate employing direct peptide assays on samples dialyzed (47) or aspirated by push-pull cannula (48,49) show that endogenous dopamine and dopamine-activated receptors affect release of peptides in a peptide-specific (and inferentially therefore, in a neuron-specific) manner. The data described in this symposium on the electrophysiological characterization of responses to peptides (Zieglgänsberger et al., *this volume*) and on the ultrastructural definition of the ligand binding sites will be critical to the many gaps that remain to be filled.

STEROID EFFECTS ON THE CNS ESTABLISH EXAMPLES OF CENTRAL VOLUME TRANSMISSION

The systems considered to this point (i.e., monoamines and peptides) operate both within the CNS and in the peripheral nervous system. Perhaps in part to avoid cross-talk between central and peripheral spheres of activity, the blood-brain barrier exists to provide a useful filter. However, as is typical for the endocrine system where release sites are at substantial distance from the response sites and in which the response sites are selectivity distributed, the well-recognized lines of neuroendocrine feedback regulation have clearly established the reality of volume transmission to the brain by lipo-soluble steroids acting on brain receptors. Neurons selectively distributed throughout the brain bear specific steroid receptors, whose distribution has begun to be well-localized by both ligand binding properties and, more directly, by *in situ* hybridization of the mRNAs for these recently cloned receptors (49). In our recent work, two aspects of this "volume transmission" would seem relevant to the present discussions.

We (50) have examined the changes in neuronal expression of oxytocin mRNA in the perinatal and mature female rat as a function of endogenous gonadal steroids. The nonapeptide oxytocin has been implicated as a neurosecretory peptide involved in labor and milk ejection, and as a centrally active neuropeptide responsible for stimulating sexual and maternal behaviors. Oxytocinergic neurons in the supraoptic (SON) and paraventricular (PVN) nuclei of the hypothalamus project to the posterior pituitary where the peptide is released into the circulation. In addition, oxytocinergic parvocellular neurons have been shown to project to many extrahypothalamic brain areas, including regions of spinal cord, brainstem, and forebrain.

Ovarian steroids, which play a central role in the control of reproductive function, influence both neurosecretory and intracerebrally projecting oxytocinergic neurons. Estrogen treatment stimulates the release of oxytocin into the peripheral circulation, induces oxytocin binding sites in the brain, influences the electrical activity of oxytocinergic neurons, and alters the pattern of oxytocinergic cells and fibers in the brain, as detected by immunocytochemistry (see ref. 49 for references). The transcriptional regulation of oxytocin mRNA by ovarian steroids during puberty appeared to us to provide a physiological opportunity to examine the regulation of this peptide during development of female reproductive biology.

In adolescent female rats, the first estrous cycle occurs at approximately 38 days postnatally. To determine whether the neural oxytocinergic system is modified during the development of sexual maturity, we examined the expression of oxytocin mRNA before and after puberty. Oxytocin mRNA levels increased between P (postnatal day) 23 and P34 and continued to increase until they reached adult levels by P50 as determined by Northern blot analysis. This increase could involve either the recruitment of new oxytocinergic neurons during the pubertal period or a net increase in the levels of oxytocin mRNA per neuron. However, by *in situ* hybridization analyses, there were no qualitative differences in the distribution of the labeled neurons in the prepubescent versus mature female animals. Labeled cells were located in the same hypothalamic nuclei, giving rise both to intra- and extra-hypothalamic circuits.

Nevertheless, marked differences in the amounts of oxytocin mRNA per neuron in all these nuclei were detected by *in situ* hybridization when we compared prepubertal and ovariectomized postpubertal females with normal postpubertal females. We interpret this as a global decrease in levels of oxytocin mRNA in all the hybridizing neurons regardless of their known or presumed targets. Neurons of the supraoptic nucleus, the paraventricular nucleus, and the anterior midline hypothalamus all contained at least twofold less oxytocin mRNA in the prepubescent and ovariectomized females than in the mature female. These results are consistent with the hypothesis that increased ovarian steroids either directly or indirectly modulate the developmental increase in neuronal oxytocin mRNA. The increase in oxytocin mRNA may, in turn, be partially responsible for the physical and behavioral transition from a sexually immature juvenile to a sexually mature adult.

Given the high likelihood that steroid hormones can regulate a feedback signal from peripheral endocrine organs, we were also drawn to the possibilities raised by the work of Beaulieu and colleagues (50–52) that the brain itself may have its own internal form of "neurosteroids." With Vatier, from the Beaulieu group, we recently examined the effects on brain neurosteroids caused by administration of centrally acting drugs (53). 3β-Hydroxy-Δ5 steroids such as pregnenolone (P), dehydroepiandrosterone (D), and their sulfate esters (PS and DS) exist in rat brain at higher concentrations than in blood and persist after removal of gonads and adrenal glands. We sought to establish the relation between blood alcohol levels (BAL) and the concentration of P, PS, D, and DS in different structures of the rat brain. Sprague-Dawley male rats (175 g) were given ethanol (16% in saline, IP) to obtain BALs of 150 mg%. Control animals were given an equal volume of saline. Animals were sacrificed 2 hr later, and the brain was quickly dissected for steroids assays. The results (N = 24–36) show an average 75% decrease of D + DS in the olfactory bulbs (OB), olfactory tubercules (OT), amygdalas (A), and hypothalamus (H). The mean concentrations of D+DS in these combined regions are 0.78 + 0.24 ng per wet weight in ethanol-injected rats and 3.24+0.44 ng per wet weight in controls. There is approximately the same depletion in each region studied (OB = average loss = 75%; OT = 80%; A = 75%; H = 60%). No significant changes were observed in the concentrations of P and PS. Thus, D seems to be a reliable marker of alcohol intoxication in the rat brain, showing a depletion and recovery comparable with that of ethanol intoxication *per se*. We interpret the data to suggest that whatever the normal functions of neurosteroids in the brain, their metabolic turnover and functional actions are clearly modified by ethanol, possibly through the formation of novel intermediate steroid catabolites, which can regulate the actions of gamma-aminobutyric acid (GABA) and perhaps other neurotransmitters (54–56). If such were the case, the regulatory actions of these neurosteroids might well represent an important opportunity to pursue volume transmission of lipid signals within the brain.

CONCLUSIONS

Absolute objectivity is difficult to retain in science because our mind tends naturally to hypotheses that are often attractive in relation to their novelty. The mismatches between ligand binding sites and ligand release sites is a valid set of perplexing phenomena in which we could draw several differing hypothetical explanations, each of which may have appeal. The questions that need to be answered are which views can be supported by data and how do these data then constrain the hypotheses. For example, one possibility that seems not to be specifically rejectable on the basis of data available now would be the idea that astrocytes also exhibit binding sites for neurotransmitters and that these glial ligand recognition sites are used to support the selectivity of interneuronal communication processes in part by maintaining the lowest possible levels of extracellular transmitter, specifically to reduce diffusion, interference, and cross-talk and to limit responses to those receptors in the immediate environment of the release sites. It may be that glia receive information in these exchanges that supports neuronal function (37) over longer time courses to fulfill some not yet well-defined "trophic" functions. Available data strongly suggest that *in vitro*, astrocytes can respond to transmitters through well-defined transductive systems to alter cellular shape, and that glia from discrete central regions can alter the appearance and other structural and chemical properties of the neurons from distant sites that inhabit their domains (57,58). For the moment, therefore, I remain open to the idea that the brain contains many aspects of intercellular regulation that remain unknown. As one dedicated to the analysis of how the integration of information is achieved through local cellular chemical events, I look forward to the harvest of ideas from conference and the research they will stimulate.

ACKNOWLEDGMENTS

My colleagues George Siggins and Steve Henriksen contributed important comments on earlier drafts of these remarks. This research was funded by Alcohol Research Center Grant 06420 and Program Project Grant NINCDS 22347. This chapter is RISC paper NP 6069.

REFERENCES

1. Agnati LF, Fuxe K, Zoli M, Ozini I, Toffano G, Ferraguti F. A correlation analysis of the regional distribution of central enkephalin and β-endorphin immunoreactive terminals and of opiate receptors in adult and old male rats. Evidence for the existence of two main type of communication in the central nervous system: the volume transmission and the wiring transmission. *Acta Physiol Scand* 1986;128:201–207.
2. Herkenham M. Mismatches between neurotransmitter and receptor localizations in brain: observations and implications. *Neuroscience* 1987;23:1–38.
3. Agnati LF, Fuxe K, Merlo Pich E, et al. Aspects of the integrative capacities of the central nervous system: evidence for "volume transmission" and its possible relevance for receptor–receptor interactions. In: Fuxe K, Agnati LF, eds. *Receptor-receptor interactions: a new intramembrane integrative mechanism*. New York: Macmillan, 1986;236–249.
4. Fuxe K, Agnati LF, Harfstrand A, et al. Principles for the hormone regulation of wiring transmission and volume transmission in the central nervous system. *Acta Physiol Scand* 1986;128: 1–53.
5. Foote SL, Bloom FE, Aston-Jones G. The nucleus locus coeruleus: new evidence of anatomical and physiological specificity. *Physiol Rev* 1983;63:844–914.
6. Bloom FE. General features of chemically identified neurons. In: Bjorklund A, Hokfelt T, eds. *Handbook of chemical neuroanatomy*, vol. 2. The Netherlands: Elsevier/North-Holland Biomedical Press, 1984;1–22.
7. Bloom FE. Chemical signalling in the spatial, temporal continuum. In: Iversen, Goodman, eds. *Fast and slow chemical signalling in the nervous system*. England: Oxford University Press, 1986;295–308.
8. Bloom FE. Neurotransmitters: past, present, and future directions. *FASEB J* 1988;2:32–41.
9. Landis SC, Bloom FE. Ultrastructural identification of noradrenergic boutons in mutant and normal mouse cerebellar cortex. *Brain Res* 1975;96:299–306.
10. Koda LY, Bloom FE. A light and electron microscopic study of noradrenergic terminals in the rat dentate gyrus. *Brain Res* 1977;120:327–335.
11. Koda LY, Wise RA, Bloom FE. Light and electron microscopic changes in the rat dentate gyrus after lesions or stimulation of the ascending locus coeruleus pathway. *Brain Res* 1978;144: 363–368.
12. Koda LY, Schulman JA, Bloom FE. Ultrastructural identification of noradrenergic terminals in rat hippocampal, unilateral destruction of the locus coeruleus with 6-hydroxydopamine. *Brain Res* 1978;145:190–195.
13. Parnavelas JG, Papadopoulos GC. The monoaminergic innervation of the cerebral cortex is not diffuse and nonspecific. *Trends Neurosci* 1989;12:315–320.
14. De Felipe J, Jones EG. A light and electron microscopic study of serotonin-immunoreactive fibers and terminals in the monkey sensory-motor cortex. *Exp Brain Res* 1988;71:171–182.
15. Molliver ME, Grzanna R, Lidov HGW, Morrison JH, Olschowka JA. Monoamine systems in the cerebral cortex. In: Chan-Palay V, Palay SI, eds. *Cytochemical methods in neuroanatomy*. New York: Alan. R. Liss, 1982;256–277.
16. Morrison JH, Foote SL, Molliver ME, Bloom FE, Lidov HGW. Noradrenergic and serotonergic fibers innervate complementary layers in monkey primary visual cortex: an immunohistochemical study. *Proc Natl Acad Sci USA* 1982;79:2401–2405.
17. Morrison JH, Foote SL. Noradrenergic and serotonergic innervation of cortical, thalamic, and tectal visual structures in old and new world monkeys. *J Comp Neurol* 1986;243:117–138.
18. deLima AD, Bloom FE, Morrison JH. Synaptic organization of serotonin-immunoreactive fibers in primary visual cortex of the macaque monkey. *J Comp Neurol* 1988;274:280–294.
19. Dahlstrom A, Fuxe K. Evidence for the existence of monoamine-containing neurons in the central nervous system. I. Demonstration of monoamines in the cell bodies of brain stem neurons. *Acta Physiol Scand* 1964;62(Suppl)232:1–55.
20. Bloom FE, Hoffer BJ, Siggins GR, Barker JL, Nicoll RA. Effects of serotonin on central neurons: microiontophoretic administration. *Fed Proc* 1972;31:97–106.
21. Hartig PR. Molecular biology of 5-HT receptors. *Trends Pharmacol Sci* 1989;10:61–69.
22. Aoki C, Zemcik BA, Strader CD, Pickel VM. Cytoplasmic loop of β-adrenergic receptors: synaptic and intracellular localization and relation to catecholaminergic neurons in the nucleus of the solitary tracts. *Brain Res* 1989;493:331–347.
23. Siggins GR, Hoffer BJ, Oliver AP, Bloom FE. Activation of a central noradrenergic project ion to cerebellum. *Nature* 1971;233:481–483.

24. Hoffer BJ, Siggins GR, Oliver AP, Bloom FE. Activation of the pathway from locus coeruleus to rat cerebellar Purkinje neurons: pharmacological evidence of noradrenergic central inhibition. *J Pharmacol Exp Ther* 1973;184:553–569.
25. Segal M, Bloom FE. The action of norepinephrine in the rat hippocampus: II. Activation of the input pathway. *Brain Res* 1974;72:99–114.
26. Segal M, Bloom FE. The action of norepinephrine in the rat hippocampus. IV. The effects of locus coeruleus stimulation on evoked hippocampal unit activity. *Brain Res* 1976;107:513–525.
27. Siggins GR, Gruol DL. Synaptic mechanisms in the vertebrate central nervous system: In: Bloom FE, ed. *Handbook of physiology, vol. on intrinsic regulatory systems of the brain*. Bethesda, MD: American Physiological Society, 1986;1–114.
28. Aston-Jones G, Bloom FE. Activity of norepinephrine-containing locus coeruleus neurons in behaving rats anticipates fluctuations in the sleep-waking cycle *J Neurosci* 1981;1:876–886.
29. Aston-Jones G, Bloom FE. Norepinephrine-containing locus coeruleus neurons in behaving rats exhibit pronounced response to non-noxious environmental stimuli. *J Neurosci* 1981;1:887–900.
30. Olpe HR, Glatt A, Laszlo J, Schellenberg A. Some electrophysiological and pharmacological properties of the cortical, noradrenergic projection of the locus coeruleus in the rat. *Brain Res* 1980;186:9–19.
31. Phillis JW, Kostopoulos GK. Activation of a noradrenergic pathway from the brain stem to rat cerebral cortex. *Gen Pharmacol* 1977;8:207–211.
32. Bloom FE. The functional significance of neurotransmitter diversity. *Am J Physiol* 1984;246:C184–C194.
33. Schultz W. Responses of midbrain dopamine neurons to behavioral trigger stimuli in the monkey. *J Neurophysiol* 1986;56:1439.
34. Schultz W, Romo R. Responses of nigrostriatal dopamine neurons to high-intensity somatosensory stimulation in the anesthetized monkey. *J Neurophysiol* 1987;57:201.
35. Nicoll RA. Neurotransmitter regulated ion channels. *Science* 1988;241:545–555.
36. Arnsten AF, Goldman-Rakic PS. Alpha-2 adrenergic mechanisms in prefrontal cortex association with cognitive decline in aged nonhuman primates. *Science* 1985;230:1273–1276.
37. Magistretti PJ, Morrison JH. Noradrenaline and vasoactive intestinal peptide-containing neuronal systems in neocortex: functional convergence with contrasting morphology. *Neuroscience* 1988;24:367–378.
38. Ferron A, Siggins GR, Bloom FE. Vasoactive intestinal polypeptide acts synergistically with norepinephrine to depress spontaneous discharge rate in cerebral cortical neurons. *Proc Natl Acad Sci USA* 1985;82:8810–8812.
39. Foote SL, Freedman R, Oliver AP. Effects of putative neurotransmitters on neuronal activity in monkey auditory cortex. *Brain Res* 1975;86:229–242.
40. Waterhouse BD, Moises HC, Woodward DJ. Noradrenergic modulation of somatosensory cortical neuronal responses to iontophoretically applied putative neurotransmitters. *Exp Neurol* 1980;69:30–49.
41. Levin BE, Craik RL, Hand PJ. The role of norepinephrine in adult rat somatosensory (SmI) cortical metabolism and plasticity. *Brain Res* 1988;443:261–271.
42. Jan LY, Jan YN. Peptidergic transmission in sympathetic ganglia of the frog. *J Physiol (Lond)* 1982;327:219–246.
43. North RA. Mechanisms of autonomic integration. In: Bloom FE, ed. *Handbook of physiology, Section I, Volume IV*. Bethesda, MD: American Physiological Society. 1986;115–153.
44. Buijs RM, Van Heerikhuize JJ. Vasopressin and oxytocin release in the brain—a synaptic event. *Brain Res* 1982;252:71–76.
45. Neumaier JF, Chavkin C. Release of endogenous opioid peptides displaces [3H]-diprenorphine binding in rat hippocampal slices. *Brain Res (in press)*.
46. Gerfen CR, Herkenham J, Thibault J. The neostriatal mosaic: II. Patch- and matrix-directed mesostriatal dopaminergic and non-dopaminergic systems. *J Neurosci* 1987;7:3915–3934.
47. Gerfen CR, Young WS III. Distribution of striatonigral and striatopallidal peptidergic neurons in both patch and matrix compartments: an *in situ* hybridization histochemistry and fluorescent retrograde tracing study. *Brain Res* 1988;460:161–167.
48. Evans RM. The steroid and thyroid hormone receptor superfamily. *Science* 1988;240:889–895.
49. Miller FD, Ozimek G, Milner RJ, Bloom FE. Regulation of neuronal oxytocin mRNA by ovarian steroids in the mature and developing hypothalamus. *Proc Natl Acad Sci USA* 1989;86:2468–2472.
50. Jung Testas I, Alliot F, Pessac B, Robel P, Beaulieu EE. Immunocytochemical localization of cytochrome P-450scc in cultured rat oligodendrocytes. *CR Acad Sci [III]*. 1989;308:165–170.
51. Hu ZY, Bourreau E, Jung Testas I, Robel P, Beaulieu EE. Neurosteroids: oligodendrocyte mitochondria convert cholesterol to pregnenolone. *Proc Natl Acad Sci USA* 1987;84:8215–8219.
52. Le Goascogne C, Robel P, Gouezou M, Sananes N, Beaulieu EE, Waterman M. Neurosteroids: cytochrome P-450scc in rat brain. *Science* 1987;237:1212–1215.
53. Vatier OC, Bloom FE. Effect of ethanol on the neurosteroids concentrations in the rat brain *(Abstr.) Soc Neurosci Abstr* 1988;14:83.13.
54. Majewska MD, Mineville JM, Vicini S. Neurosteroid pregnenolone sulfate antagonizes electrophysiological responses to GABA in neurons. *Neurosci Lett* 1988;90:279–284.

55. Turner DM, Ransom RW, Yang JS, Olsen RW. Steroid anesthetics and naturally occurring analogs modulate the gamma-aminobutyric acid receptor complex at a site distinct from barbiturates. *J Pharmacol Exp Ther* 1989;248:960–966.
56. Harrison NL, Vicini S, Barker JL. A steroid anesthetic prolongs inhibitory postsynaptic currents in cultured rat hippocampal neurons. *J Neurosci* 1987;7:604–609.
57. Barbin G, Katz DM, Chamak B, Glowinski J, Prochiantz A. Brain astrocytes express region-specific surface glycoproteins in culture. *GLIA* 1988;1:96–103.
58. Denis Donini S, Glowinski J, Prochiantz A. Glial heterogeneity may define the three-dimensional shape of mouse mesencephalic dopaminergic neurones. *Nature* 1984;307:641–643.

Volume Transmission in the Brain:
Novel Mechanisms for Neural Transmission,
edited by K. Fuxe and L.F. Agnati.
Raven Press, Ltd., New York © 1991.

3

Cell Adhesion Molecules in Neural Morphogenesis

Gerald M. Edelman and Kathryn L. Crossin

The Rockefeller University, New York, New York 10021

The fundamental processes of development leading to particular morphologies among species are under genetic control, and yet, particularly for neural structures in which the apparent specificity of patterns is so dominant, the gene products underlying pattern formation are not yet well understood. To examine this fundamental problem, it is necessary to relate both developmental genetic mechanochemical factors at several levels of organization from the gene itself and its gene products to cells, to tissues and organs, and ultimately back to the gene. In this brief survey, we will review recent evidence that cell adhesion, mediated by at least three different families of molecules, is a spatially and temporally dynamic process underlying the genetic regulation of patterns during the development of the nervous system.

Regulative development of a complex embryo from a single cell, the fertilized egg, involves the coordination of a number of primary developmental processes that are governed by complex interactions at the cell surface: cell division, differentiation, adhesion, movement, and death. These fundamental cellular processes emphasize the importance of the cellular driving forces in developmental events. Interactions among these processes lead to morphogenesis and histogenesis in a series of epigenetic sequences. For example, surface interactions between cells can lead to changes in gene expression and in cell shape, movement, and function. The events that occur subsequent to such cell interactions depend in part on the history of the cells (i.e., on the interactions they have had with other cells in the past). What is more, because these interactions depend on what cells surround any given cell, cells react differently at different places. A few years ago, topobiology was proposed as a term to describe the dynamic, interactive processes by which such "place-dependent" interactions at the cell surface lead to cell regulation (1,2). Because cell fate is critically dependent on cell interactions, the study of topobiology encompasses the study of three classes of morphoregulatory molecules: cell–cell adhesion molecules (CAMs), cell–substratum adhesion molecules (SAMs), and cell–junctional molecules (CJMs).

CELL–CELL ADHESION MOLECULES

Structure, Genetics, and Binding: N-CAM as an Example

A number of CAMs have now been identified that are related by virtue of the fact that all are homologous to immunoglobulin molecules (reviewed in refs. 3,4). Many of these are present in the nervous system.

The best studied of these CAMs to date is N-CAM, the neural CAM, first discovered as a molecule involved in the binding of neural cells to each other (5,6).

N-CAM is expressed as a number of major polypeptide chains that arise by alternative splicing of messenger ribonucleic acid (mRNA) from a single N-CAM gene (Fig. 1) (7–9). N-CAM isolated from chicken brain contains three major polypeptides that share a common extracellular sequence of 682 residues extending from their NH_2 termini to the cell membrane (7). The two larger forms, large cytoplasmic domain (ld) peptide and small cytoplasmic domain (sd) peptide, have identical transmembrane domains but the ld polypeptide contains a unique segment of 261 residues in its cytoplasmic region. This domain results from the splicing of a single exon (exon number 18, Fig. 1) in a tissue-specific fashion (8) (see below). The small surface domain (ssd) peptide terminates in a unique hydrophobic sequence and is attached to the membrane by a phosphatidylinositol-containing anchor (10,11). The extracellular

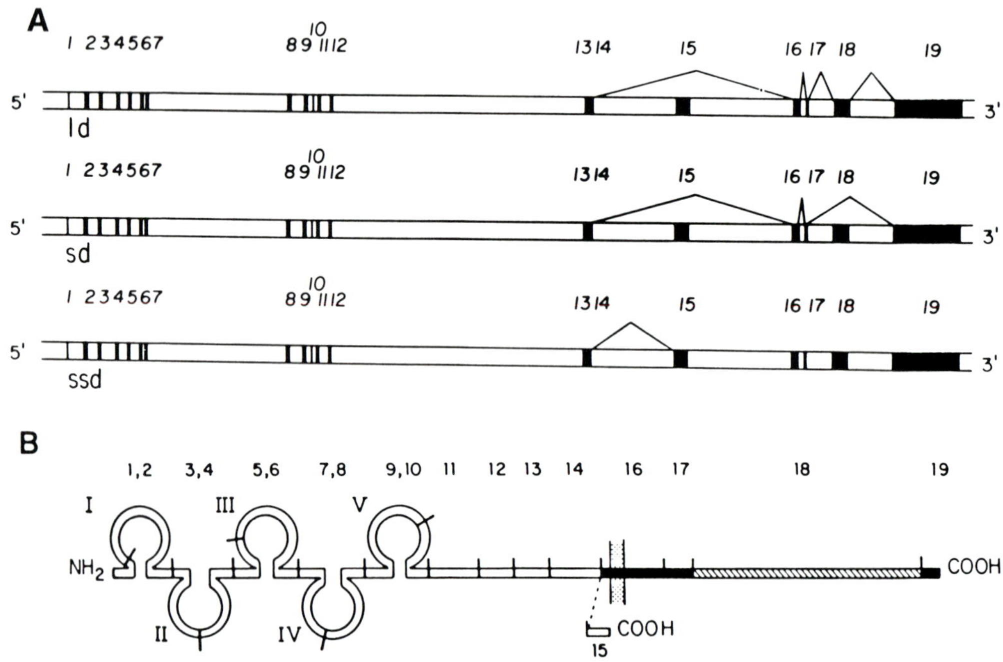

FIG. 1. A: Structure of the N-CAM gene in terms of introns and exons that code for N-CAM mRNAs from the amino terminus of the protein through the 3′ end. Segments of proteins specified by each exon are indicated on the model of N-CAM (**B**). Exons 16, 17, and 19 specify regions shared by ld and sd chains including 3′ untranslated regions of the mRNAs. Exons 15 and 18 specify regions unique to the ssd and ld chain, respectively. **B:** Model of N-CAM showing homologous loops (I–V) in the region common to all three polypeptides (*open bar*). Base of each loop corresponds to proposed intradomain disulfide bond. The cell membrane is indicated by *stippling between vertical bars,* and extracellular and intracellular regions are to the *left* and *right,* respectively. The membrane-spanning region and carboxyl-terminal segment common to ld and sd chains are indicated by the *solid bar,* whereas the cytoplasmic domain unique to the ld chain is indicated by the *hatched bar.* The unique segment of the ssd chain is indicated by the *open bar* and *dashed line* below. Numbers 1–19 correspond to the exons in the chicken N-CAM gene, and their relative boundaries in the protein are noted by transverse lines.

regions of each of the forms contain five domains, each linked by a single intrachain disulfide bond.

Each of the five domains is homologous to molecules in the immunoglobulin superfamily (7,12). In addition to these three major forms of N-CAM, several forms have been identified that contain additional inserts beyond the fifth homology region (e.g., those found in human skeletal muscle, rat brain, or chicken heart); several of these appear to be tissue-specific and also arise by alternative RNA splicing (13–15).

Attached to the fifth domain (Fig. 1) at three sites in chicken N-CAM (16) are oligosaccharides containing a carbohydrate that is unusual in vertebrates, α-2,8 polysialic acid (17,18). In the so-called embryonic (E) form of N-CAM, there is about 30 g of this carbohydrate per 100 g of polypeptide, and in the adult (A) forms, there is about one-third as much (17). The names refer to the statistical predominance of these forms at different ages: Although molecules having low amounts of the carbohydrate can be seen in early embryos and molecules with high amounts can be seen in certain locations in adult nerve, skin, and muscle, these forms are in the minority at these ages. As we discuss below, polysialic acid does not participate directly in N-CAM binding, but the presence of higher amounts of the carbohydrate does diminish binding rates (19). We would therefore expect to see the A forms in regions where stabilization of the dynamics of adhesion is required as, for example, in the adult brain; in regions of constant regeneration such as the adult olfactory bulb, we see the E form (20).

Physicochemical assays (19) that depend on measurements of rates of aggregation for N-CAM indicate that a twofold increase in surface density can result in a greater than 30-fold increase in binding rates (Table 1). Although α-2,8-polysialic acid is not directly involved in binding, changes in the amount of this negatively charged sugar, probably as a result of turnover of E forms and replacement by A forms (21), can result in three- to fourfold changes in binding rates (19) (Table 1). These data indicate that N-CAM binding is dependent on surface concentration and also on the amount of the negatively charged polysialic acid. This may indicate that at higher surface densities of CAMs, multivalent CAM structures are formed by *cis* interaction on the same cell.

TABLE 1. *Aggregation of reconstituted vesicles containing adult or embryonic forms of N-CAM*[a]

Vesicles (form of N-CAM)	N-CAM/lipid (μg/mg)	k_{agg} (units)[b]
E	17	3.5
A	17	12.2
E	14	1.5
E	17	3.5
E	19	6.4
E	28	54.0

[a]From ref. 19, with permission.
[b]One unit is the rate of aggregation (measured in nanoliters of newly formed vesicle aggregates per milliliter per minute) divided by the square of the concentration of vesicles (measured in milligrams of lipid milliliter).

Electron microscopy of rotary shadowed mixtures of the ld and sd polypeptides of N-CAM (22–24) has revealed rod-shaped molecules of uniform thickness (Fig. 2). The rods are bent at a single location and associate with one another to form rosettes with two to six branches. These rosettes were disrupted by detergent treatment. In addition the ssd form of the molecule that lacks a transmembrane and cytoplasmic domain appeared as a bent rod and never as rosettes. These results indicate that the formation of rosettes is dependent on the presence of transmembrane or cytoplasmic domains. This places the carboxyl terminal of N-CAM at the center of the rosettes and the amino-terminal binding regions as the ends of the bent rods. The images of N-CAM molecules bound to monoclonal antibodies strongly support this model: An antibody that recognizes an epitope in the amino-terminal 13 kDa of N-CAM and that

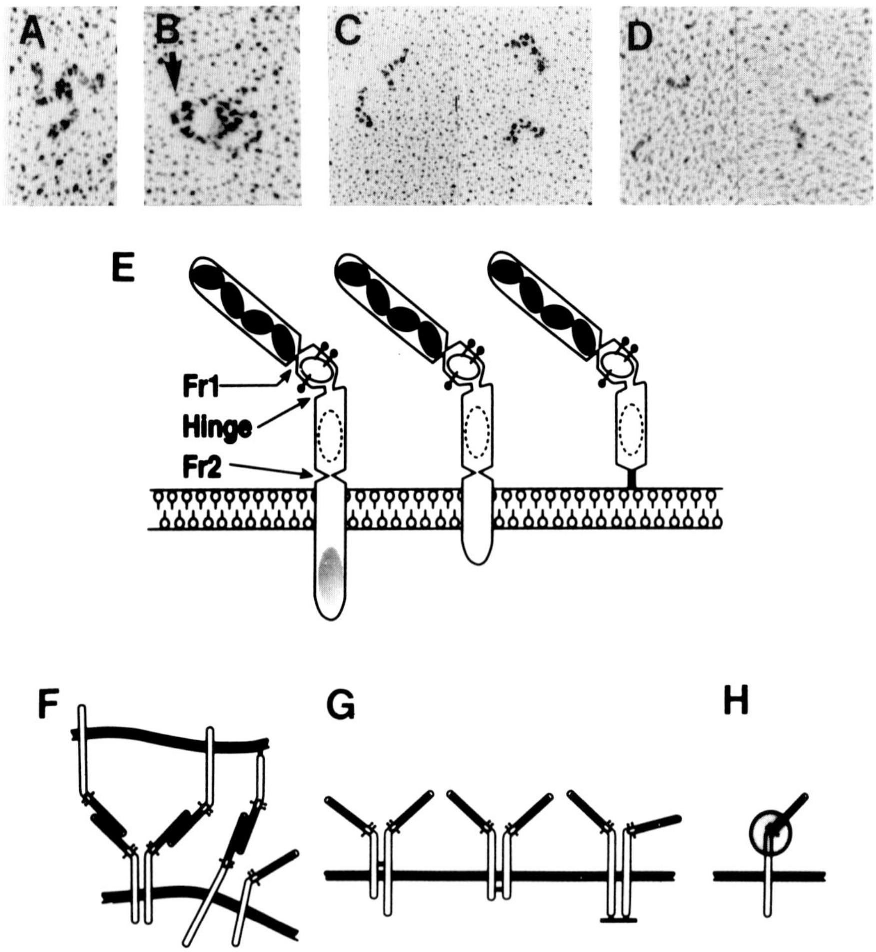

FIG. 2. Electron micrographs of CAMs. **A:** N-CAM ld/sd. **B:** N-CAM ld/sd rosettes and monoclonal antibody. **C:** N-CAM ssd. **D:** L-CAM Ft1. **E–H:** Model of N-CAM and N-CAM-mediated cell adhesion. All forms of N-CAM contain regions of sequence homology to immunoglobulin domains (*ovals*) and a homophilic binding domain (*dark color*). **E:** N-CAM on a cell membrane. The ld (to the *left*) and sd (in the *center*) forms are integral membrane proteins, whereas the ssd form (to the *right*) is attached to the membrane by a phosphatidylinositol linkage (■). ld contains a unique insert (*stippled*) in its cytoplasmic domain. All forms contain a flexible hinge 10 nm from the membrane, three attachment sites for polysialic acid (16) near the hinge, and regions of homology to fibronectin (*dashed ovals*). Sites of proteolytic cleavages that make the Fr1 and Fr2 fragments are indicated. **F:** Arrays of N-CAM molecules on one cell interact with similar arrays on apposing cells. The flexible hinge in N-CAM allows specific intermolecular contacts to be made even when cell surfaces are in motion or are not complementary in shape. **G:** N-CAM arrays may be formed by interactions between molecules on the same cell. Sites for such interaction could be extracellular (to the *left*), intracellular (in the *center*), or through cytoskeletal structures (to the *right*). **H:** Forms of the molecule may hinder intercellular adhesion by preventing formation of arrays and by restricting flexibility of the hinge region.

TABLE 2. *CAMs and related proteins*

N-CAM-like (Ig superfamily)	NS[a]	L-CAM-like	NS[a]	Others	NS[a]
N-CAM	+	L-CAM (uvomorulin, E-cadherin)	−	Cell CAM 120	−
Ng-CAM/LI	+			Neurofascin	+
Contactin/FII	+			Fasciclin I	+
MAG	+	N-cadherin (A-CAM)	+	Fasciclin III	+
Po	+				
CEA	+	P-cadherin	−		
(Amalgam)[b]	+				
(Fasciclin II)	+				
(Neuroglians I & II)	+				
I-CAM-1[c]	− (?)				
I-CAM-2[c]	− (?)				
CD-4[c]	−				

[a]NS column denotes presence (+) or absence (−) in the nervous system.
[b]Parentheses indicate that the molecule has not definitively been shown to be an adhesion molecule.
[c]Denotes molecule with known binding function but not necessarily functioning in cell-cell adhesion.

blocks N-CAM homophilic binding was shown to bind to the distal ends of the branches of rosettes. Moreover, this divalent antibody frequently cross-linked the ends of adjacent branches in the rosettes.

L-CAM, a calcium-dependent CAM, is completely different from N-CAM in size and sequence (25), but has a homologue in nervous tissue (N-cadherin) (see ref. 3) (Table 2). In electron microscopic studies, L-CAM was also seen as a rod-shaped molecule with a flexible hinge (24). The presence of such common structural traits, despite the chemical and functional differences between these two CAMs, suggests that the rod structure with a flexible hinge has an important role in specific cell–cell adhesion. The combined antibody-labeling, structural, and detergent data have led to the formulation of a model for N-CAM homophilic binding (Fig. 2E–H) that takes this into account.

Developmental Expression of Primary and Secondary CAMs

N-CAM and L-CAM appear early in development on derivatives of all three germ layers (Table 3) (3,26–29). For this reason, they are called *primary CAMs*. In contrast, Ng-CAM, a secondary CAM belonging, like N-CAM, to the Ig superfamily, appears only on postmitotic cells in derivatives of the neuroectoderm (30,31).

A systematic examination of the spatiotemporal appearance of each CAM during development (Table 3; Fig. 3) shows a characteristic expression sequence that generally becomes more restricted during development (Table 3). In the chick blastoderm, the two primary CAMs appear together on all cells; at gastrulation, however, ingressing cells stain much less for N-CAM and L-CAM. Cells in the chordamesoderm then re-express N-CAM at neural induction, and endodermal cells re-express L-CAM. During neural induction, a remarkable transition occurs in the chick epiblast: After completion of induction, the presumptive neural plate shows increased amounts of N-CAM and loses L-CAM; the surrounding somatic ectoderm loses N-CAM and shows mainly L-CAM. Thereafter, at all sites of secondary embryonic induction, N-CAM or L-CAM, or both, undergo a series of prevalence modulations that follow two main rules for these two CAMs (3,29): All epithelial-mesenchymal transformations show loss of N-CAM followed by its re-expres-

TABLE 3. *Distribution of L-CAM and N-CAM in three epochs*

N-CAM			L-CAM		
0–3-Day embryo	5–13-Day embryo	Adult	0–3 Day embryo	5–13 Day embryo	Adult
Ectoderm Upper layer Epiblast Neural plate Placodes	Nervous system	Nervous system	Ectoderm Upper layer Epiblast Presumptive epidermis Placodes	Epidermis Extraembryonic ectoderm	Skin: stratum germinativum
Mesoderm Notochord Somites Dermomyotome Somato- and splanchnopleural mesoderm Heart Mesonephric primordium	Striated muscle Adrenal cortex Gonad cortex Some mesonephric and metanephric epithelia Somato- and splanchnopleural elements Heart	[a] Testis Cardiac muscle	Mesoderm Wolffian duct	Wolffian duct Ureter Most meso- and metanephric epithelia	Epithelium of urogenital tract
			Endoderm Endophyll Hypoblast Gut primordium and buddings	Epithelium of: Digestive tract Respiratory tract Lymphoid organs Secretory glands Extraembryonic membrane	Epithelium of: Digestive tract Respiratory tract Lymphoid organs Secretory glands

[a]Distribution of N-CAM in adult tissues derived from mesoderm has not yet been rigorously established.

N-CAM (1°)	AGE: DAYS	AGE: STAGE	Ng-CAM (2°)
Blastoderm	0	1	
<u>Neural Derivatives</u> Neural plate; Neural tube; Neural crest;	2	13	
Disappears from migrating neural crest cells; Reappears in aggregating neural crest cells	4	23	Ventral neural tube; Ciliary ganglia; Dorsal root ganglia; Spinal cord: white column; Retina: ganglion cell fibers; Tectum: fiber layer under pia;
	6	29	Sympathetic ganglia
			Spinal cord: gray matter; Tectum: stratum opticum and album centrale
	8	34	
	10	36	
↓			Cerebellum: fiber tracts
	12	38	Cerebellum: molecular layer
	14	40	
	16	42	
			Cerebellum: migrating granule cells
	18	44	
E ↓ A	Birth	46	
Remains on all subsequent nervous tissues.	Adult		Diminishes in spinal cord white matter, cerebellar fiber tracts and tectal stratum opticum and album centrale.

FIG. 3. Expression sequence of two neuronal CAMs in the developing chick nervous system. A primary cell adhesion molecule (N-CAM) appears early (blastoderm stage) in both neural and non-neural derivatives (see Table 3). After birth it remains on all subsequent nervous tissues. Ng-CAM appears later during embryogenesis and is first observed on the developing spinal cord (30,31). The sequential appearance of Ng-CAM in CNS and PNS region is summarized. In the adult, Ng-CAM is limited to unmyelinated CNS regions (tracts or laminae) and PNS.

sion, and epithelial cells show other transitions in which one or the other primary CAM is lost.

Later in development, N-CAM and the secondary CAM Ng-CAM play important roles in the development and maintenance of both the central and peripheral nervous systems (Fig. 3) (31). After formation of the neural tube, neural crest cells lose N-CAM and migrate on pathways containing SAMs such as fibronectin and cytotactin, and when they reach particular sites, they differentiate and re-express N-CAM while forming ganglia (30). At about 3.5 days, as neurite extension occurs in the CNS, the secondary Ng-CAM is expressed particularly on neurites that are fasciculating (30). This molecule mediates neuron–neuron interactions essential to fasciculation as well as neuron–glia adhesion in the chick (32). It appears in a complex but regular schedule at different sites of neural development (Fig. 3) (31). Ng-CAM is the first CAM for which a putative signal is known: In culture, its synthesis in PC12 cells can be enhanced by the nerve growth factor (NGF) (33).

Obviously, a number of different CAM specificities exist in related CAMs belonging to the Ig superfamily. But additional mechanisms exist to modulate activity for a given CAM such as N-CAM. As discussed above, alternative mRNA splicing is an important mechanism for generating diversity in N-CAM polypeptides. The structurally different major polypeptides of N-CAM arise in development at different times and places, and in certain cases, the splicing events leading to the various forms can be tissue-specific. Of particular importance in the development of the nervous system is the expression of the ld polypeptide.

The ld form of N-CAM has an additional 261 amino acids encoded by a single exon (8). The splicing events that generate this form occur only in tissues of the nervous system (8,34). Moreover, the unique segment of the ld form contains phosphorylation sites (35) in addition to those shared with the cytoplasmic domain of the sd and also provides a region for possible interactions with the cytoskeleton (36,37).

In a recent study, the distribution of the mRNA for the ld form of the N-CAM was compared with mRNA for the common extracellular N-CAM domains using *in situ* hybridization (38). The mRNA for the ld form of N-CAM was present only in nervous tissues, confirming previous immunohistological and Northern blot hybridization studies. Within the nervous system, cell types could be distinguished in their capacity to express the ld mRNA. In the optic tectum (Fig. 4) and spinal cord, the ependymal cells, which line the central lumen of the entire CNS, expressed N-CAM mRNA but failed to express the ld mRNA. In the cerebellum, both Purkinje cells and internal granule cells expressed the ld mRNA.

In layered structures, such as the optic tectum (Fig. 4) and cerebellum, N-CAM mRNA was localized in cell bodies that were surrounded by fibers containing large amounts of N-CAM protein. In the tectum, fiber-rich strata such as SFP, SAC, SO, and lamina h of SGFS (39) (see Fig. 4 for abbreviations and data) were strongly positive for protein but were negative by *in situ* hybridization (Fig. 4b). In contrast, cell body–rich layers showed strong hybridization signals with both probes and lower levels of immunofluorescent staining for protein. These findings are in accord with the localization of mRNA in the soma and with the presence of high levels of protein on the fibers. In contrast, *in situ* hybridization analysis for myelin basic protein (40) and MAP2 mRNAs (41) has shown that the mRNAs encoding these proteins actually are transported to their final destination in neural processes before translation.

The observations that the N-CAM mRNA was localized in cell bodies while protein was also found at high levels in fibers suggest that the compartmentalization of the N-CAM protein occurs posttranslationally at the level of protein transport, confirming previous studies of the axonal transport of N-CAM (42). The observations of polarity modulation of N-CAM (31,43) and of the interaction of N-CAM with cytoskeletal molecules (36,37) are also consistent with the idea that the protein is synthesized in the cell body and then transported.

Modulatory Interactions and Specificity in CAM-Mediated Sorting Out of Cells

The ability to express molecules in cells that do not normally express them has provided a powerful tool to examine CAM

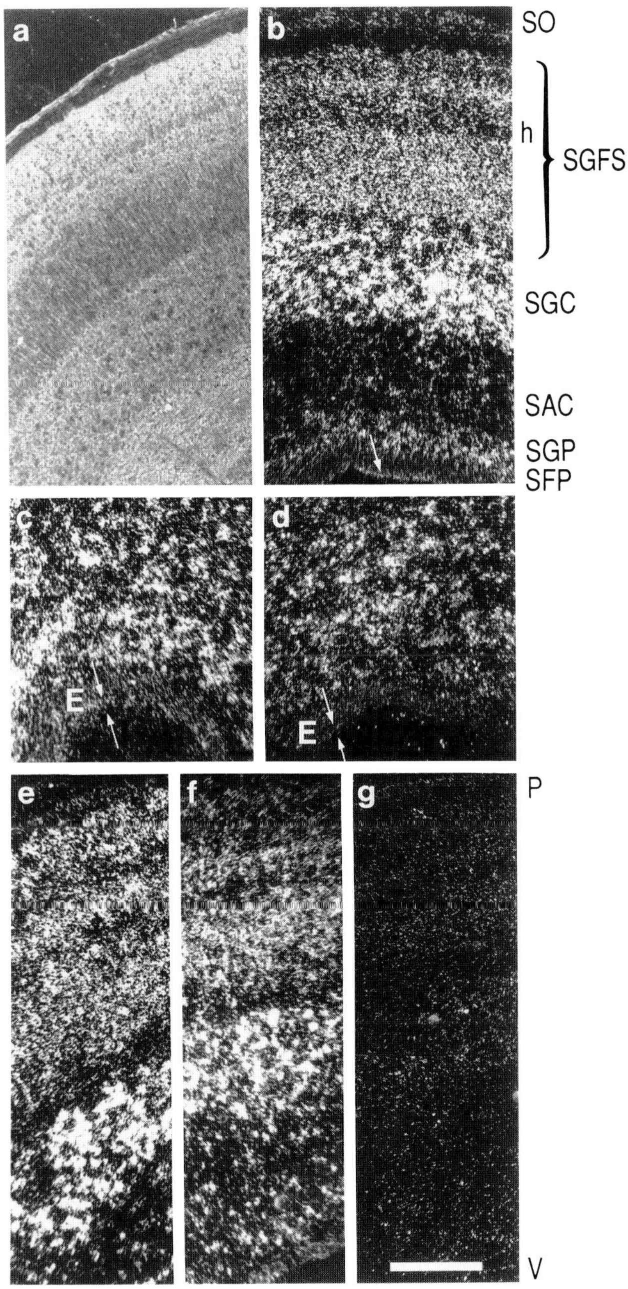

FIG. 4. Localization of total N-CAM and ld-specific mRNAs in the optic tectum. Parasagittal sections from E11 (**a** and **b**), E15 (**c** and **d**), and E17 (**e** and **g**) were stained with polyclonal anti-N-CAM, hybridized with the N-CAM antisense probe (b, c, and e), or with the ld antisense probe (d and f). *Arrows* in panels b, c, and d highlight the ependymal (E) layer. A parallel section hybridized with a total N-CAM sense probe shown in g. Bar: 180 μm for (a, b), 240 μm for (c, d), and 200 μm for (e, f, g). SO, stratum opticum; SGFS, stratum griseum et fibrosum superficiale; SGC, stratum griseum centrale; SAC, stratum album centrale; SGP, stratum griseum periventriculare; SFP, stratum fibrosum periventriculare; P, pial surface; V, ventricle. (From ref. 38, with permission.)

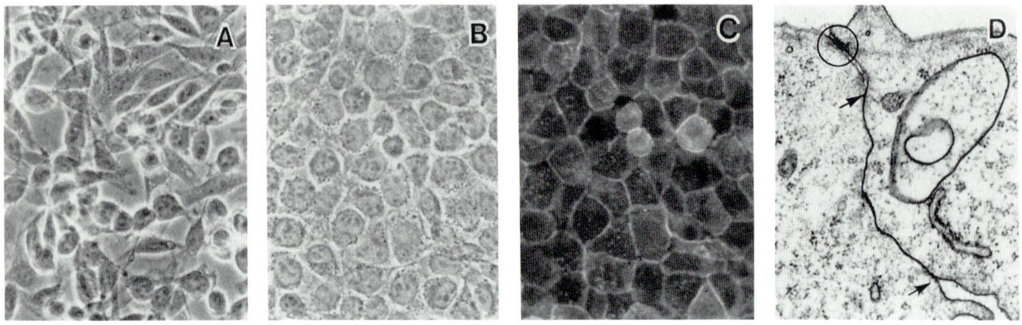

FIG. 5. Formation of epithelioid sheets and increase in cell junctions in cells transfected with L-CAM cDNA. **A:** Phase-contrast micrographs of untransfected S180 cells. **B, C:** Matched phase-contrast and fluorescence micrographs of S180 cells, transfected with chicken L-CAM cDNA (S180L) and stained with anti-L-CAM (C). At high density, S180 cells formed an epithelioid sheet (B), and L-CAM staining was brighter at boundaries of cells (C). **D:** These cell sheets expressed numerous gap junctions (*arrows*) observed by electron microscopy.

binding and its effects on cell morphology. A number of cell types have now been transfected with complementary deoxyribonucleic acids (cDNAs) for CAMs. These studies have confirmed that binding of N-CAM, L-CAM, and N-cadherin are homophilic. More recently, a system has been constructed that has established a link between CAM expression, cell morphology, and the expression of cell junctions, thus linking the functions of two families of morphoregulatory molecules (44).

The expression of L-CAM (Fig. 5) or N-cadherin in S180 mouse sarcoma cells (which do not normally produce either of these CAMs) after transfection with cDNA induced a phenotypic change in these cells from a fibroblastic to an epithelioid morphology. This alteration was accompanied by marked increases in the expression of adherens and gap junctions (Fig. 5D) (44,45). Both cell types aggregated specifically through the expressed CAM. Aggregation was inhibited by antibodies to the appropriate CAM and by cytochalasin D treatment to disrupt microfilaments. These studies provide strong evidence for the role of these two different CAMs in the reorganization of epithelia. S180L and S180cadN cells did not bind to each other, and when mixed, they separated into distinct homogeneous collectives linked internally by gap

TABLE 4. *Ability of cells expressing different levels of L-CAM and N-cadherin (cadN) to sort out from each other*[a,b]

	Cell line 1						
Cell line 2	L_L	L_H	$cadN_L$	$cadN_H$	L_LcadN_H	L_HcadN_L	L_HcadN_H
S180	+	+	+	+	+	+	+
L_L		+	+	+	+	+	+
L_H			+	+	ND	ND	−
$cadN_L$				+	+	+	+
$cadN_H$					ND	ND	−
L_LcadN_H						ND	ND
L_HcadN_L							ND

[a]From ref. 46, with permission.
[b]ND, not determined; +, cell line 1 and cell line 2 had segregated when observed 16 hr after the start of a sorting-out assay; −, no segregation.

junctions. In cells that were doubly transfected with cDNAs for L-CAM and N-cadherin, both molecules were active and accumulated at sites of cell contact.

These cell lines were used to examine the ability of CAMs to allow cell sorting into distinct cell collectives (Table 4) (46). CAM-transfected cells readily sorted from their untransfected counterparts, and L-CAM transfectants sorted from N-cadherin transfectants. Both of these results indicate that different CAM specificities are sufficient for cell sorting. More interestingly, however, was the finding that cells with different levels of the same CAM were able to sort. This suggests that qualitative and quantitative differences in the expression *in vivo* of a relatively small number of CAMs can lead to a large variety of patterns among cell collectives during tissue formation.

SUBSTRATE ADHESION MOLECULES IN NEURAL MORPHOGENESIS

The evidence suggesting that CAMs play significant roles in embryogenesis and neural histogenesis by means of modulation and regulation prompts the idea that complementary functions may be exercised by SAMs. A pair of interactive extracellular matrix molecules, cytotactin (47) [also known as myotendinous antigen (48), tenascin (49), $J1_{220/200}$ (50), or hexabrachion (51)] (reviewed in ref. 52), and its ligand, a chondroitin sulfate proteoglycan called cytotactin binding (CTB) proteoglycan (53) appears to have such a role. We concentrate here on this role in the nervous system.

Cytotactin is an extracellular matrix protein composed of structurally related polypeptides of 220, 200, and 190 kDa (47). In the nervous system the molecule is highly soluble and is made by glia but not neurons (47,53–55). It is made by various other cells including smooth-muscle cells and immature chondrocytes. The related cytotactin polypeptides arise from differential splicing of a single gene (56,57). The complete sequence analysis of cytotactin cDNAs shows extensive similarity to genes specifying three other proteins (56,57) (Fig. 6). The 5′ end encodes 13 EGF-like repeats followed by 11 fibronectin type III repeats, three of which are included by differential splicing; the 3′ end encodes a region of similarity to fibrinogen and contains a calcium binding site (Fig. 6A). These domains of the protein can be mapped onto its structure, as seen in rotary shadowed electron microscopic images. The molecule appears as a complex structure, with six arms emanating from a central core (Fig. 6B). The central core is a highly disulfide bonded region at the amino terminus of cytotactin (Fig. 6C). The arms are composed of EGF-like repeats and fibronectin type III repeats with the region of fibrinogen homology forming the terminal knob seen at the end of each arm.

Cytotactin binds CTB proteoglycan (Table 5), which is synthesized by central nervous system neurons and, in various forms, also by nonneural cells elsewhere in the body. Cytotactin also binds fibronectin (Table 5), and it has been shown to bind to fibroblasts (54). Cytotactin is involved in glia–neuron adhesion *in vitro* by a mechanism independent from Ng-CAM (47). Interaction with cytotactin causes neuronal and fibroblastic cells in culture to round up and also tends to inhibit cellular migration. In the presence of fibronectin, which generally supports cell flattening and migration, the effects of cytotactin were decreased (58,59); thus the two interactive molecules may have conjugate effects on cell behavior.

Cytotactin is notable because it appears in a series of cephalocaudal waves of expression during development (60). These waves may reflect temporal growth gradients related to macromolecular synthesis that are an important part of morphogenesis. A particularly striking example of such a wave is seen in somites. In addition, a

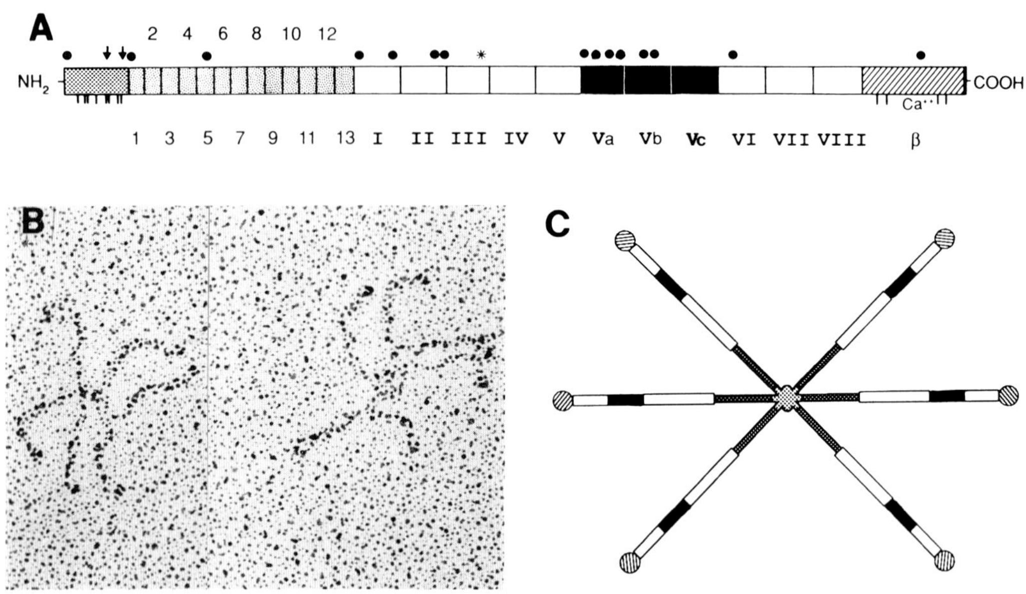

FIG. 6. Functional map of cytotactin. **A:** Schematic drawing of cytotactin polypeptides. EGFL repeats (*lightly stippled boxes*) are numbered 1–13 and type III repeats (*open boxes*) are designated by roman numerals with those in the additional insert (*black boxes*) designated Va, Vb, and Vc. The region similar to the β subunit of fibrinogen (diagonal lines and β symbol) includes a putative calcium-binding domain (Ca^{2+}) *Darkly stippled boxes* denote regions that have no extensive homology to any known protein, and they include amino-terminal 142 amino acids and the carboxyl-terminal 13 amino acids. Cysteine residues that are not in EGFL repeats (|), potential asparagine-glycosylation sites (•), potential glycosaminoglycan addition sites (↓), and the Arg-Gly-Asp sequence (*) are indicated. **B:** Electron micrograph (courtesy of Joseph W. Becker, The Rockefeller University) of chicken brain cytotactin showing two hexabrachions. **C:** Proposed model of a cytotactin hexamer. The amino-terminal regions of each polypeptide are disulfide-linked and represented as a single structure forming the core of the hexabrachion. EGFL repeats are cross-hatched; type III repeats present in both polypeptides are *white,* whereas those in the insert are *black;* the fibrinogen-like nodular region is marked with *slanted lines.* EGFL repeats are assumed to make up the thin parts of the arms and type III repeats the thicker portions. (From ref. 57, with permission.)

TABLE 5. *Intermolecular interactions among SAMs*[a,b]

Protein(s) on beads	Soluble proteins	Bead aggregates
Cytotactin only		1,660
Cytotactin binding (CTB) proteoglycan only		420
Cytotactin, CTB proteoglycan		49,500
Cytotactin, CTB proteoglycan	Anticytotactin	2,480
Cytotactin, CTB proteoglycan	Cytotactin	11,800
Cytotactin, CTB proteoglycan	CTB proteoglycan	14,700
Cytotactin, fibronectin		17,300
Cytotactin, laminin		650
CTB proteoglycan, fibronectin		320
CTB proteoglycan, laminin		610

[a]Adapted from ref. 54, with permission.

[b]In all cases, beads were coated with a single protein. Aliquots of the indicated bead preparations were mixed in the presence of bovine serum albumin and the indicated soluble proteins. A particle counter was used to quantitate the aggregation of beads during a 3-hr incubation.

correlation between cytotactin and CTB proteoglycan expression occurs in each developing somite (59).

Somites are the segmental blocks of mesoderm in vertebrate animals. They arise in development in craniocaudal sequence from the so-called segmental plate of mesoderm. After such segmentation, which appears to be correlated with N-CAM and N-cadherin activity, the SAM, cytotactin, appears adjacent to the basement membrane of the posterior portion of each somite in a cephalocaudal sequence (59). Somewhat later, a portion of each somite becomes mesenchymal, forming the sclerotome through which neural crest cells migrate to give rise to dorsal root ganglia; migration of neural crest cells occurs only through the anterior part of the sclerotome. Neural crest cells that have down-regulated N-CAM and migrated on fibronectin paths enter this region of each somite; they will later re-express N-CAM and form dorsal root ganglia.

At this time or just before, both fibronectin and CTB proteoglycan are present throughout the sclerotome. Cytotactin appears, however, in just the anterior portion (Fig. 4), and after this remarkable segmented appearance, CTB proteoglycan decreases in the anterior half and is seen only in the posterior half; fibronectin remains distributed throughout. This pattern arises even in the absence of neural crest cells. When a portion of the neural tube is removed very early, before the emigration of the crest cells, the expression of cytotactin and CTB proteoglycan is still restricted to the anterior and posterior halves of each somite. The concurrent finding that cytotactin alters neural crest cell shape and movement is consistent with the hypothesis that in the anterior region of each somite, cytotactin causes cell surface modulation of neural crest cells and that this cytotactin together with its ligand, fibronectin, differentially changes the pattern of motion of these cells.

Clearly, this example suggests a complex regulatory sequence of patterned expression of CAMs and SAMs, showing definite correlation with cell function and distribution and the primary processes of development. The presence of a network of interactive SAMs (e.g., cytotactin, fibronectin, CTB proteoglycan) in different distributions is likely to be significant in the alteration of morphogenetic patterns. Further evidence for the importance of such molecules in CNS development will be discussed below with regard to the function of CAMs and SAMs in cerebellar and cerebral cortical development.

THE MORPHOREGULATOR HYPOTHESIS: MECHANOCHEMISTRY LINKED TO DEVELOPMENTAL GENETICS

We are now in a position to consider such topobiological phenomena in terms of a theory relating CAM and SAM expression to the development of form in the embryo and the developing brain. The evidence on the structure, function, and sequences of CAM expression and that derived from perturbation experiments suggest that these molecules play a major regulatory role in morphogenesis. This conclusion provides a basis for the morphoregulator hypothesis (2,61) that states that by means of cell-surface modulation, CAMs and SAMs are key regulators of morphogenetic motion, epithelial integrity, and mesenchymal condensation. The evidence suggests that *early* control of CAM expression by regulatory genes is separate from control of expression of the molecules concerned with other cytodifferentiation events such as those governing intracellular economy (62) or regulation of cell shape. According to the morphoregulator hypothesis, one may usefully distinguish "morphoregulatory" (MR) genes regulating CAM and SAM expression from "historegulatory" (HR) genes regulating specific cytodifferentiation events.

The morphoregulator hypothesis assumes that the cell is the unit of control, and the collective (i.e., a group of neighboring cells linked by MR molecules) is the unit of signaling. Collectives are determined by cytoplasmic domains of CAMs linking them, possibly acting through different second messengers (and also by gap junctions). To coordinate various cellular processes involved in pattern formation and in tissue differentiation, three types of regulatory genes are invoked. MR genes control structural genes for MR molecules. These genes are not necessary for life of a cell but are necessary for pattern. HR genes control structural genes involving all other cytodifferentiation events. Such structural genes specify general "housekeeping" functions common to most cells and are also necessary for the emergence of histodifferentiated cells. They are required for the life and specific characteristics of any cell and constitute by far the largest of the three classes. Selector (S) genes, by affecting certain HR genes, restrict expression of sequences of cytodifferentiation at certain places and developmental times.

Given these assumptions, the morphoregulator hypothesis states that the context of epigenetic responses of cells in a collective to inductive signals is set by CAMs forming borders and response characteristics in a

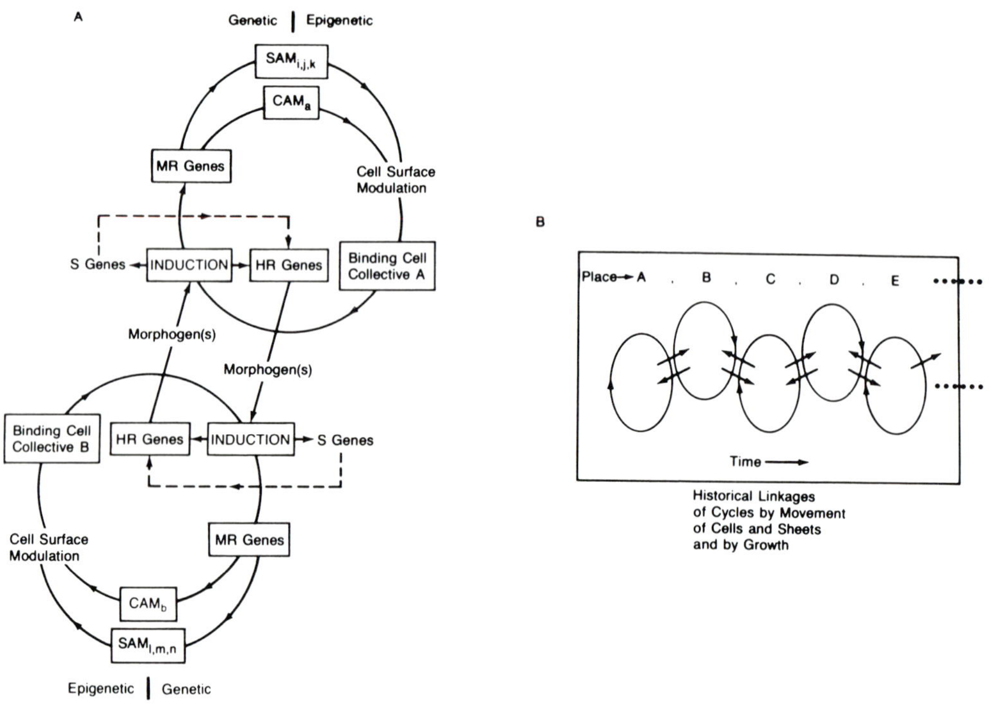

FIG. 7. The morphoregulator hypothesis. The hypothesis was proposed to explain how CAMs and SAMs create cell collectives and interact with them to determine the shape of an animal's body. **A:** In CAM cycles collectives of cells linked by CAMs make morphogens that induce cells linked by different CAMs to alter the activity of their morphoregulatory (MR) genes (i.e., genes encoding CAMs and SAMs). Morphogens also affect historegulatory (HR) genes (i.e., genes encoding tissue-specific proteins). In some cells HR genes are regulated in turn by selector (S) genes. As cell collectives interact they alter their environment; new CAM cycles (**B**) come into play, signaling to each other (*double arrows*). (From refs. 1 and 2, with permission.)

cell collective and by a network of SAMs acting by means of global modulation, resulting in a *choice* of primary processes (i.e., division, movement, death) in a given signaling situation (Fig. 7, left). The essential components linking epigenetic mechanochemical events to gene expression are the MR genes. Their action is under independent control from that of HR and S genes. The link among the genes is via inductive signal paths arising via adhesion-dominated collectives, giving rise to signals between such collectives. Such signals proceed back from the collectives to the MR, HR, and S genes to change expression of MR molecules and cytodifferentiation products. The place-dependent historical interactions of MR and HR genes via epigenetic paths provide the basis of morphological pattern by controlling temporal sequences of mitosis, movement, death, and further signaling.

It is the dynamics of expression of MR molecules interacting with morphogenetic signals over many levels of organization (Fig. 7, left) and recursively affecting cellular driving force processes that ultimately lead to form. By such means, structure at higher levels can be linked with local topobiologically conditioned signals, ensuring the sequential action of the next set of genes, whether MR or HR. Although development is irreversible, the same kinds of CAMs may be used in different contexts for regeneration and metamorphosis. Reiteration of such cycles over time during development (Fig. 7, right) would lead to rich patterns in different developmental contexts using a relatively small number of MR molecules.

The differential expression of morphoregulatory molecules in various contexts offers a potential solution to the problem of mechanochemical control of pattern during regulative development, ranging over the various levels from gene through organ and back to gene. These ideas are particularly well-illustrated in the development of the central and peripheral nervous systems and in the regeneration of the peripheral nervous system, two subjects to which we now turn.

COREGULATION AND SIGNALING LOOPS AMONG CAMS AND SAMS IN NEURAL MORPHOGENESIS AND REGENERATION

After establishment of the peripheral nervous system by neural crest cells, CAMs and SAMs play key roles in later pattern formation. Because of the readily observed regenerative effects and also the occurrence of defined interaction of cell types (i.e., Schwann cells, neurons, muscles), structures in the periphery provide an excellent opportunity to test the notion that interactions of cells in different tissues regulate CAM expression. For example, N-CAM, Ng-CAM, and cytotactin are highly concentrated at nodes of Ranvier of the adult chicken and mouse (63) (Fig. 3A–D). In contrast, unmyelinated axonal fibers were uniformly stained at low levels by specific antibodies to both CAMs but not by antibodies to cytotactin (Fig. 8). Moreover, a developmental analysis has suggested that the interaction between neurons and Schwann cells (which displayed N-CAM and Ng-CAM at their surfaces) may play a role in establishing the one-dimensional periodic pattern of nodes. At embryonic day 14, before myelination had occurred, small-caliber fibers of chick embryos showed periodic coincident accumulations of the two CAMs but not of cytotactin, with only faint labeling in the axonal membrane between accumulations. Cytotactin was found on Schwann cells and in connective tissue. By embryonic day 18, nodal accumulations of CAMs were first observed in a few medium- and large-caliber fibers. These findings are consistent with the hypothesis that surface modulation of neuronal CAMs, mediated by signals shared between neurons and glia, may be necessary for establishing and maintaining the nodes of Ranvier. Sig-

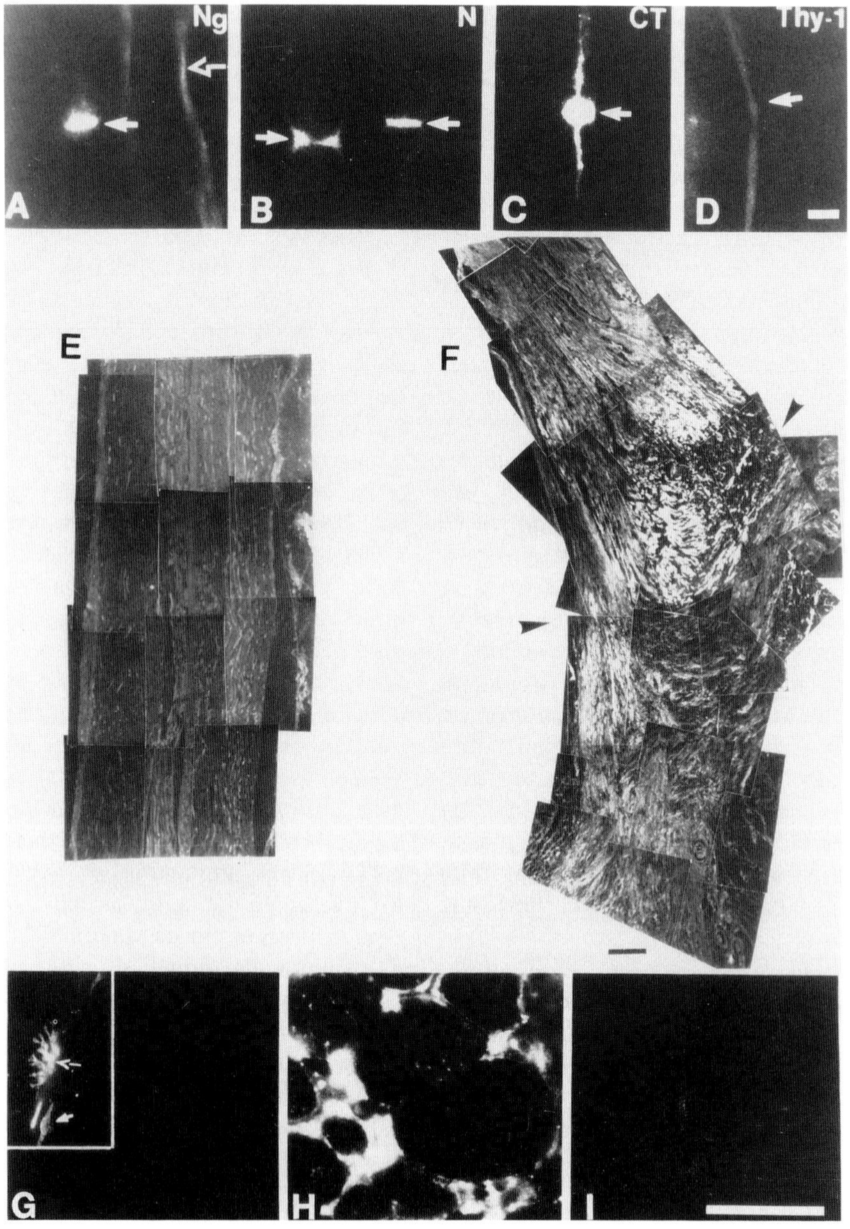

FIG. 8. A–D: Molecular staining at the node of Ranvier in whole-mount preparations of myelinated fibers of adult mice. In mouse sciatic nerve, the distribution of Ng-CAM (**A**), N-CAM (**B**), and the SAM cytotactin (**C**) overlapped with accumulations of sodium channels at nodes of Ranvier (*filled arrows*). Nonmyelinated fiber bundles (*broken arrow* in A) were also labeled with Ng-CAM antibodies. Cytotactin staining extended in a characteristic spindle-shaped pattern at what appeared to be the axon–Schwann cell interface in internodal regions. Staining intensity decreased on either side of the nodes with no staining 20 to 30 μm from the nodes. **D:** Anti-Thy-1 antibodies stained the entire axon in chicken sciatic nerves and did not accumulate exclusively at identified nodes of Ranvier (*filled arrow*). Bar = 10 μm. **E–I:** N-CAM localization in normal and regenerating neuromuscular system. Photomontages of longitudinal sections of normal sciatic nerves (**E**) and cut nerves 10 days after surgery (**F**), labeled with anti-N-CAM IgG. The *region between the black arrows* is the site of

nals of this kind may also be necessary for maintaining particular morphologies.

In support of this idea, several studies have shown that disrupted morphology or altered morphogenesis can actually lead to changes in CAM and SAM modulation patterns. For example, perturbation *in vivo* of normal cell–cell interactions during degeneration and regeneration has been shown to result in alteration of CAM expression and distribution (64–66) (Fig. 3E–G). N-CAM is present at the neuromuscular junction of striated muscles but is absent from the rest of the surface of the myofibril (64) (Fig. 8G). After cutting the sciatic nerve, the molecule appears diffusely at the cell surface and in the cytoplasm but returns to normal after regeneration (64–66) (Fig. 8H,I). These experiments indicate that early events related to regeneration can be accompanied by altered CAM modulation. More recent experiments (66,67) show that both crushing and cutting a nerve have widespread effects on CAM and SAM expression, ranging from altered anti-CAM staining in motor neurons of the spinal cord on the affected side to modulatory changes in N-CAM, Ng-CAM, and cytotactin within Schwann cells that are local to the lesion (Fig. 8E,F).

An excellent example of these principles and of the correlated activities of CAMs and SAMs is seen in cellular migration of external granule cells in the developing cerebellum to form layered structures. Recent studies (43,68) indicated that Fab′ fragments of antibodies to N-CAM have only slight effects on the migration of external granule cells in cerebellar slices *in vitro* (Fig. 9). In contrast, fragments of anti-Ng-CAM arrested most cells in the external granular layer, whereas fragments of anti-cytotactin arrested most cells in the molecular layer. Analyses of the time course combined with sequential addition of different antibodies in different orders showed that anti-Ng-CAM had a major effect in the early period of culture and a lesser effect in the second part of the culture period. Anticytotactin had essentially no effect at the earlier time but had major effects at a later period, suggesting that these molecules affected two temporally distinct processes of migration previously demonstrated in anatomical studies. The evidence suggests that Ng-CAM may play a role in both early and late events. Correlation of these effects with the spatiotemporal expression of these different CAM and SAM molecules suggested a complex complementary scheme in which temporal expression, different binding roles, and different functional effects on interactions between neurons and Bergmann glia all are required for pattern formation in the cerebellar cortex.

Recent studies have extended these studies of neural interaction and MR molecule expression to the cerebral cortex (55). The distributions of cytotactin and CTB proteoglycan were examined in the developing somatosensory cortex, in particular the so-called barrel field (Fig. 10). This is a particularly intriguing system because the cortical pattern represents a strict somatotopic mapping that results from appropriate input from the peripheral vibrissal field, giving rise to a group of axons primarily representing a single vibrissa. This results in a

the lesion (a cut). In experimental nerves, the intensity of N-CAM staining increased compared with control levels. Localization of N-CAM in normal and denervated muscle. **G:** Surfaces of muscle fibers were faintly labeled in normal chicken muscles. **Inset:** a whole-mount preparation of normal adult teased muscle fibers. The muscle surface was faintly labeled with anti-N-CAM IgG; a motor end-plate (*open arrow*) and several mononucleated cells that are probably satellite cells (*filled arrow*) were intensely labeled (**H**). 10 days after the sciatic nerve was cut, a dramatic increase occurred in the intensity of N-CAM; a normal pattern was restored after 150 days (**I**). Bar = 50 μm.

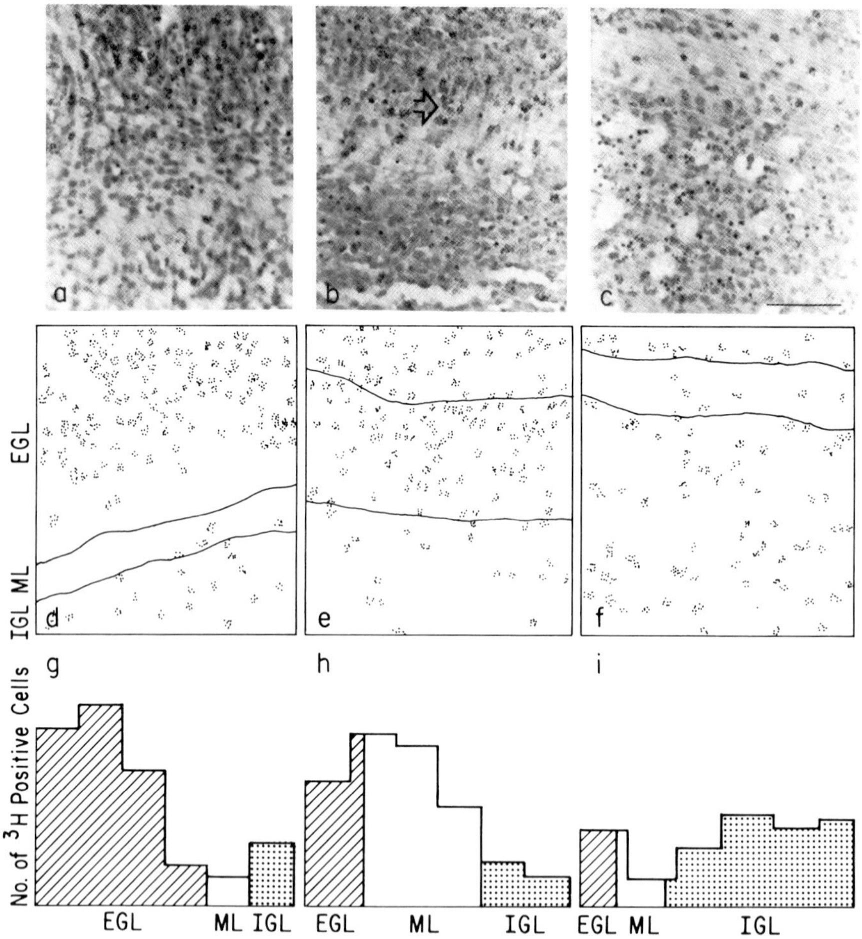

FIG. 9. Effects of anti-Ng-CAM and anti-cytotactin on the migration of tritiated cerebellar granule cells. **a–c:** Autoradiograms of paraffin sections of cerebellar explants that were pulse-labeled with tritiated thymidine for 1 hr and cultured for 3 days. **d–f:** Tracings of *a–c* to highlight the silver grains and the relative thickness of the cortical layers. **g–h:** Histograms of the number of tritiated cells versus distance along layers (*a–d, g*) anti-Ng-CAM; (*b, e, h*) anti-cytotactin; (*c, f, i*) nonimmune control. *Arrow* in *b* points to granule cells in the molecular layer (ML), which are piled up along the Bergmann glia. EGL and IGL are external and internal granule cell layer, respectively. Bar: 50 μm. (From ref. 68, with permission.)

cortical pattern of cell body–rich areas, the barrel walls, surrounding a group of thalamocortical axons and cortical neurons, one group for each vibrissa, which form the barrel hollows. Cortical neurons located in the barrel walls elaborate dendrites primarily into the hollow of one barrel and seldom into adjacent hollows. This raises the question as to what factors guide this process and why dendrites are sparser in the barrel walls than in the hollows. Recent results suggest that the expression of cytotactin and CTB proteoglycan is differentially regulated during development and that both molecules are present transiently in the walls of developing barrels (Fig. 10A,B). The cortical pattern of cellular, axonal, and dendritic distribution can be perturbed by the removal of peripheral stimulation from a vibrissa early in postnatal life. This treat-

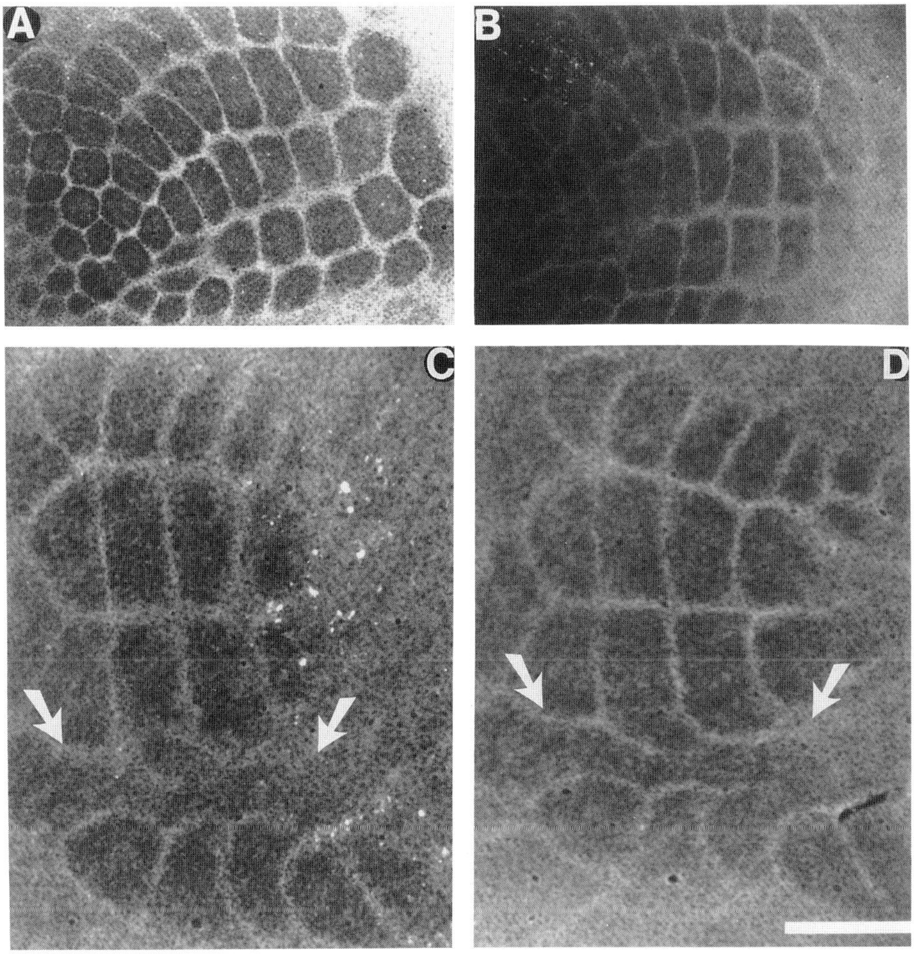

FIG. 10. Distribution of cytotactin and cytotactin binding (CTB) proteoglycan in the normal developing cortical barrel field and after whisker follicle damage. In P7 barrel cortex, antibodies to cytotactin (**A**) and to CTB proteoglycan (**B**) all strongly stain the barrel walls and are excluded from barrel hollows where thalamocortical axons are concentrated. Bar = 300 μm. Whisker follicles were electrocauterized, and whiskers were removed from row B on P1.5. Cortex was examined on P8 after this perturbation in the periphery. The expression of both cytotactin (**C**) and CTB proteoglycan (**D**) was fused into a stripe in row B of cortical barrels with no delineation between barrels (*arrows*). Bar = 300 μm (From ref. 55, with permission.)

ment results in the loss of patterned expression in the cortical row corresponding to a row of damaged vibrissae (Fig. 10C,D). Thus, activity of the thalamocortical axons plays a key role in establishing this molecular pattern as previously demonstrated for the anatomical barrel pattern. Rather than establishing a prepattern, the coordinate expression of glial cytotactin and neural CTB proteoglycan may serve to facilitate neuronal migration on radial glia and may play a role in inhibiting dendrites of cortical interneurons from traversing these molecular boundaries into adjacent barrels.

All the experiments presented above were done on *in vitro* preparations. Using agarose implants that contained anti-N-CAM antibodies placed into the tectum of

Xenopus laevis, it could be shown that severe distortion of *in vivo* map formation and regeneration ensued (69). The physiological alterations were accompanied by a sharp decrease in the density of axonal arbors in the region of the implant (70). Perhaps the most striking aspect of this experiment was the return to a more normal map pattern with time as the local concentration of anti-N-CAM antibodies diminished in the region of the implant (69).

Changes in CAM and SAM expression have also been found to be perturbed in genetic disease with altered morphology. As described above, N-CAM and Ng-CAM are colocalized at nodes of Ranvier in peripheral myelinated nerves. In two dysmyelinating mutants [i.e., trembler (+/Tr) and motor end-plate disease (med/med)], the distribution pattern of these molecules was found to be disrupted in the myelinated fibers (63). A recent study indicates that alterations in the amount of N-CAM expression are associated with a number of human myopathies (71). This is in accord with the observations on alteration of CAM expression in muscle degeneration and regeneration. These failures of regulation have a counterpart in the CNS: In the mouse mutant *staggerer,* which shows connectional defects between Purkinje cells and parallel fibers in the cerebellum and extensive cell death, E-A conversion is greatly delayed in the cerebellum (72).

The accumulated data both from interventions and from studies of disease states argue for the existence of an elaborate series of local morphology-dependent signals regulating CAM expression in various tissues. Adhesion must therefore be analyzed in a context-dependent fashion related to actual tissue and cell morphology. Although this brings concomitant increases in difficulty of analysis, the combinations of topobiological mechanisms having important effects on morphogenesis that result from such context dependence are virtually unlimited and free us from having to specify definite molecular cell addresses as the basis for neural morphogensis (73,74).

REFERENCES

1. Edelman GM. Topobiology. *Sci Am* 1989;260: 76–88.
2. Edelman GM. *Topobiology: an introduction to molecular embryology.* New York: Basic Books, 1988.
3. Edelman GM. Cell adhesion molecules in the regulation of animal form and tissue pattern. *Annu Rev Cell Biol* 1986;2:81–116.
4. Edelman GM. CAMs and Igs: Cell adhesion and the evolutionary origins of immunity. *Immun Rev* 1987;100:11–45.
5. Brackenbury R, Thiery J-P, Rutishauser U, Edelman GM. Adhesion among neural cells of the chick embryo. I. An immunological assay for molecules involved in cell-cell binding. *J Biol Chem* 1977;252:6835–6840.
6. Thiery J-P, Brackenbury R, Rutishauser U, Edelman GM. Adhesion among neural cells of the chick embryo. II. Purification and characterization of a cell adhesion molecule from neural retina. *J Biol Chem* 1977;252:6841–6845.
7. Cunningham BA, Hemperly JJ, Murray BA, Prediger EA, Brackenbury R, Edelman GM. Neural cell adhesion molecule: structure, immunoglobulin-like domains, cell surface modulation, and alternative RNA splicing. *Science* 1987;236:799–806.
8. Murray BA, Owens GC, Prediger EA, Crossin KL, Cunningham BA, Edelman GM. Cell surface modulation of the neural cell adhesion molecule resulting from alternative mRNA splicing in a tissue-specific developmental sequence. *J Cell Biol* 1986;103:1431–1439.
9. Owens GC, Edelman GM, Cunningham BA. Organization of the N-CAM gene-alternative exon usage as the basis for different membrane-associated domains. *Proc Natl Acad Sci USA* 1986; 84:294–298.
10. Hemperly JJ, Edelman GM, Cunningham BA. cDNA Clones of N-CAM lacking a membrane-spanning region consistent with evidence for membrane attachment via a phosphatidylinositol intermediate. *Proc Natl Acad Sci USA* 1986; 83:9822–9826.
11. He HT, Barbet J, Chaix JC, Goridis C. Phosphatidylinositol is involved in the membrane attachment of N-CAM-120, the smallest component of the neural cell adhesion molecule. *EMBO J* 1986; 5:2489–2494.
12. Hemperly JJ, Murray BA, Edelman GM, Cunningham BA. Sequence of a cDNA clone encoding the polysialic acid-rich and cytoplasmic domains of the neural cell adhesion molecule N-CAM. *Proc Natl Acad Sci USA* 1986;83:3037–3041.
13. Barthels D, Santoni M-J, Wille W, et al. Isola-

tion and nucleotide sequence of mouse NCAM cDNA that codes for a Mr 79 000 polypeptide without a membrane-spanning region. *EMBO J* 1987;6:907–914.
14. Dickson G, Gower HJ, Barton CH, et al. Human muscle neural cell adhesion molecule (N-CAM)—identification of a muscle specific sequence in the extracellular domain. *Cell* 1987; 50:1119–1130.
15. Prediger EA, Hoffman S, Edelman GM, Cunningham BA. Four exons encode a 93bp insert in three N-CAM mRNAs specific for chick heart and skeletal muscle. *Proc Natl Acad Sci USA* 1988;85:9616–9620.
16. Crossin KL, Edelman GM, Cunningham BA. Mapping of three carbohydrate attachment sites in embryonic and adult forms of the neural cell adhesion molecule (N-CAM). *J Cell Biol* 1984; 99:1848–1855.
17. Rothbard JB, Brackenbury R, Cunningham BA, Edelman GM. Differences in the carbohydrate structures of neural cell–adhesion molecules from adult and embryonic chicken brains. *J Biol Chem* 1982;257:11064–11069.
18. Finne J, Finne U, Deagostini-Bazin H, Goridis C. Occurrence of α2-8 linked polysialosyl units in a neural cell adhesion molecule. *Biochem Biophys Res Commun* 1983;112:482–487.
19. Hoffman S, Edelman GM. Kinetics of homophilic binding by E and A forms of the neural cell adhesion molecule. *Proc Natl Acad Sci USA* 1983;80:5762–5766.
20. Chuong C-M, Edelman GM. Alteration in neural cell adhesion molecules during development of different regions of the nervous system. *J Neurosci* 1984;4:2354–2368.
21. Friedlander DR, Brackenbury R, Edelman GM. Conversion of embryonic form to adult forms of N-CAM *in vitro* results from *de novo* synthesis of adult forms. *J Cell Biol* 1985;101:412–419.
22. Edelman GM, Hoffman S, Chuong C-M, et al. Structure and modulation of neural cell adhesion molecules in early and late embryogenesis. *Cold Spring Harbor Symp Quant Biol* 1983;58:515–526.
23. Hall AK, Rutishauser U. Visualization of neural cell adhesion molecule by electron microscopy. *J Cell Biol* 1987;104:1579–1586.
24. Becker JW, Erickson HP, Hoffman S, Cunningham BA, Edelman GM. Topology of cell adhesion molecules. *Proc Natl Acad Sci USA* 1989;86:1088–1092.
25. Gallin WJ, Edelman GM, Cunningham BA. Characterization of L-CAM, a major cell adhesion molecule from embryonic liver cells. *Proc Natl Acad Sci USA* 1983;80:1038–1042.
26. Thiery J-P, Duband J-L, Rutishauser U, Edelman GM. Cell adhesion molecules in early chick embryogenesis. *Proc Natl Acad Sci USA* 1982; 79:6737–6741.
27. Edelman GM, Gallin WJ, Delouvée A, Cunningham BA, Thiery J-P. Early epochal maps of two different cell adhesion molecules. *Proc Natl Acad Sci USA* 1983;80:4384–4388.
28. Thiery J-P, Delouvée A, Gallin WJ, Cunningham BA, Edelman GM. Ontogenetic expression of cell adhesion molecules—L-CAM is found in epithelia derived from the three primary germ layers. *Dev Biol* 1984;102:61–78.
29. Crossin KL, Chuong C-M, Edelman GM. Expression sequences of cell adhesion molecules. *Proc Natl Acad Sci USA* 1985;82:6942–6946.
30. Thiery J-P, Delouvée A, Grumet M, Edelman GM. Initial appearance and regional distribution of the neuron-glia cell adhesion molecule (Ng-CAM) in the chick embryo. *J Cell Biol* 1985; 100:442–456.
31. Daniloff J, Chuong C-M, Levi G, Edelman GM. Differential distribution of cell adhesion molecules during histogenesis of the chicken nervous system. *J Neurosci* 1986;6:739–758.
32. Grumet M, Edelman GM. Neuron-glia cell adhesion molecule interacts with neurons and astroglia via different binding mechanisms. *J Cell Biol* 1988;106:487–503.
33. Friedlander DR, Grumet M, Edelman GM. Nerve growth factor enhances expression of neuron-glia cell adhesion molecule in PC12 cells. *J Cell Biol* 1986;102:413–419.
34. Pollerberg EG, Sadoul R, Goridis C, Schachner M. Selective expression of the 180-KD component of the neural cell-adhesion molecule N-CAM during development. *J Cell Biol* 1985;101: 1921–1929.
35. Mackie KP, Sorkin BC, Nairn AC, Greengard P, Edelman GM, Cunningham BA. Identification of two protein kinases that phosphorylate the neural cell adhesion molecule, N-CAM. *J Neurosci* 1988;9:1883–1896.
36. Pollerberg GE, Schachner M, Davoust J. Differentiation state-dependent surface mobilities of two forms of the neural cell adhesion molecule. *Nature* 1986;324:462–465.
37. Pollerberg EG, Burridge K, Krebs KE, Goodman SR, Schachner M. The 180-kD component of the neural cell adhesion molecule N-CAM is involved in cell–cell contacts and cytoskeleton–membrane interactions. *Cell Tiss Res* 1987;250: 227–236.
38. Prieto AL, Crossin KL, Cunningham BA, Edelman GM. Localization of mRNA for neural cell adhesion molecule (N-CAM) polypeptides in neural and nonneural tissues by *in situ* hybridization. *Proc Natl Acad Sci USA* 1989;86:9579–9583.
39. LaVail JH, Cowan WM. The development of the chick optic tectum. I. Normal morphology and cytoarchitectonic development. *Brain Res* 1971; 28:391–419.
40. Trapp BD, Moench T, Pulley M, Barbosa E, Tennekoon G, Griffin J. Spatial segregation of mRNA encoding myelin-specific proteins. *Proc Natl Acad Sci USA* 1987;84:7773–7777.
41. Garner CC, Tucker RP, Matus A. Selective localization of messenger RNA for cytoskeletal protein MAP2 in dendrites. *Nature* 1988;336: 674–677.

42. Garner JA, Watanabe M, Rutishauser U. Rapid axonal transport of neural cell adhesion molecule. *J Neurosci* 1986;6:3242–3249.
43. Hoffman S, Friedlander DR, Chuong C-M, Grumet M, Edelman GM. Differential contributions of Ng-CAM and N-CAM to cell adhesion in different neural regions. *J Cell Biol* 1986;103: 145–158.
44. Mege R-M, Matsuzaki F, Gallin WJ, Goldberg JI, Cunningham BA, Edelman GM. Construction of epithelioid sheets by transfection of mouse sarcoma cells with cDNAs for chicken cell adhesion molecules. *Proc Natl Acad Sci USA* 1988;85:7274–7278.
45. Matsuzaki F, Mege R-M, Jaffe SH, et al. cDNAs of cell adhesion molecules of different specificity induce changes in cell shape and border formation in cultured S180 cells. *J Cell Biol* 1990;110:1239–1252.
46. Friedlander DR, Mege R-M, Cunningham BA, Edelman GM. Cell sorting-out is modulated by both the specificity and amount of different cell adhesion molecules (SAMs) expressed on cell surfaces. *Proc Natl Acad Sci USA* 1989;86: 7043–7047.
47. Grumet M, Hoffman S, Crossin KL, Edelman GM. Cytotactin, an extracellular matrix protein of neural and non-neural tissues that mediates glia–neuron interaction. *Proc Natl Acad Sci USA* 1985;82:8075–8079.
48. Chiquet M, Fambrough DM. Chick myotendinous antigen. II. A novel extracellular glycoprotein complex consisting of large disulfide-linked subunits. *J Cell Biol* 1984;98:1937–1946.
49. Chiquet-Ehrismann R, Mackie EJ, Pearson CA, Sakakura T. Tenascin: an extracellular matrix protein involved in tissue interactions during fetal development and oncogenesis. *Cell* 1986;47: 131–139.
50. Faissner A, Kruse J, Chiquet-Ehrismann R, Mackie E. The high molecular weight J1 glycoproteins are immunochemically related to tenascin. *Differentiation* 1988;37:104–114.
51. Erickson HP, Ingelsias JL. A six-armed oligomer isolated from cell surface fibronectin preparations. *Nature* 1984;311:267–269.
52. Erickson HP, Lightner VA. Hexabrachion protein (tenascin, cytotactin, brachionectin) in connective tissues, embryonic brain, and tumors. *Adv Cell Biol* 1988;2:55–90.
53. Hoffman S, Edelman GM. A proteoglycan with HNK-1 antigenic determinants is a neuron-associated ligand for cytotactin. *Proc Natl Acad Sci USA* 1987;84:2523–2527.
54. Hoffman S, Crossin KL, Edelman GM. Molecular forms, binding functions and developmental expression patterns of cytotactin and CTB proteoglycan, an interactive pair of extracellular matrix molecules. *J Cell Biol* 1988;106:519–532.
55. Crossin KL, Hoffman S, Tan S-S, Edelman GM. Cytotactin and its proteoglycan ligand mark structural and functional boundaries in somatosensory cortex of the early postnatal mouse. *Dev Biol* 1989;136:381–392.
56. Jones FS, Burgoon MP, Hoffman S, Crossin KL, Cunningham BA, Edelman GM. A cDNA clone for cytotactin contains sequences apparently homologous to EGF repeats and segments of fibronectin and fibrinogen. *Proc Natl Acad Sci USA* 1988;85:2186–2190.
57. Jones FS, Hoffman S, Cunningham BA, Edelman GM. A detailed structural model of cytotactin: protein homologies, alternative RNA splicing, and binding regions. *Proc Natl Acad Sci USA* 1989;86:1905–1909.
58. Friedlander DR, Hoffman S, Edelman GM. Functional mapping of cytotactin: proteolytic fragments active in cell-substrate adhesion. *J Cell Biol* 1988;107:2329–2340.
59. Tan SS, Crossin KL, Hoffman S, Edelman GM. Asymmetric expression in somites of cytotactin and its proteoglycan ligand is correlated with neural crest cell distribution. *Proc Natl Acad Sci USA* 1987;84:7977–7981.
60. Crossin KL, Hoffman S, Grumet M, Thiery J-P, Edelman GM. Site-restricted expression of cytotactin during development of the chicken embryo. *J Cell Biol* 1986;102:1917–1930.
61. Edelman GM. Cell adhesion and morphogenesis: the regulator hypothesis. *Proc Natl Acad Sci USA* 1984;81:1460–1464.
62. Han JH, Rall L, Rutter WJ. Selective expression of rat pancreatic genes during embryonic development. *Proc Natl Acad Sci USA* 1986;83:110–114.
63. Rieger F, Daniloff JK, Pinçon-Raymond M, Crossin KL, Grumet M, Edelman GM. Neuronal cell adhesion molecules and cytotactin are colocalized at the node of Ranvier. *J Cell Biol* 1986;103:379–391.
64. Rieger F, Grumet M, Edelman GM. N-CAM at the vertebrate neuromuscular junction. *J Cell Biol* 1985;101:285–293.
65. Covault J, Sanes JR. Neural cell adhesion molecule (N-CAM) accumulates in denervated and paralyzed skeletal muscles. *Proc Natl Acad Sci USA* 1985;82:4544–4548.
66. Daniloff JK, Levi G, Grumet M, Rieger F, Edelman GM. Altered expression of neuronal cell adhesion molecules induced by nerve injury and repair. *J Cell Biol* 1986;103:929–945.
67. Daniloff J, Crossin K, Pinçon-Raymond M, Murawsky M, Rieger F, Edelman GM. Expression of cytotactin during development and regeneration of the neuromuscular system. *J Cell Biol* 1989;108:625–636.
68. Chuong C-M, Crossin KL, Edelman GM. Sequential expression and differential functions of multiple adhesion molecules during the formation of cerebellar cortical layers. *J Cell Biol* 1987;104:331–342.
69. Fraser SE, Murray BA, Chuong C-M, Edelman GM. Alteration of the retinotectal map in *Xenopus* by antibodies to neural cell adhesion molecules. *Proc Natl Acad Sci USA* 1984;81:4222–4226.

70. Fraser SE, Carhart MS, Murray BA, Chuong C-M, Edelman GM. Alterations in the *Xenopus* retinotectal projection by antibodies to *Xenopus* N-CAM. *Dev Biol* 1988;129:217–230.
71. Walsh FS, Moore SE. Expression of cell-adhesion molecule, N-CAM, in diseases of adult human skeletal-muscle. *Neurosci Lett* 1985;59: 73–78.
72. Edelman GM, Chuong C-M. Embryonic to adult conversion of neural cell adhesion molecules in normal and staggerer mice. *Proc Natl Acad Sci USA* 1982;79:7036–7040.
73. Edelman GM. Cell surface modulation and marker multiplicity in neural patterning. *Trends Neurosci* 1984;7:78–84.
74. Sperry RW. Chemoaffinity in the orderly growth of nerve fiber patterns and connections. *Proc Natl Acad Sci USA* 1963;50:703–710.

Volume Transmission in the Brain: Novel Mechanisms for Neural Transmission, edited by K. Fuxe and L.F. Agnati. Raven Press, Ltd., New York © 1991.

4

The Meaning of Synaptic Architecture

Sanford L. Palay

Department of Anatomy, Harvard Medical School, Boston, Massachusetts

As the locus for transmission of the nerve impulse, the synapse is the preeminent intercellular relation in the nervous system. The morphological features of the synapse are well known (1), and one only needs to mention them to bring to mind the familiar image of a bulbous nerve terminal applied to the surface of a neuronal cell body or dendrite with the thin intercellular space between them. In this image, with the high resolution of electron microscopy we now see that the presynaptic terminal contains a distinctive collection of mitochondria and of vesicles clustered against the plasmalemma among projecting aggregates of fine filaments (2). On the dendritic side of the junction only a variable thicket of fine fibrils and granules marks the functional membrane.

These junctional complexes—*synaptic complexes,* as they are called—display some features common to ordinary intercellular junctions in all sorts of epithelia. First, the components belong to independent but adjacent neighboring cells. They are not sites of protoplasmic continuity; instead they are intercellular articulations, what Cajal (3), in the romantic enthusiasm of his memoirs called "protoplasmic kisses." Second, they are attached to one another by adhesive junctions, which are signaled by the dense filamentous aggregates on either side. Synapses, however, are fundamentally different from the intercellular articulations of ordinary epithelia. Although the two partners are apparently firmly stuck together, they are not merely adherent junctions, and they are not between similar and equal partners. Instead they exhibit both a morphological and a functional polarization, according to which one partner releases one or more neuroactive substances that alter the chemical and electrodynamical equilibrium of the other partner.

From the morphological arrangement of axonal terminals on the surfaces of dendrites and cell bodies, Cajal deduced the direction of transmission of the nerve impulse, although he was not acquainted with neuroactive substances or the concept of chemical transmission. Both he and Sherrington thought that the passage of the neural current into the nerve terminals induced a like phenomenon on the postsynaptic side of the articulation. This observation became the law of dynamic polarization, according to which the neural current flowed from axon terminals across the synapse to dendrites or somata. Within the nerve cell, proclaimed Cajal,

> the transmission of the nervous impulse is always from the dendritic branches and the cell body to the axon or functional process. Every neuron, then, possesses a receptor apparatus, the body and the dendritic prolongations, an apparatus of emission, the axon, and an apparatus of distribution, the terminal arborization of the nerve fiber. (3)

This polarization at the light microscope level is also evident in the electron microscope. Thus synaptic vesicles appear only on the presynaptic side of the synaptic junction, and in some synapses the thick mat of filaments attached to the cytoplasmic surface of the postsynaptic component enhances the impression of asymmetry. Thus the polarization of the synaptic functional architecture clearly points in the direction of impulse transmission. Furthermore, this junctional asymmetry conditions the central polarity of the whole neural system, for the electrical gradients and shifts in potential that take their origin from synaptic action pass indifferently in all directions in the cell, up and down dendrites, into and out of the cell body, not just toward the axon, as Cajal believed. It is not the polarity of the nerve cell that determines the direction of impulse traffic in the nervous system, but the polarity of the synaptic junction itself that drives the impulse from one cell to the next.

The transfer of the nerve impulse is accomplished by releasing neuroactive substances into the synaptic cleft, the thin intercellular space between the axon terminal and the dendrite or soma of a neighboring cell. This is primarily a merocrine secretory event, primed by the influx of calcium ions and requiring the expenditure of metabolic energy, just as in any gland. But unlike gland cells, which discharge their product into a duct that opens onto a surface of the body, the synapse is a device for precisely delivering measured amounts of neuroactive substances into the intercellular space immediately overlying the specific receptive surface for these substances.

The generally accepted view holds that the synaptic vesicles fuse with the presynaptic surface membrane and discharge their content of neuroactive substances into the synaptic cleft. The vesicular membrane is then retrieved into the internal cytoplasmic membranes by endocytosis and returned to the cell body. Alternatively the vesicle may open onto the surface, then be withdrawn, recharged, and discharged several times. In either case the vesicles are thought to be transient residents in the axonal ending and to be continuously reconstructed and passed into the ending by fast axoplasmic transport, where they are charged with their load of neuroactive substances.

The discharge of transmitter chemicals is only one aspect of the precision with which this device is constructed. The narrow interspace into which the chemicals are liberated is only 12 to 20 nm across, but it is filled with fine filaments of glycoproteins, some of which exhibit specific enzymatic activity (e.g., the acetylcholinesterase of the neuromuscular junction). Although the neuroactive substances have only a short distance to cross to reach their targets in the postsynaptic membrane, it is not to be imagined that they usually take the shortest possible path. The synaptic cleft contains multiple obstacles and the diffusing molecules must take devious paths around them. This is perhaps the price to be paid for the specificity and efficiency gained by fixing the synaptic terminal to a particular place on the surface of the postsynaptic partner where it is matched with appropriate receptors.

Once the molecules of neuroactive substances have crossed the cleft they interact with specific receptor molecules aggregated in the postsynaptic membrane. Now a new conformation results that permits the well-known influx of selected ions with a consequent change in the membrane potential and the flow of current in the postsynaptic cell. Thus, a highly restricted alteration in the membrane equilibrium is produced by a minute injection into the immediately overlying intercellular space of a small number of neuroactive molecules from a neighboring cell. What develops out of this focal alteration depends on what else is going on at the cell's surface at the same time. Its effect can be magnified by similar events in adjacent parts of the postsynaptic membrane. Or its effect can be negated by contrary events in adjacent parts. Because the alteration takes place in time, its effect can be enhanced or dampened by residues of previous events or by successive repetition. Thus, although the influence of any single synaptic discharge may be miniscule to the point of being negligible, the possibility of spatial and temporal summation raises the value of even small transient discharges. In addition, the location of the synapse on the surface of a cell and the amount of surface engaged in the synaptic interface play a role in setting the relative weight ascribable to any synaptic junction in the overall economy of the postsynaptic cell. The architecture of the nervous system takes full advantage of these features when it channels inputs from certain sources onto distal dendrites and others onto spines or proximal dendrites and still others onto somata or axon initial segments, as in the well-studied examples of the Purkinje cell in the cerebellar cortex and the pyramidal cells in the hippocampus and the visual cortex.

Although the form of the articulation between pre- and postsynaptic partners is highly variable in the nervous system, the principles of synaptic architecture remain the same, whether the articulation is with a loose arborization or in a tightly coiled glomerulus. The principles apply even for the unusual synapses (i.e., the somatodendritic, the dendrodendritic, and the axoaxonic), all variations on the central theme.

As the nerve impulse evokes an essentially merocrine response from the presynaptic partner, it should not surprise us, however, if the presynaptic terminal can also appear independently of any obvious, adjacent postsynaptic partner. The model for this arrangement is the once-puzzling neurosecretory axons in the neurohypophysis of vertebrates and in the neurohemal organs of invertebrates. These axons terminate on the basal lamina limiting the surface of the brain or in some instances in the lumen of the ventricles. They discharge their neuroactive substances into a pericapillary space whence they are carried into the bloodstream to affect distant structures, in the kidney or the mammary gland, for example. Another nonsynaptic arrangement appears in the innervation of most smooth muscle, for example, in the wall of the intestine, where *en passant* varicosities and terminals lie in the neighboring connective tissue spaces but do not attach to the muscle with differentiated postsynaptic specializations. Apparently these terminals release their neuroactive substances into the interstitial space at large. Specificity is provided, as in all synapses, by the distribution and localization of the receptors on the postsynaptic surface, but precision is not provided by the delivery system. Then the effect can be slow in onset and of long duration.

The same diffuse, nonsynaptic arrangement may also occur in the central nervous system, especially in the monoamine networks arising from the locus coeruleus and the raphe nuclei. In all these cases, topographic proximity of the pre- and postsynaptic elements is dispensed with in systems for which speed, efficiency, and precise focal interaction are not important. The effects of transmitter discharge in these systems are slow, long-lasting, and diffuse. Such effects provide a mechanism for overall tuning of the central nervous system, regulating the level of sensitivity, setting the rhythms of function, and cueing activity with the environment of the whole organism.

These apparently nonsynaptic arrangements provide an efficient way to produce generalized effects without directly and repetitiously contacting every single cell in a target area. They are comparable to sounding a fire alarm instead of ringing the doorbell of every home in an area threatened by fire. They do not violate the architectural principles of the synapse. The only difference is that the synaptic cleft is vastly enlarged.

REFERENCES

1. Peters A, Palay SL, Webster H deF. *The fine structure of the nervous system*, 3rd ed. New York: Oxford University Press, 1990.
2. Palay SL. Synapses in the central nervous system. *J Biophys Biochem Cytol* 1956;2(Suppl): 193–202.
3. Cajal SR. *Recollections of my life*. Craigie EH, transl. Cambridge, MA: MIT Press, 1989.

Volume Transmission in the Brain:
Novel Mechanisms for Neural Transmission,
edited by K. Fuxe and L.F. Agnati.
Raven Press, Ltd., New York © 1991.

5

Nonjunctional Relationships of Monoamine Axon Terminals in the Cerebral Cortex of Adult Rat

Laurent Descarries, Philippe Séguéla, and Kenneth C. Watkins

Centre de recherche en sciences neurologiques, Département de physiologie, Faculté de médecine, Université de Montréal, Montréal (Québec), Canada H3C 3J7

Like certain works of art, there are objects of scientific knowledge that come to play pivotal roles in the evolution of ideas. The monoamine systems using as transmitters the catecholamines, dopamine (DA) and norepinephrine (NE), or the indolamine serotonin (5-hydroxytryptamine, 5-HT) belong to this category. These neuronal populations were the first in mammalian brain to be structurally as well as functionally defined by their transmitter content. As such, they have prompted major transformations of the very concept of neurotransmission, allowing for the discovery of dendritic release, coexistent transmitters, and local auto- and heteroregulation of transmitter release. The first indices of "volume transmission" were also provided by the study of central monoamine neurons (for review, see ref. 1). Whether this expression is used to distinguish a broad, humoral type of electrochemical communication from the "wiring transmission" associated with synaptic connectivity (2) or, more concretely, to underline possible effects of transmitters in a relatively large sphere of influence around nerve terminals, one of its origins may be traced to the early electron microscopic descriptions of NE and 5-HT axonal varicosities in adult rat neocortex, which indicated that many of these nerve endings lacked junctional membrane specializations and hence localized postsynaptic interfaces (3,4). In addition to the possibility that these "nonsynaptic" varicosities might thus be subjected to incessant movements of translocation and/or reshaping along their parent fibers, it was proposed that their released transmitter might diffuse in surrounding tissue to reach relatively distant targets and exert desynchronized and sustained influences on vast neuronal ensembles (see also ref. 5). On the subsequent demonstration of the ubiquitous distribution of these nerve endings in cerebral cortex, it was also hypothesized that "the entire cortical volume might be within reach of their transmitter" (6), thus accounting for the diverse and widely distributed actions that were being progressively assigned to the monoamines (see ref. 7).

Numerous cytochemical, radioautographic, and immunocytochemical studies have since disclosed similar features on different monoamine innervations in other parts of the rat, cat, or monkey brain (e.g., ref. 8). However, there were also discordant views. On the basis of immunoelectron microscopic data obtained with antibodies against DA-β-hydroxylase or against formaldehyde-conjugated 5-HT (9) and then glutaraldehyde-conjugated 5-HT or NE

(10,11), the original radioautographic findings on the cortex were disputed. A recent report has even claimed that "these monoamine systems, as with other classical transmitter systems, communicate with target neurons in the cortex and elsewhere in the brain through conventional synapses" (12).

Owing to the availability of polyclonal antibodies directed against monoamine-glutaraldehyde-protein conjugates and hence compatible with optimal tissue preservation for electron microscopic analysis, we have now reinvestigated the whole issue of the relational features of NE and 5-HT axon terminals in adult rat neocortex (13,14). Moreover, we have been able to examine the cortical DA innervation, which, mainly for lack of a proper method, had previously escaped ultrastructural visualization (15). After a summary of these results, the present report briefly discusses their significance, particularly in the light of recent information on the quantified distribution of all three monoamine innervations in adult rat cortex (16–19).

IMMUNOCYTOCHEMICAL DATA ON CORTICAL DA, NE, AND 5-HT AXON TERMINALS (VARICOSITIES)

Dopamine

These cortical terminals were identified by means of peroxidase–antiperoxidase immunocytochemistry using the DA-glutaraldehyde-lysyl-protein conjugate antiserum developed by Michel Geffard (20). The tissue could thus be primarily fixed by perfusion with relatively high concentrations of glutaraldehyde. The ultrastructural features and synaptic relationships of the DA-immunostained endings were scrutinized in single and serial ultrathin sections from the deep layers of the anteromedial and suprarhinal DA fields of prefrontal cortex (15). From a total of 329 varicosities examined, 28 from the anteromedial and 50 from the suprarhinal cortex were viewed in series of 5 to 12 sections per varicosity.

As exemplified by Figs. 1 and 2, the DA-immunostained profiles from both regions averaged 0.7 μm (±0.2 SEM) in diameter, contained a mixed population of mostly small, round, and clear synaptic vesicles associated with a few larger dense-cored or fully electron-dense vesicles, and frequently exhibited synaptic contacts, which were exclusively made on dendritic branchlets and spines. These synapses were mostly of the symmetrical type (80%) and were more often seen on dendritic shafts than spines. No junctions were made on cell bodies, even though direct appositions to neuronal somata were occasionally observed.

When estimated either by stereological extrapolation from single sections or by direct observation in serial sections, the synaptic incidence of DA-immunostained varicosities was significantly greater in the anteromedial than in the suprarhinal cortex. In the longest series of thin sections, a junctional complex was observed on 93% of the immunostained varicosities from the anteromedial region, but only 56% from the suprarhinal region (Table 1). This interregional difference was statistically significant.

It was concluded that the cortical DA innervation is predominantly junctional but with significant interregional differences. Independent observations concurrently made with the same antiserum on the rat prefrontal cortex (21) and more recent ones on the rat visual cortex (22) are consistent with these findings.

Norepinephrine

The analysis of the NE innervation (14) was also carried out with PAP-immunocytochemistry using a glutaraldehyde-conjugated monoamine antibody (23). It involved the examination of more than 550 NE-immunostained varicosities from the upper

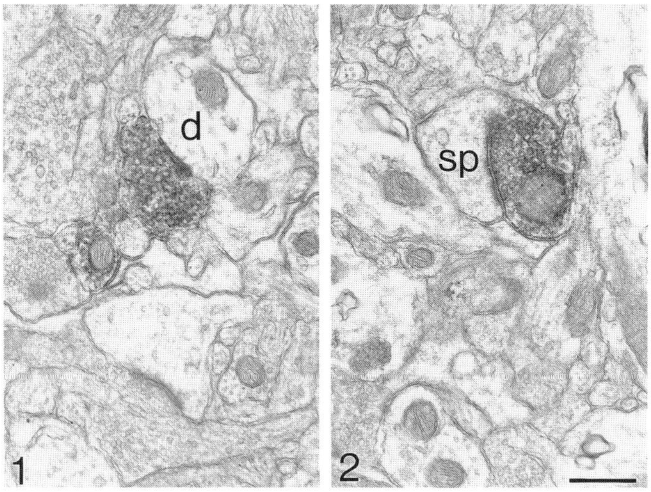

FIGS. 1 and 2. Synaptic DA-immunostained terminals (axonal varicosities) from suprarhinal cortex (single sections). **Fig. 1** is an example of a DA varicosity making a symmetrical synaptic contact with a dendritic shaft (d). **Fig. 2** shows another synaptic DA varicosity forming an asymmetrical junction on a dendritic spine (sp.) ×18,000. Calibration bar: 0.5 μm. (Reproduced from ref. 15, with permission.)

layers of the frontal (Fr1-Fr2), parietal (Par 1), and occipital (Oc1M-Oc2) neocortex [nomenclature of cortical areas according to Zilles (24)]. Most of these varicosities were identified in relatively short series of thin sections (2–5), but an appreciable number were followed in five to nine adjacent sections. Single and serial sections from 100 unlabeled axon terminals randomly selected from the same micrographs were also analyzed in each cortical region. The varicosities from both groups were then measured and scrutinized for the presence of a junctional complex indicative of synaptic differentiation. Furthermore, all cellular elements juxtaposed to their membrane were identified and counted.

The NE-immunostained profiles from all three cortical regions (e.g., Figs. 3 and 4) averaged 0.6 μm (±0.2 SEM) in diameter. In contrast to their unlabeled counterparts, they rarely showed membrane differentia-

TABLE 1. *Innervation density and synaptic incidence of monoamine terminals in different regions of adult rat neocortex*[a]

	Innervation density			Synaptic incidence		
Cortical region	DA	NE	5-HT	DA	NE	5-HT
Medial prefrontal	1.1	1.3	6.1	93%	—	—
Lateral prefrontal	0.3	1.3	7.1	56%	—	—
Frontal (motor)	0.06	1.1	4.7	—	↑	28–36%
Parietal (somatosensory)	0.05	1.0	4.4	—	17%	46%
Occipital (visual)	0.01	—	—	—	↓	37%

[a]Innervation density expressed in number (×10^6) of axonal varicosities per mm^3 of tissue (values from refs. 16,18,19). Synaptic incidence refers to frequency of junctional membrane specializations on monoamine varicosities (values from refs. 13–15). The synaptic incidence for NE represents pooled data from upper layers of frontal, parietal, and occipital cortex. In the case of 5-HT, there are no statistically significant differences between separate values from the same three regions and upper (36%) vs. lower layers (28%) of frontal cortex (interregional mean of 38%).

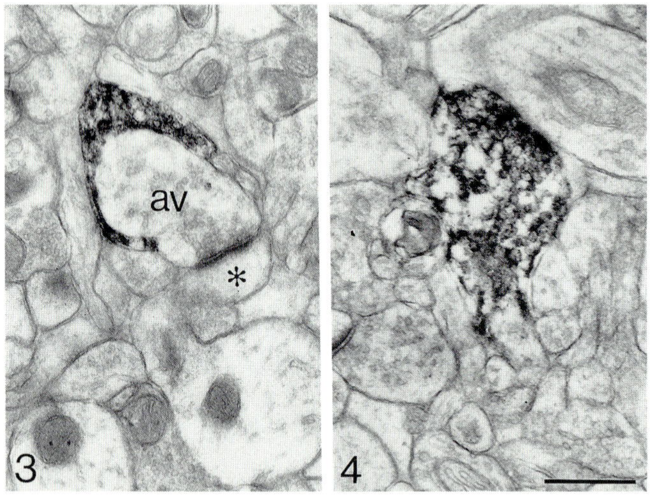

FIGS. 3 and 4. Nonjunctional NE-immunostained axonal varicosities from frontal (**Fig. 3**) and occipital (**Fig. 4**) cortex. Varicosity in Fig. 3 was examined in 10 serial thin sections, none of which exhibited any suggestion of a membrane specialization. Note how this varicosity partially engulfs another, nonimmunostained axon terminal (av), itself in synaptic contact with a dendritic spine (*). Nonjunctional varicosity in Fig. 4 was viewed in 13 serial thin sections (six others are shown in ref. 14, Fig. 4). Again, there was no junctional complex on any portion of this varicosity. ×25,000. Calibration bar: 0.5 μm.

tions characteristic of a synaptic contact (i.e., junctional complex). The rare synapses formed by NE-immunostained varicosities were invariably symmetrical and almost exclusively found on dendritic shafts. Axosomatic contacts were rare and did not show membrane differentiations. No NE varicosities were seen in close proximity to the endothelial lining of intracortical capillaries.

The proportion of cortical NE varicosities engaged in synaptic contact (i.e., synaptic incidence) was determined by plotting the frequency of visible junctions as a function of the number of thin sections available for examination. As extrapolated for whole varicosities after linear transformation (i.e., double reciprocal plot), this proportion was 17% or 26%, depending on the use of more or less stringent criteria for defining the junctional complex (Table 1). With the more stringent criteria, the analysis yielded 98% for the control population of unlabeled cortical terminals.

The low proportion of synaptic contacts established by the NE varicosities drew attention to the composition of their microenvironment (Table 2). In direct apposition with these immunostained profiles, we found a greater number of other, unlabeled axonal varicosities, but a smaller number of dendritic spines than around randomly selected unlabeled varicosities. The proportion of NE varicosities exhibiting one or more dendritic spines in their immediate microenvironment was also smaller than for the control population of unlabeled endings.

Thus, this study largely confirmed the initial radioautographic characterization of the cortical NE innervation as a mostly nonjunctional system. It nevertheless suggests that these NE varicosities might be set in a particular microenvironment often

TABLE 2. *Microenvironment of norepinephrine (NE) and serotonin (5-HT) axon terminals in three regions of adult rat neocortex*[a]

		NE (N = 300)	Non-NE (N = 300)	5-HT (N = 240)	Non-5-HT (N = 240)
Axonal varicosities	n/v	1.3	0.9	1.5	1.0
	incidence (%)	72.5	62.4	79.5	68.6
Dendritic shafts	n/v	0.7	0.8	0.9	0.7
	incidence (%)	51.7	63.4	67.8	49
Dendritic spines	n/v	0.4	0.8	0.6	0.7
	incidence (%)	37.2	63.1	43.1	53.1
Neural processes	n/v	3.6	3.0	3.4	2.8
	incidence (%)	95.6	91.9	93.3	89.5
Glia	n/v	1.4	1.5	1.4	1.4
	incidence (%)	75.8	81.2	76.2	74.1
Miscellaneous	n/v	0.3	0.3	0.3	0.3
	incidence (%)	27.2	21.1	22.2	23.8

[a]Distribution of cellular elements in direct apposition with single sectional profiles of NE and 5-HT immunostained axonal varicosities vs. nonimmunostained varicosities (non-NE and non-5-HT) randomly selected from the same sections (see text for sampling procedure). Pooled data from upper layers of frontal, parietal, and occipital cortex. Values of N refer to number of axonal profiles examined in each group. "Neural processes" include identified axonal segments together with other neuronal profiles of small diameter (axons or dendrites). "Miscellaneous" comprises a small number of nerve cell bodies, myelinated axons, and unidentified structures. n/v is the ratio n elements / n varicosities. Incidence (%) is frequency of direct apposition with one or more elements of the given type. χ^2 analysis indicated statistically significant differences between both classes of immunostained varicosities and their respective controls, and between NE vs. 5-HT varicosities ($p < 0.01$). Axonal varicosities were more numerous around NE or 5-HT varicosities than axonal varicosities in general, and there were fewer dendritic spines around NE varicosities. Moreover, both dendritic shafts and spines were more numerous around 5-HT than NE varicosities.

comprising other axonal varicosities but relatively poor in dendritic spines.

Serotonin

The ultrastructural relationships of 5-HT-immunostained terminals were also investigated in the Fr1-Fr2, Par1, and Oc1M-Oc2 cortex (13), using a glutaraldehyde-conjugated antibody (25). In the frontal region, the sampling consisted of 54 varicosities from layers I and II and 91 from layer VI; 97 each were obtained from the upper layers (I–II) of the parietal and the occipital cortex. These terminals (e.g., Figs. 5–8) were visualized in an average of 3.2 to 7.3 sections per varicosity, depending on the cortical sector examined. Their mean diameter was almost identical in the four sectors (0.7 μm ± 0.2 SEM).

The proportion making a synaptic contact was again evaluated by linear transformation of the relation between the frequency of observed junctions and the number of thin sections available for examination. In this sampling, the frequency of junctional varicosities was sufficient to obtain a high coefficient of correlation ($r > 0.95$) in each cortical sector. The synaptic incidence extrapolated for whole varicosities was thus found to range from 28% (layer VI of Fr1-Fr2) to 46% (Par1), with no statistically significant differences between the four sectors (Table 1). The interregional mean was 38%.

Synaptic cortical 5-HT terminals invariably made asymmetrical junctions, which were found either on dendritic spines or shafts but more frequently on spines, at least in the deep frontal and the upper occipital cortex. Appositions to neuronal somata were again rare and without any membrane differentiation; as with NE, there were no 5-HT varicosities in close contact with small blood vessels.

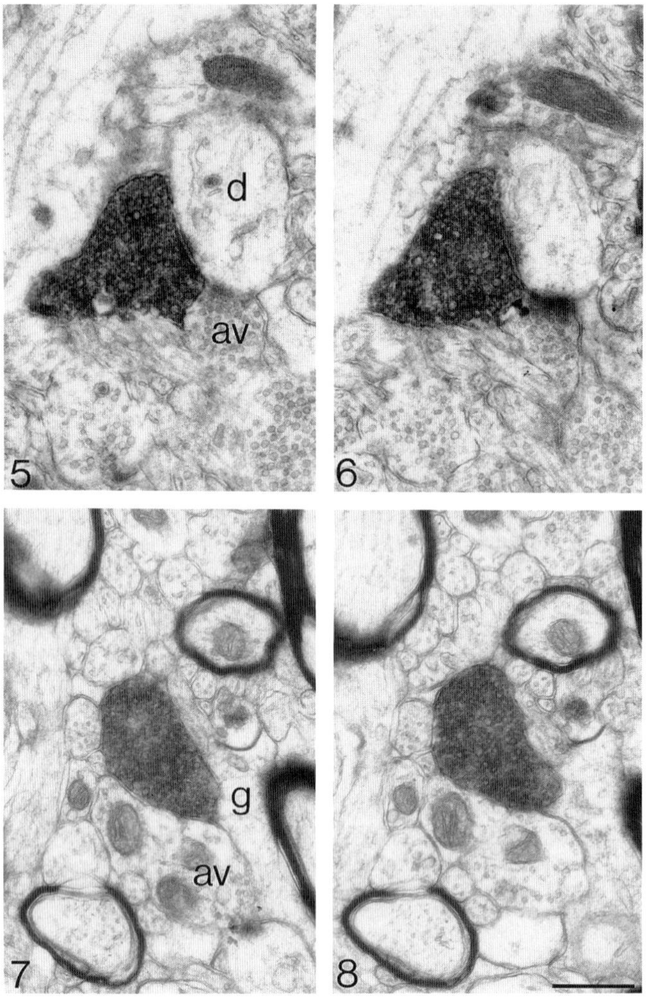

FIGS. 5–8. Adjacent thin sections across a junctional (**Figs. 5 and 6**) and a nonjunctional (**Figs. 7 and 8**) 5-HT-immunostained axonal varicosity from upper and lower layers of frontal cortex, respectively. In Figs. 5 and 6 (two of 10 serial sections), synaptic 5-HT varicosity forms an asymmetrical junction on a dendritic shaft (d), also receiving a synaptic contact from underlying unlabeled varicosity (av) directly apposed to the immunostained profile. In Figs. 7 and 8 (two of four serial sections), presumed nonsynaptic varicosity is juxtaposed to another axon terminal (av), astroglia (g), and several small neuronal processes. $\times$22,000. Calibration bar: 0.5 μm. (Figs. 5 and 7 are reproduced with permission from ref. 13.)

In all four sectors, other axonal varicosities were more frequently encountered in close apposition to 5-HT-immunostained varicosities than their randomly chosen, unlabeled counterparts (Table 2). Comparison with NE varicosities indicated a greater number of dendritic shafts as well as spines around the 5-HT terminals.

Thus, even though the synaptic incidence of cortical 5-HT varicosities appeared higher than previously evaluated by radioautography, this innervation was confirmed to be predominantly nonjunctional, as is now known to be the rule for many other 5-HT innervations in mammalian brain (for review, see ref. 8).

JUNCTIONAL FEATURES OF CORTICAL MONOAMINE TERMINALS

The above account discloses similarities and differences between the relational features of the three monoamine inputs to cerebral cortex. The morphological and/or transmitter identity of the cortical neurons synaptically innervated by the different monoamines has not yet been established. However, it is already clear that only dendritic processes are synaptically contacted by these monoamine afferents. Added to the rarity of axosomatic appositions, this preference speaks in favor of a distal character of the synaptic influences of DA, NE, and 5-HT within the cerebral cortex.

The differences in synaptic targets between the three monoamine innervations are also noteworthy. DA, NE, and 5-HT varicosities synaptically contact dendritic shafts and spines in varying proportions. The frequent DA synapses are found on spines and shafts but with some interregional differences; the rare NE synapses are exclusively found on shafts, and the occasional 5-HT synapses are shared between spines and shafts irrespective of the region considered. The structural differentiation of these contacts also varies between the three types of innervation: DA varicosities form asymmetrical (Gray's type 1) as well as symmetrical junctions (Gray's type 2), all NE varicosities make symmetrical junctions, and most if not all the occasional 5-HT junctions are asymmetrical. The functional significance of these peculiarities remains to be elucidated.

In the face of a predominantly synaptic innervation, many of the known actions of DA in cerebral cortex might be explainable on the basis of junctional relationships, particularly as these actions differ from those of NE and 5-HT (e.g., ref. 26). In the case of the NE and 5-HT varicosities, which are largely nonjunctional, our preliminary inspection of their microenvironment indicates that both are frequently apposed to other axon terminals. Such axoaxonic appositions could account for various types of presynaptic interactions with other cortical inputs of intrinsic or extrinsic origin. As regards "postsynaptic" effects, it is noteworthy that dendritic spines were less frequent around NE varicosities that other cortical terminals in general and that both dendritic shafts and spines were more commonly observed around 5-HT than NE terminals.

LACK OF SYNAPTIC JUNCTIONS ON MANY CORTICAL MONOAMINE VARICOSITIES

From the above data, it is clear that the three monoamine innervations of the cortex differ markedly in the frequency with which their axonal varicosities make synaptic junction. The DA innervation is the most junctional, the 5-HT intermediate, and the NE the least. The DA innervation also demonstrates a significant variation in synaptic incidence between two areas of the prefrontal cortex. The factors that determine the establishment and frequency of synaptic connections made by monoamine varicosities mostly remain to be identified. However, electron microscopic studies in several experimental models have already indicated that alterations of the cellular milieu can entail striking changes in the junctional incidence of these terminals. A marked increase has been documented in the case of the 5-HT innervation in the agranular cerebellum of neonatally irradiated rats (27), and a decrease in the case of the DA innervation of the striatum in weaver mice homozygotes (28). Lesser changes in synaptic incidence and targeting have also been reported for 5-HT varicosities reinnervating the dorsomedial hypothalamus of adult rat after 5,7-dihydroxytryptamine lesioning (29). In this respect, it will be of interest to examine the 5-HT hyperinnervation that occurs in adult rat neostriatum after a neonatal 6-hydroxydopamine destruction of the nigrostriatal DA system (30). In this model, it should be possible to relate eventual changes in the junctional relationships of the observed terminals (i.e., 5-HT) to the features of those that are missing (i.e., DA).

Some of the biological implications of the lack of synaptic membrane differentiations on many cortical NE or 5-HT axonal varicosities have already been discussed elsewhere (5,7). Whether this morphological characteristic is preferentially associated with some fundamental processes such as metabolic regulation (31) and/or plasticity control (32) remains to be determined. In the present context, it should be pointed out that most of the receptor subtypes that have been pharmacologically characterized for DA, NE, and 5-HT in the cerebral cortex do not show a distribution matching the regional and/or laminar densities of the respective innervations (see also refs. 16,18,19). Thus, it is obvious that many of these receptors are neither located on the monoamine varicosities themselves, on their rare synaptic interfaces, nor even on neuronal and/or glial elements in their immediate vicinity. This so-called mismatch between the topographical distribution of the cortical monoamine terminals and of their receptors certainly constitutes suggestive evidence in favor of "volume transmission" (see Herkenham, *this volume*). Presently available information does not allow relation of specific mechanisms of action of the cortical monoamines to the activation of receptor sites preferentially situated in extrajunctional as opposed to synaptic locations. Nevertheless, in view of an increasing amount of biochemical, electrophysiological, and pharmacological data, it is likely that these will be found capable of inducing different effects on activation by different concentrations in various cellular locations and notably at extrajunctional as opposed to junctional sites.

Although future studies will be needed to determine the exact localization of receptor sites for the diverse actions of cortical monoamines and to what extent these might imply junctional versus nonjunctional transmission, it must also be emphasized that the three monoamines probably exert significant effects not only on neuronal elements but also on glial cells (see Hansson, *this volume*) and microvessels (e.g., ref. 33). Interestingly, within the cerebral cortex at least, neither of these tissue constituents appear to be preferentially or tightly abutted by monoamine terminals. This simple fact constitutes another argument in favor of "volume" as opposed to "wired transmission."

RELATIONS BETWEEN SYNAPTIC INCIDENCE AND SYSTEMIC ORGANIZATION

At the theoretical level, the differing synaptic incidences of DA, NE, and 5-HT cortical terminals may be related to the degree of spatial divergence of the corresponding systems. As initially postulated in the absence of ultrastructural data (34), the cortical DA innervation is indeed largely synaptic while it is also the most circumscribed and heterogeneous in terms of regional and laminar distribution. In contrast, the predominantly nonjunctional character of both the NE and 5-HT innervations is associated with a highly collateralized, widespread, and ubiquitous topographic distribution. There seems to be a principle of coherence involved here, whereby highly divergent and dispersed afferent systems (i.e., NE and 5-HT) tend to establish rather loose relationships at the ultrastructural level as well, whereas a more convergent, focused or restricted input (i.e., DA) is also the one showing punctate and perhaps more stable, structurally differentiated synaptic connectivity. The low synaptic incidence (15%) of the histamine innervation of the cerebral cortex (35), which is widely distributed and arises from a small nucleus of the hypothalamus, appears consistent with this hypothesis. In turn, even though "volume transmission" need not to be restricted to nonjunctional systems, it would seem reasonable to consider the two widely distributed NE and 5-HT inputs as the ones characterized by transmission at a distance, multiplicity of cellular targets, and a considerable variety of transmitter actions (i.e., volume transmission), whereas the more convergent DA innervation might influence a lesser number and variety of targets for more specific and perhaps time-framed effects (i.e., wiring transmission).

ASIDE

In view of the increasing knowledge of the properties and function of the monoamine systems in brain, the absence of membrane specializations on some of these axon terminals should not be interpreted as a lack of informational capacity. Obviously, in the peripheral nervous system, neurons endowed with varicosities lacking synaptic junctions do "inform" their cellular targets. In the CNS, it remains to be demonstrated that the membrane differentiations often considered as the hallmark of synaptic transmission are more than "attachment" sites, ensuring some permanence in the structural relation between neuronal partners. Even the words "diffuse" or "nonspecific," often applied to the widely distributed, largely nonjunctional monoamine systems, should not be given a negative connotation. While avoiding quarrels on words even before their exact meaning is amenable to a clear definition, it appears more productive to envisage multiple possibilities in the modalities of information transfer, which must take place in a brain region as structurally complex and functionally powerful as the cerebral cortex.

ACKNOWLEDGMENTS

This work was supported mainly by grant MT-3544 from the Medical Research Council of Canada (L.D.), and in part by the Fonds de la recherche en santé du Québec (Equipe de recherche sur les neurotransmetteurs du cerveau). The authors are indebted to Dr. Michel Geffard (Bordeaux, France) for his generous gift of DA, NE, and 5-HT antisera. They also thank Daniel Cyr for photographic work, and Hélène Collerette for secretarial assistance.

REFERENCES

1. Vizi ES. *Non-synaptic interactions between neurons: modulation of neurochemical transmission*. New York: John Wiley, 1984.
2. Fuxe K, Agnati LF, Zoli M, Bjelke B, Zini I. Some aspects of the communicational and computational organization of the brain. *Acta Physiol Scand* 1989;135:203–216.
3. Descarries L, Beaudet A, Watkins KC. Serotonin nerve terminals in adult rat neocortex. *Brain Res* 1975;100:563–588.
4. Descarries L, Watkins KC, Lapierre Y. Noradrenergic axon terminals in the cerebral cortex of rat. III. Topometric ultrastructural analysis. *Brain Res* 1977;133:197–222.
5. Beaudet A, Descarries L. The monoamine innervation of rat cerebral cortex: synaptic and non synaptic axon terminals. *Neuroscience* 1978;3:851–860.
6. Beaudet A, Descarries L. Quantitative data on serotonin nerve terminals in adult rat neocortex. *Brain Res* 1976;11:301–309.
7. Beaudet A, Descarries L. Fine structure of monoamine axon terminals in cerebral cortex. In: Descarries L, Reader TA, Jasper HH, eds. *Monoamine innervation of cerebral cortex*. New York: Alan R Liss, 1984;77–93.
8. Soghomonian JJ, Beaudet A, Descarries L. Ultrastructural relationships of central serotonin neurons. In: Osborne NN, Hamon M, eds. *Neuronal serotonin*. Chichester: John Wiley & Sons, 1988;57–92.
9. Molliver ME, Grzanna R, Lidov HGW, Morrison JH, Olschowka JA. Monoamine systems in the cerebral cortex. In: Chan-Palay V, Palay SL, eds. *Cytochemical methods in neuroanatomy*. New York: Alan R Liss, 1982;255–277.
10. Papadopoulos GC, Parnavelas JG, Buijs RM. Light and electron microscopic immunocytochemical analysis of the serotonin innervation of the rat visual cortex. *J Neurocytol* 1987;16:883–892.
11. Papadopoulos GC, Parnavelas JG, Buijs RM. Light and electron microscopic immunocytochemical analysis of the noradrenaline innervation of the rat visual cortex. *J Neurocytol* 1989;18:1–10.
12. Parnavelas JG, Papadopoulos GC. The monoaminergic innervation of the cerebral cortex is not diffuse and nonspecific. TINS 1989;12:314–319.
13. Séguéla P, Watkins KC, Descarries L. Ultrastructural relationships of serotonin axon ter-

minals in the cerebral cortex of adult rat. *J Comp Neurol* 1989;289:129–142.
14. Séguéla P, Watkins KC, Geffard M, Descarries L. Noradrenaline axon terminals in adult rat neocortex: an immunocytochemical analysis in serial thin sections. *Neuroscience* 1990;35:249–264.
15. Séguéla P, Watkins KC, Descarries L. Ultrastructural features of dopamine axon terminals in the anteromedial and the suprarhinal cortex of adult rat. *Brain Res* 1988;442:11–22.
16. Descarries L, Lemay B, Doucet G, Berger B. Regional and laminar density of the dopamine innervation in adult rat cerebral cortex. *Neuroscience* 1987;21:807–824.
17. Doucet G, Descarries L, Audet MA, Garcia S, Berger B. Radioautographic method for quantifying regional monoamine innervations in the rat brain. Application to the cerebral cortex. *Brain Res* 1988;441:223–259.
18. Audet MA, Doucet G, Oleskevich S, Descarries L. Quantified regional and laminar distribution of the noradrenaline innervation in the anterior half of the adult rat cerebral cortex. *J Comp Neurol* 1988;274:307–318.
19. Audet MA, Descarries L, Doucet G. Quantified regional and laminar distribution of the serotonin innervation in the anterior half of adult rat cerebral cortex. *J Chem Neuroanat* 1989;2:29–44.
20. Geffard M, Buijs RM, Séguéla P, Pool CW, LeMoal M. First demonstration of highly specific and sensitive antibodies against dopamine. *Brain Res* 1984;294:161–165.
21. Van Eden CG, Hoorneman EMD, Buijs RM, Matthijssen MAH, Geffard M, Uylings HBM. Immunocytochemical localization of dopamine in the prefrontal cortex of the rat at the light and electron microscopical level. *Neuroscience* 1987;22:849–862.
22. Papadopoulos GC, Parnavelas JG, Buijs RM. Light and electron microscopic immunocytochemical analysis of the dopamine innervation of the rat visual cortex. *J Neurocytol* 1989;18:303–310.
23. Geffard M, Patel S, Dulluc J, Rock AM. Specific detection of noradrenaline in the rat brain by using antibodies. *Brain Res* 1986;363:395–400.
24. Zilles K. *The cortex of the rat. A stereotaxic atlas*. Berlin: Springer-Verlag, 1985.
25. Geffard M, Dulluc J, Rock AM. Antisera against the indole-alkylamines: tryptophan, 5-hydroxytryptophan, 5-hydroxytryptamine, 5-methoxytryptophan, and 5-methoxytryptamine tested by an enzyme-linked immunosorbent assay method. *J Neurochem* 1985;44:1221–1228.
26. Mantz J, Milla C, Glowinski J, Thierry AM. Differential effects of ascending neurons containing dopamine and noradrenaline in the contral of spontaneous activity and evoked responses in the rat prefrontal cortex. *Neuroscience* 1988; 27:517–526.
27. Beaudet A, Sotelo C. Synaptic remodeling of serotonin axon terminals in rat agranular cerebellum. *Brain Res* 1981;206:305–329.
28. Triarhou LC, Norton J, Ghetti B. Synaptic connectivity of tyrosine hydroxylase immunoreactive nerve terminals in the striatum of normal, heterozygous and homozygous weaver mutant mice. *J Neurocytol* 1988;17:221–232.
29. Frankfurt M, Beaudet A. Ultrastructural organization of regenerated serotonin axons in the dorsomedial hypothalamus of the adult rat. *J Neurocytol* 1987;16:799–809.
30. Stachowiak MK, Bruno JP, Snyder AM, Striker EM, Zigmond MJ. Apparent sprouting of striatal serotonergic terminals after dopamine-depleting brain lesions in neonatal rats. *Brain Res* 1984; 291:164–167.
31. Mobley P, Greengard P. Evidence for widespread effects of noradrenaline on axon terminals in the rat frontal cortex. *Proc Natl Acad Sci USA* 1985;82:945–947.
32. Gordon B, Allen EE, Trombley PQ. The role of norepinephrine in plasticity of visual cortex. *Prog Neurobiol* 1988;30:171–191.
33. Kalaria RN, Stockmeier CA, Harik SI. Brain microvessels are innervated by locus ceruleus noradrenergic neurons. *Neurosci Lett* 1989;97: 203–208.
34. Descarries L, Doucet G, Lemay B, Séguéla P, Watkins KC. Structural basis of cortical monoamine function. In: Avoli M, Reader TA, Dykes RW, Gloor P, eds. *Neurotransmitters and cortical function*. New York: Plenum, 1988;321–332.
35. Takagi H, Miroshima Y, Matsuyama T, Hayashi H, Watanabe T, Wada H. Histaminergic axons in the neostriatum and cerebral cortex of the rat: a correlated light and electron microscopic immunocytochemical study using histidine decarboxylase as a marker. *Brain Res* 1986;364:155–168.

Volume Transmission in the Brain: Novel Mechanisms for Neural Transmission, edited by K. Fuxe and L.F. Agnati. Raven Press, Ltd., New York © 1991.

6

Mismatches Between Neurotransmitter and Receptor Localizations: Implications for Endocrine Functions in Brain

Miles Herkenham

Section on Functional Neuroanatomy, Clinical Neuroendocrinology Branch, National Institute of Mental Health, Bethesda, Maryland 20892

Over the course of the past decade, electrophysiological, pharmacological, and molecular studies have provided clear evidence for intercellular communication mechanisms that do not adhere to the classically defined electrophysiological (1) and anatomical criteria (2) for synaptic transmission. Initially, the chemical nature of nonclassical communication was emphasized, so that attempts were made to distinguish synaptic neurotransmitters from "neuromodulators," such as peptides, hormones, and growth factors, that might be capable of acting over longer time courses (3) and greater physical distances (4). When the separation of neurochemicals into one or the other category became increasingly difficult, Schmitt (5) proposed that the term "informational substance" should encompass all neurochemicals capable of acting as intercellular messengers and that the kind of communication that works in parallel with the "hard-wired" synaptic circuitry should be called "parasynaptic." He proposed that informational substances acting in the parasynaptic mode are likely to be released at points remote from target cells, which they reach by diffusion through extracellular fluid (5,6). Later the same concept was adopted by others who provided a separate vocabulary, referring to parasynaptic communication as "volume transmission" (7).

The underlying basis for "action at a distance" (4) in the parasynaptic mode can be examined anatomically. An important test of this hypothesis is to compare the anatomical locations of release sites with locations of receptors for each of the informational substances (or simply "transmitters"). On the one hand, predominance of a close register of release sites on nerve terminals and receptors on target cells would conform to the prevailing dogma that synaptic neurotransmission, analogous to that at the neuromuscular junction, prevails within brain. On the other hand, a dissociation in the locations of transmitters and their receptors would argue that parasynaptic information transfer represents a major form of interneuronal communication.

Toward this goal, we developed an *in vitro* receptor binding and autoradiographic technique (8) to enable high-resolution light microscopic visualization of receptor localization in well-preserved tissue sections after receptor binding had been validated by studies of kinetics and structure–activity relations performed in identically prepared sections. Although receptors visualized autoradiographically at the light microscopic level cannot be assigned to particular neuronal elements, the densities, locations, and patterns in individual brain structures can be directly compared with the same aspects of nerve terminals, putatively where most transmitter release occurs, visualized by immunohistochemistry of antibodies raised against the relevant informational substances that interact with the receptors.

PERSPECTIVE

The original publication by Herkenham and Pert (9) elaborated the expectation that locations of opiate receptor distributions in brain would permit us to surmise the existence of "opiatergic" pathways, based on knowledge of tract termination patterns that conformed to receptor patterns. Some of the first reports of regional distributions of enkephalin in brain (2,10) had pointed

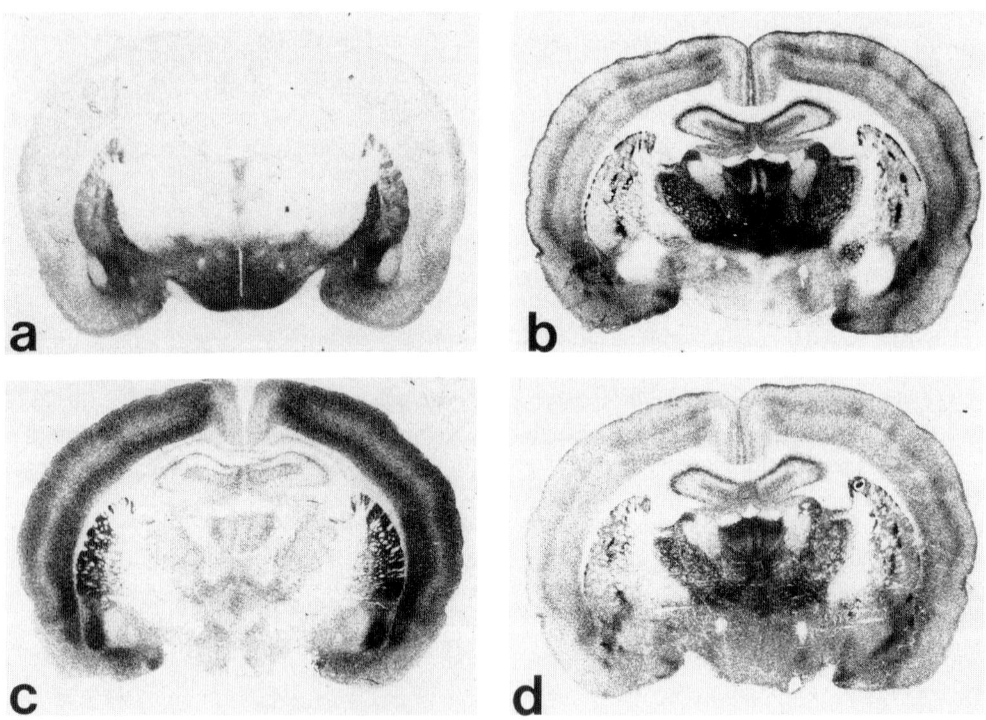

FIG. 1. Autoradiographic film images of radiolabeled coronal rat brain sections at the level of the thalamus. Distribution of enkephalin immunoreactivity (**a**) is compared with patterns of three major opiate receptor subtypes labeled in serial sections (**b–d**). Radioimmunohistochemistry on aldehyefixed sections was performed using a primary antibody against Leu-enkephalin and an ^{125}I-labeled secondary antibody (101). For receptor autoradiography, fresh frozen-cut brain sections were slide mounted and incubated *in vitro* as follows. For μ binding, slides were incubated in Tris buffer with 2nM [^{3}H]D-Ala^2, MePhe^4, Gly-ol^5-enkephalin (16). For δ binding, slides were incubated in Tris buffer with 30 nM oxymorphone and 4.8 nM [3H]D-Ala^2-D-Leu^5-enkephalin (DADLE) (102). For κ binding, sections were first placed in 50 mM MOPS buffer with 1 mM BIT and 1 mM FIT. BIT and FIT are, respectively, selective μ- and δ-opiate receptor irreversible acylating agents (103). Sections were then washed and incubated for 3 hr at 0°C in 50 mM potassium phosphate buffer containing 0.4 M NaCl, 1mg/ml BSA, and 7 nM [^{3}H]bremazocine (104). (Reproduced from ref. 16, with permission.)

out discrepancies between opioid peptide distributions and opiate receptors mapped in early autoradiography studies, but most studies placed greater emphasis on correspondences of the two markers (11–13). When greater attention was paid to anatomical detail, the disparities became clear; brain areas enriched in enkephalin-containing nerve terminals, such as the globus pallidus and hypothalamus, were remarkably opiate receptor–sparse, whereas other areas, such as the thalamus and cerebral cortex, were poor in enkephalin content but enriched in opiate receptors (Fig. 1). The term "mismatch," as it applies to the lack of register between informational substances and their receptors, was first introduced into the literature by Cuello (2), who noted the great disparity between dense enkephalin-containing terminals and sparse opiate receptors in the globus pallidus. Still, such examples were considered to be exceptional instances. The expectation that receptor patterns should match peptide patterns prevailed, and it was presumed by most workers that "explanations" for the mismatches would come forth to allow the conclusion that they were more apparent than real. Receptor localization papers published by pharmacologists in this time period typically noted that the correspondence was fairly good and that some discrepancies did exist, but usually there was little or no attempt to account for them. Interestingly, neuroanatomists publishing immunohistochemical papers localizing transmitters almost never mentioned receptor localization studies for comparison. The delegation of responsibility for consideration of this important functional aspect was given by the neuroanatomists over to the pharmacologists who were not as well-trained to recognize the anatomical disparities of the two sets of markers.

In our publication on substance P receptors (14) (Fig. 2), we argued that mismatches were the rule rather than the exception, and soon thereafter, some re-

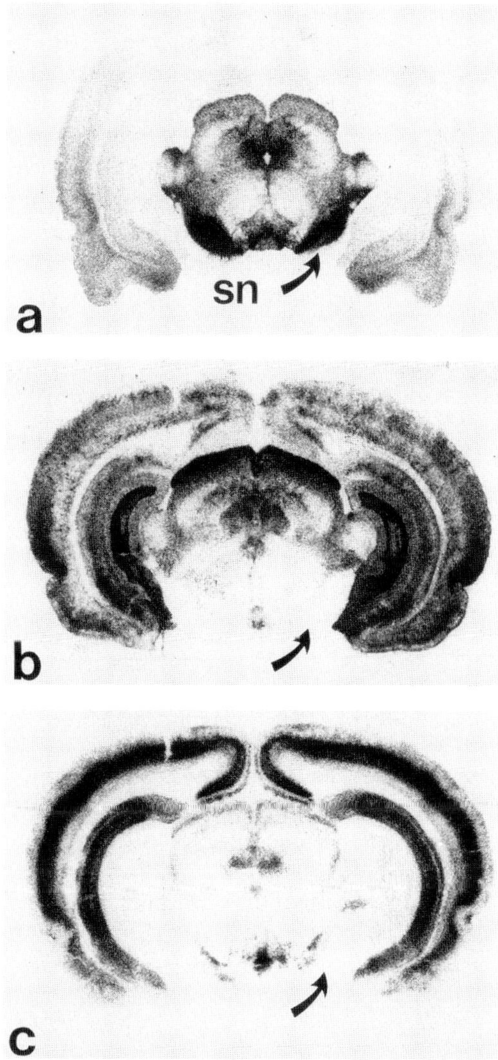

FIG. 2. Autoradiographic maps of substance P immunoreactivity (**a**) and tachykinin receptor subtypes (**b,c**) in rat brain at the level of the interpeduncular nucleus, on the ventral midline, and the substantia nigra (sn). For radioimmunohistochemistry, sections were incubated in primary antibodies against substance P, followed by incubation in ^{125}I-labeled secondary antibody (101). For receptor binding, fresh slide-mounted sections were incubated in ^{125}I-Bolton Hunter-conjugated substance P (b) (14) or eledoisin (c) (105). (Reproduced from ref. 16, with permission.)

searchers acknowledged their prevalence but regarded them as a "mismatch problem" (15). Data amassed from localization studies in other transmitter/receptor systems as well as from developmental and comparative studies provided a better appreciation of the magnitude and generality of mismatches. The definitive works, showing the extent of the mismatches and providing a thorough treatment of all possible explanations for them, appeared in the next 2 years (16–18). The conclusion drawn from the deductive process of ruling out alternative explanations was that mismatches reflect a predominance of high affinity nonsynaptic receptors that mediate parasynaptic communication. However, controversy still surrounds the implications of mismatches (19).

ANATOMICAL EVIDENCE FOR MISMATCHES: AN UPDATE

A summary of the anatomical evidence presented in the *Neuroscience* commentary (17) is given in Table 1. Figures 1 to 3 show some of the evidence provided. In Fig. 1, radioimmunohistochemical maps of opioid peptide terminal distributions and autoradiographic maps of opiate receptor distributions at one level of rat brain show that dissociations in distribution patterns are prevalent. This observation was quantified by Agnati et al. (7), who correlated the densities of enkephalin or β-endorphin immunoreactivity with densities of opiate receptors in numerous brain structures and found no significant relation.

In the tachykinin system (14), a profound mismatch is seen in the rat substantia nigra; the pars reticulata has the densest substance P terminal immunoreactivity in the rat brain and virtually undetectable receptors labeled by either iodinated substance P or eledoisin using subtype-selective binding conditions. Conversely, tachykinin receptors have dense and subtype-specific lamination patterns in cortex, where immunoreactive tachykinins are sparse (Fig. 2).

In our study of species differences in the locations of opiate receptors and opioid peptide, the distributions of μ,δ, and κ opiate receptors were mapped in the hippocampus of four rodent species (i.e., squirrel, guinea pig, rat, and hamster) and were compared with the distributions of immunoreactive dynorphin and enkephalin in each species (18,20). The results show that compared either separately or collectively, there is no consistent relation between the distributions of opioid peptides and opiate receptors—the opioid peptides are similarly distributed across species, but opiate receptors show striking species variations and no consistent relation to the peptide distributions (Fig. 3).

The anatomical evidence for mismatches in acetylcholine (ACh), glutamate, gamma-aminobutyric acid (GABA), norepinephrine, dopamine, serotonin, and 12 neuropeptide systems has been previously reviewed (16,17). Since then, autoradiographic and immunohistochemical evidence has been published to permit comparison in the oxytocin, vasopressin, and luteinizing hormone–releasing hormone (LHRH) systems, as follows.

Vasopressin

The distributions of nerve fibers and terminals containing arginine vasopressin have been presented in two papers using immunohistochemistry of antibodies raised against vasopressin or its prohormone (21,22). The distributions of receptors for vasopressin have been mapped using vasopressin or its analogues (23–25). Dense populations of vasopressin-containing cell bodies are located in the magnocellular portions of the hypothalamic paraventricular, supraoptic, and suprachiasmatic nuclei and in the bed nucleus of the stria terminalis. Immunoreactive fibers and terminals are detected in medial portions of the hypothalamus and extending dorsally and caudally into the supramammillary region, periventricular thalamus, central gray as far caudally as the isthmus region, and the nucleus of the solitary tract and the substantia gelatinosa of the spinal trigeminal and cervical horn areas. Fibers also pass laterally into the medial, central, and posterior cortical nuclei of the amygdala and rostrally into the medial preoptic areas, septum, and diagonal band of Broca. In contrast to this paramedian and limbic trajectory of vasopressinergic fibers, receptors are found in only a few of the aforementioned innervated regions, such as the posterior septum, suprachiasmatic nucleus, ventral central gray, and nucleus of the solitary tract. They are sparse or absent in most of the remaining hypothalamus, amygdala, periventricular thalamus, preoptic area, diagonal band, and caudal medulla and spinal cord. Vasopressin receptors are dense in areas not containing vasopressin innervation, such as the anterior olfactory nuclei, portions of nucleus accumbens, hippocampal formation (notably the ventral subiculum), superior colliculus, interpeduncular nucleus, area postrema, and the inferior olive. In the bed nucleus of the stria terminalis and central nucleus of the amygdala, dense receptors are confined to small regions in patterns that do not match the more dispersed terminal innervation.

Oxytocin

Like vasopressin, oxytocin-containing cells are concentrated in the magnocellular neurons of the hypothalamus and send axons along paramedian and limbic structures (26). Oxytocin receptors have been mapped in several recent studies (25,27,28), and they are concentrated in a subset of structures containing vasopressin receptors (i.e., ventral subiculum, anterior olfactory nucleus, portions of the central amygdala, bed nucleus of the stria terminalis, nucleus accumbens, and the ventromedial hypothalamic nucleus). Thus, other than in the lateral septum and part of the central nucleus, there is little overlap of sites containing both oxytocin nerve terminals and oxytocin receptors. Interestingly, a good correlation has been found between sites showing oxytocin receptor binding and neuronal sensitivity to oxytocin as determined by analysis of species differences in locations of oxytocin receptor distribution and electrophysiological responsiveness to oxytocin (29).

The development of oxytocin receptors has recently been examined (30), and the pattern of emergence dramatically underscores the independence of receptor and transmitter developmental sequelae. In early embryonic rat brain, oxytocin receptors are densely and discretely localized in patterns different from those found in the adult. The switch from the "infant pattern" to the "adult pattern" occurs postnatally around the time of puberty. The receptors in the infant pattern were examined for functionality in tests of neuronal responsiveness to applied oxytocin agonists. The transiently appearing receptors in the dorsal motor nucleus of the vagus showed the same electrophysiology and pharmacology as the adult receptors, even in a location that has no oxytocin nerve innervation.

Luteinizing Hormone–releasing Hormone

Recent work has found LHRH receptors highly concentrated in the hippocampus and the posterior pituitary of the rat but only in the pituitary of the mouse, guinea pig, hamster, rabbit, and human (31). In rat, mouse, and guinea pig, LHRH immunoreactivity is found in the preoptic area, septum, and central gray, and no immunoreactive fibers are found in the hippocampus (32–34).

TABLE 1. *Summary of anatomical evidence for mismatches in rat*[a,b]

Informational substance	High terminal density, sparse receptors	Low terminal density, dense receptors	Colocalized terminals and receptors
Peptides			
Opioids (β-endorphin, dynorphins, enkephalins)	Globus pallidus, hypothalamus, preoptic area, central nucleus of amygdala, median eminence, pontine and medullary reticular formation, hypoglossal nucleus, ventral cochlear nucleus	Olfactory bulb, striatal patches, thalamus, cerebral cortex, basolateral amygdala, olivary pretectal nucleus, presubiculum, medial terminal nucleus of accessory optic tract, tectum, vagus nerve[b]	Nucleus of solitary tract, spinal cord dorsal horn,[b] nucleus ambiguus[b]
Tachykinins (SP, SK)	Substantia nigra pars reticulata, lateral subnucleus of interpeduncular nucleus, septum, bed nucleus of stria terminalis, globus pallidus, medial nucleus of amygdala, lateral habenula, dorsal tegmentum	External plexiform layer of olfactory bulb, cerebral cortex, subfornical organ, anterodorsal nucleus, medial habenula, amygdalohippocampal area, dentate hilus, CA fields of hippocampus, parabrachial nucleus, locus ceruleus, bands in cerebellar lobules 9 and 10, inferior olive	Medial nucleus of amygdala, posterior thalamus, central gray, superior colliculus, parabrachial nucleus, dorsal horn of spinal cord
Neurotensin	Central nucleus of amygdala, ventrolateral septum, medial striatum and nucleus accumbens, bed nucleus of stria terminalis, lateral preoptic area, nucleus ambiguus	External plexiform layer of olfactory bulb, medial frontal cortex, dorsolateral septum, posterior cortical nucleus of amygdala, anterodorsal thalamic nucleus, medial habenula, suprachiasmatic, ventromedial and dorsomedial hypothalamic nuclei, molecular layer of dentate gyrus, parasubiculum, medial entorhinal area, retrosplenial cortex, superior colliculus, substantia nigra pars reticulata, cerebellum	Substantia nigra pars compacta

Somatostatin	Bed nucleus of the stria terminalis, median eminence, central nucleus of amygdala, layer I of cerebral cortex, olfactory bulb, anterior olfactory nuclei, ventromedial striatum, medial preoptic area, most of the hypothalamus (especially the ventromedial and suprachiasmatic nuclei), lateral habenula, nucleus cuneiformis, dorsal parabrachial nucleus vestibular nuclei, medial inferior olive, nucleus ambiguus	Layer V of cerebral cortex, superficial gray layer of the superior colliculus, dorsolateral septal nucleus, medial habenula, locus ceruleus, dentate gyrus, hippocampus (except the most superficial strata where some fibers are located), nucleus of lateral olfactory tract, lateral and basolateral amygdala, deep cerebellar nuclei	Substantia gelatinosa of spinal cord and of spinal trigeminal nucleus, nucleus of solitary tract, periaqueductal gray, inferior colliculus, interpeduncular nucleus
Cholecystokinin (CCK)	Anterior olfactory nucle , deep layers of olfactory tubercle and piriform cortex, bed nucleus of stria terminalis, medial preoptic area, lateral septum, medial and central nuclei of amygdala, anteromedial, reuniers and medial geniculate nuclei of the thalamus, cell layers of hippocampus and dentate gyrus, inferior colliculus, parabigeminal nucleus, ventral tegmental nucleus, nucleus of the solitary tract	Glomerular and external plexiform layers of the olfactory bulb, molecular layers of the olfactory tubercle and piriform cortex, medial striatum (excluding the most medial part), layer IV of the granular retrosplenial cortex (and also remaining cerebral cortex but with relatively lower receptor density), substantia nigra pars reticulata, anterodorsal and reticular nuclei of the thalamus, presubiculum, superficial gray layer of superior colliculus, olivary pretectal nucleus, superior olive, area postrema	Dentate hilus, posterior cortical nucleus of the amygdala, ventromedial hypothalamic nucleus, paraventricular thalamic nucleus, interpeduncular nucleus, dorsal horn of spinal cord

TABLE 1. *Continued*

Informational substance	High terminal density, sparse receptors	Low terminal density, dense receptors	Colocalized terminals and receptors
Corticotropin-releasing hormone (CRH)	Hypothalamus, central nucleus of the amygdala, rostral bed nucleus of the stria terminalis, lateral septum, substantia innominata, medial thalamus, retrorubral area, periaqueductal gray, lateral parabrachial nucleus, floor of the fourth ventricle, substantia gelatinosa of spinal cord	Main olfactory bulb, cerebral cortex, claustrum, nucleus accumbens, vertical limb of diagonal band, olfactory tubercle, caudal bed nucleus of stria terminalis, anterior cortical and basolateral nuclei of amygdala, anterodorsal thalamic nucleus, superior colliculus, interpeduncular nucleus, pontine gray, cerebellar cortex, deep cerebellar nuclei, facial, cuneate, gracile, vestibulocochlear and cochlear nuclei	Median eminence, medial amygdala, inferior olive, nucleus of the solitary tract
Vasoactive intestinal peptide (VIP)	Layers II–IV of cerebral cortex, dentate hilus, central nucleus of amygdala, ventral lateral geniculate nucleus, lateral parabrachial nucleus	Layer I of cerebral neocortex, piriform cortex, olfactory tubercle and dentate gyrus, external plexiform layer of olfactory bulb, magnocellular paraventricular and supraoptic nuclei, tuberal hypothalamus, superior colliculus, sensory relay nuclei of thalamus, interpeduncular nucleus, dorsolateral periaqueductal gray, cerebellum (molecular layer), and pineal gland. In addition, the striatum (especially the dorsomedial corner), medial habenula, lateral thalamus, and locus ceruleus	Suprachiasmatic nucleus, basolateral and medial nuclei of amygdala, bed nucleus of stria terminalis, anterior hypothalamus, midline thalamus, dorsal raphe, nucleus of solitary tract, area postrema
Thyrotropin-releasing hormone (TRH)	Septum, hypothalamus, caudate nucleus, ventral horn of the spinal cord, and dorsal motor nucleus of the vagus	Accessory olfactory bulb, amygdala, dentate gyrus, ventral hippocampus, and dorsal horn of the spinal cord	

Calcitonin gene-related peptide (CGRP)	Layer VI of frontal cortex, claustrum, septum, lateral preoptic area, dentate gyrus molecular layer, ventral pallidum, hypothalamus, medial nucleus of amygdala, gelatinosus and ventromedial thalamic nuclei, zona incerta, substantia nigra and peripeduncular area, shell of inferior colliculus, cuneiform nucleus, nuclei of lateral lemniscus, lateral parabrachial nucleus superior olive, substantia gelatinosa of spinal trigeminal nucleus, cerebellar dentate nucleus	Nucleus accumbens, olfactory tubercle, piriform cortex, dentate hilus, temporal cortex, lateral and basolateral nuclei of amygdala, rhomboid, reuniens and medial habenular nuclei of thalamus, superior colliculus, substantia nigra, pons, central part of inferior colliculus, dorsolateral tegmental nucleus, locus ceruleus, cerebellar molecular layer, inferior olive, dorsal motor nucleus of vagus, hypoglossal nucleus, medullary raphe, pia mater	Cortical area 25, anterior hippocampal area, ventral striatum, striatal fundus, insular cortex, central nucleus of amygdala (lateral part), supramammillary area, periaqueductal gray, medial geniculate nucleus, trigeminal and vestibular nuclei, nucleus of solitary tract, spinal cord
Angiotensin II	Lateral part of bed nucleus of stria terminalis, substantia innominata, central nucleus of amygdala, amygdalo-hippocampal area, lateral parabrachial nucleus	External plexiform layer of olfactory bulb, piriform cortex, lateral olfactory tract, nucleus of lateral olfactory tract, parahippocampal cortex, habenula, medial geniculate nucleus, subthalamic nucleus, superior colliculus, locus ceruleus, dorsal motor nucleus of vagus, inferior olive, cerebellar molecular layer	Anteroventral third ventricle area, paraventricular, suprachiasmatic, supraoptic and dorsomedial hypothalamic nuclei, ventrolateral septum, median eminence, medial thalamus, periaqueductal gray, nucleus of solitary tract, area postrema
Atrial natriuretic peptide (ANP)	Ventrolateral septum, central nucleus of the amygdala, raphe nuclei, periaqueductal gray, and periventricular gray areas of the third and fourth ventricles	Olfactory bulb (especially the external plexiform layer), several white matter tracts (corpus callosum, anterior commissure, lateral olfactory tract, fasciculus retroflexus, ventral hippocampal commissure, fimbria, cerebral peduncle), subfornical organ, organum vasculosum of the lamina terminalis, medial habenular "stripe," thalamic paraventricular nucleus, interpeduncular nucleus, area postrema, pia mater, choroid plexus, and ventricular ependyma	Thalamic paraventricular nucleus

TABLE 1. *Continued*

Informational substance	High terminal density, sparse receptors	Low terminal density, dense receptors	Colocalized terminals and receptors
Neuropeptide Y (NPY)	Preoptic area, hypothalamus (notably suprachiasmatic nucleus), amygdala, substantia nigra pars lateralis, periaqueductal gray, lateral parabrachial nucleus, locus ceruleus, substantia gelatinosa of spinal trigeminal nucleus, molecular layers of hippocampal CA fields and dentate gyrus	Anterior olfactory nuclei, cortical layers I–IV, lateral septum, paramedian, anterodorsal, and medial geniculate thalamic nuclei, inferior olive, area postrema, hippocampal strata oriens and radiatum	Ventral tegmental area, substantia nigra pars compacta, mammillary bodies, ventral lateral geniculate nucleus, inferior colliculus
Vasopressin	Medial hypothalamus, supramammillary region, periventricular thalamus, substantia gelatinosa of spinal trigeminal nucleus and cervical horn, medial, central, and posterior cortical nuclei of amygdala, medial preoptic area, diagonal band of Broca	Anterior olfactory nuclei, nucleus accumbens, hippocampal formation (notably ventral subiculum), superior colliculus, interpeduncular nucleus, area postrema, inferior olive	Posterior septum, bed nucleus of stria terminalis (one part), suprachiasmatic nucleus, ventral central gray, nucleus of solitary tract
Oxytocin	Medial hypothalamus, supramammillary region, periventricular thalamus, substantia gelatinosa of spinal trigeminal nucleus and cervical horn, medial, central, and posterior cortical nuclei of amygdala, medial preoptic area, diagonal band of Broca	Ventral subiculum, anterior olfactory nucleus, bed nucleus of the stria terminalis, accumbens, ventromedial hypothalamic nucleus	Lateral septum, part of central nucleus of amygdala
Luteinizing hormone–releasing hormone (LHRH)	Preoptic area, septum, median eminence, central gray, medial habenula, interpeduncular nucleus	Hippocampus	Anterior pituitary[c]

Monoamines			
Acetylcholine (ACh)	Islands of Calleja	Portions of thalamus, medial nucleus of amygdala, presubiculum, substantia nigra pars compacta	Dentate gyrus, cingulate cortex, layers I and III of retrosplenial cortex, striatum, thalamic reticular and anteroventral nuclei, posterior part of the lateral nucleus, basolateral nucleus, nucleus of the lateral olfactory tract, amygdalohippocampal area, interpeduncular nucleus, superior colliculus, brainstem cholinergic nuclei[d]
Norepinephrine (NE)	Central nucleus of amygdala, molecular layer of dentate hilus, rostral central, and intermediate subnuclei of interpeduncular nucleus, dorsomedial and paraventricular nucle of hypothalamus, anteroventral thalamic nucleus	External plexiform layer of olfactory bulb, anterior olfactory nuclei, olfactory tubercle, layer V of cerebral cortex, caudate nucleus, cerebellar molecular layer, olfactory tubercle, islands of Calleja, globus pallidus, basolateral nucleus of amygdala, mammillary bodies, thalamus, striatum, deep layers of hippocampal CA fields, dentate gyrus molecular layer, lateral subnucleus of interpeduncular nucleus, substantia nigra pars reticulata, facial, lateral reticular, and solitary tract nuclei, entorhinal cortex, and nucleus of solitary tract	Bed nucleus of stria terminalis, hypothalamus (although receptors are selectively sparse in magnocellular nuclei where adrenergic fibers are selectively dense), thalamic paraventricular nucleus, hippocampal and retrosplenial molecular layers, parabrachial nuclei, dorsal motor nucleus of vagus, inferior olive, spinal cord (but density patterns do not match)
Dopamine (DA)	Central nucleus of amygdala, hypothalamus	Somatosensory cortex, hippocampal molecular layer, anterior thalamic nuclei, medial and triangular septal nuclei, dentate gyrus, hippocampal molecular layer, medial habenula, ventroposterior thalamus, superior colliculus, cerebellum, superior and inferior colliculi, mammillary bodies, interpeduncular nucleus, cranial nerve nuclei, cerebellum, choroid plexus	Striatum, prefrontal cortex, periglomerular layer of olfactory bulb, bed nucleus of stria terminalis, globus pallidus, entopeduncular nucleus, subthalamus nucleus, lateral habenula, basolateral nucleus of amygdala, endopiriform nucleus, claustrum

TABLE 1. *Continued*

Informational substance	High terminal density, sparse receptors	Low terminal density, dense receptors	Colocalized terminals and receptors
Serotonin (5HT)	Layer III of the lateral entorhinal area, presubiculum, arcuate and suprachiasmatic nuclei	Layers I–II and IV–VI of the lateral entorhinal area, hippocampus, claustrum, lateral septum, rostral globus pallidus, thalamic intralaminar nuclei, deep cerebellar nuclei	Anterior olfactory nuclei, lateral nucleus of thalamus, subiculum, cerebral neocortex, substantia nigra, superior colliculus, facial nucleus, nucleus of solitary tract, mammillary nuclei
Amino acids			
Gamma amino butyric acid (GABA)	Subgranular zone of dentate hilus, hypothalamus, globus pallidus, reticular nucleus, ventral lateral geniculate nucleus, lateral habenula, substantia nigra, cerebellar molecular layer, deep cerebellar nuclei	Dentate gyrus molecular layer, islands of Calleja, anteromedial and posterior thalamic nuclei, cerebellar granular layer	External plexiform layer of the olfactory bulb
Glutamate		Central gray, interpeduncular nucleus, inferior colliculus, inferior olive	Olfactory bulb, neocortex and hippocampus, striatum, basolateral amygdala, thalamus, pons, red nucleus, superior colliculus, ventral cochlear nucleus, cerebellum
Glycine			Spinal cord gray matter, cranial motor nuclei, auditory, trigeminal, vestibular nuclei of medulla

[a]From ref. 17, with permission.

[b]Areas in which receptors are axonally localized as "presynaptic/nonsynaptic" receptors; [c]likely access to receptors via flow of substance in portal blood; [d]a curious match, in that these nuclei contain concentrations of ChAT-positive cell bodies and are elevated in M2 muscarinic receptor binding. In some of these nuclei, including the medial habenula and nuclei of the diagonal band, no cholinergic terminal plexi are found among the labeled cell bodies, suggesting that receptors may bind dendritically released ACh.

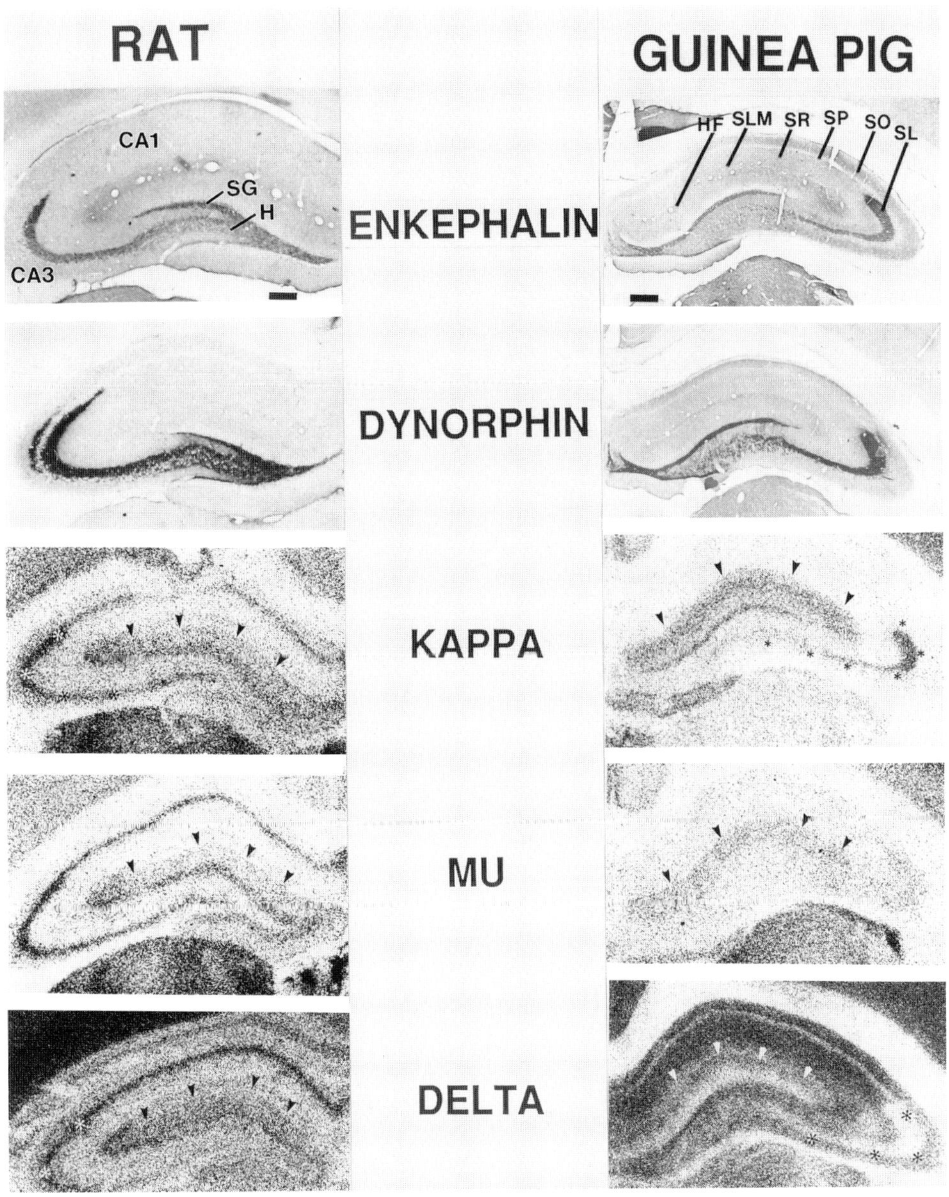

FIG. 3. Distributions of opioid peptides and receptors in hippocampus of four rodent species. Immunoreactivity was performed using avidin-biotin technique, and receptor distributions are derived from autoradiographic film images after 9-weeks exposure. For immunohistochemistry, the animals were perfused with 4% paraformaldehyde. Frozen-cut sections were incubated in antisera against dynorphin B (1:2,000) (E. Weber), Leu-enkephalin (1:1,000) (Immunonuclear), or Met-enkephalin-Arg-Gly-Leu (D. Fitzpatrick). For receptor autoradiography, fresh frozen brains were cut and thaw-mounted onto subbed slides. All slides were preincubated for 30 min in buffered isotonic saline and then incubated in subtype-selective conditions as described in the legend to Fig. 1. CA1, CA3, Corme Ammonis (Ammon's horn) fields; H, hilus; SG, stratum granulosum; HF, hippocampal fissure; SLM, stratum lacunosum moleculare; SR, stratum radiatum; SP, stratum pyramidale; SO, stratum oriens; SL, stratum lucidum. *Asterisks* mark locations of SP. *Arrowheads* mark location of HF. Calibration bar measures 450 μm for squirrel and guinea pig and 250 μm for rat and hamster. (Modified from ref. 18, with permission.)

ULTRASTRUCTURAL EVIDENCE FOR NONSYNAPTIC RECEPTORS

Recent work with antibodies to receptors is adding a great wealth of data to supplement earlier work using radiolabeled ligands at the ultrastructural level. Early electron microscopic studies emphasizing synaptic localization of receptors (35–37) have been followed by work emphasizing the prevalence of receptors not localized to synapses. The use of radiolabeled ligands to localize receptors at the ultrastructural level had been criticized because resolution is limited by radiation scatter (38). Whereas scatter may prevent designating synaptic receptors as pre- or postsynaptic, it does not impede differentiation of synaptic versus nonsynaptic receptors and has the great advantage of permitting pharmacological validation of binding. An *in vitro* receptor binding assay followed by aldehyde fixation, adapted for electron microscopic autoradiography by Hamel and Beaudet (39), provides both binding kinetics and excellent tissue preservation. By this technique, μ-opiate receptors in rat striatal patches were found to be in neuronal membranes at junctions that are predominantly nonsynaptic. Only a small percentage (7%) are located at junctional contacts satisfying the usual criteria for synapses (39,40). In rat locus ceruleus, most μ-receptor binding is localized to neuronal elements. About 50% is intracellular; the rest is localized to the outer plasma membrane, but only a few of the localized receptors reside at contacts that exhibit synaptic specializations (41). Also in rat locus ceruleus, δ receptors are intraneuronal for the most part, and less than 1% of the membrane bound label is at synaptic specializations (42).

Electron microscopic autoradiography of [^{125}I]neurotensin binding has recently been reported (43). Neurotensin receptors in the ventral tegmental area are largely confined to the outer membranes of subpopulations of neurons. In these, the vast majority of the binding overlies the outer membranes of dendrites, somata, and axons and is associated with interfaces of all types. Only about 2% of the labeling at shared interfaces is associated with synaptic contacts, the remainder being nonsynaptic.

Another receptor that lines the plasma membrane of subsets of neurons is the $GABA_A$ receptor coupled to the benzodiazepine receptor/chloride channel complex. Work on the localization of the $GABA_A$/benzodiazepine receptor is illustrative of the resolving power of immunohistochemistry. Using a receptor antibody as a marker at both the light and electron microscopic levels, a pattern of $GABA_A$ receptor distribution was found to be similar to that seen with radiolabeled ligands (44). Correspondences of receptor distributions with locations receiving dense GABAergic innervation are emphasized (44), but close inspection reveals that markers for nerve terminals and receptors may not overlap. For example, in the rat retina, presence of bands of GAD-positive terminals and $GABA_A$/benzodiazepine immunoreactive receptors in the inner plexiform layer led Richards et al. (44) to conclude that it is likely that "all benzodiazepine receptors are associated with GABAergic nerve terminals" (p. 1881). Using the same antibodies and strategy, Mariani et al. (45) examined the relations of terminals and receptors in the primate retina and came to a different conclusion: "That although both were localized to the inner plexiform layer and adjacent cell body layers, their distributions were largely non-overlapping" (p. 153). Instances of both GABAergic nerve terminals in locations lacking receptors and $GABA_A$/benzodiazepine receptors in locations lacking GABAergic terminals were found. Elsewhere in the brain, the overall patterns of GABA receptors and GABAergic nerve terminals do not provide a good match (17).

At the ultrastructural level, the $GABA_A$/benzodiazepine receptor antibody marks both pre- and postsynaptic membranes in the substantia nigra pars reticulata and also membranes not associated with synaptic contacts (44,46). In the cerebellum, the antibody is highly localized to plasma membranes of specified neuronal types, especially the granule cell, which shows immunoreactivity over most of its surface (47). Thus, much of the labeling is nonjunctional (i.e., nonsynaptic). A similar pattern of receptor distribution is found in the human hippocampus with this antibody (48). Although not examined at the ultrastructural level, receptors appear to line the entire dendritic and somal domains of the plasma membranes of selected hippocampal neurons, with large variations in overall density across hippocampal fields, resulting in patterns that have little relation to GABAergic innervation.

Recent work using antibodies to the β-adrenergic receptor in rat cerebral cortex showed that only 12.5% of the labeled processes were at postsynaptic densities. About 2% of the immunolabeling was on myelinated axons, and much of the remaining labeling was within small, unidentified processes (49). In addition, intracellular localization to cytoplasmic elements of neurons and glia was found. Of the β-adrenergic receptors found at postsynaptic sites, a curious feature emerged. In cortical synaptosomes, 12% of all postsynaptic densities were immunoreactive. However, only about 0.001% of the nerve terminals in cortex are adrenergic (50), suggesting that most of the postsynaptic adrenergic receptors are at synapses not receiving adrenergic innervation. The disparity is greater in the striatum, where the percentage of synapses containing postsynaptic β-adrenergic receptors is similarly high, but adrenergic innervation is almost absent. It appears by the morphology, localization to small distal dendrites, and abundance of the immunolabeled synaptic type that most striatal postsynaptic β-adrenergic receptors are localized to glutamatergic synapses.

In the nucleus of the solitary tract, the β-adrenergic receptor antibody has been used concomitantly with antibodies to the catecholamine synthesizing enzymes, which are presynaptic markers for adrenergic axons and terminals (51). As in cortex and striatum, much of the immunolabeling of β-adrenergic receptors is cytoplasmic, and that which is associated with the outer plasma membranes is both synaptic and nonsynaptic. In double-labeled sections, tyrosine hydroxylase immunoreactive nerve terminals and β-adrenergic receptor immunoreactive postsynaptic sites usually were not aligned. Many immunoreactive adrenergic terminals showed no synaptic specializations and were without β-adrenergic receptors. Conversely, many nonadrenergic (nonimmunoreactive) synapses did contain postsynaptic β-adrenergic receptors.

The glycine receptor has been localized by immunohistochemistry of monoclonal antibodies raised against the glycine receptor (52). This receptor is largely confined to postsynaptic sites in neurons. Such labeled postsynaptic elements appear by virtue of morphology and location to receive glycinergic innervation (53). Glycine immunoreactive nerve terminals and glycine receptor immunoreactive postsynaptic elements were found colocalized throughout the spinal cord and also in the olfactory bulb and hippocampus. In the cerebellum a discrepancy was noted as follows. Glycine immunoreactivity appears in Golgi cells of the granule layer; these neurons send axons throughout the granule layer. Glycine receptors, however, are absent in the granule layer and are scattered instead through the cerebellar molecular layer (53). In addition, glycine receptors are occasionally found at GABAergic synapses (54). Overall, however, the receptors are mainly confined to postsynaptic elements at synapses that can often be characterized as glycinergic, and both markers share the same restricted distribution in brain. The correspondence at the ultrastructural level parallels a similar finding of matches at the light microscopic level (see Table 1), suggesting that the glycine system is the only one characterized to date that meets the expectation of a match between locations of terminals and postsynaptic receptors.

MOST "PRESYNAPTIC" RECEPTORS ARE NONSYNAPTIC

Numerous autoradiographic studies have shown anterograde axonal flow of receptors (55). The pharmacological literature contains many examples of drug effects on transmitter release from nerve terminals and has designated such effects as "presynaptic." Presumably, axonally flowing receptors are destined to be inserted in terminal membranes to mediate these effects. If endogenously released transmitters bind to these presynaptic receptors in the synaptic mode, we would expect to find ultrastructural evidence for nerve terminals (ideally immunostained with the transmitter in question) making synaptic contact with other nerve terminals. However, such axoaxonic synapses between nerve terminals have almost *never* been observed, and those that are found are rare and confined to particular brain regions. The more prevalent form of axoaxonic synapse in the brain is that on the initial axon segment (axon hillock) (56). Such synapses are confined to the region of cell bodies, not terminal regions, and so do not constitute the morphological basis for presynaptic pharmacology in regions containing only axons and nerve terminals of the affected neurons. Therefore, it is likely that most physiological and pharmacological effects mediated by presynaptic receptors are, in fact, nonsynaptic. Several documented examples follow.

As summarized by Cuello (57), early work in the spinal cord showed that the substantia gelatinosa of the dorsal horn is rich in both opioid peptides and opiate receptors. Opiate receptors in the dorsal horn cord mediate narcotic analgesia. Opiates inhibit the release of substance P from primary afferents transmitting pain signals. Cutting the dorsal roots causes a drop in opiate receptor binding, indicating that some opiate receptors are presynaptically located on the terminals of primary afferents. By common expectation, therefore, enkephalin-containing axons in this region were believed to make synaptic contacts with the substance P–containing primary afferent terminals. Surprisingly, ultrastructural analysis of the region failed to find evidence for the requisite axoaxonic contacts and, instead, showed that enkephalin-positive terminals make classical axodendritic synaptic contacts with spinal cord neurons and stay "well apart" from substance P–positive terminals (57).

The striatum is a place where the presynaptic control of dopamine release from nerve terminals has been found for all the well-characterized neurotransmitters, including many peptides, monoamines, and amino acid transmitters (58). If we were to presume that the observed pharmacological effects reflect the activity of endogenous systems acting through synapses, there would have to be evidence at the ultrastructural level for axon terminals making synapses with dopamine nerve terminals. However, numerous ultrastructural studies indicate that there are no axoaxonic synapses in the striatum (59).

Opiates are one class of transmitters known to affect dopamine transmission in the striatum by acting on dopamine terminals (58). Opiate receptors are located on dopamine terminals (60). However, ultrastructural studies show that enkephalin terminals never make synaptic contacts on nerve terminals (61,62). Hamel and Beaudet (39,40) examined opiate receptor–rich striatal patches at the ultrastructural level and showed that 18% of the μ-opiate receptors are at nonjunctional (i.e., nonsynaptic) axoaxonic contacts. Perhaps endogenous opiates affect dopamine release via these nonsynaptic contacts.

Opiates applied to the cerebral cortex inhibit antidromic action potentials recorded in locus ceruleus neurons after cortical electrical stimulation. The mechanism for the opiate action would be through binding to opiate receptors on noradrenergic nerve terminals (63). Also in cortex, norepinephrine has widespread effects on nerve terminal activity, as shown by biochemical means (64). Adrenergic receptors have been shown to reside on nerve terminals in cortex (65). The above findings suggest parasynaptic mediation of the physiological effects because axoaxonic synapses on nerve terminals in cortex have never been shown at the ultrastructural level.

Nonneuronal Receptors

As summarized earlier, receptors have been localized to pial arteries, cerebral microvessels, pituitary glial cells, glial cells grown in culture, ventricular ependyma, and choroid plexus (17). These cell types do not receive classical synapses, yet in many instances the receptors on them can be shown to mediate physiological effects. One study has shown supersensitivity of choroid plexus serotonin receptors after destruction of raphe neurons, which have no direct innervation of the choroid plexus but which are known to release serotonin into the ventricles (66). In brief, functional nonneuronal receptors offer strong evidence for parasynaptic transmission.

EXPLANATIONS FOR MISMATCHES

Explanations for mismatches can be put into four categories (Table 2). (a) Typically it has been assumed that mismatches represent technical failures and/or incomplete knowledge of the full endowment of transmitters for receptors within a family of chemically related substances. (b) It has also been pointed out that we should not expect a match because transmitters and receptors represent pre- and postsynaptic elements of different neurons, which may express their features throughout their processes and not just at the sites of contact (15). (c) Biochemical possibilities for the discrepancies include occupied receptors (not available for binding by virtue of prebound endogenous ligand), spare receptors, and nonfunctional receptors. (d) Finally, we have emphasized the fact that many receptors, especially those to which the endogenous ligand binds with high affinity, are not located at synapses and may be sites of action of transmitters released from a distance (16).

TABLE 2. *Summary of categories of explanations for mismatches*

1. Technical factors
    - Limitations of immunohistochemistry
    - Receptor autoradiography artifacts
2. Transmitters and receptors distributed throughout different neurons
    - Transmitter distributed throughout the cell
    - Inability of cell to confine receptor to relevant processes
    - Receptors in axonal transit
    - Internalized receptors
3. Biochemical explanations
    - Occupied receptors
    - Low-affinity receptors
    - Receptor–effector coupling
    - Nonfunctional receptors, or "drug binding sites"
4. Nonsynaptic receptor localization and transmitter action at a distance

An effective strategy for examining the explanations for mismatches was provided in our work comparing the distributions of hippocampal opiate receptors and opioid peptides in four rodent species (18). Mismatch explanations falling in the first three categories are rendered inadequate by comparing neurotransmitter and receptor patterns in a homologous structure (i.e., the hippocampus) in guinea pig and rat (Fig. 3).

In the opiate system, three families of opioid peptides have been identified: the proenkephalin gene yielding the enkephalins and other carboxy-terminal extended forms, the prodynorphin gene providing the dynorphins, and the proopiomelanocortin gene providing β-endorphin. Extensive work has documented the presence of enkephalin and dynorphin but not β-endorphin immunoreactive systems in the rodent hippocampus. The existence of three major opiate receptor subtypes in brain is now accepted: the μ receptor exhibiting highest affinity for morphine and related alkaloids, the δ receptor binding enkephalin better than morphine, and the κ receptor having a high affinity for ethylketocyclazocine and dynorphin (67–70).

The distributions of μ-, δ-, and κ-opiate receptors in guinea pig and rat hippocampus show positive correlations (dense κ receptors and dense dynorphin terminals in the stratum lacunosum of guinea pig), negative correlations (sparse δ receptors and dense enkephalin terminals in the stratum lacunosum; conversely, dense δ receptors and sparse enkephalin in the molecular layer of CA1 of guinea pig), or no correlation (differential lamination of receptors and peptides in rat CA3). Thus, the data show that there is no consistent relation between the distributions of opioid peptides and opiate receptors; in general, the opioid peptides are similarly distributed across species, but opiate receptors show striking species variations and no consistent relation to the peptide distributions.

These results can be used to test the various explanations for the mismatches. Addressing the issues in the first category, we have used the most sensitive and optimal histological techniques currently available and have shown quantitative analysis of the receptors using previously validated binding conditions. We have used markers that recognize the predominant forms of endogenous hippocampal opioid peptides and all the major opiate receptor subtypes, so that it is unlikely that major undiscovered ligands or receptors exist that would alter the presently described relations. Even if, for example, a low-affinity synaptic receptor is discovered that is selectively localized to terminal zones, we would still have to consider the roles of the more widely distributed and mismatched high-affinity sites.

Consideration of hippocampal anatomy sheds light on explanations in the second category. Given the horizontal stratification of the receptors and the vertical alignment of hippocampal dendrites, it is clear that if receptors are localized on hippocampal neurons, they are confined to discrete portions of the dendrites. If some proportion of the receptors is located on afferent axons, it appears to be concentrated at their terminals, as the major afferent fiber tracts do not show commensurate receptor densities. Thus, cells and fibers are both capable of discrete receptor localization, making unlikely the suggestion that receptors in one hippocampal location are cytoplasmic and in transit to another location (71) or that they reflect the cell's inability to concentrate binding sites.

Transmitters, of course, are localized by immunocytochemistry throughout cell bodies, axons, and terminals, and this can be a basis for a mismatch if the sites of terminal release cannot be determined. Opioid peptide immunohistochemistry indicates that the major hippocampal terminal zone lies in the stratum lucidum, whereas scattered fibers with varicosities suggest additional release sites in the dentate and CA fields. Another source of opioid peptide–containing terminals is the perforant path (72–75), but this provides enkephalin fibers predominantly to more caudal (temporal) parts of the hippocampus than the rostral (septal) part examined in this study. Perforant path terminals are localized to the superficial parts of the hippocampal and dentate molecular layers. Thus, for the hippocampus the major sites of peptide release are known, and these taken together are the basis for the conclusion that there is no correlation between their locations and the distributions of the receptors.

In the third category, each of the various biochemical explanations for the mismatch can be put in perspective by the present results. The possibility of receptor occupation by the endogenous ligand is prevented by preincubating sections in buffer containing Na^+, which causes dissociation of endogenous agonists from the receptors (76). A more compelling argument is that we show instances of dense μ and κ receptors in the stratum lucidum, providing an excellent correspondence of dense receptors in a location where endogenous enkephalin and dynorphin are apparently released in high concentrations.

The possibility of spare receptors is an intriguing one that deserves further study. Spare receptors can be responsible for an amplification of biological response, which might be useful in locations where ligand exists in low concentrations. If spare receptors "explain" some of the hippocampal mismatches, they must exist in high relative proportions in some hippocampal regions and not others. As argued elsewhere for the substance P (16,77) and α_1-adrenergic systems (78), regional correlations between receptor density and amplitude of biological response have not documented differential distributions of spare receptors. However, this line of inquiry could be fruitful in the opiate system.

Finally, receptors are seen by some workers as either nonfunctional or "merely" drug binding sites. This is an implausible view in light of the precision with which they are distributed, the metabolic expense and synthetic machinery required to maintain them, and the proven functionality in physiological studies showing responses to drug application localized in areas, such as CA1, where the concept of nonfunctional receptors might be most appealing because of a paucity of endogenous ligand. In fact, within CA1 (Ammon's horn) of the rat, the discrete distribution of opiate receptors in stratum pyramidale and adjacent stratum oriens corresponds precisely with the DADLE (D-Ala^2-D-Leu^5-enkephalin)-sensitive sites most effective in enhancing the population spike (79). Thus, as a corollary, the pattern of electrical sensitivity to opiates applied across CA fields bears no relation to the pattern of nerve terminal distribution of opioid peptides.

The fourth explanation is based on the conclusion that most receptors are not localized to synapses and do not mediate physiologically defined synaptic transmission. Electron microscopic data directly show the anatomical basis for nonsynaptic receptors. The mismatches seen at the light microscopic level suggest further that regardless of receptor localization (e.g., a postsynaptic β-adrenergic receptor at a glutamatergic synapse would not qualify as a receptor-mediating synaptic transmission), sites of transmitter release are often far removed from sites of transmitter action, suggesting that parasynaptic volume transmission is a prevalent form of cell-to-cell communication in the brain.

Extracellular fluid and the cerebrospinal fluid (CSF) with which it is contiguous are conduits for diffusion of endogenous opioid peptides to targets close to the ventricles (e.g., the hippocampus). Ventricular administration of β-endorphin causes hippocampal seizure activity by diffusion of this endogenous ligand into the hippocampal neuropil (80). Seizure activity itself causes release of an endogenous opioid material into the CSF, and this material diffuses to distant neuronal sites to alter the threshold for subsequent seizures (81). The sources and nature of the endogenous neurochemicals and the precise targets for the opiate effects are currently unknown, and speculation about the role hippocampal opiate receptors might play in generation or inhibition of seizures is beyond the scope of the data. However, we can speculate that dense high-affinity opiate receptors in the hippocampus may be localized to participate in cell-cell communication transmitted nonsynaptically at low ligand concentrations. This mechanism is consistent with proposed roles of opiates as neuromodulators of pyramidal cell activity.

IMPLICATIONS

We have considered a number of the possible explanations for the dissociations between transmitter and receptor localizations and have concluded that the following is most likely: Namely, that receptors visualized autoradiographically are often nonsynaptically located and serve to mediate parasynaptic communication between noncontiguous cells, such as that found in the peripheral autonomic nervous system and the anterior pituitary. The anatomical evidence reviewed herein suggests that the synaptic confinement of transmitter release, receptors, and inactivating mechanism found at the neuromuscular junction is rarely found in the brain. Instead it appears that transmitter release sites and receptors can be distributed independently of each other in brain. The reason for this independence may be found in the proposal that immunohistochemistry preferentially reveals the organization of synaptic circuitry, whereas receptor autoradiography is biased toward revealing parasynaptic arrangements of information flow. Supporting this proposal are ultrastructural and high-resolution light microscopic data showing that transmitters tend to conform to expectations in that they are packaged in vesicles concentrated at synaptic junctions, whereas receptors are often nonsynaptically and occasionally nonneuronally distributed. Moreover, it seems possible that areas of high receptor density may contain primarily nonsynaptic receptors, whereas areas of low density may contain primarily synaptic receptors, making density a poor predictor of synaptic function. However, when receptor density does closely correlate with function (measured by electrophysiology, biochemistry of second messenger activation, or pharmacology), we must consider that the function measured is normally nonsynaptic.

The synaptic basis of information flow has been the major focus of neuroscience research for about the past 40 years, but recent developments point to interneuronal communication based on transmitter–receptor interactions not requiring preservation of impulse frequency coding enabled by tight physical coupling at the synaptic specialization. Nerve cells are similar to cells throughout the body in that they respond to circulating substances that bind to receptors coupled to second messengers that elicit changes in membrane potential, cell metabolism, or structure. In this functional mode neurons share many features with cells of the immune system (82). Another analogy is with the peripheral autonomic nervous system, which operates almost exclusively without synapses (83,84). Diffusion of transmitter to distant sites has been documented in the peripheral nervous system (4,85).

Within the brain, numerous hormones have peripheral sources but act centrally at discretely localized receptors via diffusion through extracellular fluids. Examples of this are the melatonin system, whose single source is the pineal gland and whose targets are receptors confined to the suprachiasmatic nucleus and median eminence (86); the insulin system, with blood-borne insulin gaining access to the CSF and widespread receptors (87–89); and the insulin-like growth factors, IGF-1 and IGF-II, with sources in the blood or choroid plexus (90) and targets widely distributed throughout brain (91,92). The generality and magnitude of mismatches suggest that all informational substances are capable of acting in hormonal fashion. Studies of chemical composition of extracellular fluid (e.g., by *in vivo* microdialysis), diffusion of informational substances, and inactivation mechanisms (e.g., reuptake, degradative enzymes) are relevant to this hypothesis but beyond the scope of the present review.

The list of cellular responses not requiring short latencies, brief durations, and tight stimulus-response coupling characteristic of synaptic transmission is steadily growing. The ability of dopamine to act through second messengers to affect the phosphorylation state of a protein (DARPP-32) and thereby alter the membrane sensitivity to another transmitter (glutamate) is one celebrated example (93). That frequency coding of synaptic transmission is not necessarily relevant in the dopamine system is underscored by the ability of drugs or transplanted grafts (94) to restore some function in a dopamine-depleted striatum. Similarly, the memory improvement seen in aged rats after augmenting the cholinergic innervation of the hippocampus by septal graft implantation (95) or in basal forebrain-lesioned rats after basal forebrain graft implantation into cerebral cortex (96) is not likely to be mediated by a frequency-coded message, even though the graft may form synaptic specializations with the host tissue. Other recently shown examples of nonclassical actions of classical neurotransmitters include the demonstration that glutamate can act as a neurotrophic factor at doses below the threshold for cell depolarization (97), that serotonin can regulate growth cone motility (98), and that nicotine acting through ACh receptors can induce rapid gene transcription in postmitotic cells (99). These events may be important for structural modifications associated with connectional rearrangements and with the establishment of long-term memory traces (100).

The parasynaptic or endocrine organization of the brain has many implications for biological psychiatry and clinical neuroendocrinology. For example, centrally acting drugs exert pervasive effects after diffusion through extracellular spaces and, thus, may mimic the mode of action of endogenous neurochemicals far more than was previously thought. In this regard, intracerebroventricular administration of peptides like corticotropin hormone–releasing hormone (CRH) sets into motion a coordinated series of behavioral and physiological responses characteristic of the stress response. Thus, endogenous neurohormones acting at disparate sites in brain seem well-suited for involvement in complex syndromes such as affective illness and anorexia nervosa, which are characterized by global changes in mood, cognition, arousal, appetite, and neuroendocrine regulation. Further work showing the kinds of operations that are mediated parasynaptically and the physiological mechanisms that underlie them should help further elucidate pathophysiological mechanisms in psychiatric illness and will constitute an important focus of our future work.

REFERENCES

1. Barker JL. Evidence for diverse cellular roles of peptides in neuronal function. *Neurosci Res Prog Bull* 1978;16:535–555.
2. Cuello AC. Nonclassical neuronal communications. *Fed Proc* 1983;42:2912–2922.
3. Bloom FE. Dynamic synaptic communication: finding the vocabulary. *Brain Res* 1973;62: 299–305.
4. Jan YN, Jan LY. A LHRH-like peptidergic neurotransmitter capable of "action at a distance" in autonomic ganglia. *Trends Neurosci* 1983;6: 320–325.
5. Schmitt FO. Molecular regulators of brain function: a new view. *Neuroscience* 1984;13: 991–1001.
6. Schmitt FO. Chemical information processing in the brain: prospect from retrospect. In: Iversen LL, Goodman E, eds. *Fast and slow chemical signalling in the nervous system*. New York: Oxford University Press, 1986;239–243.
7. Agnati LF, Fuxe K, Zoli M, Ozni I, Toffano G, Ferraguti F. A correlational analysis of the regional distribution of central enkephalin and β-endorphin immunoreactive terminals and of opiate receptors in adult and old rats. Evidence for the existence of two main types of communication in the central nervous system: the volume transmission and the wiring transmission. *Acta Physiol Scand* 1986;128:201–207.
8. Herkenham M, Pert CB. Light microscopic localization of brain opiate receptors: a general autoradiographic method which preserves tissue quality. *J Neurosci* 1982;2:1129–1149.
9. Herkenham M, Pert CB. *In vitro* autoradiography of opiate receptors in rat brain suggests loci of "opiatergic" pathways. *Proc Natl Acad Sci USA* 1980;77:5532–5536.
10. Lindberg I, Smythe SJ, Dahl JL. Regional distribution of enkephalin in bovine brain. *Brain Res* 1979;168:200–204.
11. Elde R, Hökfelt T, Johansson O, Terenius L. Immunohistochemical studies using antibodies to leucine-enkephalin: initial observations on the nervous system of the rat. *Neuroscience* 1976;1:349–351.
12. Atweh SF, Kuhar MJ. Autoradiographic localization of opiate receptors in rat brain. III. The telencephalon. *Brain Res* 1977;134:393–405.
13. Simantov R, Kuhar MJ, Uhl GR, Snyder SH. Opioid peptide enkephalin: immunohistochemical mapping in rat central nervous system. *Proc Natl Acad Sci USA* 1977;74:2167–2171.
14. Rothman RB, Herkenham M, Pert CB, Liang T, Cascieri MA. Visualization of rat brain receptors for the neuropeptide, substance P. *Brain Res* 1984;309:47–54.
15. Kuhar MJ. The mismatch problem in receptor

mapping studies. *Trends Neurosci* 1985;8: 190–191.
16. Herkenham M, McLean S. Mismatches between receptor and transmitter localizations in the brain. In: Boast CA, Snowhill EW, Altar CA, eds. *Quantitative receptor autoradiography*. New York: Alan Liss, 1986;137–171.
17. Herkenham M. Mismatches between neurotransmitter and receptor localizations in brain: observations and implications. *Neuroscience* 1987;23:1–38.
18. McLean S, Rothman RB, Jacobson AE, Rice KC, Herkenham M. Distribution of opiate receptor subtypes and enkephalin and dynorphin immunoreactivity in the hippocampus of squirrel, guinea pig, rat, and hamster. *J Comp Neurol* 1987;255:497–510.
19. Barnes DM. The receptor mismatch controversy. *Science* 1988;239:142–143.
20. Herkenham M, McLean S. The anatomical relationship of opioid peptides and opiate receptors in the hippocampus of four rodent species. *NIDA Res Monogr* 1988;82:33–47.
21. Caffé AR, van Leeuwen FW, Luiten PGM. Vasopressin cells in the medial amygdala of the rat project to the lateral septum and ventral hippocampus. *J Comp Neurol* 1987;261:237–252.
22. Dubois-Dauphin M, Zakarian S. Distribution of the C-terminal glycopeptide of the vasopressin prohormone in rat brain: an immunocytochemical study. *Neuroscience* 1987;21:903–921.
23. Brinton RE, Gehlert DR, Wamsley JK, Wan YP, Yamamura HI. Vasopressin metabolite, AVP_{4-9}, binding sites in brain: distribution distinct from that of parent peptide. *Life Sci* 1986;38:443–452.
24. van Leeuwen F. Quantitative light microscopic autoradiographic localization of binding sites labelled with [^{3}H]vasopressin antagonist $d(CH_2)_5Tyr(Me)VP$ in the rat brain, pituitary and kidney. *Neurosci Lett* 1987;80:121.
25. Tribollet E, Barberis C, Jard S, Dubois-Dauphin M, Dreifuss JJ. Localization and pharmacological characterization of high affinity binding sites for vasopressin and oxytoxin in the rat brain by light microscopic autoradiography. *Brain Res* 1988;442:105–118.
26. Sofroniew MV. Vasopressin, oxytocin and their related neurophysins. In: Björklund A, Hökfelt T, eds. *Handbook of chemical neuroanatomy*. Amsterdam: Elsevier, 1984;93–165.
27. van Leeuwen F. Light microscopic autoradiographic localization of [^{3}H]oxytocin binding sites in the rat brain, pituitary and mammary gland. *Brain Res* 1985;359:320.
28. Freund-Mercier MJ, Stoeckel ME, Palacios JM, et al. Pharmacological characteristics and anatomical distribution of [^{3}H]oxytocin-binding sites in the Wistar rat brain studied by autoradiography. *Neuroscience* 1987;20:599–614.
29. Raggenbass M, Tribollet E, Dubois-Dauphin M, Dreifuss JJ. Correlation between oxytocin neuronal sensitivity and oxytocin receptor binding: an electrophysiological and autoradiographical study comparing rat and guinea pig hippocampus. *Proc Natl Acad Sci USA* 1989;86: 750–754.
30. Tribollet E, Charpak S, Schmidt A, Dubois-Dauphin M, Dreifuss JJ. Appearance and transient expression of oxytocin receptors in fetal, infant, and peripubertal rat brain studied by autoradiography and electrophysiology. *J Neurosci* 1989;9:1764–1773.
31. Reubi JC, Palacios JM, Maurer R. Specific luteinizing-hormone-releasing hormone receptor binding sites in hippocampus and pituitary: an autoradiographic study. *Neuroscience* 1987;21: 847–856.
32. Silverman AJ, Krey LC. The luteinizing hormone-releasing hormone (LH-RH) neuronal networks of the guinea pig brain. I. Intra- and extra-hypothalamic projections. *Brain Res* 1978; 157:233–246.
33. Samson WK, McCann SM, Chud L, Dudley CA, Moss RL. Intra- and extrahypothalamic luteinizing hormone-releasing hormone distribution in the rat with special reference to mesencephalic sites which contain both LHRH and single neurons responsive to LHRH. *Neuroendocrinology* 1980;31:66–72.
34. Witkin JW, Paden CM, Silverman AJ. The luteinizing hormone-releasing hormone (LHRH) systems in the rat brain. *Neuroendocrinology* 1983;35:429–438.
35. Hunt SP, Schmidt J. The electron microscopic autoradiographic localization of α-bungarotoxin binding sites within the central nervous system of the rat. *Brain Res* 1978;142:152–159.
36. Kuhar MJ, Taylor N, Wamsley JK, Hulme EC, Birdsall NJM. Muscarinic cholinergic receptor localization in brain by electron microscopic autoradiography. *Brain Res* 1981;216:1–10.
37. Möhler H, Richards JG, Wu J-Y. Autoradiographic localization of benzodiazepine receptors in immunocytochemically identified γ-aminobutyrergic synapses. *Proc Natl Acad Sci USA* 1981;78:1935–1938.
38. Kuhar MJ, DeSouza EB, Unnerstall JR. Neurotransmitter receptor mapping by autoradiography and other methods. *Annu Rev Neurosci* 1986;9:27–59.
39. Hamel E, Beaudet A. Electron microscopic autoradiographic localization of opioid receptors in rat neostriatum. *Nature* 1984;312:155–157.
40. Hamel E, Beaudet A. Opioid receptors in rat neostriatum: radioautographic distribution at the electron microscopic level. *Brain Res* 1987;401:239–257.
41. Moyse E, Leonard K, Beaudet A. Electron microscopic localization of mu opioid receptors in the rat locus coeruleus by *in vitro* radioautography (*Abstr.*). *Soc Neurosci Abstr* 1988;14:352.
42. Pasquini F, Beaudet A, Bochet P, Rossier J, Roques BP. Electron microscopic localization of delta opioid receptors in rat brain (*Abstr.*). *Soc Neurosci Abstr* 1988;14:353.
43. Dana CD, Vial M, Leonard K, et al. Electron microscopic localization of neurotensin binding

sites in the midbrain tegmentum of the rat. I. Ventral tegmental area and interfascicular nucleus. *J Neurosci* 1989;9:2247–2257.
44. Richards JG, Schoch P, Häring P, Takacs B, Möhler H. Resolving $GAGA_A$/benzodiazepine receptors: cellular and subcellular localization in the CNS with monoclonal antibodies. *J Neurosci* 1987;7:1866–1886.
45. Mariani AP, Cosenza-Murphy D, Barker JL. GABAergic synapses and benzodiazepine receptors are not identically distributed in the primate retina. *Brain Res* 1987;415:153–157.
46. Richards JG, Glinz R, Schoch P, Möhler H. New trends in mapping benzodiazepine receptors. *Adv Biochem Psychopharmacol* 1988;45: 27–46.
47. Somogyi P, Takagi H, Richards JG, Möhler H. Subcellular localization of benzodiazepine/$GABA_A$ receptors in the cerebellum of rat, cat, and monkey using monoclonal antibodies. *J Neurosci* 1989;9:2197–2209.
48. Houser CR, Olsen RW, Richards JG, Möhler H. Immunohistochemical localization of benzodiazepine/$GABA_A$ receptors in the human hippocampal formation. *J Neurosci* 1988;8:1370–1383.
49. Aoki C, Joh TH, Pickel VM. Ultrastructural localization of β-adrenergic receptor-like immunoreactivity in the cortex and neostriatum of rat brain. *Brain Res* 1987;437:264–282.
50. Lapierre Y, Beaudet A, Demianczuk N, Descarries L. Noradrenergic axon terminals in the cerebral cortex of rat. II. Quantitative data revealed by light and electron microscope radioautography of the frontal cortex. *Brain Res* 1973;63:175–182.
51. Aoki C, Zemcik BA, Strader CD, Pickel VM. Cytoplasmic loop of β-adrenergic receptors: synaptic and intracellular localization and relation to catecholaminergic neurons in the nuclei of the solitary tracts. *Brain Res* 1989; 493:331–347.
52. Triller A, Cluzeaud F, Korn H. γ-Aminobutyric acid-containing terminals can be opposed to glycine receptors at central synapses. *J Cell Biol* 1987;104:947–956.
53. van den Pol AN, Gorcs T. Glycine and glycine receptor immunoreactivity in brain and spinal cord. *J Neurosci* 1988;8:472–492.
54. Triller A, Cluzeaud F, Betz H, Korn H. Evidence that GABAergic terminals and glycine receptors can be associated at the same central synapses. *Proc XIth Int Cong Electron Microscopy, Kyoto* 1986;2329–2330.
55. Palacios JM, Pazos A. Axonal transport of neurotransmitter receptors studied by quantitative autoradiography. In: Boast CA, Snowhill EW, Altar CA, eds. *Quantitative receptor autoradiography.* New York: Alan R. Liss, 1986;173–197.
56. Somogyi P, Freund TF, Hodgson AJ, Somogyi J, Beroukas D, Chubb IW. Identified axoaxonic cells are immunoreactive for GABA in the hippocampus and visual cortex of the cat. *Brain Res* 1985;332:143–149.
57. Cuello AC. Central distributions of opioid peptides. *Br Med Bull* 1983;39:11–16.
58. Chesselet MF. Presynaptic regulation of neurotransmitter release in the brain: facts and hypothesis. *Neuroscience* 1984;12:347–375.
59. Hattori T, Fibiger HC. On the use of lesions of afferents to localize neurotransmitter receptor sites in the striatum. *Brain Res* 1982;238:245–250.
60. Pollard H, Llorens-Cortes C, Schwartz JC. Enkephalin receptors on dopaminergic neurons in rat striatum. *Nature* 1977;268:745–747.
61. Pickel VM, Sumal KK, Beckley SC, Miller RJ, Reis DJ. Immunocytochemical localization of enkephalin in the neostriatum of rat brain: a light and electron microscopic study. *J Comp Neurol* 1980;189:721–740.
62. Somogyi P, Priestly JV, Cuello AC, Smith AD, Takagi H. Synaptic connections of enkephalin-immunoreactive nerve terminals in the neostriatum: a correlated light and electron microscopic study. *J Neurocytol* 1982;11:779–807.
63. Nakamura S, Tepper JM, Young SJ, Ling N, Groves PM. Noradrenergic terminal excitability: effects of opioids. *Neurosci Lett* 1982;30: 57–62.
64. Mobley P, Greengard P. Evidence for widespread effects of noradrenaline on axon terminals in the rat frontal cortex. *Proc Natl Acad Sci USA* 1985;82:945–947.
65. Cash R, Raisman R, Lanfumay L, Ploska A, Agid Y. Cellular localization of adrenergic receptors in rat and human brain. *Brain Res* 1986;370:127–135.
66. Conn PJ, Janowsky A, Sanders-Bush E. Denervation supersensitivity of $5HT\text{-}1_C$ receptors in rat choroid plexus. *Brain Res* 1987;400:396–398.
67. Lord JAH, Waterfield AA, Hughes J, Kosterlitz HW. Endogenous opioid peptides: multiple agonists and receptors. *Nature* 1977;267:495–499.
68. Chang K-J, Cooper BR, Hazum E, Cuatrecasas P. Multiple opiate receptors: different regional distribution in the rat brain and differential binding of opiates and opioid peptides. *Mol Pharmacol* 1979;6:91–104.
69. Chavkin C, James IF, Goldstein A. Dynorphin is a specific endogenous ligand of the kappa opioid receptor. *Science* 1982;215:413–415.
70. Corbett AD, Paterson SJ, McKnight AT, Magnan J, Kosterlitz HW. $Dynorphin_{1-8}$ and $dynorphin_{1-9}$ are ligands for the κ-subtype of opiate receptor. *Nature* 1982;299:79–81.
71. Stengaard-Pedersen K. Comparative mapping of opiate receptors and enkephalin immunoreactive nerve terminals in the rat hippocampus. *Histochemistry* 1983;79:311–333.
72. Gall C, Brecha N, Karten HJ, Chang K-J. Localization of enkephalin-like immunoreactivity to identified axonal and neuronal populations of the rat hippocampus. *J Comp Neurol* 1981; 198:335–350.
73. Stengaard-Pedersen K, Fredens K, Larsson L-I. Enkephalin and zinc in the hippocampal mossy fiber system. *Brain Res* 1981;212:230–233.

74. Tielen AM, van Leeuwen FW, Lopes daSilva FH. The localization of leucine-enkephalin immunoreactivity within the guinea pig hippocampus. *Exp Brain Res* 1982;48:288–295.
75. Fredens K, Stengaard-Pedersen K, Larsson L-I. Localization of enkephalin and cholecystokinin immunoreactivities in the perforant path terminal fields of the rat hippocampal formation. *Brain Res* 1984;304:255–263.
76. Garzón J, Sánchez-Blázquez P, Gerhart J, Loh HH, Lee NM. Dynorphin A: inhibitory effect on other opiate ligand binding in the mouse brain. *Biochem Pharmacol* 1984;33:2609–2614.
77. Mantyh PW, Pinnock RD, Downes CP, Goedert M, Hunt SP. Correlation between inositol phospholipid hydrolysis and substance P receptors in rat CNS. *Nature* 1984;309:795–797.
78. Johnson RD, Minneman KP. α_1-adrenergic receptors and stimulation of [^{3}H]inositol metabolism in rat brain: regional distribution and parallel inactivation. *Brain Res* 1985;341:7–15.
79. Dingledine R. Possible mechanisms of enkephalin action on hippocampal CA1 pyramidal neurons. *J Neurosci* 1981;1:1022–1035.
80. Henriksen SJ, Bloom FE, McCoy F, Ling N, Guillemin R. β-endorphin induces nonconvulsive limbic seizures. *Proc Natl Acad Sci USA* 1978;75:5221–5225.
81. Tortella FC, Long JB. Endogenous anticonvulsant substance in the rat cerebrospinal fluid after a generalized seizure. *Science* 1985;228: 1106–1108.
82. Pert CB, Ruff MR, Weber RJ, Herkenham M. Neuropeptides and their receptors: a psychosomatic network. *J Immunol* 1985;135:820s–826s.
83. Merrillees NCR, Burnstock G, Hollman ME. Correlation of fine structure and physiology of the innervation of smooth muscle in the guinea pig vas deferens. *J Cell Biol* 1963;19:529–550.
84. Vizi ES. *Non-synaptic interactions between neurons: modulation of neurochemical transmission*. New York: Wiley and Sons, 1984.
85. Kravitz EA, Beltz BS, Glusman S, et al. Neurohormones and lobsters: biochemistry to behavior. *Trends Neurosci* 1983;6:345–349.
86. Vanecek J, Pavlík A, Illnerová H. Hypothalamic melatonin receptor sites revealed by autoradiography. *Brain Res* 1987;435:359–362.
87. Hill JM, Lesniak MA, Pert CB, Roth J. Autoradiographic localization of insulin receptors in rat brain: prominence in olfactory and limbic areas. *Neuroscience* 1986;17:1127–1138.
88. Werther GA, Hogg A, Oldfield BJ, et al. Localization and characterization of insulin receptors in rat brain and pituitary gland using *in vitro* autoradiography and computerized densitometry. *Endocrinology* 1987;121:1562–1570.
89. Unger J, McNeill TH, Moxley RTI, White M, Moss A, Livingston JN. Distribution of insulin receptor-like immunoreactivity in the rat forebrain. *Neuroscience* 1989;31:143–157.
90. Stylianopoulou F, Herbert J, Soares MB, Efstratiadis A. Expression of the insulin-like growth factor II gene in the choroid plexus and the leptomeninges of the adult rat central nervous system. *Proc Natl Acad Sci USA* 1988; 85:141–145.
91. Bohannon NJ, Corp ES, Wilcox BJ, Figlewicz DP, Dorsa DM, Baskin DG. Localization of binding sites for insulin-like growth factor-I (IGF-I) in the rat brain by quantitative autoradiography. *Brain Res* 1988;444:205–213.
92. Lesniak MA, Hill JM, Kiess W, Rojeski M, Pert CB, Roth J. Receptors of insulin-like growth factors I and II: autoradiographic localization in rat brain and comparison to receptors for insulin. *Endocrinology* 1988;123:2089–2099.
93. Nestler EJ, Walaas I, Greengard P. Neuronal phosphoproteins: physiological and clinical implications. *Science* 1984;225:1357–1364.
94. Björklund A, Dunnett SB, Stenevi U, Lewis ME, Iverson SD. Reinnervation of the denervated striatum by substantia nigra transplants: functional consequences as revealed by pharmacological and sensorimotor testing. *Brain Res* 1980;199:307–333.
95. Gage FH, Björklund A, Stenevi U, Dunnett SB, Kelly PAT. Intrahippocampal septal grafts ameliorate learning impairments in aged rats. *Science* 1984;225:533–536.
96. Fine A, Dunnett SB, Björklund A, Iversen SD. Cholinergic ventral forebrain grafts into the neocortex improve passive avoidance memory in a rat model of Alzheimer disease. *Proc Natl Acad Sci USA* 1985;82:5227–5230.
97. Koelle GB, Sanville UJ, Thampi NS, Wall SF. L-glutamic acid, a neurotrophic factor for maintenance of acetylcholinesterase and butyrylcholinesterase in the preganglionically denervated superior cervical ganglion of the cat. *Proc Natl Acad Sci USA* 1986;83:2751–2754.
98. Haydon PG, McCobb DP, Kater SB. Serotonin selectively inhibits growth cone motility and synaptogenesis of specific identified neurons. *Science* 1984;226:561–564.
99. Greenberg ME, Ziff EB, Greene LA. Stimulaton of neuronal acetylcholine receptors induces rapid gene transcription. *Science* 1986;234: 80–83.
100. Goelet P, Castellucci VF, Schacher S, Kandel ER. The long and the short of long-term memory—a molecular framework. *Science* 1986; 322:419–422.
101. McLean S, Bannon MJ, Zamir H, Pert CB. Comparison of the substance P- and dynorphin-containing projections to the substantia nigra: a radioimmunocytochemical and biochemical study. *Brain Res* 1985;361:185–192.
102. Rothman RB, Bowen WD, Herkenham M, Jacobson AE, Rice KC, Pert CB. A quantitative study of [^{3}H]D-Ala^2-D-Leu^5-enkephalin binding to rat brain membranes: evidence that oxymorphone is a noncompetitive inhibitor of the lower affinity δ-binding site. *Mol Pharmacol* 1985;27:399–408.

103. Rice KC, Jacobson AE, Burke TR, Bajwa BS, Streaty RA, Klee WA. Irreversible ligands with high selectivity toward δ or μ opiate receptors. *Science* 1983;220:314–316.
104. Rothman RB, Bykov V, Danks JA, et al. Preparation of rat brain membranes highly enriched with opiate kappa binding sites using site-directed acylating agents: optimization of assay conditions. *Neuropeptides* 1985;6:503–516.
105. Danks JA, Rothman RB, Cascieri MA, Chicchi GG, Liang T, Herkenham M. A comparative autoradiographic study of the distributions of substance P and eledoisin binding sites in rat brain. *Brain Res* 1986;385:273–281.

Volume Transmission in the Brain: Novel Mechanisms for Neural Transmission, edited by K. Fuxe and L.F. Agnati. Raven Press, Ltd., New York © 1991.

7

Nonsynaptic Inhibitory Signal Transmission Between Axon Terminals: Physiological and Pharmacological Evidence

E. Sylvester Vizi

Institute of Experimental Medicine, Hungarian Academy of Sciences, Budapest, Hungary

The most challenging feature of the human brain is its display of individual functional diversity under the constraints imposed by uniform patterns of neuronal wiring. This means that the essence of brain function (e.g., learning, thinking) lies not in variations of neuronal circuitry (hardware) but within the chemical communication itself, which is partly wired and partly unwired.

Because synaptic communication between cells had been the most thoroughly studied form of intercellular communication, it was accepted that it is the only form of interaction between neurons and that any change in neuronal excitability is the result of electrical or synaptic transmission. Thus, the classical view of neurotransmission is that it takes place at the synapse, where the nerve ending contains the transmitter substance and comes into close contact with the effector cell containing the receptors. A nerve impulse traveling toward its target cell releases a chemical substance to transmit information, and this substance crosses the synaptic cleft and acts on the postsynaptic membrane.

In the past few years, however, neurochemical and pharmacological (cf. 1–4) and morphological (5–7) evidence has been found that some neurotransmitter modulators may be released from nonsynaptic sites, for diffusion to target cells distant from the release sites to modulate the release of another transmitter released from axon terminals. It was suggested that there is a nonsynaptic interaction between neurons (2–4). Recently Agnati et al. and Fuxe et al. (8,9) provided further evidence of this nonconventional communication and introduced a new term, "volume transmission."

NEURONES TALK TO EACH OTHER WITHOUT SYNAPTIC CONTACT

Nonsynaptic Release of Signal Transmitter

There is, as yet, no firm evidence for the nonsynaptic release of the communicator (i.e., transmitter, modulator). A critical question is: Are transmitter/modulator substances normally released from nonsynaptic boutons? There is no direct evidence on this point; indeed, it would be methodologically difficult to study. However, there is good reason to believe it. It has been known for a long time that postganglionic neurons innervating vas deferens (10), blood vessels, and smooth muscle terminate in a network of varicose areas (i.e., *boutons en passant*) that lack synaptic contact with the target cell. The transmitter is released from structurally unspecialized areas along the varicose fibers and must often travel a considerable distance before interacting with its receptors. As Descarries and co-workers (6) pointed out, nonsynaptic varicosities in the central nervous system (CNS) appear to have all the apparatus normally associated with synaptic release. Subsequently, ultrastructural examination of noradrenergic varicosities in several tissues confirmed that both large and small vesicles could undergo exocytosis in the absence of structurally specialized active zones (11). It is conceivable that a similar exocytosis occurs in monoaminergic neurons in Auerbach's plexus (12,13) and in the brain (6) where active synaptic zones are lacking. Because certain terminals in the dorsal horn of the signal cord and trigeminal nucleus caudalis are reported to lack typical synaptic zones (14), it is likely that substance P– or enkephalin-containing vesicles release their contents at nonspecialized sites. The peripheral exocytosis at a distance from the synaptic complex (15) suggests that large vesicles have a different physiological function than the small synaptic vesicles. Exocytotic release from large vesicles containing neurohormones has been also reported to occur at nonspecialized areas of the membrane in invertebrates (cf. 16–18).

The concept of nonsynaptic communication has been found mainly on neurophysiological and neurochemical studies. These studies have shown that particular neurons can influence each other by releasing chemical messengers in the apparent absence of any synaptic interconnection (cf. 4). The intercellular communication in both the peripheral nervous system and the CNS is not limited to synaptic transmission mediated by transmitters. Modulator and transmitter substances released (a) from nonsynaptic axon terminals, (b) from axon terminals making synaptic contact but acting both on synaptic and nonsynaptic target cells, or (c) from regions of the nerve cell other than the presynaptic axon terminal [i.e., dendrites, axon (19), soma] should play a physiological role in the modulation of neurochemical transmission.

In frog sympathetic ganglia it was shown (20) that leutinizing hormone–releasing hormone (LHRH) coexists with acetylcholine (ACh) in the same nerve terminals, and although the peptide is most probably released together with ACh that forms conventional morphological synaptic contacts with sympathetic neurons, it can diffuse for tens of micrometers before activating receptors. Thus, if nonsynaptic release sites are involved in nonsynaptic communication, we would expect that nonsynaptic release sites have a general occurrence in the animal kingdom.

Nonsynaptic Receptor Localization

There is an interesting difference in localization and sensitivity of receptors on effector cells where the transmitter [e.g., ACh, norepinephrine (NE), dopamine (DA)] or neuropeptide is released into a large extraneuronal space. In the case of synaptic transmission, there is a small area of the effector cell where the receptors are concentrated. The neuromuscular junction is an example of a perfect correspondence between release sites of ACh and the location of postsynaptic nicotinic receptors. However, when the transmitter release site and the target cells are widely separated from each other (i.e., 100–3000 nm), there is no specific subsynaptic arrangement, and at the recognition sites, receptors are unevenly distributed along the surface of the effector cell (i.e., presynaptic axon terminals, etc.). This morphological arrangement fits any diffusion type of transmission in which the advantage of quantal release cannot be used. There seems to be no evidence that nonquantal release can occur solely at diffuse synapses. In addition, it could be argued that the receptors distant from the synaptic sites do not represent active receptors capable of opening ion channels. However, it is also known that the sensitivity of extrasynaptic receptors is higher to chemical signals (cf. 4). In addition, it was shown that the release site/receptor matches are exceptions rather than rules (21,22). It was reported that mismatches are more numerous in the peptide systems.

The explanation for the widespread extrasynaptic distribution of the receptor complex may lie in the general role of the transmitter-gated channels. The synaptic and nonsynaptic receptors may be used under different levels of neuronal activity, providing a mechanism for adaptation (23). Somogyi et al. (23) showed in cerebellum that at low levels of neuronal activity, gamma-aminobutyric acid (GABA) acting at the synaptic junctions provides adequate inhibitory control of the neurons. The moderate amounts of GABA released can be removed by reuptake or glial uptake without reaching the extrasynaptic sites. However, at increased excitatory input, the number of channels operating at the junctions may not be adequate to keep the activity of the cell in the range of optimal sensitivity, in which case channels at extrasynaptic sites would be opened by GABA diffusing from the release sites. Consequently, the topography of GABAergic effects would primarily be regulated not by the precise placement of receptors but by the precise placement of GABA-releasing synaptic terminals and by the amount of GABA released.

The modulators exert their actions via different receptors. If there is no direct contact between two or more neurons, transmitters might reach their target cell by diffusion. This remote control of neuronal activity can be monitored only by those cells that are equipped with receptors sensitive to the modulator (i.e., discrimination) (4,22).

FUNCTIONAL INTERACTION BETWEEN NEURONS WITHOUT SYNAPTIC CONTACT

The nonconventional, nonsynaptic release of a substance and the nonsynaptic localization of receptors does not necessarily mean that there is an interaction between the release site and, for interneuronal communication, the effect cell. The substance released should be able to diffuse the distance and reach the receptor in concentration high enough to stimulate it (Table 1). It seems likely that even peptides can be used for long-distance signal transmission. An important implication of recent findings is that this release may not be the exclusive property of the axon terminal (19), and it has been suggested (1–4,8,9,24–26) that in addition to the classical synaptic form of interaction there is a nonsynaptic form of communication between neurons when the transmitter or modulator released from axon terminals may affect remote target cells. It seems likely that the released transmitter/modulator can reach surrounding target cells, neighboring axons, or axon terminals, especially when they are not insulated (e.g., in the case of the nonmyelinated nerve).

Neurochemical Evidence

There is no synaptic contact (axoaxonic synapses) between noradrenergic and cholinergic neurons in the gut; therefore, the increased ACh output from the Auerbach's plexus, at rest as well as in response to stimulation, which occurred after catecholamine depletion by reserpine and guanethidine (27,28) and after 6-hydroxy-DA (29), suggests a local permanent control by adrenergic nerves of the cholinergic function. Adrenaline, NE, and clonidine reduce the release of ACh evoked by axonal stimulation, and selective alpha$_2$-adrenoceptor antagonists prevented the inhibitory effect of

TABLE 1. *Presynaptic interaction between cholinergic and noradrenergic axon terminals in the intestinal tract*[a]

	^{3}H-NE (^{14}C-ACh) release, S_2/S_1			
			+ Physostigmine	
	^{3}H-NE	^{14}C-ACh	^{3}H-NE	^{14}C-ACh
Control	0.89±0.07	0.84±0.06	0.79±0.15	1.23±0.07
Atropine, 1 μM	0.69±0.04	1.23±0.07	3.79±0.25	5.11±0.43
CH-38083, 0.3 μM	1.66±0.07	1.68±0.03	2.12±0.19	1.95±0.12

[a]Longitudinal muscle strip attached with Auerbach's plexus (guinea pig ileum). Krebs' solution. Stimulation rate, 1 Hz (180 shocks). Physostigmine (2 μM) was present in the solution as indicated. ^{3}H-norepinephrine (NE) and ^{3}H-choline was used. Note that atropine enhanced release of acetylcholine (ACh) in the absence and presence of physostigmine, but it increased release of NE only if the cholinesterase was inhibited. This finding indicates that under physiological conditions ACh is not able to reach noradrenergic axon terminals. In contrast, NE, released from the noradrenergic axon terminals, in addition to its negative feedback modulation, tonically controls the release of ACh. Number of experiments, 4–4. S_2/S_1, ratio between the amount of transmitter released by two stimulations. Drugs were added 18 minutes prior to second (S_2) stimulation. CH-38083, alpha-2 adrenoceptor blocking agent.

agonists and in itself enhanced the release. Similarly, prior chemical sympathectomy with 6-hydroxy-DA prevented the reduction of stimulated ACh release by perivascular nerve stimulation (30). The finding that the sympathetic nerve stimulation (28) reduces the release of ACh is, in fact, the first convincing evidence that a substance released from a nonsynaptic bouton is able to reach another axon terminal.

Table 1 shows the interaction between noradrenergic and cholinergic neurons in a longitudinal muscle strip of guinea pig ileum. Both cholinergic and noradrenergic axon terminals are equipped with muscarinic and alpha$_2$ receptors. Using field stimulation both noradrenergic and cholinergic axons were stimulated; therefore any possible interaction between them can be studied. The fact that the atropine failed to enhance NE release in the absence of anticholinesterase but it did when physostigmine (Physostigmine 2 μM) was present suggests that the ACh released was not able to reach noradrenergic axon terminals. This finding indicates that the interaction between the two nerves is one-sided: NE released from the noradrenergic axon terminals controls the release of ACh through alpha$_2$-adrenoceptors. Because there is no

TABLE 2. *Differences in the synaptic and nonsynaptic inhibitory interactions between neurons*

	Inhibitory interaction between neurons	
	Synaptic	Nonsynaptic
Communication	One-to-one	One to many
Delay	Short (100–200 msec)	Long (1 sec, 1 min)
Action	Short and phasic	Long and tonic
Discrimination		Depends on presence of receptors
Sensitivity of receptors	Low	High
Concentration in vicinity of receptors	Very high (NmM)	Low (0.05 – 1 μM)[a]
Chemical agent has to	Cross the gap (5–20 nm)	Diffuse distances of some micrometers
Morphological characteristics	Axoaxonic synapse (postjunctional specializations)	Free nerve-ending, axonal varicosities (no postjunctional specializations)

[a]Data taken from refs. 27 and 33.

synaptic interaction between noradrenergic and cholinergic neurons (12,13), NE should diffuse long distances to reach cholinergic axon terminals. Acetylcholine released was possibly promptly hydrolyzed by cholinesterase (ChE) present in the extracellular space. That this was case is shown by the finding that in the presence of physostigmine, the release of NE was partly inhibited through muscarinic receptor as evidenced that atropine more than tripled the release (Table 1).

Similar interaction was observed in the CNS. It has been shown that NE reduces the release of ACh from isolated cortical slices induced either by ouabain (31) or by electrical stimulation (32). The spontaneous and evoked release of ACh was higher in those slices in which noradrenergic input was impaired in some way (2), either by phentolamine, which blocks the action of NE on alpha$_2$-adrenoceptors, or by 6-hydroxy-DA pretreatment, or by a ipsilateral locus coeruleus lesion. After a locus coeruleus lesion, the spontaneous and evoked release of ACh from slices dissected from the ipsilateral side was higher in comparison with those on the contralateral side. NE significantly reduced the resting release of ACh only in those cases in which noradrenergic control had been previously removed (2). It is suggested that the release of ACh from cholinergic neurons of the cerebral cortex is continuously controlled by NE released from nerves arising from the locus coeruleus. The removal of this inhibitory system results in an increase of ACh release. Similar action was observed on isolated human cortical slices (33); NE inhibited the release of ACh, and its effect is mediated via alpha adrenoceptors because phentolamine prevented the effect of NE. In addition, a widespread neurochemical effect of NE was observed on axon terminals in the rat frontal cortex (34). Because there is morphological contact, no axoaxonic synapsis between noradrenergic and cholinergic axon terminals (6,35) and the naked varicosity is able to release NE, which might be capable of exerting a widespread influence on vast neuronal ensembles.

In addition, there is neurochemical evidence that DA released from the nigrostriatal pathway tonically controls the release of ACh from the cholinergic interneurons (36). Removal of dopaminergic input by 6-hydroxy-DA resulted in an increase of ACh release. The interaction between dopaminergic and cholinergic neurons is nonsynaptic. A similar function of DA was described by Bjelke et al. (37), who showed that the deprivation of nonstriatal DA leads to an increase of prolactin release from the adenohypophyseal striatal transplant.

Morphological Evidence

Axon terminals devoid of synaptic membrane specialization have long been the rule in the peripheral autonomic nervous system (10). It has been suggested that in the CNS, the postsynaptic membrane specialization is not always apparent, and relative paucity or even a lack of synaptic junctions might be a feature common to all types of central monoamine terminals.

Therefore it seems likely that NE released from the varicose axon terminals reaches its target cells (e.g., the cholinergic varicosities) by diffusion and inhibits the release of ACh. This kind of interaction between two axon terminals is of physiological importance. NE released from the noradrenergic terminals inhibits the release of ACh from Auerbach's plexus (28,30), although there is no synaptic contact between them. Nevertheless, Manber and Gershon (30) provided evidence that there is also some noradrenergic-cholinergic axoaxonic contacts in the mammalian gut, but these are negligible in number.

New data on the existence and distribution of DA (38,39) and serotonin (5-HT)-containing (5) axon terminals in the cortex suggest that both DA and serotonin nerve endings have a different distribution from the axonal nerve boutons labeled with ^{3}H-

NE (cf. 7,40). Within the same region, noradrenergic terminals predominate more clearly in the superficial layers (41), whereas dopaminergic terminals appear to be mainly confined to the deeper layers (38,39). It has been suggested (7) that NE- and 5-HT-containing neurons may be ideally designed to exert sustained influences in various regions of innervation and thus modulate several integrative and/or specific functions. By contrast, a different picture began to emerge a few years ago (42). Several lines of evidence indicate that cortical monoaminergic afferents show remarkable anatomical specificity, not different from conventional synapses (cf. 42). However, Descarries and co-workers (40) showed that the overall density of noradrenergic innervation is 1.2 million varicosities/mm^3 of tissue. In the molecular layer, 1.5 to 2 times was higher the density of noradrenergic varicosities, and then progressively decreased in the underlying cortex. Only one of 1,000 terminals was noradrenergic.

PHYSIOLOGICAL IMPORTANCE

The nonconventional, nonsynaptic release of substance and nonsynaptic localization of receptors provide a possible mechanism for interneuronal communication, and it seems likely that even peptides can be used for transmitting signals.

The nonconventional interaction between neurons and the nonsynaptic modulatory role of different amines and peptides (i.e., NE, DA, 5-HT, enkephalins, etc.) in neurochemical transmission appears to be feasible: Anatomical, neurophysiological, neurochemical, and pharmacological evidence supports this suggestion. The nonsynaptic control of chemical neurotransmission by different modulators released from axonal varicosities lacking junctions might play a physiological role both in the CNS and in the neurovegetative system in shaping emotion, behavior, or learning processes or in controlling the balance between the parasympathetic and sympathetic nervous system. Presynaptic inhibition is an important form of interaction between two neurons and plays a critical role in the modulation of transmission. This type of inhibition of chemical transmission results from a decrease of transmitter/modulator release. There are two types of presynaptic modulation (Table 2):

1. Synaptic, when an inhibitory input makes synaptic contact with an axon terminal (axoaxonic synapse) or the initial segment of the cell and the inhibitory modulator released inhibits the release of the principal transmitter
2. Nonsynaptic, when a substance (modulator) is released from an axon terminal and/or varicosity devoid of synaptic characteristics, but it is able to reach the target cell.

PHARMACOLOGICAL CONSIDERATION

Widely used drugs in clinical practice have been developed or applied on the concept of presynaptic modulation of neurochemical transmission. Receptors located presynaptically or prejunctionally are the targets of different drugs having either an agonist or antagonist activity. The sensitivity of nonsynaptic receptors is higher, and they are much more accessible to drugs; a much lower concentration of drug is needed to affect them.

Drugs have difficulty reaching the receptors intrasynaptically, and, receptors being relatively insensitive, drugs are able to affect them only in high and toxic concentrations. These receptors are not target for remote (i.e., nonsynaptic) modulation or remote signal transmission. However, there are receptors (i.e., rudimentary) that are never reached by endogenous substances (i.e., transmitters, modulators, etc.) but that might be affected by relatively low concentrations of drugs.

Learning more about the molecular mechanisms that control interneuronal communication is still an important research goal because it may inevitably lead to new strategies in drug development. The drugs produced by these new approaches may or may not have therapeutic applications; these must be defined by clinical studies. Nevertheless, whatever knowledge we uncover as we pursue the study of how neurons interact will improve our overall understanding of the brain and thus contribute to the achievement of our ultimate goal. The nonsynaptic communication system has the same degree of selectivity as synaptic circuitry but possesses, in addition, a domain of versatility and plasticity in "hardwared" circuitry.

SUMMARY

The neuromodulator communication, which does not require direct synaptic contact between the nerve cell releasing the transmitter and the target cells, can be compared with a radio broadcasting system, where there is public transmission of a signal that can be picked up by any properly equipped receiver, in this case by any cell properly equipped with receptors, within range.

In addition to neurotransmitter substances acting at close range in chemical synaptic neurotransmission, there exist chemical interactions between neurons without any close synaptic contact: This interneuronal, nonsynaptic modulation of transmission, also called *volume transmission,* operates over some distance and might be regarded as a transitional form between the classical neurotransmission and the broadcasting of neuroendocrine secretion.

Brain is a wired instrument, but its neurons, besides cabled information signaling (through synapses), are able to talk to each other without synaptic contact (i.e., "wireless"). The brain is no more accepted as a mathematical but rather a geometrical system. Pellionis (43) even emphasized that the brain might be a geometrical, probably a non-Euclidean, system.

REFERENCES

1. Vizi ES. Presynaptic modulation of neurochemical transmission. *Prog Neurobiol* 1979;12:181–290.
2. Vizi ES. Non-synaptic modulation of transmitter release: pharmacological implication. *TIPS* 1980; 1:172–175.
3. Vizi ES. Non-synaptic intercellular communication: presynaptic inhibition. *Acta Biol Acad Sci Hung* 1982;33:331–351.
4. Vizi ES. Non-synaptic interactions between neurons: modulation of neurochemical transmission. In: *Pharmacological and clinical aspects.* Chichester: Wiley, 1984.
5. Descarries L, Beaudet A, Watkins KC. Serotonin nerve terminals in adult rat neocortex. *Brain Res* 1975;100:563–588.
6. Descarries L, Watkins KC, Lapierry Y. Noradrenergic axon terminals in the cerebral cortex of the rat. III Topometric ultrastructural analyses. *Brain Res* 1977;133:197–222.
7. Beaudet A, Descarries L. The monoamine innervation of rat cerebral cortex synaptic and nonsynaptic axon terminals. *Neuroscience* 1978; 3:851–860.
8. Agnati LF, Fuxe K, Zoli M, Zini I, Toffano G, Ferraguti F. A correlation analysis of the regional distribution of central enkephaline and beta-endorphin immunoreactive terminals and of opiate receptors in adult and old male rats. Evidence for the existence of two main types of communications in the central nervous system: the volume transmission and the wiring transmission. *Acta Physiol Scand* 1986;128:201–207.
9. Fuxe K, Agnati LF, Cintra A, et al. Studies on central D1 receptors role in volume transmission, neuroendocrine regulation and release of noradrenaline. In: Goldstein M, Fuxe K, Tabachnick I, eds. *Advances in experimental medicine and biology,* vol 235. New York, London: Plenum Press, 1988;83–119.
10. Merrillees NCR, Burnstock G, Hollman ME. Correlation of fine structure and physiology of the innervation of smooth muscle in the guinea pig vas deferens. *J Cell Biol* 1963;19:529–550.
11. Thureson-Klein A. Exocytosis from large and small dense cored vesicles in noradrenergic nerve terminals. *Neuroscience* 1983;10:245–252.
12. Gordon-Weeks PR. Are there noradrenergic synaptes in Auerbach's plexus? *Scand J Gastroenterol* 1981;16:11–182.
13. Gordon-Weeks PR. Noradrenergic and non-noradrenergic nerves containing small granular vesicles in Auerbach's plexus of the guinea pig. *Neuroscience* 1982;11:2925–2936.
14. Glazer EJ, Basbaum AI. Opioid neurons and pain modulation: an ultrastructural analysis of

enkephalin in cat superficial dorsal horn. *Neuroscience* 1983;10:357–376.

15. Zhu PC, Thureson-Klein A, Klein RL. Exocytosis from large dense cored vesicles outside the active synaptic zones of terminals within the trigeminal subnucleus caudalis: a possible mechanism for neuropeptide release. *Neuroscience* 1986;19:43–54.
16. Golding DW, Bayraktaroglu E. Exocytosis of secretory granules—a probable mechanism for the release of neuromodulators in invertebrate neuropiles. *Experientia* 1984;40:1277–1280.
17. Buma P, Roubos EW. Calcium dynamics, exocytosis and membrane turnover in the ovulation hormone releasing caudo-dorsal cells of Lymnea stagnalis. *Cell Tiss Res* 1983;233:143–159.
18. Buma P, Roubos EW, Buijs RM. Ultrastructural demonstration of exocytosis of neural neuroendocrine and endocrine secretions with an *in vitro* tannic acid (TARI) method. *Histochemistry* 1984;80:247–256.
19. Vizi ES, Gyires K, Somogyi GT, Ungvari G. Evidence that transmitter can be released from regions of the nerve cell other than presynaptic axon terminal axonal release of acetylcholine without modulation. *Neuroscience* 1983;10:967–972.
20. Jan YN, Jan LY. A LHRH-like peptidergic neurotransmitter capable of "action at a distance" in autonomic ganglia. *TINS* 1983;320–325.
21. Herkenham M, McLean S. Mismatches between receptor and transmitter localizations in the brain. In: Boast CA, Snowhill EW, Altar CA, eds. *Quantitative receptor autoradiography*. New York: Alan R. Liss, 1986;137–171.
22. Herkenham M. Mismatches between neurotransmitters and receptors. *Neuroscience* 1987; 23:38.
23. Somogyi P, Takagi H, Pichrads JG, Mohler H. Subcellular localization of benzodiazepine (GABA) receptors in the cerebellum of rat, cat and monkey using monoclone antibodies. *J Neurosci* 1989;9:2197–2209.
24. Dismukes K. New look at the aminergic nervous system. *Nature* 1977;269:557–558.
25. Cuello AC. Non-classical neuronal communications. *Fed Proc* 1983;42:2912–2922.
26. Schmitt FO. Molecular regulators of brain function: a new view. *Neuroscience* 1984;13:991–1001.
27. Paton WDM, Vizi ES. The inhibitory action of noradrenaline and adrenaline on acetylcholine output by guinea-pig ileum longitudinal muscle strip. *Br J Pharmacol* 1969;35:10–28.
28. Vizi ES, Knoll J. The effects of sympathetic nerve stimulation and guanethidine on parasympathetic acetylcholine release. *J Pharmacol Pharm* 1971;23:916–929.
29. Knoll J, Vizi ES. Presynaptic inhibition of acetylcholine release by endogenous and exogenous noradrenaline at high rate of stimulation. *Br J Pharmacol* 1970;40:554–555.
30. Manber L, Gershon MD. A reciprocal adrenergic-cholinergic axo-axonic synapse in the mammalian gut. *Am J Physiol* 1979;236:738–745.
31. Vizi ES. Stimulation by inhibition of (Na^+, K^+-Mg^{2+}) activated ATPase of acetylcholine release in cortical slices from rat brain. *J Physiol* 1972;226:95–117.
32. Beani L, Bianchi C, Gracomelli A, Tamberi F. Noradrenaline inhibition of acetylcholine release from guinea-pig brain. *Eur J Pharmacol* 1978;48:179–183.
33. Vizi ES, Pásztor E. Release of acetylcholine from isolated human cortical slices: inhibitory effect of norepinephrine and phenytoin. *Exp Neurol* 1981;73:144–153.
34. Mobley D, Greengard P. Evidence for widespread effects of noradrenaline on axon terminals in the rat frontal cortex. *Proc Natl Acad Sci* 1985;82:945–947.
35. Audet MA, Descarries L, Doucet G. Quantified regional and laminar distribution of the serotonin innervation in the anterior half of adult rat cerebral cortex. *J Chem Neuroanat* 1989;2:29–44.
36. Vizi ES, Hársing LG, Knoll J. Presynaptic inhibition leading to disinhibition of acetylcholine release from interneurons of the caudate nucleus: effects of dopamine, beta-endorphin and D-Ala_2-Pro_5 enkephalinamide. *Neuroscience* 1977; 2:953–961.
37. Bjelke B, Fuxe K, Agnati LF. Increased diffusion of prolactin-like material into the brain neuropil from homologous adenohypophyseal transplants in the rat neostriatum after a 6-OH-dopamine induced degeneration of the neostriatal dopamine neurons. *Neurosci Lett* 1989.
38. Hökfelt T, Fuxe K, Johansson O, Ljungdahl A. Pharmacohistochemical evidence of the existence of dopamine nerve terminals in the limbic cortex. *Eur J Pharmacol* 1974;25:98–112.
39. Descarries L, Lemay B, Doucet G, Berger B. Regional and laminar density of the dopamine innervation in adult rat cerebral cortex. *Neuroscience* 1987;3:807–824.
40. Audet MA, Doucet G, Oleskevich S, Descarries L. Quantified regional and laminar distribution of the noradrenaline innervation in the anterior half of the adult rat cerebral cortex. *J Chem Neuroanat* 1988;274:307–318.
41. Fuxe K, Hamberger B, Hökfelt T. Distribution of noradrenaline nerve terminals in cortical areas of the rat. *Brain Res* 1968;8:125–131.
42. Parnavelas JG, Papadopulos GC. The monoaminergic innervation of the cerebral cortex is not diffuse and non-specific. *TINS* 1989;12:219–315.
43. Pellionis AJ. Neural geometry: towards a fractal model of neurons. In: Cotterill RM, ed. *Models of brain function*. Cambridge University Press 1989;1–12.

Volume Transmission in the Brain: Novel Mechanisms for Neural Transmission,
edited by K. Fuxe and L.F. Agnati.
Raven Press, Ltd., New York © 1991.

8

D_2 Dopamine Receptors: Autoreceptors and Receptors in a Prolactin-Secreting Pituitary Tumor

*Menek Goldstein, *Kazuho Harada, *Jow Y. Lew, and **Emanuel Meller

**Neurochemistry Research Laboratories and **Millhauser Laboratories, New York University Medical Center, New York, New York 10016*

Dopamine (DA) synthesis, release, and impulse flow in nigrostriatal DA neurons are regulated by inhibitory autoreceptors on cell bodies and nerve terminals (1,2). DA released from terminals most likely diffuses readily and, under physiological conditions, the concentration of DA at autoreceptor sites is probably low. When the concentration of the transmitter is low at receptor sites, the presence of receptor reserve amplifies the transmitter-elicited biological response (3,4). In recent studies we have investigated the presence of spare receptors for DA agonists at autoreceptors regulating synthesis and release of DA (5). In this chapter we will review some of these studies. In addition, we will describe some of our studies on the molecular properties and regulation of the D_2 DA receptors in the rat prolactin-secreting 7315a anterior pituitary tumor.

REGULATION OF TYROSINE HYDROXYLASE ACTIVITY IN STRIATAL SYNAPTOSOMES BY DA AUTORECEPTORS

Tyrosine hydroxylase (TH) activity in striatal synaptosomes is inhibited by apomorphine (Apo) and by other DA agonists, and this inhibition is partially reversed by D_2 DA antagonists (6). These findings indicate that TH activity in striatal synaptosomal preparations is regulated by DA autoreceptors. Because DA autoreceptors exhibit the pharmacological profile of D_2 DA receptors (7) and because postsynaptic D_2 DA receptors are linked to G_i/G_o proteins, we have investigated whether the G proteins are involved in the DA autoreceptor-mediated inhibition of TH activity by DA agonists. The results presented in Table 1 show that inactivation of G_i/G_o proteins in striatal synaptosomal preparations by pertussis toxin (PTX) attenuates the Apo-elicited inhibition of TH. The inhibition of TH by low concentrations of Apo was completely blocked, whereas that by high concentrations was only partially blocked by PTX. This is in agreement with the findings that the inhibition of TH by low concentrations of Apo is mediated by DA autoreceptors, whereas that by high concentrations is mediated not only by DA autoreceptors but also by direct inhibition of enzyme activity (8). It is also evident from the results presented in Table 1 that the D_2 antagonist sulpiride attenuates the Apo-elicited inhibition of TH activity, and treatment of synapto-

TABLE 1. *The effects of PTX and of (−) sulpiride on APO-induced inhibition of striatal synaptosomal TH activity*[a]

Apomorphine (M)	Sulpiride (M)	% inhibition by Apo	
		Untreated	PTX-treated
0	0	—	—
5×10^{-8}	—	14.0	1.5
5×10^{-8}	10^{-4}	0	0
1×10^{-7}	—	27.7	8.5
1×10^{-7}	10^{-4}	6.1	7.8
2×10^{-7}	—	50.0	39.1
2×10^{-7}	10^{-4}	27.9	31.1

[a]Results are the means from four experiments ± SE of 0.5–3%.

somes with PTX abolishes the effects elicited by sulpiride.

To determine the extent of inactivation of G proteins by PTX in synaptosomal preparations, we measured its effect on the back adenosine diphosphate (ADP) ribosylation of G_i/G_o proteins. Figure 1 shows that incubation of synaptosomes with PTX resulted in partial inactivation of the G_i/G_o proteins. ADP ribosylation with [^{32}P]nicotinamide-adenine dinucleotide (NAD) was decreased for both G_i and G_o after incubation of synaptosomes with PTX (62% and 48% of control values of G_o and G_i, respectively). Thus, partial inactivation of G_i/G_o proteins in striatal synaptosomes attenuated the Apo-elicited inhibition of TH mediated by DA autoreceptors.

REGULATION OF STRIATAL DOPA SYNTHESIS *IN VIVO* BY DA AUTORECEPTORS: RECEPTOR RESERVE FOR DA AGONISTS

To determine the receptor reserve for DA agonists at DA autoreceptors regulating the synthesis of 3,4-dihydroxyphenylalanine (DOPA) we have combined the GBL procedure with the partial irreversible receptor inactivation procedure (9,10). DA receptors were inactivated by the irreversible DA antagonist *N*-ethoxycarbonyl-2-ethoxy-1,2-dihydroquinoline (EEDQ) (10), and the effects of DA agonists on GBL-enhanced striatal DOPA synthesis were studied. Although EEDQ inactivates both pre- and postsynaptic DA receptors, the effects of DA agonists on DA autoreceptors can be selectively studied because after GBL

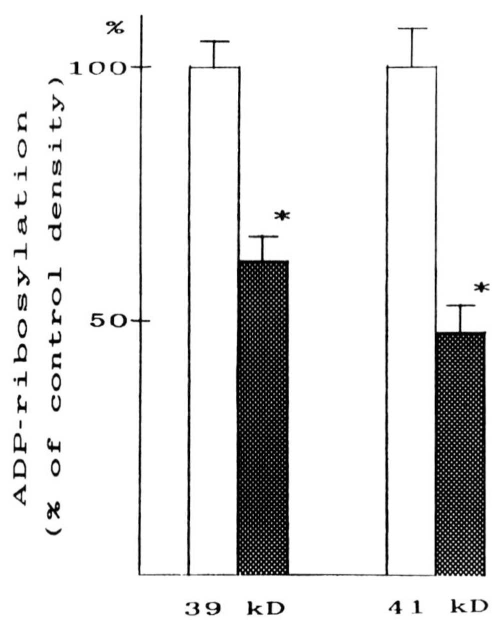

FIG. 1. Effect of pertussis toxin (PTX) (16 μg/ml/60 min) on ADP-back ribosylation of G proteins in striatal synaptosomes. Autoradiograms were evaluated by scanning with a densitometer. Ordinate indicates ADP-ribosylation as % of control density (mean ± SE from four samples). *Open and shaded columns* represent vehicle and PTX-treated groups, respectively. *Significantly different from control: $p < 0.005$.

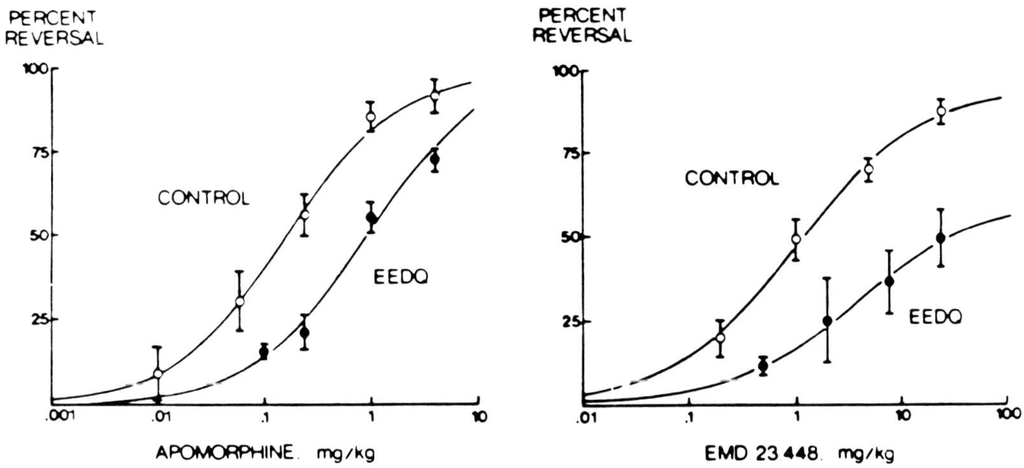

FIG. 2. Dose-response curves for apomorphine and indolyl-3-butylamine (EMD) 23,448 reversal of GBL-elicited L-DOPA elevation after treatment with vehicle or EEDQ (6 mg/kg). Each point is the mean ± SE of 4–6 experiments.

administration the influence of the postsynaptic DA receptors is eliminated. The results presented in Fig. 2 show that the DA agonists Apo and indolyl-3-butylamine (EMD) 23-448 dose-dependently reversed the elevation of L-DOPA accumulation induced by GBL and that after partial inactivation of DA receptors with EEDQ the dose-response curves for both agonists were shifted to the right. The maximum response elicited by Apo was not diminished after partial inactivation with EEDQ, whereas that elicited by the partial DA agonist EMD 23-448 was reduced. Similar results were obtained with other DA agonists (5). Partial inactivation of DA receptors with EEDQ produced a rightward shift in the dose-response curves and no significant change in the maximal response for full agonists [e.g., Apo, *N*-propylapomorphine (NPA)] but a reduced maximal response for partial agonists [e.g., EMD 23-448, (+)3-PPP]. These results demonstrate that a large receptor reserve exists for DA agonists at striatal autoreceptors regulating the synthesis of DOPA. Calculation of response as a function of receptor occupancy showed that 50% of maximal response occurred at 3.8%, 18%, and 30% occupancy for the full DA agonist NPA and for the partial DA agonists EMD 23-448 and (+)3-PPP, respectively (5). The maximal response elicited by the weak partial DA agonist (−)3-PPP was only about 50%, yet this required occupancy of virtually all the receptors (5). Relative efficacies were calculated from the ratio of the fraction of the receptors occupied by the reference full agonist NPA to that occupied by each of the tested agonists (Table

TABLE 2. *Relative efficacies of DA agonists to reverse GBL-elicited striatal L-DOPA accumulation*[a]

Agonist	Relative efficacy
NPA	1.00
Quinpirole (LY 171555)	0.61
(+) PHNO	0.29
EMD 23-448	0.19
EMD 49980	0.16
LY 163502	0.12
(+)-3-PPP	0.12
(−)-3-PPP	0.05

[a]The relative efficacy is expressed as the ratio of the fraction of receptors recognized to be occupied by NPA over that fraction required to be occupied by each of the other agonists at the 50% response level. NPA, *R*-(−)-*N*-*n*-propylnorapomorphine; EMD 23-448, indolyl-3-butylamine; 3-PPP, 3-(3-hydroxyphenyl)*N*-*n*-propylpiperidine; (+)PHNO, (+)-4-propyl-9-hydroxynaphoxazine.

TABLE 3. *Purification of rat striatal and pituitary tumor (7315a) D_2 DA receptors on wheat germ agglutinin (WGA)-agarose columns*[a]

	^{3}H-Spiroperidol binding		
	Total activity (fmol)	Spiroperidol activity (fmol/mg)	% Recovery
Striatum			
Solubilized fraction	4,800 ± 450	230 ± 20	
WGA nonadsorbed fraction	1,440 ± 250	190 ± 15	
N-acetyl-glucosamine eluate (0.1 M)	1,950 ± 220	1,150 ± 130	40
Pituitary tumor			
Solubilized fraction	2,000 ± 280	30.5	
WGA nonadsorbed fraction	1,750 ± 250	21.0	
N-acetyl-glucosamine eluate (0.1 M)	200 ± 25	78.0	10

[a]These results are the means from three experiments ± SEM.

2). The data in Table 3 show that the recently described D_2 DA agonists quinpirole and PHNO have lower efficacy than NPA and that the DA autoreceptor selective agonists EMD 23-448 and (+)3-PPP require a larger receptor occupancy than NPA to produce the same response. These differences in efficacy may explain the selective action of these DA agonists at DA autoreceptors and at postsynaptic supersensitive DA receptors.

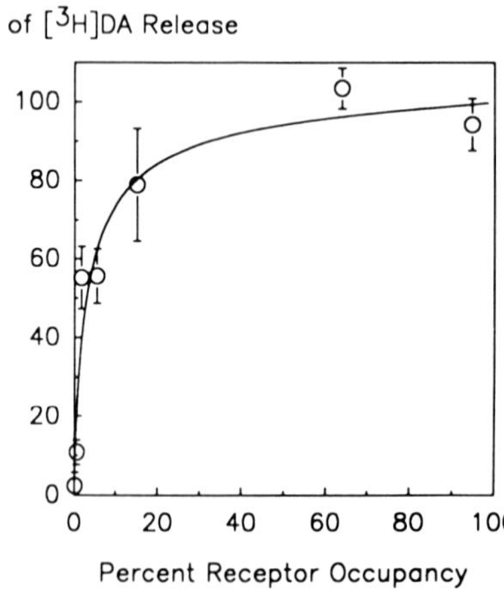

FIG. 3. Percent receptor occupancy as function of the percentage of maximum inhibition of ^{3}H-dopamine (DA) release produced by apomorphine (11).

MODULATION OF DA RELEASE FROM STRIATAL SLICES: RECEPTOR RESERVE FOR DA AGONISTS

To determine whether a DA receptor reserve exists at DA autoreceptors modulating the release of DA we have investigated the effect of Apo on the electrically stimulated release of ^{3}H-DA from striatal slices. The electrically evoked release of ^{3}H-DA was reduced by Apo in a dose-dependent manner (11). Partial irreversible inactivation of striatal DA receptors with EEDQ shifted the dose-response curve for Apo to the right and reduced the maximum response only slightly. A steep hyperbolic relation was obtained when receptor occupancy was plotted against percent of maximum inhibition of ^{3}H-DA release (Fig. 3). The maximum response plateaued at 40% to 50% receptor occupancy, suggesting a receptor reserve of 50% to 60% of Apo-elicited inhibition of ^{3}H-DA release.

D_2 DA RECEPTORS IN THE 7315A PROLACTIN-SECRETING TUMOR

It has been shown that D_2 DA receptors are present in the rat prolactin 7315a anterior pituitary tumor (12) and that the DA receptors in the tumor are negatively linked to adenylate cyclase (13). Because DA does

TABLE 4. *Purification of D_2 DA receptors from rat tumors (7315a) on fluphenazine Sepharose (Fpz-Seph) affinity column*

	³H-Spiroperidol binding			
Preparation	Total (mg protein)	Total activity (fmol)	Sp. activity (fmol/mg)	Folds purification
Solubilized fraction	1,230	35,793	29.1	1
Polyethylenglycol (10%) ppt	535	22,192	41.5	1.4
Fpz-Seph fraction[a]	0.043	710	16,703	574

[a]Receptor binding activity was assayed after reconstitution with phospholipid.

not regulate prolactin release in the tumor cells as it does in normal anterior pituitary cells, we have investigated whether the molecular properties of the tumor receptor differ from the brain receptor. The solubilized striatal D_2 DA receptor binds to wheat germ agglutinin (WGA) agarose columns (14), and we have therefore investigated whether the D_2 DA receptors solubilized from the tumors are adsorbed by this lectin. The results presented in Table 3 show that the D_2 DA receptors solubilized from the rat tumors, unlike the striatal receptors, do not bind to WGA-agarose columns. However, the 3-(3-cholamidopropyl) dimethylammonio-2-hydroxy-1-propanesulfonate (CHAPSO)-solubilized D_2 DA receptor from the tumors was partially purified on a fluphenazine-Sepharose affinity column (Table 4). The ligand binding subunit of the D_2 DA receptors was identified in the tumor and in the striatum by photoaffinity labeling with N-(p-azido-m[^{125}I] iodophenethyl) spiperone ([^{125}I]-NAPS) and subsequent sodium dodecyl sulfate (SDS) electrophoresis. In membrane preparations obtained from the tumor, the major labeled binding protein had a molecular weight (MW) of approximately 32 to 34 kDa and a minor component of approximately 92 to 94 kDa, whereas in the striatum relative amounts of these two proteins were reversed (Fig. 4).

To determine whether the DA receptors in the tumor were sensitive to chronic treatment with neuroleptics, we treated the tumor-bearing animals for 21 days with haloperidol. The results presented in Table 4 show that chronic treatment with haloperidol had no significant effect on the K_D for ³H-spiroperidol (³H-Spi) but increased the maximal binding (B_{max}) for ³H-Spi in both the tumor and the striatum; the percent in-

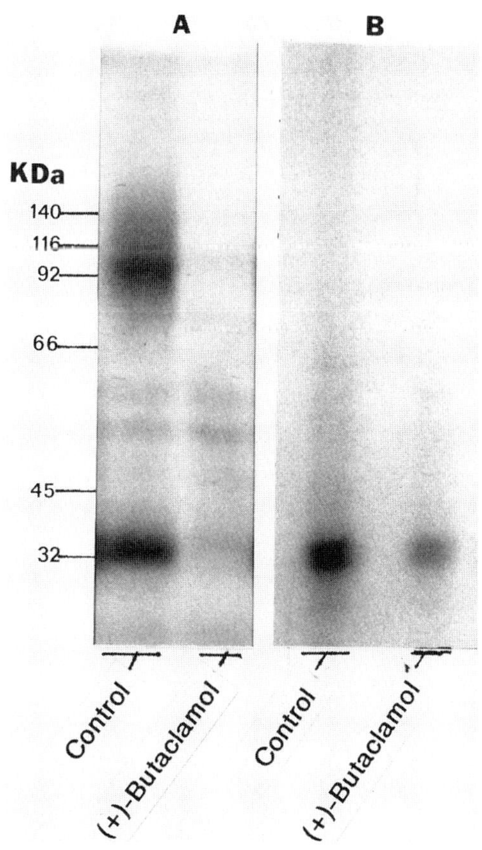

FIG. 4. Sodium dodecyl sulfate-polyacrylamide gel electrophoresis (SDS-PAGE) pattern of the labeling of rat striatal (**A**) and 7315a tumor membranes (**B**) with [^{125}I]N_3-NAPS. Membranes were incubated in presence and absence of 10^{-5} M butaclamol.

crease was much greater in the tumor than in the striatum.

DISCUSSION

The finding of a large receptor reserve for DA agonists at autoreceptors regulating synthesis and release of DA (5,11), but not at striatal postsynaptic DA receptors regulating cholinergic activity (15), raises the question whether receptor reserve is a phenomenon associated with presynaptic receptors (autoreceptors) and whether these reserves are of physiological significance. We have shown that presynaptic α_2 adrenoreceptors, which regulate the release of norepinephrine (NE) from cortical nerve endings (16), and somatodendritic 5-HT 1A autoreceptors, which regulate 5-HT synthesis in cortex and hippocampus (17), also have a large receptor reserve for the corresponding agonists. These studies support the idea that receptor reserve may be a phenomenon associated primarily with neurotransmitter autoreceptors. The function of autoreceptors might provide an explanation for the physiological significance of receptor reserve. The concentration of the released transmitter at autoreceptor sites is probably variable and dependent on the activity states of the neurons. Thus, receptor reserve could serve as a buffer, which provides an increased modulatory role for responses elicited by fluctuating concentrations of the transmitter at the autoreceptor sites.

The findings that PTX attenuates Apo-elicited inhibition of synaptic TH activity (18) and Apo-elicited inhibition of GBL-stimulated striatal DOPA synthesis *in vivo* indicates that DA autoreceptors regulating synthesis of the transmitter are linked to the effector by G_i/G_o proteins. It is noteworthy that synaptosomal DA autoreceptors regulating release of DA are apparently not linked to PTX-sensitive G proteins (19). Our data indicate that partial inactivation of G_i/G_o proteins in striatal synaptosomes by PTX abolishes DA autoreceptor-regulated synthesis of the transmitter. If it is assumed that PTX inactivates the entire pool of G_i/G_o proteins in synaptosomes to the same extent, the concentration of G proteins may be critical for autoreceptor function.

The molecular characteristics of D_2 DA receptors in the prolactin-secreting tumor were compared with striatal D_2 DA receptors. In our studies on purification of DA receptors we have found that solubilized striatal D_2 DA receptors are completely adsorbed on WGA (14), whereas only a small proportion of the tumor receptor is adsorbed on this lectin. This was the first indication that tumor receptors are either not glycosylated or glycosylated differently than striatal receptors. Photoaffinity labeling of the D_2 DA receptor binding units with $[^{125}I]N_3$-NAPS revealed differences between tumor and striatal D_2 receptors. Receptor binding in the tumor was mainly associated with the 32- to 34-kDa protein, whereas that in the striatum with the 92- to 94-kDa protein, and to a smaller extent, with the 32- to 34-kDa protein (20). Deglycosylation of the 92- to 94-kDa and 32- to 34-kDa D_2 DA receptor binding proteins results in the formation of a 44- and 23-kDa protein, respectively (21). These findings indicate that the 32-kDa binding unit is not a proteolytic degradation product of the larger binding unit and that the 32-kDa binding protein in the tumor represents a different molecular entity than the 92 KDa binding protein in striatum.

Although the D_2 DA receptor is refractory to DA agonist inhibition of prolactin release (12), it is responsive to chronic treatment with neuroleptics as evidenced by an increase in the maximal binding site of ^{3}H-Spi and the levels of D_2 DA receptor messenger ribonucleic acid (mRNA) (20). The regulatory effect of neuroleptic treatment may reflect the blockade of the response to DA derived from systemic circulation. Because the D_2 DA receptor density is low in naive animals, the blockade of receptors elicited by chronic treatment with haloperidol is probably efficient and may explain the relatively high increase in both D_2 DA receptor binding density and in D_2 DA receptor mRNA. Alternatively, it is possible that the D_2 DA receptor subtype in the tumor is more sensitive to "up-regulation" by neuroleptics than the subtype in striatum. Based on our studies we suggest that the defect in regulation of prolactin release by DA in the tumor is not due to a defect at a postreceptor site, but rather to the molecular properties of the D_2 DA receptor subtype.

SUMMARY

Dopamine receptors regulating synthesis of the transmitter were characterized in striatal synaptosomes. Partial inactivation of G proteins with PTX attenuates the Apo-elicited inhibition of TH mediated by DA autoreceptors, suggesting that the concentration of G proteins is critical for DA autoreceptor function. A large receptor reserve exists for DA agonists at autoreceptors regulating synthesis and release of DA. The data obtained in our studies support the hypothesis that receptor reserve for agonists may be a phenomenon associated with transmitter autoreceptors.

D_2 DA receptors in a prolactin-secreting pituitary were also characterized, and the regulation by neuroleptics was investigated. The major D_2 DA receptor binding protein in the tumor is associated with a 32- to 34-kDa unit but in the striatum with a 92- to 94-kDa unit. The D_2 DA receptor in the tumor is more sensitive to "up-regulation" by neuroleptics than the striatal receptor. The defect in regulation of prolactin release in the tumor might not be due to a defect at the postreceptor site but rather to the molecular properties of the D_2 DA receptor subtype.

ACKNOWLEDGMENTS

These studies were supported in part by grants from the National Institute of Mental Health (NIMH 02717) and the National Institutes of Health (NINCDS 06801).

REFERENCES

1. Carlsson A, Kehr W, Lindquist M, Magnusson T, Atack CV. Regulation of monoamine metabolism in the central nervous system. *Pharmacol Rev* 1972;24:371–384.
2. Bronaugh RL, Goldstein M. The effect of various chlorpromazine derivatives on the apomorphine-elicited inhibition of synaptosomal tyrosine hydroxylase activity. *Psychopharmacol Commun* 1975;1:201–208.
3. Stephenson RP. A modification of receptor theory. *Br J Pharmacol* 1956;11:379–393.
4. Furchgott RF, Bursztyn P. Comparison of dissociation constants and of relative efficacies of selected agonists acting on parasympathetic receptors. *Ann NY Acad Sci* 1967;144:882–899.
5. Meller E, Bohmaker K, Namba Y, Friedhoff AJ, Goldstein M. Relationship between receptor occupancy and response at striatal dopamine autoreceptors. *Mol Pharmacol* 1987;31:592–596.
6. Goldstein M, Bronaugh RL, Ohashi T, Ebstein B. Reversal of apomorphine elicited inhibition of synaptosomal tyrosine hydroxylase activity by neuroleptics. In: Usdin E, Hamburg DA, Barchas JD, eds. *Neuroregulators and psychiatric disorders*. New York: Oxford University Press, 1977:538.
7. Goldstein M. Regulatory mechanisms of dopamine biosynthesis at the tyrosine hydroxylase step. *Ann NY Acad Sci* 1984;430:1–5.
8. Goldstein M, Freedman LS, Backstrom T. The inhibition of catecholamine biosynthesis by apomorphine. *J Pharm Pharmacol* 1970;22:715–717.
9. Roth R. Dopamine autoreceptors: pharmacology, function and comparison with post-synaptic dopamine receptors. *Commun Psychopharmacol* 1979;3:429–445.
10. Meller E, Bohmaker K, Goldstein M, Friedhoff AJ. Inactivation of D_1 and D_2 dopamine receptors by *N*-ethoxycarbonyl-2-ethoxy-1,2-dihydroquinoline *in vivo:* selective protection by neuroleptics. *J Pharmacol Exp Ther* 1985;233:656–662.
11. Yokoo H, Goldstein M, Meller E. Receptor reserve at striatal dopamine receptors modulating the release of [^{3}H]dopamine. *Eur J Pharmacol* 1988;155:323–327.
12. Cronin MJ, Valdenegro CA, Perkins SN, MacLeod RM. The 7315a pituitary tumor is refractory to dopaminergic inhibition of prolactin release but contains dopamine receptors. *Endocrinology* 1981;109:2160–2166.
13. Lin C, McGonigle P, Molinoff P. Characterization of D-2 dopamine receptors in a tumor of the rat anterior pituitary gland. *J Pharmacol Exp Ther* 1987;242:950–956.
14. Lew JY, Goldstein M. Solubilization and characterization of striatal dopamine receptors. *J Neurochem* 1984;42:1298–1305.

15. Meller E, Enz A, Goldstein M. Absence of receptor reserve at striatal dopamine receptors regulating cholinergic neuronal activity. *Eur J Pharmacol* 1988;155:151–154.
16. Adler CH, Meller E, Goldstein M. Receptor reserve at the alpha-2 adrenergic receptor in the rat cerebral cortex. *J Pharmacol Exp Ther* 1986;240:508–515.
17. Meller E, Goldstein M, Bohmaker K. Receptor reserve for 5-hydroxytryptamine$_{1A}$-mediated inhibition of serotonin synthesis: possible relationship to anxiolytic properties of 5-hydroxytryptamine$_{1A}$ agonists. *Mol Pharmacol* 1990;37:231–237.
18. Harada K, Meller E, Goldstein M. The effects of pertussis toxin on inhibition of synaptosomal tyrosine hydroxylase activity by apomorphine. *Eur J Pharmacol* 1990 (*in press*).
19. Bowyer JF, Weiner N. K^+ channel and adenylate cyclase involvement in regulation of Ca^{2+}-evoked release of [^{3}H]dopamine from synaptosomes. *J Pharmacol Exp Ther* 1989;248:514–520.
20. Lew JY, Zawadzka H, Feigenblum D, et al. Regulation and molecular characterization of the D_2 dopamine receptors in the 7315a prolactin-secreting pituitary tumor. *Eur J Pharmacol* 1990 (*in press*).
21. Clagget-Dame M, McKelvy JF. N-linked oligosaccharides are responsible for rat striatal dopamine D_2 receptor heterogeneity. *Arch Biochem Biophys* 1989;274:145–154.

Volume Transmission in the Brain: Novel Mechanisms for Neural Transmission,
edited by K. Fuxe and L.F. Agnati.
Raven Press, Ltd., New York © 1991.

9

On the Existence of Volume Transmission in the Central Neuropeptide Y Neuronal Systems. Studies on Transmitter Receptor Mismatches and on Biological Effects of Neuropeptide Y Fragments

*Kjell Fuxe, **Luigi F. Agnati, *J. A. Aguirre, *B. Bjelke, *B. Tinner, **E. Merlo Pich, and †P. Eneroth

**Department of Histology and Neurobiology, Karolinska Institute, Stockholm, Sweden; **Department of Human Physiology, University of Modena, Modena, Italy; and †Department of Applied Chemistry, Huddinge Hospital, Huddinge, Sweden*

Recently it has been possible to demonstrate that central neuropeptide Y (NPY) neuronal systems are characterized by the existence of topological and functional transmitter receptor mismatches (see refs. 1–3). On the basis *inter alia* of these mismatch phenomena, we have suggested the existence of volume transmission processes in the NPY neurons (i.e., the existence of a chemical communication in these systems, which involves the diffusion of NPY and its fragments in extracellular fluid pathways to reach distant high-affinity NPY receptors). However, other types of phenomena may also contribute to the existence of transmitter receptor mismatches in the brain as summarized in Table 1. In previous work we have also obtained indications for the existence of volume transmission processes within the central β-endorphin and enkephalin immunoreactive (IR) neurons, within the renin angiotensin system in the brain (4–8; Bunnemann et al, *this volume*), and within the central dopamine (DA) pathways (9).

In the present chapter we will summarize the evidence for the existence of transmitter receptor mismatches in the central NPY neurons and also present evidence that the high-affinity NPY receptors in the mismatch areas represent biologically active NPY receptors. Finally, we will, in agreement with previous findings (10,11), present observations demonstrating marked biological activity of porcine (p) NPY13-36 in neuroendocrine and cardiovascular studies involving intraventricular injections of the NPY fragment. These results give further evidence for the existence of volume transmission in the central NPY neurons and also support the view of the existence of both Y1 and Y2 receptors for NPY in the brain (12).

TRANSMITTER RECEPTOR MISMATCHES IN CENTRAL NEUROPEPTIDE Y NEURON SYSTEMS

In these immunocytochemical and receptor autoradiographical studies, specific

TABLE 1. *Possible explanations for the transmitter-receptor mismatch phenomenon*

Volume transmission
Low-affinity receptors
Receptor subtypes
Transmitter postrelease processing/receptors for transmitter fragments
Isoreceptor interconversion
Uncoupled receptors
Intracytoplasmic receptors (receptor synthesis, transport, and metabolism)
Different neuronal localization (e.g., autoreceptors)

pathogen-free Sprague-Dawley rats have been used.

In the immunocytochemical analysis, several types of antibodies against pNPY-conjugated bovin serum albumin have been used and found to give similar results (13–15). In these papers the antisera have been extensively characterized. After transcardiac perfusion with a 0.2 M sodium phosphate buffer, pH 7.4, containing 4% (wt/vol) paraformaldehyde, serial coronal 20- to 30-μm-thick sections were cut using a vibratome or a cryotome at various rostrocaudal levels from the most rostral part of the brain to the caudal medulla oblongata. The sections were processed according to the indirect immunoperoxidase procedure of Sternberger (16) or the avidin biotin method (17) using a 3,3′-diaminobenzidine as a chromophor to demonstrate IR NPY. To relate the NPY IR neurons to the cytoarchitectonics, counterstaining with cresylviolet was performed.

Semiquantitative evaluation of IR NPY nerve terminal systems was obtained by means of computer-assisted image analysis based on previously published methods (2,18). The optical density of IR NPY was shown to be linearly correlated with the contents of IR NPY as measured by means of radioimmunoassays (see also ref. 1). In this way it has been possible to describe the distribution of the NPY nerve terminal systems at various rostrocaudal levels of the brain and especially at the upper pons level, the upper mesencephalic level, the mid-hypothalamic level, and the striatal level (Bregma −8.2, −5.8, −2.8, and +12.7 mm, respectively) (20).

In separate brains but at the same rostrocaudal level, autoradiographical analysis has been performed of the high-affinity ^{125}I-NPY binding sites (0.5 nM). The procedure of Young and Kuhar (21) was adopted for the quantitative receptor autoradiography (22–25). After transcardiac perfusion with saline, coronal sections (14 μm thick) were cut in a Leitz cryostat at the various rostrocaudal levels indicated above. ^{125}I-pNPY was used in a concentration close to the K_D value (0.1–0.5 nM) and the binding procedure was performed under equilibrium conditions. The unspecific binding was defined as the binding in the presence of 10^{-6} M pNPY. The slides with the sections were placed in X-ray cassettes together with a sheet of 3H-Ultrofilm (LKB, Bromma, Sweden), and the exposure time for ^{125}I-pNPY was 24 to 36 hr. Quantitative receptor autoradiography was performed as previously described (24). For the microdensitometrical measurements, a Zeiss Kontron IBAS image analyzer was used in combination with a Bosch video camera attached to a macroobjective. Boundaries of the measuring fields were related to the cytoarchitectonic features of the sections. The area was delineated, and the average gray tone value measured for total and unspecific binding and the respective background values. The relation between the absorption values of the standard spots and the respective radioactivity made is possible to obtain a standard curve, which could be used to convert the values for total and unspecific binding to femtomoles per square millimeter. These values were subsequently converted to femtomoles per microgram protein by determination of the protein content in the gray matter of the section (see ref. 24).

By means of neuroanatomical landmarks, it was possible to obtain immunocytochemical preparations and receptor autoradiograms from the same rostrocaudal levels so that the relation of the high-affinity NPY binding and the degree of IR NPY could be semiquantitatively estimated. We will now photographically document some of the receptor transmitter matches and mismatches that have been obtained in this analysis (3).

NPY Transmitter-Receptor Matches and Mismatches in the Cerebral Cortex

It is evident in Fig. 1 that the NPY IR nerve terminals form a plexus evenly distributed over all the layers of the cerebral cortex, which probably arise from axons belonging to NPY IR nerve cell bodies also found within all layers of the cerebral cortex. Instead, the high-affinity NPY binding sites are predominantly concentrated to the outer layers of the cerebral cortex as seen in Figs. 1 and 2. These results open up the possibility that the plexus of NPY IR nerve terminals within the outer layer may operate via high-affinity NPY receptors, which are mainly located outside the synaptic regions and reached by NPY via diffusion in the extracellular space (i.e., volume transmission). Instead, in the basal layers the plexus of NPY IR nerve terminals may operate via low-affinity NPY receptors, which could mainly represent synaptic receptors. It is also possible that in these basal layers the NPY is more rapidly fragmented into smaller NPY peptides, which may act via high-affinity receptors for NPY fragments not labeled by ^{125}I-pNPY1-36. It must be emphasized that NPY is synthesized and stored exclusively within neurons (26,27). Thus, a glial origin of NPY for, e.g., the NPY 1-36 binding sites in the outer layers of the cerebral cortex, can be excluded.

NPY Transmitter-Receptor Mismatches in the Septal Area

This area is characterized by a marked transmitter-receptor mismatch as seen in Fig. 2, where a high density of high-affinity ^{125}I-NPY binding sites is present all over the lateral part of the septal area, having little association with NPY IR nerve terminals. In high magnification, the diffusely scattered NPY IR nerve terminals can be observed, but these cannot be visualized in the low magnification shown in Fig. 2. Based on the concept of volume transmission, it seems possible that the high-affinity NPY receptors in the septal area may be reached by NPY peptides diffusing *via* extracellular pathways and originating both from the scattered NPY IR terminals present within the septal region itself as well as from the cerebrospinal fluid (CSF) of the surrounding lateral ventricle. NPY may also diffuse into the lateral septal region from the surrounding strongly NPY IR nerve terminals of the shell nucleus of the nucleus accumbens and from the lambdoid zone of the septal region surrounding the medial septal nucleus.

NPY Transmitter-Receptor Mismatches in the Nucleus Accumbens and the Bed Nucleus of the Striae Terminalis

As illustrated in Fig. 2, a high density of strongly NPY IR nerve terminals exists within the bed nucleus of the striae terminalis and within the nucleus accumbens, especially the shell part. However, these nerve terminal plexa are not associated with high-affinity ^{125}I-pNPY binding sites. These results may imply that these strongly NPY IR networks are mainly associated with low-affinity NPY binding sites. Thus, sufficiently high concentrations of NPY may be released from these terminals to activate low-affinity synaptic NPY receptors.

NPY Transmitter-Receptor Mismatches in the Hypothalamus and the Thalamus

Also in these regions, marked topological mismatches are demonstrated, as seen in Fig. 3. Thus, within the thalamus a high density of high-affinity NPY binding sites exists within the midline thalamic nuclei and in the laterodorsal thalamic nuclei, and all these areas contain few NPY IR nerve terminal networks. Instead, within the thalamus the NPY IR nerve terminals are mainly concentrated to the paraventricular thalamic nucleus and to a certain degree to the habenular complex, which contain few high-affinity NPY binding sites. These re-

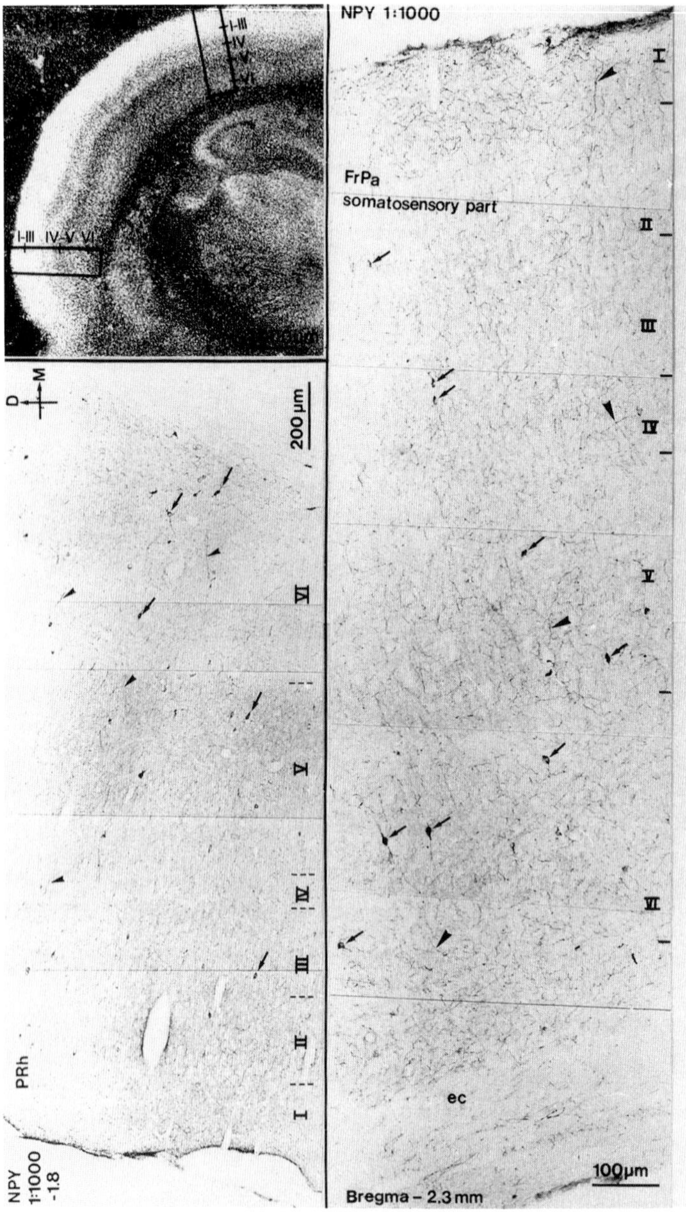

FIG. 1. Autoradiogram of ^{125}I-NPY (0.5 nM) binding and corresponding NPY IR in frontoparietal and perirhinal cortex at Bregma level -2.3 mm. A strong labeling is found in cortical layers 2 to 3, and a medium to weak labeling is found between layers 4 and 5 and in layer 6. Homogeneous distribution of NPY IR terminals are spread through all layers in both frontoparietal and perirhinal cortex. NPY IR cell bodies are also scattered through all layers in both areas. *Arrows* indicate NPY IR cell somas, and *arrowheads* indicate terminals.

sults again underline the possibility that the high-affinity NPY receptors in this case within the thalamus may be reached by NPY peptides diffusing via the extracellular fluid. These NPY peptides may originate from the scattered NPY IR nerve terminals found within the midline thalamic nuclei and in the laterodorsal thalamic nuclei. It cannot even be excluded that NPY may diffuse from the paraventricular thalamic nuclei and the hypothalamic nuclei and reach some of the high-affinity NPY receptors in the thalamic midline region.

In contrast, within most hypothalamic nuclei, high densities of strongly NPY IR terminals can be demonstrated (e.g., within the dorsomedial hypothalamic nucleus, the arcuate nucleus, and the internal layer of the median eminence) (see Fig. 3). However, as seen in Fig. 3, these NPY IR nerve terminal networks are not associated with high-affinity NPY receptors and may there-

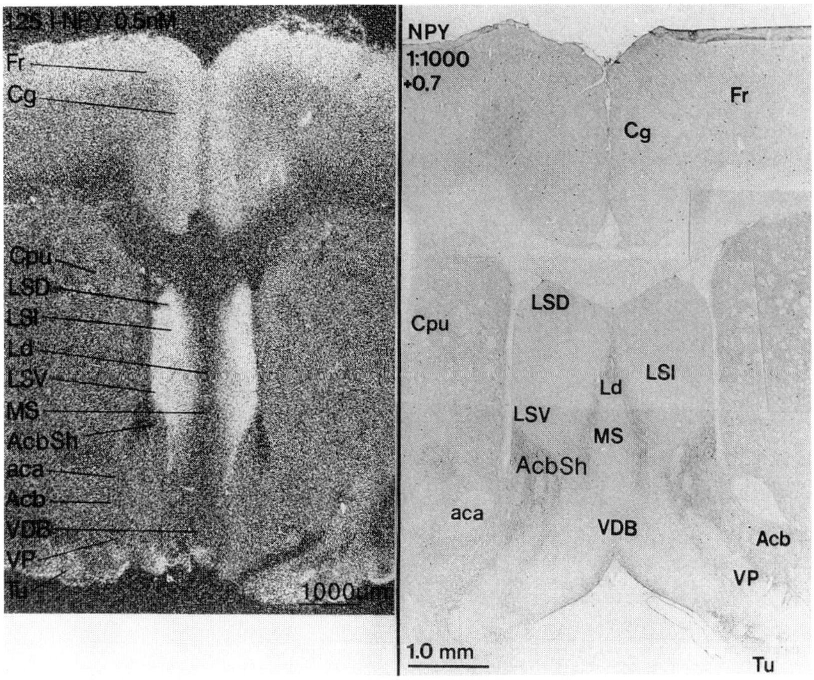

FIG. 2. Autoradiogram of ^{125}I-NPY (0.5 nM) binding and corresponding NPY IR from the Bregma level +0.7 mm. NPY IR neurons are scattered in the region of cortex, Cpu, BST and Acb. In cortex and Cpu an even distribution of weak NPY IR terminal plexa is present, whereas a higher density of strong NPY IR terminals is present in the region of BST, Ld and Tu. Hardly any terminals are visualized in LSI, LSD and few terminals are present in MS and VDB. Brighter tones in the autoradiogram indicate higher binding level. More intense labeling is present in cortical layers 1–3 and a high degree of binding is located in LSD, LSI, and LSV, whereas a low binding is present in BST. Thus, mismatch is present in the region of LSD, LSI, and LSV, where low NPY IR and high binding is present, whereas in the region of BST high NPY IR and low binding is visualized. Fr, frontal cortex; Cg, cingulum; AcbSH, nucleus accumbens, shell; aca, anterior commissure, anterior part; VP, ventral pallidum.

fore be associated with NPY receptors of the low-affinity type. These low-affinity NPY receptors are thus activated by the high concentrations of NPY, which may be reached in the synaptic cleft of these NPY IR terminals containing high amounts of IR. It should also be considered, however, that many of the strongly NPY IR hypothalamic nerve terminals described above may be associated with cleavage enzymes forming a special NPY fragment. This fragment may activate a receptor that cannot be labeled by ^{125}I-pNPY but may have a high affinity for the special NPY fragment formed in these areas. Finally, it should also be considered that within the areas with large numbers of high-affinity NPY receptors associated with few NPY IR nerve terminal networks, there may exist nerve terminal networks containing NPY- and peptide YY (PYY)-related peptides, which do not cross-react with the NPY antibody used. The antiserum used exhibits less than 1% cross-activity with PYY (see refs. 15–28). However, so far PYY IR neurons have only been demonstrated in the lower brainstem and spinal cord (29), having no clear association with high-affinity ^{125}I-NPY binding sites.

NPY Transmitter-Receptor Matches and Mismatches in the Midbrain

In Fig. 4, the analysis has been focused on the ventral component of the medial ge-

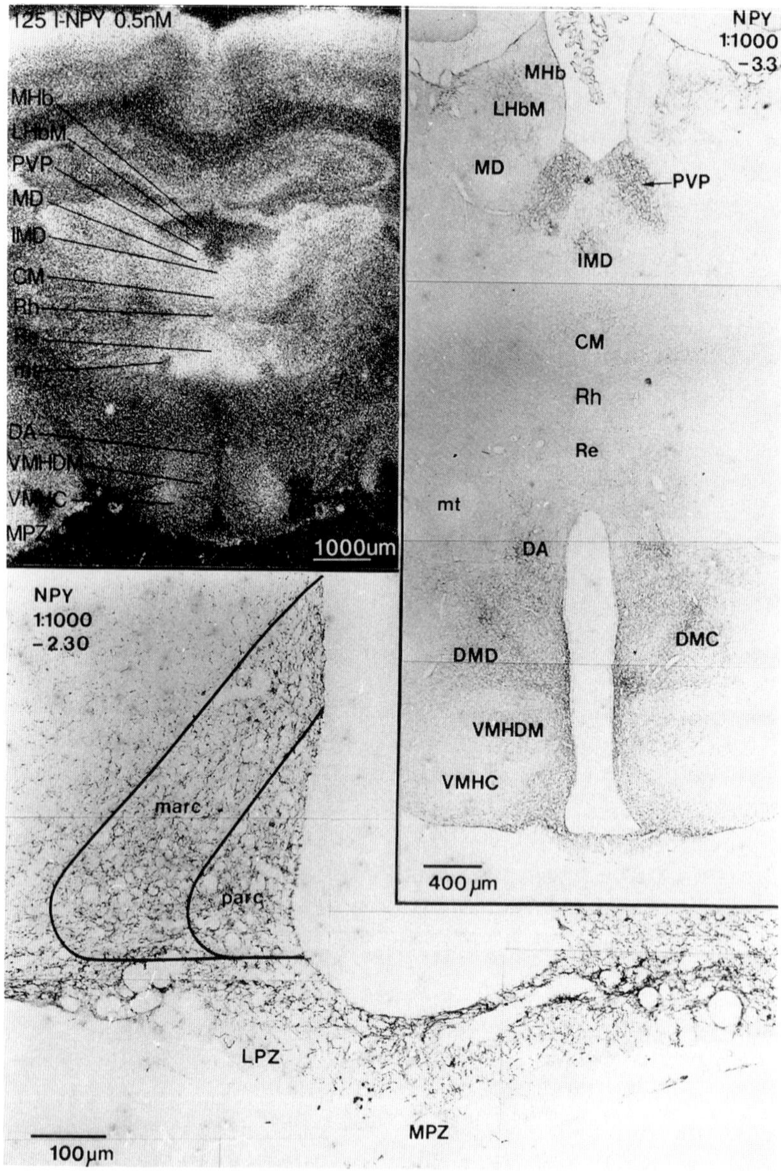

FIG. 3. Autoradiogram of ^{125}I-NPY (0.5 nM) binding and corresponding NPY IR in the area of thalamus and hypothalamus at Bregma level -2.3 mm. Strong NPY binding is present in PVP, MD, IMD, CM and Re, whereas weak NPY binding is present in MHb, LHbM, VMHDM, VMHC and MPZ. A strong NPY IR is seen in PVP, DA, DMC and DMD, whereas moderate NPY IR is present in LHbM, VMHDM, marc, Parc, SEL, and few if any NPY IR terminals are present in MHb, MD, CM, Rh, Re, and VMHC. Mismatches with strong binding and few if any NPY IR terminals are seen in MD, CM, Rh, Re, and VMHC. mt, mammillothalmic tract.

niculate body, which is shown to contain a moderate density of high-affinity NPY binding sites but few NPY IR nerve terminal networks. Instead, it is surrounded by strongly NPY IR nerve terminal networks along its dorsal and ventral border. With the low magnification used, the sparse plexus of NPY IR terminals found within the ventral part of the medial geniculate body is not visualized. Thus, it seems possible that also

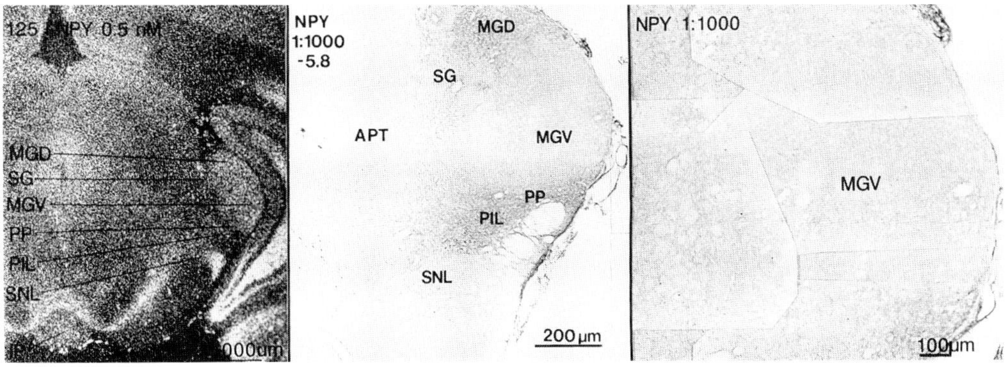

FIG. 4. Autoradiograms of ^{125}I-NPY (0.5 nM) binding and corresponding distribution of NPY IR at Bregma level -5.8 mm. Moderate to high labeling is present throughout the medial geniculate body; the pars compacta and ventral tegmental area, and strong binding is present in the dorsolateral part of substantia nigra pars compacta. NPY IR terminal plexa are of moderate density in the MGD, SG, PIL and PP. Thus, mismatch is evident in region of MGV and in the dorsolateral part of substantia nigra pars compacta, where high binding but low IR is visualized. APT, anterior pretectal nucleus; SNL, substantia nigra, lateral part.

within this nucleus the high-affinity NPY receptors may be reached by NPY peptides diffusing from surrounding nerve terminal networks to reach the high-affinity NPY receptors via the local extracellular fluid pathways. It should be noted that the high-affinity NPY binding sites present within the lateral part of the substantia nigra also are associated with only a sparse plexus of high-affinity NPY IR nerve terminal networks (Fig. 4). Again strong NPY IR is present in the area immediately dorsal to these binding sites (posterior intralaminar thalamic nucleus), from which NPY peptides may diffuse to reach the lateral substantia nigra. Instead, within the zona compacta and within the ventromedial midbrain (A10 area), where the DA cell bodies are located, there is a match in the distribution of high-affinity NPY binding sites and NPY IR nerve terminal networks, both showing a moderate density.

FUNCTIONAL MISMATCHES IN THE CENTRAL NPY IR NEURONAL SYSTEMS

By means of computer-assisted microdensitometry and morphometry (see ref. 19), it has been possible to demonstrate that there exists a negative and highly significant correlation between the distribution of total NPY IR and high-affinity NPY receptors (Fig. 5) (see refs. 1,2). This analysis has been based on the analysis of 27 brain areas. It became clear that the NPY neuronal features of many areas are the mirror image of one another (see also above). The functional mismatch is characterized by a situation in which no correlation exists between the density distribution of the pre- and postsynaptic markers of the chemical transmission analyzed, or where there exists an inverse relation between the density distributions of the pre- and postsynaptic markers. The latter case is in fact demonstrated in Fig. 5. In contrast to the topological mismatches, the functional mismatch is not very dependent on the sensitivity of the procedures used but is mainly dependent on the reliability of the immunocytochemistry and the receptor autoradiographical techniques involved. It must be emphasized that the reliability of the techniques used for demonstration of NPY IR and ^{125}I-NPY binding has been demonstrated, because these techniques correlate to a high degree with results obtained by means of radioimmunoassay and biochemical receptor binding assays, respectively (see ref. 1). It is obvious that the

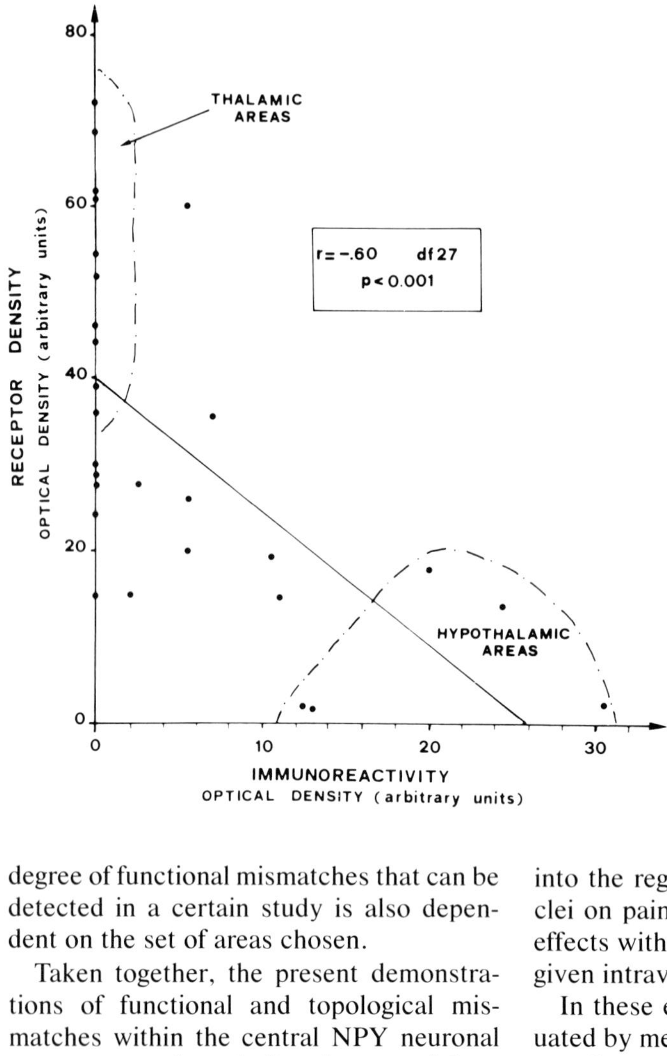

FIG. 5. Correlation between density of NPY IR and relative density of NPY binding sites in various brain regions of the rat. The quantification of NPY IR and receptor autoradiography was carried out as previously described. For each sampled region, optical density value of the NPY IR is reported on the abscissa, and respective optical density value of ^{125}I-pNPY binding is reported on the ordinate.

degree of functional mismatches that can be detected in a certain study is also dependent on the set of areas chosen.

Taken together, the present demonstrations of functional and topological mismatches within the central NPY neuronal systems strongly underline the potential existence of volume transmission, which may also involve NPY fragments (see below).

Finally, the existence of low-affinity synaptic NPY receptors has been postulated, linked mainly to wiring transmission.

BIOLOGICAL ROLE OF THE HIGH-AFFINITY NPY RECEPTORS IN THE MISMATCH REGION OF THE MIDLINE THALAMUS

In view of the role of the midline thalamic nuclei in the regulation of pain and arousal mechanisms (30) we have evaluated the effects of stereotactic microinjections of NPY into the region of the midline thalamic nuclei on pain threshold and compared these effects with those obtained when NPY was given intraventricularly (see ref. 31).

In these experiments we have first evaluated by means of quantitative receptor autoradiography the diffusion of ^{125}I-NPY (1.2 nmol) after the injection of 1 μl of a saline solution containing the iodinated radioligand. The injection was made via an intrathalamic cannula, and the rats were decapitated 30 min after the injection. As seen in Fig. 6, ^{125}I-NPY strongly labeled the midline thalamic area but did not reach into the lateral thalamus nor adjacent brain regions, such as the hypothalamus. Thus, it seems likely that with this technique it is possible selectively to affect the high-affinity NPY receptors within the midline thalamus where a topological mismatch exists (see above) because few NPY IR nerve terminals are found in this region.

Spontaneously hypertensive rats (SHR)

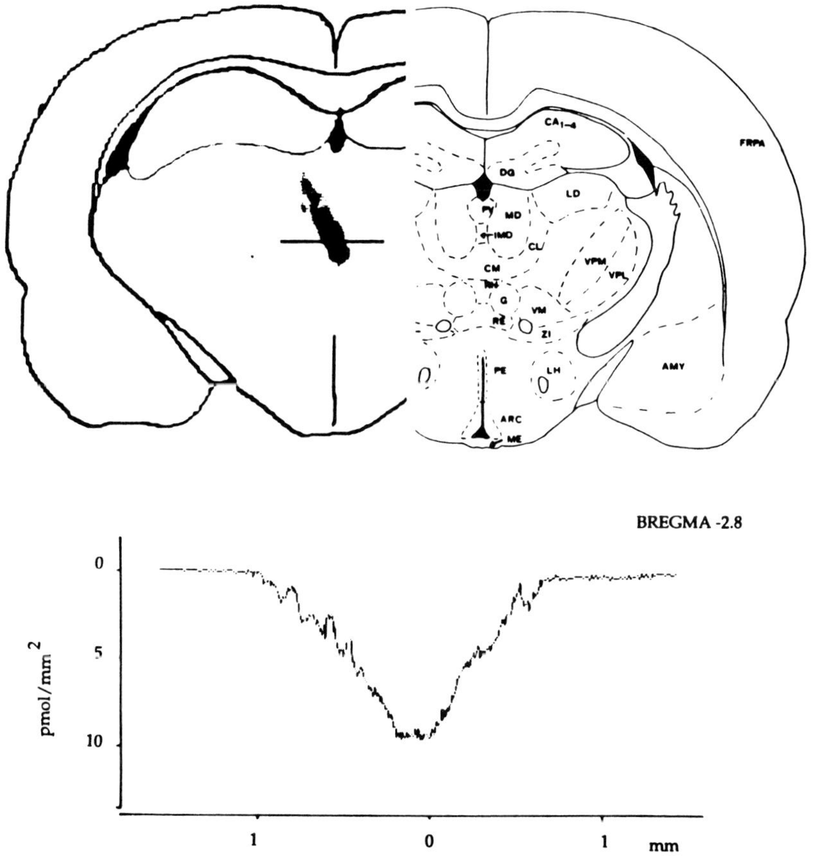

FIG. 6. ^{125}I-NPY (SA = 2,000 Ci/mmol, 1.2 nmol/μl, injected 30 min before killing in freely moving animals) diffusion after intrathalamic injection in an experimental rat (placement of cannula: Bregma: -2.8 mm, according to ref. 20). *Upper left panel* shows the darkening of the radiographical plate after 2 days of exposure. The line along which the scanning reported in the lower panel was performed is indicated. *Upper right panel* shows landmarks of anatomical areas (from ref. 20). *Lower panel* shows scanning of the gray tones on the radiographical plate. By means of the proper standard curve, the concentration of NPY present at any distance from the site of injection could be calculated. AMY, amygdala; ARC, nuc. arcuatus; CA1–4, hippocampal fields 1–4; CL, centrolateral thalamic nucleus; CM, central medial thalamic nucleus; DG, dentate gyrus; FRPA, frontoparietal cortex; G, nuclear gelatinosus; IMD, intermediodorsal thalamic nucleus; LD, laterodorsal thalamic nucleus; LH, lateral hypothalamus; ME, median eminence; MD, mediodorsal thalamic nucleus; PE, periventricular hypothalamic nucleus; PV, paraventrical thalamic nucleus; RE, reuniens nucleus; RH, rhomboid nucleus; VM, ventromedial thalamic nucleus; VPL, ventroposterolateral thalamic nucleus; VPM, ventroposteromedial thalamic nucleus; ZI, zona incerta.

and normotensive male Wistar-Kyoto rats (WKR) were implanted with chronic cannulas in the lateral ventricle (icv) or in the midline thalamic nuclei (A-2.8; V4.0; L0.0 from Bregma) (20). The pain threshold was measured by using a hot-plate procedure (55°C), and EEG recordings of the sleep–wakefulness cycle were made. Two doses of NPY (0.25 nmol and 1.25 nmol) were used, which were given either icv or directly into the thalamus.

The results are shown in Figs. 7 and 8. The basal values of the hot-plate latency were considerably higher in SHR than in WKR. Intraventricular injections of NPY in the high dose but not in the low dose increased the pain threshold in the SHR but not in WKR. The effects became significant

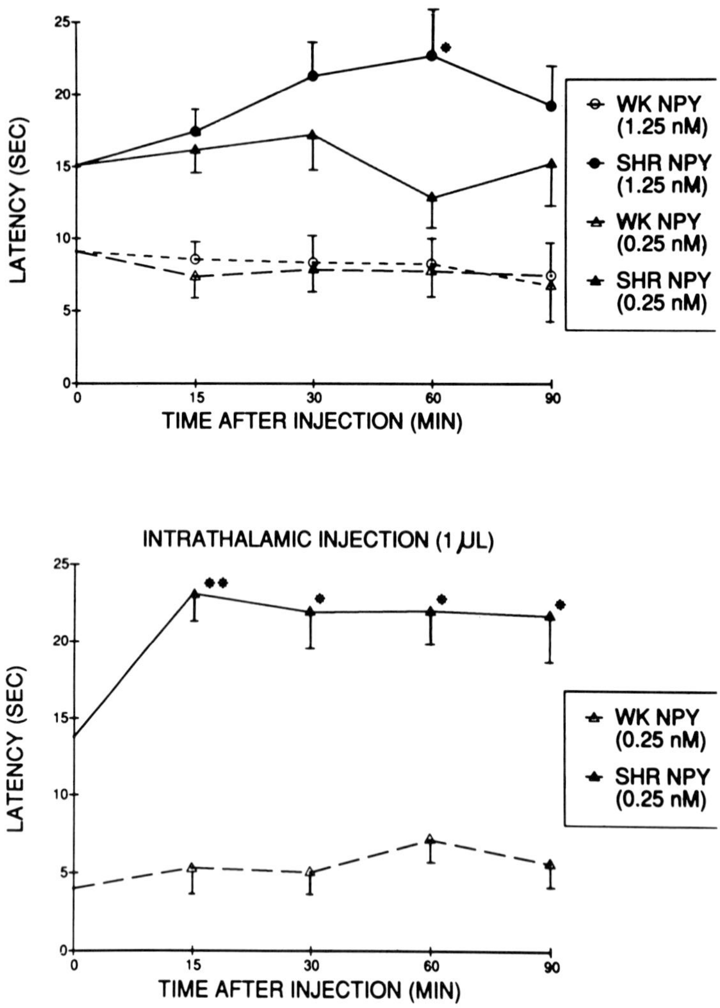

FIG. 7. Effects of microinjections of NPY on hot-plate latencies in SHR and WKR rats. Microinjections were performed icv or intrathalamic via chronically implanted cannulas. Values are expressed as means + SEM. $n = 7–10$. Statistical analysis was carried out by means of ANOVA for repeated measurements followed by Dunnet test (*$p < 0.05$, **$p < 0.01$, versus basal mean values).

after 1 hr. In contrast, the direct injections of NPY into the midline thalamus resulted in a significant increase in hot-plate latency after 15 min (first tested time), and both doses were effective. Again, no effects could be observed in WKR. Furthermore, NPY1-36 produced a selective reduction of slow wave sleep (SWS) time in the SHR, probably caused by increased formation of the NPY fragments producing arousal.

The present results indicate that, in fact, the high-affinity NPY receptors present in mismatch areas, such as the midline thalamus, can, on activation, produce biological actions and that the high-affinity NPY receptors may be reached by NPY diffusing in the extracellular space. In line with this hypothesis, it was found that intraventricularly injected NPY, which had to diffuse within the extracellular fluid to reach areas controlling pain such as intrathalamic nuclei, only produced effects on pain threshold with a higher dose and only after a long latency. In contrast, intrathalamic injections of NPY lead to an increase in pain threshold already at a dose six times lower than that used for intraventricular injections and with a short latency (<15 min). These results indicate that the high-affinity

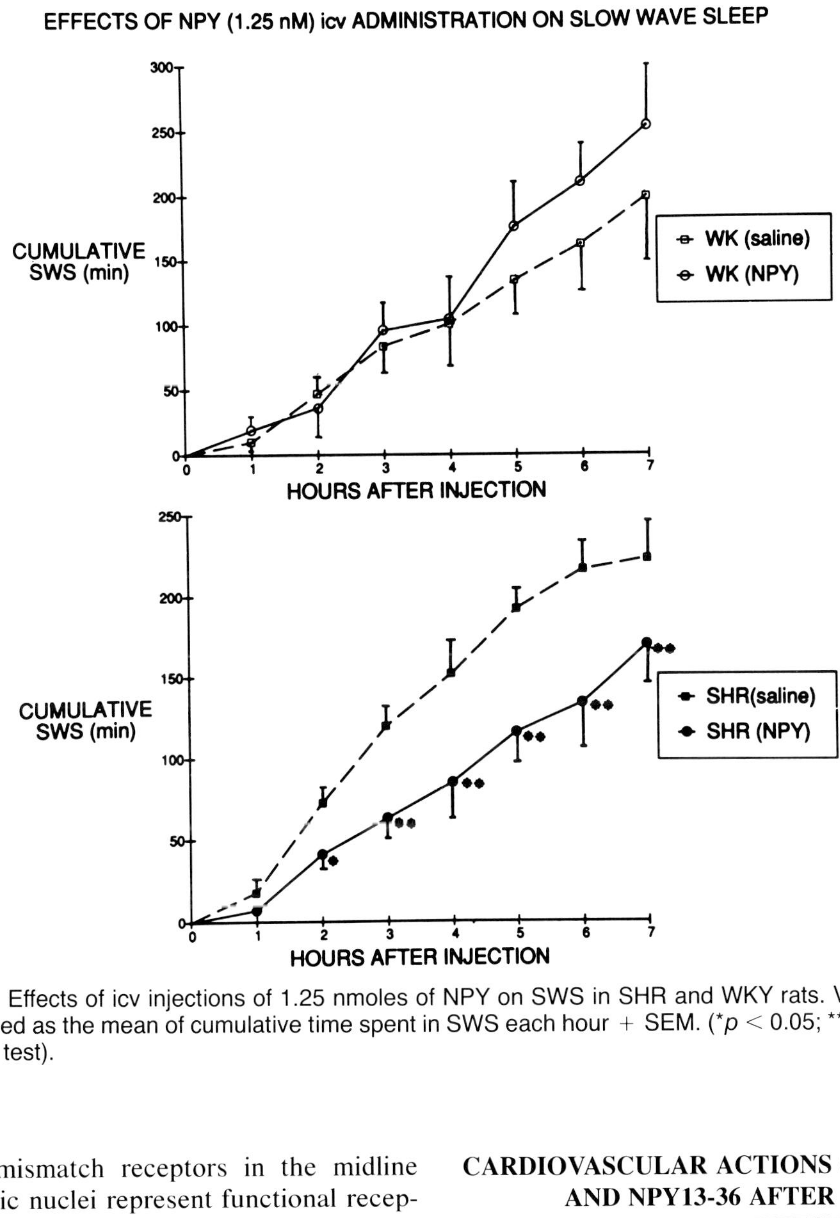

FIG. 8. Effects of icv injections of 1.25 nmoles of NPY on SWS in SHR and WKY rats. Values are expressed as the mean of cumulative time spent in SWS each hour + SEM. (* $p < 0.05$; ** $p < 0.01$, Tukey's test).

NPY mismatch receptors in the midline thalamic nuclei represent functional receptors and that they can be reached by long-distance diffusion in the extracellular fluid. Thus, volume transmission can play an important role in central NPY communication, especially within thalamic midline nuclei. The results, in fact, open up the possibility that even far away located networks of NPY terminals and CSF NPY via diffusion can reach midline thalamic circuits containing large numbers of high-affinity NPY receptors.

CARDIOVASCULAR ACTIONS OF NPY AND NPY13-36 AFTER INTRAVENTRICULAR INJECTION IN THE RAT. POSSIBLE ROLE OF VOLUME TRANSMISSION IN EPINEPHRINE/NPY COSTORING CARDIOVASCULAR NEURONS OF THE MEDULLA OBLONGATA

It seems likely that a large peptide such as NPY may be split by peptidases in the extracellular fluid into biologically active fragments to reach the mismatch high-affin-

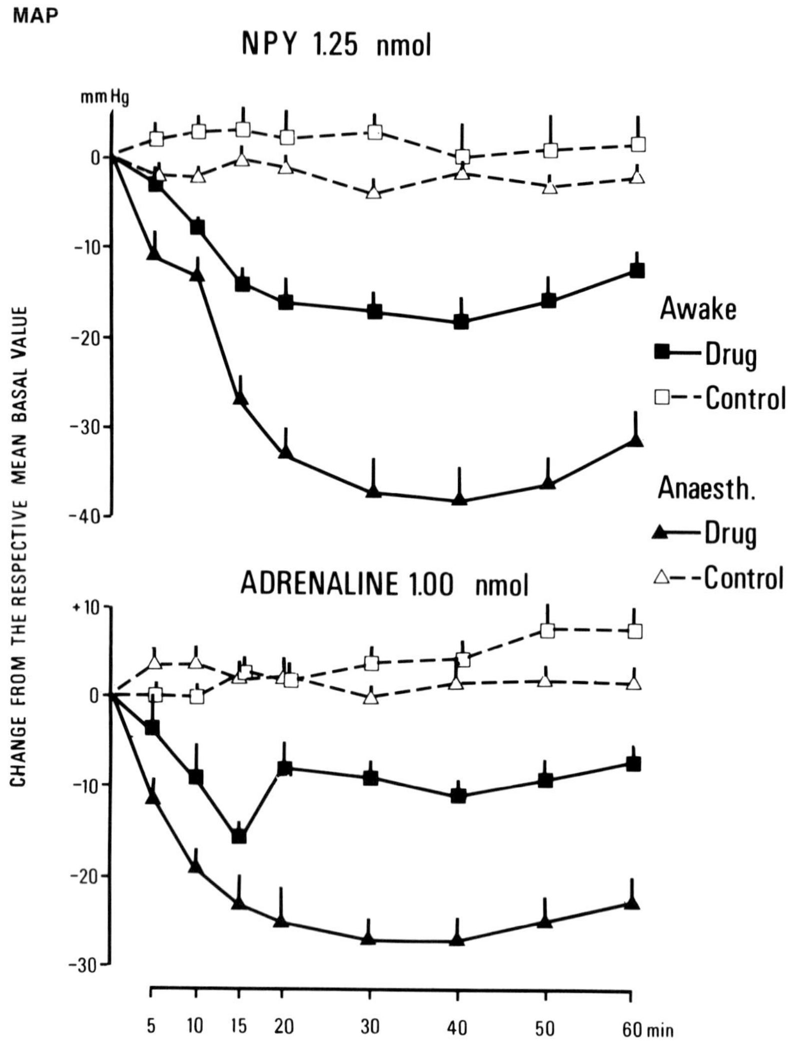

FIG. 9. Comparison of the effect of central administration of maximal doses of NPY and epinephrine in the awake freely moving and in α-chloralose anesthetized male rats. Difference in the hypotensive action is marked between awake and anesthetized animals when considering blood pressure both for NPY and epinephrine, whereas almost a similar response in heart rate is observed especially for NPY. Note sustained peak action of NPY in awake animals, which is not observed after epinephrine administration in awake rats. Means ± SEM are shown. MAP, mean arterial blood pressure.

ity NPY receptors (see refs. 2,32–37). It is, therefore, of substantial interest to evaluate if high-affinity NPY receptors in mismatch areas mainly represent high-affinity NPY receptors of the Y2 type (i.e., if those receptors possess a high affinity for long C-terminal fragments, such as NPY13-36) (see ref. 12). These aspects have been evaluated by studying the cardiovascular actions of NPY and NPY fragments and their interaction with epinephrine.

As seen in Fig. 9, epinephrine on intraventricular injection produces in the awake unrestrained male rat a substantially less prolonged lowering of the mean arterial blood pressure than NPY (both given in maximal doses). Thus, unlike in the NPY treated animals, the vasodepressor action of epinephrine is not maintained around the peak reduction level but is partially restored after 20 min. This difference appears to be the major difference between the vasodepressor responses induced by epinephrine and NPY. We have suggested that the cardiovascular responses by NPY and epinephrine are at least in part mediated via effects on the nucleus tractus solitarius (nTS) (38), where the baroreceptor reflex afferents terminate, and which contains high amounts of high-affinity α_2 agonist binding sites and high amounts of high-affinity ^{125}I-NPY binding sites (see refs. 39,40). Also within the nTS there exists a receptor transmitter mismatch in the NPY neuronal systems (25,41). Thus, the high-affinity NPY receptors are predominantly found within the dorsal cardiovascular region of the nTS, whereas the NPY IR is predominantly found within ventral parts of nTS and within the dorsal motor nucleus of the vagus (see ref. 41). Such a mismatch is not found for the catecholamine (CA) nerve terminal networks and the α_2-adrenergic receptors within the nTS. Based on the above morphological and functional observations, we propose that the coexistence of classical transmitters and neuropeptides such as epinephrine/NPY within central nerve terminal networks may reflect the need of these neurons to operate via both a short-term mode (wiring transmission) and a long-term mode (volume transmission), allowing long-lasting responses and widespread influences on adjacent networks.

In view of the postulated role of volume transmission in the mediation of the cardiovascular actions of NPY within the dorsal medulla, it became of interest to analyze the effects of NPY fragments on cardiovascular function after central administration of the NPY fragments. So far we have analyzed the actions of pNPY13-36 on cardiovascular functions after intraventricular injections in the awake unrestrained male rat. Previous work has indicated that the NPY fragment 13-36 only binds to the Y2 receptors for NPY (see ref. 12). Furthermore, this NPY fragment has no ability to pro-

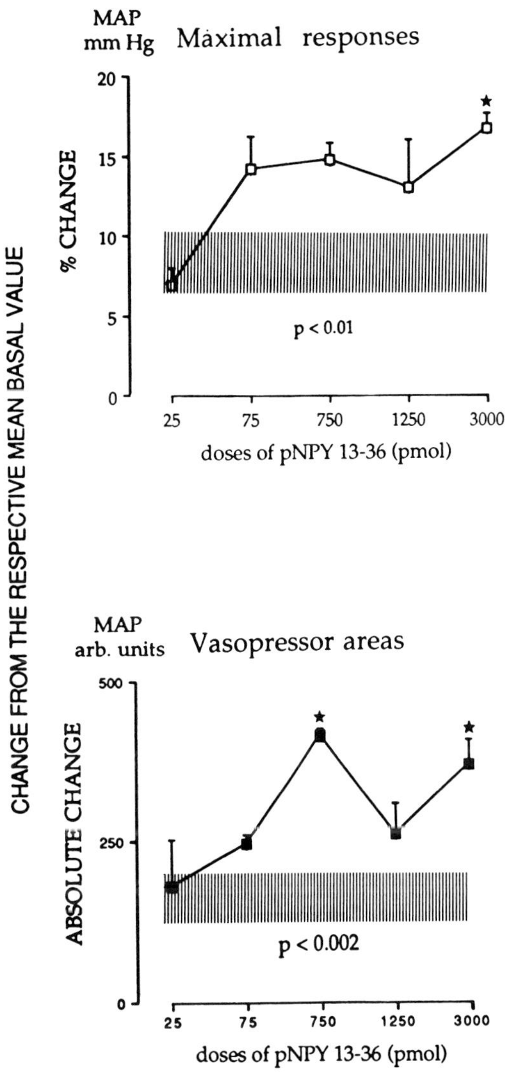

FIG. 10. Cardiovascular effects of different doses of a pNPY fragment (13-36) administered i.v.t. in the awake unrestrained male rat. Peak actions (maximal responses) and overall effects (vasopressor areas) on mean arterial blood pressure (MAP, in mm Hg) are expressed as means $\pm$ SEM. The Jonchere-Terpstra test for ordered alternatives to evaluate dose-related changes ($p < 0.01$ and $p < 0.002$) and a one-way ANOVA followed by the least square difference (LSD) test (* $p < 0.05$) were used. Basal values for MAP for each group receiving a dose were as follows: pNPY 25 pmol, 98 ± 5; pNPY 75 pmol, 92 ± 3; pNPY 750 pmol, 109 ± 5; pNPY 1,250 pmol, 100 ± 5; pNPY 3,000 pmol, 100 ± 5. The MAP basal mean value in the group treated with solvent (mock cerebrospinal fluid) alone (*shaded area*) was 104 ± 3. Each dose is represented by six experiments.

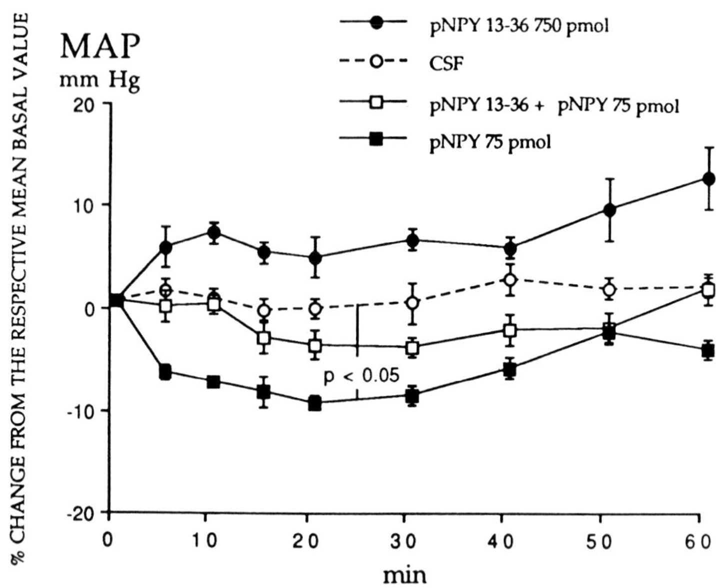

FIG. 11. Comparison of the time course of central effects of pNPY13-36 and pNPY1-36 given i.v.t. on mean arterial blood pressure (MAP) in awake freely moving male rats. Values are expressed as means ± SEM along a time axis (60 min). The dose corresponding to maximal overall action of the pNPY fragment (pNPY13-36, 750 pmol) was administered alone or simultaneously with a dose of pNPY1-36 (75 pmol) close to its ED_{50} value for the overall hypotensive action (78 pmol) in the awake rat. MAP curve for pNPY (75 pmol) alone is also shown. A one-way ANOVA followed by the LSD test ($p < 0.05$) was used (46). $n = 6$. Hypotensive action of pNPY given i.v.t. is slightly counteracted by pNPY13-36. Note hypertensive action of the fragment (pNPY13-36) on MAP in comparison with pNPY1-36 hypotensive action. MAP (mmHg) basal values: CSF, 104 ± 3; pNPY13-36 (750 pmol), 109 ± 5; pNPY13-36 (750 pmol) + pNPY (75 pmol), 92 ± 2; pNPY (75 pmol), 93 ± 2.

duce a lowering of blood pressure nor of heart rate (38,40). These results indicate that the lowering of blood pressure by NPY is mediated via Y1 receptors.

As seen in Fig. 10, it has, however, now been possible to demonstrate that the NPY fragment 13-36 on intraventricular transfusion (i.v.t.) can produce a dose-related increase of mean arterial blood pressure as seen both from the maximal response and from the overall response (i.e., area value). These results are of substantial interest because they indicate that released NPY1-36 may be converted into NPY13-36, which via activation of Y2 receptors may act to counteract the vasodepressor response induced by NPY1-36 at the Y1 receptors (Fig. 10). Thus, in volume transmission within the NPY neuronal systems, fragments can be formed, which via activation of a special subset of NPY receptors (Y2 receptors) may counteract the response of the parent peptide. These results well illustrate the existence of "chemical networks" (i.e., formation of a cascade of active fragments from a parent molecule) within volume transmission (see refs. 27,32–37,40). In Fig. 11 it is in fact demonstrated that NPY13-36 can counteract the lowering of mean arterial blood pressure induced by NPY1-36, using a dose of 75 pmol of NPY1-36, which is around the ED_{50} value for lowering of mean arterial blood pressure (28,38). These results then, in fact, indicate that the Y2 receptor activation via the NPY fragment 13-36 can at least in the high dose used (750 pmol) significantly counteract the hypotensive effects of NPY in a dose around the ED_{50} value. It may be speculated that an increased breakdown of NPY1-36 may take place in hypertension development of central origin, leading to formation of NPY fragments with pressor actions. Such an event may also explain why NPY1-36 only affected pain threshold in SHR but not in WKR (Fig. 7). This hypothesis also offers an explanation for the ability of NPY1-36 exclusively to increase arousal in the Sprague-Dawley rat (Fig. 8) (31,42). We therefore introduce the hypothesis that increased formation of long C-terminal NPY fragments leading to increased Y2 receptor activation is responsible for the altered responses to NPY in SHR, a neurochemical change that may contribute to the hypertension development. Such events can also explain the reduced potency of NPY1-36 to reduce arterial blood pressure in SHR (43). Taken together, these results certainly indicate that disturbances in the chemical networks in volume transmission may lead to pathophysiological consequences. Thus, peptide fragments can produce unique biological effects via subtypes of peptide receptors, some of which may contribute to a counteraction of the action elicited by the full peptide. It may be pointed out that in the β-endorphin and adrenocorticotropic hormone (ACTH) IR neurons it seems as if it is the ACTH peptides that fulfill an antagonistic role in these arcuate neuron systems by counteracting the biological effects of β-endorphin on DA neurons in the median eminence and prolactin secretion (6,33,35, 44).

EFFECTS OF CENTRAL ADMINISTRATION OF NPY AND OF THE NPY FRAGMENT 13-36 ON NEUROENDOCRINE FUNCTION IN THE MALE RAT. RELEVANCE FOR VOLUME TRANSMISSION

In previous work we have found that i.v.t. injections of NPY 1–36 in low doses of 7.5 to 25 pmol/rat lead to inhibitory effects on corticosterone and aldosterone secretion and to increases of prolactin secretion in the normal male rat as evaluated at the 15- to 60-min time interval (see Fig. 12) (14,27,45,46). With high doses of NPY 1–36 ranging from 250 to 750 pmol inhibitory effects were observed on serum thyroid-stimulating hormone, prolactin, and growth hormone (GH) levels, while serum levels of corticosterone, adrenocorticotropin, and aldosterone were increased. After tyrosine hydroxylase (TH) inhibition, the serum corticosterone levels were increased and were no longer influenced by various doses of NPY 1–36 (45). Based on the mapping of high-affinity ^{125}I-NPY binding sites, it seems likely that within the hypothalamus the NPY IR nerve terminals are mainly linked to low-affinity synaptic NPY receptors (see above). It is therefore postulated that the neuroendocrine actions of NPY1-36 found in the high-dose range (250–1,250 pmol) were related to activation of these types of NPY receptors within the hypothalamus and the preoptic area (27,46). Instead, the neuroendocrine effects found with the lowest i.v.t. doses of NPY may have been produced by activation of high-affinity NPY receptors within the hypothalamus and preoptic area, although as described above, they are only present in low densities in these areas.

To test the hypothesis that Y2 NPY receptors may have a neuroendocrine function we have also studied the effects of central administration of NPY13-36 on serum levels of hypophyseal hormones in the awake unrestrained male rat. Again, various doses of NPY13-36 were administered and the effects evaluated at the 1-hr time interval. NPY13-36 is known to have effects only on Y2 receptors (see ref. 12). The results obtained are summarized in Figs. 13–15. It is shown that pNPY13-36 produces a U-shaped dose-response curve with a marked and significant reduction of serum corticosterone levels in a dose of 75 pmol/rat (Fig. 13). These effects have disappeared at the highest dose studied (1.25 nmol). In the presence of the α-methyltyrosine methylester (αMT), a TH inhibitor, these inhibitory effects on corticosterone levels can no longer be observed. These results indicate at least a permissive role of catecholamines for the inhibitory effects of

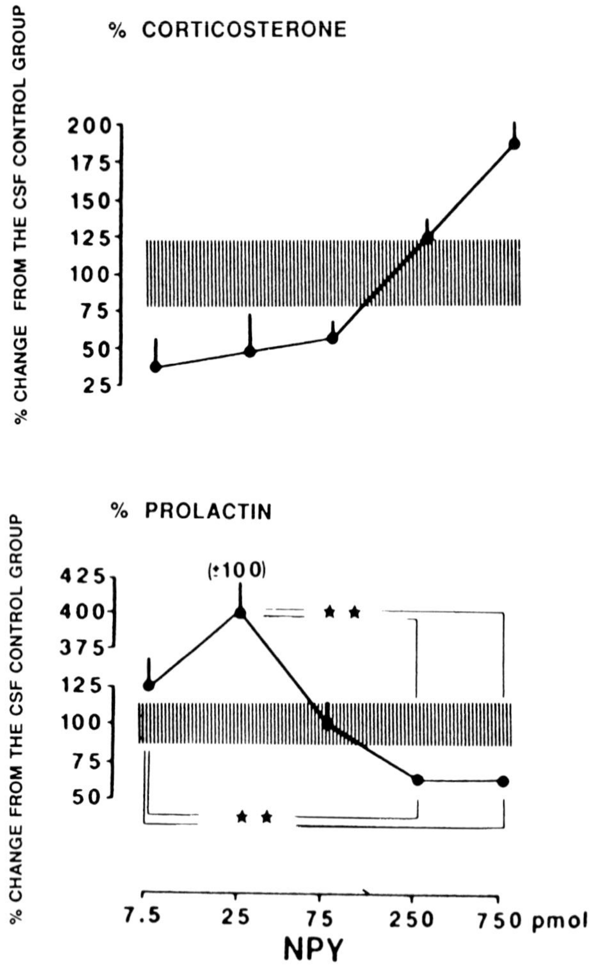

FIG. 12. Effects of intraventricular administration of different doses of pNPY 1–36 on serum corticosterone and prolactin levels in the awake unrestrained male rat. Means ± SEM are shown and given in percent of the respective CSF alone group mean value (*striped area* represents SEM limits). A treatment vs. control nonparametric procedure was used (** $p < 0.01$), $n = 6$ for all groups (see ref. 27).

the NPY fragment. Furthermore, the Y2 NPY receptors appear to be involved in the inhibitory control of corticosterone secretion, probably via an effect on corticotropin-releasing factor (CRF) neurons located within the paraventricular hypothalamic nucleus richly innervated by NPY IR terminals. The fact that the effects of the NPY fragment vanished with the highest dose (1.25 nmol) used may be related to the possibility that the NPY fragment in the highest doses can act also on low-affinity Y2 receptors. This may then lead to excitatory effects on the CRF neurons. Thus, the results obtained on corticosterone secretion would favor a major role of high- and low-affinity NPY receptors of the Y2 type, in the control of CRF release from the median eminence. It cannot be excluded, however, that in the highest doses the NPY fragment may also activate Y1 NPY receptors.

The results obtained on the serum prolactin levels with the NPY fragment 13-36 show that in the higher dose range (150–1,250 pmol), the NPY fragment can significantly reduce the serum prolactin levels at the 1-hr time intervals (Fig. 14). These inhibitory effects on serum prolactin levels by the NPY fragment were abolished by treatment with TH inhibitor αMT, suggesting an involvement of DA in the inhibitory effects of the NPY fragment on prolactin secretion as has previously been demonstrated for NPY1-36 (see refs. 14,45) (see below). In contrast, in the lower doses used, the reverse appears to be true. Thus,

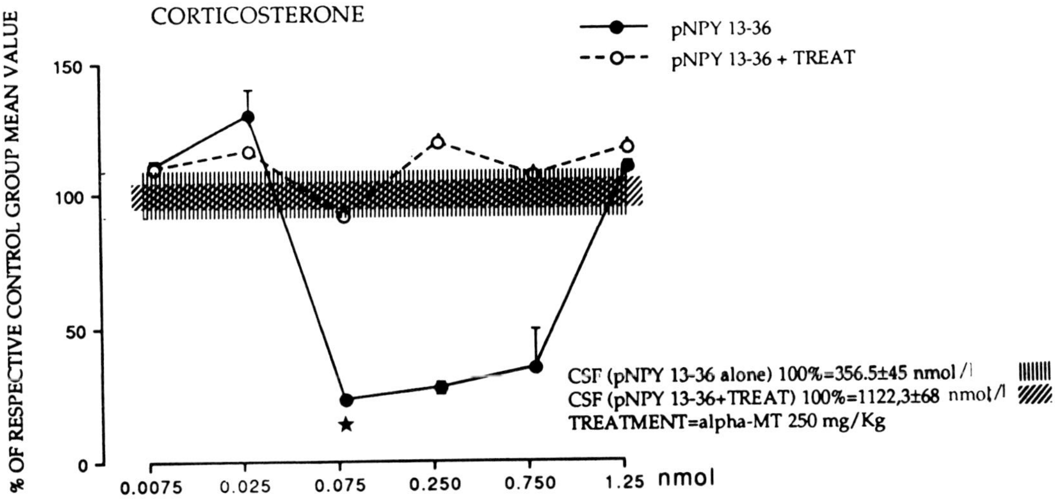

FIG. 13. Studies on effects of various doses of pNPY13-36 given i.v.t. in the absence or presence of tyrosine hydroxylase inhibition (αMT) on serum corticosterone levels in the awake unrestrained male rat. The pNPY fragment 13-36 was given i.v.t. immediately after the intraperitoneal administration of αMT (1 hr before killing). *Striped areas* represent SEM limits for respective CSF control groups. Values are given in percent of the CSF alone and the CSF + αMT treated group mean value. Present treatment vs. control (nonparametric) test was used (* $p < 0.05$); $n = 6$ rats (see ref. 27).

in the absence of the TH inhibitor, no significant effects were observed on serum prolactin levels, whereas in the presence of αMT, marked and significant inhibitory effects were observed with the lowest doses of pNPY13-36 (7.5 and 25 pmol). Thus, there appear to exist both high- and low-affinity Y2 NPY receptors in control of prolactin secretion, but the inhibitory action of the high-affinity Y2 receptors can only be unmasked in the presence of a TH inhibitor, [e.g., because of the removal of an excit-

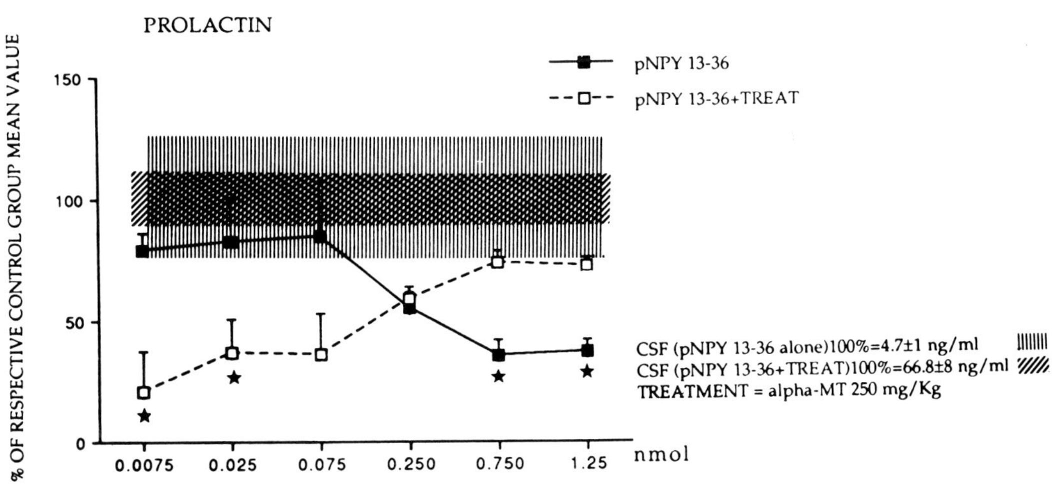

FIG. 14. Studies on effects of various doses of pNPY 13-36 given i.v.t. in absence or presence of tyrosine hydroxylase inhibition on serum prolactin levels in the awake unrestrained male rat. Values are given in percent of the CSF alone and CSF + αMT treated group mean values (*striped area*), respectively. Means ± SEM; n, 6 rats. The treatment versus control (nonparametric) test was used (*$p < 0.05$).

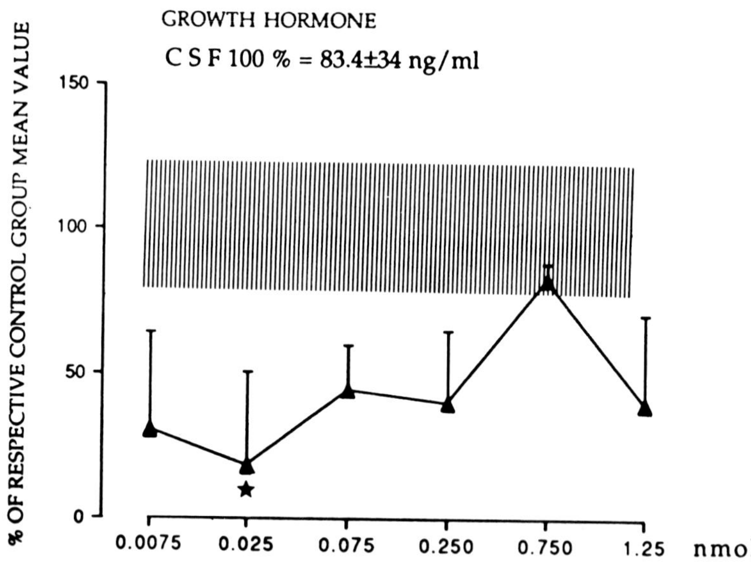

FIG. 15. Effects of various doses of pNPY13-36 given i.v.t. on the serum growth hormone levels in the awake freely moving rat. Means ± SEM are shown. $n = 6$ rats. *Striped area* represents SEM limits for the CSF alone group. Program treatment vs. control (nonparametric) test was used in statistical analysis (* $p < 0.05$) (see ref. 27).

atory epinephrine (NE) influence on prolactin secretion] (see refs. 47,48). It seems likely based on the present observations that the inhibitory effects of low doses of pNPY13-36 on serum prolactin levels in the presence of a TH inhibitor probably does not involve a DAergic mechanism (see below and refs. 9,27). It is presently unclear if the failure to demonstrate an increase of prolactin serum levels with low doses of NPY fragment intact rats is related to the fact that the time interval chosen (1 hr) is too long to demonstrate a possible increase of prolactin secretion. This is a likely possibility in view of the results obtained on median eminence DA turnover with the NPY fragment 13-36 (see below) and in view of the fact that NPY1-36 mainly increases prolactin secretion at the 15-min time interval (46).

In the present experiments, the NPY fragment 13-36 was also found to inhibit significantly GH secretion in the low dose range (25 pmol), whereas no significant inhibitory effects were observed in the higher dose range (Fig. 15). This is in contrast to the results obtained using the parent NPY peptide 1-36, where inhibitory effects were observed in the highest dose range (250–750 pmol), but not in the low dose range. Thus, high-affinity NPY receptors of the Y2 type appear to be involved in the inhibitory control of GH secretion, probably via effects at the level of the GH-releasing hormone IR cell bodies in the arcuate nucleus richly innervated by NPY IR terminals (see refs. 14,27,49).

In conclusion, the present results underline an important role of high-affinity Y2 receptors in the neuroendocrine regulation of corticosterone, GH, and prolactin secretion. These results also favor the role of volume transmission in the hypothalamic and preoptic NPY IR nerve terminal networks despite the low densities of high-affinity ^{125}I-NPY 1–36 binding sites demonstrated and may be related to the existence of NPY fragment receptors, which cannot bind ^{125}I-NPY 1-36 with a high affinity. In addition, the present results indicate a biological role of putative low-affinity NPY receptors of the Y2 type in the control of corticosterone and prolactin secretion. These receptors are postulated to be linked to

NPY synapses and thus to wiring transmission. However, it remains to be demonstrated whether the NPY fragment is selective for Y2 receptors also in high doses. In conclusion, both wiring and volume transmission appear to have an important role in mediating via low- and high-affinity NPY receptors, respectively, the effects of central NPY transmission on neuroendocrine regulation. Furthermore, long C-terminal NPY fragments may be involved via high-affinity Y2 receptors in volume transmission, leading to neuroendocrine actions.

EFFECTS OF CENTRAL ADMINISTRATION OF NPY 1–36 AND THE NPY FRAGMENT 13-36 ON REGIONAL CA UTILIZATION IN THE AWAKE UNRESTRAINED MALE RAT. RELEVANCE FOR VOLUME TRANSMISSION

In previous work we have demonstrated that NPY 1–36 given i.v.t. produces a biphasic change in CA utilization in the tuberoinfundibular DA neurons and in the

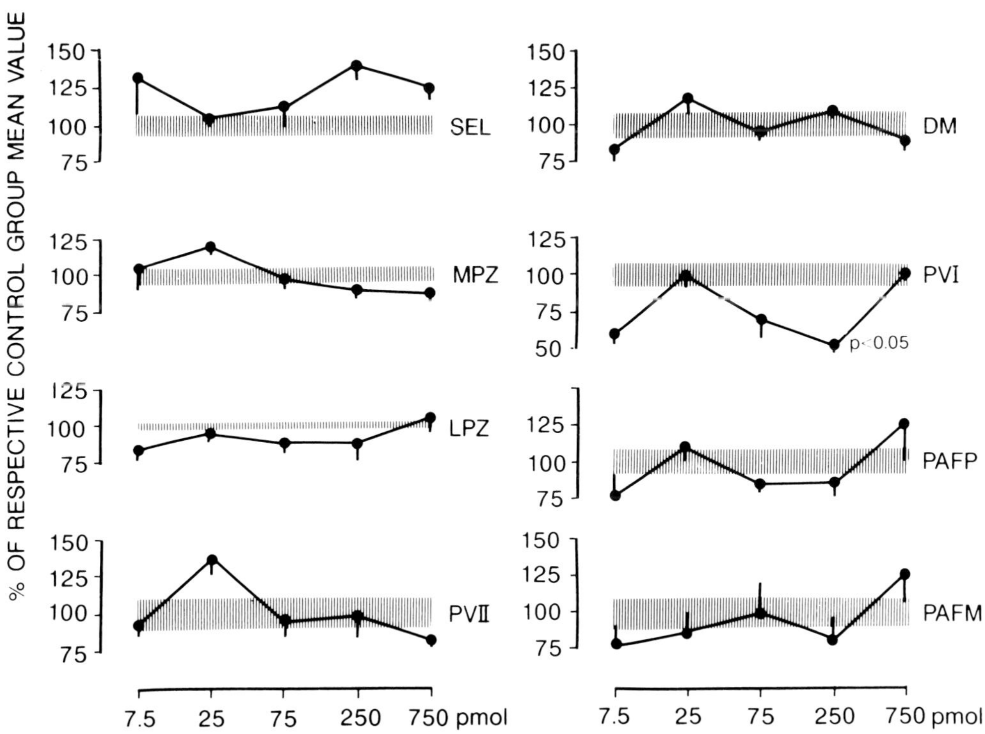

FIG. 16. Effects of i.v.t. injections of various doses (13–36) of the pNPY fragment on catecholamine (CA) levels 1 hr after injection in discrete CA nerve terminal systems of the hypothalamus in the male rat. Means ± SEM are shown. $n = 6$ in each group. Treatment vs. control (nonparametric) test was used for statistics (see ref. 27). The CA levels in the group treated with solvent alone (*shaded area*) were as follows: SEL 100%: 99 ± 9 nmol/g; MPZ 100%: 203 ± 11 nmol/g; LPZ 100%: 593 ± 22 nmol/g; PV II 100%: 106 ± 12 nmol/g; DM 100%: 246 ± 24 nmol/g; PV 100%: 175 ± 29 nmol/g; PA FP 100%: 285 ± 28 nmol/g; PA FM 100%: 189 ± 17 nmol/g. SEL, supependymal layer of the median eminence; MPZ, medial palisade zone of the median eminence; LPZ, lateral palisade zone of the median eminence; PVII, posterior periventricular hypothalamic nucleus; DM, dorsomedial hypothalamic nucleus; PVI, anterior periventricular hypothalamic region; PAFP, parvocellular part of the paraventricular hypothalamic nucleus; PAFM, magnocellular part of the paraventricular hypothalamic nucleus.

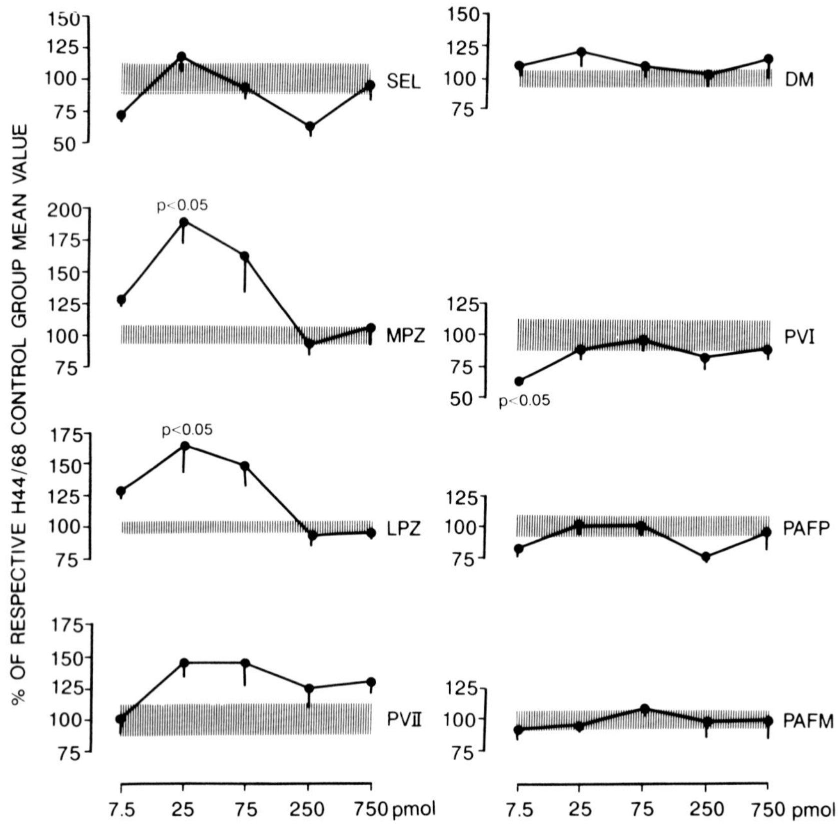

FIG. 17. Effects of i.v.t. injections of various doses in the male rat of pNPY13-36 on CA disappearance in discrete CA nerve terminal systems of hypothalamus after TH inhibition (αMT 250 mg/kg, i.p. 1 hr). pNPY13-36 was given immediately after αMT. Means ± SEM are shown. $n = 6$ in each group. Treatment vs. control (nonparametric) test was used (see ref. 24). CA levels in the group treated with solvent alone (*shaded area*) were as follows: SEL 100%, 94±14 nmol/g; MPZ 100%; 102±8 nmol/g; LPZ 100%, 277±18 nmol/g; PV II 100%, 40±5 nmol/g; DM 100%, 196±17 nmol/g; PV I 100%, 83±11 nmol/g; PA FP 100%, 228±20 nmol/g; PA FM 100%, 132±11 nmol/g. SEL, supependymal layer of the median eminence; MPZ, medial palisade zone of the median eminence; LPZ, lateral palisade zone of the median eminence; PVII, posterior periventricular hypothalamic nucleus; DM, dorsomedial hypothalamic nucleus; PVI, anterior periventricular hypothalamic region; PAFP, parvocellular part of the paraventricular hypothalamic nucleus; PAFM, magnocellular part of the paraventricular hypothalamic nucleus.

FIG. 18A. Effects of the pNPY fragment (1,250 pmol) 13-36 on catecholamine (CA) levels in various types of CA nerve terminal systems of the hypothalamus. Means ± SEM are shown in percent of respective group mean value (n, 6 in each group). Solvent used was mock CSF. The 100% values represent the following absolute concentrations of CA expressed in nmol/g of tissue: SEL, 99±9; MPZ, 203±11; LPZ, 593±22; PV II, 106 ±12; DM, 246±24; PV I, 175±29; PA FP, 285±28; PA FM, 89±17. Mann-Whitney U-test. $^*p < 0.05$ (see ref. 45). SEL, supependymal layer of the median eminence; MPZ, medial palisade zone of the median eminence; LPZ, lateral palisade zone of the median eminence; PVII, posterior periventricular hypothalamic nucleus; DM, dorsomedial hypothalamic nucleus; PVI, anterior periventricular hypothalamic region; PAFP, parvocellular part of the paraventricular hypothalamic nucleus; PAFM, magnocellular part of the paraventricular hypothalamic nucleus. **FIG. 18B.** Effects of the pNPY fragment 13-36 at a high dose (1,250 pmol) on the H 44/68 (αMT) induced catecholamine (CA) disappearance in various hypothalamic CA nerve terminal sys-

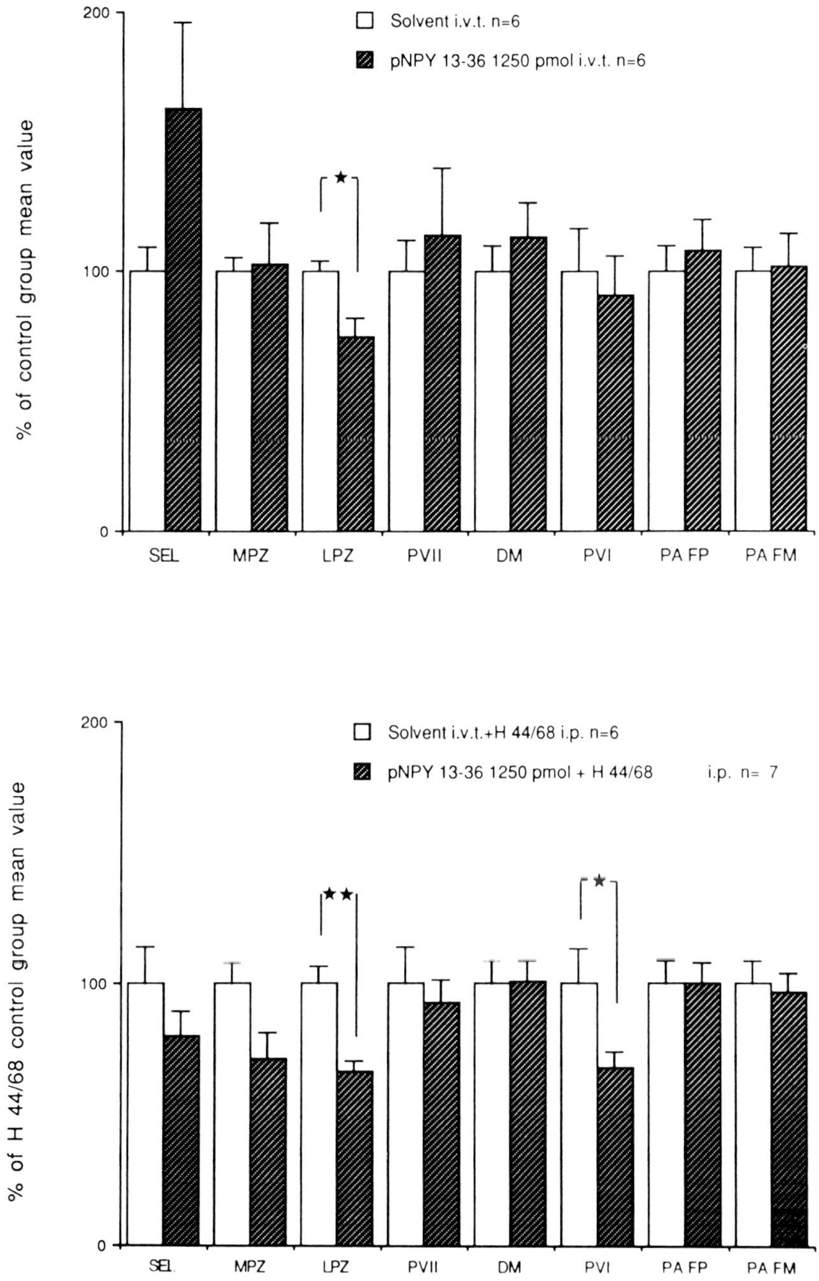

tems. Means ± SEM are shown in percent of the respective H 44/68 group mean value. H 44/68 (250 mg/kg i.p.) was administered immediately before the intraventricular injection of the pNPY fragment. Mann-Whitney U-test. *$p < 0.05$; **$p < 0.01$ (see ref. 45). The 100% values represent the following absolute concentrations of CA expressed in nmol/g of tissue: SEL, 94 ± 14; MPZ, 102 ± 8; LPZ, 277 ± 18; PV II, 40 ± 5; DM, 196 ± 17; PV I, 83 ± 11; PA FP, 228 ± 20; PA FM, 132 ± 11. SEL, supependymal layer of the median eminence; MPZ, medial palisade zone of the median eminence; LPZ, lateral palisade zone of the median eminence; PVII, posterior periventricular hypothalamic nucleus; DM, dorsomedial hypothalamic nucleus; PVI, anterior periventricular hypothalamic region; PAFP, parvocellular part of the paraventricular hypothalamic nucleus; PAFM, magnocellular part of the paraventricular hypothalamic nucleus.

norepinephrine terminals in various hypothalamic areas (14,45). Thus, the lower doses appear to inhibit CA utilization, whereas the higher doses appear to enhance CA utilization. The present results are summarized in Figs. 16 and 17 and demonstrate the influence of i.v.t. injections of NPY13-36 on regional hypothalamic CA levels and utilization in the awake unrestrained male rat. The CA levels in any region studied are uninfluenced by the various doses tested at the 1-hr time interval (Fig. 16). A major result was obtained in the analysis of regional hypothalamic CA utilization, using the TH inhibition method involving αMT (H44/68). Thus, it was demonstrated that in a low dose of 25 pmol/rat NPY13-36 could produce a significant reduction of the CA depletion induced by H44/68 in the medial palisade zone and in the lateral palisade zone of the median eminence (Fig. 17). These results indicate that, in fact, NPY13-36 can produce a reduction in DA utilization in these regions, because DA is the major CA found (see refs. 45,50). With increasing doses of NPY13-36, however, this effect on DA utilization disappears. Thus, an inverse U-shaped dose-response curve is demonstrated in the external layer of the median eminence. Finally, with the highest dose used of the NPY fragment 13-36 (i.e., 1,250 pmol/rat), a significant increase of DA utilization is observed in the medial and lateral palisade zone (see Fig. 18).

The results obtained on regional CA utilization with i.v.t. injections of the NPY fragment 13-36 suggests that both the high- and low-affinity type of NPY Y2 receptors are involved in the control of the tuberoinfundibular DA neurons. Previous work indicates that these Y2 receptors may control the DA nerve cells in the arcuate nucleus indirectly via the β-endorphin neurons (see refs. 6,27). The biphasic changes in DA utilization induced by both pNPY13-36 and pNPY1-36 within the medial palisade zone may explain the biphasic changes in prolactin secretion demonstrated after i.v.t. injections of pNPY13-36 and pNPY1-36 (45). Thus, the NPY regulation of the tuberoinfundibular DA neurons may involve both volume transmission (high-affinity Y2 receptors) and wiring transmission (low-affinity Y2 receptors). In fact, low densities of high-affinity ^{125}I-NPY1-36 binding sites have been demonstrated in the arcuate nucleus.

In conclusion, neuroendocrine regulation by NPY appears to involve in part Y2 receptors of the high- and low-affinity type, whereas cardiovascular regulation by NPY mainly involves high-affinity NPY receptors of the Y1 type (the depressor function) but also high-affinity NPY receptors of the Y2 type (pressor function). Thus, NPY transmission in central autonomic centers appears to operate via both wiring (low-affinity receptors) and volume transmission (high-affinity receptors).

CONCLUDING REMARKS

The present morphological and functional studies on transmitter-receptor mismatches in the central NPY neurons and the functional studies on the NPY fragment 13-36 suggest a role of volume transmission in central NPY transmission operating via high-affinity Y1 and Y2 receptors and involving chemical networks with formation of NPY fragments, such as NPY13-36.

ACKNOWLEDGMENTS

This work has been supported by a grant (O4X-715) from the Swedish Medical Research Council. J.A.A. was supported by Ministerio de Educacion y Ciencia (Spain). This chapter is dedicated to Prof. Mario Tiengo, Chairman of the Brain Clinic, University of Milano, for his 70th year.

List of Abbreviations

A9 dopamine cell group A9
A10 dopamine cell group A10
aca anterior commissure, anterior part
Acb nucleus accumbens
AcbC nucleus accumbens, core
AcbSh nucleus accumbens, shell
ACo anterior cortical amygdaloid nucleus
APT anterior pretectal nucleus
Aq aqueduct (Sylvius)

BL basolateral amygdaloid nucleus
BLV basolateral amygdaloid nucleus, ventral part
BMA basomedial amygdaloid nucleus, anterior part
BST bed nucleus of the stria terminalis

CA1–3 fields CA1-3 of Ammon's horn
CB cell bridges of the ventral striatum
Ce central amygdaloid nucleus
CG central gray
cg cingulum
CGD central gray, dorsal part
CIC central nucleus of the inferior colliculus
Cl claustrum
CM central medial thalamic nucleus
CPu caudate putamen
csc commissure of the superior colliculus

DA dorsal hypothalamic area
DCIC dorsal cortex of the inferior colliculus
DG dentate gyrus
Dk nucleus of Darkschewitsch
DMC dorsomedial hypothalamic nucleus, compact part
DMD dorsomedial hypothalamic nucleus, diffuse part

ec external capsule
En endopiriform nucleus
EW Edinger-Westphal nucleus

Fr frontal cortex
FrPa frontoparietal cortex
FStr fundus striati

gcc genu of the corpus callosum
GrDG granular layer of the dentate gyrus

ICjM islands of Calleja
IMD intermediodorsal thalamic area
IP interpeduncular nucleus

Ld lambdoid septal zone
LHbM lateral habenular nucleus, medial part
LMol lacunosum moleculare layer of the hippocampus
lo lateral olfactory tract
LPZ lateral palisade zone
LSD lateral septal nucleus, dorsal part
LSI lateral septal nucleus, intermediate part
LSV lateral septal nucleus, ventral part
LV lateral ventricle

marc magnocellular part of the arcuate nucleus
MD mediodorsal thalamic nucleus
MeA medial amygdaloid nucleus, anterior part
MF mossy fibers
MGD medial geniculate nucleus, dorsal part
MGV medial geniculate nucleus, ventral part
MHb medial habenular nucleus
Mol molecular layer of the dentate gyrus
MPZ medial palisade zone
MS medial septal nucleus
mt mammillothalamic tract

opt optic tract
Or oriens layer of the hippocampus

Parc parvocellular part of the arcuate nucleus
PIL posterior intralaminar thalamic nucleus
PIR piriform cortex
PoDG polymorph layer of the dentate gyrus
PP peripenduncular nucleus
PrH prepositus hypoglossal nucleus
PVP paraventricular thalamic nucleus, posterior part
Py pyramidal cell layer of the hippocampus

Rad stratum radiatum of the hippocampus
Re reuniens thalamic nucleus
Rh rhomboid thalamic nucleus

SG suprageniculate thalamic nucleus
SN substantia nigra
SNL substantia nigra, lateral part
SuG superficial gray layer of the superior colliculus

Tu olfactory tubercle

VDB nucleus of the vertical limb of the diagonal band
VMHC ventromedial hypothalamic nucleus, central part
VMHDM ventromedial hypothalamic nucleus, dorsomedial part
VP ventral pallidum

Zo zonal layer of the superior colliculus

REFERENCES

1. Agnati LF, Zoli M, Merlo Pich E, et al. NPY receptors and their interactions with other transmitter systems. In: Mutt V, Fuxe K, Hökfelt T, Lundberg JM, eds. *Neuropeptide Y*. New York: Raven Press, 1989;103–114.
2. Zoli M, Agnati LF, Fuxe K, Bjelke B. Demonstration of NPY transmitter receptor mismatches in the central nervous system of the male rat. *Acta Physiol Scand* 1989;135:201–202.
3. Bjelke B, Zoli M, Tinner B, Agnati LF, Fuxe K. On the relationship of NPY immunoreactive nerve terminal networks of high affinity NPY receptors in the male rat brain. Evidence for transmitter receptor mismatches. (*submitted*).
4. Agnati LF, Fuxe K, Zoli M, Zini I, Toffano G, Ferraguti F. A correlation analysis of the regional distribution of central enkephalin and β-endorphin immunoreactive terminals and of opiate receptors in adult and old male rats. Evidence for the existence of two main types of communication in the central nervous system: the volume transmission and the wiring transmission. *Acta Physiol Scand* 1986;128:201–207.
5. Agnati LF, Fuxe K, Zoli M, et al. Morphofunctional evidence for a alteration in central cardiovascular regulation in aging especially in relation to stress induced alterations in cardiovascular function. In: Vezzadini P, Facchini A, Labó G, eds. *Neuroendocrine system and aging* (Topics in aging research in Europe; 8). Rijswijk: Eurage; 1986:37–50.
6. Fuxe K, Agnati LF, Zoli M, et al. The opioid peptide systems: their organization and role in volume transmission and neuroendocrine regulation. In: Illes P, Farsang C, eds. *Regulatory roles of opioid peptides*. New York: VCH, Weinheim, 1988;33–68.
7. Fuxe K, Bunnemann B, Aronsson M, et al. Pre- and postsynaptic features of the central angiotensin systems. Indications of a role of angiotensin peptides in volume transmission and for interactions with central monoamine neurons. *Clin Exp Hyperten* 1988;A10(Suppl 1):143–168.
8. Härfstrand A, Fuxe K, Kalia M, Agnati LF, Zini I, Terenius L. On a possible role of neuropeptide Y, peptide YY, and opioid peptide in the regulation of central respiratory rhythm in the rat. In: von Euler C, Lagercrantz H, eds. *Neurobiology of the control of breathing*. New York: Raven Press, 1986;175–186.
9. Fuxe K, Cintra A, Agnati LF, Härfstrand A, Goldstein M. Studies on the relationship of tyrosine hydroxylase, dopamine and cyclic AMP–regulated phosphoprotein-32 immunoreactive neuronal structures and D1 receptor antagonistic binding sites in various brain regions of the male rat—mismatches indicate a role of D1 receptors in volume transmission. *Neurochem Int* 1988;13:179–197.
10. Danho W, Triscari J, Vincent G, Nakajima T, Taylor J, Kaiser ET. Synthesis and biological evaluation of pNPY fragments. *Int J Peptide Protein Res* 1988;32:496–505.
11. Heilig M, Wahlestedt C, Widerlöv E. Neuropeptide Y (NPY)-induced suppression of activity in the rat: evidence for NPY receptor heterogeneity and for interaction with α-adrenoceptors. *Eur J Pharmacol* 1988;157:205–213.
12. Sheikh S, Håkanson R, Schwartz TW. Y_1 and Y_2 receptors for neuropeptide Y. *FEB* 1989;245 (1,2):209–214.
13. Lundberg JM, Terenius L, Hökfelt T, Tatemoto K. Comparative immunohistochemical and biochemical analysis of pancreatic polypeptide-like peptides with special reference to presence of neuropeptide Y in central and peripheral neurons. *J Neurosci* 1984;4(9):2376–2386.
14. Härfstrand A, Fuxe K, Agnati LF, et al. Studies on neuropeptide Y–catecholamine interactions in the hypothalamus and in the forebrain of the male rat. Relationship to neuroendocrine function. *Neurochem Int* 1986;8/3;355–376.
15. Everitt B, Hökfelt T, Terenius L, Tatemoto K, Mutt V, Goldstein M. Differential co-existence of NPY (NPY)-like immunoreactivity with CA in the central nervous system of the rat. *Neuroscience* 1984;11l:443–462.
16. Sternberger LA. *Immunocytochemistry*, 2nd ed. Chichester: John Wiley & Sons, 1979.
17. Hsu SM, Raine L, Fanger H. A comparative study of the peroxidase-antiperoxidase method and an avidin-biotin complex method for studying polypeptide hormones with radioimmunoassay antibodies. *Am J Clin Pathol* 1981;75:734–738.
18. Agnati LF, Fuxe K, Zoli M, et al. Morphometrical and microdensidometrical studies on phenylethanolamine-*N*-methyltransferase- and neuropeptide Y–immunoreactive neurons in the rostral medulla oblongata of the adult and old male rat. *Neuroscience* 1988;26:461–478.
19. Zoli M, Cintra A, Zini I, et al. Nerve cell clusters in dorsal striatum and nucleus accumbens of the male rat demonstrated by glucocorticoid receptor immunoreactivity. *J Chem Neuroanat* 1990 (*in press*).
20. Paxinos G, Watson C. *The rat brain in stereotaxic coordinates*. New York: Academic Press, 1982.
21. Young WS, Kuhar MJ. Noradrenergic α_1 and α_2 receptors: autoradiographic visualization. *Eur J Pharmacol* 1979;59:317–319.
22. Fuxe K, Calza L, Benfenati F, Zini I, Agnati LF. Quantitative autoradiographic localization of (^{3}H)imipramine binding sites in the brain of the rat: relationship to ascending 5-hydroxytryptamine neuron systems. *Proc Natl Acad Sci USA* 1983;80:3836–3840.
23. Agnati LF, Fuxe K, Benfenati F, Calza L, Fabbri L. Computer-assisted morphometry in immunohistochemistry and receptor autoradiography: impact on neuroendocrine control. In: Rodbard D, Forti G, eds. *Computers in endocrinology*, vol 14. New York: Raven Press, 1984; 171–185.

24. Benfenati F, Cimino M, Agnati LF, Fuxe K. Quantitative autoradiography of central neurotransmitter receptors: methodological and statistical aspects with special reference to computer-assisted image analysis. *Acta Physiol Scand* 1986;128:129–146.
25. Härfstrand A, Fuxe K, Agnati LF, Benfenati F, Goldstein M. Receptor autoradiographical evidence for high densities of ^{125}I-neuropeptide Y binding sites in the nucleus tractus solitarius of the normal male rat. *Acta Physiol Scand* 1986; 128:195–200.
26. Chronwall BM. Anatomical distribution of NPY and NPY messenger RNA in rat brain. In: Mutt V, Fuxe K, Hökfelt T, Lundberg JM, eds. *Neuropeptide Y.* New York: Raven Press, 1989;51–59.
27. Fuxe K, Agnati LF, Härfstrand A, et al. Studies on the neurochemical mechanisms underlying the neuroendocrine actions of neuropeptide Y. In: Mutt V, Fuxe K, Hökfelt T, Lundberg JM, eds. *Neuropeptide Y.* New York: Raven Press, 1989;115–135.
28. Härfstrand A, Fuxe K, Terenius L, Kalia M. Neuropeptide Y–immunoreactive perikarya and nerve terminals in the rat medulla oblongata: relationship to cytoarchitecture and catecholaminergic cell groups. *J Comp Neurol* 1987;260:20–35.
29. Broomé M, Hökfelt T, Terenius L. Peptide YY (PYY)-immunoreactive neurons in the lower brain stem and spinal cord of rat. *Acta Physiol Scand* 1985;125:349–352.
30. Albe-Fessard D, Berkley KJ, Kruger L, Ralston HJ, Willis WD. Diencephalic mechanisms of pain sensation. *Brain Res Rev* 1985;356/3:217–296.
31. Merlo Pich E, Zoli M, Zini I, et al. Effects of central administration of neuropeptide Y on vigilance and pain threshold in spontaneously hypertensive rats. In: Lipton, Tunks, Zoppi, eds. *Advances in pain research and therapy.* New York: Raven Press, 1990;13:55–63.
32. Agnati LF, Andersson K, Fuxe K, Locatelli V, Mutt V, Hökfelt T. Differential actions of CCK-39, CCK-8 and CCK-7 on pre- and postsynaptic dopamine mechanisms in various parts of the forebrain of the rat. *Neurosci Lett* 1980;suppl 5:S204.
33. Agnati LF, Fuxe K, Blake C, et al. Morphometrical characterization of the ACTH-endorphin and dopamine neurons systems of the mediobasal hypothalamus of the rat. Effects of neonatal monosodium glutamate treatment. *INSERM* 1983; 110:87–102.
34. DeWied D, Jolles J. Neuropeptides derived from proopiocortin: behavioural, physiological and neurochemical aspects. *Physiol Rev* 1982;62: 976–1059.
35. Fuxe K, Andersson K, Locatelli V, et al. Neuropeptides and central catecholamine systems: interactions in neuroendocrine and central cardiovascular regulation. In: Costa E, Trabucchi M, eds. *Neural peptides and neuronal communication.* New York: Raven Press, 1980;37–50.
36. Fuxe K, Agnati LF, Härfstrand A. Morphofunctional studies in the dorsal cardiovascular region of the medulla oblongata. Focus on receptor–receptor interactions in cotransmission. In: Hökfelt T, Fuxe K, Pernow B, eds. *Progress in brain research,* vol 28. Amsterdam: Elsevier Science Publishers, 1986;303–320.
37. Fuxe K, Härfstrand A, Agnati LF, et al. Physiological and biochemical characteristics of epinephrine/neuropeptide Y–costoring neurons in the medulla oblongata of the rat, with particular reference to vasodepressor functions within the nucleus tractus solitarii. In: Stolk JM, U'Prichard, Fuxe K, eds. *Epinephrine in the central nervous system.* New York, Oxford: Oxford University Press, 1988;346–365.
38. Härfstrand A. Brain neuropeptide Y mechanisms. Basic aspects and involvement in cardiovascular and neuroendocrine regulation. *Acta Physiol Scand* 1987;suppl 565:131.
39. Fuxe K, Härfstrand A, Agnati LF, et al. Evidence for a vasodepressor role of neuropeptide Y/adrenaline costoring neurons of the dorsal subnuclei of the nucleus tractus solitarius and for their interactions with CCK peptides and their regulation by calcitonin gene–related peptide and substance P immunoreactive nerve terminal networks. In: Buckley JP, Ferrario CM, eds. *Brain peptides and catecholamines in cardiovascular regulation.* New York: Raven Press, 1987;27–42.
40. Fuxe K, Härfstrand A, Agnati LF, von Euler G, Svensson T, Fredholm B. On the role of NPY in central cardiovascular regulation. In: Mutt V, Fuxe K, Hökfelt T, Lundberg JM, eds. *Neuropeptide Y.* New York: Raven Press, 1989;201–214.
41. Fuxe K, Härfstrand A, Agnati LF, et al. Central catecholamine–neuropeptide Y interactions at the pre- and postsynaptic level in cardiovascular centers. *J Cardiovasc Pharmacol* 1987;10(12): 1–13.
42. Zini I, Merlo Pich E, Fuxe K, et al. Actions of centrally administered neuropeptide Y on EEG activity in different rat strains and in different phases of their circadian cycle. *Acta Physiol Scand* 1984;122:71–77.
43. Härfstrand A, Fuxe K, Agnati LF, et al. Studies on neuropeptide Y/catecholamine interactions in central cardiovascular regulation in the α-chloralose anaesthetized rat. Evidence for a possible new way of activating the α_2 adrenergic transmission line. *Clin Exp Hypertens* 1984;A6 (10,11):1947–1950.
44. Fuxe K, Sedvall G. Pharmacological and clinical implications of chemical neurotransmission: chairmen's overview. In: Stjärne L, Hedqvist P, Lagercranz H, Wennmalm Å, eds. *Chemical neurotransmission.* London: Academic Press, 1981; 541–548.
45. Härfstrand A, Eneroth P, Agnati LF, Fuxe K. Further studies on the effects of central admin-

istration of neuropeptide Y on neuroendocrine function in the male rat: relationship to hypothalamic catecholamines. *Regul Pept* 1987;17:167–179.

46. Fuxe K, Agnati LF, Härfstrand A, et al. On the role of neuropeptide Y in information handling in the central nervous system in normal and physiolpathological states: focus on volume transmission and neuropeptide Y/α-2 receptor interactions. *Ann NY Acad Sci* 1990;579:28–79.
47. McCann SM, Krulich L, Ojeda SR, Negro-Vilar A, Vijayan E. Neurotransmitters in the control of anterior pituitary function. In: Fuxe K, Hökfelt T, Luft R, eds. *Central regulation of the endocrine system*. New York: Plenum Press, 1979; 329–347.
48. Andersson K, Fuxe K, Eneroth P, Nyberg P, Roos P. Rat prolactin and hypothalamic catecholamine nerve terminal systems. Evidence for rapid and discrete increases in dopamine and noradrenaline turnover in the hypophysectomized male rat. *Eur J Pharmacol* 1981;76:261–265.
49. McCann SM, Rettori V, Milenkovic L, Riedel M, Aquila C, McDonald JK. The role of neuropeptide Y (NPY) in control of anterior pituitary hormone release in the rat. In: Mutt V, Fuxe K, Hökfelt T, Lundberg JM, eds. *Neuropeptide Y*. New York: Raven Press, 1989;215–227.
50. Löfström A, Jonsson G, Fuxe K. Microfluorimetric quantitation of CA fluorescence in rat median eminence. I. Aspects on the distribution of dopamine and noradrenaline nerve terminals. *J Histochem Cytochem* 1976;24:415–429.

Volume Transmission in the Brain: Novel Mechanisms for Neural Transmission, edited by K. Fuxe and L.F. Agnati.
Raven Press, Ltd., New York © 1991.

10

The Brain Renin-Angiotensin System and Its Possible Involvement in Volume Transmission

*B. Bunnemann, *Kjell Fuxe, *B. Bjelke, and **D. Ganten

**Department of Histology and Neurobiology, Karolinska Institute, Stockholm, Sweden; and **Department of Pharmacology and German Institute for High Blood Pressure Research, University of Heidelberg, Heidelberg, West Germany*

Since about 50 years ago (1,2) the circulating renin-angiotensin system (RAS) with its effector peptide angiotensin II (ANG II) is well known as one of the most important hormonal systems in the regulation of blood pressure and body fluid homeostasis. The aspartyl protease renin—synthetized and released into the blood mainly by the juxtaglomerular cells of the kidney—cleaves the precursor molecule angiotensinogen (AOGEN) originating in the liver to the decapeptide (ANG I), which has no known biological activity. ANG I is then converted by the angiotensin converting enzyme (ACE) to the octapeptide ANG II acting on high-affinity ANG receptors, which may also be structurally related to the *mas*-oncogene (3). The circulating ANG II exerts several well-described functions such as constriction of the vascular smooth muscle, aldosterone release in the adrenal glands, and sodium retention in the kidney, thereby increasing blood pressure. Also ANG II fragments such as ANG III (4–8), ANG-(3-8) (9), and ANG-(1-7) (10) have been found to have biological functions that may be mediated by angiotensin fragment–specific receptors.

During recent years evidence has accumulated that besides the circulating RAS, endogenous RAS might exist in a variety of tissues (e.g., arterial wall, kidney, heart, adrenal glands) and in the brain. Ganten et al. (11) and Fischer-Ferraro et al. (12) gave the first reports about central renin activity and during the following years all the components of the RAS have been demonstrated in the brain (13–29) by different techniques. The demonstration of the genetic message for renin and the precursor molecule AOGEN (30–33) provided the final proof for the existence of an independent brain-RAS.

After Bickerton and Buckley (34) had shown that ANG II not only acts on the periphery but also increases blood pressure after injection into the brain, vast efforts were made to investigate the role of ANG II in the central nervous system. The stimulation of brain ANG II receptors leads to a complex pattern of cardiovascular and endocrine effects and changes in behavior (35–37). An increase in blood pressure, changes in drinking behavior and influences on the release of pituitary hormones have been observed. ANG II causes the release of adrenocorticotropic hormone (ACTH), β-endorphin, and luteinizing hormone (LH) from the anterior pituitary and inhibits prolactin secretion via a central action. The release of vasopressin and oxytocin from neurosecretory neurons by ANG II has been described. Furthermore, it stimulates thirst and salt appetite and increases natriuresis in the kidney. Today ANG II is regarded not only as a circulating hormone that plays an important role in blood pressure control but also as a neuropeptide acting as a neurotransmitter (13,14,37,38).

Agnati and Fuxe (39,40) have recently suggested the involvement of two main types of transmission in the central nervous system. The "wiring transmission" working via synapses and the "volume transmission" using the extracellular fluid pathways. The latter includes not only the diffusion of humoral factors through the extracellular fluid to their specific receptors but also the diffusion of electrotonic signals along those pathways. Volume transmission seems to be responsible for long-lasting regulation effects in the brain. This chapter summarizes the present data on the distribution of the components of the RAS in the rat brain. By relating them to one another we will point out several mismatch regions in the brain that strongly suggest the involvement also of volume transmission mechanisms in the organization of the brain-RAS (41).

METHODOLOGIES USED IN THE STUDIES OF THE BRAIN-RAS

Ribonucleic Acid Probes

Two rat AOGEN complementary cDNA fragments were used for RNA probe synthesis. One 291-bp-long Bam H I/Pvu II fragment subcloned into the Bam H I site of the vector pGEM 4, representing positions 133–424 and one 712-bp-long Bam H I fragment subcloned into the Bam H I site of the vector pSPT 18, representing positions 424–1136 of the AOGEN messenger RNA (mRNA).

For the *mas* RNA probe synthesis, we used a 615-bp-long Bam H I/Hinc II rat *mas*-cDNA fragment subcloned into the vector pGEM 4 and representing positions 258–873 of the *mas*-mRNA.

The RNA probes were synthesized and labeled with ^{35}S-α-UTP (Dupont NEN, Boston, MA) by either SP 6 or T 7 *in vitro* transcription, purified on Nensorb cartridges (Dupont NEN) and used directly for hybridization. The specific activity of the probes was about 5×10^8 dpm/μg RNA.

All RNA probes have been checked before use by denaturing formaldehyde gel electrophoresis.

In situ Hybridization

Adult male Sprague-Dawley rats (300 g body weight) were anesthetized with sodium pentobarbital (60 mg/kg body weight) and perfused transcardially with 200 ml of ice-cold 0.9% saline. The brains were immediately removed and snap-frozen with CO_2. Coronal sections (10 μm thick) were taken on a cryostat, thaw mounted onto poly-L-lysine-coated slides, and fixed for 15 min in 4% paraformaldehyde in PBS, pH 7.0. After three short washes in PBS, they were stored in 70% ethanol at 4°C until further use.

After a deproteination with 0.1 M HCl for 10 min and an acetylation step in 0.1 M triethanolamin/0.25% acetic anhydride for 20 min, the slides were dehydrated in graded ethanol and air dried. Then 150 μl prehybridization buffer (50% deionized formamide, 50 mM Tris-HCl pH 7.6, 25 mM EDTA pH 8.0, 20 mM NaCl, 0.25 mg/ml yeast tRNA, 2.5× Denhardt's solution) was applied per section, and the slides were incubated in a humidified chamber for 2 hr at 37°C.

The prehybridization buffer was drained off the slides, and subsequently, 10 μl hybridization buffer (50% deionized formamide, 20 mM Tris-HCl pH 7.6, 1 mM EDTA pH 8.0, 0.3 M NaCl, 0.1 M dithiothreitol, 0.5 mg/ml yeast tRNA, 0.1 mg/ml poly-A, 1× Denhardt's solution, 10% dextransulphate) was applied per section. This buffer also contained the ^{35}S-α-UTP labeled RNA probes at the appropriate concentrations. For hybridization with AOGEN-RNA 10 pg/μl were used. The concentration in the experiments with the *mas*-RNA was 60 pg/μl. The sections were covered with siliconized coverslips and incubated in a humidified chamber at 37°C for 16 to 18 hr.

To remove the coverslips, the slides were immersed in 2× SSC at 48°C for 30 min. After that step, the slides were washed in 1× SSC/50% formamide at 48°C for 6 hr. The washing solution was changed every 90 min. After two short washes in 1× SSC for 10 min each, the slides were dehydrated in graded ethanol and air dried. The slides were exposed to ^{3}H-LKB-Ultrofilm at −70°C or covered with Kodak NTB 2 photographic emulsion. Counterstaining of sections was performed with cresyl violet.

Hybridization with the respective ^{35}S-α-UTP labeled mRNA at the same specific activity and concentration as the cRNA served as control in all experiments.

Glial Fibrillary Acidic Protein Immunocytochemistry in Combination with *in situ* Hybridization

Immunocytochemistry was performed by means of the biotin-avidin immunoperoxidase procedure using the Vectastain kit (Vector, Burlingame) and 3-3′ diaminobenzidine tetrahydrochloride (DAB) as chromogen. For the combination of *in situ* hybridization and immunocytochemistry, chrome-gelatine-coated slides were used. After the final washing step in 1× SSC, the slides were transferred to PBS, omitting the graded ethanol. Then 100 μl of primary antibody [cow glial fibrillary acidic protein (GFAP) 1:300, Dakopatts, Denmark] was applied per section and incubated overnight at 4°C in a humidified chamber. After washing 3× 10 min in PBS, the sections were incubated with 100 μl/section of the secondary antibody (antirabbit IgG, 1:200, Vector) for 1 hr at room temperature (RT). The slides were then washed 3× 10 min in PBS and incubated for 45 min with 100 μl/section Vectastain at RT. After 3 washes for 10 min each in 0.05 M Tris-HCl pH 7.4, the slides were stained with DAB for approximately 1 min, and the reaction was stopped in 0.05 M Tris-HCl pH 7.4. After dehydration in graded ethanol, the slides were air dried and dipped in Kodak NTB 2 photographic emulsion. As neuronal marker, counterstaining with cresyl violet was used.

ANG II Immunocytochemistry and ANG II Receptor Autoradiography

The procedures for the immunocytochemical detection of ANG II-immunoreactivity (IR) and the method for the [^{125}I]-ANG II receptor autoradiography have been described elsewhere (41).

DISTRIBUTION OF AOGEN-mRNA

Only a few studies have been performed on the distribution and cellular localization of AOGEN-mRNA in the central nervous system using *in situ* hybridization with radiolabeled oligonucleotides (42) also combined with immunocytochemistry for GFAP (43). In our laboratory we mapped out the distribution of AOGEN-mRNA in the rat brain by means of *in situ* hybridization using radiolabeled RNA probes. AOGEN-mRNA turned out to be widely distributed throughout the whole brain with highest expression levels in several discrete areas. To be able to relate this expression pattern to the other components of the RAS, we measured the gray values of the film autoradiograms using a SAS Biovision image analyzing system (Milan, Italy). The specific mean gray values have been obtained by subtracting the control values from the specific values and dividing by the film background.

Cerebral Cortex

The cerebral cortex and also several other regions (e.g., the anterior olfactory nucleus, the nucleus accumbens, the caudate putamen, and the reticular formation of the medulla oblongata) exhibited a uniform signal slightly above background that seemed to be due to a few scattered cells expressing the AOGEN gene in these areas.

Septum

At rostral levels a strong labeling was mainly present in the medial septal nucleus and the nuclei of the band of Broca. The lateral septum was devoid of a detectable signal.

Basal Ganglia

The basal ganglia showed a moderate to low labeling in the globus pallidus, the ventral pallidum, and the substantia innominata. A higher level was measured in the reticular part of the substantia nigra. No significant signal was detectable in the caudate putamen.

Preoptic Region

The expression in the nuclei of the preoptic region (i.e., anteromedial, anteroventral, and median preoptic nuclei) were within the highest levels measured. The medial preoptic nucleus, the medial and lateral preoptic areas, and the magnocellular preoptic nucleus showed moderate to high signals (Fig. 1a).

Hippocampal Formation

The hippocampal formation exhibited predominantly unspecific labeling, but with the emulsion technique, weak labeling over single scattered glial cells was observed in all fields.

Amygdaloid Cortex

In the amygdala, a moderate to high signal was present in the medial amygdaloid nucleus (Fig. 2c), whereas the rest of the amygdala showed only moderate labeling. The central amygdaloid nucleus was devoid of AOGEN-mRNA. A moderate to low signal was present in the bed nucleus of the stria terminalis.

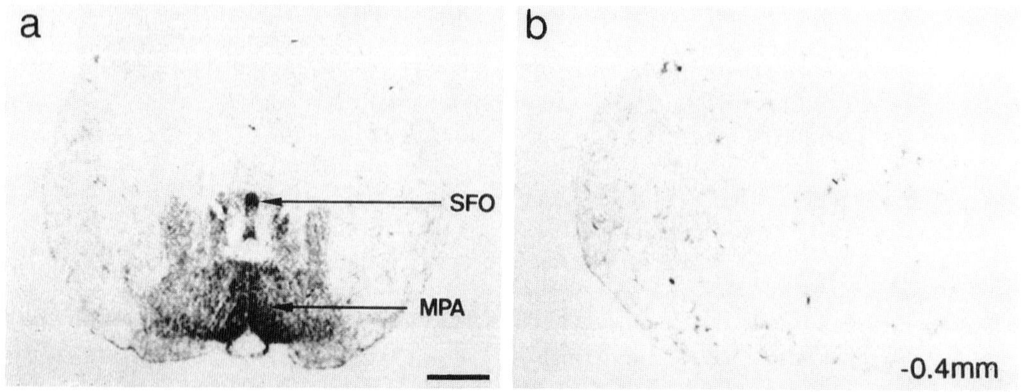

FIG. 1. Autoradiograms of adjacent coronal sections at the level of the preoptic region hybridized with ^{35}S-α-(UTP) labeled AOGEN-cRNA (**a**) and ^{35}S-α-UTP labeled AOGEN-mRNA as control (**b**). Note strong labeling in the subfornical organ (SFO) and the preoptic area. (Bar = 2mm.) MPA, medial preoptic area.

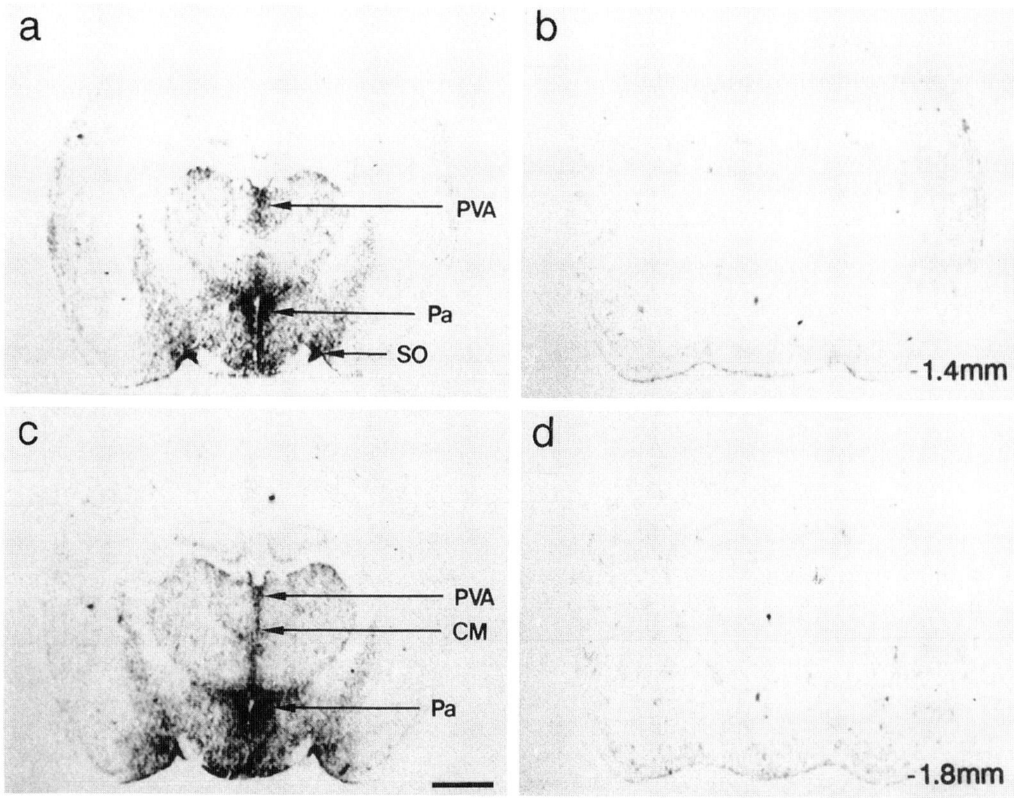

FIG. 2. Autoradiograms of adjacent coronal sections hybridized with ^{35}S-α-UTP labeled AOGEN-cRNA (**a,c**) and ^{35}S-α-UTP labeled AOGEN-mRNA as control (**b,d**). Note strong labeling in the hypothalamus compared with the thalamus. (Bar = 2mm.) PVA, paraventricular thalamic nucleus, anterior; Pa, paraventricular hypothalamic nucleus; SO, supraoptic nucleus; CM, centromedial thalamic nucleus.

Hypothalamus

The hypothalamus is one of the brain regions containing the highest levels of AOGEN-mRNA (Fig. 2a,c). A strong specific signal was present in the supraoptic region and in the supraoptic and suprachiasmatic nuclei, followed by the paraventricular hypothalamic, the periventricular hypothalamic, and the arcuate nuclei. High amounts of AOGEN-mRNA were also detected in the posterior part of the anterior hypothalamic area, the dorso- and ventromedial hypothalamic nuclei, and the retrochiasmatic area. The rest of the anterior, posterior, and lateral hypothalamic areas and the corpus mammillare showed a moderate signal.

Thalamus

In the thalamus, considerably lower amounts of AOGEN-mRNA were present (Fig. 2a,c). High expression levels were demonstrated only in the olivary pretectal nucleus, the medial geniculate body, and the paraventricular thalamic nucleus. The laterodorsal thalamic nuclei, the reticular thalamic nucleus, and the ventral nuclear group exhibited low labeling. The rest of the thalamus showed a rather uniform moderate signal with somewhat higher levels in

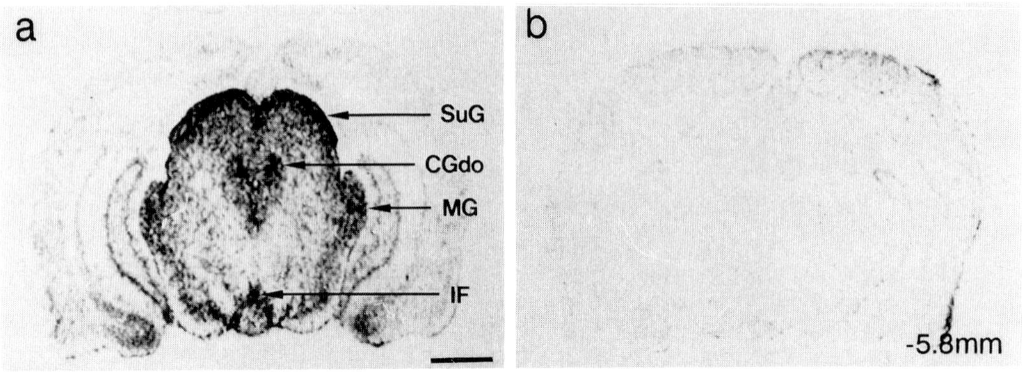

FIG. 3. Autoradiograms of adjacent coronal midbrain sections hybridized with ^{35}S-α-UTP labeled AOGEN-cRNA (**a**) and ^{35}S-α-UTP labeled AOGEN-mRNA as control (**b**). Note strong signal in the superficial gray layer of the superior collicle (SuG) and the dorsal part of the central periaquaeductal gray (CGdo). IF, interfascicular nucleus; MG, medial geniculate nucleus. (Bar = 2mm.)

the subthalamic nucleus, the habenulae, and some midline nuclei (i.e., intermediodorsal, centromedial, reuniens, and rhomboideus nuclei).

Midbrain

In the midbrain, high amounts of AOGEN-mRNA were present in the interfascicular nucleus and the dorsal part of the periaqueductal gray. High signal was exhibited by the paranigral nucleus, the Edinger-Westphal nucleus, and the zonal and superficial gray layers of the superior collicle, whereas other layers showed only moderate levels (Fig. 3a). The rest of the central gray, the dorsal and median raphe nuclei, and the interpeduncular nucleus also contained moderate amounts of AOGEN-mRNA.

Pons

The highest expression in the pons was present in the locus ceruleus followed by

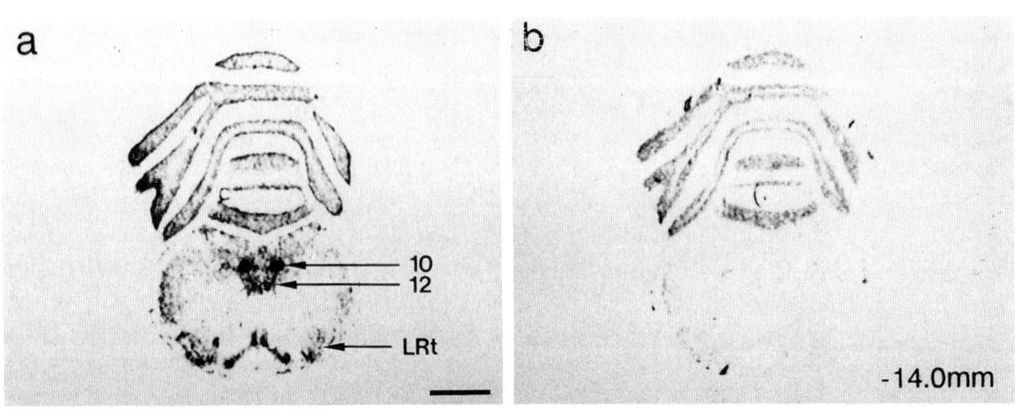

FIG. 4. Autoradiograms of adjacent coronal sections of the medulla hybridized with ^{35}S-α-UTP labeled AOGEN-cRNA (**a**) and ^{35}S-α-UTP labeled AOGEN-mRNA as control (**b**). Note unspecific binding in the granular layer of the cerebellar cortex. 10, dorsal motor nucleus of the vagus; 12, hypoglossal nucleus; LRt, lateral reticular nucleus. (Bar = 2mm.)

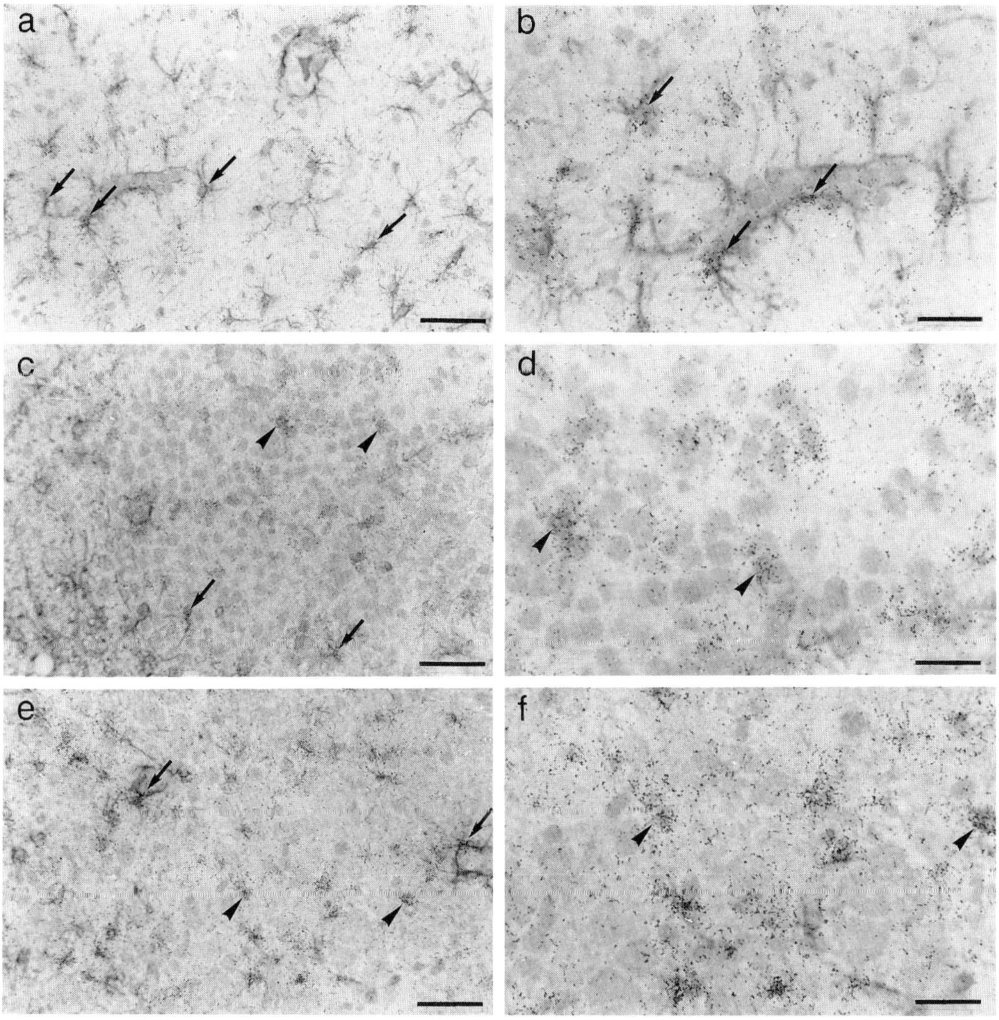

FIG. 5. Photomicrographs of emulsion-coated sections hybridized with ^{35}S-α-UTP labeled AOGEN-cRNA, subsequently immunostained for GFAP and counterstained with cresyl violet (neuronal marker). **a:** Globus pallidus (Bregma -0.9 mm), labeling is only present over glial cells (*arrows*); **b:** same area in higher magnification, note labeling in the glial cells (*arrows*) surrounding the blood vessel. **c:** Suprachiasmatic nucleus (Bregma -0.9 mm), note presence of labeling over both glia (*arrows*) and neurons (*arrowheads*); **d:** same area in higher magnification, note labeled neurons (*arrowheads*). **e:** Paraventricular hypothalamic nucleus (Bregma -1.8 mm), labeling is present over glia (*arrows*) and neurons (*arrowheads*); **f:** same area in higher magnification, note labeled neurons (*arrowheads*). Bars in a, c, e = 50 μm; bars in b, d, f = 25 μm.

the mesencephalic trigeminal nucleus and the central gray of the pons, whereas only moderate to low signals were present in various other regions (i.e., pontine nuclei, sensory trigeminal nucleus, and rostral preolivary region).

Medulla Oblongata

High expression could be demonstrated throughout the medulla oblongata (Fig. 4a). The highest levels were detected in the dorsal motor nucleus of the vagus, the prepositus hypoglossal nucleus, the inferior olive, and the medial part of the nucleus of the solitary tract. The external cuneate nucleus, the medial vestibular nucleus, and the hypoglossal nucleus also contained high amounts of AOGEN-mRNA. Moderate levels were present in the raphe pallidus and in the facial and the lateral reticular nuclei. The spinal trigeminal nucleus showed a low labeling.

Cerebellum

In the cerebellar cortex, high levels of AOGEN-mRNA were detected in the Purkinje cell layer, whereas no significant values could be measured in the other layers (Fig. 4a).

Circumventricular Organs

Among the circumventricular organs and among all other brain structures, the subfornical organ showed the highest expression but interestingly only in its very rostral part (Fig. 1a). The rest of the subfornical organ contained only low to very low amounts of AOGEN-mRNA. A moderate to high signal was observed in the area postrema, whereas a very low signal was present in the median eminence.

Cellular Localization of AOGEN-mRNA

To demonstrate the cellular localization of the AOGEN-mRNA, we performed a combination of *in situ* hybridization and immunocytochemistry for GFAP, because it has been previously reported that AOGEN is synthesized in glial cells (41). Our results confirmed and extended those findings. Most labeling was clearly related to astrocytes, but in at least several areas, the signal was present over neurons (suprachiasmatic nucleus Fig. 5c,d; paraventricular hypothalamic nucleus Fig. 5e,f; lateral hypothalamic area, superficial gray layer of the superior collicle Fig. 6c,d).

DISTRIBUTION OF AOGEN IMMUNOREACTIVITY

Also relatively few studies describe the immunocytochemical localization of AOGEN (28,29,44,45). In contrast to the cellular localization of AOGEN-mRNA, which is mainly present in astrocytes, AOGEN immunoreactivity (AOGEN-IR) was demonstrated in both neuroglial as well as in neuronal perikarya. AOGEN-IR has been mapped out in rat brain using antisera to rat AOGEN as a primary antibody and streptavidin-biotin-linked peroxidase and DAB as a chromogenic detection system (29).

Cerebral Cortex

In cortical regions, few, but strongly stained, neuronal and glial cell bodies were observed.

Septum

In the forebrain, AOGEN-IR was demonstrated in neuroglial cells of the medial septal nucleus and the nuclei of the band of Broca. In the lateral septal nucleus, some scattered glial cells showed a less intense staining. No neuronal staining was observed in this area.

Basal Ganglia

Nearly all parts of the globus pallidus, the caudate putamen, the ventral pallidum, and the substantia innominata showed a light glial staining, whereas moderate to high AOGEN-IR was observed in scattered neuronal perikarya. However, in the globus pallidus the labeling for AOGEN-mRNA was exclusively over glial cells. This discrepancy remains to be explained but could involve the release of AOGEN from glia and neuronal uptake. Also in the internal capsule, moderate to highly immunoreactive glial cell bodies were present. The substantia nigra was completely devoid of any staining.

Preoptic Region

Of all the brain areas examined the preoptic region and the hypothalamus showed the strongest glial staining. In the preoptic area, strongly immunoreactive glial cells were present in the medial preoptic nucleus and in the medial and lateral preoptic area extending to the nucleus of the horizontal limb of the diagonal band. No neuronal-IR was demonstrated in this area.

Hippocampal Formation

In the hippocampal formation, only neuronal staining at moderate to low levels was present in the pyramidal layer of all regions.

Amygdaloid Cortex

The amygdala showed a moderate level of glial AOGEN-IR in all subnuclei, whereas neuronal staining was present primarily in the basolateral and basomedial part. In the bed nucleus of the stria terminalis, a moderate to low glial staining but no neuronal IR was observed.

Hypothalamus

Heavy staining was found in the anterior and lateral hypothalamic area and in the suprachiasmatic and the peri- and paraventricular hypothalamic nuclei. High levels of AOGEN-IR were also present in the arcuate nucleus, the dorso- and ventromedial hypothalamic nuclei, and the dorsal hypothalamic area. The mammillary area showed a moderate staining. The only neuronal AOGEN-IR in the hypothalamus was present at moderate levels in the supraoptic nucleus. All other parts were devoid of AOGEN-IR neurons.

However, it should be pointed out that in the suprachiasmatic nucleus (Fig. 5c,d), we have been able to observe indications of AOGEN-mRNA levels over neuronal profiles and also that in the supraoptic nucleus the labeling obtained with the AOGEN-cRNA was mainly located over astroglial profiles. Also within the paraventricular hypothalamic nucleus, there seemed to exist a labeling over some neuronal perikarya.

Thalamus

Throughout all thalamic structures, a low to moderate neuroglial staining was observed in agreement with our *in situ* hybridization analysis. The neuronal IR was moderate and scattered, however, more concentrated in the ventrolateral nuclei (e.g., reticular nucleus, zona incerta, ventroposterior and ventrolateral thalamic nuclei). Other areas such as the medial geniculate body showed moderate astroglial and neuronal IR, except the reuniens and ventromedial thalamic and habenular nuclei, which were totally devoid of neuronal staining but still contained low amounts of neuroglial AOGEN-IR.

Midbrain

In the midbrain, a strong neuroglial staining was observed in the periaqueductal gray, the interpeduncular and dorsal raphe nuclei, and the inferior and superior colliculi. Neuronal IR in these areas was present in the central gray and the inferior colliculus. The interpeduncular nucleus and the dorsal raphe nucleus were devoid of neuronal staining. The superior colliculus contained moderate glial staining and some scattered AOGEN-IR neurons in agreement with the present analysis on AOGEN-mRNA levels (Fig. 6c,d). Throughout the deep mesencephalic nucleus, scattered

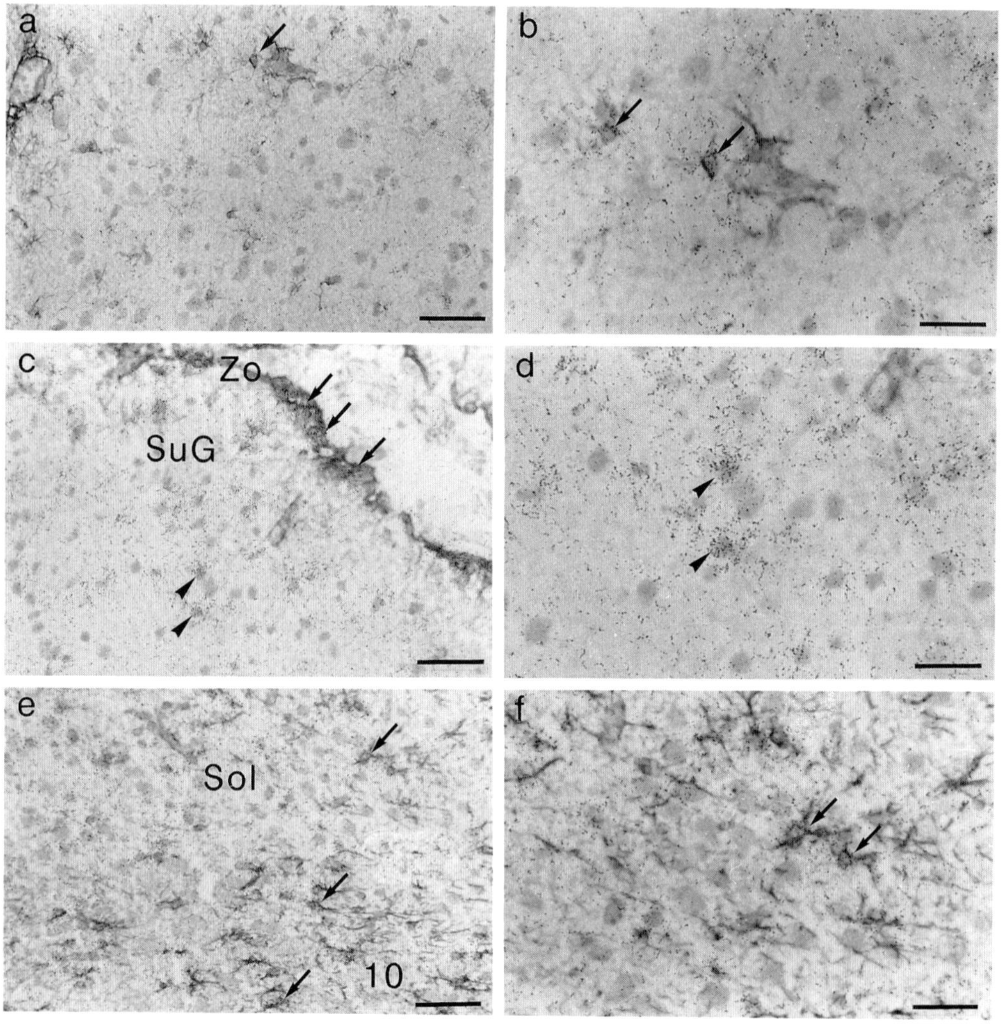

FIG. 6. Photomicrographs of emulsion-coated sections hybridized with ^{35}S-α-UTP labeled AOGEN-cRNA, subsequently immunostained for GFAP and counterstained with cresyl violet (neuronal marker). **a:** Medial geniculate body (Bregma -5.8 mm), silver grains are only located over glial cells (*arrows*). **b:** Same area in higher magnification. **c:** Superficial gray layer of the superior collicle (SuG) (Bregma -5.8 mm), note strong glial labeling (*arrows*) in the zonal layer (Zo) and the labeled neurons in the superficial gray layer (*arrowheads*). **d:** Same area in higher magnification, note labeled neurons (*arrowheads*). **e:** Nucleus of the solitary tract (Sol) and dorsal motor nucleus of the vagus (10) (Bregma -14.0 mm), strong signal is only present over glial cells; **f:** same area in higher magnification. Bars in a, c, e = 50 μm; bars in b, d, f = 25 μm.

moderately stained neurons and glial cells were present.

Pons

In the pons, moderate to low levels of only neuroglial AOGEN-IR were present in the pontine nuclei, the central pontine gray, the sensory trigeminal nucleus, and the superior olive.

It should be noted that within the locus ceruleus, no AOGEN-IR could be demonstrated in neurons or glial cells despite our demonstration of high levels of AOGEN-mRNA within the astroglia of the locus ceruleus and the mesencephalic trigeminal nucleus.

Medulla Oblongata

The medulla was rich in AOGEN-IR structures. The nucleus of the solitary tract, the inferior olive, and the prepositus hypoglossal nucleus contained a great number of strongly immunoreactive astrocytes in agreement with the present findings on the localization of AOGEN-mRNA (Fig. 6e,f). Less amounts of glial IR were detected in the hypoglossal nucleus, the dorsal motor nucleus of the vagus, and the facial nucleus. The rest of the medulla possessed an even density of low to moderately stained glial cells. Neuronal staining in moderate to high amounts was restricted to dorsolateral regions like all portions of the vestibular nucleus, the external cuneate and cuneate nuclei, and the spinal trigeminal nucleus.

Within the substantia gelatinosa of the nucleus tractus nervi trigemini, it was possible to observe labeling over both astroglial and neuronal profiles after the incubation within the AOGEN-cRNA probe. Some scattered AOGEN-IR neurons and glial cells were present in the reticular formation.

Cerebellum

The cerebellar cortex showed AOGEN-IR in a number of Purkinje cells and in a lower amount in neurons and glial cells of the granular layer. The molecular layer was devoid of staining.

Nevertheless, in the *in situ* hybridization analysis, the labeling found within the Purkinje cell layer did not appear to be located over the Purkinje cell bodies but was probably located over the surrounding glia.

Circumventricular Organs

No immunoreactive cells have been reported to exist in the circumventricular organs.

DISTRIBUTION OF RENIN

Renin-mRNA has been demonstrated in brain tissue by liquid hybridization assays and by Northern blotting analysis (32,33,46). The regional and cellular localization of renin expression in the central nervous system, however, still remains to be investigated.

The presence of renin itself in the central nervous system has been described for the first time by Ganten et al. (11) by measuring the generated ANG I in homogenized brain tissue incubated with AOGEN. Until now, not much is known about the anatomical localization of renin in the rat brain. By means of radioimmunoassays, renin activity has been shown to be present in the frontal cortex, the striatum, the amygdala, the hypothalamus, and the cerebellum. The medulla contained the highest levels of renin activity (47). Fuxe et al. (16,20,21) reported on the localization of renin in the para- and periventricular nuclei and the supraoptic nucleus of the hypothalamus as shown immunocytochemically. Renin-IR was also present in the Purkinje cells of the cerebellar cortex.

DISTRIBUTION OF ANG-CONVERTING ENZYME

The distribution of ACE in the rat central nervous system has been demonstrated in several papers. In these studies, either a [^{125}I]-labeled ACE inhibitor has been used for the localization and autoradiographic quantification (26,48) or the enzymatic activity of ACE in dissected brain nuclei has been measured *in vitro* (17,23). Also immunocytochemical attempts have been made to localize ACE in the brain (49).

Cerebral Cortex

The cortex exhibited an overall very low binding of the ACE inhibitor.

Septum

The medial septal nucleus as well as the nuclei of the band of Broca displayed a slight binding. In the lateral septum, no labeling was detected.

Basal Ganglia

The reticular part of the substantia nigra is the most densely labeled structure in the brain, followed by the caudate putamen. The globus pallidus exhibited a high binding, and also the accumbens nucleus showed a dense labeling as did the entopeduncular nucleus. The ventral pallidum contained moderate amounts of ACE, whereas the substantia innominata was almost without labeling.

Preoptic Region

A moderate labeling was present in the median and medial preoptic areas, whereas the lateral preoptic area was without binding.

Hippocampal Formation

In the hippocampal formation, a high density of ACE was present in the molecular layer of the dentate gyrus. The rest of the hippocampus was without binding.

Amygdaloid Cortex

The amygdala contained moderate amounts of ACE with the basolateral part as the most densely labeled nucleus. The lateral part was without activity. The central, basomedial, and medial amygdaloid nuclei showed moderate amounts of ACE. The bed nucleus of the stria terminalis was only slightly labeled.

Hypothalamus

In agreement with the AOGEN-mRNA distribution, the supraoptic and paraventricular hypothalamic nuclei showed high levels of ACE activity, whereas the suprachiasmatic nucleus displayed an even lower binding than the adjacent anterior hypothalamic area. Other hypothalamic areas (the lateral hypothalamic area, the periventricular hypothalamic, the arcuate, the ventro-, and dorsomedial hypothalamic nuclei, and the mammillary body) were moderately labeled.

Thalamus

In the thalamus, a moderate to low labeling was present in the anterior nuclear group, the reticular nucleus, paraventricular thalamic, parential, rhomboid, and reuniens nuclei. The habenular nuclei showed a moderate binding. The interanteromedial nucleus was nearly without ACE. Low labeling was detected in the mediodorsal nucleus, and the laterodorsal thalamic nuclei are without binding. The ventral nuclei (ventrolateral, ventromedial, ventroposteriolateral, and ventroposteriomedial thalamic nuclei) all displayed low ACE activity. A moderate labeling was also present in the lateral geniculate body but only in its ventral portion; the dorsal part was almost negative, as was the medial geniculate body.

Midbrain

In midbrain regions, the central gray and the interfascicular and the paranigral nuclei displayed a moderate labeling. The superior collicle showed a low density of binding, except its intermediate gray and white layers, which showed a slightly higher activity. The dorsal and medial raphe nuclei exhibited low binding.

Pons

A moderate binding was present in the locus ceruleus and the parabrachial and the pontine nuclei. The cochlear nuclei and the superior olive showed only low labeling.

Medulla Oblongata

Almost all regions in the medulla (nucleus of the solitary tract, motor trigeminal, facial, hypoglossal, cuneate, medial vestibular, prepositus hypoglossal, and lateral reticular nuclei) showed low binding. Only the superior vestibular and the inferior olive exhibited moderate signals.

Cerebellum

In the cerebellar cortex, a high density binding was present in the Purkinje cell layer, whereas the molecular layer was only moderately labeled.

Circumventricular Organs

Among the periventricular structures, the subfornical organ, the vascular organ of the lamina terminalis, and the median eminence exhibited high labeling. The area postrema in contrast showed only a low binding. The choroid plexus of all ventricles as well as their ependymal walls contained high densities of ACE.

DISTRIBUTION OF ANG II-IR

A number of studies described the distribution of ANG II-IR in the rat brain (13,14,18–20,25,41,50,51) using different immunohistochemical approaches. ANG II-IR neurons were mainly present within the hypothalamus and the subfornical organ with projections to other hypothalamic nuclei, the median eminence, the subfornical area, the preoptic region, and the cardiovascular areas of the medulla oblongata and the spinal cord (i.e., nucleus of the solitary tract, sympathetic lateral column). However, in contrast to AOGEN, no neuroglial localization of ANG II has been reported.

Cerebral Cortex

No ANG II-IR nerve cell bodies could be demonstrated in cortical regions, but scattered ANG II-IR fibers were reported.

Septum

Nearly all parts of the septum contained at least a few stained fibers. The highest concentration was shown in the lateral part of the lateral septal nucleus. The rest of the lateral septal nucleus and the medial septal nucleus contained a scattered distribution of ANG II-IR nerve fibers.

Basal Ganglia

The striatum showed a scattered fiber distribution, except the ventral caudal part of the caudate putamen, which contained a higher amount of stained nerve fibers. The substantia innominata displayed a moderate density of fibers and contained immunoreactive nerve cell bodies in its caudal part. In the substantia nigra, moderate amounts of fibers were present in the ventral part of the pars reticulata and in and around the pars compacta.

Preoptic Region

Angiotensin II-IR nerve cell bodies were demonstrated in the medial and lateral preoptic area and in the periventricular preoptic nucleus, which also contained a strong fiber plexus. High amounts of ANG II-IR fibers were also found in the median preoptic nucleus, whereas the medial and lateral preoptic areas and the medial preoptic nucleus contained a low density of fibers.

Hippocampal Formation

At least a few ANG II-IR fibers were detected in all fields of the hippocampus.

Amygdaloid Cortex

The central nucleus of the amygdala is one of the brain structures with the highest densities of ANG II-IR fibers (Fig. 7a). The medial nucleus contained some ANG II-IR nerve cell bodies as did the bed nucleus of the stria terminalis. Both nuclei also showed moderate amounts of stained fibers. The other parts of the amygdala were without ANG II-IR nerve fibers.

Hypothalamus

The highest number of ANG II-IR nerve cell bodies in the brain was present in the neurosecretory nuclei of the hypothalamus (paraventricular hypothalamic Fig. 7b and supraoptic nuclei) and the lateral hypothalamus. Lower amounts of cells were found in the suprachiasmatic nucleus, the anterior, lateral, and posterior hypothalamic area, the supraoptic region, the arcuate nucleus, and the corpus mammillare. A fiber plexus was present in the same areas except the anterior hypothalamic area and the ventromedial hypothalamic nucleus, which were almost without fibers. The highest amounts of ANG II-IR fibers were demonstrated in the lateral hypothalamus, the para- and periventricular hypothalamic nuclei, and the posterior hypothalamic area. A moderate fiber density was present in the dorsomedial hypothalamic nucleus.

Thalamus

Only a weak staining was observed in thalamic regions. ANG II-IR nerve fibers were mainly restricted to the paraventricular thalamic nucleus, the reuniens nucleus, the centromedial nucleus, and the zona incerta. All other regions of the thalamus including the medial geniculate body (Fig. 7c) did not contain any detectable ANG II-IR.

Midbrain

Only little ANG II-IR is present in the midbrain compared with other regions. ANG II-IR cell bodies and fibers were found in the peripeduncular nucleus (Fig. 7c) and interpeduncular nucleus. ANG II-IR fibers were scattered throughout the central periaqueductal gray and the paranigral, interfascicular, dorsal raphe, and pedunculopontine nuclei. Only few fibers were present in the superficial gray layer of the superior colliculus.

Pons

The only region containing high amounts of ANG II-IR in the pons is the lateral parabrachial nucleus. Nerve cells as well as fibers have been demonstrated in this area. Other areas such as the locus ceruleus, the medial parabrachial nucleus, and the central pontine gray showed only few scattered ANG II-IR fibers.

Medulla Oblongata

Angiotensin II-IR in nerve cells of the medulla was restricted to one group in the medial part of the nucleus of the solitary tract. This nucleus also contained high densities of ANG II-IR fibers as did the dorsal motor nucleus of the vagus (Fig. 7d) and the substantia gelatinosa of the spinal trigeminal nucleus. A diffuse ANG II-IR was also ob-

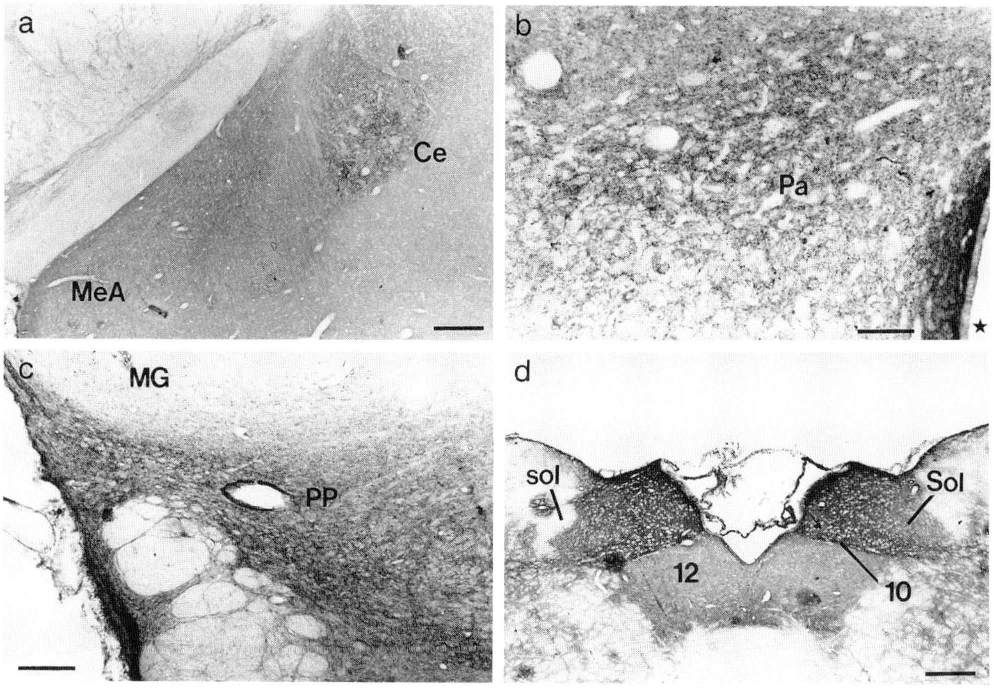

FIG. 7. Photomicrographs of coronal sections of the rat brain incubated with ANG II antibodies (1:700). **a:** Amygdaloid cortex, note strong ANG II-IR in the central amygdaloid nucleus (Ce) and slight staining in the medial amygdaloid nucleus (MeA). **b:** Hypothalamus, a dense plexus of ANG II-IR nerve terminals is present in the paraventricular nucleus of the hypothalamus. **c:** Midbrain, note absence of ANG II-IR in the medial geniculate body (MG), whereas strong ANG II-IR is present in the adjacent peripeduncular nucleus (PP). **d:** Medulla oblongata, strong ANG II-IR is present in the nucleus of the solitary tract (Sol) and the dorsal motor nucleus of the vagus (10), note absence of ANG II-IR in the hypoglossal nucleus (12). Pa, paraventricular hypothalamic nucleus; sol, solitary tract. Bars = 250 μm (a, d), 100 μm (c), and 50 μm (b).

served in the inferior olive. All other regions were without detectable staining.

Cerebellum

No ANG II-IR cell bodies or nerve fibers could be demonstrated in the cerebellar cortex.

Circumventricular Organs

The subfornical organ contained large numbers of ANG II-IR nerve cell bodies and a moderate to dense plexus of fibers. A high density of ANG II-IR terminals was present in the external layer of the median eminence. The area postrema showed staining in the form of a fine fiber plexus.

ANGIOTENSIN III

The C-terminal heptapeptide ANG-(2-8) (ANG III) has been reported in recent studies to have a higher potency to affect the functions of the central nervous system than ANG II (4–8) itself. It has even been suggested to be the most active form of brain ANG (4). The anatomical distribution of this peptide, however, is unknown.

DISTRIBUTION OF ANG-(1-7)

Schiavone and co-workers (10) have been able to show that an active fragment is formed from ANG II, namely, ANG-(1-7), which possesses the same potency as ANG II in releasing vasopressin. This N-terminal heptapeptide ANG-(1-7) has been mapped out in the rat brain using immunocytochemistry and was demonstrated to be present in neuronal perikarya as well as in nerve fibers (52). Immunoreactive staining was mainly confined to certain areas of the tel- and diencephalon, whereas ANG II-IR has been demonstrated in distinct nuclei of all parts of the central nervous system.

Angiotensin-(1-7) immunoreactive nerve cell bodies were present in the paraventricular hypothalamic nucleus, the supraoptic and suprachiasmatic nuclei, the posteriolateral part of the bed nucleus of the stria terminalis, and the substantia innominata.

A dense plexus of ANG-(1-7) immunoreactive nerve fibers was located in the internal layer of the median eminence. Only scattered ANG-(1-7) positive fibers were demonstrated in the substantia innominata, the periventricular hypothalamic nucleus, and the ventral part of the retrochiasmatic area.

DISTRIBUTION OF ANG II BINDING SITES

In a number of studies, ANG II binding sites were mapped out by using receptor autoradiography with either [^{125}I]-ANG II (27,41,53,54) or ^{125}I-labeled ANG II analogues (24,55–57). These studies revealed a large number and a distinct distribution of ANG II binding sites in the rat brain. However, with these analyses, only high-affinity binding sites are detected.

Cerebral Cortex

Low levels of ANG II binding sites were present in cortical regions.

Septum

The triangular septal nucleus contained the highest amounts of ANG II receptors followed by the lateral septum, where moderate densities of binding sites were found. The medial septal nucleus showed no binding.

Basal Ganglia

The basal ganglia were nearly without ANG II binding sites.

Preoptic Region

Moderate densities of binding sites were present in the median preoptic area and the median preoptic nucleus.

Hippocampal Formation

Only low specific binding was present in the hippocampal formation but with somewhat higher densities in the caudal part.

Amygdaloid Cortex

The amygdala showed a moderate to high density of ANG II binding sites in their medial part, whereas other parts exhibited no binding.

Hypothalamus

In the hypothalamus, high binding was observed in the paraventricular hypothalamic, periventricular hypothalamic, supraoptic, and suprachiasmatic nuclei. Moderate levels were present in the dorso- and ventromedial hypothalamus and in the mammillary nuclei.

Thalamus

Among thalamic areas, the subthalamic nucleus contained substantial amounts of ANG II binding sites. Other regions such as the habenular nuclei, the medial geniculate body, and several midline nuclei (e.g., rhomboideus and reuniens) showed a moderate labeling.

Midbrain

The binding in the midbrain was confined to a few discrete nuclei. High densities were present in the superficial gray layer of the superior colliculus, and moderate binding was shown in the central periaqueductal gray.

Pons

The locus ceruleus in the pons displayed high binding of [^{125}I]-ANG II.

Medulla Oblongata

High densities of ANG II binding sites were present in several nuclei of the medulla oblongata (i.e., the nucleus of the solitary tract, the dorsal motor nucleus of the vagus, and the inferior olive). The hypoglossal nucleus was without any binding. The substantia gelatinosa of the spinal trigeminal nucleus showed a moderate labeling.

Cerebellum

The molecular layer of the cerebellum displayed a moderate binding. No specific binding was detected in the other layers.

Circumventricular Organs

Among the periventricular organs, the subfornical organs, the organum vasculosum of the lamina terminalis, and the median eminence showed high amounts of ANG II binding sites, whereas the area postrema exhibited only moderate levels, as did the choroid plexus.

mas ONCOGENE

Recently Jackson and co-workers (3) described the gene product of the *mas* oncogene to be acting as a functional ANG receptor. In another study, the distribution of *mas* expression in the rat brain has been examined by means of RNase protection assays (58). The *mas* oncogene turned out to be expressed in the hippocampal formation and the cerebral cortex.

Together with Dr. M. Hanley, we have mapped out the expression of this gene in the rat brain using *in situ* hybridization. The highest levels of *mas* gene expression have been found in the CA 3 and CA 4 region of the hippocampus, the olfactory bulb, and the olfactory tubercle. Rather low expression levels were present in the outer and basal layers of the cerebral cortex and in the basomedial amygdaloid nucleus. The CA 1 and CA 2 regions of the hippocampus displayed a low labeling. Also, arterial walls and the ependymal cells of the cerebral ventricles showed low hybridization signals. The distribution of the *mas*-mRNA was not related to the distribution of high-affinity ANG II binding sites detected by receptor autoradiography.

MISMATCH REGIONS

As described in a great number of studies and summarized in this chapter, the components of the RAS are widely distributed throughout all levels of the rat brain. They are present not only in areas important in cardiovascular control, but also in functionally different regions, suggesting also other functions of the RAS in the rat brain.

Few attempts have been made to correlate all the components of the brain RAS

	AOGEN-mRNA	ANG II-IR	ANG II-Rec
AOGEN-mRNA		r=0.34	r=0.63**
ANG II-IR	r=0.34		r=0.37
ANG II-Rec	r=0.63**	r=0.37	

FIG. 8. Matrix of the correlation of regional distribution of AOGEN-mRNA, ANG II-IR and ANG II binding sites (ANG II-Rec) in 19 areas of the rat brain.

with one another (41). We have performed correlations between the distribution of AOGEN-mRNA, ANG II-IR, and ANG II receptors for 19 areas for which semiquantitative data were available (Fig. 8). Interestingly, it turned out that there is no correlation between AOGEN-mRNA and ANG II-IR and between ANG II-IR and ANG II receptors, but a significant correlation between the AOGEN-mRNA and the ANG II receptors (Fig. 9). In Table 1, the regional distributions of the RAS compo-

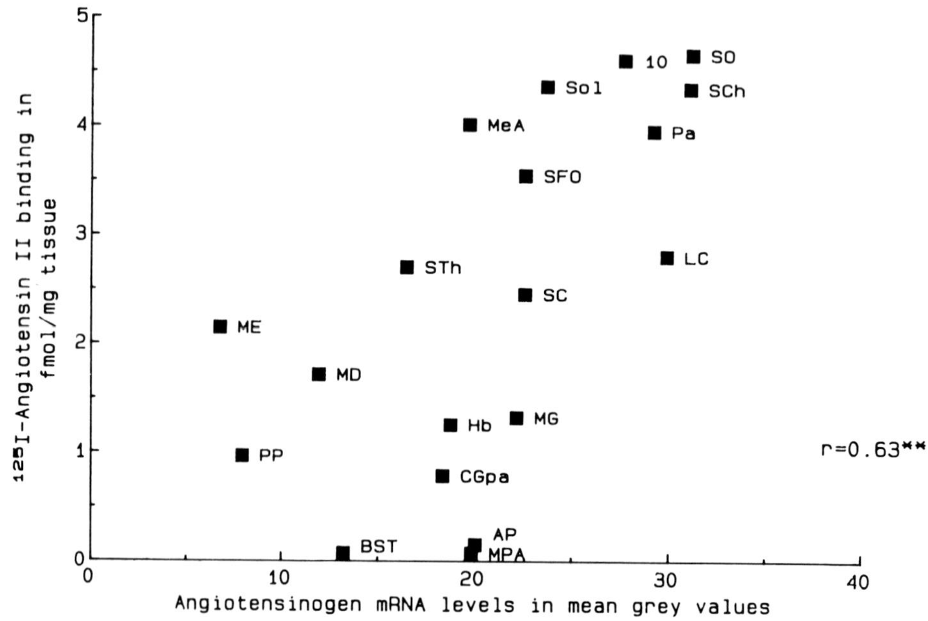

FIG. 9. Correlation between the regional AOGEN-mRNA levels and ANG II binding sites in the rat brain. 10, dorsal motor nucleus of the vagus; SO, supraoptic nucleus; SCh, suprachiasmatic nucleus; Sol, nucleus of the solitary tract; MeA, medial amygdaloid nucleus; Pa, paraventricular hypothalamic nucleus; SFO, subfornical organ; STh, subthalamic nucleus; LC, locus ceruleus; SC, superior colliculus; ME, median eminence; MD, mediodorsal thalamic nucleus; Hb, habenulae; MG, medial geniculate nucleus; PP, peripeduncular nucleus; CGpa, central gray, periaqueductal; AP, area postrema; MPA, medial preoptic area; BST, bed nucleus of the stria terminalis.

TABLE 1. *Comparison between the distribution of AOGEN-mRNA, AOGEN-IR* [29], *ACE (23,26,47), ANG II-IR (25,39), and ANG II binding sites (ANG II-Rec) (24,27,39) in the rat brain*[a]

Area	AOGEN-mRNA	AOGEN-IR	ACE	ANG II-IR	ANG II-Rec
Telencephalon					
Cerebral Cortex	+	+	+	+	+
Basal Ganglia					
Substantia nigra	+ +	+	+ + +	+ +	n.d.
Ventral pallidum	+ +	+	+ +	+	n.d.
Globus pallidus	+ +	+	+ + +	+	+
Substantia innominata	+ +	+	+	+ +	n.d.
Caudate putamen	+	+	+ + +	+ +	+
Accumbens nucleus	n.d.	+	+ +	+	+
Septum					
Nucleus of the horizontal limb of a diagonal band	+ +	+ +	+	+	n.d.
Nucleus of the vertical limb of the diagonal band	+ +	+ +	+	+	n.d.
Medial septal nucleus	+ +	+ +	+	+	n.d.
Lateral septum	n.d.	+	n.d.	+ +	+ +
Preoptic Region					
Anteromedial preoptic nucleus	+ + +	+ + +	n.d.	n.d.	n.d.
Anteroventral preoptic nucleus	+ + +	+ + +	n.d.	n.d.	+ +
Median preoptic nucleus	+ + +	n.d.	+ +	+ + +	+ +
Medial preoptic area	+ +	+ + +	+ +	+	+ +
Medial preoptic nucleus	+ +	+ +	n.d.	+	n.d.
Lateral preoptic area	+ +	+ +	n.d.	+	n.d.
Magnocellular preoptic nucleus	+ +	+ +	n.d.	n.d.	n.d.
Hippocampal Formation	+	+	+ + +	+	+
Amygdaloid Cortex					
Medial amygdaloid nucleus	+ +	+	+ +	+	+ +
Bed nucleus of the stria terminalus	+ +	+	+	+	n.d.
Central amygdaloid nucleus	n.d.	+	+ +	+ + +	n.d.
Basolateral amygdaloid nucleus	n.d.	+	+ + +	n.d.	n.d.
Basomedial amygdaloid nucleus	+ +	+	+ +	n.d.	n.d.
Diencephalon					
Hypothalamus					
Supraoptic region	+ + +	+ +	n.d.	+ + +	n.d.
Supraoptic nucleus	+ + +	+ + +	+ + +	+ + +	+ + +
Suprachiasmatic nucleus	+ + +	+ + +	+	+ +	+ + +
Paraventricular hypothalamic nucleus	+ + +	+ + +	+ + +	+ + +	+ + +
Arcuate hypothalamic nucleus	+ + +	+ + +	+ +	+ +	n.d.
Periventricular hypothalamic nucleus	+ + +	+ + +	+ +	+ + +	+ + +
Dorsomedial hypothalamic nucleus	+ + +	+ + +	+ +	+	+ +
Ventromedial hypothalamic nucleus	+ + +	+ + +	+ +	+	+ +
Retrichiasmatic area	+ + +	+ +	+ +	n.d.	n.d.
Dorsal hypothalamic area	+ +	n.d.	n.d.	n.d.	n.d.
Anterior hypothalamic area	+ +	+ + +	+	+	n.d.
Mammillary nuclei	+ +	+ +	+ +	+ +	+ +
Lateral hypothalamic area	+ +	+ + +	+ +	+ +	n.d.
Thalamus					
Olivary pretectal nucleus	+ + +	n.d.	n.d.	n.d.	n.d.
Medial geniculate nucleus	+ + +	+ +	n.d.	n.d.	+ +
Paraventricular thalamic nucleus	+ + +	+	+ +	+	n.d.
Lateral habenulae	+ +	+	n.d.	n.d.	+ +
Medial habenulae	+ +	+	+ +	n.d.	+ +
Subthalamic nucleus	+ +	n.d.	n.d.	n.d.	+ + +

TABLE 1. *Continued*

Area	AOGEN-mRNA	AOGEN-IR	ACE	ANG II-IR	ANG II-Rec
Thalamus *Continued*					
Reuniens nucleus/ Rhomboideus nucleus	+ +	+ +	+ +	+	+ +
Zona incerta	+ +	+ +	+ +	+	n.d.
Centromedial thalamic nucleus	+ +	+ +	+ +	+	n.d.
Anteroventral thalamic nucleus	+ +	+	+ +	+	n.d.
Mediodorsal thalamic nucleus	+ +	+	+	n.d.	+ +
Lateral geniculate nucleus	+ +	+ +	+ +	n.d.	+
Laterodorsal thalamic nuclei	+	+	n.d.	n.d.	+
Ventral posteriolateral thalamic nucleus/Ventral posteriomedial thalamic nucleus	+	+	+	+	+
Reticular thalamic nucleus	+	+	+ +	n.d.	n.d.
Ventromedial thalamic nucleus/ Ventrolateral thalamic nucleus	+	+	+	+	n.d.
Mesencephalon					
Interfascicular nucleus	+ + +	n.d.	+ +	+	n.d.
Central gray dorsal part	+ + +	+ + +	+ +	+	+
Paranigral nucleus	+ + +	n.d.	+ +	+	n.d.
Zonal layer of the superior colliculus	+ + +	n.d.	n.d.	n.d.	n.d.
Edinger-Westphal nucleus	+ + +	n.d.	n.d.	n.d.	+
Superficial gray layer of the superior colliculus	+ + +	+ +	+	n.d.	+ + +
Central gray	+ +	+ + +	+ +	+	+
Dorsal hypothalamic area	+ +	+ + +	+	+	n.d.
Superior colliculus	+ +	+ +	+	n.d.	+ +
Median raphe nucleus	+ +	n.d.	+	n.d.	n.d.
Interpeduncular nucleus	+ +	+ + +	+	+	n.d.
Peripeduncular nucleus	n.d.	n.d.	n.d.	+	n.d.
Pons					
Locus coeruleus	+ + +	n.d.	+ +	+	+ + +
Mesencephalic trigeminal nucleus	+ + +	n.d.	n.d.	n.d.	n.d.
Central gray	+ + +	+ +	+ +	+	n.d.
Pontine nuclei	+ +	+ +	+ +	n.d.	n.d.
Paramedian raphe nucleus	+ +	n.d.	n.d.	n.d.	n.d.
Inferior colliculus	+ +	+ +	+ +	n.d.	n.d.
Rostral preolivary region	+ +	n.d.	n.d.	n.d.	n.d.
Principal sensory trigeminal nucleus	+	+	n.d.	n.d.	n.d.
Superior olive/preolivary region	+	+	+	n.d.	n.d.
Dorsal cochlear nucleus/ Ventroposterior cochlear nucleus	+ +	n.d.	+	n.d.	n.d.
Medulla Oblongata					
Dorsal motor nucleus of the vagus	+ + +	+ +	+	+ +	+ + +
Prepositus hypoglossal nucleus	+ + +	+ +	+	n.d.	n.d.
Inferior olive	+ + +	+ +	+ +	+ +	+ +
Nucleus of the solitary tract	+ + +	+ +	+	+ +	+ + +
External cuneate nucleus	+ + +	+	+	n.d.	n.d.
Medial vestibular nucleus	+ + +	+	+	n.d.	n.d.
Hypoglossal nucleus	+ + +	+	+	n.d.	n.d.
Facial nucleus	+ +	+	+	n.d.	n.d.
Spinal trigeminal nucleus	+	+	+	+ +	+ +
Reticular formation	+	+	+	n.d.	n.d.
Cerebellum					
Purkinje cell layer of the cerebellar cortex	+ + +	+	+ + +	n.d.	n.d.
Granular layer of the cerebellar cortex	+	+	+ +	n.d.	n.d.

TABLE 1. *Continued*

Area	AOGEN-mRNA	AOGEN-IR	ACE	ANG II-IR	ANG II-Rec
Cerebellum *Continued*					
Molecular layer of the cerebellar cortex	n.d.	n.d.	+ +	n.d.	+ +
Circumventricular Organs					
Subfornical organ	+ + +	n.d.	+ + +	+ +	+ + +
Area postrema	+ +	n.d.	+	+ +	+ +
Median eminence	+	n.d.	+ + +	+ +	+ + +
Choroid plexus	n.d.	n.d.	+ + +	n.d.	+ +
Organum vasculosum of the lamina terminalis	n.d.	n.d.	+ + +	+ +	+ + +

+, low; + +, moderate; + + +, high–very high; n.d., not detected.

nents are listed and compared. It can clearly be seen that there exist a number of regions with topological mismatches, where one or more components of the system are lacking.

In the neurosecretory nuclei of the hypothalamus (i.e., paraventricular hypothalamic nucleus and supraoptic nucleus), all the components of the RAS have been demonstrated in high amounts (Fig. 10a). The same is true for the periventricular hypothalamic nucleus. The suprachiasmatic and the dorso- and ventromedial hypothalamic nuclei also contain all components but to various degrees.

One of the mismatch regions is the amygdaloid cortex (Fig. 10a). High ANG II-IR is present in its central part together with moderate levels of AOGEN-IR and ACE, whereas no AOGEN-mRNA and no high affinity ANG II receptors have been detected in this subnucleus. In the medial amygdaloid nucleus, we find moderate amounts of AOGEN-mRNA, ACE, and high-affinity ANG II binding but only few sparsely distributed ANG II-IR nerve fibers. Another mismatch is present in the bed nucleus of the stria terminalis, which contains all the components but has a relative lack of high-affinity ANG II receptors. In the medial geniculate body (Fig. 10b), in contrast, we find another type of mismatch with high levels of AOGEN-mRNA and moderate amounts of AOGEN-IR and high-affinity ANG II receptors, but no presence of ACE and few scattered ANG II-IR fibers. At the same rostrocaudal level, high AOGEN expression is present in the superficial gray layer of the superior collicle, where also high densities of high-affinity ANG II binding sites were demonstrated. However, only little ACE activity and only some scattered ANG II-IR nerve fibers are present in this area. Another interesting nucleus at this rostrocaudal level is the interpeduncular nucleus where we can find moderate levels of AOGEN-mRNA, AOGEN-IR, and low amounts of ACE and ANG II-IR, but no high-affinity ANG II receptors. However, a perfect match region is the central periaqueductal gray where all the components of the RAS are present.

Mismatch regions can also be observed in the medulla oblongata (Fig. 10c). All components of the RAS are present in the nucleus of the solitary tract, the dorsal motor nucleus of the vagus, and the substantia gelatinosa of the spinal trigeminal nucleus. The inferior olive contains high amounts of AOGEN-mRNA and moderate amounts of AOGEN-IR, ACE, ANG II-IR fibers, and high-affinity ANG II binding sites. Mismatches are present in the prepositus hypoglossal nucleus, the hypoglossal nucleus, the external cuneate nucleus, and the medial vestibular nucleus; moderate to high levels of AOGEN-mRNA are associated with low amounts of AOGEN-IR and ACE. No ANG II-IR and no high-affinity ANG II binding sites are present in these areas.

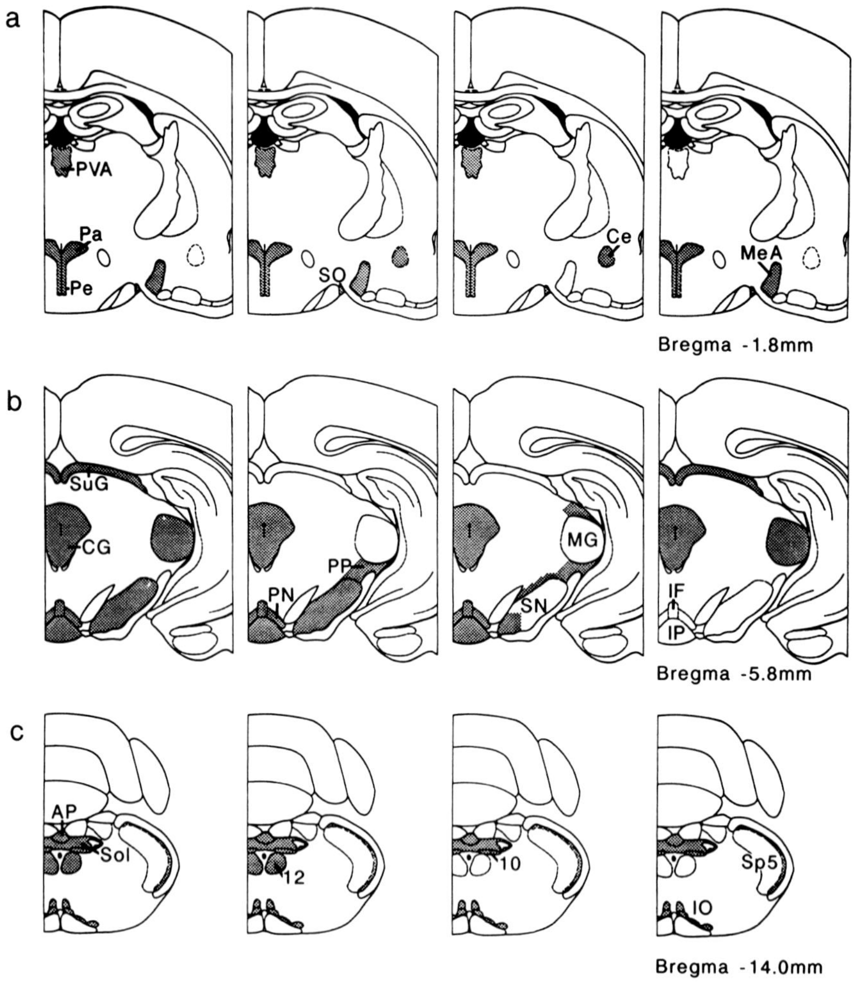

FIG. 10. Schematic examples of topological mismatches in the brain = RAS. **a:** Hypothalamus and amygdaloid cortex: for better overview the distribution is shown only in indicated areas. For simplicity slight ANG II-IR in the medial amygdaloid nucleus was considered as negative. **b:** Midbrain; **c:** medulla oblongata. *Shading* of area indicates presence of parameter in question. PVA, paraventricular thalamic nucleus, anterior; Pa, paraventricular hypothalamic nucleus; Pe, periventricular hypothalamic nucleus; SO, supraoptic nucleus; Ce, central amygdaloid nucleus; MeA, medial amygdaloid nucleus; IF, interfascicular nucleus; IP, interpeduncular nucleus; MG, medial geniculate nucleus; SN, substantia nigra; PP, peripenducular nucleus; PN, paranigral nucleus; SuG, superficial gray layer of the superior colliculus; CG, central gray; AP, area postrema; Sol, nucleus of the solitary tract; 12, hypoglossal nucleus; 10, dorsal motor nucleus of the vagus; Sp5, spinal trigeminal nucleus levels (62).

DISCUSSION

In conclusion we can state the existence of three major types of mismatches in the brain-RAS (Table 2):

1. AOGEN-mRNA levels, AOGEN-IR, and high-affinity ANG II receptors are present, but ACE and/or ANG II is not detected in this area.

ACE is not the only enzyme able to generate ANG II. Cathepsin G and tonin have been shown to generate ANG II directly from AOGEN (59). The presence of one of these enzymes in areas lacking only ACE remains to be demonstrated. ANG II could therefore still be formed in these regions.

Angiotensinogen may be released from glia and cleaved in the extracellular space to ANG II, which is then acting on its specific high-affinity receptors present in the mismatch area. Another possibility is that ANG II originates in adjacent brain areas and diffuses into the mismatch region using the extracellular fluid pathways. It is, for example, true that in the brain areas surrounding the medial geniculate body, a plexus of ANG II-IR nerve fibers was observed (i.e., peripeduncular nucleus). ANG II could be released from nerve terminals in these areas and diffuse into the medial geniculate body to activate the high-affinity ANG II receptors.

In both cases the angiotensin peptides formed use extracellular pathways to reach high-affinity ANG II receptors, which thus would represent volume transmission processes.

2. There are no detectable ANG II binding sites in the presence of all other components of the brain-RAS.

The lack of high-affinity ANG II receptors in areas where all other components are present could imply that ANG II is acting in this region via low-affinity binding sites, which could not be detected with the current receptor autoradiographic methods. One kind of low-affinity binding site could be the *mas* gene product because no high-affinity ANG II binding was detected in areas expressing the *mas* gene.

Recent evidence for a physiological role of angiotensin fragments like ANG III (4–8), ANG-(3-8) (9), and ANG-(1-7) (10) in the brain also suggests the possible existence of high-affinity binding sites for these peptides different from the high-affinity ANG II receptor. Such receptors would also not be detectable with current techniques. This seems to be true at least for ANG-(3-8) because it has a 1,000-fold lower affinity to high-affinity ANG II receptors than ANG II itself (60) and because it improved processes related to learning and memory in rats with the same potency as ANG II (9). This effect must be independent of ANG II receptors. ANG II might be processed in these areas intra- and/or extracellularly to biologically active ANG fragments acting on their own highly specific receptors.

These processes could involve both wiring and volume transmission mechanisms.

3. AOGEN-mRNA is associated with AOGEN-IR and in some cases also with moderate to low amounts of ACE. ANG II-IR and ANG II receptors are not detected.

ANG-converting enzyme can metabolize a variety of peptides besides ANG I such as bradykinin, enkephalins, and neurotensin (61). Thus ACE might have different functions in areas where there is no association with the other components of the RAS.

In the present type of mismatch demonstrating only the presence of AOGEN-mRNA and AOGEN-IR, it seems likely that ANG II formation does not take place. AOGEN could have another as yet unknown function in these mismatch regions.

When comparing the cellular localization of the AOGEN expression, which is mainly located in astroglia, with the cellular location of AOGEN-IR (i.e., glia and neurons) and ANG II-IR, which is exclusively pres-

TABLE 2. *Match regions and three major types of topological mismatches in the brain-renin angiotensin system (RAS) of the rat*

No Mismatch	
Basal ganglia	Globus pallidus
	Caudate putamen
Preoptic region	Median preoptic area
Hypothalamus	Supraoptic nucleus
	Suprachiasmatic nucleus
	Paraventricular hypothalamic nucleus
	Periventricular hypothalamic nucleus
	Dorsomedial hypothalamic nucleus
	Ventromedial hypothalamic nucleus
	Mammillary nuclei
Midbrain	Central periaqueductal gray
Medulla oblongata	Dorsal motor nucleus of the vagus
	Nucleus of the solitary tract
	Spinal trigeminal nucleus (substantia gelatinosa)
	Inferior olive
No Detectable Angiotensin II Immunoreactivity	
Amygdaloid cortex	Medial amygdaloid nucleus
Thalamus	Mediodorsal thalamic nucleus
	Lateral habenula
	Medial habenula
	Subthalamic nucleus
	Medial geniculate body
	Lateral geniculate body
	Laterodorsal thalamic nuclei
Midbrain	Superficial gray layer of the superior colliculus
No Detectable Angiotensin II Receptors	
Basal Ganglia	Substantia nigra
	Ventral pallidum
	Substantia innominata
Septum	Nuclei of the band of Broca
	Medial septal nucleus
Amygdaloid cortex	Bed nucleus of the stria terminalis
	Central amygdaloid nucleus
Hypothalamus	Supraoptic region
	Nucleus arcuatus
	Anterior hypothalamic area
	Lateral hypothalamic area
Thalamus	Paraventricular thalamic nucleus
	Zona incerta
	Centromedial thalamic nucleus
	Anteroventral thalamic nucleus
Midbrain	Dorsal raphe nucleus
	Interpeduncular nucleus
Pons	Central pontine gray
AOGEN-mRNA and AOGEN-IR alone	
Preoptic region	Anteromedial preoptic nucleus
	Magnocellular preoptic nucleus
Medulla oblongata	Prepositus hypoglossal nucleus
	External cuneate nucleus
	Medial vestibular nucleus
	Hypoglossal nucleus
	Facial nucleus
Cerebellum	Purkinje cell layer of the cerebellar cortex

ent in neurons, it becomes obvious that there may exist an extracellular transport of AOGEN from glial cells to neurons. The further processing of AOGEN to ANG II will then take place within the nerve cells.

Another possibility is that AOGEN is released by glial cells and processed extracellularly to ANG I and ANG II. ANG II may then diffuse to reach high-affinity ANG II receptors or be processed to smaller biologically active fragments, which have their own specific binding sites.

It should be remembered that in some regions, AOGEN is in fact expressed in neurons, and in this case AOGEN will be processed within the same cellular compartment.

We have now also evidence that in several brain regions such as the nucleus tractus solitarius, angiotensin peptides can be released from both nerve terminals and astroglia. These two types of angiotensin peptide signals may be in part selectively detected if we assume that the synaptic receptor for ANG II is a low-affinity receptor. Thus it seems reasonable that only neuronally released ANG II reaches sufficient concentrations to activate the low-affinity receptors. In contrast the high-affinity ANG II receptors in the area may be

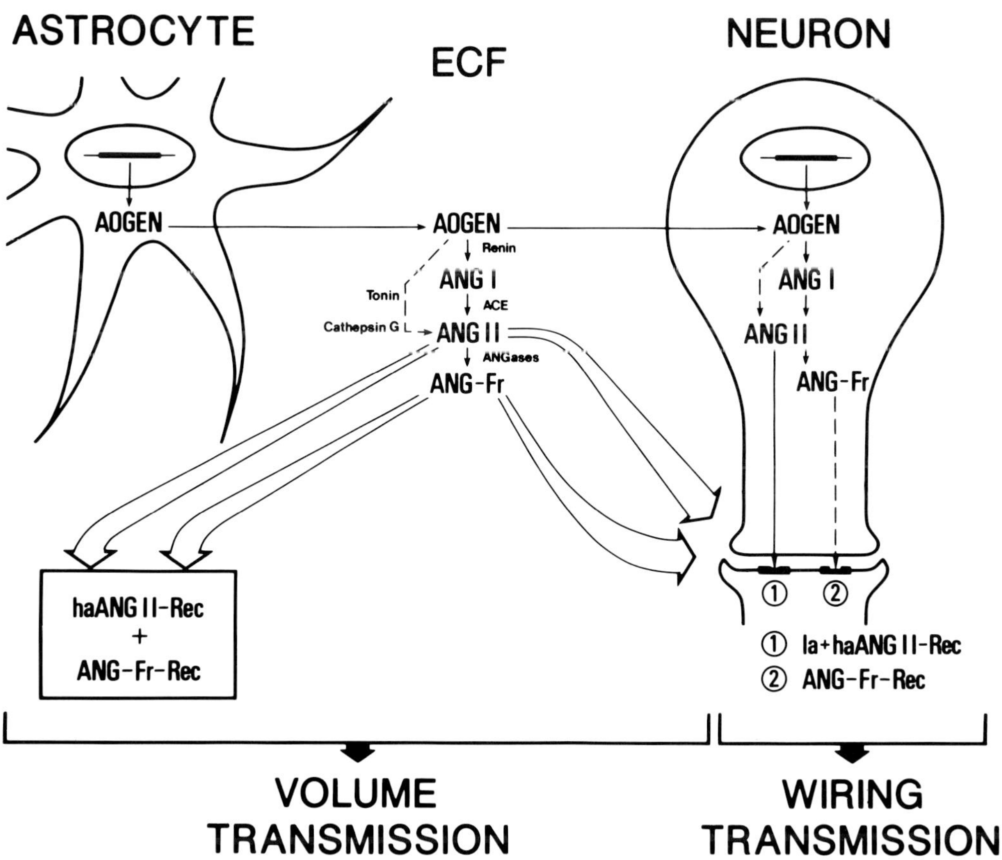

FIG. 11. Scheme of possible transmission pathways in the organization of the brain-RAS. (ANGases, angiotensinases; ANG-Fr, angiotensin fragments; ANG-Fr-Rec, angiotensin fragment receptors; ECF, extracellular fluid; Ia + haANG II-Rec = low-affinity and high-affinity ANG II receptors, AOGEN, angiotensinogen; ACE, ANG-converting enzyme). The scheme illustrates that the brain-RAS may use both wiring and volume (ECF pathways) transmission in chemical communication.

reached by ANG peptides originating both from glia and from leaking nerve endings followed by a diffusion in the extracellular space.

In conclusion, the present evidence suggests the existence of both wiring transmission and volume transmission in the brain-RAS (Fig. 11) based on the correlation of the distribution of its components (Table 2).

The brain-RAS may operate in the following three modes:

1. In some areas ANG II or ANG fragments are released from nerve endings to activate synaptic receptors (wiring transmission) or diffuse to adjacent regions to reach their specific receptors (volume transmission).
2. In some areas, AOGEN is released from glial cells into the extracellular space where it is processed to ANG II or to active ANG fragments, which act on their specific receptors (volume transmission).
3. In some areas, AOGEN is not related to the formation of ANG peptides but might have other still unknown functions.

ACKNOWLEDGMENTS

This study was supported by a grant of the Swedish Medical Research Council (04X715), the Deutsche Forschungsgemeinschaft (SFB 317), and NIH grant 1 RO1 HL 35821-01.

REFERENCES

1. Braun-Menendez E, Fasciolo JC, Leloir LF, Munoz JM. The substance causing renal hypertension. *J Physiol* 1940;98:283–298.
2. Page IH, Helmer OM. A crystalline pressor substance (angiotonin) resulting from the reaction between renin and reninactivator. *J Exp Med* 1940;71:29–42.
3. Jackson TR, Blair LAC, Marshall J, Goedert M, Hanley MR. The *mas* oncogene encodes an angiotensin receptor. *Nature* 1988;335:437–440.
4. Harding JW, Felix D. Angiotensin-sensitive neurons in the rat paraventricular nucleus: relative potencies of angiotensin II and angiotensin III. *Brain Res* 1987;410:130–134.
5. Wright JW, Roberts KA, Harding JW. Drinking to intracerebroventricularly infused angiotensin II, III, IV in the SHR. *Peptides* 1988;9:979–984.
6. Wright JW, Quirk WS, Hanesworth JM, Harding JW. Influence of aminopeptidase inhibitors on brain angiotensin metabolism and drinking in rats. *Brain Res* 1988;441:215–220.
7. Sullivan MJ, Harding JW, Wright JW. Differential effects of aminopeptidase inhibitors on angiotensin-induced pressor responses. *Brain Res* 1988;456:249–253.
8. Lin KS, Chan JYH, Chan SHH. Reduction in baroreceptor reflex response by angiotensin III and its modification by Ile^7-angiotensin III and bestatin in the rat. *Neurosci Lett* 1988;90:172–176.
9. Braszko JJ, Kupryszewski G, Witczuk B, Wisniewski K. Angiotensin II-(3-8)-hexapeptide affects motor activity, performance of passive avoidance and a conditioned avoidance response in rats. *Neuroscience* 1988;27:777–783.
10. Schiavone MT, Santos RAS, Brosnihan KB, Khosla MC, Ferrario CM. Release of vasopressin from the rat hypothalamo-neurohypophysial system by angiotensin-(1-7) heptapeptide. *Proc Natl Acad Sci USA* 1988;85:4095–4098.
11. Ganten D, Merquez-Julio A, Granger P, et al. Renin in dog brain. *Am J Physiol* 1971;221:1733–1737.
12. Fischer-Ferraro C, Nahmod VE, Goldstein DJ, Finkielman S. Angiotensin and renin in rat and dog brain. *J Exp Med* 1970;133:353–361.
13. Fuxe K, Ganten D, Hökfelt T, Bolme P. Immunohistochemical evidence for the existence of angiotensin II containing nerve terminals in the brain and spinal cord of the rat. *Neurosci Lett* 1976;2:229–234.
14. Ganten D, Fuxe K, Philips MI, Mann JFE, Ganten U. The brain isorenin-angiotensin system: biochemistry, localization and possible role in drinking and blood pressure regulation. In: Ganong WF, Martini L, eds. *Frontiers in neuroendocrinology*, vol 5. New York: Raven Press, 1978; 61–99.
15. Lewicki JA, Fallon JH, Printz MP. Regional distribution of angiotensinogen in the rat brain. *Brain Res* 1978;158:359–371.
16. Fuxe K, Ganten D, Hökfelt T, et al. Renin-like immunocytochemical activity in the rat and mouse brain. *Neurosci Lett* 1980;18:245–250.
17. Fuxe K, Ganten D, Köhler C, Schull B, Speck G. Evidence for differential localization of angiotensin I converting enzyme and renin in the corpus striatum of the rat. *Acta Physiol Scand* 1980;110:321–323.
18. Fuxe K, Andersson K, Ganten D, Hökfelt T, Eneroth P. Evidence for the existence of an angiotensin II-like immunoreactive central neuron system and its interaction with the central catecholamine pathways. In: Gross F, Vogel G, eds. *Enzymatic release of vasoactive peptides*. New York: Raven Press, 1980;161–170.
19. Kilcoyne MM, Hoffman DL, Zimmerman EA.

Immunocytochemical localization of angiotensin II and vasopressin in rat hypothalamus: evidence for the production in the same neuron. *Clin Sci* 1980;59:57s–60s.
20. Fuxe K, Ganten D, Andersson K, et al. Immunocytochemical demonstration of angiotensin II and renin-like immunoreactive nerve cells in the hypothalamus. Angiotensin peptides as comodulators in vasopressin and oxytocin neurons and their regulation of various types of central catecholamine nerve terminal systems. *Exp Brain Res* 1982;Suppl 4:208–232.
21. Fuxe K, Agnati LF, Ganten D, et al. Morphometric evaluation of renin-like and oxytocin-like immunoreactivity in nerve cells of the paraventricular hypothalamic nucleus of the rat. *Neurosci Lett* 1982;33:19–24.
22. Calza L, Fuxe K, Agnati LF, et al. Presence of renin-like immunoreactive nerve cells of the paraventricular and supraoptic nuclei in the rat hypothalamus. *Acta Physiol Scand* 1982;116: 313–316.
23. Saavedra JM, Fernandez-Pardal J, Chevillard C. Angiotensin-converting enzyme in discrete areas of the rat forebrain and pituitary gland. *Brain Res* 1982;245:317–325.
24. Mendelsohn FAO, Quirion R, Saavedra JM, Aguilera G, Catt KJ. Autoradiographic localization of angiotensin II receptors in rat brain. *Proc Natl Acad Sci USA* 1984;81:1575–1579.
25. Lind RW, Swanson LW, Ganten D. Organization of angiotensin II immunoreactive cells and fibers in the rat central nervous system. *Neuroendocrinology* 1985;40:2–24.
26. Chai SY, Mendelsohn FAO, Paxinos G. Angiotensin-converting enzyme in rat brain visualized by quantitative *in vitro* audioradiography. *Neuroscience* 1987;20:615–627.
27. Gehlert DR, Speth RC, Wamsley JK. Distribution of ^{125}I-angiotensin II binding sites in the rat brain: a quantitative autoradiographic study. *Neuroscience* 1986;18:837–856.
28. Imboden H, Harding JW, Hilgenfeldt U, Celio MR, Felix D. Localization of angiotensinogen in multiple cell types of rat brain. *Brain Res* 1987; 410:74–77.
29. Thomas WG, Sernia C. Immunocytochemical localization of angiotensinogen in the rat brain. *Neuroscience* 1988;25:319–341.
30. Lynch KR, Simnad VI, Ben-Ari ET, Garrison JC. Localization of preangiotensinogen messenger RNA sequences in the rat brain. *Hypertension* 1986;8:540–543.
31. Dzau VJ, Ingelfinger J, Pratt RE, Ellison KE. Identification of renin and angiotensinogen messenger RNA sequences in mouse and rat brains. *Hypertension* 1986;8:544–548.
32. Hellmann W, Suzuki F, Okhubo H, Nakanishi S, Ludwig G, Ganten D. Angiotensinogen expression in extrahepatic tissues: application of a solution hybridization assay. *Naunyn Schmiedebergs Arch Pharmacol* 1988;338:327–331.
33. Suzuki F, Hellmann W, Paul M, et al. Renin gene expression in rat tissues: a new quantitative assay method for rat renin mRNA using synthetic cRNA. *Clin Exp Hypertens* 1988;A10:345–359.
34. Bickerton RK, Buckley JP. Evidence for a central mechanism in angiotensin-induced hypertension. *Proc Soc Exp Biol* 1961;106:834–836.
35. Fitzsimons JT. Angiotensin stimulation of the central nervous system. *Rev Physiol Biochem Pharmacol* 1980;87:118–146.
36. Unger T, Badoer E, Ganten D, Lang RE, Rettig R. Brain angiotensin: pathways and pharmacology. *Circulation* 1988;77:I40–I54.
37. Phillips MI. Functions of angiotensin in the central nervous system. *Annu Rev Physiol* 1987;49:413–435.
38. Brosnihan KB, Diz DI, Schiavone MT, Averill DB, Ferrario CM. Approaches to establishing ANG II as a neurotransmitter. In: Buckley JB, Ferrario CM, eds. *Brain peptides and catecholamines in cardiovascular regulation*. New York: Raven Press, 1987;313–329.
39. Agnati LF, Fuxe K, Zoli M, et al. Aspects on the information handling by the central nervous system: focus on cotransmission in the aged rat brain. In: Hökfelt T, Fuxe K, Pernow B, eds. *Progress in brain research*, vol 68. Amsterdam: Elsevier, 1986;291–301.
40. Agnati LF, Fuxe K, Zoli M, Zini J, Toffano G, Ferraguti F. A correlation analysis of the regional distribution of central enkephalin and β-endorphin immunoreactive terminals and of opiate receptors in adult and old male rats. Evidence for the existence of two main types of communication in the central nervous system: the volume transmission and the wiring transmission. *Acta Physiol Scand* 1986;128:201–207.
41. Fuxe K, Bunnemann B, Aronsson M, et al. Pre- and postsynaptic features of the central angiotensin systems. Indications for a role of angiotensin peptides in volume transmission and for interactions with central monoamine neurons. *Clin Exp Hypertens* 1988;A 10 [Suppl. 1]:143–168.
42. Lynch KR, Hawelu-Johnson CL, Guyenet PG. Localization of brain angiotensinogen mRNA by hybridization histochemistry. *Mol Brain Res* 1987;2:149–158.
43. Stornetta RL, Hawelu-Johnson CL, Guyenet PG, Lynch KR. Astrocytes synthesize angiotensinogen in brain. *Science* 1988;242:1444–1446.
44. Healy DP, Printz MP. Distribution of immunoreactive angiotensin II, angtiotensin I, angiotensinogen, and renin in the central nervous system of intact and nephrectomized rats. *Hypertension* 1984;6[Suppl. 1]:I130–I136.
45. Deschepper CF, Bouhnik J, Ganong WF. Colocalization of angiotensinogen and glial fibrillary acidic protein in astrocytes in the rat brain. *Brain Res* 1986;374:195–198.
46. Paul M, Wagner D, Metzger R, et al. Quantification of renin mRNA in various mouse tissues by a novel solution hybridization assay. *J Hypertens* 1988;6:247–252.
47. Genain CP, Van Loon GR, Kotchen TA. Distribution of renin activity and angiotensinogen in the rat brain. *J Clin Invest* 1985;76:1939–1945.
48. Correa FMA, Plunkett LM, Saavedra JM. Quan-

titative distribution of angiotensin-converting enzyme (kinase II) in discrete areas of the rat brain by autoradiography with computerized microdensitometry. *Brain Res* 1986;375:259–266.

49. Brownfield MS, Reid IA, Ganten D, Ganong WF. Differential distribution of immunoreactive angiotensin and angiotensin-converting enzyme in rat brain. *Neuroscience* 1982;7:1759–1769.
50. Lind RW, Swanson LW, Ganten D. Angiotensin II immunoreactivity in the neural afferents and efferents of the subfornical organ of the rat. *Brain Res* 1984;321:209–215.
51. Lind RW, Swanson LW, Bruhn TO, Ganten D. The distribution of angiotensin II-immunoreactive cells and fibers in the paraventriculo-hypophysial system of the rat. *Brain Res* 1985;338: 81–89.
52. Block CH, Santos RAS, Brosnihan KB, Ferrario CM. Immunocytochemical localization of angiotensin-(1-7) in the rat forebrain. *Peptides* 1989; 9:1395–1401.
53. Gehlert DR, Speth RC, Healy DP, Wamsley JK. Autoradiographic localization of angiotensin II receptors in the rat brainstem. *Life Sci* 1984;34: 1565–1571.
54. Healy DP, Printz MP. Localization of angiotensin II binding sites in rat septum by autoradiography. *Neurosci Lett* 1984;44:167–172.
55. Israel A, Correa FMA, Niwa M, Saavedra JM. Quantitative determination of angiotensin II binding sites in rat brain and pituitary gland by autoradiography. *Brain Res* 1984;322:341–345.
56. Speth RC, Wamsley JK, Wright JW, Abhold RH, Harding JW. Characterization of putative sites of action of angiotensin II in the rat central nervous system by *in vitro* receptor autoradiography with ^{125}I-sarcosine^{1}isoleucine^{8} angiotensin II. In: Buckley JP, Ferrario CM, eds. *Brain peptides and catecholamines in cardiovascular regulation*. New York: Raven Press, 1987;257–271.
57. Healy DP, Rettig R, Nguyen T, Printz MP. Quantitative autoradiography of angiotensin II receptors in the rat solitary-vagal area: effects of nodose gangliectomy or sinoaortic denervation. *Brain Res* 1989;484:1–12.
58. Young D, O'Neill K, Jessell T, Wigler M. Characterization of the rat *mas* oncogene and its high-level expression in the hippocampus and the cerebral cortex of the rat brain. *Proc Natl Acad Sci USA* 1988;85:5339–5342.
59. Ferrario CM, Schiavone MT. The renin angiotensin system: importance in physiology and pathology. *Cleve Clin J Med* 1989;56:439–446.
60. Speth RC, Harik SI. Angiotensin II receptor binding sites in brain microvessels. *Proc Natl Acad Sci USA* 1985;82:6340–6343.
61. Erdös EG, Skidgel RA. The unusual substrate specifity and the distribution of human angiotensin I-converting enzyme: special lecture. *Hypertension* 1986;8[Suppl 1]:I34–I37.
62. Paxinos G, Watson C. *The rat brain in stereotaxic coordinates*. 2nd ed. Australia: Academic Press, 1986.

Volume Transmission in the Brain: Novel Mechanisms for Neural Transmission, edited by K. Fuxe and L.F. Agnati. Raven Press, Ltd., New York © 1991.

11

5-HT Systems in a Lower Vertebrate Model: Ultrastructure, Distribution, and Synaptic and Cellular Mechanisms

Johan Christenson, Peter Wallén, Lennart Brodin, and Sten Grillner

Nobelinstitute for Neurophysiology, Karolinska Institutet, Stockholm, Sweden

The lamprey central nervous system has proved a useful model for studying the cellular basis of central nervous mechanisms in vertebrates (1). A special interest has been devoted to the underlying mechanisms controlling locomotion, which is due to an activation of spinal networks generating the motor pattern. The initiation of locomotion depends largely on activation of excitatory amino acid (EAA) receptors, but the neuronal activity is modulated by different putative neurotransmitters in the spinal cord, including 5-hydroxytryptamine (serotonin; 5-HT). 5-HT is distributed both within the brainstem and spinal cord as well as in the periphery. 5-HT exerts specific physiological effects at the cellular level as well as on the network level responsible for the pattern generation. The functional organization of the different 5-HT systems in brainstem–spinal cord will be reviewed. The spinal 5-HT varicosities do not form synaptic specializations ultrastructurally and may thus use a paracrine mode of operation (i.e., volume transmission). The 5-HT acts postsynaptically by a depression of calcium-dependent potassium channels responsible for the frequency regulation in several types of interneurons and motoneurons in the spinal cord. These 5-HT effects cause marked changes in the network generating locomotion, and it is shown by computer simulation that these behavioral effects can be explained solely by the effects on the after-hyperpolarization (AHP).

5-HT IMMUNOREACTIVE NEURONS AND FIBERS WITHIN THE BRAINSTEM

5-HT immunoreactivity (ir) is observed throughout the brain and brainstem. The 5-HT-ir neurons innervate locally within the brainstem and/or send their processes down the spinal cord. Recently, 5-HT-ir neurons and fibers have been found with ir varicosities in close apposition on cell bodies of reticulospinal (RS) neurons of the excitatory amino type (2). This innervation was observed in the mesencephalic reticular nucleus (mrn), anterior (arrn), middle (mrrn), and posterior rhombencephalic nucleus (prrn). In the arrn and mrrn, the 5-HT-ir fibers are densely packed in "plexus-like" formations with close appositions onto cell bodies including the large descending Müller cells.

DESCENDING FIBERS FROM 5-HT-IR NEURONS IN THE BRAINSTEM

Using retrograde transport of the fluorescent tracer Fast blue in combination with 5-HT immunohistochemistry reveals the ex-

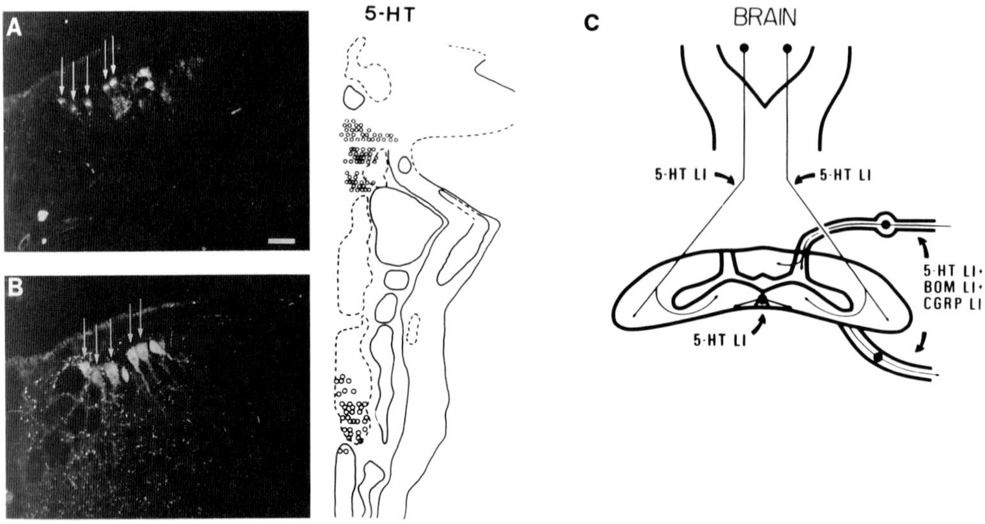

FIG. 1. Fluorescent neurons in lamprey brainstem after injection of Fast blue in the rostral spinal cord and after 5-HT immunohistochemistry. **A:** Micrograph including posterior reticular nucleus of the rhombencephalon after spinal Fast blue injection. **B:** Same section after incubation with 5-HT antiserum followed by FITC—conjugated antiserum. *Arrows* indicate double-labeled cells. Cells that are stained with both Fast blue and 5-HT-ir are indicated in schematic drawing of the brainstem. **C:** Schematic drawing of 5-HT distribution in the spinal cord including descending fibers from the brainstem, intraspinal 5-HT-ir neurons, and an afferent 5-HT-ir system. LI,; BOM,; CGRP, calcitonin-gene related protein. (Modified from ref. 3, with permission.)

istence of two groups of RS 5-HT neurons in the brainstem that project to the spinal cord (3) (Fig. 1). One group is located within the posterior and another within the anterior reticular nucleus of the prrn and arrn, respectively. The cells are medium-sized (20–40 μm in diameter), and their dendrites usually have a ventral orientation. In these areas, the double labeled cells (cells that were labeled with both Fast blue and 5-HT-ir) represent roughly 10% to 20% of all Fast blue labeled cells in these two regions of the rhombencephalon.

In the animals with a spinal hemisection, a decreased, asymmetric 5-HT labeling of the brainstem is observed, with the majority of the labeled cells contralateral to the side of the lesion. Lesion studies show that descending 5-HT fibers supply the ipsilateral and ventromedial columns of the spinal cord but not the dorsal horn.

INTRASPINAL 5-HT NEURONS AND FIBERS

The lamprey spinal cord also contains intrinsic 5-HT neurons close to the central canal (4–8). These cells are small (8–12 μm in diameter), and they have bilateral processes giving rise to the bilateral dense 5-HT-ir plexus in the ventromedial area of the spinal cord. Intraspinal 5-HT neurons have been observed in many different species such as reptiles and nonteleost fishes (9) but also in mammals such as the rat (10) and monkey (11). "Primitive" species such as elasmobranchs (12), teleosts, and reptiles appear to have larger numbers of intraspinal neurons, which may become progressively fewer in number during phylogenesis.

In a study of possible target neurons for 5-HT terminals in lamprey, different classes of spinal neurons were injected with Lucifer yellow and subsequently processed for 5-HT immunohistochemistry (13). Close appositions of 5-HT boutons (≤1 μm) were observed on both motoneurons and lateral interneurons, mainly in the ventromedial spinal cord.

AFFERENT 5-HT SYSTEM

5-HT-ir is observed in small cell bodies in the spinal ganglia, which project to the dorsal horn and dorsal column (14). This system in addition co-contains both calcitonin-gene related peptide (CGRP) and bombesin-ir. However, in contrast to what has been described in other species, no descending serotonergic innervation could be observed to the dorsal horn in lamprey.

As for many other species such as monkey, cat (15), rat (16), guinea pig (17), and chicken (18), the lamprey have 5-HT systems within the brainstem that innervate neuronal structures in the brainstem as well as descending 5-HT fibers to the spinal cord.

The descending fibers in the lamprey have been observed in "close apposition" to spinal neurons as well as the 5-HT varicosities of the intraspinal neurons. The question whether these close appositions form point-to-point synapses can only be answered by using the electron microscope.

IMMUNOHISTOCHEMISTRY AND ELECTRON MICROSCOPY—LACK OF SYNAPTIC SPECIALIZATION IN 5-HT VARICOSITIES

Spinal ir cell bodies and fibers were studied using the peroxidase–antiperoxidase (PAP) immunohistochemical technique (19–21) and subsequent analysis of ultrathin serial sections (for methods, see ref. 22). The distribution of cell bodies, nerve fibers, and terminals corresponds well with the pattern described with immunofluorescence techniques. The staining is strongest in the varicosities in the ventromedial plexus, but in general, the fine structural characteristics of 5-HT-ir varicosities are identical in all regions. The PAP-reaction product is found scattered throughout the cytoplasm, with a predilection for the large granular vesicles. In many instances, the 5-HT-ir varicosities are positioned in close apposition to other structures, and yet there are no signs of synaptic specializations such as synaptic clefts or pre- and postsynaptic membrane thickenings. The analysis of synaptic complex formations reveals that no synaptic specializations are found in conjunction with any of the 5-HT-ir varicosities, which could be followed through their entire length (Fig. 2). In the same sections, a number of synaptic complexes are found. We conclude that 5-HT-ir varicosities lack synaptic specializations, at least of any conventional type, in the areas of the spinal cord studied. In fact, these varicosities appear distributed in the tissues without any clear preference regarding surrounding profiles such as axons, dendrites, and/or other unidentified neuronal elements.

The distribution of 5-HT fibers in the vertebrate spinal cord has been extensively investigated. The descending 5-HT fibers in the spinal cord have been studied light microscopically by using aldehyde-induced fluorescence or immunofluorescence techniques (23,24). Ultrastructural studies have been performed in the rat (25), the guinea pig, the cat, and the chicken. 5-HT-ir varicosities with or without synaptic complex formations have been reported in all four species, although the percentage of 5-HT-ir terminals that form synaptic junctions differs.

SUBCELLULAR DISTRIBUTION OF 5-HT LOCATED IN SMALL AND LARGE VESICLES

In the electron microscope, the 5-HT varicosities are heavily labeled with dense pre-

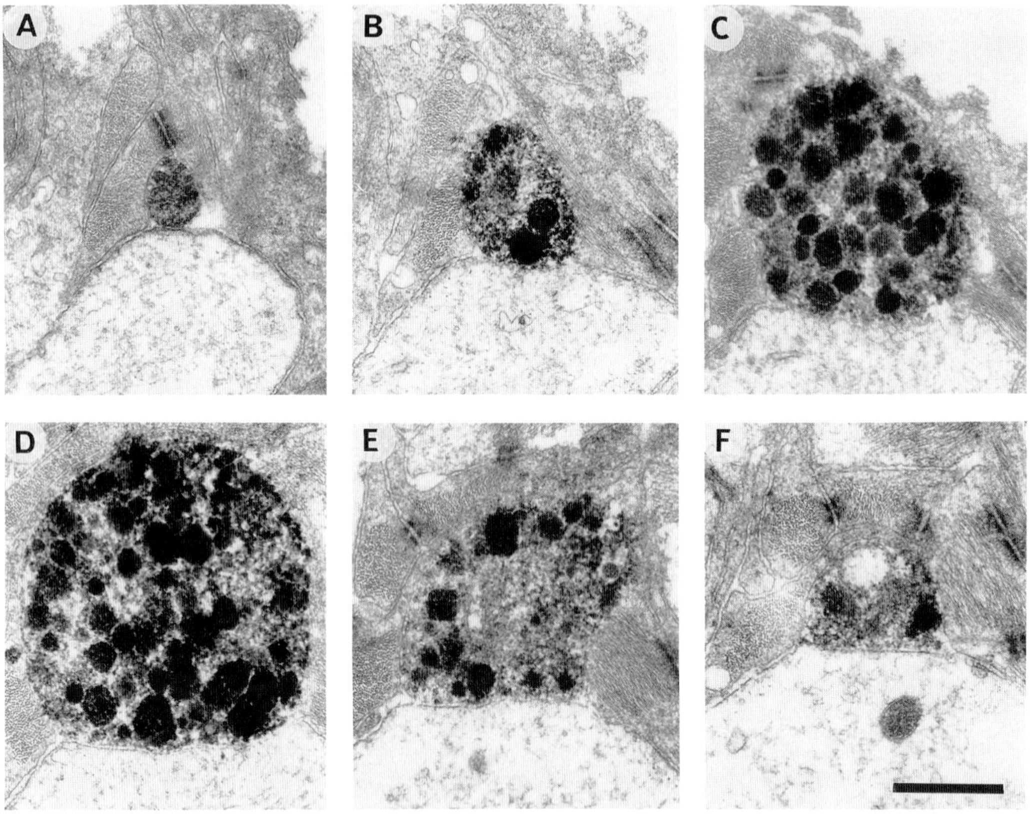

FIG. 2A–F. Electron micrograph showing a serially sectioned 5-HT immunoreactive varicosity. Although varicosity is in close relation to an unidentified neuronal structure, no synaptic junction is formed. Note the higher content of large granular vesicles in the middle part of the varicosity. Bar = 0.5 μm. (From ref. 22, with permission.)

cipitates over large granular vesicles. However, also the space around these large vesicles shows reaction product, as described above. Comparison with tissue not processed for 5-HT-ir showed that the varicosities contained both large dense-cored vesicles and small agranular vesicles. It would thus appear likely that both types of vesicles would contain 5-HT.

The subcellular distribution of 5-HT was therefore studied using ultracentrifugation on continuous density gradients. The distribution of 5-HT showed two maxima, one corresponding to 0.30 to 0.40 M sucrose and one to 0.85 to 1.05 M sucrose density. These results indicate that 5-HT may be distributed in more than one subcellular compartment apart from the cytosol, possibly corresponding to small and large synaptic vesicles. Studies on the rat spinal cord suggest a similar storage of 5-HT in small vesicles and large dense cored vesicles (26).

ENDOGENOUS RELEASE OF 5-HT MODULATES THE FICTIVE LOCOMOTOR PATTERN

Bath application of 5-HT during fictive locomotion in the *in vitro* preparation of the lamprey spinal cord will markedly influence the burst pattern (27). The rate of bursting is slowed down, and both intensity and duration of the burst discharges will increase.

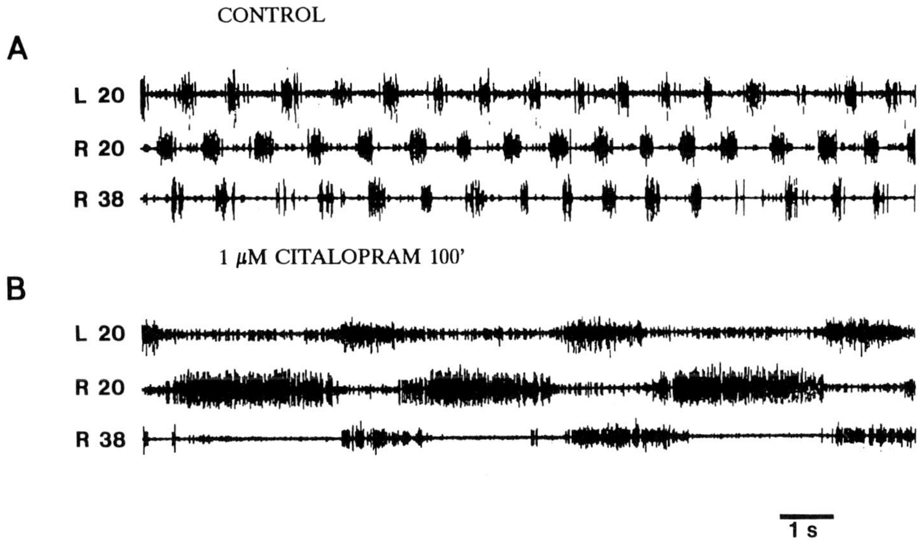

FIG. 3. A: Ventral root recordings from left and right side of spinal segment 20 and right ventral root recording from the more caudal spinal segment 38 during NMDA (100 μM) induced "fictive locomotion." Phase lag 0.23 ± 0.06 between R 20 and R 38. **B:** Ventral root recordings after reperfusion of 5-HT uptake blocker citalopram (1 μM) after 100 min. During reperfusion, the 5-HT is accumulated in the perfusate, and a clear effect caused by increased 5-HT concentration is seen after 70–105 min. Locomotor frequency is decreased from 1.1 to 0.2 Hz. Strength of the ventral root bursts is increased, and intersegmental coordination has changed so that R 38 is bursting at the same time as L 20. Phase lag 0.51 ± 0.05 between R 20 and R 38. (From ref. 28, with permission.)

To test whether an endogenous release of 5-HT occurs and whether it has a similar effect, the 5-HT uptake system was blocked. This treatment markedly affected similar effects on the fictive burst pattern in a similar manner (28) (Fig. 3). This finding suggests that an endogenous release of 5-HT from the spinal cord fibers, and hence a natural activation of them, may modulate the locomotor activity.

CELLULAR MECHANISMS UNDERLYING THE EFFECTS OF 5-HT ON FICTIVE LOCOMOTION

It is thus clear that 5-HT will in some way influence the activity of the neurons responsible for generating the locomotor rhythm. To investigate how individual neurons will be affected, intracellular studies have been performed in which a release of 5-HT was mimicked by local droplet application near the ventral surface of the spinal cord, while recording from a neuron (29,30). The membrane input resistance was found to remain unaffected after 5-HT application, and no effect was observed with regard to the shape or the threshold of the action potential. Thus, the resting membrane properties as well as the mechanisms underlying the action potential appear not to be influenced by 5-HT. In contrast, when investigating the AHP after the action potential, a marked effect was revealed. Within seconds after the 5-HT application, the amplitude of the AHP was substantially reduced (Fig. 4). Whenever the early and late phases of the AHP in these cells could be distinguished, it was seen that only the late phase was reduced by 5-HT (Fig. 4A,B). Subsequent experiments have shown that the 5-HT has the same effect on the AHP on RS neurons.

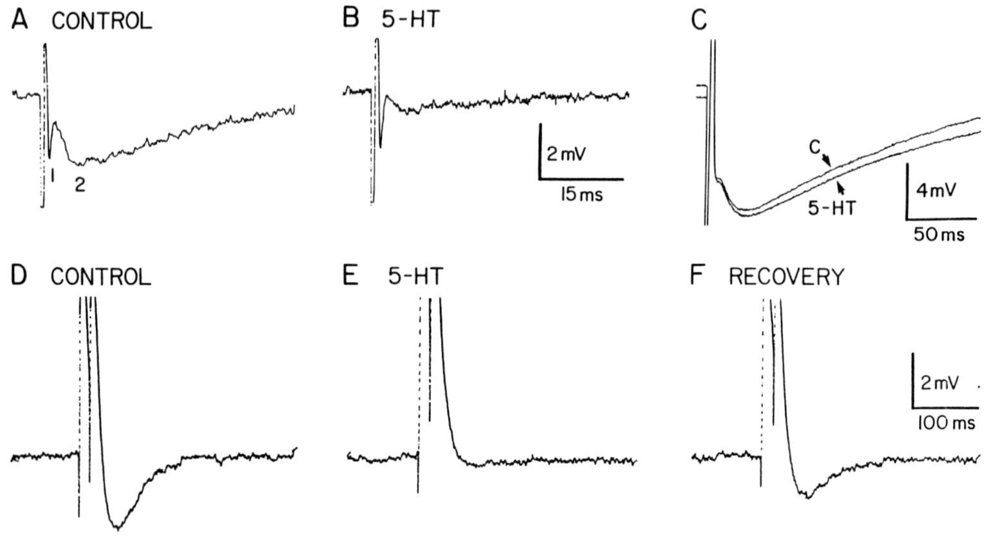

FIG. 4. Local application of 5-HT affects the late phase of AHP. **A:** Control record from a motoneuron before 5-HT, illustrating early (1) and late (2) phases of AHP after an action potential elicited by a brief depolarizing current pulse. **B:** After application of 5-HT to the ventral surface of the spinal cord near the medial dendrites of the motoneuron, the late phase of AHP was markedly reduced, whereas the early phase remained unaffected. **C:** Two superimposed computer averaged records of AHPs, elicited in an edge cell. The average AHP (n = 18) after 5–HT application did not differ from control average AHP. **D:** AHP in a motoneuron following two action potentials elicited by two current pulses at a 15 ms interval. **E.** One minute after 5–HT application. **F:** Partial recovery of the AHP 30 minutes after 5–HT. (Modified from ref. 30, with permission.)

The detailed neuronal mechanisms underlying this specific effect of 5-HT on the AHP have recently been investigated. As in several other types of CNS neurons, the AHP in these lamprey neurons is associated with an increase in membrane conductance (31). Local application of 5-HT markedly reduces this conductance increase, which indicates that the ionic currents underlying the AHP are being blocked by 5-HT. The late phase of the AHP is in general attributed to the activation of calcium-dependent potassium channels (32,33), and it was found that this holds true also for the lamprey neurons studied here.

A possible mechanism by which 5-HT could reduce the amplitude of the AHP could be a blockade of the entry of calcium during the action potential. The other alternative is that 5-HT could act later in the chain of events and, directly or indirectly, prevent the activation of the calcium-dependent potassium channels. When 5-HT was droplet-applied in the presence of tetraethylammonium (TEA), which blocks the "early" potassium channels and leads to increased calcium entry during a much prolonged action potential, the spike remained prolonged (Fig. 5A,C), and thus the entry of calcium does not appear to be blocked by 5-HT. Still, the AHP was depressed after 5-HT application (Fig. $5D_1$). From these experiments, we may thus conclude that 5-HT appears to exert its action after calcium has entered the cell.

The possibility that 5-HT might act via a second-messenger system, as is the case for the corresponding effect of norepinephrine and acetylcholine on hippocampal neurons (34–37), was tested with regard both to cyclic adenosine monophosphate (AMP), cyclic guanosine monophosphate (GMP), diacylglycerol, and arachidonic acid. No evidence was obtained for an involvement of any of these second-messenger systems nor for a mediation via a pertussis toxin–sensitive G protein (38). It thus appears likely that the 5-HT effect on the late phase

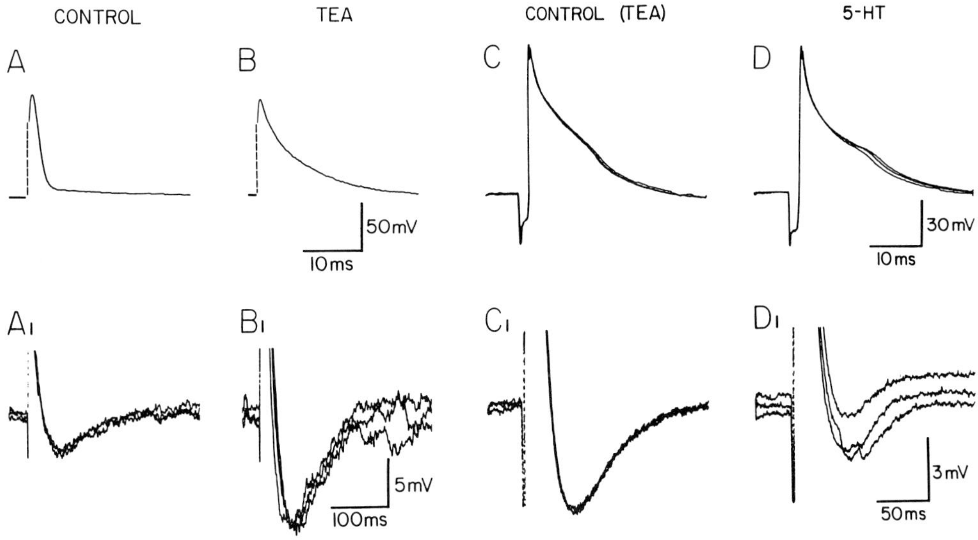

FIG. 5. A,B: Effect of tetraethylammonium (TEA) on AHP. **A, A_1:** In normal saline, action potentials were elicited in a motoneuron (V_m, −56 mV) by brief depolarizing current pulses (*bar* and *dashed line* in A). The two phases of the AHP could not be distinguished in this case (A_1). After bath application of TEA (**B, B_1**), the spike was prolonged, and the size of AHP was increased. Three sweeps were superimposed in A_1 and B_1. The cell spontaneously hyperpolarized slightly between A and B. **C–D_1:** Local application of 5-HT in presence of TEA. **C:** Prolonged action potential in TEA. **C_1:** AHP of TEA action potential. **D:** After local ejection of 5-HT, the duration of the action potential was little affected, but the AHP was reduced by more than one-half (D_1). V_m level was constant (−52 mV) between C and D, but the potential fluctuated somewhat around this value during later part of recording session (evident in D_1). C–D_1: Three sweeps superimposed. Calibrations in B, B_1, D, and D_1 also apply to A, A_1, C, and C_1, respectively. (From ref. 30, with permission.)

of the AHP results from a direct action on the calcium-dependent potassium channels.

FIRING PROPERTIES OF THE INDIVIDUAL NEURON ARE INFLUENCED BY 5-HT

The size of the AHP is a major factor determining the spike frequency characteristics of the neuron (30,40). 5-HT would therefore be expected to influence the spike response to a given excitatory input. Figure 6 shows that the response to a long, depolarizing stimulus pulse is markedly altered after 5-HT application—the number of spikes is more than doubled. Note also the reduction in size of the late AHP (arrows). The same excitatory input thus results in a more powerful spike response when the neuron is under the influence of 5-HT. The input–output relation, or the "gain" of the neuron, is altered by 5-HT, and the response to excitatory input in each neuron that is affected by 5-HT will become potentiated.

EFFECTS ON THE SPINAL GENERATOR CIRCUITRY—SIMULATION STUDIES

A release of 5-HT will potentiate the spike response of both motoneurons and interneurons with dendrites in the region of the 5-HT fibers. In addition to motoneurons, lateral interneurons and CC-interneurons, both of which are inhibitory premotor interneurons, are influenced by 5-HT as described above (i.e., the late phase of the AHP is markedly reduced and the frequency response characteristics are potentiated). Both these interneuron types are part of the circuitry proposed as a model

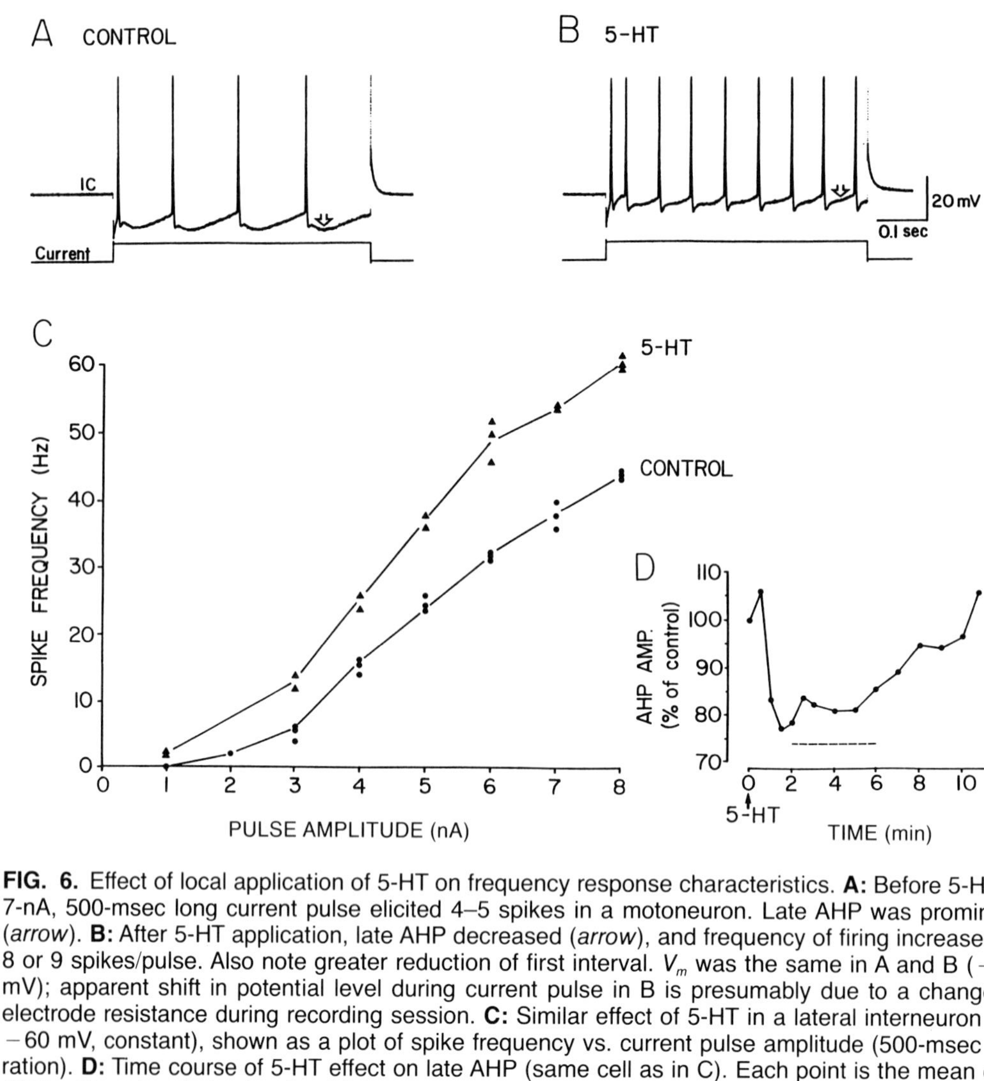

FIG. 6. Effect of local application of 5-HT on frequency response characteristics. **A:** Before 5-HT, a 7-nA, 500-msec long current pulse elicited 4–5 spikes in a motoneuron. Late AHP was prominent (*arrow*). **B:** After 5-HT application, late AHP decreased (*arrow*), and frequency of firing increased to 8 or 9 spikes/pulse. Also note greater reduction of first interval. V_m was the same in A and B (−60 mV); apparent shift in potential level during current pulse in B is presumably due to a change in electrode resistance during recording session. **C:** Similar effect of 5-HT in a lateral interneuron (V_m −60 mV, constant), shown as a plot of spike frequency vs. current pulse amplitude (500-msec duration). **D:** Time course of 5-HT effect on late AHP (same cell as in C). Each point is the mean of 7 AHPs. Slow depolarization (7 mV) began 10 sec before 5-HT application, probably causing the increase in AHP seen just after 5-HT pulse and in recovery. Data in C were taken between 2 and 6 min after 5-HT application (*dashed line*). (From ref. 30, with permission.)

neuronal network for the generation of the locomotor rhythm in the lamprey (41,42). CC-interneurons (having a crossed, caudally projecting axon) inhibit the network neurons on the contralateral side of the spinal cord, thereby assuring alternating activity. Lateral interneurons inhibit CC-interneurons on the same side, terminating their burst activity and thereby allowing the contralateral side to become active. Both types of inhibitory interneurons are excited by local excitatory interneurons, which are also an integral part of the model network (43).

5-HT will increase the excitability in both excitatory and inhibitory neurons. Furthermore, the AHP is likely to play an important role in the circuitry as a burst-terminating factor. To investigate the possible influence of these factors, experiments were carried out with a computer simulation of the rhythm-generating network of the lamprey spinal cord based on experi-

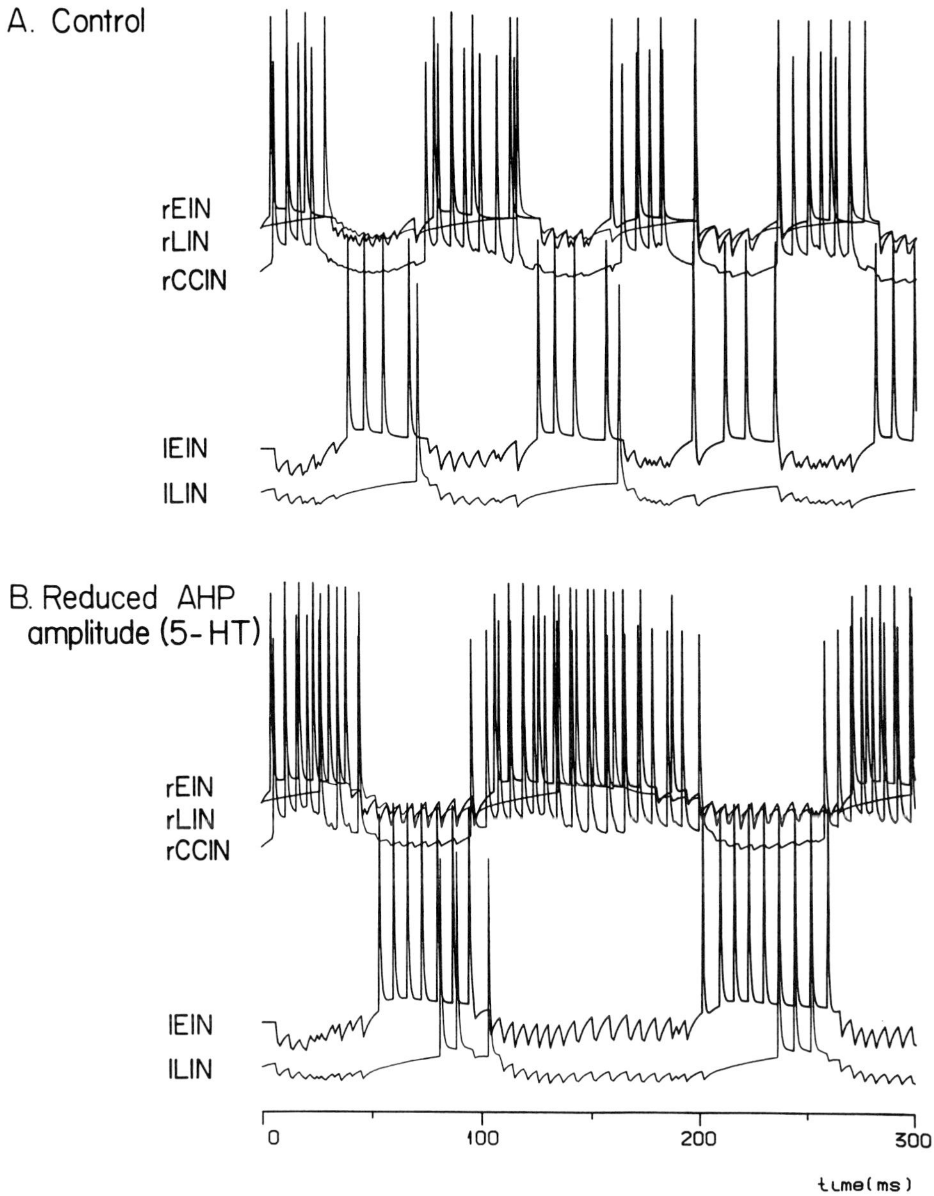

FIG. 7. Simulated test of the role of the postspike after-hyperpolarization (AHP) in burst generation. **A:** Control simulation. **B:** Simulation with identical parameters except for a reduced postspike AHP. Note prolonged burst duration and increase in firing frequency of all neurons. Physiological effect of 5-HT is to reduce AHP and produce the motor pattern in B. rEIN,; rLIN,; CC-interneuron (rCCIN); excitatory interneuron (lEIN); lateral interneuron (lLIN) (r, right side; l, left side). (From ref. 42, with permission.)

mentally established neuronal properties and connectivity. When a release of 5-HT was simulated by reducing the size of the AHP in the neurons of the network model, the burst rate decreased, and bursts were prolonged and more spikes were fired at a higher frequency (Fig. 7). These effects are analogous to those found experimentally with 5-HT uptake blockade or bath application of 5-HT (cf above). It thus appears that the modulatory effects of 5-HT on the locomotor burst pattern can be explained by the reduction of the AHP in the neurons of the generator circuitry. Also the motoneurons will be influenced by a release of 5-HT, which will further increase spike frequency and burst intensity. The effects on the motoneurons will, however, not contribute to the alteration in burst rate, because the motoneurons themselves are not part of the generator circuitry (44).

CONCLUDING REMARKS

5-HT varicosities in all parts of the spinal cord lack synaptic specializations, and yet 5-HT is released in the spinal cord as shown by using 5-HT uptake blockers. Moreover, the 5-HT released produces the same behavioral effects as bath-applied 5-HT. Most likely, the 5-HT released from the varicosities will affect neurons located in the vicinity, thus having a "paracrine" mode of operation (45,46). This would correspond to what has been referred to as "volume transmission."

5-HT may thus influence neurons of the lamprey spinal cord in a specific manner—the amplitude of the late phase of the AHP may be markedly attenuated, without any influence on passive membrane properties or on the action potential *per se*. The size of the late AHP appears to be modulated through a direct action on the underlying calcium-dependent potassium channels.

As suggested by computer simulation experiments, the reduction in size of the AHP will lead to a slowing down of the rhythmic activity, with an increased intensity and duration of the burst discharges. This is in good agreement with experimental findings with either externally applied 5-HT or after blocking 5-HT uptake to increase endogenous levels. Taken together, these results thus suggest that the 5-HT systems of the lamprey CNS may modulate the efferent rhythmic activity underlying locomotion by simply regulating the amplitude of the AHP in the different neuron types involved in motor activity. It may here be noted that 5-HT does not affect the AHP in dorsal cells and edge cells, both of which are primary sensory neurons.

Regulation of the AHP amplitude appears to be a simple and straightforward strategy by which a modulatory control of the motor activity can be accomplished. Whenever the external demands require that the locomotor movements be altered, perhaps when the resistance to the movements increases as during burrowing, activation of the 5-HT fibers will change the firing characteristics of the involved neurons by influencing one particular ionic conductance, the calcium-dependent potassium channels responsible for the late phase of the AHP. Further experimentation is needed to clarify how and under what circumstances the 5-HT fibers are naturally activated, and the specific roles of the different 5-HT systems also remain to be investigated.

REFERENCES

1. Grillner S, Wallén P, Dale N, Brodin L, Buchanan J, Hill R. Transmitters, membrane properties and network circuitry in the control of locomotion in lamprey. *Trends Neurosci* 1987;10: 34–41.
2. Di Prisco GV, Dubuc R, Wallén P, Grillner S. Effects of 5-HT on reticulospinal neurons and distribution of 5-HT immunoreactive fibers in the lamprey brainstem. *Neuroscience* 1989;15, part 1:159.3.
3. Brodin L, Buchanan JT, Hökfelt T, Grillner S. A spinal projection of 5-hydroxytryptamine neurons in the lamprey brainstem; evidence from combined retrograde tracing and immunohistochemistry. *Neurosci Lett* 1986;67:53–57.
4. Baumgarten HG. Biogenic monoamines in the cyclostome and lower vertebrate brain. *Prog Histochem Cytochem* 1972;4:1–90.
5. Van Dongen PAM, Hökfelt T, Grillner S, et al. Immunohistochemical demonstration of some putative neurotransmitters in the lamprey spinal cord and spinal ganglia: 5-hydroxytryptamine-, tachykinin-, and neuropeptide-Y-immunoreactive neurons and fibers. *J Comp Neurol* 1985; 234:501–522.
6. Van Dongen PAM, Hökfelt T, Grillner S, Rehfeld J, Verhofstad AAJ. A cholecystokinin-like peptide is present in 5-hydroxytryptamine neu-

rons in the spinal cord of the lamprey. *Acta Physiol Scand* 1985;125:557–560.
7. Harris-Warrick RM, McPhee JC, Philler JA. Distribution of serotonergic neurons and processes in the lamprey spinal cord. *Neuroscience* 1985;14:1127–1140.
8. Honma S. Presence of monoaminergic neurons in the spinal cord and intestine of the lamprey. *Arch Histol Jpn* 1970;32:383–393.
9. Parent A, Northcutt RG. The monoamine containing neurons in the brain of the garfish, *Lepisosteus osseus*. *Brain Res Bull* 1982;9:182–204.
10. Newton BW, Maley BE, Hamill RW. Immunohistochemical demonstration of serotonin neurons in autonomic regions of the rat spinal cord. *Brain Res* 1986;376:155–163.
11. LaMotte CC, Johns DR, de Lanerolle NC. Immunohistochemical evidence of indolamine neurons in monkey spinal cord. *J Comp Neurol* 1982;206:359–370.
12. Ritchie TC, Roos LJ, Williams BJ, Leonard RB. The descending and intrinsic serotonergic innervation of an elasmobranch spinal cord. *J Comp Neurol* 1984;224:395–496.
13. Van Dongen PAM, Hökfelt T, Grillner S, Verhofstad AA, Steinbusch HW. Possible target neurons of 5-hydroxytryptamine fibers in the lamprey spinal cord: immunohistochemistry combined with intracellular staining with Lucifer yellow. *J Comp Neurol* 1985;234:523–535.
14. Brodin L, Buchanan JT, Hökfelt T, et al. Cholecystokinin (CCK)-like peptides in the brainstem and spinal cord of the lamprey: an immunohistochemical study with particular reference to the relation with 5-hydroxytryptamine (5-HT), bombesin and calcitonin gene related peptide (CGRP) immunoreactivities. *J Comp Neurol* 1988;271:1 18.
15. Ulfhake B, Arvidsson U, Cullheim S, et al. An ultrastructural study of 5-hydroxytryptamine-, thyrotropin-releasing hormone- and substance P-immunoreactive axonal boutons in the motor nucleus of spinal cord segments L7-S1 in the adult cat. *Neuroscience* 1987;23:917–929.
16. Descarriers L, Beaudet A, Watkins KC. Serotonin nerve terminals in adult rat neocortex. *Brain Res* 1975;100:563–588.
17. Chiba T, Masako S. Synaptic structure of the monoamine and peptide nerve terminals in the intermediolateral nucleus of the guinea pig thoracic spinal cord. *J Comp Neurol* 1987;262: 242–255.
18. Atsumi S, Sakamoto H, Yokota S, Fuiwara T. Substance P and 5-hydroxytryptamine immunoreactive presynaptic boutons on presumed α-motoneurons in the chicken ventral horn. *Arch Histol Jpn* 1985;48:159–172.
19. Steinbusch HWM, Verhofstad AAJ, Joosten HWJ. Localization of serotonin in the central nervous system by immunohistochemistry: description of a specific and sensitive technique and some applications. *Neuroscience* 1978;3: 811–819.
20. Hanker JS, Yates PE, Metz CB, Rustioni A. A new specific, sensitive and noncarcinogenic reagent for the demonstration of horesradish peroxidase. *J Histochem* 1977;9:789–792.
21. Somogyi P, Takagi H. A role on the use of picric acid-paraformaldehyde-glutaraldehyde fixative for correlated light and electron microscopic immunohistochemistry. *Neuroscience* 1982;7:1779–1783.
22. Christenson J, Cullheim S, Grillner S, Hökfelt T. 5-HT varicosities in the lamprey spinal cord have no synaptic specializations. *Brain Res.* 1990;512:201–209.
23. Dahlström A, Fuxe K. Evidence for the existence of monoamine-containing neurons in the central nervous system. II. Experimentally induced changes in the intraneuronal amine levels of bulbospinal neuron systems. *Acta Physiol Scand* 1965;64,(suppl. 247):7–36.
24. Hökfelt T. *In vitro* studies on central and peripheral monoamine neurons at the ultrastructural level. *Z Zellforsch Mikrosk Anat* 1968;91:1–74.
25. Maxwell DJ, Leranth C, Verhofstad AAJ. Fine structure of serotonin-containing axons in the marginal zone of the rat spinal cord. *Brain Res* 1983;266:253–259.
26. Fried G, Franck J, Brodin E. Differential distribution of 5-hydroxytryptamine and substance P in synaptosomal vesicles of rat ventral spinal cord. *Neurosci Lett* 1988;91:315–320.
27. Harris-Warrick RM, Cohen AH. Serotonin modulates the central pattern generator for locomotion in the isolated lamprey spinal cord. *J Exp Biol* 1985;116:27–46.
28. Christenson J, Franck J, Grillner S. Increase in endogenous 5-hydroxytryptamine levels modulates the central network underlying locomotion in the lamprey spinal cord. *Neurosci Lett* 1989;100:188–192.
29. Van Dongen PA, Grillner S, Hökfelt T. 5-Hydroxytryptamine (serotonin) causes a reduction in the afterhyperpolarization following the action potential in lamprey motoneurons and premotor interneurons. *Brain Res* 1986;366:320–325.
30. Wallén P, Buchanan JT, Grillner S, Hill RH, Christenson J, Hökfelt T. The effects of 5-hydroxytryptamine on the afterhyperpolarization, spike frequency regulation and oscillatory membrane properties in lamprey spinal cord neurons. *J Neurophysiol* 1989;61:4:759–768.
31. Gustafsson B. Afterhyperpolarisation and the control of repetitive firing in spinal neurons of the cat. *Acta Physiol Scand* 1974; Suppl. 416.
32. Meech RW. Calcium-dependent potassium activation in nervous tissues. *Annu Rev Biophys Bioeng* 1978;7:1–18.
33. Hill RH, Århem P, Grillner S. Ionic mechanisms of 3 types of functionally different neurons in the lamprey spinal cord. *Brain Res* 1985;358:40–52.
34. Madison DV, Nicoll RA. Noradrenaline blocks accommodation of pyramidal cell discharge in the hippocampus. *Nature* 1982;299:636–638.
35. Madison DV, Nicoll RA. Actions of noradrenaline recorded intracellularly in rat hippocampal CA1 pyramidal neurones, *in vitro*. *J Physiol* 1986;372:221–244.

36. Madison DV, Nicoll RA. Cyclic adenosine 3′,5′-monophosphate mediates beta-receptor actions of noradrenaline in rat hippocampal pyramidal cells. *J Physiol* 1986;372:245–260.
37. Malenka RC, Madison DV, Andrade R, Nicoll RA. Phorbol esters mimic some cholinergic actions in hippocampal pyramidal neurons. *J Neurosci* 1986;6:475–480.
38. Dunlap K, Holz GG, Rane SG. G proteins as regulators of ion channel function. *Trends Neurosci* 1987;10:241–244.
39. Kernell D. The limits of firing frequency in cat lumbosacral motoneurones possessing different time course of afterhyperpolarization. *Acta Physiol Scand* 1965;65:87–100.
40. Baldissera F, Gustafsson B. Firing behaviour of a neurone model based on the afterhyperpolarization conductance time course and summation. Adaptation and steady state firing. *Acta Physiol Scand* 1974;92:27–47.
41. Grillner S, Buchanan JT, Lansner A. Simulation of the segmental burst generating network for locomotion in lamprey. *Neurosci Lett* 1988;89: 31–35.
42. Grillner S, Lansner A, Wallén P, et al. The segmental burst-generating network used in lamprey locomotion: experiments and simulations. In: Goldbeter A, ed. *Cell to cell signalling: from experiments to theoretical models*. New York: Academic Press, 1989;77–88.
43. Buchanan JT, Grillner S. Newly identified "glutamate interneurons" and their role in locomotion in the lamprey spinal cord. *Science* 1987; 236:312–314.
44. Wallén P, Lansner A. Do the motoneurones constitute a part of the spinal network generating the swimming rhythm in the lamprey? *J Exp Biol* 1984;113:493–497.
45. Beaudet A, Descarries L. The monoamine innervation of rat cerebral cortex: synaptic and nonsynaptic axonal terminals. *Neuroscience* 1978;3:851–860.
46. Schmitt FO. Molecular regulators of brain function: a new view. *Neuroscience* 1984;13:991–1001.

Volume Transmission in the Brain: Novel Mechanisms for Neural Transmission,
edited by K. Fuxe and L.F. Agnati.
Raven Press, Ltd., New York © 1991.

12

Transient Neuropeptides and Receptors in the Human Inferior Olive of the Medulla Oblongata and Dentate Nucleus of the Cerebellum and in Intracranial Tumors

Pierre Mailleux, Monique Verslijpe, and Jean-Jacques Vanderhaeghen

Laboratory of Neuropathology and Neuropeptide Research, Erasme and Brugmann Hospitals, Université Libre de Bruxelles, Brussels, Belgium

The central theme of this book is the role of volume transmission in brain function. The evaluation of the presence of this mode of transmission in the human brain is difficult, but morphological studies of the distribution of the neuropeptides and their receptors help in such studies.

In this chapter, we report two situations in humans in which neuropeptides may be involved in functions other than classical neurotransmission. The first example is the transient presence of neuropeptides and receptors in the human inferior olive of the medulla oblongata and dentate nucleus of the cerebellum during critical periods of their development.

The second example is the distribution of several neuropeptides receptors and endopeptidase 24.11 in human meningiomas.

TRANSIENT NEUROPEPTIDES AND RECEPTORS IN THE HUMAN INFERIOR OLIVE OF THE MEDULLA OBLONGATA AND DENTATE NUCLEUS OF THE CEREBELLUM DURING DEVELOPMENT

Transient Neurotensin 1-13 Nerve Terminals and Receptors

Using immunohistochemistry, radioimmunoassay, and gel filtration, a transient rich network of neurotensin (NT) 1-13 nerve terminals is present in the human inferior olive of the medulla oblongata (Fig. 1), mainly during the first year of age (1). The absence of large molecular forms of NT in this terminal nerve fiber-rich region is probably due to progressive processing of

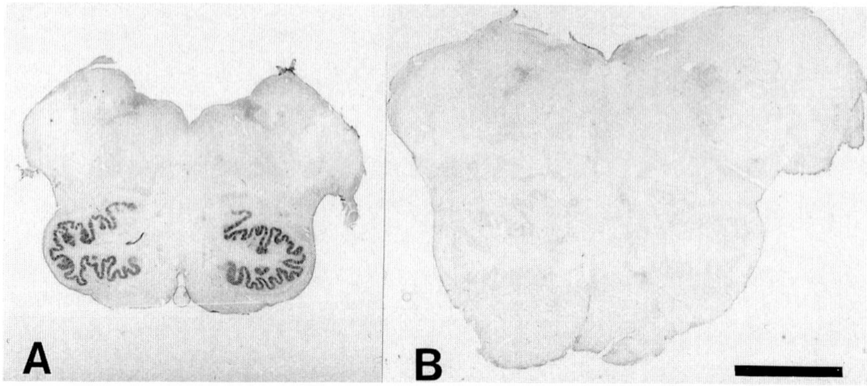

FIG. 1. Rich neurotensin 1-13 innervation (visible in *black*) is present transiently in the infant principal olive (**A**), mainly during the first year of age, disappears in children after 8 years, and remains absent in the adult (**B**). 35-μm cryostat sections—PAP technique of Sternberger—NT antisera (1). Bar: 5 mm.

NT during axonal transport (2). NT terminals are present exclusively in the principal inferior olive, and they are totally absent in the accessory dorsal and medial olives (Fig. 2). The network is transient, appearing at 22 weeks of gestation (Fig. 4), in the dorsal lamella of the principal olive and extending to the whole of the principal olive in the infants, although remaining at all ages more concentrated in the dorsal lamella. It is still absent in the medial part of the ventral lamella of the principal olive at birth; it is at a maximum in infants of 3 months, when 582 pmol/g wet weight is detected, and dis-

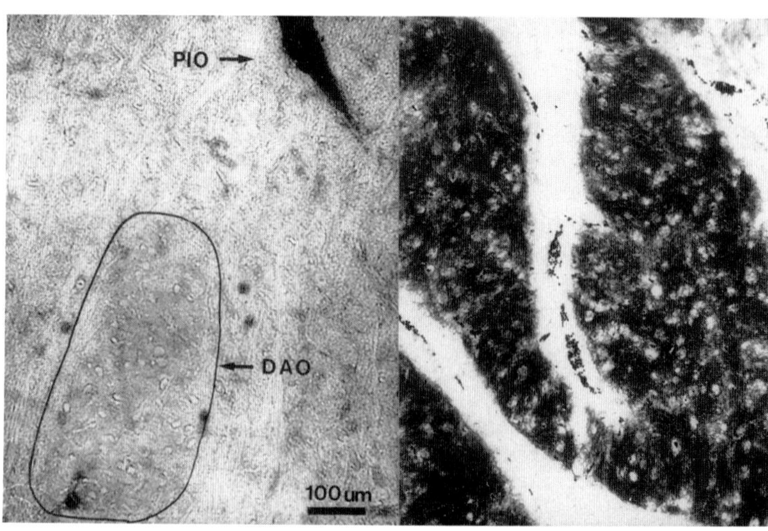

FIG. 2. At high magnification, neurotensin 1-13 nerve terminals (visible in *black*) are exclusively found in the infant principal olive (PIO), delimiting small empty cells, which correspond to unlabeled olivary neurons (**B**), and they are totally absent from accessory olives (e.g., DAO in **A**). 35-μm cryostat sections—PAP technique of Sternberger—NTantisera (1).

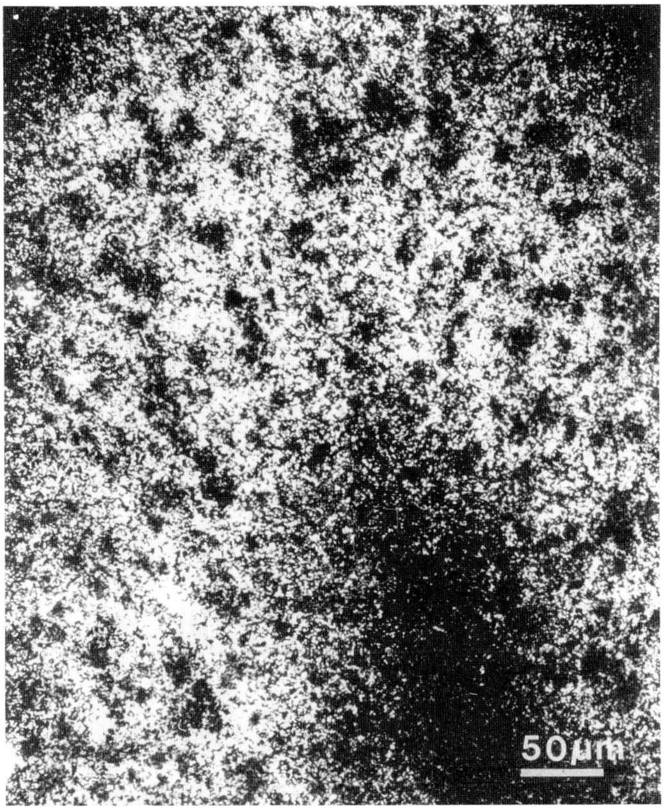

FIG. 3. At high magnification (*dark field*) of an olivary fetal convolution, numerous fibers rich in neurotensin 1-13 binding sites (visible in *white*) delimit small empty cells (visible in *black*), which correspond to olivary neuronal cell bodies. 20-μm cryostat section previously incubated with ^{125}I-Tyr^{3}-NT, postfixed in glutaraldehyde and exposed to emulsion for 6 weeks.

appears in children after 8 years of age, remaining absent in the young and old adult. In the cat, transient NT nerve terminals are also found in the kitten principal olive (3). They are absent in the rat inferior olive.

The distribution of NT nerve terminals does not match but follows the disappearance of NT receptors, which are abundant in the fetal inferior olive (4). On the fetal olivary membranes, NT binds with high affinity to a single population of sites (K_d = 561 pM), which are levocabastine-insensitive and possess a B_{max} value of 71 fmol/mg protein. At the light microscopic level, the receptors are located on afferent fibers of the olivary neurons (Fig. 3). In the 15-week-old fetuses, receptors are located in the whole inferior olive. In the fetuses of 25 weeks and of 33 weeks, receptors are concentrated in the ventral lamella of the principal olive and in the accessory olives (Fig. 4). In the newborns, receptors are still found in the medial part of the ventral lamella of the principal olive. The receptors are absent in the infants after 1 month and remain absent in the adults.

Transient Substance P Nerve Terminals

Using immunohistochemistry, transient substance P (SP) immunoreactive nerve terminals appear in the h nucleus of the medial accessory olive in the 20-week fetus, per-

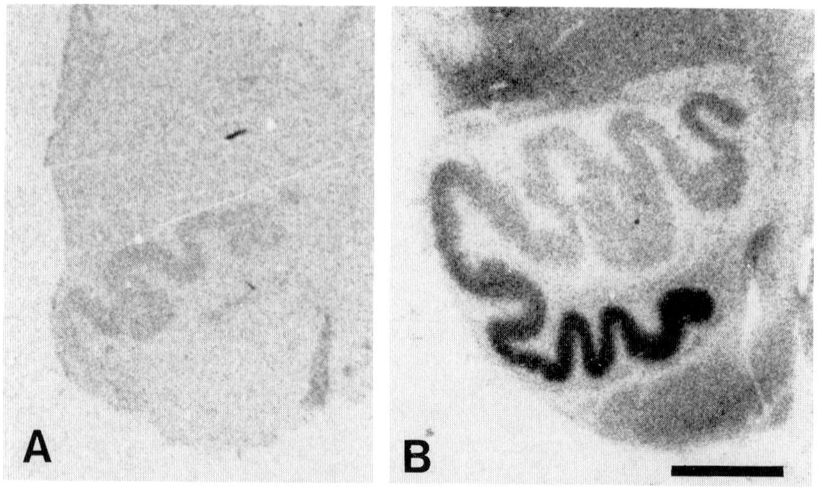

FIG. 4. A mismatch exists in the fetal principal olive between distribution of neurotensin 1-13 terminals, which are located in dorsal lamella (**A**), and of receptors, which are located in ventral lamella (**B**). 20-μm cryostat section—PAP technique of Sternberger—NTantisera (1) in part A. 20-μm cryostat section previously incubated with ^{125}I-Tyr^{3}-NT and exposed to film for 1 week in B. Bar: 1 mm.

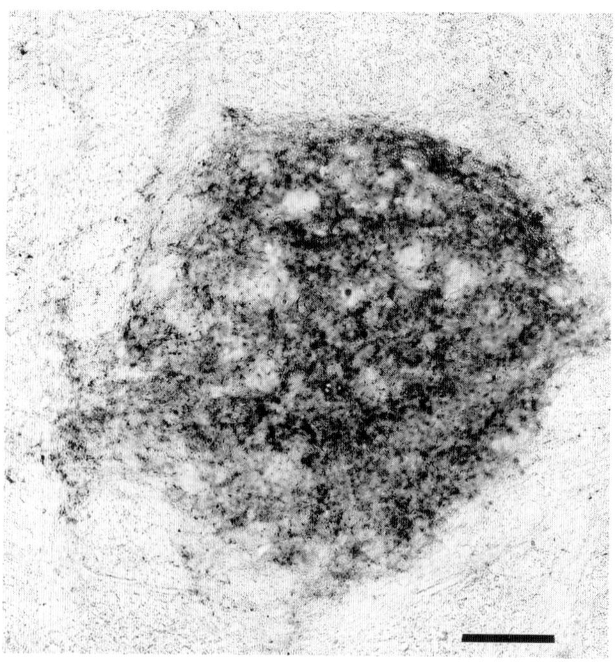

FIG. 5. At high magnification, transient substance P nerve terminals (visible in *black*) are present in abundance in infant h nucleus of neo-olive, which disappears in the adult. 35-μm cryostat sections—PAP technique of Sternberger—substance P antiserum (18). Bar: 50 μm.

sist at birth and during the first year of age (Fig. 5), and disappear in the adult.

Cholecystokinin, Substance P, and Methinone-Enkephalin Nerve Terminals and Receptors

Using immunohistochemistry, cholecystokinin (CCK), SP, and methionine-enkephalin (Met-Enk) immunoreactive nerve terminals are present in the human inferior olive during development. They are exclusively present in the accessory parts and are always absent in the principal olive. They appear in the caudal pole of the fetal and infant accessory olives before extending also to the rostral levels in the adult. At all ages, they are more concentrated in the a and b nuclei of the medial accessory olive and in the lateral part of the dorsal accessory olive.

Receptors for CCK, SP, and Met-Enk are also present in abundance in the whole fetal inferior olive before the arrival of the corresponding neuropeptidergic nerve terminals. They are still present in the infant and in the adult. They are more concentrated in the infant than in the adult. For example, in the case of the CCK receptors, which are of the central type, the B_{max} are, respectively, 15 fmol/mg protein and 3 fmol/mg protein.

Transient Cholecystokinin Receptors in the Human Dentate Nucleus of the Cerebellum

Using autoradiography and *in vitro* membrane binding assay, transient CCK receptors of the central type are found in the fetal dentate nucleus of the cerebellum and disappear in the infant and adult (Fig. 6).

Discussion

The human olivary nucleus is comprised of the neo-olive, which comprises the principal olive and the h nucleus, and of the two accessory medial and dorsal olives (5). CCK, SP, and Met-Enk nerve terminals are the first to appear in the fetal human accessory olives. The neurons at their origin are probably located in the spinal cord, because during development of the monkey inferior olive (6), the first afferent fibers to the olive are from the spinal cord, and they innervate first the caudal levels of the accessory olives before extending to the rostral levels. The NT nerve terminals, which appear later in the fetal dorsal lamella of the principal olive, should originate in neurons in the parvocellular red nucleus, because during development, the rubral terminals appear after the spinal ones, and they first innervate the dorsal lamella of the principal olive before extending, to a more limited degree, to the ventral lamella. The afferents of the h nucleus are not known. Because this nucleus is exclusively present in the human, the transient SP nerve terminals could originate in cortical neurons.

The finding of transient neuropeptidergic nerve terminals has been reported in the human neo-olive. Transient populations of neurons containing neuropeptides exist in the cat cortex during fetal and early postnatal development (7), and a large population of neurons containing CCK and NT has been reported, respectively, in the entorhinal cortex (8) and the subiculum (9) of the human newborn.

The ontogeny of the receptors in the inferior olive indicates that there is a down-regulation of receptors in concert with an increased concentration of neuropeptides. However, with NT, it seems that the situation in the principal olive differs from that in the accessory olives. In the principal olive, especially in the dorsal lamella, NT released from the terminals may act on the receptors in close proximity. In contrast, NT terminals are never found in the accessory olives nor in the ventral lamella of the fetal principal olive, so NT may be acting at a distance from its site of release. The transient presence of CCK receptors in the

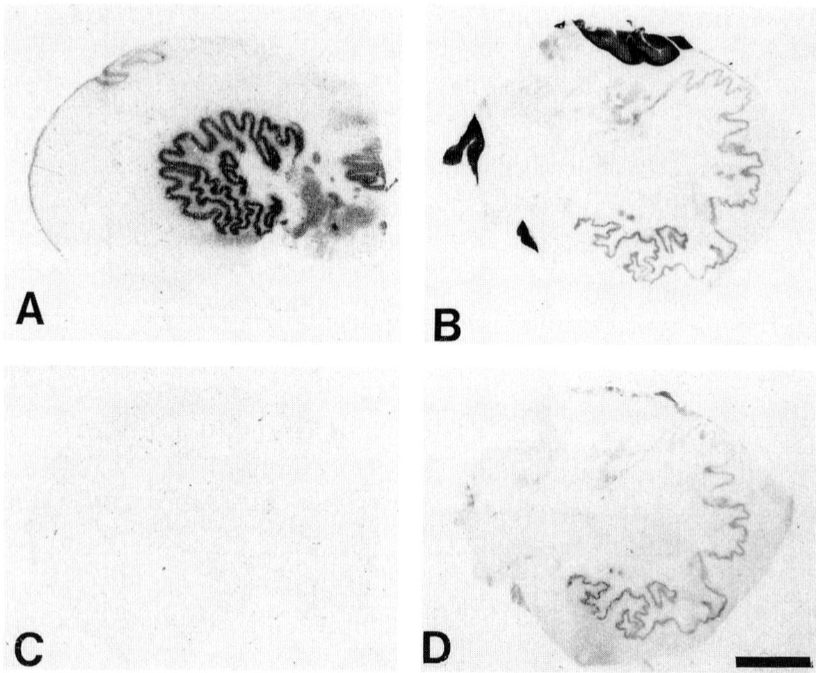

FIG. 6. Cholecystokinin receptors of the central type are present transiently in fetal dentate nucleus of cerebellum (**A**) and disappear in infant and adult nucleus (**B**). Nonspecific labeling is high in the adult nucleus (**D**) in comparison with the one found in the fetal nucleus (**C**) and in a small portion of the adult cerebellar cortex visible at the border of D. 20-μm cryostat sections previously incubated with ^{125}I-Bolton and Hunter CCK-8 and exposed to film for 1 week. Bar: 5mm.

fetal dentate nucleus of the cerebellum, which completes its maturation *in utero* (10), may also be suggestive of CCK acting at a distance, because no CCK innervation is detected in this nucleus.

In conclusion, the transient existence of both neuropeptide nerve terminals and receptors, coupled with the differences observed in their distribution, raises the possibility of paracrine distal and trophic actions of neuropeptides in the developing human inferior olive and, indirectly, in the cerebellum because the axons of the olivary neurons are the only source of the excitatory climbing fibers to the Purkinje cells (11) of the cerebellum, which completes its growth essentially during the first postnatal year (12).

NEUROPEPTIDES RECEPTORS AND ENDOPEPTIDASE 24.11 IN HUMAN MENINGIOMAS

Neuropeptides Receptors

Some neuropeptides receptors are present on the meningothelial parts of the human meningiomas (Fig. 7). Using *in vitro* membrane binding assay and autoradiography, specific high-affinity binding sites for (ig) NT are found in the human meningiomas (19) and share common properties with the NT G-coupled receptor. Because the carboxy-terminal hexapeptide sequence of NT binds to the same site as NT but with higher affinity, they are levocabastine-insensitive and their number is reduced by

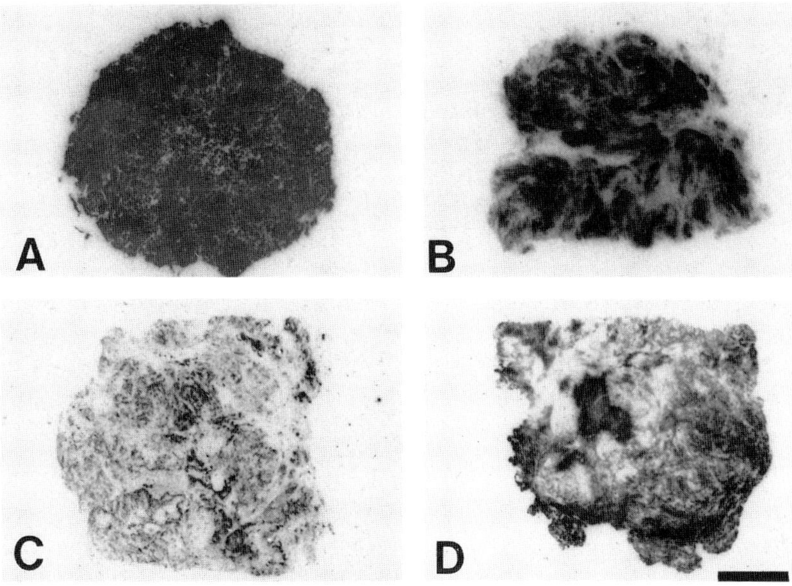

FIG. 7. Neurotensin 1-13 (**A**), cholecystokinin (**B**), and neuropeptide Y (**C**) receptors are exclusively present in meningothelial parts of human meningiomas, in contrast to membrane-bound endopeptidase 24.11 (**D**), which is exclusively found in fibroblastic parts of human meningiomas. Because the majority of meningiomas are of transitional type, which bridges both histological components, labeling for different markers is often heterogeneous. 20-μm cryostat sections previously incubated with ^{125}I-Tyr^{3}-NT, ^{125}I-Bolton and Hunter CCK-8, ^{125}I-PYY, and the selective inhibitor of endopeptidase 24.11, ^{3}H-HACBO-Gly, and exposed to film for 1 week (A–C) and 3 months (D). Bar: 1 cm.

guanosine triphosphate (GTP) and sodium. Receptors for CCK of the peripheral type are also present (20), in contrast with the presence of CCK receptors of the central type on the small cell lung carcinoma (20). Receptors for neuropeptide Y and SRIF (13) are also present on the meningothelial parts of the meningiomas. Receptors for tachykinins are found around blood vessels; there are no opioid receptors and no galanine receptors.

Endopeptidase 24.11

Functional membrane-bound endopeptidase 24.11 (E-24.11) is detected in the fibroblastic parts of human meningiomas (19), using binding to its active site of selective inhibitors (Fig. 7) and selective enzymatic assay.

Discussion

These neuropeptides could influence the growth of the meningiomas, because SRIF inhibits endothelial growth factor (EGF)-stimulated growth of meningiomas in culture (14). Because no neuropeptides are detectable in the meningiomas, an autocrine action is improbable. Neuropeptides synthesized in the brain may eventually reach the meningiomas via the cerebral spinal fluid. Neuropeptides in the peripheral circulation may also reach the meningiomas, because, for example, it has been demonstrated using spect that peripherally injected radioiodinated SRIF selectively accumulates in the human meningiomas (15).

Neuropeptides, for which receptors are found in the meningiomas, are substrates of E-24.11, and E-24.11 could be a modulator (16) of the trophic actions of neuropeptides on the meningiomas, for example, by regulating their access to the receptors on the adjacent meningothelial cells. Because E-24.11 was recently cloned and shown to be identical to the so-called common acute lymphoblastic leukemia antigen (CALLA), trophic actions of the enzyme are possible.

Because there is no barrier to the uptake by the meningiomas of the neuropeptides present in the blood, peripheral injections of neuropeptides could be clinically performed for the diagnosis of small meningiomas using spect and for the treatment of surgically inaccessible meningiomas. In the majority of the cases, the diagnosis of meningiomas with classical radiology is easy, and the excision of benign meningiomas is successful.

Meningiomas presumably arise from the arachnoidal cell of the leptomeninx. The arachnoidal cell has both meningiothelial (endotheloid) and fibroblastic properties (17). This duality is reflected in the neoplastic state with the statistical prevalence of the transitional meningiomas, which are composed of both meningiothelial and fibroblastic parts. The presence of neuropeptides receptors on the meningothelial cells and of E-24.11 on the fibroblastic cells is an indicator of such duality and is useful for the classification of the meningiomas, whose histological heterogeneity could already be a sign of malignancy.

CONCLUSIONS

To understand fully neurological and psychiatric diseases, particularly those of a degenerative and metabolic type, which are peculiar to the human species, it is necessary to study neuropeptides and their receptors in the human brain. Similarly, the large interspecies variation observed in such neuropeptides and receptors limits the degree to which they can be compared, again necessitating human studies.

The transient presence of neuropeptide nerve terminals and receptors during critical periods of the development of the human brain and the presence of neuropeptide receptors on tumors suggest that in addition to their participation in neurotransmission, neuropeptides may be involved in other functions and are excellent candidates for volume transmission.

ACKNOWLEDGMENTS

This work has been partly supported by the Belgian Fund For Medical Scientific Research (FRSM 3.4574.90), the neurobiology program of the Queen Elisabeth Medical Foundation (FMRE 89-91), and the Hoguet Foundation. P.M. is research assistant of the Belgian National Fund for Scientific Research (FNRS 89-91).

REFERENCES

1. Mailleux P, Vanderhaeghen J-J. Transient neurotensin in the human inferior olive during development. *Brain Res* 1988;456:199–203.
2. Liston DR, Vanderhaeghen J-J, Rossier J. Presence in brain of syn-enkephalin, a proenkephalin-immunoreactive protein which does not contain enkephalin. *Nature (Lond)* 1983;302:62–64.
3. Mailleux P, Schiffmann S, Halleux P, Lotstra F, Vanderhaeghen J-J. Transient neurotensin in the cat inferior olive during development. *Neurochem Int* 1989;14:159–161.
4. Mailleux P, Pelaprat D, Vanderhaeghen J-J. Transient neurotensin high-affinity binding sites in the human inferior olive during development. *Brain Res* 1990;508:345–348.
5. Olzewski J, Baxter D. *Cytoarchitecture of the human brainstem,* 2nd ed. Karger, 1982.
6. Robertson LT, Stotler WA. The structure and connections of the developing inferior olivary nucleus of the rhesus monkey. *J Comp Neurol* 1979;158:167–190.
7. Chun JJM, Nakamura MJ, Shatz CJ. Transient cells of the developing mammalian telencephalon and peptide-immunoreactive neurons. *Nature (Lond)* 1987;325:617–620.
8. Lotstra F, Vanderhaeghen J-J. High concentra-

tions of cholecystokinin neurons in the newborn human entorhinal cortex. *Neurosci Lett* 1987; 80:191–196.
9. Lotstra F, Mailleux P, Schiffmann S, Vierendeels G, Vanderhaeghen J-J. Neurotensin containing neurones in the human hippocampus of the adult and during development. *Neurochem Int* 1989;14:143–151.
10. Mihajlovic P, Zecevic N. Development of the human dentate nucleus. *Hum Neurobiol* 1986;5: 189–197.
11. Desclin JC. Histological evidence supporting the inferior olive as the major source of cerebellar climbing fibers in the rat. *Brain Res* 1974; 77:365–384.
12. Zecevic N, Rakic P. Differentiation of Purkinje cells and their relationship to other components of developing cerebellar cortex in man. *J Comp Neurol* 1980;167:27–48.
13. Reubi JC, Maurer R, Klijn JGM, et al. High incidence of somatostatin receptors in human meningiomas; biochemical characterization. *J Clin Endocrinol Metab* 1986;63:433–438.
14. Reubi JC, Horisberger U, Lang W, Koper JW, Braakman R, Lamberts SWJ. Coincidence of EGF receptors and somatostatin receptors in meningiomas but inverse, differentiation-dependent relationship in glial tumours. *Am J Pathol* 1989;134:337–344.
15. Krenning EP, Breemann WAP, Kooij PPM, et al. Localisation of endocrine-related tumours with radioiodinated analogue of somatostatin. *Lancet* 1989;4:242–244.
16. Kenny AJ, O'Hare MJ, Gusterson BA. Cell-surface peptidases as modulators of growth and differentiation. *Lancet* 1989;785–787.
17. Kartenbeck J, Schwechheimer K, Moll R, Franke WW. Attachment of vimentin filaments to desmosomal plaques in human meningiomal cells and arachnoidal tissue. *J Cell Biol* 1984; 98:1072–1081.
18. Lotstra F, Mailleux P, Vanderhaeghen J-J. Substance P neurons in the human hippocampus: an immunohistochemical analysis in the infant and adult. *J Chem Neuroanat* 1988;1:11–123.
19. Mailleux P, Przedborski S, Beaumont A, Verslijpe M, Depierreux M, Levivier M, Kitabgi P, Roques B, Vanderhaeghen J-J. Neurotensin high–affinity binding sites and endopeptidase 24.11 are present respectively in meningothelial and in the fibroblastic components of human meningiomas. *Peptides*. (*In Press*).
20. Mailleux P, Vanderhaeghen J-J. Cholecystokinin receptors of A type in the human dorsal medulla oblongata and meningiomas, and of B type in small cell lung carcinomas. *Neuroscience Letters*. (*In press*).

Volume Transmission in the Brain: Novel Mechanisms for Neural Transmission, edited by K. Fuxe and L.F. Agnati. Raven Press, Ltd., New York © 1991.

13

Small Synaptic Vesicles and Large Dense-Core Vesicles: Secretory Organelles Involved in Two Modes of Neuronal Signaling

M. Matteoli, A. Thomas Reetz, and P. De Camilli

Department of Cell Biology, Yale University School of Medicine, New Haven, Connecticut 06510

The property to secrete signal molecules in a regulated way into the extracellular space is the key feature of endocrine cells and neurons. Endocrine cells secrete molecules (hormones) that have a relatively slow action and that act on target cells, which can be distant. Hormones reach these cells by simple diffusion (paracrine secretion) or by the blood stream (*bona fide* endocrine secretion). Neurons, instead, are regarded as cells highly specialized for fast, point-to-point signaling. They secrete molecules (neurotransmitters) that have a rapid action on target sites localized in direct proximity of the sites of release.

During the past few years, however, evidence has accumulated indicating that in the nervous system at least two modes of neuronal signaling coexist. The first type is the conventional point-to-point signaling, which is also called "wire transmission" because it relies on neuronal processes (dendrites and axons) interconnected with each other. It is characterized by high speed and fidelity. In "wire transmission," pathways of signaling are primarily defined by the cytoarchitectural organization of neuronal networks.

The other type of signaling is mediated by substances that diffuse in the extracellular space of the nervous system to act on distant target cells. This type of signaling has been defined as "volume transmission" to emphasize the fact that, in principle, these substances can produce their effects throughout the volume of the nervous system. In this mode of intercellular communication, pathways of signaling are defined primarily by the chemistry of the signal molecules and by the cellular localization of their receptors, rather than by the topological proximity of release and target sites.

Wire transmission and volume transmission are based to a large extent on different classes of signal molecules, which are released by different mechanisms and operate on different types of receptors. Wire transmission is mediated by nonpeptide molecules, the so-called "classical" neurotransmitters, and its structural basis is the synapse. Classical neurotransmitters are stored in and secreted from typical 50 nm-diameter synaptic vesicles (referred henceforth as small synaptic vesicles, SSVs) (Fig. 1), which are a key morphological feature of nerve endings and are clustered in direct proximity of the synaptic junction (1). During a single synaptic event a high, but also extremely transient, increase of neurotransmitter in the synaptic cleft occurs. The neurotransmitter is then rapidly cleared by enzymatic machineries, dilution,

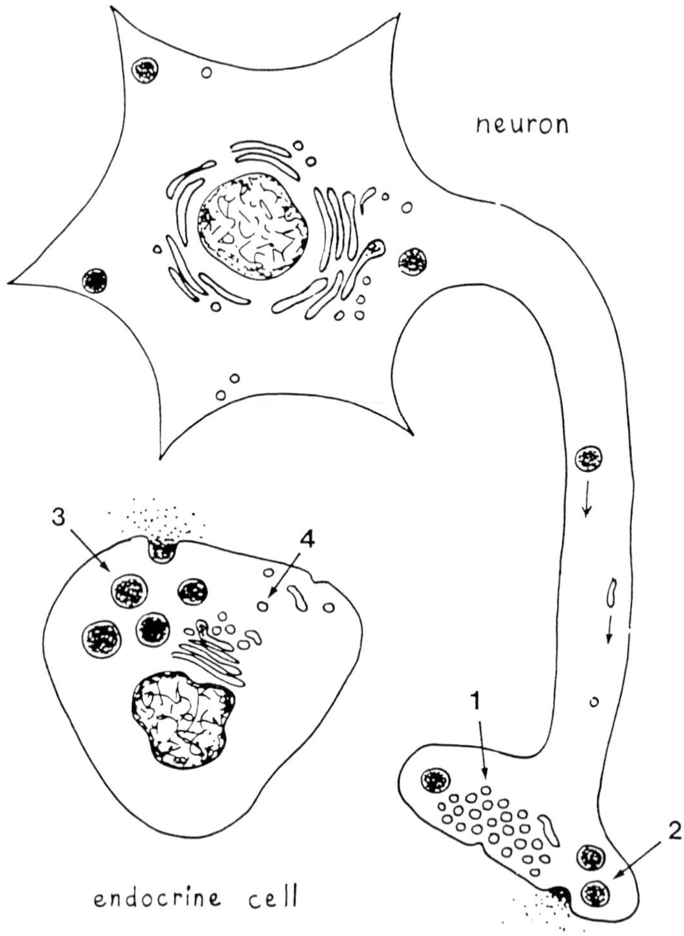

FIG. 1. Schematic drawing illustrating coexistence of small synaptic vesicles (SSVs)(1) and large dense-core vesicles (LDCVs)(2) in nerve endings and relation of these organelles to two distinct types of vesicles of endocrine cells. SSVs, which are clustered and undergo exocytosis at active zones of synapses, have a dominant role in wire transmission. LDCVs have a dominant role in wire transmission. LDCVs are equivalent organelles in neurons of secretory granules of endocrine cells (3). SSVs are related to a microvesicle population in endocrine cells (4) of still unknown function (synaptic-like microvesicles, SLMVS).

or re-uptake mechanisms. Secreted neurotransmitter binds to postsynaptic receptors that are clustered in the patch of target cell plasmalemma facing the synaptic junction. These receptors are ligand-gated channels that rapidly convert chemical signals into electrical signals (2–7). The relatively low affinity these receptors display for their ligands explains why activation of a single nerve terminal results in the activation of only its postsynaptic site, even though synapses that use the same neurotransmitter are located nearby (4). This is turn explains why relatively few neurotransmitter molecules are sufficient for the myriads of synapses present in the brain.

Volume transmission is primarily mediated by peptide molecules, which are often the same molecules that can be secreted by endocrine cells and act as hormones. These molecules are stored in and released from a distinct population of secretory organelles of nerve cells, the so-called large dense-core vesicles (LDCVs) (Fig. 1) (8,9). These molecules have longer half-lives outside the cells than classical neurotransmitters. In addition, they act in general on G protein–coupled receptors, which operate on a time scale longer than that of ligand-gated channels (10). Their main function is to produce modulatory changes on target neurons with a mechanism similar to that used by peptide hormones on their target cells. Peptides bind to their receptors with high affinity. This accounts for their property to act at long distances from release sites, where their concentration is low. The large number of different peptide neurotransmitters is explained by the need to define pathways of signaling by the chemistry of the signal, rather than by the topological proximity of release and target sites.

Although the concept of wire and volume transmission is useful to emphasize the distinct ways by which neurons can exchange information, they represent opposite ends of a spectrum that includes many different ways of signaling. For example, activation of ligand-gated membrane channels can produce intracellular modulatory signals in addition to fast electric events (11). Amines are nonpeptide neurotransmitters and yet produce most of their effects by acting on G protein–coupled receptors (12). Amino acid neurotransmitters, as well as acetylcholine, may bind to G protein–linked receptors in addition to ligand-gated channels (13). These receptors may mediate some action of classical neurotransmitters away from synapses. Thus, the same neurotransmitter may produce multiple effects. Conversely, the concentration of peptides in the extracellular medium is undoubtedly much higher in proximity of release sites. Thus, connectivity is also likely to play an important role in the mechanism of action of peptides. Finally, a unique mode of neuronal signaling is that used by nerve terminals of the neurohypophysis, which release peptides into blood capillaries. Peptides released by these neurons act as *bona fide* hormones (14,15).

In this short chapter we will review work from our and other laboratories on the two secretory organelles that play key roles in these two types of neuronal signaling.

COEXISTENCE OF SSVs AND LDCVs IN NERVE ENDINGS

After the discovery that classical and peptide neurotransmitters can coexist in the same nerve ending, it has become clear that these two classes of substances are stored in and released from two different types of secretory vesicles (16) (Fig. 1). SSVs contain nonpeptide neurotransmitters only (8,16). These organelles undergo cycles of exo–endocytosis in nerve endings, and at each cycle they can be reloaded with neurotransmitter because all the cellular machinery necessary to synthesize and load classical neurotransmitter into vesicles is present in the nerve terminal itself (17). LDCVs contain peptides, although they may also contain nonpeptide neurotransmitters (e.g., amines) (8,9,16, 18,19). They are assembled in the Golgi complex region of the cell, and their membranes need to return to the Golgi complex area to be reloaded with content (15).

Coexistence of SSVs and LDCVs has been shown in a large number of neurons, and most likely these two organelles coexist in all neurons. However, their relative proportion varies from nerve terminal to nerve terminal, and their number is generally low in comparison with the number of SSVs. Neuropeptide-containing LDCVs are present at particularly high concentration in certain interneurons and in most nerve terminals of hypothalamic neurons (14). In other nerve terminals (e.g., those of the neuromuscular junction), they may represent 1% or less of the total vesicle population (19,20). These differences may reflect specific functions of nerve terminals. Relatively slow modulatory signals may play an important role in the hypothalamus, a region of the brain involved in autonomic functions. Conversely, fast signaling is required at the motor end plate.

Changes in the relative proportion of SSVs and LDCVs were also observed at a given synapse depending on its functional state. During development and regeneration the relative number of peptide-containing LDCVs in nerve terminals has been found to be higher than in mature nerve terminals (21,22). Developmental changes in the expression of a neuroactive peptide were recently investigated in mammalian motor neurons and motor nerve terminals. It was found that calcitonin gene-related peptide (CGRP), a neuroactive peptide contained in LDCVs of motor nerve terminals (19), is expressed at high levels during the first weeks of postnatal development but is down-regulated in the adult animal (23). A decrease in the number of LDCVs (relative to SSVs) in motor nerve terminals concomitant with maturation was also observed (S. Balbi, M. Matteoli, C. Sala, and G. Fumagalli, *unpublished observations*). CGRP expression was also found to be up-regulated during motor nerve regeneration (23). One may speculate that changes in the relative proportion of LDCVs may be related to changes in the relative role of trophic signals with maturation of the synapse.

SSVs AND LDCVs UNDERGO EXOCYTOSIS AT DIFFERENT SITES

Small synaptic vesicles and LDCVs have a different intracellular distribution and undergo exocytosis at distinct sites of the neuronal surface. Electron microscopy studies carried out at a number of synapses have shown that SSVs are generally organized in clusters closely opposite to the active zones (i.e., the part of the plasma membrane that faces the synaptic cleft). These are also the sites at which exocytosis of SSVs takes place (9,17,24). Quick-freezing/deep-etching studies have suggested that SSV clusters are represented by SSVs bound to a dense meshwork of actin filaments, which are connected to the active zones by their barbed end (25,26). The formation of these clusters is strictly controlled by the interaction of a nerve terminal with its postsynaptic site (27). Thus, exocytotic sites for SSVs are tightly defined by the connections of a neuron. LDCVs are in general excluded from SSV clusters (28), and their exocytosis takes place preferentially away from active zones (9).

DIFFERENTIAL EXOCYTOSIS OF LDCVs AND SSVs

Pharmacological studies have suggested that the relative proportion of classical and peptide neurotransmitters released from nerve endings may vary with the conditions of stimulations. In particular, high frequencies of electrical stimulation are more effective for the release of peptides (29,30).

A direct demonstration that exocytosis of LDCVs and SSVs from the same nerve ending can be dissociated has been provided by experiments carried out at the frog neuromuscular junction involving α-latrotoxin. α-Latrotoxin is the main toxin of black widow spider venom and is an agent known to trigger a massive, Ca^{2+}-independent exocytosis of SSVs from the nerve terminal (31,32). When applied to neuromuscular preparations at low concentrations in the absence of extracellular Ca^{2+} or at high concentrations in the presence of extracellular Ca^{2+}, it also blocks recycling of SSV membranes, thus leading to a complete depletion of SSVs from the terminal (31,33). This massive depletion is not accompanied by any significant decrease in the number of LDCVs within the same nerve terminals (Fig. 2) (19). Although α-latrotroxin is not a physiological stimulus, these results are consistent with the hypothesis that the ex-

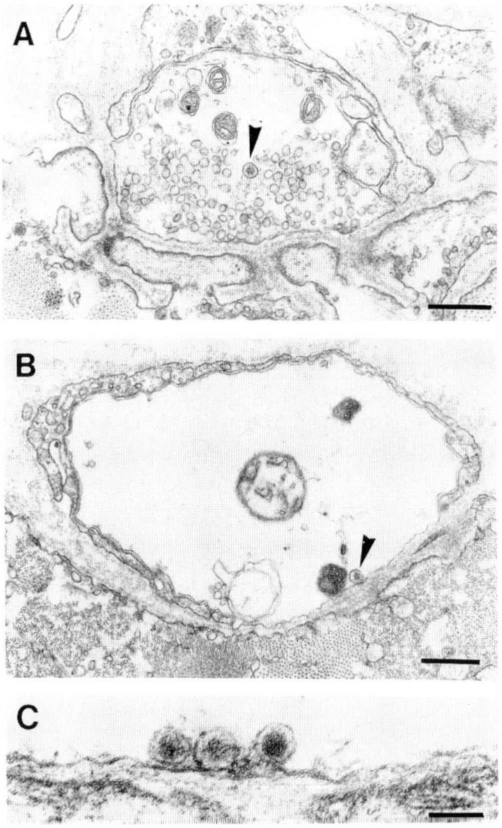

FIG. 2. Differential effect of α-latrotoxin on exocytosis of LDCVs and SSVs at frog neuromuscular junction. **A** shows a control nerve terminal where LDCVs (*arrowhead*) represent about 1% of total vesicle content. **B** and **C** show nerve terminals incubated for 1 hr in a Ca^{2+}-free solution in the presence of α-latrotoxin. In these terminals, all SSVs, but not LDCVs, have disappeared as a result of massive exocytosis. A single LDCV (*arrowhead*) surrounded by a SSV-free cytoplasm is visible in B. Three LDCVs clustered in front of the synaptic cleft are visible in C. (Calibration bars = A, 380 μm; B, 582 μm; C, 185 μm. (From ref. 19, with permission.)

ocytotic machineries for SSVs and LDCVs are at least partially different.

A difference in the stimulus-secretion coupling for SSVs and LDCVs was also suggested by studies on the subtypes of Ca^{2+} channels involved in neurotransmitter release. These studies have raised the possibility that different types of Ca^{2+} channels play a dominant role in the release of classical and peptide neurotransmitters (34,35). Sites of exocytosis for SSVs and LDCVs may have different topological relations to different types of Ca^{2+} channels. In addition, exocytotic machineries for SSVs and LDCVs may have a different sensitivity to cytosolic Ca^{2+}. It has been proposed that the intracellular signal for exocytosis of SSVs is represented by a prominent (hundreds of micromoles), but also extremely transient (hundreds of microseconds), increase in the concentration of cytosolic Ca^{2+} that takes place at active zones during an action potential. A rise in cytosolic Ca^{2+} of lower magnitude, but of longer duration, may instead be responsible for LDCV exocytosis (36). Such a rise may follow a train of closely spaced action potentials.

SSVs AND LDCVs HAVE A DISTINCT MEMBRANE COMPOSITION

Recently, major advancements have been made in our understanding of the protein composition of SSV membranes, and SSVs are perhaps the best characterized secretory organelles. Several major membrane proteins of SSVs, which appear to have a general function in their life cycle (i.e., a function not specific for a given neurotransmitter), have been identified. For some of them the primary amino acid sequence is known.

The most thoroughly investigated SSV proteins are the synapsins. These represent a family of four similar proteins, synapsin Ia, Ib, IIa, and IIb, which are the products of alternative splicing of primary transcripts from two distinct genes. Synapsin is present in all nerve terminals, although the relative ratio of the four synapsin isoforms varies from nerve terminal to nerve terminal (37,38). Both synapsin I and synapsin II are substrates for endogenous phosphorylation mediated by Ca^{2+}/calmodulin dependent- and by cyclic adenosine monophosphate (cAMP)-dependent protein kinases. They are peripheral membrane proteins localized at the cytoplasmic surface of the vesicle membrane (37). The interaction of synapsin I with the vesicle surface has been thoroughly studied and found to involve both hydrophilic and hydrophobic interactions (39). Synapsin I has also been found to bind cytoskeletal elements and, with particular high affinity, actin (40). It is therefore thought to act as a connecting link between the vesicle surface and the actin-based cytomatrix adjacent to the active zone. Phosphorylation of synapsin I affects the binding of the protein to both synaptic vesicles and actin (40–42). In addition, physiological and pharmacological manipulations, which promote or facilitate the release of neurotransmitters from nerve endings, also result in an increase of phosphorylation of synapsin I (37). Synapsin I may therefore control one or more steps leading to SSV exocytosis. The properties of synapsin II have not been studied in detail so far. The primary sequence similarity between synapsin I and synapsin II and their similar localization on SSVs suggest that the two proteins have similar functions (38).

The other best characterized SSV membrane proteins are synaptophysin (also referred to as p38) and synaptobrevin (also referred to as p18 and as VAMP1/2). Synaptophysin is an intrinsic membrane glycoprotein (total of 307 amino acids) that spans the membrane four times. Both the N-terminus and the C-terminus of the protein are located in the cytoplasm. The C-terminus is tyrosine-phosphorylated and contains 10 copies of an unusual pentapeptide repeat (43–47). It has been reported to form ion channels when inserted in lipid bilayers and to bind Ca^{2+} at its cytoplasmic tail (48,49). Because of the latter property, it may play a role in Ca^{2+}-dependent exocytosis of SSVs.

Synaptobrevin is a protein of 116 amino acids in mammals, with a single putative transmembrane domain and the C-terminus within the lumen of the vesicle (50–52). It is the only SSV protein that has already been conclusively identified in *Drosophila* (51). It can exist in at least two isoforms (product of two different genes), which are differentially expressed in the nervous system (52).

Other intrinsic proteins of SSV membranes, for which the primary sequence is still unknown, include p29, p65, and SV2. p29 is a protein that shares some important similarity with synaptophysin, such as acidic pH, epitopes recognized by monoclonal antibodies, and the property to be tyrosine-phosphorylated (53). p65 (54) is a calmodulin-binding protein (55) and is probably identical to a ganglioside-binding protein found on SSVs (56). SV2 is a protein of approximately 100,000 daltons, which is highly glycosylated (57). Finally, SSVs have been found to contain a proton adenosine triphosphatase (ATPase) of the vacuolar type (58) and to be enriched in a RAS-related low molecular weight G protein: rab3 (59).

In contrast to this large body of information available on the composition of SSVs, little is known on proteins of LDCV membranes. LDCVs of the brain represent a minor and heterogeneous vesicle population. They have been obtained in sufficient quantity and purity only from highly specialized organs that are highly enriched in them [e.g., the posterior pituitary (60) and the bovine splenic nerve (61)]. Studies on these vesicles have suggested important similarities with granules of endocrine cells, for example, those of chromaffin cells, which are the secretory granules most thoroughly characterized (62). Identified membrane components of at least some LDCVs include peptide α-amidase (63), cytochrome b561 (64), dopamine β-hydroxylase (65), and a proton pump of the vacuolar type (66).

Although some compositional overlap appears to exist between the membranes of SSVs and LDCVs [e.g., a proton pump is present on both organelles (58,66)], clearly the two membranes are different. Some of

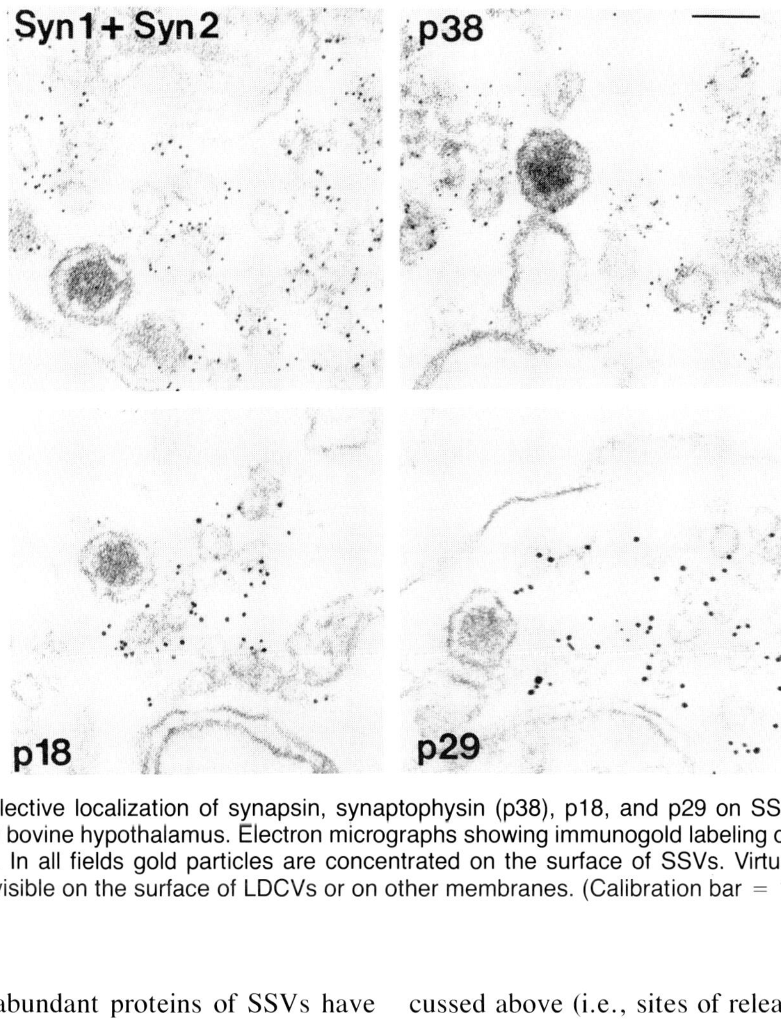

FIG. 3. Selective localization of synapsin, synaptophysin (p38), p18, and p29 on SSVs in nerve terminals of bovine hypothalamus. Electron micrographs showing immunogold labeling of lysed synaptosomes. In all fields gold particles are concentrated on the surface of SSVs. Virtually no gold labeling is visible on the surface of LDCVs or on other membranes. (Calibration bar = 100 nm.)

the most abundant proteins of SSVs have not been found at significant concentration in the membranes of LDCVs. This applies to at least synapsin, synaptophysin, synaptobrevin, and p29 (Fig. 3) (28,50,53,67, 68). Such findings suggest that SSVs and LDCVs are organelles with a distinct biogenesis and speak against the hypothesis, first proposed for nerve terminals of the neurohypophysis (69) and of sympathetic neurons (70) and subsequently generalized to all neurons, that SSVs are formed in nerve endings as a byproduct of exocytosis of LDCVs. In addition, the existence of major differences in the membrane composition of SSVs and LDCVs demonstrates a biochemical basis for the different functional properties of SSVs and LDCVs discussed above (i.e., sites of release, regulation of release, type of content, etc.). For example, the selective association of synapsin I with SSVs may be responsible for the selective association of SSVs with active zones.

SSVs AND LDCVs ARE RELATED TO TWO DISTINCT ORGANELLES OF ENDOCRINE CELLS

Even before the discovery that a large number of peptide hormones are also used by nerve cells as neurotransmitters, the existence of special biochemical similarities between neurons and the endocrine cells that are specialized for the regulated secretion of peptide hormones and amines was known (14,71,72). The discovery that many, and perhaps all, neurons secrete peptide hormones via LDCVs (16) has revealed an even closer relation between neurons and endocrine cells. LDCVs may be seen as the equivalent organelles of secretory granules of endocrine cells (73,74). LDCVs and endocrine granules are similar in morphology, content, biogenesis, and probably

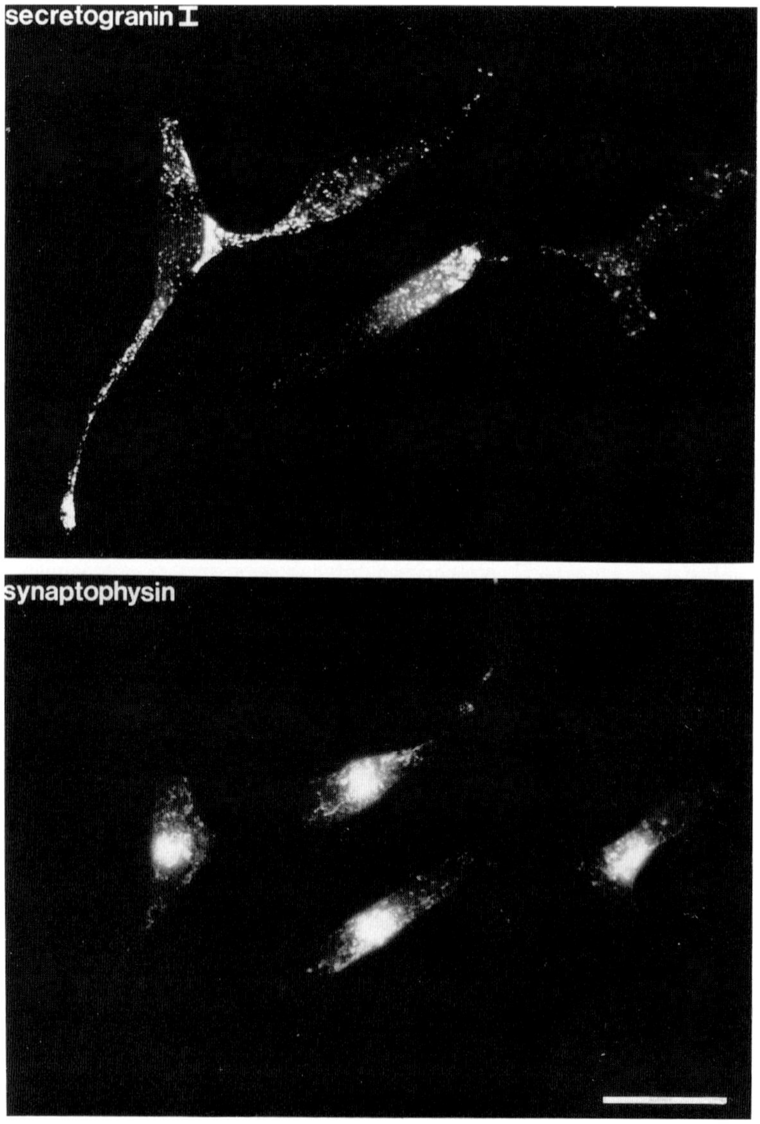

FIG. 4. Double-immunofluorescence staining of four insulinoma cells (RINm5F) showing that synaptophysin, a major protein of SSVs, does not colocalize with secretory granules. Secretory granules are visualized with an antibody directed against a secretory protein, secretogranin I (chromogranin B), costored with insulin in the granules. Puncta of secretogranin I immunoreactivity, which represent individual secretory granules, are particularly concentrated in the distal portion of cell processes. Synaptophysin immunoreactivity, which reflects distribution of synaptic-like microvesicles (SLMVs), is particularly concentrated in the area of Golgi complex. (Calibration bar = 20 μm.)

also the stimulus-secretion coupling involved in their exocytosis (74).

A further relationship between neurons and endocrine cells has been suggested by the finding that a number of major membrane proteins of SSVs are also present at significant concentration in at least some endocrine cells (Fig. 4). These include synaptophysin, synaptobrevin, p29, p65, and SV2 (44,50,53,54,57,67). The subcellular distribution of two of these proteins, synaptophysin and p29, in endocrine cells has been thoroughly investigated. These studies have shown that both proteins are not present at detectable concentration in the membranes of secretory granules (for an al-

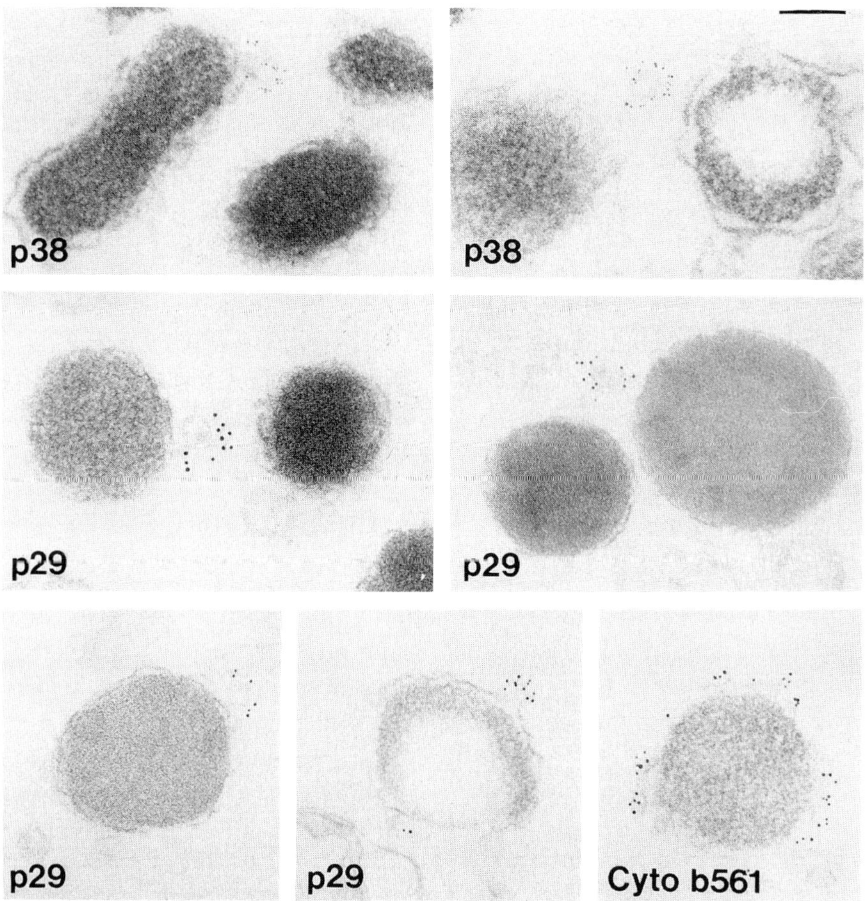

FIG. 5. Gallery of electron micrographs illustrating localization of synaptophysin (p38), p29, and cytochrome b561 (cyto b561) in subcellular organelles of endocrine cells. Immunogold labeling of agarose-embedded homogenates of bovine anterior pituitary (**A–D**) and adrenal medulla (**E–G**). Both synaptophysin and p29 are selectively localized on microvesicles (synaptic-like microvesicles, SLMVs), and no immunoreactivity for these proteins is visible on secretory granules. In F a p29-positive small membrane evagination of a vacuole with a partially dense core is shown. This structure may represent an immature granule. If so, the figure suggests that separation of SLMV proteins from secretory granule membranes takes place (or is completed) during secretory granule maturation. Intense labeling of secretory granule membranes after cyto b561 immunolabeling indicates that their surface is freely accessible to antibodies and represents a positive control for their lack of synaptophysin and p29 immunoreactivity. (Calibration bar = 200 nm.) (From refs. 53 and 67, with permission.)

ternative view, see ref. 75) and are instead localized on a population of pleomorphic microvesicles (referred to as synaptic-like microvesicles, SLMVs) with the same size range of neuronal SSVs (Figs. 4, 5, 6A and B) (53,67). Some evidence suggests that synaptobrevin as well is localized on SLMVs (53). SLMVs are scattered throughout the cytoplasm and are particularly concentrated in the Golgi complex area. Presence of a few SLMV membranous profiles in continuity with the plasmalemma (i.e., omega-shaped invaginations positive for synaptophysin) and of SLMVs labeled by extracellular tracers after exposure of cells to such tracers (Fig. 6A and B) indicates that SLMVs undergo exo–endocytosis. Finally, the rapid labeling of SLMVs with extracellular tracers (minutes/hours) despite the relatively long half-life of synaptophysin (days) suggests that SLMVs undergo several cycles of exo–endocytosis (67,76). Since the latter property is a key property of SSVs, SLMVs appear to share with SSVs not only biochemical, but also functional, similarities. It will be of interest to determine whether SLMVs contain small nonpeptide molecules (e.g., classical neurotransmitters), and whether SLMVs are involved in some form of intercellular signaling as their neuronal counterparts.

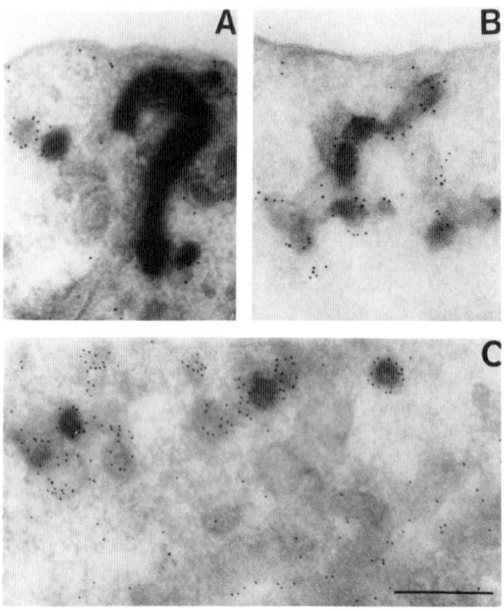

FIG. 6. Electron micrograph demonstrating localization of synaptophysin on small endocytic vesicles in undifferentiated PC12 cells (**A** and **B**) and in CHO cells transfected with synaptophysin (complementary deoxyribonucleic acid) (cDNA) (**C**). Immunogold labeling of cells that have been exposed for 1 hr to an extracellular tracer (horseradish peroxidase) before fixation to label endocytic compartments (for experimental details see ref. 76). In both cell types immunogold selectively decorates the surface of round or elongated microvesicles. Some of them are also positive for extracellular tracer as shown by their content of dark peroxidase reaction product. Larger peroxidase-positive vacuoles, probably late endosomes or lysosomes, are unlabeled by immungold. Note lack of labeling of the plasmalemma. (Calibration bar = 200 nm.) (From ref. 76, with permission.)

These findings suggest that SSVs represent a neuron-specific adaptation of endocrine SLMVs. Synapsin, which is expressed at significant concentration only by neuron tissues (37), may represent one of the key features of this neuron-specific adaptation. As mentioned above, synapsin may be involved in the clustering of SSVs at release sites. It is therefore of interest that this clustering is typical of neuronal SSVs and is not seen in the case of endocrine SLMVs. It remains to be seen whether SLMVs are in turn related to some organelle present in all cells. So far, proteins immunoreactive with antibodies directed against major SSV proteins have been identified only in neurons and endocrine cells. This does not exclude the possibility that SSVs and SLMVs might be distantly related to recycling organelles present in all cells. If so, the identification and study of such organelles and the comparison of their properties with those of SSVs and SLMVs may help to advance our knowledge on neuronal secretion.

With the goal of identifying such organelles, we have investigated the fate of synaptophysin when expressed by cDNA transfection in a fibroblastic cell line, that normally does not express this protein (CHO cells). The working hypothesis behind this experimental approach was that if an organelle related to SLMVs and SSVs exists in all cells, synaptophysin may be targeted to that organelle and be a tool to identify it. Results of these experiments have shown that posttranslational modification and half-life of synaptophysin were remarkably similar in transfected CHO cells and in PC12 cells (an endocrine cell line that naturally expresses the protein). In addition, in transfected CHO cells, synaptophysin was localized in a population of microvesicles similar to SMLVs of PC12 cells in morphology, dimensions, intracellular distribution, and property to become labeled by extracellular tracers (Fig. 6). Moreover, synaptophysin-positive microvesicles of transfected CHO cells appeared to coincide with the microvesicles that recycle the transferrin receptor (76,77). These findings raise the possibility that SLMVs and SSVs represent the endocrine- and the neuron-specific adaptation, respectively, of a recycling pathway expressed by all cells.

SUMMARY AND CONCLUSIONS

Wire transmission and volume transmission represent two distinct ways of interneuronal signaling. The latter is closely related to typical endocrine signaling, the former is a neuron-specific type of signaling that is specially designed to allow exchange of information with high speed and great spatial precision. One of the cellular basis for the coexistence of these two modes of interneuronal communication is the presence in neurons of two distinct secretory pathways, which use different organelles with different contents, different sites of exocytosis, and different stimulus-secretion couplings: SSVs and LDCVs. LDCVs are the equivalent organelles of secretory granules of endocrine cells and play a dominant role in volume transmission. SSVs represent the neuronal adaptation of SLMVs of endocrine cells (and possibly of recycling organelles present in all cells) and play a dominant role in wire transmission.

ACNOWLEDGMENTS

The work carried out in our laboratory and reviewed in this manuscript was supported in parts by grants from MDA, the Klingenstein Foundation, and the Diabetes Research and Education Foundation and by funds from Fidia Research Foundation to P.D.C.

REFERENCES

1. Peters A, Palay SL, Webster HDF. *The fine structure of the nervous system. The neurons and supporting cells*. W.B. Saunders, Philadelphia, 1976.
2. Fischbach GD, Frank E, Jessell TM, Rubin LL, Schuetze SM. Accumulation of acetylcholine receptors and acetylcholinesterase at newly formed nerve-muscle synapses. *Pharmacol Rev* 1979;30:411–428.
3. Triller A, Cluzeaud F, Pfeiffer F, Betz H, Korn H. Distribution of glycine receptors at central synapses: an immunoelectron microscopic study. *J Cell Biol* 1985;101:683–688.
4. Changeux JP. The acetylcholine receptor. An allosteric protein. *Science* 1984;225:1335–1345.
5. Grenningloh G, Rienitz A, Scmitt B, et al. The strychnine-binding subunit of the glycine receptor shows homology with nicotinioc acetylcholine receptors. *Nature* 1987;328:215–220.
6. Schofield PR, Darlison MG, Fujita N, et al. Sequence and functional expression of the GABAa receptor shows a ligand-gated receptor superfamily. *Nature* 1987;328:221–227.
7. Stevens C. Channel families in the brain. *Nature* 1987;328:198–199.
8. Klein R, Lagercrantz H, Zimmermann H. *Neurotransmitter vesicles*. New York: Academic Press, 1982;1–375.
9. Thureson Klein A, Klein RL. Exocytosis from neuronal large dense-cored vesicles. *Int Rev Cytol* 1990;121:67–126.
10. Gilman AG. G proteins and regulation of adenyl cyclase. *JAMA* 1989;262:1819–1825.
11. Mayer M. Two channels reduced to one. *Nature* 1987;325:480–481.
12. Kebabian JB, Caine DB. Multiple receptors for dopamine. *Nature* 1979;277:93–96.
13. Sugiyama H, Ito I, Hirono C. A new type of glutamate receptor linked to inositol phospholipid metabolism. *Nature* 1987;325:531–533.

14. Scharrer E, Scharrer B. Hormones produced by neurosecretory cells. *Recent Prog Horm Res* 1954;10:183–240.
15. Brownstein MJ, Russell JT, Gainer H. Synthesis, transport and release of posterior pituitary hormones. *Science* 1980;207:373–378.
16. Hokfelt T, Millhorn D, Seroogy K, et al. Coexistence of peptides with classical neurotransmitters. *Experientia* 1987;43:768–780.
17. Ceccarelli B, Hurlbut WP, Mauro A. Turnover of transmitter and synaptic vesicles at the frog neuromuscular junction. *J Cell Biol* 1973;57: 499–524.
18. Coulter HD. Vesicular localization of immunoreactive (Met5) enkephalin in the globus pallidus. *Proc Natl Acad Sci USA* 1988;85:7028–7032.
19. Matteoli M, Haimann C, Torri-Tarelli F, Polak JM, Ceccarelli B, De Camilli P. Differential effect of α-latrotoxin on exocytosis of ACh-containing small synaptic vesicles and CGRP-containing large dense core vesicles at the frog neuromuscular junction. *Proc Natl Acad Sci USA* 1988;85:7366–7370.
20. Lynch K. Stimulation-induced reduction of large dense core vesicle numbers in cholinergic motor nerve endings. *Brain Res* 1980;194:249–254.
21. Van Arsdall CB, Lentz TL. Neurons: secretory activity during limb regeneration and induction in the newt. *Science* 1968;162:1296–1297.
22. Pellegrino de Iraldi A, De Robertis E. The neurotubular system of the axon and the origin of granulated and non-granulated vesicles in regenerating nerves. *Z Zellforschung* 1968;87:330–344.
23. Matteoli M, Balbi S, Sala C, et al. Developmentally regulated expression of calcitonin gene-related peptide at the mammalian neuromuscular junction *J Mol Neurosci* (in press).
24. Torri-Tarelli F, Grohovaz F, Fesce R, Ceccarelli B. Temporal coincidence between synaptic vesicle fusion and quantal secretion of acetylcholine. *J Cell Biol* 1985;101:1386–1399.
25. Landis DMD, Hall AK, Weinstein LA, Reese TS. The organization of cytoplasm at the presynaptic active zone of a central nervous system synapse. *Neuron* 1988;1:201–209.
26. Hirokawa N, Sobue K, Kanda K, Harada A, Yorifuji H. The cytoskeletal architecture of the presynaptic terminal and molecular structure of synapsin. I. *J Cell Biol* 1989;108:111–126.
27. Lindsley TA, De Camilli P, Banker GA. The influence of cell-cell contact on the distribution of synapsin I in hippocampal neurons in culture (*Abstr.*) *Neurosci Abstr* 1987;13:318.
28. Navone F, Greengard P, De Camilli P. Synapsin I in nerve terminals: selective association with small synaptic vesicles. *Science* 1984;226:1209–1211.
29. Andersson PO, Bloom SR, Edwards AV, Järhult J. Effects of stimulation of the chorda tympani in bursts on submaxillary responses in the cat. *J Physiol* 1982;322:469–483.
30. Lundberg JM, Hokfelt T. Coexistence of peptides and classical neurotransmitters. *Trends Neurosci* 1983;6:325–333.
31. Hurlbut WP, Ceccarelli B. In: Ceccarelli/Clementi eds. *Advances in cytopharmacology*, vol 3. New York: Raven, 1979;87–115.
32. Meldolesi J, Scheer H, Madeddu L, Wanke L. On the mechanism of action of alpha-latrotoxin, the presynaptic stimulatory toxin of the black widow spider venom. *Trends Pharmacol Sci* 1986;7:151–155.
33. Valtorta F, Jahn R, Fesce R, Greengard P, Ceccarelli B. Synaptophysin (p38) at the frog neuromuscular junction: its incorporation into the axolemmma and recycling after intense quantal secretion. *J Cell Biol* 1988;107:2717–2727.
34. Rane S, Holz GG, Dunlap K. Dihydropiridine inhibition of neuronal calcium current and substance P release. *Pflugers Arch* 1987;409:361–366.
35. Hirning LD, Fox AP, McCleskey EW, et al. Dominant role of N-type Ca^{++} channels in evoked release of norepinephrine from sympathetic neurons. *Science* 1988;239:57–61.
36. Smith S, Augustine GJ. Calcium ions, active zones and synaptic transmitter release. *Trends Neurosci* 1988;11:458–465.
37. De Camilli P, Greengard P. Commentary. Synapsin I: a synaptic vesicle associated neuronal phosphoprotein. *Biochem Pharmacol* 1986;35: 4349–4357.
38. Südhof TC, Czernik AJ, Kao M, et al. The synapsins: mosaics of shared and unique domains in a family of synaptic vesicle phosphoprotein. *Science* 1989;245:1474–1480.
39. Benfenati F, Bahler M, Jahn R, Greengard P. Interaction of synapsin I with small synaptic vesicles: distinct sites in synapsin I bind to vesicle phospholipids and vesicle proteins. *J Cell Biol* 1989;108:1863–1872.
40. Bähler M, Greengard P. Synapsin I bundles F-actin in a phosphorylation-dependent manner. *Nature* 1987;326:704–707.
41. Huttner WB, Schlieber W, Greengard P, De Camilli P. Synapsin I (protein I), a nerve terminal specific phosphoprotein. III. Its association with synaptic vesicles studied in a highly purified synaptic vesicles preparation. *J Cell Biol* 1983;96:1374–1388.
42. Schlieber WB, Jahn R, Doucet JP, Rothlein J, Greengard P. Characterization of synapsin I binding to small synaptic vesicles. *J Biol Chem* 1986;261:8383–8390.
43. Jahn R, Schlieber W, Ouimet C, Greengard P. A 38,000 dalton membrane protein (p38) present in synaptic vesicles. *Proc Natl Acad Sci* 1985; 82:4137–4141.
44. Wiedenmann B, Franke WW. Identification and localization of synaptophysin, an integral membrane glycoprotein of M 38,000 characteristic of presynaptic vesicles. *Cell* 1985;41:1017–1028.
45. Südhof TC, Lottspeich F, Greengard P, Mehl E, Jahn R. A synaptic vesicle protein with a novel cytoplasmatic domain and four transmembrane regions. *Science* 1987;238:1142–1144.
46. Leube RE, Kaiser P, Seiter A, et al. Synaptophysin: molecular organization and mRNA expression as determined from cloned cDNA. *EMBO J* 1987;6:3261–3268.

47. Pang D, Wang J, Valtorta F, Benfenati F, Greengard P. Protein tyrosine-phosphorylation in synaptic vesicles. *Proc Natl Acad Sci USA* 1988; 85:762–766.
48. Thomas L, Hartung K, Langosch D, et al. Identification of synaptophysin as a hexameric channel protein of the synaptic vesicle membrane. *Science* 1988;242:1050–1053.
49. Rehm H, Wiedenmann B, Betz H. Molecular characterization of synaptophysin, a major calcium-binding protein of the synaptic vesicle membrane. *EMBO J* 1986;5:535–541.
50. Baumert M, Maycox PR, Navone F, De Camilli P, Jahn R. Synaptobrevin: an integral membrane protein of 18,000 Dalton present in small synaptic vesicles of rat brain. *EMBO J* 1989;8:379–384.
51. Südhof TC, Baumert M, Perin MS, Jahn R. A synaptic vesicle membrane protein is conserved from mammals to *Drosophila*. *Neuron* 1989a; 245:1474–1480.
52. Elferink LA, Trimble WS, Scheller RH. Two vesicle-associated membrane protein genes are differentially expressed in the rat central nervous system. *J Biol Chem* 1989;264:11061–11064.
53. Baumert M, Takei K, Hartinger J, et al. P29, a novel tyrosine-phosphorylated membrane protein present in small clear vesicles of neurons and endocrine cells. *J Cell Biol* 1990;110:1285–1294.
54. Matthew WD, Tsavaler L, Reichardt LF. Identification of a synaptic vesicle specific membrane protein with a wide distribution in neuronal and neurosecretory tissues. *J Cell Biol* 1981;91:257–269.
55. Fournier S, Trifaro JM. A similar calmodulin-binding protein expressed in chromaffin, synaptic and neurohypophyscal secretory vesicles. *J Neurochem* 1988;50:27–37.
56. Popoli M, Mengano A. A hemagluttinin specific for sialic acid(s) in rat brain synaptic vesicle enriched fractions. *Neurochem Res* 1988;13:63–67.
57. Buckley K, Kelly RB. Identification of a transmembrane glycoprotein specific for secretory vesicles of neural and endocrine cells. *J Cell Biol* 1985;100:1284–1294.
58. Hell JW, Maycox PR, Stadler H, Jahn R. Uptake of GABA by rat brain synaptic vesicles isolated by a new procedure. *EMBO J* 1988;7:3023–3036.
59. Fischer V, Mollard G, Mignery GA, et al. Rab3 is a small GTP-binding protein exclusively localized to synaptic vesicles. *Proc Natl Acad Sci* 1990;87:1988–1992.
60. Russell JT. The secretory vesicle in processing and secretion of neuropeptides. *Curr Top Membranes Transport* 1987;31,277–312.
61. Lagergrantz H. On the composition and function of large dense cored vesicles in sympathetic nerves. *Neuroscience* 1976;1:81–92.
62. Winkler H, Westhead E. The molecular organization of adrenal chromaffin granules. *Neuroscience* 1980;5:1803–1823.
63. Eipper BA, Mains RE. Peptide alpha-amidation. *Annu Rev Physiol* 1988;50:333–344.
64. Kent UM, Fleming PJ. Purified cytochrome b561 catalyzes transmembrane electron transfer for dopamine beta-hydroxylase and peptidyl glycine alpha-amidating monoxygenase activities in reconstituted systems. *J Biol Chem* 1987;262: 8174–8178.
65. Lamoroux A, Vigny A, Francon Bignet N, et al. The primary structure of human dopamine-beta-hydroxylase. Insight into the relationship between the soluble and the membrane bound forms of the enzyme. *EMBO J* 1987;6:3931–3937.
66. Rudnick G. ATP-driven H^+ pumping into intracellular organelles. *Annu Rev Physiol* 1986; 48:403–413.
67. Navone F, Jahn R, Di Gioia G, Stukenbrok H, Greengard P, De Camilli P. Protein p38: an integral membrane protein specific for small vesicles of neurons and neuroendocrine cells. *J Cell Biol* 1986;103:2511–2527.
68. Navone F, DiGioia G, Browning M, Greengard P, De Camilli P. Microvesicles of the neurohypophysis are biochemically related to small synaptic vesicles of presynaptic nerve terminals. *J Cell Biol* 1989;109:3415–3434.
69. Douglas WW, Nagasawa J, Schulz R. Electron microscopic studies on the mechanism of secretion of posterior pituitary hormones and significance of microvesicles (synaptic vesicles): evidence of secretion by exocytosis and formation of microvesicles as a by-product of this process. *Mem Soc Endocrinol* 1970;19:353–377.
70. Smith AD. Mechanisms involved in the release of noradrenaline from sympathetic nerve. *Br Med Bull* 1972;29;123–129.
71. Pearse AGE, Takov TT. Embryology of the diffuse neuroendocrine system and its relationship to the common peptides. *Fed Proc* 1979;38: 2288–2294.
72. Schmechel D, Marangos PJ, Brightman M. Neurone-specific enolase is a molecular marker for peripheral and central neuroendocrine cells. *Nature* 1978;276:834–836.
73. Rosa P, Hille A, Lee RVH, Zanini A, De Camilli P, Huttner W. Secretogranins I and II: two tyrosine-sulphated secretory proteins common to a variety of cells secreting peptides by the regulated pathway. *J Cell Biol* 1985;101:1999–2011.
74. De Camilli P, Jahn R. Pathways to regulated exocytosis in neurons. *Annu Rev Physiol* 1990;52:625–645.
75. Lowe AW, Madeddu L, Kelly RB. Endocrine secretory granules and neuronal synaptic vesicles have three integral membrane proteins in common. *J Cell Biol* 1988;106:51–59.
76. Johnston PA, Cameron PL, Stukenbrok H, Jahn R, De Camilli P, Sudhof TC. Synaptophysin expressed in CHO cells is targeted to microvesicles analogous to those of PC12 cells. *EMBO J* 1989;8:2863–2872.
77. Cameron PL, Johnston PA, Stukenbrok H, Jahn R, Sudhof TC, De Camilli P. Synaptophysin when expressed in CHO cells is targeted to the microvesicles which recycle transferrin receptors. (*Abstr.*) *Cell Biol Abstr* 1989;29:612.

Volume Transmission in the Brain: Novel Mechanisms for Neural Transmission,
edited by K. Fuxe and L.F. Agnati.
Raven Press, Ltd., New York © 1991.

14

Steroid Actions on Hypothalamic Neurons with Special Reference to Estrogen Control of Luteinizing Hormone–Releasing Hormone Biosynthesis and Release

George Fink, Roberta Rosie, and Elspeth Thomson

MRC Brain Metabolism Unit, University Department of Pharmacology, Edinburgh, Scotland

Steroid action on neurons may be classified as transient, long-term reversible, or permanent (1). Permanent actions are exemplified by the effect of steroids on sexual differentiation of brain. Long-term reversible effects are exemplified by stimulatory action of gonadal steroids on the vasopressinergic neurons of the bed nucleus of the stria terminalis (2,3) and by the way in which estrogen treatment lowers by three to fourfold the dose of alphaxalone required to anesthetize male rats (4). Transient actions can be further subdivided into those that occur within milliseconds to minutes [e.g., the action of the steroid anesthetic alphaxalone (4,5) and the inhibition of pulsatile luteinizing hormone (LH) and LH-releasing hormone (LHRH) release by estrogen (6)] and those that take hours. The latter, termed "intermediate," are exemplified by negative feedback actions of adrenocortical steroids and the stimulatory effects of gonadal steroids on LHRH release and motor and mating behavior (1).

Underlying the mechanism of sexual differentiation of the brain are key principles regarding neuronal determination, proliferation, and degeneration and synapse formation. Sexual differentiation of the brain involves massive morphological changes such as the fourfold greater size of the sexually dimorphic nucleus of the preoptic area in the male compared with the female rat (7). More subtle changes such as numbers of synapses (8) and sex differences in brain proteins (9,10) are also seen. However, the functional significance of these differences and the mechanism by which these sex differences in brain morphology and protein synthesis are produced by gonadal steroids remain to be elucidated. The long-term reversible and the intermediate effects of gonadal steroids also involve protein synthesis.

Whether protein synthesis is also involved in the fast transient action of steroids remains to be determined. Derivatives of progesterone do have direct effects on ion flux across membrane channels by way of GABA receptors (5), an effect that is likely to be direct on the cell membrane. However, because genomic transcription and translation can occur within minutes, changes in electrophysiological activity that occur minutes (or longer) after steroid challenge may also involve protein synthesis.

Here attention will be focused on a unique type of "intermediate" steroid action, the "positive feedback" of estrogen, which results in the release and increased biosynthesis of LHRH. The latter is a decapeptide that mediates the neural control of gonadotrophin biosynthesis and release and the "induction" and maintenance of pituitary gonadotrophs. The surge of LHRH, triggered by estradiol-17β, results in the surge of LH that causes ovulation. This is a *positive feedback* cascade in the true sense of the term because the cascade is terminated by the destruction (rupture of the ovarian follicle) of a major component of the feedback system. In the rat and man, the spontaneous preovulatory increase in estradiol-17β secretion by the ovary also stimulates the increased synthesis and release of prolactin, and the possible interactions between the gonadotrophin and prolactin control systems will be briefly discussed.

Our understanding of steroid action on hypothalamic pituitary function is still mainly at a physiological level. Molecular pharmacological studies designed to estab-

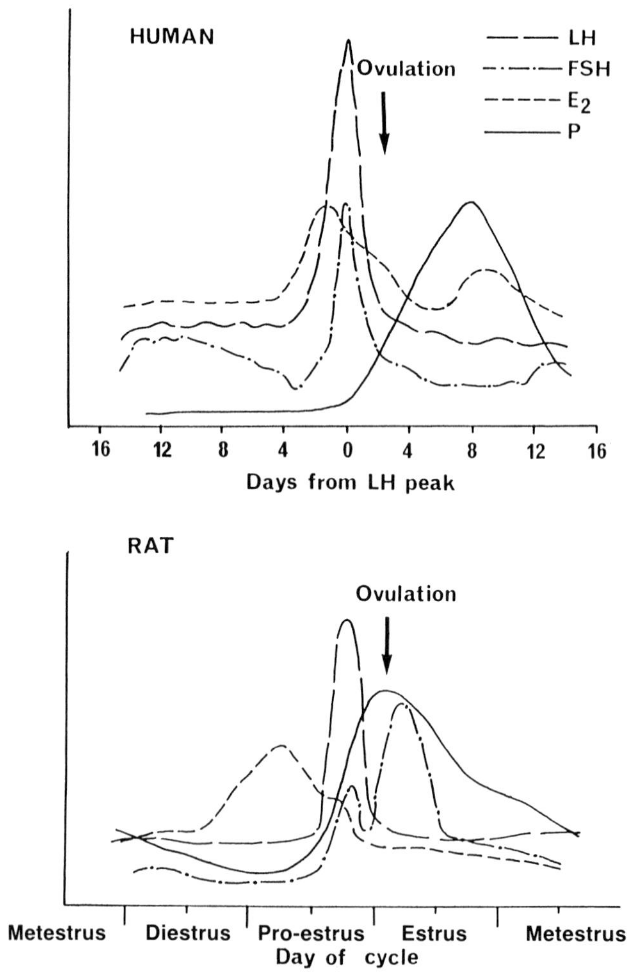

FIG. 1. Schematic diagram of key hormonal changes during human menstrual cycle and rat estrous cycle. Ovulation is preceded by a surge of luteinizing hormone (LH), which is triggered by a surge of estradiol-17β (E_2) and accompanied and followed by a surge of progesterone (P). Most of P is secreted by the ovary in response to LH, but in the rat a small amount of P is secreted by the adrenal gland just before onset of the LH surge. FSH, follicle-stimulating hormone. (From ref. 1, with permission.)

lish the mechanism of steroid action will provide important principles related to the neurotransmitter mechanisms, synaptic and "volume," that control the biosynthesis, release, and action of neurohormones that mediate the hypothalamic control of the anterior pituitary gland.

ESTROGEN TRIGGERS THE SPONTANEOUS OVULATORY GONADOTROPHIN SURGE: POSITIVE FEEDBACK CASCADE

The LH surge is the most dramatic spontaneous signal emitted by the hypothalamic-pituitary system and depends on resonance between neural and ovarian oscillators. The surge is arguably one of the most important events in the mammal because, without it, reproduction would not occur, and it is also unique in that it is produced by one of the few positive feedback cascades that occurs under physiological conditions. The mechanism of the LH surge has been most extensively studied in the rat, which, allowing for a much shorter cycle length, has hormonal profiles that resemble those that occur during the menstrual cycle in women (Fig. 1). The positive feedback cascade that triggers the LH surge has been reviewed in detail (11,12). Briefly, the cascade (Fig. 2) that leads to the LH surge is initiated by the increased secretion of estradiol-17β, which facilitates the expression of a neural signal for LH release in the form of a surge of LHRH and an increase in the responsiveness of the pituitary gonadotrophs to LHRH. The LH surge reaches a peak when increased concentrations of LHRH in hypophysial portal blood act on an anterior pituitary gland whose responsiveness has been further augmented by the priming effect of LHRH and by progesterone secreted during the LH surge. The probable role of the priming effect of LHRH, the capacity of LHRH to increase pituitary responsiveness to itself, is to coordinate increasing pituitary responsiveness with increasing concentrations of

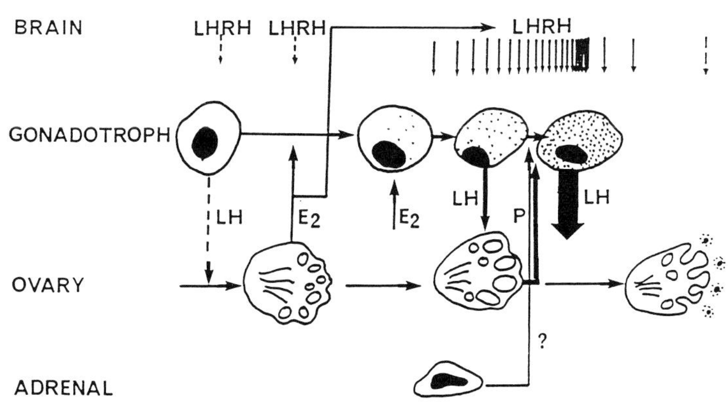

FIG. 2. Schematic diagram that shows the cascade of events that produce spontaneous ovulatory LH surge in the rat. Increase in plasma concentrations of estradiol-17β (E_2: the ovarian signal) increases responsiveness of pituitary gonadotrophs (increased stippling) to LHRH and also triggers the surge of LHRH. Pituitary responsiveness to LHRH is further increased by progesterone (P) secreted from the ovary in response to LH released during the early part of the LH surge and by the priming effect of LHRH, the unique capacity of the decapeptide to increase pituitary responsiveness to itself. The priming effect of LHRH coordinates surges of LHRH with increasing pituitary responsiveness so that the two events reach a peak at the same time. Conditions are thereby made optimal for a massive surge of LH. This cascade, which represents a form of positive feedback, is terminated by destruction of a major component of the system in the form of the rupture of ovarian follicles (ovulation). (From ref. 1, with permission.)

LHRH in hypophysial portal vessel blood so that both reach a peak at about the same time.

THE ESTROGEN-TRIGGERED SURGE OF LHRH

The LHRH neurons of the hypothalamus constitute the final common pathway for the neural signal. The neural signal occurs daily in the rat but is only expressed as a surge of LHRH in the presence of high plasma concentrations of estradiol. Depending on dosage, progesterone was found to have either no effect or to reduce the facilitatory effect of estradiol on LHRH release (13), and this suggests that the facilitatory effect of progesterone on the LH surge is mainly exerted at the level of the pituitary gland. The LHRH surge in hypophysial portal blood can only be detected under certain anesthetics, alphaxalone in the rat (14) and phencyclidine hydrochloride in the rhesus monkey (15).

The neural signal for LHRH release is probably generated in the anterior hypothalamus, which encompasses the medial preoptic area (MPOA) and suprachiasmatic nuclei and contains the cell bodies of most of the hypothalamic LHRH neurons. This view is supported by studies on the effects of electrical stimulation of the brain and hypothalamic deafferentation, the fact that estrogen implants in the preoptic area trigger the LH surge (16), and the fact that rats in which the major connections between hippocampus and amygdala and the hypothalamus have been severed show regular estrous cycles (17).

MECHANISM OF ACTION OF ESTROGEN IN TRIGGERING THE LHRH SURGE

As assessed by measurements of the LHRH release into portal blood estrogen significantly potentiates the responsiveness of the LHRH release system to electrical stimulation of the preoptic area (18,19). However, immunocytochemical/autoradiographic studies have shown that few, if any, LHRH neurons contain estrogen receptors (20). This suggests that estrogen does not have a direct action on LHRH neurons but rather triggers the LHRH surge by an action on other types of neurons such as noradrenergic (NA), dopaminergic (DA), and opioid, which might terminate on and affect the activity of LHRH neurons (21–28).

Most NA and adrenergic (A) nerve terminals in the hypothalamus are derived from ascending projections from nuclei in the midbrain and hindbrain. The precise role of these ascending tracts is not established (29). Overall, present data suggest that the ascending NA neurons stimulate LHRH release, but this action is modulatory and not mandatory for the occurrence of the ovulatory surge of LH (24). The NA control of pulsatile LH release differs from that of the surge release in that stimulation of the dorsal, but not the ventral, NA tract inhibits pulsatile LH release (29).

The most prominent group of DA neurons are the tuberoinfundibular group (so-called TIDA neurons), which have their cell bodies in the arcuate and ventromedial nuclei and project to the primary plexus of hypophysial portal vessels where they form close axoaxonic relations with the terminals of LHRH neurons. Although there are no structures that resemble classical synapses between the LHRH and TIDA terminals, it is generally assumed that DA released at TIDA terminals could inhibit LHRH release by a direct action on LHRH terminals. This action, comparable to presynaptic inhibition in other parts of the CNS, is an example of volume transmission in the CNS. There is experimental evidence for decreased DA turnover in the arcuate nucleus during the afternoon of proestrus (22–24), and estrogen significantly reduces the rate of transcription of the tyrosine hydroxylase gene in rat arcuate nucleus (25).

Other DA nerve terminals in the hypothalamus that could also influence the activity of LHRH neurons are derived from cell groups in the zona incerta as well as the midbrain and hindbrain (24,27).

Enkephalin and endorphin-containing neurons are situated in several hypothalamic nuclei but predominate in the region of the arcuate nucleus (28,30). As in the case of DA neurons, it is conceivable that the endorphin-containing neurons in the arcuate nucleus could influence LHRH release by an axoaxonic mechanism as well as by way of their terminals on the soma and/or dendrites of the LHRH neurons. Opioid binding sites are present on LHRH terminals in the medial basal hypothalamus (31), and the injection of anti-β-endorphin or antidynorphin serum into the arcuate nucleus produced a significant release of LH in immature rats (32). Estrogen treatment significantly reduces the amount of pro-opiomelanocortin (POMC) messenger ribonucleic acid (mRNA) in the arcuate nucleus (30). These, together with other data, suggest that estrogen could stimulate the LHRH surge by reducing POMC synthesis, thereby reducing the amount of β-endorphin available for release onto LHRH terminals where it is presumed that the opioid exerts an inhibitory effect.

Thus, present data suggest that the LHRH/LH surge is stimulated by central NA and inhibited by central opioid and DA mechanisms and that the stimulatory effect of estrogen on LHRH release could be brought about by stimulation of the facilitatory mechanisms, inhibition of the inhibitory mechanisms, or both. Figure 3 illustrates several possible modes for the way in which monoaminergic and opioid neurons could modulate LHRH release and biosynthesis (see below). This scheme, it must be stressed, is necessarily oversimplified and does not take account of the facts that other transmitters such as GABA (33–38) or neuropeptides such as neuropeptide Y (39–41) may influence the activity of LHRH neurons and/or the gonadotrophs.

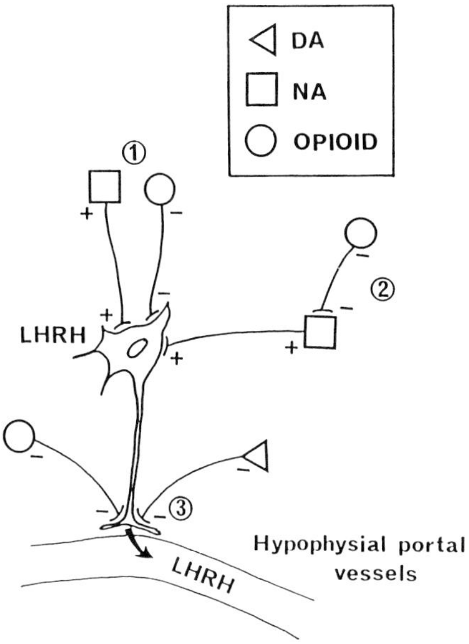

FIG. 3. Schematic diagram of how estrogen could trigger LHRH surge. LHRH neurons have no estrogen receptors, and therefore estrogen probably triggers the LHRH surge by acting on "interneurons" that modulate activity of LHRH neurons. The + and − signs indicate stimulatory and inhibitory effects, respectively, of estrogen at the cell bodies and of the modulatory neurons at synapses or points of contact. Three of many different possible mechanisms are shown. *Scheme 1* shows that estrogen could stimulate noradrenergic neurons, which stimulate LHRH neurons, or inhibit opioid neurons, which inhibit LHRH neurons. *Scheme 2* shows that estrogen could inhibit opioid neurons, which could inhibit noradrenergic neurons: this disinhibition of noradrenergic neurons would effect an increase in firing of LHRH neurons. *Scheme 3* shows that estrogen could inhibit activity of arcuate dopaminergic and opioid neurons, some of whose terminals in the median eminence are juxtaposed to and may have a "presynaptic" inhibitory effect on terminals of LHRH neurons. Again, LHRH release would occur as a consequence of disinhibition of LHRH release. It must be stressed that this diagram is highly oversimplified and does not take into account other types of neurotransmitters, especially GABA and neuropeptide Y (see text), or other types of circuits by which the activity of LHRH neurons could be altered. DA, dopamine; NA, norepinephrine. (From ref. 1, with permission.)

SIGNIFICANCE OF LATENCY BETWEEN ONSET OF ESTRADIOL AND LHRH/LH SURGES

Elevated plasma concentrations of estrogen take about 26 to 28 hr to trigger the surges of LHRH and LH in the rat and at least 36 hr in the rhesus monkey (1,42). This latency between the onset of the estradiol and LHRH surges would be more than sufficient for RNA and protein synthesis and even for structural changes in neurons or neuronal connectivity. This latency is of the order also required for estrogen to increase the firing of hypothalamic units in the rat (43–45) and rhesus monkey (46). Estrogen also takes a relatively long time to increase the firing rate of hypothalamic units involved in mating behavior (47) and units in the neostriatum (48). The mechanism underlying the latency of estrogen action on neurons is not known but will offer an important clue as to how steroids affect neuronal connectivity and function within the organized and relatively precise neuronal system that underlies the LHRH surge and mating.

CHANGES IN PITUITARY RESPONSIVENESS TO LUTEINIZING HORMONE–RELEASING HORMONE

Pituitary responsiveness to LHRH increases significantly before and during the spontaneous LH surge in women, monkeys, sheep, hamsters, and rats (1,42). Because the spontaneous surge of LHRH (above) is far too small by itself to produce the ovulatory surge of LH, the massive (20–50-fold) increase in pituitary responsiveness to LHRH is crucial for the LH surge (49). The precise mechanism by which estrogen and progesterone increase pituitary responsiveness to LHRH remains to be determined. However, the key points (reviewed in detail in ref. 50) are as follows:

1. There is no molecular biological evidence that estrogen stimulates LH biosynthesis (51), and therefore the most likely direct effect of estrogen on the pituitary gonadotrophs is to initiate a cascade of intracellular events that increase the "sensitivity" of the gonadotrophs to LHRH. Studies with antiestrogens as well as steroid manipulation show that this action of estrogen is mandatory for increased pituitary responsiveness to LHRH (52).
2. Studies using dispersed rat pituitary cells or rats in which the pituitary stalk has been cut show that estrogen and progesterone can facilitate pituitary responsiveness by a direct action on the gonadotrophs.
3. The nuclear and total concentrations of estradiol-receptor complexes in the pituitary are greater at proestrus than at other times of the cycle, and progesterone receptors in the pituitary are estrogen-dependent.
4. As well as a direct action on pituitary gonadotrophs, several data suggest that at least part of estradiol's action in increasing pituitary responsiveness to LHRH may depend on "normal" or increased concentrations of LHRH in hypophysial portal blood, which increases pituitary responsiveness by way of the priming effect of LHRH. This bears on the fact that there is a marked sex difference in the effect of estrogen on the LH response to LHRH (53,54), which was puzzling because there is good evidence that in the rat, sexual differentiation of the hypothalamic-pituitary-ovarian system is due to changes predominantly in the brain (55). However, if at least part of the action of estrogen in increasing pituitary responsiveness to LHRH requires normal function of the hypothalamic LHRH system, this could explain why there is a marked sex difference in the effect of estrogen on pituitary responsiveness to LHRH.

POSSIBLE RELATIONSHIP BETWEEN ESTROGEN EFFECTS ON THE BRAIN AND THE PRIMING EFFECT OF LHRH ON THE PITUITARY

There are major differences between the priming and the releasing action of LHRH, the most prominent of which is that the priming but not the releasing action of LHRH involves the synthesis of a new protein and depends on the integrity of microfilaments (56). The new protein synthesized as part of LHRH priming has a relative molecular mass of about 70 kD (56) and comigrates, on two-dimensional gel electrophoresis, with an estrogen-stimulated protein extracted from the hypothalamic ventromedial nucleus (57). The precise nature and function of this protein has yet to be determined, but it could provide a common mechanism for the action of estradiol-17β in the brain [e.g., involved in estrogen-induced lordosis behavior (58)] and pituitary [margination of secretory granules in pituitary gonadotrophs (59,60, and below)] and may provide a common point of action of estradiol and LHRH.

Ultrastructural morphometry showed that LHRH priming is associated with a change in orientation and length of the microfilaments and a significant migration of gonadotroph secretory granules to a "marginal" zone close to the plasmalemma from where the granules can easily be released as a consequence of a second exposure to a secretagogue such as K^+ depolarization, Ca^{2+} ionophores, or LHRH (59,60). This migration of secretory granules, presumably caused by changes in the contractile elements of the cell, may be an important factor in the increase in the readily releasable pool of LH that occurs before and during the spontaneous ovulatory LH surge (61). Evidence in support of this hypothesis comes from studies in testicular feminized mice in which there is spontaneous, massive margination of secretory granules in the gonadotrophs and in which the response to LHRH is remarkably high (62).

It remains to be determined whether estrogen has similar effects as LHRH on the cytoskeleton and the movement of secretory granules and whether the pituitary gonadotroph model is applicable to neurotransmitter transport and release in neurons.

ESTROGEN STIMULATION OF LHRH BIOSYNTHESIS

We have recently investigated whether estrogen in its positive feedback mode stimulates the biosynthesis as well as the release of LHRH (63). The small amount of LHRH mRNA in hypothalamus (as assessed by immunocytochemistry, there are only about 500 LHRH-containing neurons in the rat hypothalamus) made it necessary for us to resort to *in situ* hybridization. We carried out the study in an experimental rat preparation that has been shown to be a robust model for the positive feedback action of estradiol-17β in triggering the LH and LHRH surge (13,52,54,64). Briefly, female rats that showed regular 4-day estrous cycles were ovariectomized under halothane anesthesia between 0900 hr and 1100 hr of diestrus. Immediately after ovariectomy, four rats were injected subcutaneously with 10 μg estradiol benzoate (EB) in 0.2 ml arachis oil and another four with 0.2 ml arachis oil alone. Animals were killed by cervical dislocation between 1600 hr and 1700 hr of the following day (presumptive proestrus) just before or during the early part of the expected LHRH and LH surges.

Figure 4 shows that the numbers and distribution of cells expressing LHRH mRNA closely resembled those of previous immunohistochemical and *in situ* hybridization studies (65–67). The highest concentrations of cells were located within the MPOA and, to a lesser extent, within the lateral POA and the diagonal band of Broca. There was no difference in the numbers and distribution of LHRH mRNA expressing cells in

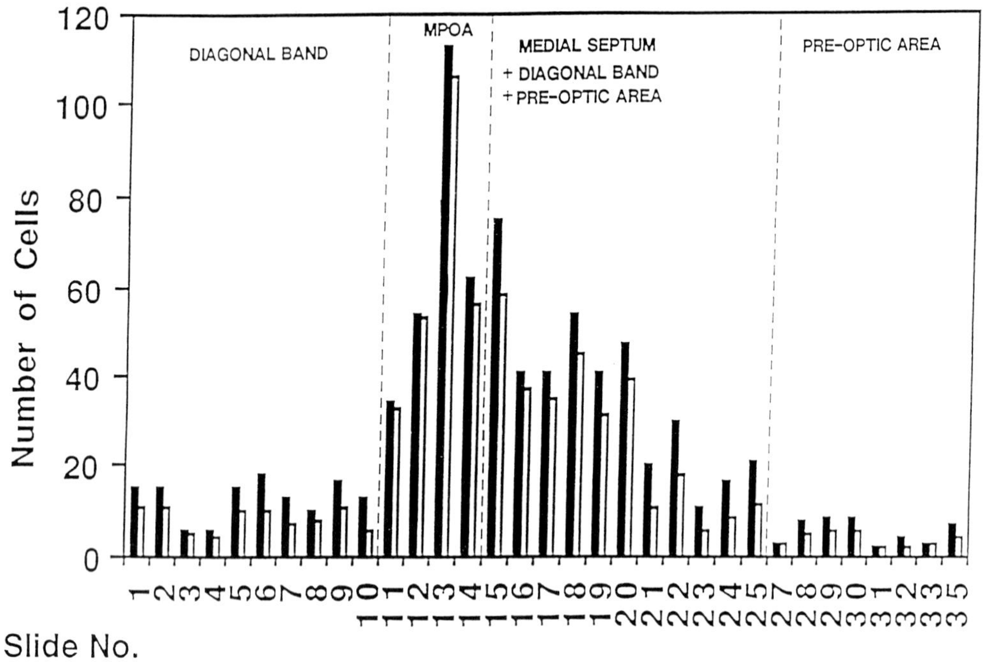

FIG. 4. *In situ* hybridization for LHRH mRNA in rat brain. Total cell count and anatomical distribution of LHRH mRNA expressing cells within the hypothalamus of the adult female rat, showing that the cells were situated mainly in the diagonal band of Broca, medial preoptic area (MPOA), lateral preoptic area, and medial septum. *Solid bars* represent the gross number of cells per slide, and *open bars* represent the net number (discounts parts of cells counted on adjacent sections) of cells per slide. The LHRH mRNA was detected using a 30 base oligonucleotide probe complementary to the sequence for the LHRH decapeptide domain. The location of cells was assessed with the aid of a stereotactic atlas (82). (For details, see ref. 63, from which this figure is reproduced with permission.)

EB compared with oil-treated control animals. No positive cells were detected when sections were pretreated with RNase A or when the LHRH probe was replaced with the complementary sense strand oligonucleotide.

To determine whether there was a difference between LHRH mRNA expression in EB compared with oil-treated rats, the numbers of silver grains were counted in 40 to 50 cells from each of the oil- and EB-treated brains. These cells were selected randomly in nine to 12 brain sections at the peak of LHRH cell concentration (i.e., mainly the MPOA) (Fig. 4). Brain sections and the anatomical locations of cells for grain counting in EB- and oil-treated brains were carefully matched. Figure 5 shows that in each of four pairs of experimentally matched brains significantly ($p < 0.001$; chi-square) more cells had higher grain counts in the EB-treated compared with oil-treated rats. The grain counts per cell were also significantly greater ($p < 0.002$; Mann-Whitney U test) in EB-treated compared with oil-treated rats (Figs. 5 and 6). The plasma LH concentrations showed that as expected, EB had triggered the LH surge (Fig. 5).

These results show for the first time that as well as stimulating the release of LHRH, estradiol in its positive feedback mode also stimulates the synthesis of LHRH mRNA. There are no reports of a comprehensive survey of LHRH mRNA expression in all hypothalamic LHRH neurons through the

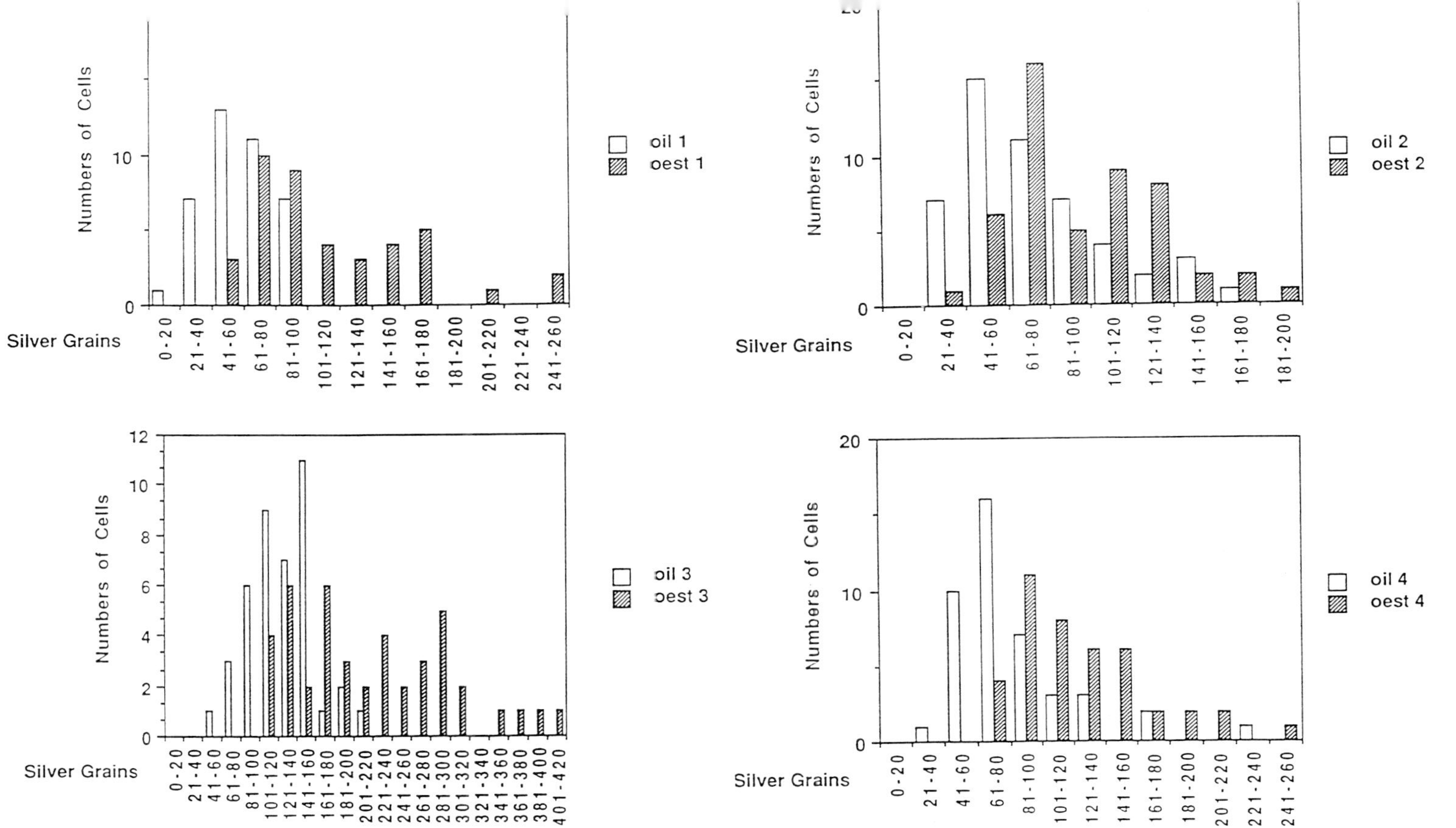

FIG. 5. *In situ* hybridization for LHRH mRNA in rat brain. Frequency distribution histograms of cells expressing LHRH mRNA in four estradiol benzoate (oest)-treated animals and four oil-treated animals. Each panel shows the distribution of LHRH mRNA-containing cells in the MPOA from one oil- treated (*open bars*) and one oest-treated (*hatched bars*) animal in which the brains were processed together. Abscissa shows numbers of silver grains per cell, and the ordinate, numbers of cells expressing LHRH mRNA per frequency interval. Mean ($\pm$ SEM) grain counts per cell was 138 $\pm$ 32 in oest compared with 86 $\pm$ 16 in oil-treated rats. Plasma LH concentrations, in trunk blood, ranged from 2.7 to 3.8 ng NIH-LH-S18/ml in oil-treated rats; in oest-treated rats, values were 12.0, 17.0, 2.0, and 25.5 ng/ml for rats 1 to 4, respectively. Increases in plasma LH in three of the four oest-treated rats confirmed our expectation that in this preparation, estradiol was acting in its positive feedback mode. Animal 3 was probably killed just before the onset of LH surge: its vaginal smear was as heavily cornified as that of the other oest-treated animals. (From ref. 63, with permission.)

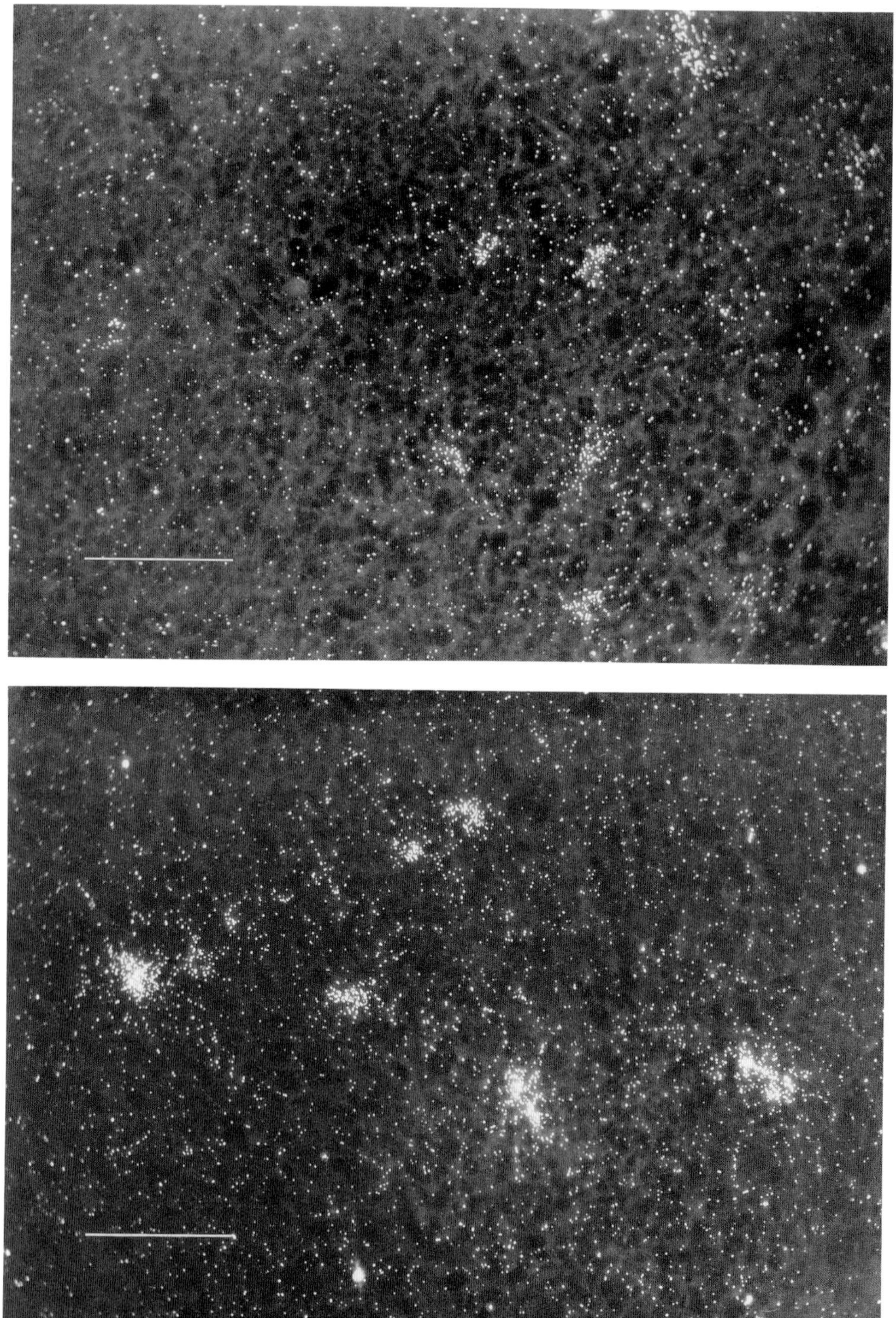

A

B

FIG. 6. *In situ* hybridization for LHRH mRNA. Dark-field autoradiographs of matched coronal sections through the MPOA from an oil- (**a**) and an EB-treated (**b**) rat. Note density of silver grains is much greater in the EB-treated compared with the oil-treated animal. Bars indicate 100 μm.

estrous cycle. However, LHRH mRNA concentrations of neurons at the level of the organum vasculosum of lamina terminalis increased twofold between the morning and evening of proestrus (68). Our findings suggest that this proestrous increase in LHRH mRNA is due to the preovulatory surge of estradiol-17β. Cellular LHRH mRNA concentrations were also increased in ovariectomized rats treated with estrogen for 7 days (66). Although plasma LH concentrations were not measured in the latter study, it is likely that the experimental paradigm used resulted in daily surges of LH (and LHRH) (6) and that, therefore, estrogen was also acting in its positive feedback mode. For reasons given above in relation to LHRH release, the effect of estradiol on LHRH mRNA synthesis is likely to be mediated by "interneurons" that do have estradiol receptors and that modulate LHRH neurons. Detailed pharmacological studies will determine which type(s) of interneurons mediate the action of estrogen in stimulating LHRH mRNA synthesis.

STEROID-INDUCED PROLACTIN SURGE

In the rat and man a spontaneous surge of prolactin occurs coincidentally with the spontaneous proestrous surge of LH. In contrast to the LH surge, several hypothalamic factors may mediate the neural control of prolactin release.

Early studies by Everett and Nikitovitch-Winer on pituitary grafts to the kidney capsule (69) suggested that the CNS exerted an inhibitory effect on prolactin secretion. So far, DA remains the favorite and most convincing candidate for a prolactin-inhibiting factor (PIF) (70). There is also pharmacological evidence for prolactin-releasing factors, the favorite candidates being thyrotropin-releasing factor (TRH), vasoactive intestinal peptide, and PHI (70). However, although DA agonists and antagonists and TRH are used extensively in humans to stimulate or inhibit prolactin release, robust evidence that these factors are involved in the physiological control of prolactin release awaits further studies (50,71). Thus, for example, no DA could be detected in sheep hypophysial portal blood, and the concentrations of DA in rat portal blood are not inversely related (as would be expected if DA is *the* PIF) to the peripheral plasma concentrations of prolactin (71). The inhibition of prolactin by DA may involve DA released from nerve terminals in the distal pituitary stalk or neurointermediate lobe and/or conveyed to the pars distalis by the short portal vessels (70,72).

The control of prolactin release is further complicated by the fact that estrogen stimulates prolactin secretion by stimulating prolactin mRNA synthesis in prolactotrophs (73,74). The stimulatory action of estrogen on pituitary prolactin secretion overrides inhibition of prolactin secretion by DA and its agonists (75). In addition to stimulating prolactin synthesis and release by a direct action on the prolactotrophs, estrogen also augments the prolactin response to TRH (76). These data could be interpreted to mean that the spontaneous proestrous surge of prolactin is mainly due to the stimulatory action of estrogen on the pituitary gland. However, this simple explanation is unlikely to be correct. Rather, for reasons detailed previously (50), it seems likely that the spontaneous prolactin surge is a consequence of the activation of central prolactin stimulating systems and/or the inhibition of central prolactin-inhibiting systems by estradiol and possibly progesterone.

THEORETICAL MODEL OF THE CONTROL OF THE LHRH/LH AND PROLACTIN SURGES AND LHRH BIOSYNTHESIS

The data reviewed above enable us to develop working models that provide the basis for more sophisticated molecular pharmacological studies designed to elucidate the circuits and transmitter mechanisms involved in the estrogen-induced LHRH/LH and prolactin surges and LHRH biosynthesis. Several different model systems of varying complexity are possible. Figures 7 and 8 show a relatively simple and parsimonious model, which allows for the fact that estrogen, and possibly progesterone, may activate the same surge generator "module," which then triggers the LH and prolactin surges through "subsystems" specific for each hormone. The steroids also act at the anterior pituitary gland to increase pituitary responsiveness to LHRH. Similar increases in gain may also occur in the neural components of the surge generator and the LH and prolactin subsystems, a point illustrated by the estrogen-induced increase in responsiveness of MPOA-LHRH neurons to electrical stimulation (18,19) and the estrogen-induced increase in firing of certain central neurons (43–46). The precise neurotransmitters and pathways involved in the systems are unknown. However, there is reasonable evidence that implicates serotonergic, NA, opioid, and/or gamma-aminobutyric acid (GABA) neurons in the surge command system. Within the subsystems for LH and prolactin release it is con-

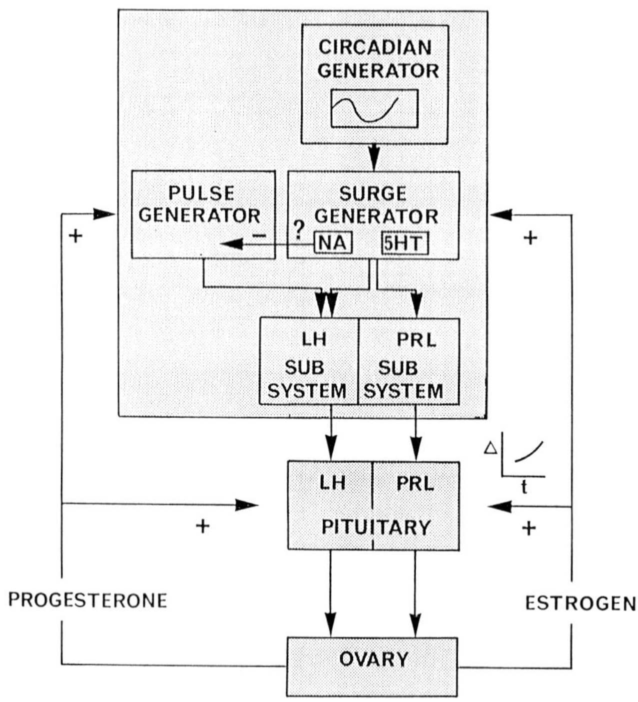

FIG. 7. Highly simplified, hypothetical model of main components in the system that controls gonadotrophin and prolactin surges. In the CNS there is a surge generator that drives LH and prolactin (PRL) control subsystems. Activity of the surge generator is influenced to a large extent by a circadian rhythm generator. The surge generator is probably linked to the pulse generator, and pathways that are involved in generating the surge may switch off the pulse generator (e.g., ascending noradrenergic systems). Estrogen and progesterone trigger the surge generator, but the time at which this generator fires is determined by the activity of the circadian generator. Estradiol and, to a lesser extent for LH, progesterone, activate the surge generator. Steroids also increase the responsiveness of the anterior pituitary gland to LHRH (for details see text). NA, norepinephrine; 5-HT, serotonin. (From ref. 1, with permission.)

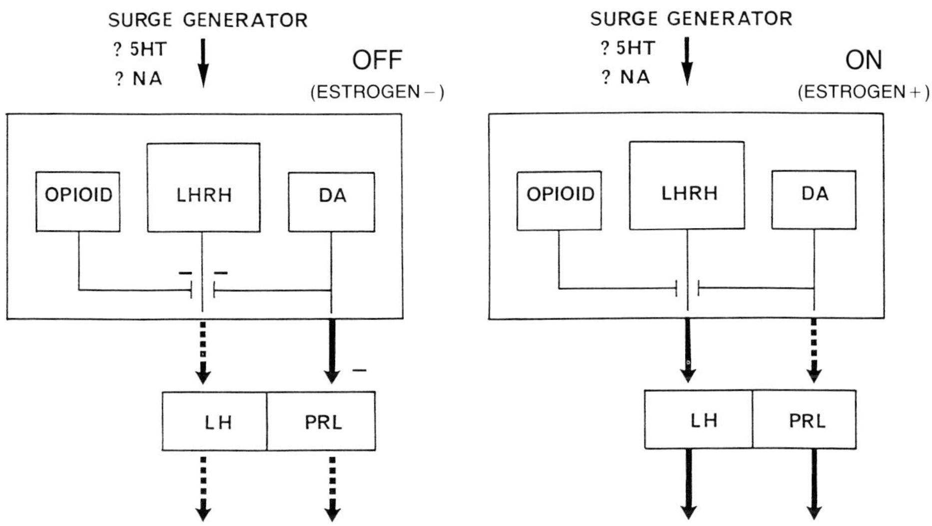

FIG. 8. A highly simplified diagram illustrating the hypothesis for how estradiol could, by inhibiting dopamine (DA) neurons, produce both a surge of LH and prolactin (PRL). Note that DA control of LHRH release has a back-up in the form of the opioid system, which may add precision to the LHRH surge. Other possible mechanisms are shown in Fig. 3. 5-HT, serotonin; NA, norepinephrine. (From ref. 1, with permission.)

ceivable that at least one unitary mechanism may operate (Fig. 8). That is, estrogen could reduce the activity of the DA neurons in the arcuate nucleus, which would have the effect of disinhibiting LHRH neurons and reducing the inhibition of the prolactotrophs. As a consequence, a surge of LH and prolactin would occur. Similarly, inhibition by estradiol-17β of the arcuate POMC neurons, which also inhibit LHRH release, would also disinhibit the LHRH neurons, resulting in increased LHRH release. Collateral evidence for the importance of disinhibition as a mechanism for the LH surge is provided by data from [^{14}C]2-deoxyglucose studies in which we found that the relative metabolic activity of brain areas involved in neuroendocrine control (specifically, the arcuate nucleus, preoptic and anterior hypothalamic areas, the median eminence, and amygdala) decreased rather than increased around the time of the proestrous surge of LH (77).

Figure 7 also shows that the control system would have to allow for a drive from the circadian oscillator and a switch that changes the output from the pulsatile mode, which operates during basal secretion of the gonadotrophins, to a surge mode. Finally, the control system would have to incorporate a mechanism that allows a switch from the spontaneous to the reflex (in response to copulation) surge mode. Such a switch occurs in female rats exposed to constant light or treated neonatally with androgens (androgenized rats) (78). The surge generator cannot operate in rats exposed to constant light or in androgenized rats, but the pulse generator continues to operate at a reduced frequency and amplitude (78). The switch from the spontaneous to reflex surge mode also involves a switch from estradiol to progesterone as the main trigger for the LH surge (78). This is possibly because the brain-pituitary system in the rats exposed to constant light and in androgenized rats has been "sensitized" to progesterone by the relatively high plasma estradiol concentrations in these preparations.

Thus, the key to unraveling the neural mechanisms responsible for generating the estrogen-induced LH and prolactin surges and LHRH biosynthesis is to determine which type(s) of neuron is the primary target for estradiol-17β. Clearly, estrogen is likely to affect a variety of processes in a variety of neuronal cell types, and therefore, to establish the mechanism involved, strict attention will have to be paid to the latency of action and concentrations at which estrogen exerts its effects. For this purpose the model used in our studies on LHRH biosynthesis (above, and ref. 63) and release (13,64) would seem to be physiologically appropriate.

Which parts of the LH and prolactin surge control system involve volume or synaptic transmission also awaits further studies. However, the absence of axoaxonic synapses in the median eminence (79,80) and the close juxtaposition of LHRH, DA, and opioid terminals suggest that the DA/opioid "presynaptic" clamp of LHRH release (Fig. 8) may mainly depend on volume transmission; transmitter actions on LHRH perikarya and dendrites could involve both synaptic (38,81) and volume transmission.

CONCLUSIONS

Steroid action on central neurons is one of the best examples of volume transmission, in that gonadal and adrenocortical steroids are secreted remote from their specific receptors in target CNS neurons. Steroids must reach their target neurons by way of the extracellular brain space.

Gonadal steroids exert potent irreversible effects on brain differentiation as well as long-term reversible effects. At the other end of the scale, some steroids exert rapid (10^{-3}–10^{2} sec) effects on neuronal activity. Here attention is focused on steroid action, intermediate in type, which has a latency of hours and decay time of minutes to hours. The examples used are the preovulatory surges of LH and prolactin. These are precise, spontaneous, neuroendocrine signals and are triggered by a surge of estradiol-17β. The LH surge is the result of a positive feedback cascade in which estradiol-17β, secreted as a result of follicle-stimulating hormone (FSH) and LH stimulation of the ovary, triggers a surge of LHRH and initiates an increase in pituitary responsiveness to LHRH. Pituitary responsiveness is further enhanced by the priming effect of LHRH, the capacity of LHRH to increase pituitary responsiveness to itself. The surge of estradiol-17β also stimulates the synthesis of LHRH mRNA in hypothalamic neurons. LHRH neurons contain few, if any, estradiol receptors, and so the action of estrogen, both on LHRH release and biosynthesis, is likely to be mediated by one or more "interneurons." The mechanism of the estrogen-induced prolactin surge is less well understood because the precise nature of the neurohormones that mediate neural control of prolactin secretion has yet to be determined. A relatively simple, parsimonious model is proposed to explain the mechanism of the LH and prolactin surge generators. One component of this model, a DA and opioid clamp of LHRH neurons at a "presynaptic" level, would involve volume transmission of DA and/or opioids from their respective nerve terminals to receptors on LHRH nerve terminals in the median eminence. The LH/prolactin surge generator provides an excellent model for studying the potent neuromodulation by steroids of volume and synaptic transmission in the CNS.

ACKNOWLEDGMENTS

We thank Ann Bradley for preparing the typescript. Some parts of the review are based on updated versions of previous articles by the authors in the Journal of Endocrinology (Journal of Endocrinology Ltd), the Journal of Steroid Biochemistry (Pergamon Press), and the Quarterly Journal of Physiology (Cambridge University Press).

REFERENCES

1. Fink G. The GW Harris lecture. Steroid control of brain and pituitary function. *Q J Exp Physiol* 1988;73:257–293.
2. de Vries GJ, Buijs RM, van Leeuwen FW. Sex differences in vasopressin and other neurotransmitter systems in the brain. In: de Vries GJ, de Bruin JPC, Uylings HBM, Corner MA, eds. *Progress in brain research,* vol. 1. Amsterdam: Elsevier Science Publishers BV, 1984;185–203.
3. Mayes CR, Watts AG, McQueen JK, Fink G, Charlton HM. Gonadal steroids influence neurophysin-II distribution in the forebrains of normal and mutant mice. *Neuroscience* 1987;25: 1013–1022.
4. Fink G, Sarkar DK, Dow RC, et al. Sex differences in response to alphaxalone anaesthesia may be oestrogen dependent. *Nature* 1982; 298:270–272.
5. Lambert JJ, Peters JA, Cottrell GA. Actions of synthetic and endogenous steroids on the $GABA_A$ receptors. *TIPS* 1987;8:224–227.
6. Sarkar DK, Fink G. Luteinizing hormone releasing factor in pituitary stalk plasma from long-term ovariectomized rats: effects of steroids. *J Endocrinol* 1980;86:511–524.
7. Gorski RA. Critical role for the medial preoptic area in the sexual differentiation of the brain. In: de Vries GJ, de Bruin JPC, Uylings HBM, Corner MA, eds. *Progress in brain research,* vol 1. Amsterdam: Elsevier Science Publishers BV, 1984;129–145.
8. Raisman G, Field PM. Sexual dimorphism in the neuropil of the preoptic area of the rat and its dependence on neonatal androgen. *Brain Res* 1973,54.1–29.
9. Stanley HF, Fink G. Synthesis of specific brain proteins is influenced by testosterone at mRNA level in the neonatal rat. *Brain Res* 1986; 370:223–231.
10. Stanley HF, Borthwick NM, Fink G. Brain protein changes during development and sexual differentiation in the rat. *Brain Res* 1986;370:215–222.
11. Fink G. Neuroendocrine control of gonadotrophin secretion. *Br Med Bull* 1979;35:155–160.
12. Fink G. Gonadotropin secretion and its control. In: Knobil E, Neill J, eds. *The physiology of reproduction,* vols 1 and 2. New York: Raven Press, 1988;1349–1377.
13. Fink G, Sarkar DK, Chiappa SA. Gonadotrophin releasing hormone surge during pro–estrus: role of steroid hormones. *J Endocrinol* 1977;75:46–47.
14. Sherwood NM, Chiappa SA, Sarkar DK, Fink G. Gonadotropin-releasing hormone (GnRH) in pituitary stalk blood from proestrous rats: effects of anesthetics and relationship between stored and released GnRH and luteinizing hormone. *Endocrinology* 1980;107:1410–1417.
15. Neill JD, Patton JM, Dailey RA, Tsou RC, Tindall GT. Luteinizing hormone releasing hormone (LHRH) in pituitary stalk blood of rhesus monkeys: relationship to level of LH release. *Endocrinology* 1977;101:430–434.
16. Goodman RL. The site of the positive feedback action of estradiol in the rat. *Endocrinology* 1978;102:151–159.
17. Chaippa SA, Fink G, Sherwood NM. Immunoreactive luteinizing hormone releasing factor (LRF) in pituitary stalk plasma from female rats: effects of stimulating diencephalon, hippocampus and amygdala. *J Physiol (Lond)* 1977;267: 625–640.
18. Sherwood NM, Chiappa SA, Fink G. Immunoreactive luteinizing hormone releasing factor in pituitary stalk blood from female rats; sex steroid modulation of response to electrical stimulation of preoptic area or median eminence. *J Endocrinol* 1976;70:501–511.
19. Fink G, Jamieson MG. Immunoreactive luteinizing hormone releasing factor in rat pituitary stalk blood: effects of electrical stimulation of the medial preoptic area. *J Endocrinol* 1976;68: 71–87.
20. Shivers BD, Harlan RE, Morell JI, Pfaff DW. Absence of oestradiol concentration in cell nuclei of LHRH-immunoreactive neurones. *Nature* 1983;304:345–347.
21. Sarkar DK, Fink G. Gonadotropin-releasing hormone surge: possible modulation through postsynaptic α-adrenoreceptors and two pharmacologically distinct dopamine receptors. *Endocrinology* 1981;108:862–867.
22. Barraclough CA, Wise PM, Selmanoff MK. A role of hypothalamic catecholamines in the regulation of gonadotropin secretion. *Recent Prog Horm Res* 1984;40:487–521.
23. Kalra SP, Kalra PS. Neural regulation of luteinizing hormone secretion in the rat. *Endocrinol Rev* 1983;4:311–351.
24. Fink G, Stanley HF, Watts AG. Central nervous control of sex and gonadotropin release: peptide and nonpeptide transmitter interactions. In: Krieger D, Brownstein M, Martin J, eds. *Brain peptides* New York: John Wiley & Sons, 1983; 413–435.
25. Blum M, McEwen BS, Roberts JL. Transcriptional analysis of tyrosine hydroxylase gene expression in the tuberoinfundibular dopaminergic neurons of the rat arcuate nucleus after estrogen treatment. *J Biol Chem* 1987;262:817–821.
26. Akabori A, Barraclough CA. Effects of morphine on luteinizing hormone secretion and catecholamine turnover in the hypothalamus of estrogen-treated rats. *Brain Res* 1986;362:221–226.
27. MacKenzie FJ, Hunter AJ, Daly C, Wilson CA. Evidence that the dopaminergic incerto-hypothalamic tract has a stimulatory effect on ovulation and gonadotrophin release. *Neuroendocrinology* 1984;39:289–295.
28. Ferin M, Van Vugt D, Wardlaw S. The hypothalamic control of the menstrual cycle and the role of endogenous opioid peptides. *Recent Prog Horm Res* 1984;40:441–480.
29. Bergen H, Leung PCK. Electrical stimulation of ventral versus dorsal mesencephalic segmental

areas in the conscious rat: effects on luteinizing hormone release. *Neuroendocrinology* 1978;46: 117–124.

30. Roberts JL, Wilcox JN, Blum M. The regulation of proopiomelanocortin gene expression by estrogen in the rat hypothalamus. In: Fink G, Harmar AJ, McKerns KW, eds. *Neuroendocrine molecular biology*. New York: Plenum Press, 1986;261–270.
31. Rostene WH, Drouva SV, Pollard H, Sokoloff P, Pattou E, Kordon C. Further evidence for the existence of opiate binding sites on neurosecretory LHRH mediobasal hypothalamic terminals. *Eur J Pharmacol* 1982;80:139–141.
32. Schulz R, Wilhelm A, Pirke KM, Gramsch C, Herz A. β-Endorphin and dynorphin control serum luteinizing hormone level in immature rats. *Nature* 1981;294:757–759.
33. Mitchell R, Grieve G, Dow RC, Fink G. Endogenous GABA receptor ligands in hypophysial portal blood. *Neuroendocrinology* 1983;37:169–176.
34. Jennes L, Stumpf WE, Tappaz ML. Anatomical relationships of dopaminergic and GABAergic systems with the GnRH-systems in the septohypothalamic area. *Exp Brain Res* 1983;50:91–99.
35. Fuchs E, Mansky T, Stock K-W, Vijayan E, Wuttke W. Involvement of catecholamines and glutamate in GABAergic mechanism regulatory to luteinizing hormone and prolactin secretion. *Neuroendocrinology* 1984;38:484–489.
36. Tappaz ML, Bosler O, Paut L, Berod A. Glutamate decarboxylase-immunoreactive boutons in synaptic contacts with hypothalamic dopaminergic cells: a light and electron microscopy study combining immunocytochemistry and radioautography. *Neuroscience* 1985;16:111–122.
37. Leranth CS, Sakamoto H, MacLusky NJ, Shanabrough M, Naftolin F. Application of avidin-ferritin and peroxidase as contrasting electron-dense markers for simultaneous electron microscopic immunocytochemical labelling of glutamic acid decarboxylase and tyrosine hydroxylase in the rat arcuate nucleus. *Histochemistry* 1985;82:165–168.
38. Leranth C, MacLusky NJ, Sakamoto H, Shanabrough M, Naftolin F. Glutamic acid decarboxylase-containing axons synapse on LHRH neurons in the rat medial preoptic area. *Neuroendocrinology* 1985;40:536–539.
39. McCann SM, Samson WK, Aguila MC, et al. The role of brain peptides in the control of anterior pituitary hormone secretion. In: Fink G, Harmar AJ, McKerns KW, eds. *Neuroendocrine molecular biology*, New York: Plenum Press, 1986;101–111.
40. Crowley WR, Kalra SP. Neuropeptide Y stimulates the release of luteinizing-releasing hormone from medial basal hypothalamus *in vitro:* modulation by ovarian hormones. *Neuroendocrinology* 1987;46:97–103.
41. Sutton SW, Mitsugi N, Plotsky PM, Sarkar D. Neuropeptide Y (NPY): a possible role in the initiation of puberty. *Endocrinology* 1988;123: 2152–2154.
42. Knobil E. The neuroendocrine control of the menstrual cycle. *Prog Horm Res* 1986;36:53–88.
43. Cross BA. Unit responses in the hypothalamus. In: Ganong WF, Martini L, eds. *Frontiers in neuroendocrinology*. Oxford: Oxford University Press, 1973;133–171.
44. Hayward JN. Functional and morphological aspects of hypothalamic neurons. *Physiol Rev* 1977;57:574–658.
45. Dyer RG. An electrophysiological dissection of the hypothalamic regions which regulate the preovulatory secretion of luteinizing hormone in the rat. *J Physiol (Lond)* 1973;234:421–442.
46. Yeomen RR, Terasawa E. An increase in single unit activity of the medial basal hypothalamus occurs during the progesterone-induced luteinizing hormone surge in the female rhesus monkey. *Endocrinology* 1984;115:2445–2452.
47. Pfaff DW. Impact of estrogens on hypothalamic nerve cells: ultrastructural, chemical and electrical effects. *Recent Prog Horm Res* 1983;39: 127–175.
48. Tansey EM, Arbuthnott GW, Fink G, Whale D. Oestradiol-17β increases the firing rate of antidromically identified neurones of the rat neostriatum. *Neuroendocrinology* 1983;37:106–110.
49. Fink G, Aiyer M, Chiappa S, et al. Gonadotropin-releasing hormone: release into hypophyseal portal blood and mechanism of action. In: McKerns K, ed. *Hormonally active brain peptides: structure and function*. New York: Plenum Press, 1982;397–426.
50. Fink G. Oestrogen and progesterone interactions in the control of gonadotrophin and prolactin secretion. *J Steroid Biochem* 1988;30:169–178.
51. Stanley HF, Lyons V, Obonsawin MC, et al. Regulation of pituitary α-subunit β-luteinizing hormone and prolactin messenger ribonucleic acid by gonadotropin-releasing hormone and estradiol in hypogonadal mice. *Mol Endocrinol* 1989;2:1302–1310.
52. Aiyer MS, Fink G. The role of sex steroid hormones in modulating the responsiveness of the anterior pituitary gland to luteinizing hormone releasing factor in the female rat. *J Endocrinol* 1974;62:553–572.
53. Turgeon JL, Barraclough CA. Further studies of the hypothalamo-hypophysial-gonadal axis of the androgen-sterilized rat. In: Forrest MG, Bertrand J, eds. *Endocrinologie sexuelle de la periode perinatale*, Paris: INSERM, 1974;339–356.
54. Fink G, Henderson SR. Steroids and pituitary responsiveness in female, androgenized female and male rats. *J Endocrinol* 1977;73:157–164.
55. Harris GW. Hormonal differentiation of the developing central nervous system with respect to patterns of endocrine function. *Philos Trans R Soc Lond B* 1970;259:165–177.
56. Curtis A, Lyons V, Fink G. The priming effect of LH-releasing hormone: effects of cold and in-

volvement of new protein synthesis. *J Endocrinol* 1985;105:163–168.

57. Mobbs C, Fink G, Johnson M, Welch W, Pfaff D. Similarity of an estrogen-induced protein and a luteinizing hormone releasing hormone-induced protein. *Mol Cell Endocrinol* 1989;62: 297–306.
58. Mobbs CV, Harlan RE, Burrous MR, Pfaff DW. An estradiol-induced protein synthesized in the ventral medial hypothalamus and transported to the midbrain central gray. *J Neurosci* 1988;8: 113–118.
59. Lewis CE, Morris JF, Fink G. The role of microfilaments in the priming effect of LH-releasing hormone: an ultrastructural study using cytocholasin B. *J Endocrinol* 1985;106:211–218.
60. Lewis CE, Morris JF, Fink G, Johnson M. Changes in the granule population of gonadotrophs of hypogonadal (hpg) and normal female mice associated with the priming effect of LH-releasing hormone *in vitro*. *J Endocrinol* 1986;109:35–44.
61. Pickering AJMC, Fink G. Variation in size of the "readily releasable pool" of luteinizing hormone during the oestrous cycle of the rat. *J Endocrinol* 1979;83:53–59.
62. Morris JF, Lewis CE, Fink G. LHRH priming in gonadotrophs: a model system for the analysis of neuroendocrine mechanisms at the cellular level. In: Fink G, Harmar AJ, McKerns KW, eds. *Neuroendocrine molecular biology*. New York: Plenum Press, 1986;341–352.
63. Rosie R, Thomson E, Fink G. Oestrogen positive feedback stimulates the synthesis of LHRH mRNA in neurones of the rostral diencephalon of the rat. *J Endocrinol* 1989;124:285–289.
64. Sarkar DK, Fink G. Effect of gonadal steroids on output of luteinizing hormone releasing factor into pituitary stalk blood in the female rat. *J Endocrinol* 1979;80:303–313.
65. Shivers BD, Harlan RE, Hejtmancik JF, Conn PM, Pfaff DW. Localization of cells containing LHRH-like mRNA in rat forebrain using *in situ* hybridization. *Endocrinology* 1986;118:883–885.
66. Pfaff DW. Gene expression in hypothalamic neurons: luteinizing hormone releasing hormone. *J Neurosci Res* 1986;16:109–115.
67. Ronnekleiv OK, Naylor BR, Bond DT, Adelman JP. Combined immunohistochemistry for gonadotropin-releasing hormone (GnRH) and pro-GnRH and *in situ* hybridization for GnRH messenger ribonucleic acid in rat brain. *Mol Endocrinol* 1989;3:363–371.
68. Zoeller RT, Young WS III. Changes in cellular levels of messenger ribonucleic acid encoding gonadotropin-releasing hormone in the anterior hypothalamus of female rats during the estrous cycle. *Endocrinology* 1988;123:1688–1689.
69. Everett JW, Nikitovitch-Winer M. Physiology of the pituitary gland as affected by transplantation or stalk transection. In: Nalbandov AV, ed. *Advances in neuroendocrinology*. Urbana: University of Illinois Press, 1963;289–303.
70. Neill JD. Prolactin secretion and its control. In: Knobil E, Neill J, eds. *The physiology of reproduction*, vols 1 and 2. New York: Raven Press, 1988;1379–1390.
71. Thomas GB, Cummins JT, Smythe GA, et al. Concentrations of dopamine and noradrenaline in hypophysial portal blood in the sheep and the rat. *J Endocrinol* 1989;121:141–147.
72. Ben-Jonathan N. Dopamine: a prolactin-inhibiting hormone. *Endocr Rev* 1985;6:564–589.
73. Maurer RA. Estradiol regulates the transcription of the prolactin gene. *J Biol Chem* 1982;257: 2133–2136.
74. Stanley HF, Curtis A, Sheward WJ, Roberts JL, Fink G. Prolactin messenger ribonucleic acid levels in normal and hypogonadal mouse pituitary gland. *Endocrinology* 1986;119:2422–2426.
75. de Greef WJ, Klootwijk W, Karels B, Visser TJ. Levels of dopamine and thyrotrophin-releasing hormone in hypophysial stalk blood during an oestrogen-stimulated surge of prolactin in the ovariectomized rat. *J Endocrinol* 1985;105:107–112.
76. Willoughby JO, Pederick H, Jervois P, Menadue M, Judd SJ. Responses to prolactin secretagogues in oestrogen treated rats suggest that the defect in prolactin regulation produced by oestrogen is at the level of the pituitary gland. *J Endocrinol* 1985;104:447–452.
77. McQueen JK, Fink G. Changes in local cerebral glucose utilization associated with the spontaneous ovulatory surge of luteinizing hormone in the rat. *Neuroendocrinology* 1988;47:551–555.
78. Watts AG, Fink G. Pulsatile luteinizing hormone release, and the inhibitory effect of estradiol-17β in gonadectomized male and female rats: effect of neonatal androgen or exposure to constant light. *Endocrinology* 1984;115:2251–2259.
79. Fink G, Smith GC. Ultrastructural features of the developing hypothalamo-hypophysial axis in the rat. *Z Zellforsch Mikrosk Anat* 1971;119: 208–226.
80. Page RB. The anatomy of the hypothalamo-hypophyseal complex. In: Knobil E, Neill J, eds. *The physiology of reproduction*, vols 1 and 2. New York: Raven Press, 1988;1161–1233.
81. Silverman A-J. The gonadotropin-releasing hormone (GnRH) neuronal system: immunocytochemistry. In: Knobil E, Neill J, eds. *The physiology of reproduction*, vols 1 and 2. New York: Raven Press, 1988;1283–1304.
82. Pellegrino LJ, Pellegrino AS, Suchman AJ. *A stereotaxic atlas of the rat brain*. New York: Plenum Press, 1979.

Volume Transmission in the Brain: Novel Mechanisms for Neural Transmission, edited by K. Fuxe and L.F. Agnati. Raven Press, Ltd., New York © 1991.

15

Brain Corticosteroid Receptors and Neurotransmission

*E. Ronald de Kloet and **Marian Joëls

**Rudolf Magnus Institute for Pharmacology and **Institute for Molecular Biology and Medical Biotechnology, Utrecht, The Netherlands*

Corticosteroids secreted by the adrenal cortex readily enter the brain and bind to intracellular receptor proteins. The steroid receptor complex subsequently alters the expression of cell-specific genes. These genes encode for the synthesis of neuropeptide precursors, enzymes involved in transmitter synthesis and intermediary metabolism, and proteins underlying transmitter signaling. These, together with the steroids, regulate metabolism and have a pronounced influence on electrical and chemical communication of nerve cells (1,2). The steroids, therefore, may be considered as a clear example of the theme of this book: volume transmission. Much detailed information has become available on the genomic actions of corticosteroids. The receptor-mediated genomic actions are slow in onset but long-lasting and may persist hours after removal of the steroids. The modulation of gene expression and axonal transport of gene products are responsible for this delay in the actions of steroids.

More recently direct actions of steroid hormones with pre- and postsynaptic neurotransmitter receptors have been described, too (3,4). These actions are supposedly mediated via membrane rather than intracellular steroid receptors. The most prominent example of this direct steroid action on neurotransmission is the binding of A-ring-reduced metabolites of progesterone and deoxycorticosterone to the gamma-aminobutyric acid (GABA) a-benzodiazepine-chloride channel receptor complex, which underlies their potent sedative, hypnotic, and anesthetic effects (5). The steroids have a barbiturate-like action in that they potentiate GABA-evoked conductances by prolongation of chloride channel open time. Alternatively, other adrenal steroids have antagonistic activity in the action of GABA. The latter steroids include the naturally occurring glucocorticoids, corticosterone and cortisol, as well as pregnenolone sulfate, a (neuro)steroid that can be synthesized from cholesterol in oligodendrocytes (4,6). The direct actions of steroids are fast in onset, of short duration, and occur parallel to the changing levels in steroid hormone concentration. Control of neurosteroid biosynthesis and secretion is still poorly understood, whereas also the neurosteroid concept requires further solid substantiation.

In this communication we will focus on the properties of central intracellular rather than membrane corticosteroid receptors and their role in neurotransmission. These receptors can be distinguished in mineralocorticoid (MR) and glucocorticoid (GR) receptors. In the CA1 pyramidal neurons of the hippocampus the MR and GR appear to be colocalized. Using intracellular electrophysiological recording techniques, we find that in these neurons MR-mediated actions transiently facilitate excitability but that excitation is slowly and progressively overruled by a long-lasting suppressive action exerted via GR.

MINERALOCORTICOID RECEPTORS

Twenty years ago McEwen and colleagues discovered that ^{3}H-corticosterone, when injected into adrenalectomized (ADX) rats, was retained in large amounts in cell nuclei of the hippocampus (7). Radioautography revealed predominant localization in the cell nuclei of the CA1, CA2, and CA3 pyramidal neurons and the granular neurons of the dentate gyrus. Retention of ^{3}H-corticosterone also occurred in the lateral septum, amygdaloid nuclei, and motor and sensory neurons in the brainstem. Interestingly, ^{3}H-aldosterone was also retained in the same neurons, whereas both steroids were equally effective in competing for each other's uptake and retention. In addition, ^{3}H-aldosterone retention was also observed in circumventricular tissues and the anteroventral parts of the hypothalamus. The uptake of both steroids was not affected by dexamethasone, whereas the distribution of tracer amounts of this synthetic steroid in the brain also differed from the pattern observed for corticosterone or aldosterone (8).

In vitro binding of the steroids to soluble receptor fractions in cytosol prepared from hippocampus tissue revealed that the corticosteroid did bind to a heterogeneous pool of receptors (see Table 1). Scatchard plots of ^{3}H-aldosterone binding were curvilinear and could be linearized after inclusion of "pure" glucocorticoids (9). Such glucocorticoids, as with prototype RU28362, do not display any affinity to the classical MRs in the kidney nor do they have any mineralocorticoid effect on the mineral balance. Aldosterone and corticosterone did bind competitively and with similar affinity to cytosol receptors extracted from hippocampal tissue, provided RU28362 was included in the medium. Accordingly, the group of Funder and our laboratory reached the conclusion that the corticosteroid receptors incapable of binding RU28362 actually represented MRs that lacked the ability to discriminate between aldosterone and the naturally occurring glucocorticoid of the rat, corticosterone (9–11). Because the latter steroid circulates in a 10- to 100-fold higher concentration and we had been able to define corticosterone-specific effects on the brain, the receptors were indicated originally as "corticosterone-preferring receptors" or "type 1" receptors (1,11,12). However, because we now know that the receptor genes in brain and kidney are indistinguishable, it is perhaps better to refer to these receptors as MRs (13,14).

TABLE 1. *Relative binding affinity (IC50) of steroids for MR and GR in cytosol of rat hippocampus*[a]

	MR (IC50, nM)	GR (IC50, nM)
Corticosterone	0.5	5.0
Cortisol	2.1	19.7
Aldosterone	2.5	26.0
Spirondactone	4.9	88.0
RU28362	>1,000.0	2.1
RU38486	>1,000.0	5.0

[a] IC50 values were determined according to methods described in ref. 10. MR, incubation with 2 nM 3H-corticosterone in the presence of a 100-fold excess of RU28362; GR, incubation with 3 nM 3H-RU28362; RU28362, 11,17-dihydroxy-6-methyl-17α-(1-propynyl)androsta 1,4,6-triene-3-one; RU38486, 17β-hydroxy-11β-(4-dimethylaminophenyl) 17α-(1-propynyl)-estra-4,9-diene-3-one.

ALDOSTERONE-SELECTIVE AND NONSELECTIVE MINERALOCORTICOID RECEPTORS

Why certain tissues such as kidney and parotid glands express aldosterone-selective receptors *in vivo* while others in brain and heart do not remained a mystery until recently. It has now become apparent that two factors are of prime importance for the cellular specificity conferring mechanism for aldosterone. These factors are corticosteroid binding globulin (CBG or transcortin) as well as the enzyme, 11β-hydroxysteroiddehydrogenase (11β-OHSD). Both the steroid plasma binding protein and the enzyme are present intracellularly in tissues, such as the kidney, that are aldosterone-selective. CBG does not bind aldosterone but displays high affinity toward corticosterone. Accordingly, it has been hypothesized that intracellular CBG competes with the receptors for binding of corticosterone. CBG could therefore function as a buffer toward changing corticosterone levels, which is bypassed by aldosterone (15,16).

In mammalian cells the enzyme 11β-OHSD acts in concert with 11 oxo-reductase to catalyze the reversible conversion of corticosterone/cortisol to their 11 oxo metabolites, which show low affinity to MRs. Considerable enzyme activity is present in kidney medulla and papillary cells, parotid gland, liver, testis, and even the anterior pituitary. The enzyme is blocked by glycyrrhctinic acid (an active component of licorice) with a Ki in the nanomolar range when tested in broken cell preparations. *In vivo* administration of this licorice component also blocks enzyme 11β-OHSD. Under such conditions a considerable retention is observed of ^{3}H-corticosterone in the kidney to an extent comparable with ^{3}H-aldosterone. These observations suggest, therefore, that aldosterone–selectivity of MR is conferred by 11β-OHSD (17–19).

Excess licorice consumption increases blood pressure and shows in its pathophysiology many of the symptoms found with a syndrome of apparent mineralocorticoid excess (20). This syndrome is observed in children who have a deficiency in 11β-OHSD. In animal experiments we have recently found that daily administration of glycyrrhizic acid (the glucuronide of glycyrrhetinic acid) results in elevated blood pressure (D. VandenBerg, *unpublished data*). Removal of the adrenals reduced blood pressure. This observation is consistent with the hypothesis that the mineralocorticoid-like effects of licorice require an intact adrenocortical system. Accordingly, these effects are not due to a direct interaction of the compound with MR, but probably due to excessive stimulation of MR by endogenous corticosterone or cortisol (17–19).

The hippocampus seems to lack 11β-OHSD. This observation is consistent with a postulated role of the enzyme in aldosterone selectivity of MR-containing cells. MRs in neurons, therefore, may respond to aldosterone as well as corticosterone. However, with the exception of low A.M. basal resting levels of circulating corticosterone (<1 μg%), the latter steroid exceeds aldosterone concentrations with a factor 10 to 100, whereas the affinity *in vitro* is in the same range. The central MR are, thus, predominantly exposed to corticosterone (10).

GLUCOCORTICOID RECEPTORS

Radioautography of brain sections of rats treated with the highly selective synthetic glucocorticoid agonist (RU28362) showed a completely different binding pattern in the brain. This steroid was retained in particular by the paraventricular nucleus (PVN), supraoptic nucleus (SON), and arcuate nucleus (AN). These brain areas are apparently well-endowed with GRs (21). The findings were subsequently confirmed with immunocytochemistry using a monoclonal antibody raised against the liver GR. In virtually all brain cells, we found the GR. The highest density of GR occurs, however, in neurons of the PVN, SON, AN, thalamic nuclei, all nuclei of the ascending aminergic pathways, limbic (i.e., septal, hippocampal CA1, CA2, dentate gyrus, and amygdaloid) neurons, and the neocortex (22,23). In some cases, extensive knowledge has been gained on the neurotransmitter content of the receptor-containing neurons. GR are, for example, present in parvocellular neurons of the PVN. These neurons synthesize vasopressin and corticotrophin-releasing factor (CRF), among other neuropeptides. GR are also abundantly present in dopaminergic, adrenergic/noradrenergic, and serotonergic neurons in the brainstem. Peptides from the tachykinin (neuropeptide Y) and opioid family (POMC, proenkephalin) are also found in these neurons, and they are likely to be under the control of glucocorticoids (22).

In cytosol binding experiments involving punched brain nuclei, essentially the same regional distribution of GR was observed (10). The affinity of corticosterone to GR was approximately 10-fold lower than to MR. The hippocampus contains the highest amounts of both receptors. Both radioautography and immunocytochemistry suggest that MR binding sites and GR proteins are present particularly in CA1, CA2, and dentate gyrus neurons; GRs were to a lesser extent observed in the CA3 pyramidal neurons.

COLOCALIZATION OF GR AND MR IN HIPPOCAMPAL NEURONS

Meanwhile, the steroid receptors have been cloned, and their primary structure has been resolved (14). MR and GR have a 94% homology in the deoxyribonucleic acid (DNA) binding domain and a 57% homology in the steroid binding domain, respectively. Recent studies, furthermore, suggest that the steroid receptors are bound to heat shock proteins (HSP) in the cell nuclear compartment until the steroid hormone binds to the receptor and promotes dissociation of the HSP receptor complex (24, 25). Transcription is initiated on anchoring of the steroid receptor "Zn fingers" in the responsive elements of the DNA (14,25).

Using *in situ* hybridization with complementary ribonucleic acid (cRNA) probes, van Eekelen et al. (26) observed that limbic neurons express both MR- and GR-mRNA in large amounts. GR-mRNA was expressed in neurons as well as glial cells widely distributed in the brain (27). In complete agreement with immunocytochemical observations the PVN, SON, AN, and the ascending aminergic neurons as well as cortical and thalamic neurons were richly endowed with GR message. MR gene expression is much more discrete and resembles in distribution the previous radioligand binding studies with tracer amounts of ^{3}H-aldosterone and ^{3}H-corticosterone. Thus, the MR message was particularly high in neurons of the hippocampal formation, lateral septal, amygdala, olfactory nucleus, layer II in the cortex, and in brainstem sensory and motor neurons. The amounts of MR- and GR-mRNA were about equal in the CA1 and CA2 pyramidal neurons of the hippocampus and the granular neurons of the dentate gyrus. CA3 neurons expressed much more MR than GR (26–28).

These findings raise questions on the biological significance of GR and MR that (1) bind to corticosterone with a different affinity, (2) are coexpressed in high quantities in the same neurons, and (3) recognize the same or closely related hormone responsive elements. A clue to the significance comes from studies on the extent of receptor occupancy as a function of the circulating steroid level (see Fig. 1). When graded doses of corticosterone are administered to ADX rats, hippocampal MRs are already half maximally occupied with a dose of 1 μg/100 g body weight. This dose does not lead to detectable occupancy of the hippocampal GR within the 1-hr interval after administration. ED_{50} of GR is only reached at a dose of 60 μg/100 g body weight. Likewise, basal AM (<1 μg%) plasma corticosterone level already induces a 75% MR occupancy but occupies only about 15% of GR. Substantial GR occupany is reached after stress and at the circadian peak (10). As will be pointed out below, these occupancy studies of MR and GR have been fundamental for our hypothesis on the function of both receptors in control of neuronal activity.

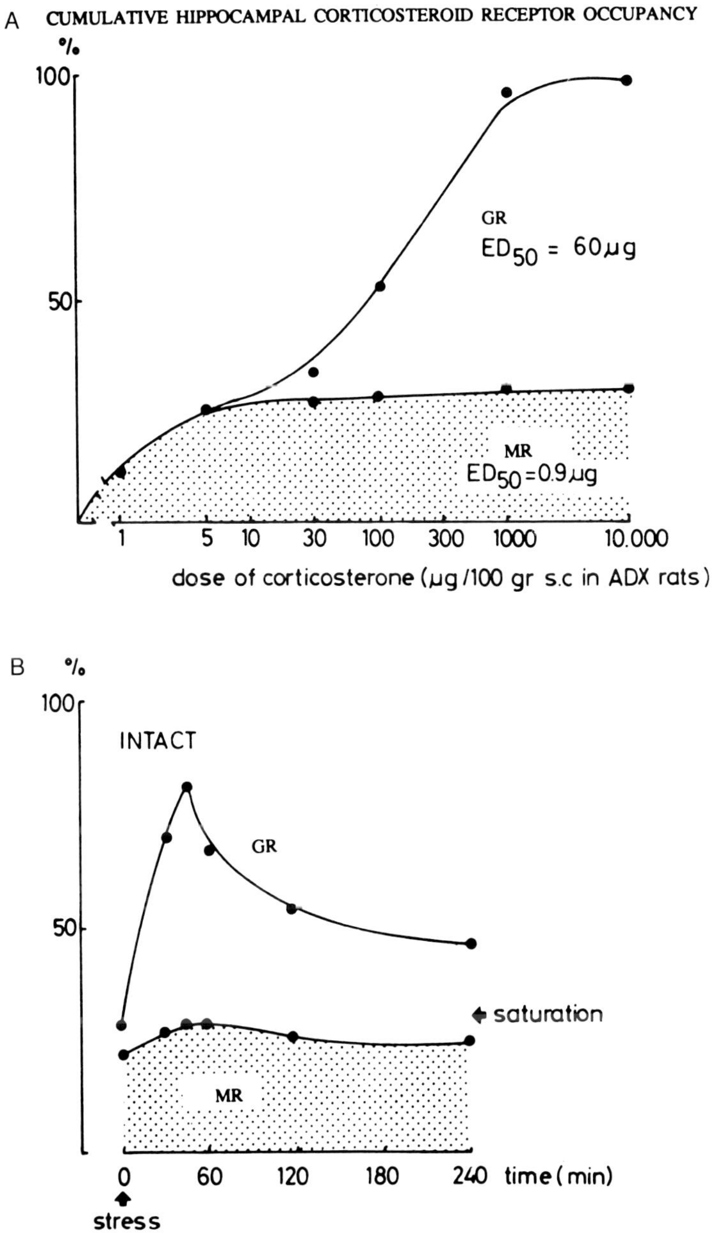

FIG. 1. Cumulative occupancy of MR and GR in hippocampal cytosol, from adrenalectomized (ADX) rats (**A**) by exogenously administered corticosterone and (**B**) after exposure of rats from basal AM resting state to novel environment (new cage). ED_{50} = effective dose of steroid in microgram per 100 gram body weight required to occupy 50% of the receptor. 100% = full occupancy of MR + GR. (Data taken from ref. 10 and ref. 29.)

IMPLICATIONS OF CORTICOSTEROID RECEPTOR DIVERSITY IN HIPPOCAMPUS

Recent studies of Arrizza et al. (28) show that interaction of MR and GR with the same hormone responsive elements is at least a potential possibility. Using a *cis/trans* cotransfection system in cultured kidney CV-1 cells, which lack these receptors, the response characteristics to recombinant human MR (hMR) and hGR expression vectors were investigated. The data suggest that MR can be a weak activator of the typical GR responsive promoter mouse mammary tumor virus (MMTV). MR has only 5% to 10% of the activational capacity of GR. Therefore, when expressed in the same cell, MR may only partially activate GR-responsive gene networks. A specific MR responsive promoter may exist as well. On the basis of these observations, Evans (30) recently proposed that when MR and GR are coexpressed, the receptors may coordinatively exert a synergistic or competitive action in transcriptional control, i.e., act as a binary hormone response system.

On a different level of analysis, our studies point to a well-defined role of central MR and GR in control of molecular, cellular, and behavioral events. These studies were guided by the topography, binding specificity, and affinity of central MR and GR. In these studies, we have either subjected ADX animals to steroid replacement therapy, or we have applied receptor-specific (ant) agonists locally in the brain. The findings have provided a new concept of MR and GR function in the brain (i.e., tonic activating influences of corticosterone via MR versus a feedback action via GR aimed to restore disturbances in homeostasis induced by stress) (12).

In the next sections we will focus on the neuronal response to corticosterone in the hippocampal CA1 cell field. The experiments support that the excitability of these cells is regulated by corticosterone via a receptor-mediated genomic mechanism. Corticosterone effects are biphasic and involve a relatively fast facilitation of excitation via MR and subsequent suppression via GR.

ELECTROPHYSIOLOGICAL INVESTIGATION OF STEROID RECEPTOR–MEDIATED EFFECTS

Electrophysiological studies during the past decades have supplied evidence that steroids, once they have reached their target cells in the brain, can directly or indirectly affect neuronal activity. Thus, unidentified hippocampal cells in hypophysectomized rats displayed a suppressed spontaneous firing rate after peripheral injection of corticosterone (31). In contrast, intraperitoneal injection of cortisol in freely moving rats resulted in either decreased or increased activity in 50% of hippocampal single units (32); no changes were observed with iontophorized corticosterone in urethane anesthetized animals (33). Field potentials were either increased or decreased by steroid treatment, partly depending on the concentration that was used (34–36). Similarly, steroids induced variable effects on electrical activity of septal and preoptic single units of the rat (37).

Although steroids thus appear to affect neuronal activity, these studies have not yielded a general consensus with respect to the underlying mechanism. The incompatibility of some of the observations may have resulted from one or more of the following reasons: (1) Some of the experiments in which corticosterone was given were performed in (tissue from) intact rather than ADX animals. Therefore we can assume, in retrospect, that probably all the MRs and a large proportion of the GRs were occupied by the circulating endogenous ligand and not accessible for the exogenously applied steroid. (2) At the time that these experiments were carried out, it was not yet appreciated that corticosteroids can bind to two different intracellular receptor types in the brain. In so far as the concentration of the exogenously added steroids can be estimated (after peripheral injection in intact animals or with superfusion in *in vitro* studies), it seems likely that in most cases the steroids must have fully occupied both the MR and the GR population, thereby not only inducing mixed agonistic effects but probably masking the more subtle effects on the MR (see below). Only since the existence of two instead of one corticosteroid receptor types in the brain has been appreciated and selective ligands for these receptors have become available has it become possible to study the MR- and GR-mediated changes in electrical activity separately (3). Most of the steroid effects were studied extracellularly. Although this technique allows the registration of direct steroid-induced changes in spontaneous activity, more subtle effects of the steroids such as permissive or conditional (e.g., voltage-dependent) effects might remain unnoticed. The latter actions can be better studied with intracellular recording techniques in brain slices, where direct control over the electrical characteristics and the ionic environment of the neuron under study is possible.

Given these considerations, we set out to study the effect of selective activation of the MR and GR on electrical activity in the brain (38). We studied the steroid effects with intracellular recording techniques in pyramidal cells of the rat CA1 hippocampal area, where both types of receptors are present. Selective MR- and GR-mediated effects were studied in slices from ADX rats. Ligands for the GR and MR were superfused (for 20 min) approximately 1 to 2 hr before impalement of the neuron, for two reasons: (1) Biochemical studies have shown that genomic effects mediated by the GR can develop in this time span (39,40), and (2) possible membrane effects, which have been reported for high concentrations of corticosterone, have disappeared by this time and will therefore no longer interfere with the delayed, genomic actions (41,42).

Neurons recorded before and after exposure of the tissue to selective MR or GR ligands were compared for their general membrane properties such as resting membrane potential, input resistance, spike size, and spontaneous activity. In addition, we investigated the characteristics of the afterhyperpolarization (AHP), which can be recorded in pyramidal cells after a brief (50 msec) depolarizing current pulse (43–45). The AHP, which occurs directly after the termination of the current pulse, is due to the activation (and subsequent inactivation) of a Ca^{2+}-dependent K^+ conductance (46). If the duration of the depolarizing pulse is prolonged (e.g., to 500 msec), activation of the Ca^{2+}-dependent K^+ conductance starts already during the depolarization and results in a slowing down of the cellular firing rate, a phenomenon that is known as accommodation. Together, the accommodation and the AHP represent an intrinsic mechanism by which hippocampal and other cells are protected against excessive excitation (47).

GR-MEDIATED EFFECTS ON ELECTRICAL ACTIVITY

The most apparent effect of a brief exposure of hippocampal tissue to high concentrations of corticosterone (1–2 hr before recording) turned out to be a marked increase in the amplitude and duration of the AHP after a brief (50 msec) depolarizing pulse (38) (Fig. 2). The increase in amplitude was observed over the whole range of applied current steps (0.2–1.0 nA). The number of spikes associated with the brief depolarizing response, however, did not differ from the number of spikes obtained in neurons recorded before steroid exposure. If the duration of the depolarizing step was prolonged to 500 msec, thus allowing not only activation but probably also partial inactivation of the Ca^{2+}-dependent K^+ conductance underlying the AHP, neurons exposed to the steroid still displayed an increased AHP amplitude, although the differences were somewhat less pronounced

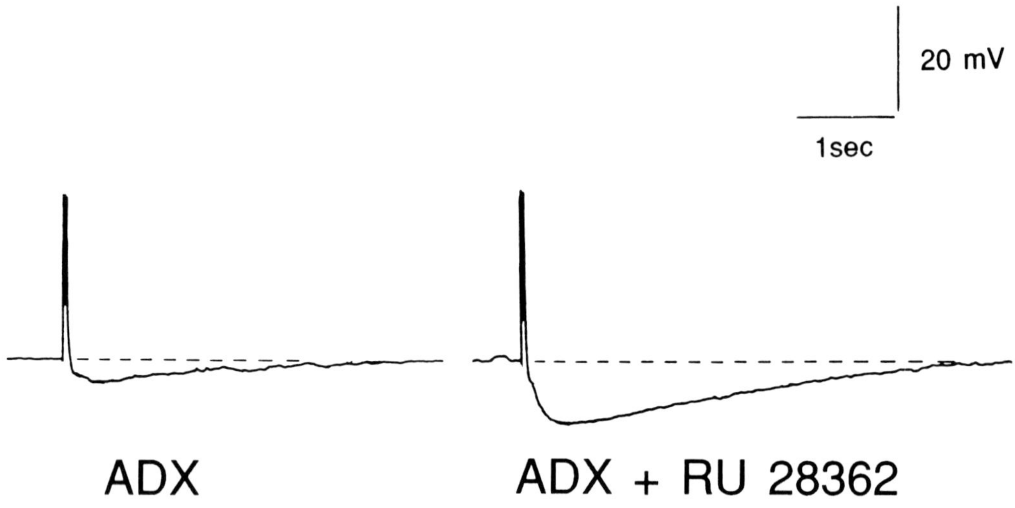

FIG. 2. Afterhyperpolarizations obtained in two CA1 neurons that were consecutively recorded in the same hippocampal slice, with the same recording electrode. The slice was prepared from a rat that was ADX 7 days before the experiment. The afterhyperpolarization on the left was obtained in a neuron before the slice was perfused with steroids; the signal on the right was from a neuron approximately 1 hr after the slice was treated with 100 nM RU28362 for 20 min.

than those observed with the short pulse. Interestingly, these changes in AHP occurred in the absence of consistent differences in resting membrane potential or resistance between the neurons recorded before or after steroid treatment.

This delayed increase in the AHP amplitude by the steroids was already fully expressed with 30 nM corticosterone, a concentration that under similar *in vitro* conditions, occupies nearly all the MR and GR sites. If corticosterone (1 μM) was tested in the presence of RU38486, a GR antagonist, no steroid effects on the AHP amplitude were observed, indicating that the increase in amplitude is mediated by the GR. Accordingly, superfusion with RU28362 (1 nM–1 μM), a selective GR agonist, resulted in a marked increase in the AHP amplitude. In conclusion, selective activation of the GR in hippocampal pyramidal cells leads to a marked increase in the amplitude and duration of the AHP, in the absence of apparent changes in other tested membrane properties. Both the receptor specificity and the delay in onset of this effect are compatible with a genomic action of the steroid hormone. The functional implication of this GR-mediated effect may be that a temporarily increased excitatory input to these neurons is successfully suppressed by high levels of circulating steroids, with a time delay of 1 to 2 hr. Such an action of the steroids is in line with the proposed role of the GR in the feedback control of the stress response (12).

MR-MEDIATED EFFECTS ON ELECTRICAL ACTIVITY

Binding characteristics of corticosterone to the MR or GR revealed that occupation of the MR precedes the occupation of the GR. For this reason we selected a slightly different protocol for the study of MR-mediated effects in which we continuously followed the passive membrane properties of pyramidal cells and their AHP characteristics before, during (20 min), and at least 20 min after application of the ligands for the MR (48). The data indicate that 1 nM corticosterone, which probably occupies most of the MR but only a small part of the GR population, evokes a decrease in pyramidal cell accommodation (500 msec pulse), starting approximately 15 min after onset of the steroid perfusion and reaching a maximal effect 10 to 20 min after offset (i.e., 30–40 min after first exposure of the tissue to the steroid). The decrease in accommodation is accompanied by a small decrease in the AHP amplitude. The resting membrane potential was on average slightly (1–2 mV) depolarized, whereas no changes in input resistance were observed. Similar effects have been observed for 1 nM aldosterone. Some of the CA1 pyramidal cells (10–20%) did not respond to corticosterone or aldosterone with any apparent change in accommodation or AHP amplitude. Spironolactone, a MR antagonist, was able to prevent the corticosterone-induced effects on accommodation and AHP. Thus, selective activation of the MR seems to result in a decrease of the accommodation and AHP amplitude, an effect that is opposite to that of GR-mediated processes. In addition, MR-mediated events seem to reach their maximal effect with a shorter delay (~30–40 min) than GR mediated effects (~90 min).

MIXED EFFECTS OF CORTICOSTERONE ON THE MR AND GR

We were particularly interested in the effects of the endogenous ligand corticosterone, a mixed agonist for the MR and GR. At a low concentration (1 nM), corticosterone initially decreased the pyramidal cell accommodation and AHP, reaching a maximal effect approximately 30 to 40 min after the steroid treatment was started (i.e., 10–20 min after washout of the steroid was commenced). However, approximately 60 min later, the accommodation and AHP showed a small increase compared with the group of ADX neurons tested before steroid treatment. This observation suggests that although with 1 nM corticosterone only a small proportion of the GR are occupied, the GR-mediated effects can overcome the MR-mediated actions. Similarly, when 30 nM corticosterone was perfused, resulting in a near-complete occupation of both receptor types, a small decrease in AHP amplitude and accommodation was observed 30–40 min after steroid perfusion was started; yet, approximately 60 min later there was a marked increase in these parameters compared with the values obtained before steroid treatment, approaching the levels obtained after such an interval with the selective glucocorticoid agonist. Based on these observations, we assume that the GR-mediated effects on electrical activity can "overrule" the MR-mediated effects, provided that the time delay necessary for a full expression of the GR-mediated events is allowed.

Insight into these mixed agonistic effects of corticosterone is important for understanding the differences in electrical properties that exist between hippocampal slices from ADX versus sham-operated rats. According to the biochemical data, we can assume that when slices from sham-operated rats are prepared, most of the MR and part of the GR are occupied by the endogenous ligand. Because the natural source for the endogenous ligand, the adrenals, is no longer present *in vitro,* receptor-occupation and steroid-mediated effects will probably decline during the course of the experiment. As long as both MR- and GR-mediated effects are present in the slices from sham-operated rats, we can expect that the cellular accommodation and AHP are large in comparison to the values obtained in slices from ADX rats. However, if the GR-mediated action expires (e.g., at the end of the experiment), no differences or even a decrease in AHP/accommodation might occur in comparison to the ADX animals. Taken together, neurons recorded in slices from sham-operated rats are expected to yield a stronger accommodation and AHP than neurons from ADX

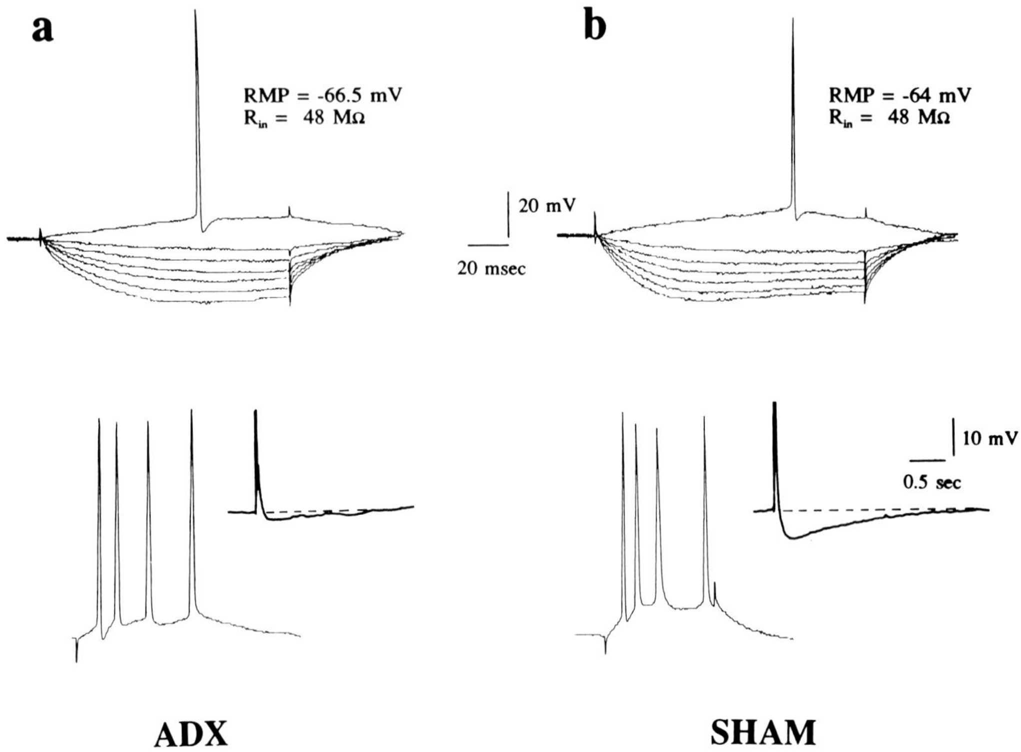

FIG. 3. a: Voltage deflections in response to depolarizing (0.1 nA) and hyperpolarizing (−0.1–0.6 nA) current steps, recorded in a neuron that was recorded in a slice from a 7-day ADX rat (*top*). The neuron responded with four spikes to a 0.5-nA depolarizing current pulse of 50 msec duration (*bottom*). The *insert* shows the afterhyperpolarization that followed the depolarizing current step. The neuron depicted in **b,** which was recorded in a slice from a sham-operated rat, displayed similar current-voltage properties, resting membrane potential (RMP), input resistance (Rin), and number of spikes evoked by a short (50-msec) depolarizing (0.5 nA) current pulse as the neuron in A but showed a markedly larger amplitude of the afterhyperpolarization.

rats, although the difference is likely to be smaller than observed for ADX neurons before and after glucocorticoid treatment. In fact, this expected difference turned out to be true in experiments from our group and from the group of Landfield (38,49) (Fig. 3).

CONCLUDING REMARKS

Corticosteroids appear to have a pronounced effect on the electrical and chemical mechanisms underlying neuronal communication. The direction of the effects exerted by the corticosteroids depends on the type of intracellular receptor that is stimulated as well as on the time that elapses between the steroid hormone stimulus and the recording of the neuronal activity. Thus, stimulation by MR ligands results in an enhancement of neuronal excitability, which develops within 15 min of steroid application and reaches a maximal effect after 30 to 40 min. GR agonists suppress excitability with a longer time delay (i.e., suppression reaches a maximal level after 90 min).

Corticosterone activates and suppresses excitability sequentially *in vitro* in the hippocampus neurons, where MR and GR are colocalized. A transient facilitation of excitability is slowly and progressively overruled by the suppressive GR action. However, GR-mediated suppression is amplified further as a function of the corticosterone concentrations. Over 30 nM corticosterone, the MR response does not develop at all. Under such circumstances, corticosterone acts reminiscent to the effect of the selective synthetic glucocorticoid analogues. The findings are consistent with the hypothesis on the "tonic and feedback action" of corticosterone exerted via MR and GR, respectively (1,12). This hypothesis was based on the concentration and timing of receptor-specific opposing actions of corticosterone on behavioral and neurotransmitter events. The findings are also partly in line with the "binary hormone response hypothesis" postulated by Evans (30) in which MR and GR are thought to control coordinatively the same set of genes. The working model of Evans is based on studies with a cell system transfected with MR and GR genes that act synergistically. In this line of reasoning, the high- (MR) and low- (GR) affinity receptor could provide a continuum of control and thus expanding the range of the physiological response. In contrast, our studies measuring the response of hippocampal neurons containing endogenous MR as well as GR mediate rather competitive than synergistic effects of corticosterone that can be distinguished temporally and in direction.

The nature of the MR-mediated and GR-mediated responses suggests a critical role for the Ca^{2+}-dependent K^{+} conductance in cellular homeostasis. This conductance is also subject to regulation by several transmitters and second messengers (47) [e.g., the cyclic adenosine monophosphate (AMP)-mediated β-adrenergic response] (50–52). Interestingly, neuronal excitation evoked by application of norepinephrine is suppressed by glucocorticoids via a GR-mediated mechanism (38). In line with this observation are data showing that glucocorticoids suppress norepinephrine-stimulated cyclic AMP formation in hippocampus and cerebral cortex (53–55). These corticosteroid–amine interactions can be extended to behavioral observations. ADX rats show retention deficits in conditioned avoidance behavior, which are reversed by subcutaneous administration of catecholamines. Glucocorticoids decrease the sensitivity to the amines (56), which is consistent with the glucocorticoid-dependent suppression in norepinephrine-stimulated cyclic AMP generation and norepinephrine-induced neuronal excitation. Taken together these findings suggest that we are beginning to recognize similar features of corticosteroid–transmitter interactions converging in control of neuronal activity as well as in control of animal behavior.

The characteristics of the MR and GR control exerted on cellular excitability fits well with the concept of corticosteroid physiology. Selye (5) already noted that that mineralocorticoids and glucocorticoids have opposing actions (e.g., prophlogistic and antiphlogistic actions, respectively). Tausk (57) and Munck et al. (58) postulated that glucocorticoid actions via GR are aimed to dampen cellular and organ responses to stress. Our studies demonstrate that this function of GR can be applied to hippocampal neurons and extended with the opposing MR-enhanced neuronal excitability. Thus, both receptor systems exert gene-mediated control of state and efficacy of limbic circuits in response to changing levels of circulating corticosterone caused by stress and circadian periodicity.

REFERENCES

1. de Kloet ER, Ratka A, Reul JMHM, Sutanto W, van Eekelen JAM. Corticosteroid receptor types in brain: regulation and putative function. *Ann NY Acad Sci* 1987;512:351–361.
2. McEwen BS, de Kloet ER, Rostene W. Adrenal steroid receptors and actions in the nervous system. *Physiol Rev* 1986;66:1121–1188.
3. Lambert JJ, Peters JA, Cottrell GA. Actions of synthetic and endogenous steroids on the $GABA_a$ receptor. *Trends Pharmacol Sci* 1987;8:224–227.
4. Majewska MD, Harrison NL, Schwartz RD, Barker JL, Paul SM. Steroid hormone metabo-

lites are barbiturate-like modulators of the GABA receptor. *Science* 1986;232:1004–1007.

5. Selye H. *The story of the adaptation syndrome*. Montreal, Canada: Acta, Medical Publishers, 1952.
6. Hu ZY, Bourreau E, Jung-Testas I, et al. Neurosteroids: oligodendrocyte mitochondria convert cholesterol to pregnenolone. *Proc Natl Acad Sci* 1987;84:8215–8219.
7. McEwen BS, Weiss JM, Schwartz LS. Selective retention of corticosterone by limbic structures in rat brain. *Nature* 1968;220:911–912.
8. de Kloet ER, Wallach G, McEwen BS. Difference in binding of corticosterone and dexamethasone to rat brain and pituitary. *Endocrinology* 1975;96:598–611.
9. Veldhuis HD, van Koppen C, Van Ittersum M, et al. Specificity of the adrenal steroid receptor system in rat hippocampus. *Endocrinology* 1982; 110:2044–2053.
10. Reul JMHM, de Kloet ER. Two receptor systems for corticosterone in rat brain: microdistribution and differential occupation. *Endocrinology* 1985;117:2505–2512.
11. Funder JW. Adrenocortical receptors in the brain. In: Ganong WF, Martini L, eds. *Frontiers in Neuroendocrinology*. New York: Raven Press, 1986;10:169–189.
12. de Kloet ER, Reul JMHM. Feedback action and tonic influence of corticosteroids on brain function: a concept arising from the heterogeneity of brain receptor systems. *Psychoneuroendocrinology* 1987;12:83–105.
13. Arriza JL, Weinberger C, Cerelli G, et al. Cloning of human mineralocorticoid receptor complementary DNA: structural and functional kinship with the glucocorticoid receptor. *Science* 1987;237:268–275.
14. Evans RM. The steroid and thyroid hormone receptor superfamily. *Science* 1988;240:889–895.
15. de Kloet ER, Burbach JPH, Mulder GH. Localization and role of transcortin-like molecules in the anterior pituitary. *Mol Cell Endocrinol* 1977;7:261–273.
16. Reventos J, Hammond GL, Minto NA, et al. Tissue distribution and hormonal regulation of corticosteroid binding globulin (CBG) mRNA in rhesus monkey. 8th International Congress of Endocrinology 1988; abstract 13.21-109.
17. Funder JW, Pearce PT, Smith R, et al. Mineralocorticoid action: target tissue specificity is enzyme, not receptor, mediated. *Science* 1988;242: 583–586.
18. Edwards CRW, Stewart PM, Burt D, et al. Localization of 11β-hydroxysteroid dehydrogenase: tissue specific protector of the mineralocorticoid receptor. *Lancet* 2:986–989.
19. Monder C, Stewart PM, Lakshi V, et al. Licorice inhibits corticosteroid 11β-dehydrogenase of rat kidney and liver: *in vivo* and *in vitro* studies. *Endocrinology* 1989;125:1046–1053.
20. Ulick S, Levine LS, Gunczler P, et al. A syndrome of apparent mineralocorticoid excess associated with defects in the peripheral metabolism of cortisol. *J Clin Endocrinol Metab* 1979;49:757–776.
21. Sutanto W, Van Eekelen JAM, Reul JMHM, et al. Species specific topography of corticosteroid receptor types in rat and hamster brain. *Neuroendocrinology* 1988;47:398–404.
22. Fuxe K, Wikstrom AC, Okret S, et al. Mapping of glucocorticoid receptor immunoreactive neurons in the rat tel- and diencephalon using a monoclonal antibody against rat liver glucocorticoid receptors. *Endocrinology* 1985;117:1803–1812.
23. Van Eekelen JAM, Kiss JZ, Westphal HM, de Kloet ER. Immunocytochemical study on the intracellular localization of the type 2 glucocorticoid receptor in the rat brain. *Brain Res* 1987;436:120–128.
24. Baulieu EE, Binart N, Cadepond F, et al. Do receptor-associated nuclear proteins explain earliest steps of steroid hormone function? In: Carlstedt-Duke J, Eriksson H, Gustafsson JA, eds. *The steroid/thyroid hormone receptor family and gene regulation*. Basel: Birkhauser Verlag, 1989: 301–318.
25. Gustafsson J, Carlstedt-Duke J, Poellinger L, et al. Biochemistry, molecular biology and physiology of the glucocorticoid receptors. *Endocr Rev* 197;8:185–234.
26. Van Eekelen JAM, Jiang W, de Kloet ER, et al. Distribution of the mineralocorticoid and the glucocorticoid receptor mRNAs in the rat hippocampus. *J Neurosci Res* 1988;21:88–94.
27. Aronsson M, Fuxe K, Dong Y, et al. Localization of glucocorticoid receptor mRNA in the male rat brain *in situ* hybridization. *Proc Natl Acad Sci* 1988;85:9331–9335.
28. Arriza JL, Simerly RB, Swanson LW, et al. Neuronal mineralocorticoid as a mediator of glucocorticoid response. *Neuron* 1988;1:887–900.
29. Reul JMHM, Van den Bosch F, de Kloet ER. Differential response of type 1 and type 2 corticosteroid receptors to changes in plasma steroid level and circadian rhythmicity. *Neuroendocrinology* 1987;45:407–412.
30. Evans RM. A molecular framework for the actions of glucocorticoid hormones in the nervous system. *Neuron* 1989;2:1105–1112.
31. Pfaff DW, Silva MTA, Weiss JM. Telemetered recording of hormone effects on hippocampal neurons. *Science* 1971;171:394–395.
32. Dafny N, Philips MI, Taylor AN, Gilman S. Dose effects of cortisol on single unit activity in hypothalamus reticular formation and hippocampus of freely behaving rats correlated with plasma steroid levels. *Brain Res* 1973;59:257–272.
33. Ben Barak Y, Gutnick MJ, Feldman S. Iontophoretically applied corticosteroids do not affect the firing of hippocampal neurons. *Neuroendocrinology* 1977;23:248–256.
34. Reiheld CT, Teyler TJ. Effects of corticosterone on the electrophysiology of hippocampal CA1 pyramidal cells *in vitro*. *Brain Res Bull* 1984;12: 349–353.

35. Vidal C, Jordan W, Zieglgansberger W. Corticosterone reduces the excitability of hippocampal pyramidal cells *in vitro*. *Brain Res* 1986; 383:54–59.
36. Rey M, Carlier E, Soumireu-Mourat B. Effects of corticosterone on hippocampal slice electrophysiology in normal and adrenalectomized BALB/c mice. *Neuroendocrinology* 1987;46: 424–429.
37. Saphier D. Cortisol alters firing rate and synaptic responses of limbic forebrain units. *Brain Res Bull* 1987;19:519–524.
38. Joëls M, de Kloet ER. Effects of glucocorticoids and norepinephrine on the excitability in the hippocampus. *Science* 1989;245:1502–1505.
39. Mosher KM, Young DA, Munck A. Evidence for irreversible, actinomycin D-sensitive, and temperature-sensitive steps following binding of cortisol to glucocorticoid receptors and preceding effects on glucose metabolism in rat thymus cells. *J Biol Chem* 1971;246:654–659.
40. Nichols NR, Verner SP, Masters JN, May PC, Millar SL, Finch CE. Rapid corticosterone-induced changes in gene expression in rat hippocampus display type II glucocorticoid receptor specificity. *Mol Endocrinol* 1988;2:284–299.
41. Aldenhoff JB, Hennig S. Corticosterone hyperpolarizes pyramidal neurons of the hippocampus *in vitro*. (*Abstr.*) *Soc Neurosci Abstr* 1988;14: 443.
42. Hua S-Y, Chen Y-Z. Membrane receptor-mediated electrophysiological effects of glucocorticoid on mammalian neurons. *Endocrinology* 1989;124:687–691.
43. Hotson JR, Prince DA. A calcium-activated hyperpolarization follows repetitive firing in hippocampal neurons. *J Neurophysiol* 1980;43:409–419.
44. Gustafsson B, Wigstrom H. Evidence for two types of afterhyperpolarization in CA1 pyramidal cells in the hippocampus. *Brain Res* 1981; 206:462–468.
45. Madison DV, Nicoll RA. Control of the repetitive discharge of rat CA1 pyramidal neurones *in vitro*. *J Physiol (Lond)* 1984;354:319–331.
46. Lancaster B, Adams PR. Calcium-dependent current generating the afterhyperpolarization of hippocampal neurons. *J Neurophysiol* 1986;55: 1268–1282.
47. Nicoll RA. The coupling of neurotransmitter receptors to ion channels in the brain. *Science* 1988;241:545–551.
48. Joëls M, de Kloet ER. Mineralocorticoid receptor-mediated effects on membrane properties of rat CA1 pyramidal neurons *in vitro*. *Proc Natl Acad Sci* 1990;87:4495–4498.
49. Kerr DS, Landfield PW. A corticosteroid-sensitive component of the hippocampal calcium-dependent afterhyperpolarization increases with aging (*Abstr.*) *Soc Neurosci Abstr* 1988;14:1271.
50. Madison DV, Nicoll RA. Noradrenaline blocks accommodation of pyramidal cell discharge in the hippocampus. *Nature* 1982;299:636–638.
51. Haas HL, Konnerth A. Histamine and noradrenaline decrease calcium-activated potassium conductance in hippocampal pyramidal cells. *Nature* 1983;302:432–434.
52. Madison DV, Nicoll RA. Actions of noradrenaline recorded intracellularly in rat hippocampal CA1 pyramidal neurones, *in vitro*. *J Physiol (Lond)* 1986;372:221–244.
53. Mobley PL, Sulser F. Adrenal corticoids regulate sensitivity of noradrenaline receptor–coupled adenylate-cyclase in brain. *Nature* 1980; 286:608–609.
54. Roberts DCS, Bloom FE. Adrenal steroid-induced changes in β-adrenergic receptor binding in rat hippocampus. *Eur J Pharmacol* 1981;74: 37–41.
55. Stone EA, McEwen BS, Herrera AS, Carr KD. Regulation of α and β components of noradrenergic cyclic AMP response in cortical slices. *Eur J Pharmacol* 1987;141:347–356.
56. Borrell J, de Kloet ER, Bohus B. Corticosterone decreases the efficacy of adrenaline to affect passive avoidance retention of adrenalectomized rats. *Life Sci* 1984;34:99–105.
57. Tausk M. Hat die Nebenniere tatsachlich eine Verteidigungsfunction. In: *Das Hormon*, vol 3. The Netherlands: Organon, 1951;1.
58. Munck A, Guyre PM, Holbrook NJ. Physiological functions of glucocorticoids in stress and their relation to pharmacological actions. *Endoc Rev* 1984;5:25–44.

Volume Transmission in the Brain: Novel Mechanisms for Neural Transmission, edited by K. Fuxe and L.F. Agnati.
Raven Press, Ltd., New York © 1991.

16

Peptide Expression Patterns in Neuroendocrine Tumors of Transgenic Mice

*†Guido Rindi, *Enrico Solcia, and †Julia M. Polak

**Department of Human Pathology, University of Pavia, Anatomic Pathology Unit, Pavia, Italy; and †Department of Histochemistry, Royal Postgraduate Medical School, Hammersmith Hospital, London, England*

Tumors producing regulatory peptides or amines are usually grouped in a special category, named *neuroendocrine tumors*, independently of their origin from endocrine cells, neurons, or related progenitor cells (1,2). They are characterized by the nature of their neuroendocrine products (3), which may cause hyperfunctional syndromes (4). Among these, some, such as the so-called vasoactive intestinal peptide (VIP)oma syndrome caused by excessive production of the neuropeptide VIP, are caused by both endocrine and neurogenic tumors (5). In addition, neuroendocrine tumors share a number of morphologic markers normally expressed by neurons, endocrine cells, or both (2,6). To investigate *in vivo* functional changes induced by regulatory peptide overproduction, animals bearing well-characterized neuroendocrine tumors would be of help. Tumors induced in transgenic mice may offer an efficient way to obtain such tumors.

TRANSGENIC MICE AS SOURCES OF NEUROENDOCRINE TUMORS

By transgenic technology, a recent gene manipulation technique (7–9), it has been rendered possible to obtain strains of mice that stably inherit the property of developing functioning tumors of neuroendocrine organs (10–19). Indeed, it is now theoretically possible to obtain the growth of cells specialized in the production of a given peptide, by using transgenes made up of the regulatory sequences of the peptide gene aligned as to transcribe sequence coding for transforming oncoprotein(s). The early region of the simian virus 40 (SV40) encoding for the potent oncoprotein large T antigen has been largely used (20,21). If heritable formation of neuroendocrine neoplasms results, obtaining growths of neuroendocrine cells producing the expected peptide would give important information on the structure and functions of the gene under study and, aside, would provide strains of mice with a nonphysiologic, excessive source of such peptide. This implies that mice programmed to rise the level of a predetermined peptide in their circulating fluids are available for study during the whole process of tumorigenesis, from early preneoplastic changes to progressive growth. Complete characterization of the secretory products of neuroendocrine tumors developing in transgenic mice represents the first step to enrole these strains of mice as possibly useful tools for functional studies. In this article we review the relevant morphofunctional features of neuroendocrine tumors developing in four transgenic mice harboring peptide-promoted hybrid oncogenes.

INSULIN/SV40 TRANSGENIC MICE

In this pioneering transgenic experiment, Hanahan (11) used hybrid oncogenes composed of the rat insulin II gene 5′ flanking deoxyribonucleic acid (DNA) linked to the SV40 early region (Tag). Resulting transgenic mice proved to die prematurely while developing heritable tumors of the pancreas between 70 and 140 days of age. Sudden death was shown to be caused by severe hypoglycemia, and transgenic mice needed lifelong hyperglucose diet to grow. Pancreas tumors were composed of proliferating insulin-producing B cells showing beta granules at ultrastructural analysis (12). In addition, Power et al. (13), in a subsequent analysis of INS/SV40 advanced tumors, identified a minor pancreatic-polypeptide subpopulation in 30% of examined growths. The whole endocrine pancreas was largely remodeled by the presence of hyperplastic–dysplastic islets: Atypical cells formed the large majority of B cells in each islet. The oncoprotein large T antigen (Tag) was expressed in all islet B cells, besides in tumor cells. The development of pancreatic tumors seemed to result from Tag-promoted proliferation of B cells, and a possible sequence from hyperplasia to neoplasia was suggested (11).

It was concluded that 520 bp of 5′ flanking rat insulin II gene DNA was capable of directing site-specifically to pancreatic insulin-producing B cells the expression of SV40 large T antigen.

ARGININ VASOPRESSIN/SV40 TRANSGENIC MICE

Murphy et al. (15) developed this model (15). The transgene in use was made up of 1.25 kb of the 5′ flanking DNA, deriving from the bovine Arg-vasopressin gene (AVP) linked to the early region of SV40. This transgenic experiment aimed at assessing the regulatory properties of the putative promoter of the vasopressin gene. Of six founder transgenic mice, one line (VT-C) was derived that showed an heritable pathology leading transgenic mice to death between 12 and 20 weeks. VT-C mice developed neuroendocrine tumors of anterior pituitary and pancreas, but none of these growths expressed Arg-vasopressin (15). We performed extensive analysis of VT-C mice phenotype (16). Tumors of the pancreas were composed of insulin-producing B cells with a significant pancreatic polypeptide (PP)-immunoreactive subpopulation (in 25% of tumors). Pituitary tumors were composed of poorly granulated cells, which failed to react to the tests performed for the hormones normally produced in anterior pituitary (see Fig. 1) (15,16). Although no direct evidence of endocrine imbalance was at that time monitored, it was reasonable that the cause of premature death of VT-C mice was related to acute hyperinsulinemia and consequent lethal hypoglycemia or to acute phenomena caused by large intracranial growths (i.e., lethal endocranial hypertension caused by rupture of the abnormal blood vessels). By time course histological analysis (16), it was possible to establish that the endocrine pancreas underwent a large remodeling, which started early after birth: Hyperplastic-dysplastic changes were detected in mature B cells single or organized in foci inside well-formed islets. All the atypical cells expressed insulin and were positive for large T antigen, the transforming oncoprotein encoded by the transgene. In pituitaries, similar lesions appeared later, had cytologic features of severe anaplasia, and the elapsing time between detection of early lesions and development of large neoplasms was significantly shorter than in endocrine pancreas. It was concluded that the transgene under study was capable of inducing transformation in two separate endocrine organs, that the target cells were insulin-producing B cells in the pancreas and an

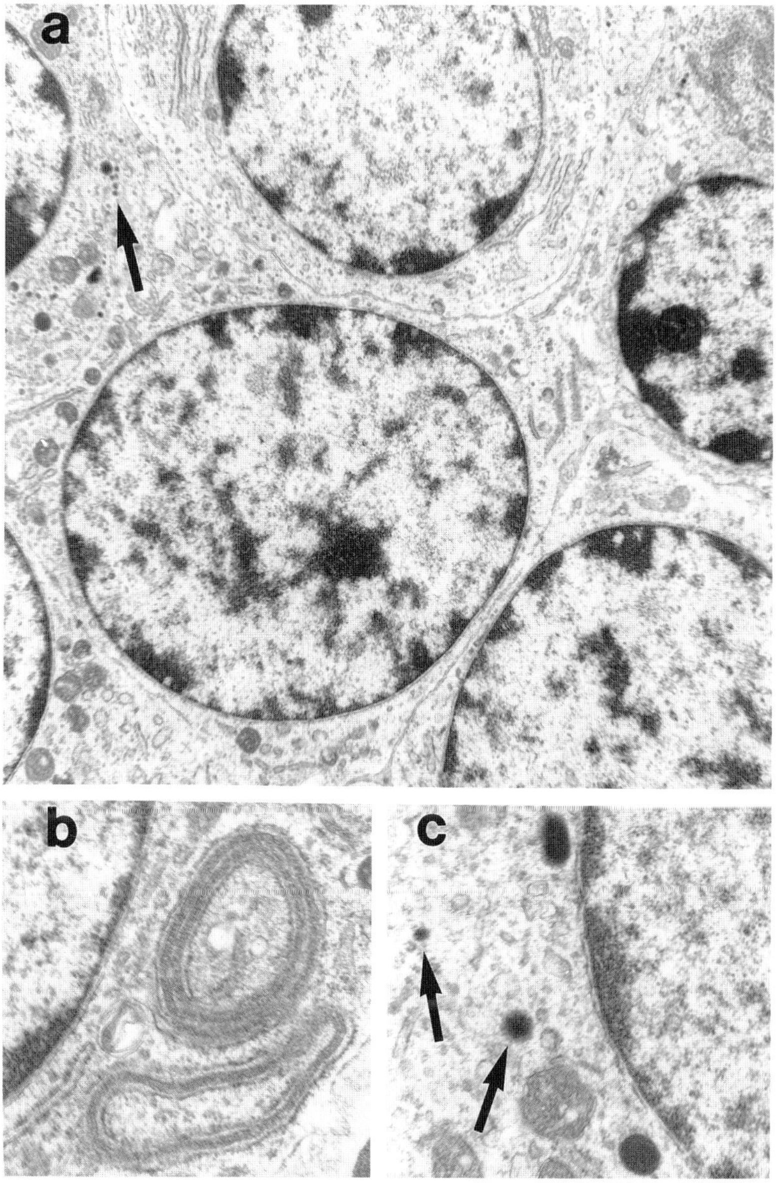

FIG. 1. Arginin-vasopressin/SV40 transgenic mice, pituitary tumor. **a:** ×8,000, uranyl acetate/lead citrate: general view of the ultrastructure of neoplastic elements. Note variable size of nuclei, prominent rough endoplasmic reticulum, and the few, small granules (*upper left corner, arrow*). **b:** ×24,000 uranyl acetate/lead citrate: whorls of rough endoplasmic reticulum. **c:** ×32,000, uranyl acetate/lead citrate: electron-dense granules of different size and shape (*arrows*).

undetermined endocrine element in the anterior pituitary, and that the proliferative processes had different paces in the two target organs. In VT-C mice, the hypotalamic vasopressinergic system appeared not to be affected by the transgene: By immunohistochemistry, no expression of the transgene could be detected in those areas of the central nervous system where Arg-vasopressin is normally produced, nor was the synthesis of vasopressin perturbed (15).

The above transgenic experiment indicated that the vasopressin sequences used to build the transgene were not sufficient to direct the expression of the SV40 early region (Tag) to the hypotalamic vasopressinergic neurons. It was hypothesized that further regulatory sequences could be needed for site-specific expression, as suggested by complex regulatory patterns elucidated in other transgenic experiments (14,22–25). Moreover, possible effects caused by chromosomal integration site could not be ruled out (15). Further genetic experiments appear necessary to assess the above hypothesis and to elucidate the regulatory system of the vasopressin gene. Nevertheless, VT-C transgenic mice constitute an intriguing model for the study of multiple endocrine neoplasia syndrome type I (MEN I) (26,27). It is an attractive observation that *mature* islet B cells were target of the transforming process (16), showing a slowly progressive, multifocal impairment of a B-cell subpopulation. This gives morphological support to the hypothesis of an early, progressive alteration of the hormone secretion needed to preserve glucose homeostasis. Functional studies are required to monitor these biochemical modifications.

As no hormone content has been detected in AVP/SV40 pituitary tumor cells, these lesions are likely "nonfunctioning," a type of neuroendocrine tumor known to arise in human anterior pituitary and whose origin still requires clarification (28–30).

GLUCAGON/SV40 TRANSGENIC MICE

Efrat et al. (17) produced these transgenic mice by using an hybrid oncogene made by the rat preproglucagon gene 5′ flanking DNA linked to the SV40 early region (Tag). Resulting GLU/Tag transgenic mice heritably developed tumors of pancreas at 9 to 12 months of age. Pancreatic growths were isolated growths composed of glucagon-producing A cells, showing typical alpha-granules at the ultrastructural level. All tumor cells coexpressed large T antigen, as did proliferating A cells, forming peripheral mantles in rare hyperplastic islets. A sequence from focal hyperplasia to neoplasia was proposed to monitor tumor development in A cells. Tag expression was also detected in neurons of those areas of the central nervous system (CNS) where preproglucagon gene is normally expressed and in other unexpected sites of CNS. Unexpected Tag-positive neurons failed to colocalize glucagon or glucagon-like peptide (GLP)I, the parts of preproglucagon peptide (31) they were tested for.

It was concluded that 500 bp of 5′ flanking rat preproglucagon gene DNA was capable of directing the expression of Tag site-specifically to pancreatic glucagon-producing A cells and to CNS neurons partly coexpressing the preproglucagon molecule. The fact that, despite Tag expression, no appreciable proliferation of neurons could be identified, was attributed to transformation resistance properties inherent to neurons and their lack of proliferative capacity.

INSULIN-SV40/INSULIN-PyST TRANSGENIC MICE

Grant et al. (18) developed transgenic mice carrying two hybrid oncogenes made up by the rat insulin II gene promoter (RIP) (11) aligned so as to transcribe either the early region of SV40 (Tag) or the polyoma

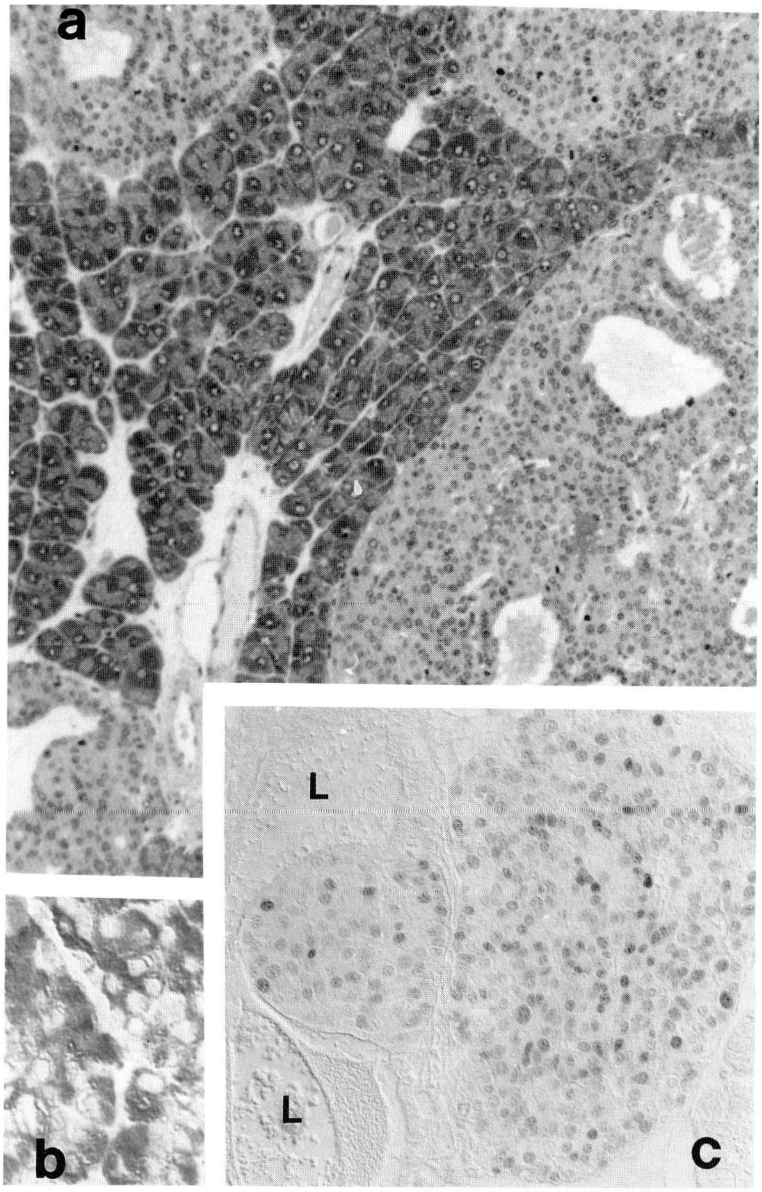

FIG. 2. Insulin/SV40-insulin/PyST double transgenic mice pancreas. **a:** ×100, hematoxilin and eosin: general view of pancreas showing large neuroendocrine tumors (*upper and lower right corners*) and two dysplastic islets (*upper and lower left corners*). The exocrine parenchima (*center* of the micrograph) is compressed by the expanding endocrine growths. **b:** ×500, peroxidase-antiperoxidase method, Nomarsky optic: insulin-immunoreactive tumor cells. **c:** ×250, avidin-biotinylated peroxidase complex (ABC) method, Nomarsky optic: large T-antigen immunoreactive nuclei of B cells in a large, dysplastic islet (*right*) and in an adjacent small islet bulging into the lumen of a ductule. L, lumen of the duct.

virus small T antigen coding sequences (PyST) (20). Double transgenic mice were obtained by crossing two lineages of single transgenic mice, i.e., RIP1/Tag2 mice [one of the strains obtained in the above-described insulin/SV40 experiment (11)] crossed with RIP2/PyST1 strain. RIP2/PyST1 transgenic mice were produced with a transgene made with the insulin promoter (RIP) (11) linked to the polyoma virus small T-antigen cDNA (PyST). The latter experiment was planned to study the effect of the polyoma oncoprotein small T antigen on B cells; however, no pathology resulted in the three strains of transgenic mice obtained (for details see ref. 18).

RIP1Tag2/RIP2PyST1 double transgenic mice proved to develop the expected insulinomas of the pancreas, as well as highly metastatic neuroendocrine tumors of the small intestine (18). Tumors of the pancreas appeared to retain the morpho-functional aspects previously described (11–13), being composed of insulin-producing B cells (Fig. 2a,b) with a significant PP-immunoreactive subpopulation in 42% of analyzed neoplasms. Neuroendocrine tumors of the small intestine did not express insulin but were mainly composed of secretin-producing S-type cells (19). Other gastrointestinal hormones [i.e., glicentin, gastrin, glucose-dependent insulinotropic polypeptide (GIP), neurotensin, somatostatin, and PP] were also detected in minor subpopulations of intestinal tumors, either as scattered elements (e.g., somatostatin-immunoreactive cells) or organized in larger clones of neoplastic cells (e.g., glicentin-immunoreactive cells) (19). In addition, by radioimmunoassay of plasma, high levels of insulin and secretin immunoreactants were detected, confirming the secretory properties of *both* pancreatic and intestinal growths (19). SV40 large T-antigen expression was detected in all B cells of nontumor pancreatic islets and in some mucosal secretin cells of the small intestine (19). It was concluded that the insulin promoter under study, besides inducing transformation in pancreatic B cell, is also capable of promoting tumorigenesis in a separate endocrine cell, the intestinal secretin cell. This phenomenon suggests the existence of a possible relation between B cells and S cells, which underlies the capability of the transgene to express in both these sites.

The above transgenic model is a unique source of double hyperproduction of hormones and is the first recorded model for endocrine tumorigenesis in mammalian intestine. Similar to INS/SV40 mice but at variance with the previously discussed VT-C mice, the progressive derangement of the endocrine pancreas appears to embrace diffusely the whole B-cell population. The oncoprotein large T antigen was detected in almost all B cells, leading to a diffuse pattern of hyperplastic-dysplastic-adenomatous lesions. Interestingly, nuclear Tag-immunoreactivity was also detected in a few B cells composing neoforming islets budding off from ducts and in single insulin-immunoreactive elements of some pancreatic duct (see Fig. 2c). The diffuse hyperplastic–dysplastic changes of B cells may contribute with B-cell tumors to the severely impaired glucose metabolism of RIP1Tag2/RIP2PyST1 transgenic mice, which require lifelong high glucose diet. Which effect(s) may play the high levels of circulating secretin in the mechanisms related to food digestion and absorption remains to be elucidated.

CONCLUDING REMARKS

Peptide Expression in Neuroendocrine Tumors of Transgenic Mice

We have reviewed four transgenic mice models in which peptide-promoted oncoprotein was successfully used to induce oncogenesis in endocrine organs. The pattern of peptide expression in neuroendocrine tumors of transgenic mice appeared to be specifically determined by the transgene used in INS/SV40 (11–13) and GLU/SV40 mice

(17), whereas in AVP/SV40 (15,16) and partly in INS/SV40-INS/PyST mice (18,19), inappropriate endocrine cells were transformed with subsequent production of unexpected peptides. Nevertheless, in neuroendocrine tumors of INS/SV40 (13) and GLU/SV40, minor clonal subpopulations of cells expressing peptides inappropriate to the promoter sequences could be detected. In any case, the peptide pattern of all the above-discussed neuroendocrine tumors appeared to reflect the endocrine component of the organ in which the tumors developed. Heterogeneity of peptide expression is a frequent finding in human neuroendocrine tumors (1,2,5,32,33). This phenomenon could find explanation considering the known plasticity of endocrine cells when transformed (34–36).

Unexpected Cellular Targets of Hormone-Promoted Transgenes

Discrepancies in respect to the expected hormone production as targeted by the hybrid oncogene appear to be frequent. Also the brief (520 bp) insulin promoter, whose site-specificity has been largely confirmed by a number of transgenic experiments (11,37–39), in RIP1Tag2/RIP2PyST1 mice (and in RIP1Tag2 mice with low frequency of intestinal tumors growth) is capable of targeting transformation to an unexpected endocrine element (18,19). Grant et al. (18) suggested that some effect of chromosomal location of the transgenes integration might be responsible for this. Hypothetically, chromosomal integration site effects might potentiate the proliferative pressure induced by large T antigen, thus dramatically enhancing the otherwise low promoting capabilities of the insulin promoter in the secretin cell. Interestingly, in RIP1Tag2/RIP2PyST1 intestinal tumors, a high proliferative rate was coupled with high degrees of anaplasia and loss of immunohistochemically detectable endocrine markers in association with tumor progression (18,19). Analogue morphological features have been detailed in pituitary tumors of VT-C mice (15,16), perhaps supporting a transgene chromosomal site integration effect(s) as a cooperating factor for the severe transforming process that takes place in that organ.

Possible Routes of Transformation in Neuroendocrine Cells

From the above observations it is possible to speculate that endocrine cells, whatever their specific product, behave similarly when under comparable proliferative pressure, leading to neoplasia via two possible distinct routes: a slow-rate mechanism, which requires long-term action of transforming factor(s) and is associated with hyperplastic-dysplastic changes (the mechanism typical of pancreatic growths in all models discussed); and a high-rate mechanism, in which transformation takes place suddenly with rapid gain of all the peculiar features of neoplasia (i.e., high mitotic rate, loss of intercellular adherency, progressive loss of differentiation markers, etc.) and is not preceded or associated with slow-growing preneoplastic lesions [mechanism observed in the intestine of RIP1Tag2/RIP2PyST1 mice (18,19) and in the pituitary of VT-C mice (15,16)].

Transgenic mice harboring hormone-promoted oncoproteins proved to be a unique tool for research, allowing us to gain important information in the fields of genetics and pathology. The use of transgenic mice for functional studies on derangements of neuroendocrine balances is suggested.

SUMMARY

Transgenic technology made it possible to promote proliferation in peptide-producing neuroendocrine cells. Relevant morpho-functional features of neuroendocrine tumors developing in four transgenic mice harboring peptide-promoted hybrid oncogenes were reviewed and discussed. Insulin/SV40 early region, coding for the oncoprotein large T antigen (Tag), Arg-vasopressin/SV40 Tag, glucagon/SV40 Tag single transgenic mice, and insulin/SV40 Tag-insulin PyST double transgenic mice all develop heritable neuroendocrine tumors. Peptide production in neuroendocrine tumors of pancreas, intestine, and anterior pituitary is analyzed in respect to transgene-induced peptide expression. The need of complete characterization of peptide expression is stressed in view of possible application of transgenic strains for functional studies.

ACKNOWLEDGMENTS

GR is supported by a postdoctoral fellowship from IRCCS Policlinico San Matteo, Italy; this work was supported in part by the Italian Research Council (Oncology Project), Italy, and by the Cancer Research Campaign, UK.

REFERENCES

1. Polak JM, Bloom SR. *Endocrine tumors.* Edinburgh, London, Melbourne, New York: Churchill Livingstone, 1985.
2. Solcia E, Capella C, Buffa R, Tenti P, Rindi G, Cornaggia M. Antigenic markers of neuroendocrine tumors: their diagnostic and prognostic value. In: Fenoglio-Preiser CM, Weinstein RS, Kaufmann N, eds. *New concepts in neoplasia as applied to diagnostic pathology.* Int. Acad. of Pathol. Monograph. Baltimore, London, Los Angeles, Sydney: Williams & Wilkins, 1986;242–261.
3. Solcia E, Usellini L, Buffa R, et al. Endocrine cells producing regulatory peptides. *Experientia* 1987;43:834–850.
4. Ch'ng JLC, Polak JM, Bloom SR. Endocrine syndromes. In: Polak JM, Bloom SR, eds. *Endocrine tumors.* Edinburgh, London, Melbourne, New York: Churchill Livingstone, 1985; 264–280.
5. Solcia E, Capella C, Riva C, Rindi G, Polak JM. The morphology and neuroendocrine profile of pancreatic epithelial VIPomas and extrapancreatic VIP-producing neurogenic tumors. *Ann NY Acad Sci* 1988;527:508–527.
6. Bishop AE, Power RF, Polak JM. Markers for neuroendocrine differentiation. *Pathol Res Pract* 988;183:119–128.
7. Gordon JW, Scangos GA, Plotkin DJ, Barbosa JA, Ruddle FH. Genetic transformation of mouse embryos by microinjection of purified DNA. *Proc Natl Acad Sci USA* 1980;77:7380–7384.
8. Hogan B, Costantini F, Lacy PE. *Manipulating the mouse embryo: a laboratory manual.* Cold Spring Harbor, New York, 1986.
9. Murphy D, Hanson J. The production of transgenic mice by the microinjection of fertilized one-cell eggs. In: Glover, ed. *DNA cloning: a practical approach.* London: IRL Press, 1988; 213–248.
10. Rindi G, Solcia E, Polak JM. Transgenic mouse models and peptide producing endocrine tumours: morpho-functional aspects. In: Polak JM, ed. *Regulatory peptides.* Basel, Boston, Berlin: Birkhauser Verlag, 1989;210–219.
11. Hanahan D. Heritable formation of pancreatic B-cell tumors in transgenic mice expressing recombinant insulin simian virus 40 oncogenes. *Nature* 1985;315:115–122.
12. Holm RF, Varndell I, Power RF, et al. Ultrastructure and electron immunocytochemistry of insulin-producing B-cell tumors from transgenic mice: comparison with counterpart human tumor. *Ultrastruct Pathol* 1988;12:547–559.
13. Power RF, Holm RF, Bishop AE, et al. Transgenic mouse model: a new approach for the investigation of endocrine pancreatic B-cell growth. *Gut* 1987;28:121–129.
14. Messing A, Chen HY, Palmiter RD, Brinster RL. Peripheral neuropathies, hepatocellular carcinomas and islet cell adenomas in transgenic mice. *Nature* 1985;316:461–463.
15. Murphy D, Bishop AE, Rindi G, et al: Mice transgenic for a vasopressin hybrid oncogen develop tumors of the endocrine pancreas and the anterior pituitary. A possible model for human multiple endocrine neoplasia syndrome type I. *Am J Pathol* 1987;129:552–566.
16. Rindi G, Bishop AE, Murphy D, Solcia E, Hogan B, Polak JM. A morphological analysis of endocrine tumor genesis in pancreas and anterior pituitary of AVP/SV40 transgenic mice. *Virchows Arch A* 1988;412:255–266.
17. Efrat S, Teitelman G, Anwar M, Ruggiero D, Hanahan D. Glucagon gene regulatory region directs oncoprotein expression to neurons and pancreatic alpha-cells. *Neuron* 1988;1:605–613.
18. Grant SGN, Seidman I, Hanahan D, Bautch VL. Early invasiveness characterizes metastatic carcinoid tumors in transgenic mice. (*submitted*)
19. Rindi G, Grant SGN, Yiangos Y, et al. Development of neuroendocrine tumors in the gastrointestinal tract of transgenic mice: heterogeneity of hormone expression. *Am J Pathol* 1990;136:1349–1363.
20. Tooze J. *Molecular biology of tumor viruses, part 2: DNA tumor viruses,* rev. 2nd ed. Cold Spring Harbor, New York, 1982.
21. Rigby PWJ, Lane DE. Structure and function of simian virus 40 large T-antigen. In: Klein G, ed. *Advances in viral oncology.* New York: Raven Press, 1983;31–57.

22. Hammer RE, Krumlauf R, Camper S, Brinster RL, Tilghman JM. Diversity of alpha-fetoprotein gene expression is generated by a combination of separate enhancer elements. *Science* 1987; 235:53–58.
23. Krumlauf R, Hammer RE, Brinster R, Chapman VM, Tilghman JM. Regulated expression of alpha-fetoprotein genes in transgenic mice. *Cold Spring Harbor Symp Quant Biol* 1985;50:371–378.
24. Palmiter RD, Brinster RL, Hammer RE, et al. Dramatic growth of mice that develop from egg microinjected with metallothionein-growth hormone fusion genes. *Nature* 1985;316:461–463.
25. Palmiter RD, Chen HY, Messing H, Brinster RL. SV40 enhancer and large T-antigen are instrumental in development of choroid tumors in transgenic mice. *Nature* 1985;316:457–460.
26. Wermer P. Genetic aspects of adenomatosis of endocrine glands. *Am J Med* 1954;16:363–371.
27. De Lellis RA, Dayal Y, Tischler AS, Lee AK, Wolfe JH. Multiple endocrine neoplasis (MEN) syndromes: cellular origins and interrelationships. *Int Rev Exp Pathol* 1986;28:163–215.
28. Kovacs K, Horvath E, Ryan N, Ezrin C. Null cell adenoma of the human pituitary. *Virchows Arch (Pathol Anat)* 1980;387:165–174.
29. Mashiter K, Adams EF, Van Noorden S. Secretion of LH, FSH and PRL shown by cell culture and immunocytochemistry of human functionless pituitary adenomas. *Clin Endocrinol(OXF)* 1981;15:103–112.
30. Capella C, Buffa R, Usellini L, et al. Alpha and beta subunits of glycoprotein hormones in argyrophil pituitary tumors with small granule cells. *Ultrastruct Pathol* 1983;4:35–50.
31. Conlon JM. Proglucagon-derived peptides: nomenclature, biosynthetic relationship and physiological roles. *Diabetologia* 1988;31:538–566.
32. Kloppel G, Heitz PU. Pancreatic endocrine tumors. *Pathol Res Pract* 1988;183:155–168.
33. Pilato FP, D'Adda T, Banchini E, Bordi C. Nonrandom expression of polypeptide hormones in pancreatic endocrine tumors. *Cancer* 1988;61: 1815–1820.
34. Oie HK, Gazdar AF, Minna JD, Weir GC, Baylin SB. Clonal analysis of insulin and somatostatin secretion and L-DOPA decarboxylase expression by a rat islet cell tumor. *Endocrinology* 1983;112:1070–1075.
35. Tischler AS. Gut peptide production by tumor cell lines in culture. *Scand J Gastroenterol Suppl* 1983;18:33–43.
36. Madsen OD, Larsson LI, Rehfeld JF, et al. Cloned cell lines from a transplantable islet cell tumor are heterogeneous and express cholecystokinin in addition to islet hormones. *J Cell Biol* 1986;103:2025–2034.
37. Sarvetnick N, Liggit D, Pitts SL, Hanse SE, Stewart TA. Insulin-dependent diabetes mellitus induced in transgenic mice by ectopic expression of class II MHC and interferon-gamma. *Cell* 1988;52:773–782.
38. Allison J, Campbell IL, Morahan G, Mandel TE, Harrison LC, Miller JFAP. Diabetes in transgenic mice resulting from overexpression of class I histocompatibility molecules in pancreatic B cells. *Nature* 1988;333:529–533.
39. Lo D, Burkly LC, Widera G, et al. Diabetes and tolerance in transgenic mice expressing class II MCII molecules in pancreatic beta cells. *Cell* 1988;53:159–168.

Volume Transmission in the Brain:
Novel Mechanisms for Neural Transmission,
edited by K. Fuxe and L.F. Agnati.
Raven Press, Ltd., New York © 1991.

17

Neuronotrophic Factors in CNS Degeneration and Regeneration and Their Modulation

Stephen D. Skaper, Guido Vantini, Gino Toffano, and Alberta Leon

Fidia Research Laboratories, Abano Terme, Italy

The causes of some of the most common and devastating diseases of the nervous system remain unknown, as well as effective strategies for their treatment. Prominent on this list are Alzheimer's disease, amyotrophic lateral sclerosis, and Parkinson's disease. Each of these conditions is classified as a neurodegenerative disorder of undefined origin. One interesting feature is that all three diseases reflect a pathological change within a relatively limited neuronal network in the central (CNS) or peripheral nervous systems (PNS).

In this chapter, we will discuss some aspects of current neurobiological research that relates to these problems. The focus will be on special protein molecules that, directly or indirectly, control the maintenance, survival, and repair capacity of distinct populations of nerve cells. The actions of these chemicals are subject to physiological modulations by other biological agents and, therefore, may be interwoven with various conditions of neurological relevance. After briefly surveying the various protein factors active toward nervous tissue, some aspects of their participation in CNS repair processes will be described. We will conclude with a consideration of how specific pharmacological agents may find use in the potentiation of endogenous neuronotrophic actions, in turn facilitating the natural ability of the brain to repair itself.

NERVE GROWTH FACTOR

A neuron is subject to a variety of influences from its surroundings, which can include the extracellular fluid and matrix, and nonneuronal cells (e.g., glia) in the immediate vicinity. The behaviors of neurons in physiological and pathological situations are sensitive to a particular group of proteins presented from this microenvironment, molecules termed *neuronotrophic factors*. The classical example of such macromolecules is nerve growth factor (NGF), a target-derived neuronotrophic factor that in the PNS is essential for the development and maintenance of sympathetic neurons and neural crest-derived sensory neurons (1,2). During development of the nervous system, the limiting amount of NGF present in the embryo supports only a fraction of the NGF-responsive neurons, with a large number being lost because of this limited trophic support (3,4). In the PNS, the amounts of NGF protein and its messenger ribonucleic acid (mRNA) show, in most cases, a good correlation with the density of sympathetic innervation (5,6). In the brain, cholinergic target areas synthesize NGF mRNA and protein, with the NGF being retrogradely transported to support cholinergic neurons of the basal forebrain nuclei (7–9).

Nerve growth factor effects are mediated

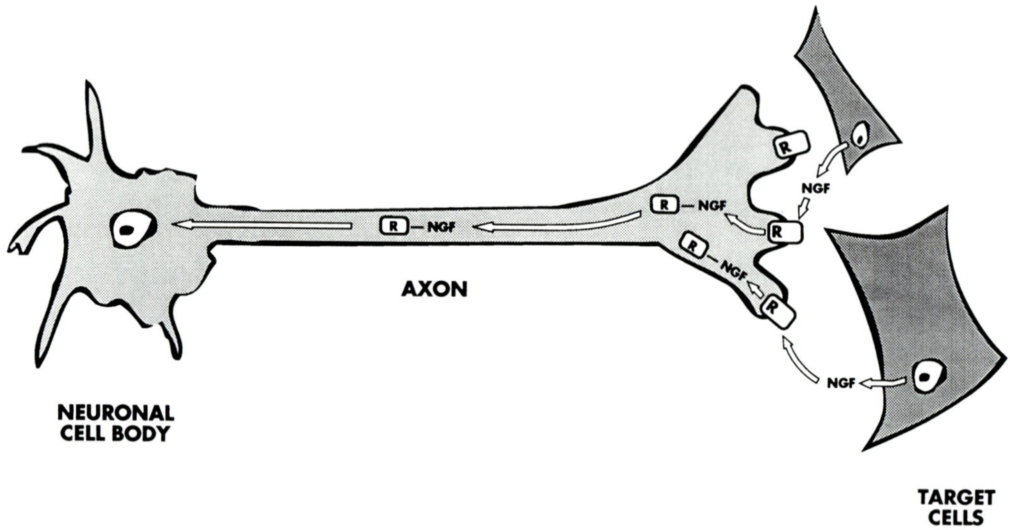

FIG. 1. Schematic representation of neuronotrophic action of NGF (NGF) R–NGF, receptors.

by interaction with specific receptors that, in the PNS, have been located on the surface of sensory and sympathetic neurons (10,11). The receptor for NGF has also been identified on central cholinergic neurons (12–15). After binding to its receptor, the NGF is internalized by receptor-mediated endocytosis and is transported retrogradely from the nerve terminal to the cell body (16,17), where NGF induces a variety of metabolic changes in the responsive neuron. Figure 1 schematically illustrates these events.

It is important to realize that the actions of NGF are not restricted to neurons only during certain periods of their development but that central and peripheral neurons in the adult animal have the capacity to respond to NGF. These investigations with NGF have led to the recognition of new families of protein trophic factors capable of influencing the growth, regeneration, and cellular functions of nervous tissue.

OTHER NEURONOTROPHIC FACTORS

During nervous system development, neurons are produced in numbers that far exceed what will normally be found in the adult. This reduction occurs at a time when nerve fibers are extending to their target territories to connect with their postsynaptic elements. Developmental neuronal death can be manipulated by experimentally changing the amount of target tissue in the embryo (18). This phenomenon has been clearly demonstrated in the case of NGF (3,4).

The identification, purification, and characterization of new neuronotrophic factors is being pursued in a number of laboratories, although these studies have yet to reach the level of biological and molecular information already obtained for NGF. Although beyond the scope of this chapter, we will briefly summarize what is known concerning other protein molecules having neuronotrophic actions.

Brain-derived neurotrophic factor (BDNF) is a macromolecule purified from pig brain (19), which promotes the survival of neural crest-derived chicken embryo dorsal root ganglion neurons and neural placode-derived nodose ganglion neurons, but not sympathetic or ciliary ganglion neurons (20). In the CNS, BDNF is active toward retinal ganglion neurons *in vitro* (21).

Investigations to identify trophic factors for cholinergic neurons of the ciliary ganglion led to the isolation of a ciliary neuronotrophic factor (CNTF) from the ciliary ganglion neuron-innervated tissues of the chicken embryo (22). CNTF proteins from adult rat sciatic nerve (23) and bovine heart (24) have also been purified. Interestingly, both the avian and mammalian CNTFs, although originally described for their activity on parasympathetic neurons, also support the survival of sensory and sympathetic neurons (20,21).

Another group of protein factors of relevance are the traditional "growth factors," like the fibroblast growth factors (FGF) and epidermal growth factor (EGF). Originally identified for their ability to promote cell proliferation, more recent evidence indicates that such polypeptide growth factors also display trophic activity for various CNS and PNS neurons *in vitro* (25–29). These growth factors and/or their receptors may also occur in the brain (30).

NEURONOTROPHIC FACTORS IN THE ADULT NERVOUS SYSTEM

Neuronotrophic factors like NGF appear to have an important role for central neurons, with adult CNS neurons continuing to depend on an adequate supply of endogenous trophic molecules. Dysfunction or degeneration in the adult brain could thus result from a neuronotrophic deficit. Such deficits may reflect an increased need by the neuron above and beyond the normal supply or from a reduced production, transport, or capacity of the neuron to use its endogenous supply. This hypothesis carries potentially important clinical implications, in that administration of the appropriate neuronotrophic factor could be used to interrupt or reduce the effects of a neurodegenerative process or even promote the regeneration of damaged brain neurons.

The notion that neuronotrophic factors are involved in adult CNS function derives from a number of observations. Trophic activities are present in brain extracts and increase after a traumatic lesion, particularly in the vicinity of the lesion (31,32). In the case of NGF, strong evidence supports its role for central cholinergic neurons, principally in the septum, the nucleus basalis of Meynert and in the corpus striatum. As already mentioned, NGF and its mRNA are present in highest amounts in the hippocampus and neocortex (7–9,33), the innervation territories of the cholinergic neurons of the septum and nucleus basalis, respectively, with NGF undergoing retrograde transport to the cell bodies of these same neurons (16,17). Receptors for NGF (12–15) and the associated mRNA (33–36) have also been localized to central cholinergic neurons. *In vitro,* these cholinergic neurons respond to exogenous NGF with an increase in choline acetyltransferase (ChAT) activity (37) and survival and neurite outgrowth (38,39). Intraventricular administration of NGF leads to up-regulation of NGF receptor mRNA expression (36,40) and receptor immunopositivity (36), whereas antibodies to NGF given by the same route reduce ChAT activity (41). In nonlesioned adult rats, NGF infusion is able to elevate ChAT activity in the hippocampus and septum (42) and ChAT immunostaining in striatal cholinergic interneurons (43); in the neonate, ChAT activity in striatum (44) and septum (Fig. 2) is also increased.

The cholinergic septo-hippocampal system has proven useful in evaluating *in vivo* the neuronotrophic hypothesis for the CNS. Cholinergic neurons of the medial septum project their axons to the hippocampus via a well-defined area, the fimbria. A unilateral fimbrial axotomy results in a lesion that presumably interrupts the retrograde flow of hippocampal-derived NGF to the above cholinergic neurons, thereby imposing on them a trophic deficit. The injury induced by this lesion can be reduced by the exogenous administration of NGF into the lateral ventricles adjacent to the axotomized neurons (45–48). Collectively,

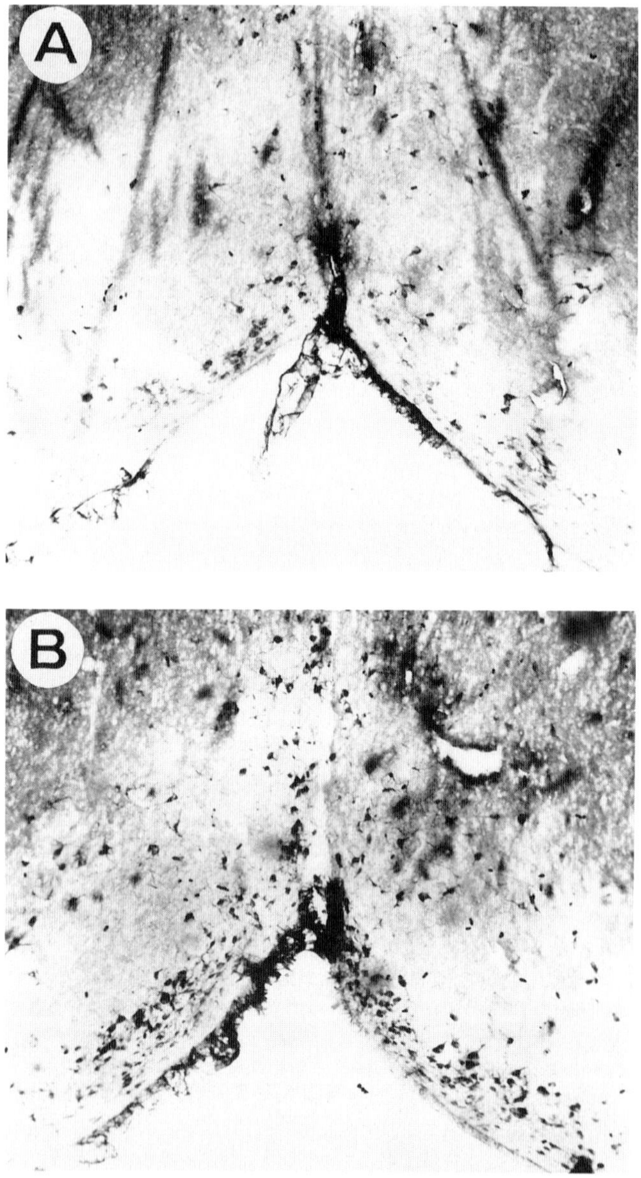

FIG. 2. Effect of intraventricular NGF administration on ChAT immunostaining in neonatal rat septum. Coronal sections through the septal complex of 9-day-old rats were stained to visualize ChAT-immunoreactive neurons. **A:** control; **B:** NGF-treated (daily, from postnatal days 2 to 8).

these findings strongly support a function for NGF as a retrograde trophic messenger for such cholinergic neurons.

In addition to basal forebrain neurons, other CNS populations may be responsive to NGF and to other trophic factors. NGF has been reported to protect adult rat retinal ganglion cells from degeneration after optic nerve section (49). Basic FGF (bFGF) may also be active against this CNS neuronal injury paradigm (50), as well as in the septo-hippocampal model for cholinergic neuron dysfunction (51). Thus, like NGF, the trophic effects of bFGF first described *in vitro* are confirmed at the *in vivo* level.

DEGENERATION IN CNS

A number of pathological states in the CNS appear to target particular populations of neurons. For example, Alzheimer's disease is characterized by a deficit of central cholinergic neurons; Parkinson's disease affects primarily dopaminergic neurons within the substantia nigra; amyotrophic lateral sclerosis strikes motor neurons. The very selectivity of the neurons involved in these degenerative situations suggests the action of a specific chemical substance (or substances) for each affected neuronal population and, in particular, the possibility of a neuronotrophic insufficiency (52,53). The ability of NGF to prevent the dysfunction and/or death of adult cholinergic neurons (45–48) and to stimulate their cholinergic behaviors in the *intact* adult animal (42,43) has provoked speculation that the availability of a neuronotrophic factor for these neurons (perhaps NGF) may well control their functional performance and survival. A trophic deficit, conceivably, could lead to cholinergic cell deterioration and memory deficits—a situation known to occur in Alzheimer's disease (53). In cognitively impaired aged rats, exogenous NGF is reported to ameliorate both spatial memory impairment (54,55) and cholinergic neuron atrophy (54).

Brain aging may also reflect a trophic deficit. Several similarities can be found between Alzheimer's disease and the apparently progressive reduction in the regenerative capacity of neural systems with age. Most affected appear to be cholinergic neurons of the nucleus basalis of Meynert and of the septum—the same areas as in Alzheimer's (56). Interestingly, decreased levels of NGF and its mRNA have been described in the hippocampus of the aged rat brain (57) as well as a loss of dendritic NGF receptor immunoreactivity in cholinergic nuclei (58). Trophic factors may have a role here also, as already hypothesized for Alzheimer's disease (53). The septohippocampal model for CNS regeneration (45–48) and reduction of cholinergic neuron atrophy in aged rats by NGF (54) can provide useful paradigms for further exploration of this question.

GANGLIOSIDES AND THE MODULATION OF NEURONOTROPHIC FACTOR SIGNALS

Gangliosides comprise a family of naturally occurring sialic acid–containing glycosphingolipids (59) found within all vertebrate cell membranes, with the greatest concentration in mammals being in CNS gray matter, especially in the region of synaptic terminals (60). These molecules are believed to participate in a variety of neurobiological functions related to the maturation and repair of neuronal tissue. Modifications of the ganglioside profile take place during mammalian brain development (61); changes also occur in aging and in some neuropathologies (62). Antiganglioside GM1 antibodies interfere with neurite regeneration *in vitro* and axonal elongation *in vivo* (63) and produce long-lasting alterations in CNS circuitry maturation and behavioral processes in developing animals (64).

Studies with a number of neuronal cell types *in vitro* have demonstrated a clear capacity of exogenous gangliosides, mainly GM1, to enhance *de novo* neurite formation or neurite regeneration in the presence of the appropriate trophic signal (65–68). The ability of GM1 to potentiate neuronotrophic actions is not limited to NGF but is also applicable to NGF-unlike trophic activities (68,69). GM1, added to cultures of fetal mouse mesencephalic cells, stimulates the biochemical development and survival of the dopaminergic and gamma-aminobutyric acid (GABA)ergic neuronal populations (70).

These types of culture experiments emphasize two important aspects of ganglioside action. In cells with a neuronotrophic factor requirement for survival and/or neurite growth, ganglioside will improve the morphological or biochemical actions of the trophic molecule but cannot replace the latter. Gangliosides thus seem able to modulate the plastic responses of neurons to microenvironmental signals. The neuritogenic response to ganglioside is the result of an earlier onset of neurite regeneration and not a permanent increase in the number of neurite-positive cells (66,68).

In addition to enhancing *in vitro* neuronal cell responsiveness to exogenous trophic signals, GM1 improves recovery processes consequent to CNS injury in adult rodents (cf. refs. 71,72 for recent reviews). The ability of GM1 to act *in vivo* depends on the degree of the lesion, suggesting the need for a minimum level of trophic support. Because trophic signals increase in the brain after lesion (31,32), the *in vivo* actions of GM1 may reflect a potentiation of endogenous trophic molecules that undergo a lesion-dependent rise. Although a critical component in the brain's self-repair capacity, these latter activities may be insufficient when needed most.

Recent experimental evidence supports this concept of ganglioside potentiation of *in vivo* neuronotrophic signals. The biochemical and morphological changes that lesioned cholinergic neurons undergo are prevented by intraventricular NGF infusion, with NGF and GM1 in combination giving synergistic effects (73–75). GM1 also potentiates the NGF-induced increase of ChAT activity in cultured septal neurons

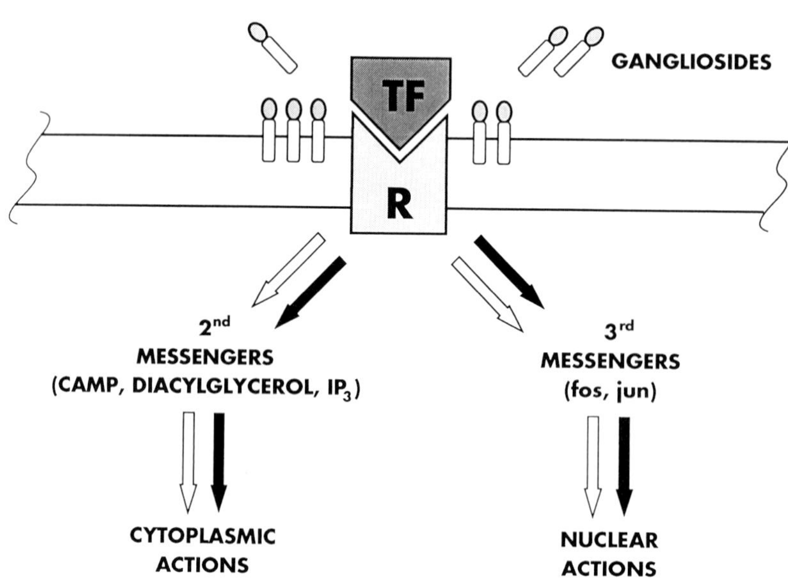

FIG. 3. One hypothetical view of how exogeneous gangliosides could function in relation to neuronotrophic signals. Trophic factor (TF) binds to specific membrane receptors (R), leading to transduction of biochemical signals that appear as "second" or "third" messengers in the cytosol. These molecules, in turn, can affect various cytoplasmic or nuclear processes. Insertion of gangliosides into the plasma membrane could, in principle, serve to amplify these messenger systems (*heavy arrows*), thereby leading to a potentiation of trophic signal(s). CAMP, cyclic adenosine monophosphate; IP_3, inositol trisphosphate; fos, proto–oncogene; jun, proto–oncogene.

(73). In the PNS, GM1 facilitates NGF actions after vinblastine-induced sympathectomy in newborn rats (76). Figure 3 presents one hypothetical interpretation of how exogenous gangliosides may act in relation to trophic signals.

CONCLUDING REMARKS

The information reviewed here argues strongly for the view that in adult life, as well as during development, nerve cells will continue to depend on neuronotrophic signals for their maintenance and function. A trophic deficit may well be one of the basic causes of axonal degeneration and neuronal death that occur after an acute brain insult or, in particular, chronic neuropathological situations or aging. Thus, an increase in trophic availability or efficacy could be directed at preventing or delaying neuronal losses and their accompanying behavioral deficits. The apparent need to apply a trophic protein, or genetically modified cells (77), at the immediate vicinity of a lesion can present practical difficulties in a clinical setting. However, the application of gangliosides to this problem offers an alternative, attractive approach. The data discussed here point to the pharmacological potential of ganglioside for improving the recovery of nerve cell dysfunction or impairment caused by insufficiency of NGF or other neuronotrophic molecules.

REFERENCES

1. Levi-Montalcini R. Nerve growth factor: its mode of action on sensory and sympathetic neurons. *Harvey Lect* 1966;60:217–259.
2. Levi-Montalcini R. The nerve growth factor thirty five years later. *Science* 1987;237:1154–1162.
3. Hamburger V, Brunso-Bechtold JK, Yip JW. Neuronal death in the spinal ganglia of the chick embryo and its reduction by nerve growth factor. *J Neurosci* 1981;1:60–71.
4. Hamburger V, Yip JW. Reduction of experimentally induced neuronal death in spinal ganglia of the chick embryo by nerve growth factor. *J Neurosci* 1984;4:767–774.
5. Heumann R, Korsching S, Scott S, Thoenen H. Relationship between levels of nerve growth factor (NGF) and its messenger RNA in sympathetic ganglia and peripheral target tissues. *EMBO J* 1984;3:3183–3189.
6. Shelton DL, Reichardt LF. Expression of the β-nerve growth factor gene correlates with the density of sympathetic innervation in effector organs. *Proc Natl Acad Sci USA* 1984;81:7951–7955.
7. Korsching S, Auburger G, Heumann R, Scott J, Thoenen H. Levels of nerve growth factor and its mRNA in the central nervous system of the rat correlate with cholinergic innervation. *EMBO J* 1985;4:1389–1393.
8. Shelton DL, Reichardt LF. Studies on the expression of the β-nerve growth factor (NGF) gene in the central nervous system: level and regional distribution of NGF mRNA suggest that NGF functions as a trophic factor for several distinct populations of neurons. *Proc Natl Acad Sci USA* 1986;83:2714–2718.
9. Whittemore SR, Ebendal T, Lärkfors L, et al. Developmental and regional expression of nerve growth factor messenger RNA and protein in the rat central nervous system. *Proc Natl Acad Sci USA* 1986;83:817–821.
10. Banerjee SP, Snyder SH, Cuatrecasas P, Greene LA. Binding of nerve growth factor in superior cervical ganglia. *Proc Natl Acad Sci USA* 1973;70:2519–2523.
11. Herrup K, Shooter EM. Properties of the β-nerve growth factor receptor of avian dorsal root ganglia. *Proc Natl Acad Sci USA* 1973;70:3884–3888.
12. Taniuchi M, Schweitzer JB, Johnson EM Jr. Nerve growth factor receptor molecules in rat brain. *Proc Natl Acad Sci USA* 1986;83:1950–1954.
13. Hefti F, Hartikka J, Salvatierra A, Weiner WJ, Mash DC. Localization of nerve growth factor receptors in cholinergic neurons of the human basal forebrain. *Neurosci Lett* 1986;69:37–41.
14. Schatteman GC, Gibbs L, Lanahan AA, Claude P, Bothwell M. Expression of NGF receptor in the developing and adult primate central nervous system. *J Neurosci* 1988;8:860–873.
15. Large TH, Weskamp G, Helder JC, et al. Structure and developmental expression of the nerve growth factor receptor in the chicken central nervous system. *Neuron* 1989;2:1123–1134.
16. Stoeckel K, Paravicini U, Thoenen H. Specificity of the retrograde axonal transport of nerve growth factor. *Brain Res* 1974;76:413–421.
17. Johnson EM Jr, Taniuchi M, Clark HB, et al. Demonstration of the retrograde transport of nerve growth factor receptor in the peripheral and central nervous system. *J Neurosci* 1987;7:923–929.
18. Hamburger V, Oppenheim RM. Naturally occurring neuronal death in vertebrates. *Neurosci Comment* 1982;1:39–55.
19. Barde YA, Edgar D, Thoenen H. Purification of a new neurotrophic factor from mammalian brain. *EMBO J* 1982;1:549–553.

20. Lindsay RM, Thoenen H, Barde YA. Placode and neural crest-derived sensory neurons are responsive at early developmental stages to brain-derived neurotrophic factor. *Dev Biol* 1985; 112:319–328.
21. Johnson JE, Barde YA, Schwab M, Thoenen H. Brain-derived neurotrophic factor supports the survival of cultured rat retinal ganglion cells. *J Neurosci* 1986;6:3031–3038.
22. Barbin G, Manthorpe M, Varon S. Purification of the chick eye ciliary neuronotrophic factor (CNTF). *J Neurochem* 1984;43:1468–1478.
23. Manthorpe M, Skaper SD, Williams LR, Varon S. Purification of adult rat sciatic nerve ciliary neuronotrophic factor. *Brain Res* 1986;367:282–286.
24. Watters DJ, Hendry IA. Purification of a ciliary neurotrophic factor from bovine heart. *J Neurochem* 1987;49:705–713.
25. Walicke P, Cowan WM, Ueno N, Baird A, Guillemin R. Fibroblast growth factor promotes survival of dissociated hippocampal neurons and enhances neurite extension. *Proc Natl Acad Sci USA* 1986;83:3012–3016.
26. Unsicker K, Reichert-Preibsch H, Schmidt R, Pettmann B, Labourdette G, Sensenbrenner M. Astroglial and fibroblast growth factors have neuronotrophic functions for cultured peripheral and central nervous system neurons. *Proc Natl Acad Sci USA* 1987;84:5459–5463.
27. Hatten ME, Lynch N, Rydel RE, et al. *In vitro* neurite extension by granule neurons is dependent upon astroglial-derived fibroblast growth factor. *Dev Biol* 1988;125:280–289.
28. Ferrari G, Minozzi MC, Toffano G, Leon A, Skaper SD. Basic fibroblast growth factor promotes the survival and development of mesencephalic neurons in culture. *Dev Biol* 1989;133: 140–147.
29. Morrison RS, Kornblum HI, Leslie FM, Bradshaw RA. Trophic stimulation of cultured neurons from neonatal rat brain by epidermal growth factor. *Science* 1987;238:72–75.
30. Herschman HR. Polypeptide growth factors and the CNS. *Trends Neurosci* 1986;9:53–57.
31. Nieto-Sampedro M, Manthorpe M, Barbin G, Varon S, Cotman C. Injury-induced neuronotrophic activity in adult rat brain: correlation with survival of delayed implants in the wound cavity. *J Neurosci* 1983;3:2219–2229.
32. Gasser UE, Weskamp G, Otten U, Dravid AR. Time course of the elevation of nerve growth factor (NGF) content in the hippocampus and septum following lesion of the septohippocampal pathway in rats. *Brain Res* 1986;376:351–356.
33. Lu B, Buck CR, Dreyfus CF, Black IB. Expression of NGF and NGF receptor mRNAs in the developing brain: evidence for local delivery and action of NGF. *Exp Neurol* 1989;104:191–199.
34. Ernfors P, Hallböök F, Ebendal T, et al. Developmental and regional expression of β-nerve growth factor receptor mRNA in the chick and rat. *Neuron* 1988;1:983–996.
35. Buck CR, Martinez HJ, Chao MV, Black IB. Differential expression of the nerve growth factor receptor gene in multiple brain areas. *Dev Brain Res* 1988;44:259–268.
36. Cavicchioli L, Flanigan TP, Vantini G, et al. NGF amplifies expression of NGF receptor messenger RNA in forebrain cholinergic neurons of rats. *Eur J Neurosci* 1989;1:258–262.
37. Hefti F, Hartikka J, Eckenstein F, Gnahn H, Heumann R, Schwab M. Nerve growth factor (NGF) increases choline acetyltransferase but not survival or fiber outgrowth of cultured fetal septal cholinergic neurons. *Neuroscience* 1985; 114:55–68.
38. Hartikka J, Hefti F. Development of septal cholinergic neurons in culture: plating density and glial cells modulate effects of NGF on survival, fiber growth, and expression of transmitter-specific enzymes. *J Neurosci* 1988;8:2967–2985.
39. Arimatsu Y, Miyamoto M, Tsukui H, Hatanaka H. Nerve growth factor promotes survival of retrogradely labeled hippocampus-projecting neurons in the rat basal forebrain *in vitro*. *Dev Brain Res* 1989;45:297–301.
40. Higgins GA, Koh S, Chen KS, Gage FH. NGF induction of NGF receptor gene expression and cholinergic neuronal hypertrophy within the basal forebrain of the adult rat. *Neuron* 1989; 3:247–256.
41. Vantini G, Schiavo N, Di Martino A, et al. Evidence for a physiological role of nerve growth factor in CNS of neonatal rats. *Neuron* 1989;3:267–273.
42. Fusco M, Oderfeld-Nowak B, Vantini G, et al. NGF affects uninjured, adult rat septohippocampal cholinergic neurons. *Neuroscience* 1989;33:47–52.
43. Hagg T, Hagg F, Vahlsing HL, Manthorpe M, Varon S. Nerve growth factor effects on cholinergic neurons of neostriatum and nucleus accumbens in the adult rat. *Neuroscience* 1989; 30:95–103.
44. Mobley WC, Rutkowski JL, Tennekoon GI, Buchanan K, Johnston MV. Choline acetyltransferase in striatum of neonatal rats increased by nerve growth factor. *Science* 1985;229:284–287.
45. Hefti F. Nerve growth factor promotes survival of septal cholinergic neurons after fimbrial transections. *J Neurosci* 1986;6:2155–2162.
46. Williams LR, Varon S, Peterson GM, et al. Continuous infusion of nerve growth factor prevents forebrain neuronal death after fimbria-fornix transection. *Proc Natl Acad Sci USA* 1986;83: 9231–9235.
47. Kromer LF. Nerve growth factor treatment after brain injury prevents neuronal death. *Science* 1987;235:214–216.
48. Montero CN, Hefti F. Rescue of lesioned septal cholinergic neurons by nerve growth factor: specificity and requirement for chronic treatment. *J Neurosci* 1988;8:2986–2999.
49. Carmignoto G, Maffei L, Candeo P, Canella R, Comelli C. Effects of NGF on the survival of rat retinal ganglion cells following optic nerve section. *J Neurosci* 1989;9:1263–1272.

50. Sievers J, Hausmann B, Unsicker K, Berry M. Fibroblast growth factors promote the survival of adult rat retinal ganglion cells after section of the optic nerve. *Neurosci Lett* 1987;76:157–162.
51. Anderson KJ, Dam D, Lee S, Cotman CW. Basic fibroblast growth factor prevents death of lesioned cholinergic neurons *in vivo*. *Nature* 1988;332:360–361.
52. Appel SH. A unifying hypothesis for the cause of amyotrophic lateral sclerosis, parkinsonism, and Alzheimer disease. *Ann Neurol* 1981;10: 499–505.
53. Hefti F, Weiner WJ. Nerve growth factor and Alzheimer's disease. *Ann Neurol* 1986;20:275–281.
54. Fischer W, Wictorin K, Björkund A, Williams LR, Varon S, Gage FH. Amelioration of cholinergic neuron atrophy and spatial memory impairment in aged rats by nerve growth factor. *Nature* 1987;329:65–68.
55. Mandel RJ, Gage FH, Thal LJ. Spatial learning in rats: correlation with cortical choline acetyltransferase and improvement with NGF following NBM damage. *Exp Neurol* 1989;104:208–217.
56. Whitehouse PJ, Price DL, Struble RG, Clark AW, Coyle JT, DeLong MR. Alzheimer's disease and senile dementia: loss of neurons in the basal forebrain. *Science* 1982;215:1237–1239.
57. Lärkfors L, Ebendal T, Whittemore SR, Persson H, Hoffer B, Olson L. Decreased level of nerve growth factor (NGF) and its messenger RNA in the aged rat brain. *Mol Brain Res* 1987;3:55–60.
58. Gomez-Pinella F, Cotman CW, Nieto-Sampedro M. NGF receptor immunoreactivity in aged rat brain. *Brain Res* 1989;479:255–262.
59. Svennerholm L. Biological significance of gangliosides. In: Dreyfus H, Massarelli R, Freysz L, Rebel G, eds. *Cellular and pathological aspects of glyconjugate metabolism*, vol 126. Paris: INSERM, 1984;21–44.
60. Ledeen RW. Gangliosides. In: Lajtha A, ed. *Handbook of neurochemistry*, vol 3. New York: Plenum Press, 1983;41–90.
61. Suzuki K. The pattern of mammalian brain gangliosides. III. Regional and developmental differences. *J Neurochem* 1965;12:969–979.
62. Suzuki K. Gangliosides and disease: a review. *Adv Exp Med Biol* 1984;174:407–420.
63. Sparrow JR, McGuiness C, Schwartz M, Grafstein B. Antibodies to ganglioside inhibit goldfish optic nerve regeneration *in vivo*. *J Neurosci Res* 1984;12:233–243.
64. Kasarskis EJ, Karpiak SE, Rapport MM, Yu RK, Bass NH. Abnormal maturation of cerebral cortex and behavioral deficit in adult rats after neonatal administration of antibodies to ganglioside. *Dev Brain Res* 1981;1:25–35.
65. Roisen FJ, Bartfeld H, Nagale R, Yorke G. Ganglioside stimulation of axonal sprouting *in vitro*. *Science* 1981;214:577–578.
66. Katoh-Semba R, Skaper SD, Varon S. Interaction of GM1 with PC12 pheochromocytoma cells: serum and NGF-dependent effects on neuritic growth (and proliferation). *J Neurosci Res* 1984;12:299–310.
67. Leon A, Benvegnù D, Dal Toso R, et al. Dorsal root ganglia and nerve growth factor: a model for understanding the mechanisms of GM1 effects on neuronal repair. *J Neurosci Res* 1984;12:277–287.
68. Skaper SD, Katoh-Semba R, Varon S. GM1 ganglioside accelerates neurite outgrowth from primary peripheral and central neurons under selected culture conditions. *Dev Brain Res* 1985; 23:19–26.
69. Spoerri PE, Roisen FJ. Ganglioside potentiation of NGF-independent agents on sensory ganglia. *Neurosci Lett* 1988;90:21–26.
70. Leon A, Dal Toso R, Presti D, et al. Development and survival of neurons in dissociated fetal mesencephalic serum-free cultures: II. Modulatory effects of gangliosides. *J Neurosci* 1988;8: 746–753.
71. Stein DG, Sabel BA. *Pharmacological approaches to the treatment of brain and spinal cord injury*. New York: Plenum Press, 1988.
72. Tettamanti G, Ledeen RW, Sandhoff K, Nagai Y, Toffano G. *Gangliosides and neuronal plasticity*. Fidia Research Series, vol 6. Padova: Liviana Press, 1986.
73. Cuello AC, Garofalo L, Kenigsberg RL, Maysinger D. Gangliosides potentiate *in vivo* and *in vitro* effects of nerve growth factor on central cholinergic neurons. *Proc Natl Acad Sci USA* 1989;86:2056–2060.
74. DiPatre PL, Casamenti F, Cenni A, Pepeu G. Interaction between nerve growth factor and GM1 monosialoganglioside in preventing cortical choline acetyltransferase and high affinity choline uptake decrease after lesion of the nucleus basalis. *Brain Res* 1989;480:219–224.
75. Casamenti F, Di Patre PL, Milan F, Petrelli L, Pepeu G. Effects of nerve growth factor and GM1 ganglioside on the number and size of cholinergic neurons in rats with unilateral lesion of the nucleus basalis. *Neurosci Lett* 1989;103:87–91.
76. Vantini G, Fusco M, Bigon E, Leon A. GM1 ganglioside potentiates the effect of nerve growth factor in preventing vinblastine-induced sympathectomy in newborn rats. *Brain Res* 1988;448:252–258.
77. Rosenberg MB, Friedman T, Robertson RC, et al. Grafting genetically modified cells to the damaged brain: restorative effects of NGF expression. *Science* 1988;242:1575–1578.

Volume Transmission in the Brain: Novel Mechanisms for Neural Transmission, edited by K. Fuxe and L.F. Agnati. Raven Press, Ltd., New York © 1991.

18

Neuron–Astroglia Interactions in Physiopathological Conditions: Possible Role of Volume Transmission

*R. Grimaldi, *I. Zini, *M. Zoli, *E. Merlo Pich, *F. Ferraguti, **P. Davalli, †A. Toffano, ††Kjell Fuxe, and *Luigi F. Agnati

**Institutes of Human Physiology and **Chemical Biology, University of Modena, Italy; †Fidia Research Laboratories, Abano Terme, Italy; and ††Department of Histology and Neurobiology, Karolinska Institutet, Stockholm, Sweden*

Polyamines (putrescine, spermidine, and spermine) are ubiquitous organic cations heterogeneously distributed in the central nervous system (CNS) (1); these compounds can be released in the extracellular space and also be actively taken up by nerve cells (2); they can also diffuse from the CNS into the cerebrospinal fluid (3). Polyamines are known to play a fundamental role in the regulation of cellular growth processes, including pre- and postnatal growth, and cell regeneration (4–6). In previous studies it has been shown that chemical, mechanical, and ischemic lesions of the CNS cause activation of the polyamine biosynthesis with a huge increase in ornithine decarboxylase (ODC) activity and putrescine accumulation (7–11). In models of mechanical, toxic, and ischemic brain injury, astroglial reaction has also been observed (12–16).

The aim of the present study was to assess whether the polyamine biosynthesis activation and release and the astroglial reaction induced by a localized mechanical lesion or by an ischemic insult are causally linked phenomena.

MATERIALS AND METHODS

Animals

Male-specific pathogen-free Sprague-Dawley rats (Charles River, Italy) of 250 to 275 g body weight are used. They were given food pellets and water *ad libitum* and kept under standardized temperature, humidity, and lighting conditions (light on at 8:00 AM and off at 8:00 PM).

Model of Mechanical Lesion

Rats were mounted in a stereotaxic frame and anesthetized with a mixture of air and halothane throughout the experimental period. The rat rectal temperature was maintained between 36.5 and 37.5°C. A probe for microdialysis (Carnegie Medicin, Sweden) was inserted into the striatum at the following coordinates: Ant + 0.1 mm from bregma, Lat 2.5 mm from midline, and Depth −5.3 mm from dura mater according to the Paxinos and Watson atlas (17). The probe was connected to a microinfusion pump (CMA/100 Carnegie Medicin, Sweden) and continuously perfused for 5 hr at 2 μl/min with Ringer's solution (148 mM Na^+, 2.4 mM Ca^{2+}, 4 mM K^+, and 156 mM Cl^-) or Ringer's solution + putrescine 20 nmol/ml. Two groups of animals were considered: saline- and α-difluoromethylornithine (αDFMO)-treated rats. The αDFMO (generous gift of Merrel Dow, Strasbourg, France), a suicide inhibitor of ODC, was administered subcutaneously at a dose of 400 mg/kg, 2 hr before and 3 hr after the probe implantation. In a first set of experiments, the rats (n = 4/group) were sacrificed by intracardiac perfusion 24 hr after probe implantation, and the brain dissected out and processed for immunocytochemistry. In a second set of experiments (n = 8/group), ODC activity and polyamine content in the tissue and in the extracellular fluid (ECF) were evaluated via the microdialysis probe. Samples of 50 μl were collected for polyamine determination during the perfusion period. Then the rats were sacrificed by decapitation, and the brain regions dissected out and processed for ODC activity and polyamine evaluation. Some animals (n = 4) perfused with Ringer's solution for 5 hr were used for *in situ* hybridization to evaluate ODC messenger ribonucleic acid (mRNA) content.

Model of Transient Forebrain Ischemia

A transient forebrain ischemia was induced by means of 30′ occlusion of common carotid arteries in rats with prior permanent occlusion of vertebral arteries (18). After 4 hr of reperfusion, the rats (n = 4/group) were killed by intracardiac perfusion, and the brain processed for immunocytochemistry. The experiment was performed in saline- or αDFMO-treated rats (400 mg/kg s.c. 3 hr before and 1 hr after carotid occlusion).

Immunocytochemical Procedure

The indirect immunoperoxidase procedure using the avidin-biotine technique (19) was used. Fifty-μm-thick sections were cut at striatal level [from bregma 1.7 mm to bregma 0.3 mm according to the Paxinos and Watson atlas (17)]. The sections were incubated with the primary rabbit antiserum against the glial fibrillary acidic protein (GFAP) (Dako, Denmark, lot no. 015; dilution 1:300), a specific class of intermediate filament present in the cytoplasm of differentiated astrocytes (20,21). The immunocytochemical procedure has been described elsewhere (22).

In situ Hybridization

In striatum, ODC mRNA content was determined by *in situ* hybridization histochemistry using a ^{35}S-radiolabeled complementary ribonucleic acid (cRNA) probe transcribed from a complementary deoxyribonucleic acid (cDNA) cloned in a SP6-T7 promoter-containing vector. The rats were intracardiacally perfused with 50 ml ice-cold saline, and the brains were dissected out, frozen, and cut in the cryostat (20-μm-thick sections) throughout the entire striatum. Details on the histochemical procedure have been described elsewhere (23). The specificity of the hybridization was assessed by hybridizing adjacent sections with a sense-labeled RNA probe of the same specific activity as the antisense RNA. The hybridized sections were exposed to autoradiographical plates (LKB ^3H-ultrofilm).

Computer-Assisted Image Analysis

Semiquantitative microdensitometry of immunocytochemical preparations was performed by means of an automatic image analyzer (IBAS I-II, Zeiss Kontron, Munich, West Germany) according to Agnati et al. (22). Briefly, the specific profiles were interactively discriminated from the background, and the gray tone distribution of the specific profile population was evaluated. Two parameters were obtained: the field area (FA) (i.e., the area covered by specific profiles), and the mean gray tone value of the specific profiles (X). An overall evaluation of the intensity of immunostaining (TIR = total immunoreactivity) could be obtained by multiplying FA × X. In the case of *in situ* hybridization histochemistry, a relative quantitation of the mRNA content was obtained according to the procedure illustrated in Ferraguti et al. (23).

Biochemical Evaluations

The activity of ODC was determined by quantitating the release of $^{14}CO_2$ from (1-^{14}C)ornithine according to the method of Jänne and William-Ashman (24). Polyamines were assayed by HPLC with fluorescence detection after precolumn derivatization with dansyl chloride (25).

The recovery of polyamines through the microdialysis probe, as assessed by *in vitro* test, was about 10% to 20%.

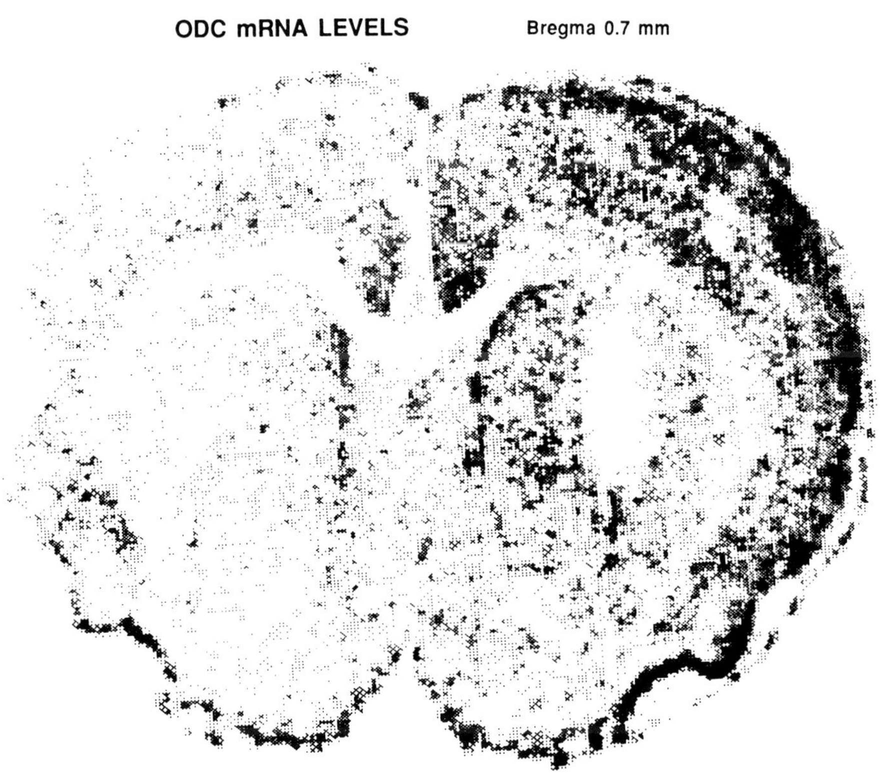

FIG. 1. Histochemical evaluation of ODC mRNA content in the striatum implanted with a microdialysis probe as evaluated by *in situ* hybridization. Anatomical level: bregma 0.7 mm according to Paxinos and Watson atlas (17). For procedure details see text.

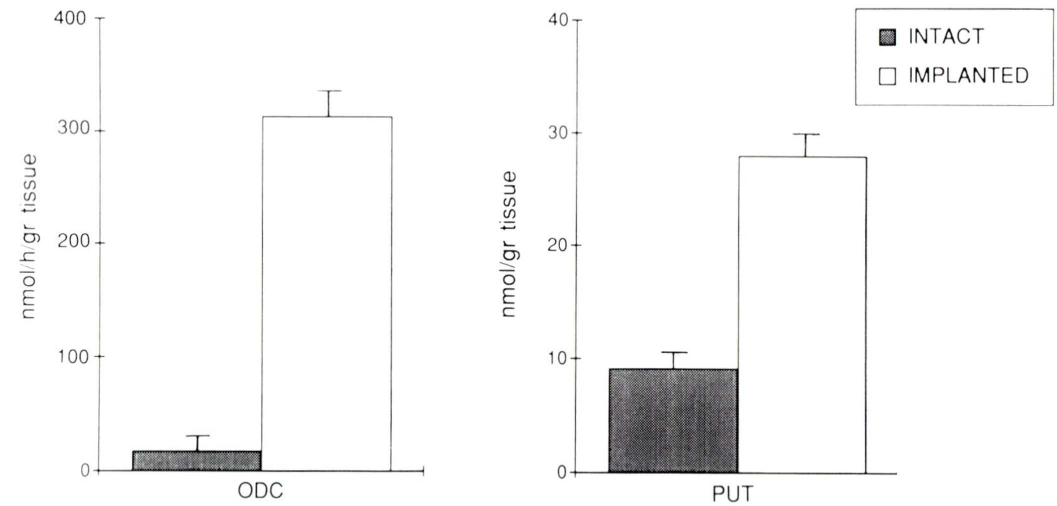

FIG. 2. Effects of microdialysis probe implantation in tissue ODC activity and putrescine (PUT) in rat striatum. Means ± SEM (n = 8/group). ODC activity is expressed as nmol/hr/g tissue; putrescine levels are expressed as nmol/g tissue. For procedure details see text.

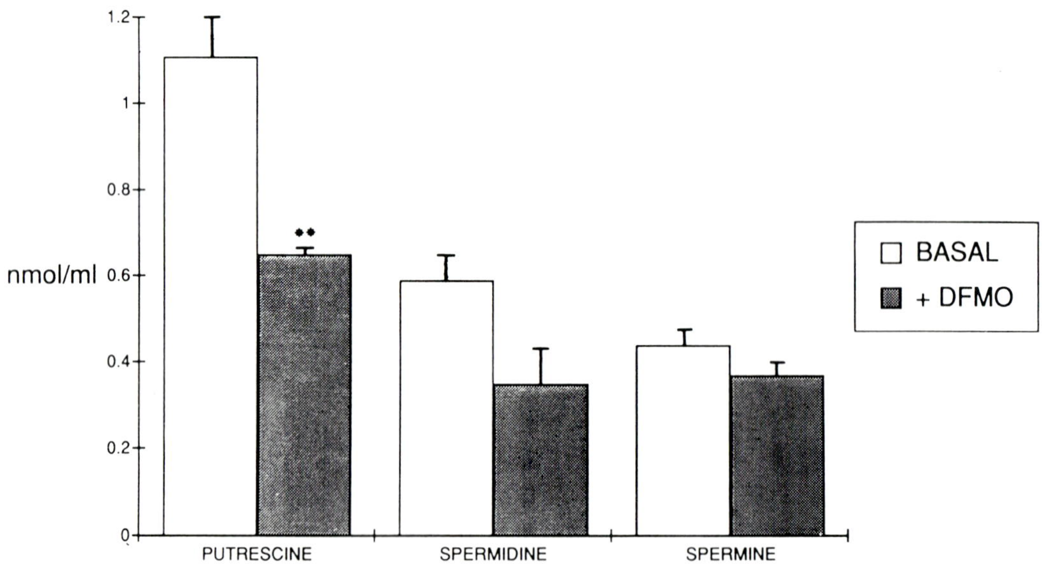

FIG. 3. Effects of αDFMO on polyamine levels in extracellular fluid of rat striatum as measured by microdialysis technique. Means ± SEM (n = 8/group). For further details see text. Statistical analysis was performed according to nonparametric Mann-Whitney U test; $*p < 0.01$.

RESULTS

An increase of ODC mRNA levels was detected in the probe implanted striatum (Fig. 1). The semiquantitative microdensitometric evaluation of the entire striatum showed an increase of about 400% (data not shown). A similar increase of ODC activity and an accumulation of putrescine in the tissue (Fig. 2) were observed in the striatum implanted with the microdialysis probe. The evaluation of polyamine content in the perfusate showed a 50% reduction of putrescine recovery in the αDFMO-treated rats (Fig. 3). In contrast, the increase of putrescine content in the tissue was insensitive to the αDFMO treatment (data not shown). An activation of astroglial cells as assessed by immunocytochemistry for GFAP was detected in the implanted striatum [see Fig. 4A (control) and B (lesioned), and for quantitative comparisons Fig. 5]. In the αDFMO-treated rats, a decrease of GFAP-immunoreactivity (IR) was observed [see Fig. 4B (lesioned) and C (αDFMO-treated), and for quantitative comparisons

FIG. 4. Binary discriminated images of GFAP-IR (black, specific staining; white, nonspecific staining) in (**A**) striatum of control rat; (**B**) striatum of probe-implanted rat perfused with Ringer's solution; (**C**) striatum of probe-implanted rat treated with αDFMO and perfused with Ringer's solution; (**D**) striatum of probe-implanted rat treated with αDFMO and perfused with Ringer's solution + putrescine (20 nmol/ml). For further procedure details see text. Calibration bar = 50 μm.

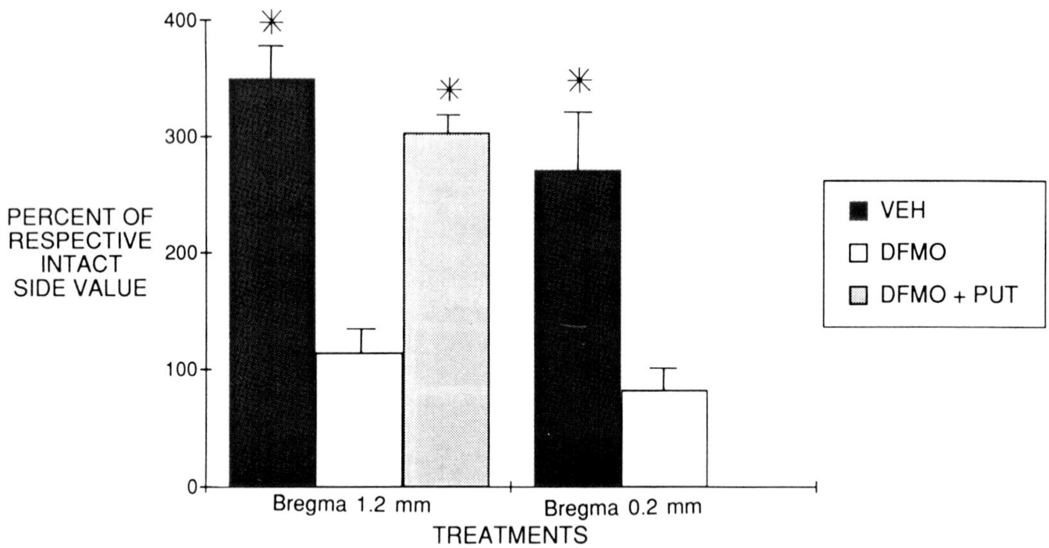

FIG. 5. Effects of αDFMO on GFAP-IR increase induced by probe implantation in striatum of rat perfused with Ringer's solution and with Ringer's solution + putrescine (PUT) (20 nmol/ml). Total immunoreactivity values (means ± SEM, $n = 4$/group) are reported as percent changes of respective intact side values. For further procedure details see text. Anatomical levels according to Paxinos and Watson atlas (17) are given. Statistical analysis was performed according to nonparametric Mann-Whitney U test; * $p < 0.01$.

Fig. 5]. Putrescine (20 nmol/ml), locally administered via the microdialysis probe, was able to restore the astroglial response in the αDFMO-treated rats [see Fig. 4C (αDFMO-treated) and D (αDFMO-treated + local perfusion with putrescine) and for quantitative comparisons Fig. 5]. This effect was clearly detectable 1 mm apart from the site of the probe implantation. A 40% to 50% reduction of the GFAP immunoreactivity increase induced by 30′ transient forebrain ischemia was observed in ischemic rats treated with αDFMO and sacrificed 4 hr after reperfusion (Fig. 6).

DISCUSSION

The microdialysis probe implantation (in these experimental conditions) causes a minor lesion of the tissue with a relatively modest GFAP-IR increase (26,27) and, in the meantime, allows both the recovery of chemical signals (e.g., polyamines) released in the extracellular space by the lesioned cells and the local administration of exogenous substances. The present study gives support to the hypothesis that the astroglial response, evaluated by means of GFAP-IR, is causally linked to an increase in polyamine biosynthesis and to putrescine diffusion in the ECF. This has been proved for the mechanical lesion. In fact, αDFMO pretreatment reduces both the putrescine levels in the ECF and the GFAP-IR response. However, the local administration of putrescine in concentrations of the same order of magnitude as the ECF levels (10–15 nmol/ml) can restore the GFAP-IR response in αDFMO-treated rats.

Circumstantial evidence for a causal link between ODC activation and GFAP response has also been obtained in ischemia. In fact, αDFMO pretreatment can cause a marked reduction in the GFAP response observed in the postischemic period (16).

The present data indicate that the release of putrescine and its diffusion at a long dis-

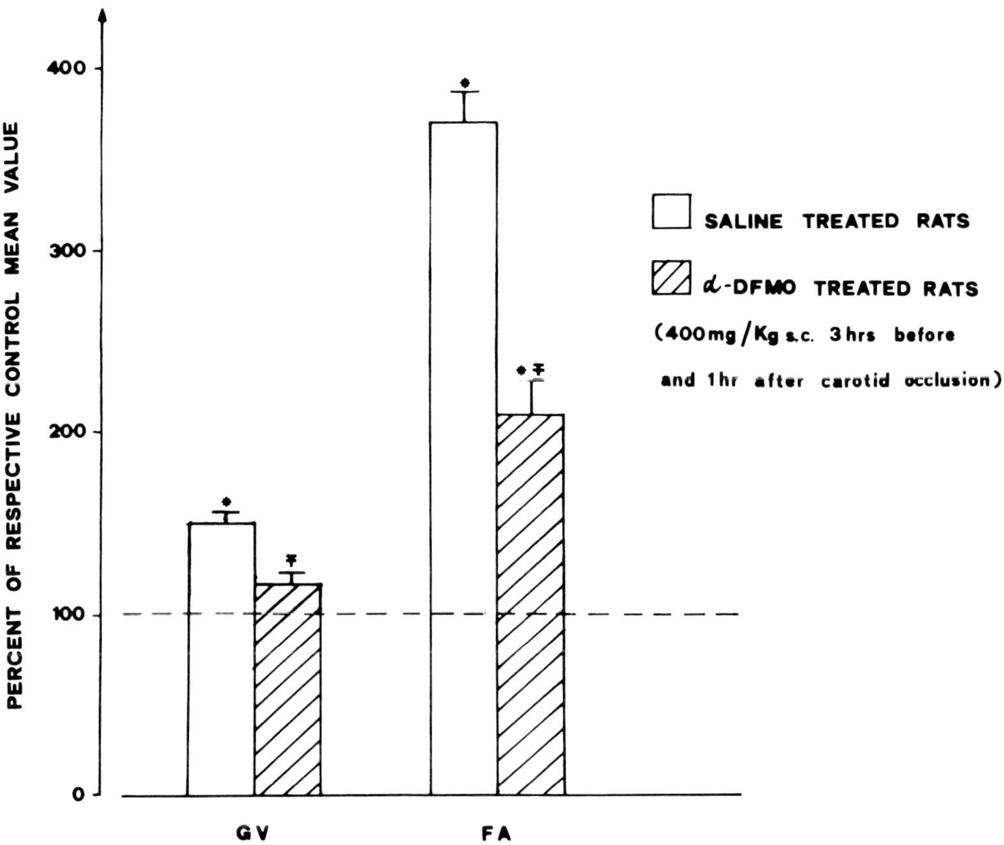

FIG. 6. Effects of αDFMO treatment on GFAP-IR increase induced by 30′ transient forebrain ischemia in striatum of the male rat. Animals were sacrificed 4 hr after reperfusion. Gray values (GV) and field area values (FA) are reported as percent changes of respective control mean value (means + SEM, $n = 4$/group). For further procedure details see text. Statistical analysis has been carried out using nonparametric Mann-Whitney U test; * $p < 0.01$ versus control rats, ‡ $p < 0.05$ αDFMO- versus saline-treated rats.

tance to affect astroglial cell metabolism can play an important role in the reaction of the neural tissue to an injury. In fact, in the implanted striatum of αDFMO-treated rats the local administration of putrescine was able to activate the astroglial response more than 1 mm apart from the site of probe implantation. This evidence favors the existence of a volume transmission mode (28,29) for this trophic signal. Furthermore, long-distance action of lesion signals can, at least in part, explain the concerted response of an entire brain area (e.g., the striatum) to, for example, an ischemic insult. It is presently not known whether the increased putrescine levels observed after brain injury are produced by neurons or glial cells. However, by means of immunocytochemistry, ODC-IR has been shown mainly in neurons but also in activated glial cells (30,31). Moreover, Petito and Babiak (32) suggested that the increase of ODC activity observed after brain ischemia may, at least in part, take place in the proliferating glial cells. It has been shown that the GFAP-IR increase is a marker of astroglial reaction to various types of brain injury, both toxic and mechanical (12–15). The increase of GFAP-IR starts a few hours after the insult, reaches the peak at 3 to 7 days, and declines after 1 to 2 months. Many biological signals have been suggested to trigger the astroglial response to neural damage such as macrophage- and/or microglia-released factors (33,34), serum factors and hormones entering the brain after blood-brain barrier breakdown (35), putrescine and other trophic factors such as prostaglandins and fibroblast growth factor (36, 37).

The physiological role of the astrocytic reaction observed after brain injury is not well known. It has been shown that astrocytes can produce neurotrophic factors such as nerve growth factor (38,39) and that the gliosis process correlates well with the induction of neurotrophic activity by neural damage (38). In fact, the highest *in vitro* trophic activity, obtained after a brain injury, has been demonstrated to reside within the tissue surrounding the lesion, a region rich in gliotic reactions (38,40). Thus, if the glial reaction can be considered as a trophic response to a brain damage, the polyamine system should represent one of the factors mediating this response.

However, it has been recently suggested that the changes in polyamine metabolism after cerebral ischemia may contribute to the ischemic cell injury (11). According to this view, it has been shown that putrescine is involved in the Ca^{2+}-mediated aspartate release from synaptosomes (41); it can contribute to the breakdown of the blood-brain barrier after cold injury (42) and induce electrical hyperactivity in the cerebral cortex after intraventricular injection (43). All these effects seem to play a role in the ischemic cell injury.

However, our previous data (7,8) on the effects of mechanical lesions at mesodiencephalic level point to a positive trophic role of the polyamine system in these experimental conditions. Also preliminary data (Grimaldi et al., *in preparation*) in ischemic animals treated with αDFMO indicate a trophic role for polyamines. In fact, a worsening of the ischemic lesion has been observed in αDFMO-treated rats. In particular, malacic tissue was present in hippocampus and striatum with a higher degree of anatomical disarrangement and microglial infiltration. Moreover, the analysis of Nissl stained neuronal cells in the CA3 hippocampal field, an area that usually shows little ischemic damage, demonstrated a significant reduction of neuron number in αDFMO-treated compared with saline-treated ischemic animals. Thus, it seems that an optimal range of polyamine concentration is required in the ECF for trophic support of brain tissue. Both reductions and increases may be harmful.

In conclusion, the present results demonstrate that putrescine via volume transmission may importantly participate in neuron–glia interactions and influence far away–located astroglia cell populations via long-distance diffusion in the ECF, events that take place especially after brain injury, allowing trophic support of entire brain areas.

ACKNOWLEDGMENTS

This work has been supported by CNR-I and MPI grants and by a grant from the Swedish Medical Research Council. This chapter is dedicated to Prof. Mario Tiengo, Institute of Physiopathology and Therapy of Pain, University of Milan, on his 68th birthday, for his contribution to the study of pain in Italy.

REFERENCES

1. Seiler N. Polyamine metabolism and function in the brain. *Neurochem Int* 1981;3:95–110.
2. Seiler N, Deckardt K. Uptake of polyamines and related compounds into nerve endings. In: Campbell RA, et al., eds. *Advances in polyamine research*. vol 2. New York: Raven Press, 1978;101–107.
3. Marton LJ, Heby O, Levin VA, Lubich WP, Crafts DC, Wilson CB. The relationship of polyamines in cerebrospinal fluid to the presence of central nervous system tumors. *Cancer Res* 1976;36:937–977.
4. Pegg AE. Recent advances in the biochemistry

of polyamines in eukaryotes. *Biochem J* 1986; 234:249–262.

5. Jänne J, Pösö H, Raina A. Polyamines in rapid growth and cancer. *Biochem Biophys Acta* 1978;473:241–293.
6. Pegg AE, McCann PP. Polyamine metabolism and function. *Am J Physiol* 1982;243:C212–C221.
7. Agnati LF, Fuxe K, Davalli P, Zini I, Corti A, Zoli M. Striatal ornithine decarboxylase activity following neurotoxic and mechanical lesions of the mesostriatal dopamine system of the male rat. *Acta Physiol Scand* 1985;125:173–175.
8. Agnati LF, Fuxe K, Zoli M, Davalli P, Corti A, Zini I, Toffano G. Effects of neurotoxic and mechanical lesions of the mesostriatal dopamine pathway on striatal polyamine levels in the rat: modulation by chronic ganglioside GM1 treatment. *Neurosci Lett* 1985;61:339–344.
9. Dienel GA, Cruz NF. Induction of brain ornithine decarboxylase during recovery from metabolic, mechanical, thermal or chemical injury. *J Neurochem* 1984;42:1053–1061.
10. Paschen W, Schmidt-Kastner R, Djuricic B, Meese C, Linn F, Hossmann KA. Polyamine changes in reversible cerebral ischemia. *J Neurochem* 1987;49:35–37.
11. Paschen W, Röhn G, Meese CO, Djuricic B, Schmidt-Kastner R. Polyamine metabolism in reversible cerebral ischemia: effect of α-difluoromethylornithine. *Brain Res* 1988;453:9–16.
12. Brock TO, O'Callaghan JP. Quantitative changes in the synaptic vesicle proteins, synapsin I and p38, and the astrocyte specific protein, glial fibrillary acidic protein, are associated with chemical-induced injury to the rat central nervous system. *J Neurosci* 1987;7:931–942.
13. Ludwin SK. Reaction of oligodendrocytes to trauma and implantation. A combined autoradiographic and immunohistochemical study. *Lab Invest* 1985;52:20–30.
14. Bignami A, Dahl D, Rueger DG. Glial fibrillary acidic (GFA) protein in normal neuronal cells and in pathological conditions. In: Federoff S, Hertz L, eds. *Advances in cellular neurobiology*, vol 1. New York: Academic Press, 1980;285–310.
15. Mathewson AJ, Berry M. Observations on astrocyte response to a cerebral stab wound in adult rats. *Brain Res* 1985;327:61–69.
16. Grimaldi R, Zini I, Zoli M, et al. Related responses between dopamine synapses and astrocytes in the rat striatum after global cerebral ischemia: a morphometrical and microdensitometrical study. In: Hartmann A, Kuschinsky W, eds. *Cerebral ischemia and calcium*. Berlin, Heidelberg, New York: Springer-Verlag, 1989; 105–110.
17. Paxinos G, Watson C. *The rat brain in stereotaxic coordinates*. New York: Academic Press, 1982.
18. Pulsinelli WA, Brierly JB. A new model of bilateral hemispheric ischemia on the unanesthetized rat. *Stroke* 1979;10:267–272.
19. Hsu SM, Raine L, Fanger H. Use of avidin-biotin-peroxidase complex (ABC) in immunoperoxidase techniques: a comparison between ABC and unlabeled antibody (PAP procedures). *J Histochem Cytochem* 1981;29:577–580.
20. Bignami A, Dahl D. Differentiation of astrocytes in the cerebellar cortex and the pyramidal tracts of the newborn rat. An immunofluorescence study with antibodies to a protein specific to astrocytes. *Brain Res* 1973;49:393–402.
21. Eng LF. Glial fibrillary acidic protein (GFAP): the major protein of glial intermediate filaments in differentiated astrocytes. *J Neuroimmunol* 1985;8:203–214.
22. Agnati LF, Fuxe K, Zoli M, et al. Morphometrical and microdensitometrical studies on phenylethanolamine-*N*-methyltransferase and neuropeptide Y immunoreactive neurons in the rostral medulla oblongata of the adult and old rats. *Neuroscience* 1988;26:461–478.
23. Ferraguti F, Zoli M, Aronsson M, et al. Distribution of glutamic acid decarboxylase messenger RNA containing nerve cell populations of the male rat brain. *J Chem Neuroanat* 1990 (*in press*).
24. Jänne J, Williams-Ashman HG. On the purification of L-ornithine decarboxylase from rat prostate and effects of thiol compounds on the enzyme. *J Biol Chem* 1971;246:1725–1732.
25. Desiderio MA, Davalli P, Perin A. Simultaneous determination of gamma-aminobutyric acid and polyamines by high-performance liquid chromatography. *J Chromatogr* 1987;419:285–290.
26. Benveniste H, Diemer NH. Cellular reactions to implantation of a microdialysis tube in the rat hippocampus. *Acta Neuropathol* 1987;74:234–238.
27. Ruggeri M, Zoli M, Grimaldi R, et al. Experimental and theoretical aspects on neural plasticity in the central nervous system. III. Methodological aspects on the microdialysis technique. *Neurochem Int* 1990,16(4).427–436.
28. Agnati LF, Fuxe K, Zoli M, et al. Aspects on the information handling by the central nervous system: focus on cotransmission in the aged rat brain. In: Hökfelt T, Fuxe K, and Pernow B, eds. *Coexistence of neuroactive substances. A new principle in synaptic transmission. Prog Brain Res*, vol. 68, Amsterdam: Elsevier, 1986; 291–301.
29. Fuxe K, Agnati LF, Zoli M, Bjelke B, Zini I. Some aspects on the communicational and computational organization of the brain. *Acta Physiol Scand* 1989;135:203–216.
30. Cintra A, Fuxe K, Agnati LF, et al. Evidence for the existence of ornithine decarboxylase-immunoreactive neurons in the rat brain. *Neurosci Lett* 1987;76:268–274.
31. Nemeth G, Cintra A, Fuxe K, et al. Immunocytochemical studies on some brain regions in a rat model of incomplete cerebral ischemia. Relationship to behaviour and aging. In: Hartmann A, Kuschinsky W, eds. *Cerebral ischemia and calcium*. Berlin, Heidelberg, New York: Springer-Verlag, 1989;83–96.
32. Petito CK, Babiak T. Early proliferative changes in astrocytes in postischemic noninfarcted rat brain. *Ann Neurol* 1982;11:510–518.
33. Giulian D, Baker TJ. Peptides released by ame-

boid microglia regulate astroglial proliferation. *J Cell Biol* 1985;101:2411–2415.
34. Giulian D, Allen R, Baker A, Tomozawa Y. Brain peptides and glial growth. I. Glia-promoting factors as regulators of gliogenesis in the developing and injured central nervous system. *J Cell Biol* 1986;102:803–811.
35. Bologa L, Cole R, Chiappelli F, Saneto RP, De Vellis J. Expression of glial fibrillary acidic protein by differentiated astrocytes is regulated by serum antagonistic factors. *Brain Res* 1988; 457:295–302.
36. Morrison RS, de Vellis J, Lee YL, Bradshaw RA, Eng LF. Hormone and growth factors induce the synthesis of glial fibrillary acidic protein in rat brain astrocytes. *J Neurosci Res* 1985;14:167–176.
37. Chiu FC, Goldman JE. Regulation of glial fibrillary acidic protein (GFAP) expression in CNS development and in pathological states. *J Neuroimmunol* 1985;8:283–292.
38. Nieto-Sampedro M, Manthrope M, Barbin G, Varon S, Cotman CW. Injury-induced neuronotrophic activity in adult rat brain: correlation with survival of delayed implants in the wound cavity. *J Neurosci* 1983;3:2219–2229.
39. Needels DL, Nieto-Sampedro M, Cotman CW. Induction of a neurite-promoting factor in rat brain following injury or deafferentation. *Neuroscience* 1986;18:517–526.
40. Nieto-Sampedro M, Lewis ER, Cotman CW, et al. Brain injury causes a time-dependent increase in neuronotrophic activity at the lesion site. *Science* 1982;217:860–861.
41. Bondy SC, Walker CH. Polyamines contribute to calcium-stimulated release of aspartate from brain particulate fractions. *Brain Res* 1986; 371:96–100.
42. Koenig H, Goldstone AD, Lu CY. Blood-brain barrier breakdown in brain edema following cold injury is mediated by microvascular polyamines. *Biochem Biophys Res Commun* 1983;116:1039.
43. Nisticò G, Ientile R, Rotiroti D, Di Giorgio RM. Gaba depletion and behavioural changes produced by intraventricular putrescine in chicks. *Biochem Pharmacol* 1980;29:954–957.

Volume Transmission in the Brain: Novel Mechanisms for Neural Transmission, edited by K. Fuxe and L.F. Agnati. Raven Press, Ltd., New York © 1991.

19

Transmitter Receptors on Astroglial Cells

Elisabeth Hansson

Institute of Neurobiology, University of Göteborg, Göteborg, Sweden

It is well-established that astrocytes from the central nervous system (CNS) in primary culture possess receptors for a variety of neurotransmitter substances (1,2). Much research has been focused on those astrocyte receptors linked to adenylate cyclase with cyclic adenosine monophosphate (AMP) as second messenger (3). The astrocytes respond to norepinephrine (NE) and isoproterenol with an increase in the intracellular levels of cyclic AMP (4,5). Beta-adrenoceptors are present in high concentrations in various rat brain regions. However, the contribution of each β-adrenoceptor subtype varies markedly depending on the brain region examined (6). In addition to the β adrenoceptors, astrocytes in culture also possess α_2 adrenoceptors, which mediate an inhibition of cyclic AMP accumulation in the cells (5). The occurrence of dopamine (D_1) receptors on astrocytes of the striatum with cyclic AMP as second messenger has been previously demonstrated (7–9). Another transduction mechanism that has attracted attention recently employs an inositol lipid as a precursor to generate the second messengers diacylglycerol, which activates protein kinase C, and inositol 1,4,5-trisphosphate, which mobilizes Ca^{2+} (10,11). Stimulation of α_1 adrenoceptors or serotonin (5-HT_2) receptors induces an accumulation of radiolabeled inositol phosphates in astrocytes preincubated with [^{3}H]-myo-inositol (2,12).

Receptors are highly flexible molecules and sensitive to ionic and lipid environment but also to the presence of other proteins and particularly to other receptors and second messengers. The interaction between signal substances coexisting in CNS neurons has been studied by Fuxe and Agnati (13), and coexistence of classical neurotransmitters with peptide neurotransmitters has also been demonstrated (14). Concerning astrocytes, this field is just in its beginning (15). Astrocytes in culture express many receptor sets and complex receptor interactions. One function of the adrenoceptors may be to regulate uptake of amino acid neurotransmitters (16). This suggests that astrocytes may be an integral regulatory component in the local control of neuronal activity in the CNS.

OCCURRENCE OF ASTROCYTES IN THE MATURE NERVOUS SYSTEM

Astrocytes constitute as much as 50% of the cell number in the cerebral cortex of higher mammals (17). They have long processes that surround vessels and synapses, form a continuous subpial and subependymal layer, surround neuronal surfaces, and form a syncytium with other astrocytes by gap junctions (18). Astrocytes actively participate in the formation of the blood-brain barrier and thus "know about" toxic substances and drugs reaching the nervous system via the blood. In addition, astrocytes participate in the regulation of the neuronal activity. Because of the complexity of the interactions of neural cells, model systems need to be used to provide possibilities for evaluating metabolic, pharmacological, and physical reactions at the cellular level.

PRIMARY CULTURES AS A MODEL SYSTEM TO STUDY ASTROCYTES

Techniques for maintaining highly enriched astrocytes in dissociated primary culture have been successfully developed during the past 20 years (19). These preparations consist of actively proliferating cells isolated from prenatal, neonatal, or postnatal animals, and by different means, they can be enriched in one specific cell type. The primary cultures originate from cells, tissues, or organs taken directly from the organism and are regarded as primary until subcultivated. Such cultures provide the option of choosing brain region and age of the animal at seeding. However, it should be emphasized that cell and tissue culture is a model system, and whatever is discovered in culture must also be demonstrated *in vivo*.

It is important to characterize the primary cell cultures, using morphological, biochemical, and electrophysiological properties for various cell types expected to be in the culture, as cells other than those desired can survive. This is even more important when evaluating experimental results from cell cultures obtained from various brain regions to ensure similar cellular proportions in the various cultures (20).

Using astroglial cell primary cultures, two morphologically distinct astroglial cell types have been observed (21,22). The predominant of these were polygonally shaped flat cells that formed a confluent bed layer (Fig. 1). These cells have been called *type 1 astrocytes*. A second cell type that exhib-

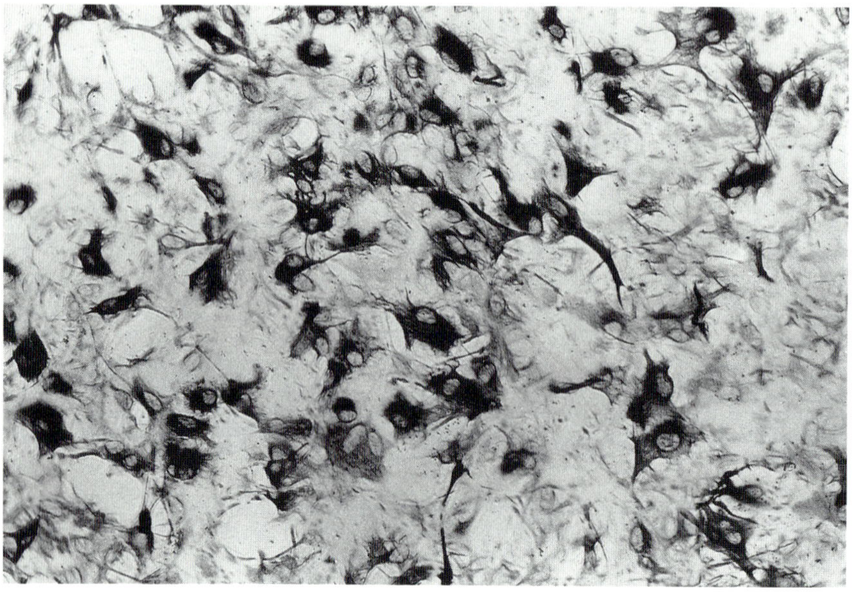

FIG. 1. Primary astroglial cell culture grown for 2 weeks from newborn rat cerebral cortex was stained with antibodies against glial fibrillary acidic protein (GFAp). A peroxidase antiperoxidase (PAP) kit (BioGenex Labs, Dublin, CA) was used, and primary antibodies for GFAp were diluted 1:10.

ited a central cell body with numerous short processes grew on top of the bed layer cells. These cells are in a stage of plasticity and can be directed to express either oligodendroglial or astroglial cell markers. These astroglial cells have been called *type 2* (22). As other cells than astrocytes may survive in the culture, it is important to evaluate if such cells are present by using specific markers for the expected cells. Cells with phagocytic properties were considered to be macrophages. Also a few number of endothelial cells as well as small phase-dark cells, considered to be oligoblasts, have been detected. Ependymal cells with beating cilia and mesenchymal-like undifferentiated cells were demonstrated. Neurons will not survive in this type of primary culture from newborn rat brain (4).

It is important to understand that the results obtained using cultured cells reflect the properties of astrocytes obtained from immature animals and maintained in the absence of their normal cellular and chemical milieu. It seems quite likely that the properties expressed by cultured astrocytes do reflect the potential of astrocytes in brain and that these properties may be differentially expressed either by different groups of astrocytes in brain or at different stages of development. New techniques must be developed to bridge the gap between *in vitro* studies and *in vivo* studies if we are to establish definitely that astrocytes in brain exhibit the variety of properties that have been shown to be associated with cultured astrocytes.

TRANSMITTER RECEPTORS AND SECOND MESSENGERS

Beta Adrenoceptors

It has been known for some time that astroglial β adrenoceptors in primary cultures are coupled to adenylate cyclase, which leads to increases in cyclic AMP (3,5,23). A number of investigators have examined the subtypes of β adrenoceptors present in brain, β_1 and β_2. Astrocyte receptors are of the β_1 subtype (24), whereas fibroblasts possess β_2 receptors. There is no evidence that neurons prepared from rat fetal cerebral cortex exhibit β receptors (25). This stresses the need for well-defined cultures when investigating receptor populations.

The β-receptor binding site density associated with the astrocytes appeared to peak with cultures prepared from neonatal animals (23,26). However, primary cultures prepared from rats older than 2 days contained a relatively large number of fibroblasts (26), and therefore both β-receptor subtypes can be expressed. The maximal stimulation of intracellular cyclic AMP accumulation by β agonists is enhanced when astrocytes reach confluency in culture (23). An important question that can be addressed here is whether there is a heterogeneity among astrocytes with respect to the expression of receptors. A more prominent cyclic AMP accumulation is seen in astrocytes derived from cerebral cortex and striatum than in astrocytes from brainstem (23). Also McCarthy's group has indicated a heterogeneity among cultured astrocytes with respect to receptor expression (27). They have used iodinated receptor ligands selective for either the β adrenoceptor or the α adrenoceptor. The results indicate that essentially all polygonal (type 1) astrocytes exhibit β adrenoceptors, whereas process-bearing (type 2) astrocytes do not appear to express this receptor. Furthermore, α adrenoceptors seem to be restricted to a subpopulation of type 1 astrocytes. The results strongly suggest that there are multiple populations of astrocytes that can be distinguished by their expression of receptors and therefore their ability to respond to specific neuronal signals. Another important point is that noncultured astrocytes from mature brain exhibit β receptors (27), showing that astrocytes express at least β receptors even *in vivo*.

Beta-receptor stimulation and the resulting increases in cyclic AMP lead to phosphorylation of intracellular proteins (28,29), changes in cell shape (30), changes in glycogen metabolism (31), stimulation of release of taurine (32), and a stimulation of the active uptake of gamma-aminobutyric acid (GABA) into astrocytes (16).

Alpha$_2$ Receptors

Astrocytes also possess α_2 adrenoceptors, which mediate an inhibition of cyclic AMP accumulation in the cells (5,33,34). Because α_2 adrenoceptors inhibit the cyclic AMP production induced by isoproterenol, a specific β-adrenoceptor agonist, it is likely that both α_2 and β adrenoceptors are expressed on the same cell (34). This observation raises the problem of the physiological significance. Similarly, it is not impossible that α_2 and β adrenoceptors are localized on different cell surface domains of the astrocyte (35). Whatever the physiological significance of the presence of both β and α_2 receptors on the same cell, it is clear that both receptors will contribute to the control of intracellular cyclic AMP levels. This will be discussed below.

Dopamine Receptors

Recent work indicates the presence of D_1 receptors on astrocytes from striatum, positively coupled to adenylate cyclase (7). It has been speculated that this D_1-activated accumulation of cyclic AMP may be the result of the stimulation of β adrenoceptors (3). Using selective D_1 agonists and antagonists, the presence of D_1 receptors with cyclic AMP as second messenger could be demonstrated in a primary astroglial culture from striatum (8). This strengthens the hypothesis of heterogeneity among the brain astrocytes as D_1 receptors were not identified in astroglial cultures from cerebral cortex, hippocampus, or brainstem. Hösli and Hösli (9) have also described both D_1 and D_2 binding sites on astrocytes in explant cultures from rat striatum and spinal cord, using autoradiography.

Alpha$_1$ Receptors

The phosphoinositide cascade, which results in mobilization and entry of calcium and activation of protein kinase C, is triggered by activation of phospholipase C–coupled receptors. The development of the methodology by Berridge and colleagues (36), using lithium-dependent accumulation of labeled inositol phosphates to measure phosphoinositide (PI) hydrolysis, markedly increases the sensitivity of detection and has facilitated the identification of cell surface receptors coupled to phospholipase C in several cell types. There is strong evidence that the α_1-receptor subtype is linked to stimulation of inositol phospholipid metabolism on astrocytes grown in primary culture (37,38).

5-HT Receptors

The presence of high-affinity 5-HT receptors on astrocytes in primary culture was first reported by Hertz et al. (39), using a radioligand binding assay. At present pharmacological characterization has divided the receptor into three major subtypes termed 5-HT$_1$, 5-HT$_2$, and 5-HT$_3$. The 5-HT$_1$ receptor has five different subclasses, i.e., 5-HT$_{1A}$, 5-HT$_{1B}$, 5-HT$_{1C}$, 5-HT$_{1D}$, and 5-HT$_{1E}$ (40). Evidence has been provided for the coupling of 5-HT$_{1A}$, 5-HT$_{1B}$, and 5-HT$_{1D}$ to adenylate cyclase (41–43) and 5-HT$_{1C}$ and 5-HT$_2$, receptors to phosphatidylinositol metabolism (44,45). The work by Pearce et al. (38) demonstrated a stimulation of PI hydrolysis, achieved at a 5-HT concentration of 100 μM in primary astrocyte cultures. This concentration was more suggestive of the activity of a 5-HT$_2$ receptor than any subtype of the 5-HT$_1$ (38), as the 5-HT$_1$ receptors bind 5-HT at the nanomolar concentrations and the 5-HT$_2$ receptor binds 5-HT at the micromolar concentrations. Recently, we have found 5-HT_2 receptors on astrocytes in primary culture from cerebral cortex, striatum, hippocampus, and brainstem with a stimulation of the turnover of the PIs (12).

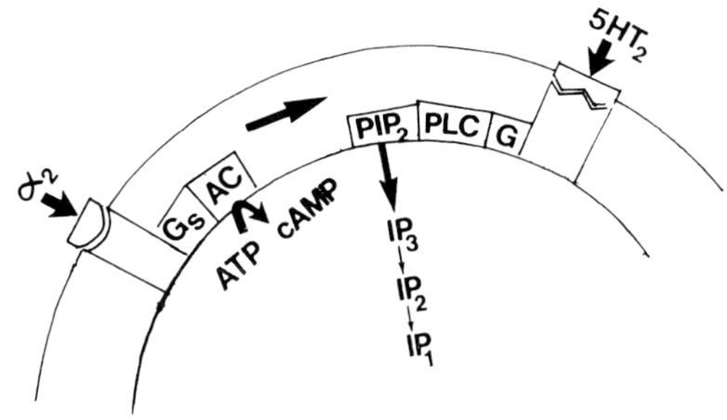

FIG. 3. Interactions between the 5-HT_2 receptor and the α_2 adrenoceptor in a cultured astrocyte. In the presence of the α_2 agonist clonidine and 5-HT there was a potentiation of the IP formation. G_s, guanine nucleotide stimulatory protein; G, guanine nucleotide protein; AC, adenylate cyclase; cAMP, cyclic adenosine monophosphate; PIP_2, phosphatidylinositol 4,5-bisphosphate; PLC, phospholipase C; IP_3, inositol - 1,4,5 - trisphosphate; IP_2, inositol - 1,4 - bisphosphate; IP_1, inositol - 4 - monophosphate; ATP, adenosine triphosphate.

COMMUNICATION BETWEEN DIFFERENT RECEPTORS AT THE LEVEL OF THE SECOND MESSENGER

Progress in receptor research has yielded an intriguing picture of the cell surface signal transduction elements. All this information provides a rich perspective for considering the questions of how receptor function is regulated and by what mechanisms might the activation of one receptor regulate the behavior of a second receptor. There are many levels at which receptors may be controlled, both with respect to receptor number and with respect to overall receptor activity. Taking the possible mechanisms into account that may be involved in regulating receptors, one must consider not only the ligand-binding domain of the receptor, but also other receptor domains that may be involved in overall receptor function.

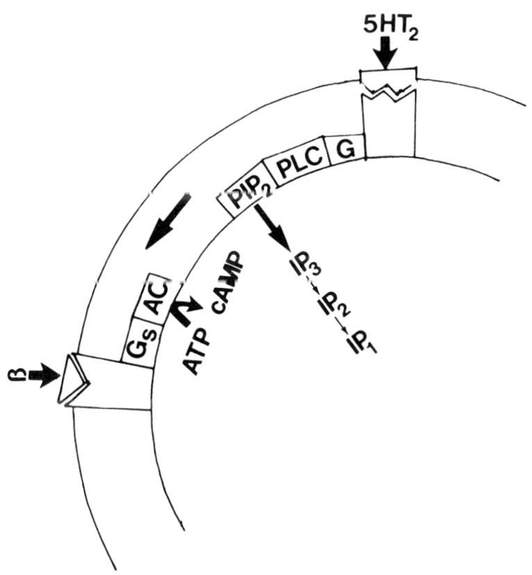

FIG. 2. Interactions between the 5-HT_2 receptor, stimulating formation of IP_3, and the β adrenoceptor, stimulating formation of cyclic AMP, in a cultured astrocyte. In the presence of 5-HT and the β agonist isoproterenol, there was a potentiation of cyclic AMP formation. G_s, guanine nucleotide stimulatory protein; G, guanine nucleotide protein; AC, adenylate cyclase; cAMP, cyclic adenosine monophosphate; PIP_2, phosphatidylinositol 4,5-bisphosphate; PLC, phospholipase C; IP_3, inositol - 1,4,5 - trisphosphate; IP_2, inositol - 1,4 - bisphosphate; IP_1, inositol - 4 - monophosphate; ATP, adenosine triphosphate. For details see ref. 37.

Research of the past years has provided many examples of coexistence in CNS neurons of classical neurotransmitters and peptide neurotransmitters (13,14). In astrocytes, this field is just in the beginning. There may be formations of mosaics of receptors on the cell membrane of astrocytes, and the second messengers of these receptors may communicate with each other through intramembrane signals. An interaction between the β-adrenergic agonist isoproterenol and vasoactive intestinal peptide (VIP) has been obtained on astrocytes (46,47). The demonstration of a synergistic effect of VIP and NE on an intracellular event strongly suggests that the two neurotransmitters may act, at least in part, on identical target cells (48). Synergistic interactions have also been observed between NE and some neuropeptides (49).

5-HT, which had no effect on the formation of cyclic AMP, potentiated the β-adrenergic stimulation of adenylate cyclase induced by isoproterenol. This effect was inhibited by ketanserin, indicating that the potentiation was mediated via the 5-HT_2 receptor (Fig. 2) (37). We have also observed that the amount of inositol phosphates (IP) accumulated was substantially increased when 5-HT was added together with adrenergic agonists that stimulate phospholipase C. The inositol monophosphate (IP_1) response after simultaneous stimulation with phenylephrine (an α_1 agonist) and 5-HT was found to be significantly larger than the sum of the individual phenylephrine and 5-HT effects (Fig. 4). Clonidine, an α_2 agonist, known to lower cyclic AMP (33), markedly potentiated the 5-HT-stimulated formation of IP_1 (Fig. 3). However, the β-receptor agonist isoproterenol did not potentiate the 5-HT-stimulated IP_1 formation (37). Our observation that adrenergic and serotonergic receptors on astrocytes interact are in line with the concept of receptor–receptor cross-talk as a means of processing extracellular information (50–52). Protein kinase C can phosphorylate the adenylate cyclase–coupled β-adrenergic receptor, and protein kinase A, the α_1-adrenergic receptor (53). The physiological regulatory significance is less clear; these reactions could plan an important amplification role in tissues in which a physiological or biochemical response of the target cell is reciprocally controlled by agonists, using two distinct signal transduction pathways.

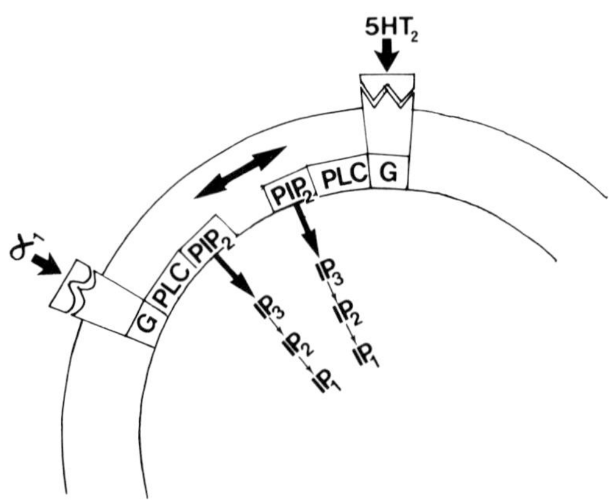

FIG. 4. Interactions between 5-HT_2 and α_1 adrenoceptors, both receptors stimulating the formation of IP_3 in a cultured astrocyte. In presence of both agonists there was potentiation of IP formation. G, guanine nucleotide protein; PIP_2, phosphatidylinositol 4,5-bisphosphate; PLC, phospholipase C; IP_3, inositol-1,4,5-triphosphate; IP_2, inositol-1,4-bisphosphate; IP_1, inositol-4-monophosphate.

RECEPTORS REGULATING THE CARRIER-MEDIATED UPTAKE OF NEUROTRANSMITTERS

Chemical neurotransmission is the predominant mode of information transfer between neurons and between neurons and astrocytes in the CNS. As evidence has already been obtained, at least in aplysia, ion channels in their nerve cell membranes can be controlled by receptors on the membrane that communicate with their channel by means of a second-messenger system, such as cyclic AMP and the cyclic AMP–dependent protein kinase. The possibility of a monoamine control over uptake carriers in astrocytes was therefore examined (54,55).

The evidence that the glutamate and GABA carriers and their ion channels in astrocytes can be controlled by α_1 and β receptors (Figs. 5, 6) (16) suggests that these adrenoceptors also regulate the intracellular enzymatic control of glutamic oxaloacetate transaminase (GOT), glutamine synthetase (GS), and γ-aminobutyric acid α-ketoglutarate transaminase (GABA-T). The receptors seem also to regulate the intracellular control of the amino acid pool

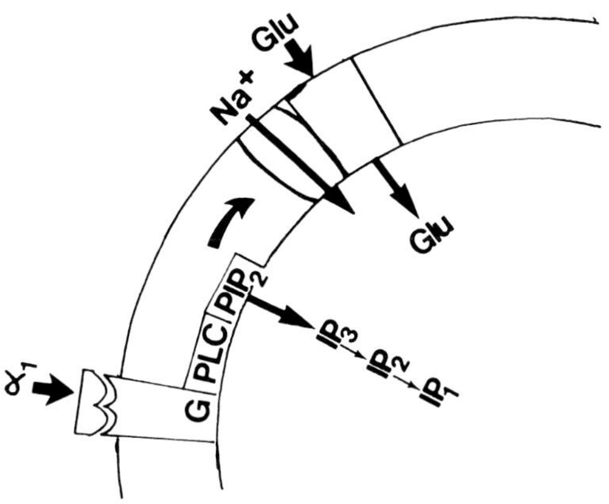

FIG. 5. α_1-receptor regulation of the glutamate (Glu) uptake carrier in a cultured astrocyte. G, guanine nucleotide protein; PLC, phospholipase C; PIP_2, phosphatidylinositol 4,5-bisphosphate; IP_1, inositol-4-monophosphate; IP_2, inositol-1,4-bisphosphate; IP_3, inositol-1,4,5-triphosphate. For details see ref. 16.

and transmitter release (32,56). Astrocytes in culture express many receptor sets and complex receptor interactions. We have recently observed that 5-HT regulates the carrier mediated taurine uptake in form of an inhibition (*unpublished data*). It seems

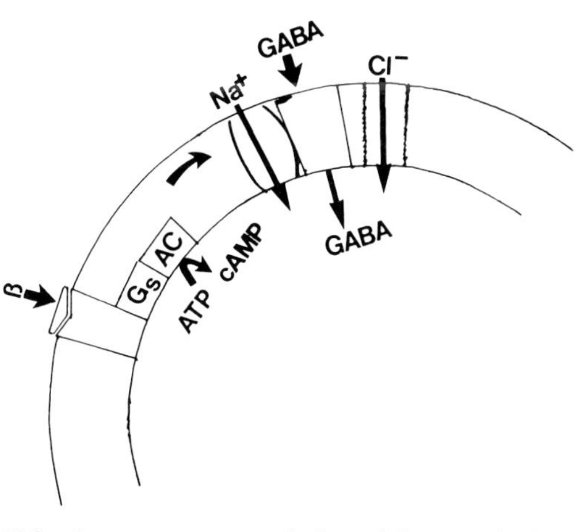

FIG. 6. β-receptor regulation of the γ-aminobutyric acid (GABA) uptake carrier in a cultured astrocyte. ATP, adrenosine triphosphate; cAMP, cyclic adenosine monophosphate; G_s, guanine nucleotide stimulatory protein; AC, adenylate cyclase. For details see ref. 16.

that the β adrenoceptor also regulates the taurine uptake (*unpublished data*).

FUTURE VIEWS

Cells respond to changes in their environment by means of specific receptors that detect incoming signals and translate the information into a form that can be recognized by intracellular effector systems. Astrocytes are known to participate in many physiological and pathophysiological events in the nervous system. They may be of importance in the pathogenesis of Parkinson disease and Huntington's chorea, as well as symptoms ensuing from metal and solvent toxicity. They also take part in the immunologic system. This participation is partly due to the ability of astrocytes to recognize and respond to specific signals. Astrocytes exhibit a wide variety of both membrane and intracellular receptors that regulate intracellular processes. Therefore, astrocyte receptors might be one way of learning how to restore changes in neuronal activity caused by different substances. Receptors are flexible molecules, which are sensitive not only to the ionic and lipid environment, but also to the presence of other proteins and particularly to other receptors and second messengers. As coexistence is established between different signal substances within the CNS, recent research indicating the presence of receptor–receptor interactions in astrocytes suggests new perspectives on information transfer dynamics within the CNS. The consequences of such interactions might be a modulation of second-messenger activity. Another possibility is that receptor–receptor interaction serves to adjust the number of receptors after, for example, up- or down-regulation of receptors. Also the presence of receptor regulation of amino acid neurotransmitter transport over the cell membrane has significant implications, as it suggests that astrocytes may be an integral regulatory component in the local control of neuronal activity in the CNS. In this context it might be that astroglial transmitter receptors in addition of mediating information along specific pathways might also be targets for volume transmission.

ACKNOWLEDGMENTS

This work was supported by grants from the Swedish Medical Research Council (project No. 12X-06812), Torsten and Ragnar Söderberg's Foundation, O.E. and Edla Johansson's Foundation, and Åke Wiberg's Foundation.

REFERENCES

1. Kimelberg HK, ed. Glial cell receptors. New York: Raven Press, 1988.
2. Murphy S, Pearce B. Functional receptors for neurotransmitters on astroglial cell. *Neuroscience* 1987;22:381–394.
3. Van Calker D, Hamprecht B. Effects on neurohormones on glial cells. In: Fedoroff S, Hertz L, eds. *Advances in cellular neurobiology,* vol 1. New York: Academic Press, 1980;32–67.
4. Hansson E, Rönnbäck L, Persson LI, et al. Cellular composition of primary cultures from cerebral cortex, striatum, hippocampus, brain stem and cerebellum. *Brain Res* 1984;300:9–18.
5. McCarthy KD, de Vellis J. Alpha-adrenergic receptor modulation of beta-adrenergic, adenosine and prostaglandin E_1 increased adenosine 3′:5′-cyclic monphosphate levels in primary cultures of glia. *J Cyclic Nucleotide Res* 1978;4:15–26.
6. Minneman KP, Wolfe BB, Pittman RN, Molinoff PB. Beta-adrenergic receptor subtypes in rat brain. In: Segawa T, ed. *Molecular pharmacology of neurotransmitter receptors.* New York: Raven Press, 1983;73–81.
7. Hansson E, Rönnbäck L, Sellström Å. Is there a "dopaminergic glial cell"? *Neurochem Res* 1984;9:679–689.
8. Hansson E, Rönnbäck L. Neurons from substantia nigra increase the efficacy and potency of second messenger arising from striatal astroglia dopamine receptor. *GLIA* 1988;1:393–397.
9. Hösli E, Hösli L. Binding sites for ^{3}H-dopamine and dopamine-antagonists on cultured astrocytes of rat striatum and spinal cord: an autoradiographic study. *Neurosci Lett* 1986;65:177–182.
10. Berridge MJ. Inositol trisphosphate and diacylglycerol as second messengers. *Biochem J* 1984;220:345–360.
11. Berridge MJ. Intracellular signalling through inositol trisphosphate and diacylglycerol. *Biol Chem Hoppe-Seyler* 1986;367:447–456.
12. Hansson E, Simonsson P, Alling C. 5-Hydroxytrytamine stimulates the formation of inositol phosphate in astrocytes from different regions of the brain. *Neuropharmacology* 1987;26:1377–1382.
13. Fuxe K, Agnati LF. Receptor–receptor interaction in the central nervous system. A new integrative mechanism in synapses. *Med Res Rev* 1985;5:441–482.
14. Lundberg JM, Hökfelt T. Coexistence of peptides and classical neurotransmitters. *Trends Neurosci* 1983;6:325–333.
15. Hansson E. Co-existence between receptors, carriers, and second messengers on astrocytes grown in primary cultures. *Neurochem Res* 1989;14:811–910.
16. Hansson E, Rönnbäck L. Regulation of glutamate and GABA transport by adrenoceptors in primary astroglial cell cultures. *Life Sci* 1989; 44:27–34.
17. Pope A. Neuroglia: quantitative aspects. In: Schoffeniels E, Franck G, Hertz L, and Tower DB, eds. *Dynamic properties of glia cells.* London: Pergamon Press, 1978;13–20.
18. Peters A, Palay SL, Webster H de F, eds. *The fine structure of the nervous system: the neurons and supporting cells.* Philadelphia: Saunders, 1976.
19. Booher J, Sensenbrenner M. Growth and cultivation of dissociated neurons and glial cells from embryonic chick, rat and human brain in flask cultures. *Neurobiology* 1972;2:97–105.
20. Hansson E. Astroglia from defined brain regions as studied with primary cultures. *Prog Neurobiol* 1988;30:369–397.
21. McCarthy KD, de Vellis J. Preparation of separate astroglial and oligodendroglial cell cultures from rat cerebral tissue. *J Cell Biol* 1980;85:890–902.
22. Raff MC, Abney ER, Cohen J, Lindsay R, Noble M. Two types of astrocytes in cultures of developing rat white matter: differences in morphology, surface gangliosides, and growth characteristics. *J Neurosci* 1983;3:1289–1300.
23. Hansson E. Primary cultures from defined brain areas; effects of seeding time on the development of β-adrenergic and dopamine stimulated cAMP activity during cultivation. *Dev Brain Res* 1985;21:187–192.
24. Trimmer PA, Evans T, Smith MM, Harden TK, McCarthy KD. Combination of immunocytochemistry and radioligand receptor assay to identify β-adrenergic receptor subtypes on astroglia *in vitro. J Neurosci* 1984;4:1598–1606.
25. Harden TK, McCarthy KD. Identification of the β-adrenergic receptor subtype on astroglia purified from rat brain. *J Pharmacol Exp Ther* 1982;222:600–605.
26. Trimmer PA, McCarthy KD. Immunocytochemically defined astroglia from fetal, newborn and young adult rats express β-adrenergic receptors *in vitro. Brain Res* 1986;27:151–165.
27. McCarthy KD, Salm A, Lerea LS. Astroglial receptors and their regulation of intermediate fil-

ament protein phosphorylation. In: Kimelberg HK, ed. *Glial cell receptors*. New York: Raven Press, 1988;1–22.
28. Browning ET, Sanders MM. Vimentin: a phosphoprotein under hormonal regulation. *J Cell Biol* 1981;90:803–808.
29. McCarthy KD, Prime J, Harmon T, Pollenz R. Receptor-mediated phosphorylation of astroglial intermediate filament proteins in cultured astroglia. *J Neurochem* 1985;44:723–730.
30. Narumi S, Kimelberg HK, Bourke RS. Effects of norepinephrine on the morphology and some enzyme activities of primary monolayer cultures from rat brain. *J Neurochem* 1978;31:1479–1490.
31. Cummins CJ, Lust WD, Passonneau JV. Regulation of glycogenolysis in transformed astrocytes *in vitro*. *J Neurochem* 1983;40:137–144.
32. Shain WG, Madelian V, Martin DL, Kimelberg HK, Perrone M, Lepore R. Activation of β-adrenergic receptors stimulated release of an inhibitory transmitter from astrocytes. *J Neurochem* 1986;46:1298–1303.
33. Hauser KL. Adrenergic receptors in glioblast cultures from newborn rat cortex. *Experientia* 1977;33:805.
34. Van Calker D, Muller M, Hamprecht B. Adrenergic α- and β-receptors on the same cell type in primary cultures of perinatal mouse brain. *J Neurochem* 1978;30:713–718.
35. Bockaert J, Ebersolt C. α-Adrenergic receptors on glial cells. In: Kimelberg HK, ed. *Glia cell receptors*. New York: Raven Press, 1988;35–51.
36. Berridge MJ, Downes CP, Hanley MR. Lithium amplifies agonist-dependent phosphatidylinositol responses in brain and salivary glands. *Biochem J* 1982;206:587–595.
37. Hansson E, Simonsson P, Alling C. Interactions between cyclic AMP and inositol phosphate transduction systems in astrocytes in primary culture. *Neuropharmacology* 1990;29:591–598.
38. Pearce B, Cambray-Deakin M, Morrow C, Grimble J, Murphy S. Activation of muscarinic and of α_1-adrenergic receptors on astrocytes results in the accumulation of inositol phosphates. *J Neurochem* 1985;45:1534–1540.
39. Hertz L, Baldwin F, Schousboe A. Serotonin receptors on astrocytes in primary cultures. Effects of methysergide and fluoxetine. *Can J Physiol Pharmacol* 1979;57:223–226.
40. Conn P-J, Sanders-Bush E. Central serotonin receptors: effector systems, physiological roles and regulation. *Psychopharmacology* 1987;92: 267–277.
41. De Vivo M, Maayani S. Characterization of 5-hydroxytryptamine 1A-receptor-mediated inhibition of forskolin-stimulated adenylate cyclase activity in guinea-pig and rat hippocampal membranes. *J Pharmacol Exp Ther* 1986;238:248–253.
42. Herrick-Davis K, Titeler M, Leonhardt S, Struble R, Price D. Serotonin 5-HT_{1D} receptors in human prefrontal cortex and caudate: interaction with a GTP binding protein. *J Neurochem* 1988;51:1906–1912.
43. Murphy TJ, Bylund DB. 5-HT_{1B} receptors are negatively coupled to adenylate cyclase in OK cells (*Abstr.*) *Fed Am Soc Exp Biol* 1988;2A1397.
44. Conn P-J, Sanders-Bush E. Selective 5-HT_2 antagonists inhibit serotonin stimulated phosphatidylinositol metabolism in cerebral cortex. *Neuropharmacology* 1984;8:993–996.
45. Conn P-J, Sanders-Bush E, Hoffman BJ, Hartig PR. A unique serotonin receptor in choroid plexus is linked to phosphatidylinositol turnover. *Proc Natl Acad Sci* 1986;83:4086–4088.
46. Chneiweiss H, Glowinski J, Prémont J. Vasoactive intestinal polypeptide receptors linked to an adenylate cyclase, and their relationship with biogenic amine- and somatostatin-sensitive adenylate cyclases on central neuronal and glial cells in primary cultures. *J Neurochem* 1985; 44:779–786.
47. Hansson E, Rönnbäck L. Interaction between catecholamines and vasoactive intestinal peptide in cultured astrocytes. *Neuropharmacology* 1988;27:295–300.
48. Rougon G, Noble M, Mudge AW. Neuropeptides modulate the beta-adrenergic response of purified astrocytes *in vitro*. *Nature* 1983;305:715–717.
49. Niehoff DL, Mudge AW. Somatostatin alters β-adrenergic receptor effector coupling in cultured rat astrocytes. *EMBO J* 1985;4:317–321.
50. Enna SJ, Karbon EW. Receptor regulation: evidence for a relationship between phospholipid metabolism and neurotransmitter receptor-mediated cAMP formation in brain. *Trends Pharmacol Sci* 1987;8:21–24.
51. Lichtstein D, Rodbard D. A second look at the second messenger hypothesis. *Life Sci* 1987;40: 2041–2051.
52. Mooibroek MJ, Wang JH. Integration of signal-transduction processes. *Biochem Cell Biol* 1988; 66:557–566.
53. Yoshimasa T, Sibley DR, Bouvier M, Lefkowitz RJ, Caron MG. Cross-talk between second messenger generating systems: phorbol esters induce phosphorylation of the catalytic unit of adenylate cyclase activity and enhancement of its activity. *Nature* 1987;327:67–70.
54. Levitan IB, Adams WB. Cyclic AMP modulation of a specific ion channel in an identified nerve cell: possible role for protein phosphorylation. *Adv Cyclic Nucleotide Res* 1981;14:647–653.
55. Siegelbaum SA, Camardo JS, Kandel ER. Serotonin and cyclic AMP close single K^+ channels in Alypsia sensory neurons. *Nature* 1982;299: 413–417.
56. Lehmann A, Hansson E. Kainate-induced stimulation of amino acid release from primary astroglial cultures of the rat hippocampus. *Int J Neurochem* 1988;13:557–561.

Volume Transmission in the Brain: Novel Mechanisms for Neural Transmission, edited by K. Fuxe and L.F. Agnati. Raven Press, Ltd., New York © 1991.

20

Lymphokines, Monokines, and Glial Cells

Jean E. Merrill

Department of Neurology, Reed Neurological Research Center, UCLA School of Medicine, Los Angeles, California 90024

Although the interactions between the central nervous system (CNS) and immune system were well-documented more than 100 years ago (1), the molecular biology and biochemistry of shared cell surface molecules and of homologies/identities in circulating intercellular proteins have only been elucidated in the past decade (2). Indeed, recognition that regular trafficking of cells from the bone marrow to CNS occurs during embryogenesis, fetal development, and in normal adult brains is also recent (3–7). The function of shared receptors and their ligands in the CNS and immune systems can be examined by illustrating similarities and differences in the biological events in the two systems resulting from signal transduction or cell–cell interactions (Tables 1 and 2). Cells other than glial cells produce and respond to lymphokines and monokines. A summary of these findings appears in Table 2, although no discussion of these findings will be provided here.

MACROGLIAL CELLS

The study of astrocytes helps to illustrate themes shared by or duplications between the nervous and immune systems: autocrine factor production, morphological and functional mimicry of unrelated cells, production of proteins belonging to supergene families, ability to be stimulated by both lymphokines and neurotransmitters, and mechanisms for autofeedback. The first factors described in the field of neural–immune interactions were lymphocyte-derived factors that had both stimulating and inhibitory activities *in vitro* on astrocytes or their malignant counterparts (8–12).

Astrocytes and microglia function in similar ways when activated, in that induction of major histocompatibility (MHC) class II antigens occur on both cells, allowing for antigen presentation. The ability of both these cell subtypes to become Ia (HLA-DR) positive when induced by trauma or disease has implications for perpetuation of nervous system interaction with the immune system and the potential for them to influence CNS disease processes such as bacterial infections (meningitis), viral infections [acquired immune deficiency syndrome (AIDS), encephalitis], autoimmune diseases [multiple sclerosis (MS)], and parasitic infections (cerebral malaria).

In normal adult brain tissue, MHC class II antigens are exceedingly rare if present at all. They are occasionally seen on scattered microglia but not on astrocytes and, even with disease, never on oligodendroglia or Schwann cells in peripheral nerve tissue (13–17). In disease processes such as experimental autoimmune encephalomyelitis (EAE) in animals and MS, subacute sclerosing panencephalitis (SSPE), brain tumors, and Alzheimer's in humans, there is

TABLE 1. *Cytokine effects on glial cells*[a]

Glial cell target	Cytokine	Cytokine effect	In vivo	In vitro	Reference
Astrocytes and astrocytomas	Interleukin (IL) 1	Proliferation	X	X	37–43,49
		Induction of IL6		X	69
		Induction of GMCSF mRNA		X	110
		Differentiation		X	42
		Induction of PGE		X	53
	Interleukin 6	Production of NGF-like factor		X	69
	Tumor necrosis factor α	Proliferation		X	38,43
		Induction of IL6		X	69
		Up-regulation of MHC class II	X	X	32
	Interferon γ	Up-regulation of MHC Class II		X	27–29,34
	Transforming growth factor β	Inhibition of proliferation	X	X	70–72
		Inhibition of differentiation		X	36
Microglia	Interleukin 1	Induction of IL6		X	69
	Tumor necrosis factor	Induction of IL6		X	69
	Macrophage-colony stimulating factor	Induction of IL6		X	69
	Multicolony stimulating factor	Proliferation	X		68
	Granulocyte-macrophage colony stimulating factor	Proliferation	X	X	67,68
	Interleukin 3	Proliferation		X	66,67
	Interferon γ	Up-regulation MHC class II		X	67
		Increased cytotoxicity		X	67
Oligodendroglia precursors	Interleukin 1	Proliferation		X	94
	Interleukin 2	Inhibition of proliferation		X	94
Oligodendrogliomas	Interleukin 1	Proliferation		X	92
	Interleukin 2	Proliferation		X	90
Mature oligodendrocytes	Interleukin 2	Proliferation		X	88
		Increase in MBP		X	87,88
		Up-regulation MBP mRNA		X	87
		Up-regulation IL2R		X	95
	Tumor necrosis factor α	Inhibition of oligodendrocyte survival & myelin production		X	62,63
	Interferon γ	Induction MHC class I		X	15

[a]GMCSF, granulocyte-macrophage colony stimulating factor; PGE, prostaglandin E; NGF, nerve growth factor; IL, Interleukin; MHC, major histocompatibility; MBP, myelin basic protein.

an up-regulation of Ia on microglia and a *de novo* induction of Ia on astrocytes (13,14,16–22). In MS, astrocytes at the lesion edge and well out into the adjacent, normal-appearing white matter become DR positive (Fig. 1) (14–17). Endothelial cells in blood vessels of affected white matter areas in MS also become DR positive (15,16).

Bacteria and viruses have been shown to induce Ia on astrocytes *in vitro* and *in vivo* (23–26). Although IFNγ can induce Ia in a short period (24–48 hr) (27–29), viral induction of Ia that is independent of IFNγ requires several days (24,30). Tumor necrosis factor-alpha (TNFα) also can act synergistically with infectious agents to induce MHC class I and II antigens on astrocytes (31,32). This infers the need for two signals in the induction of Ia and implies a safety mechanism within the CNS to prevent unwanted MHC class II expression. Both IFNγ and TNFα have been demonstrated to be associated with astrocytes, especially

TABLE 2. *Cytokine effects on nonglial cells*[a]

Nonglial cell target	Cytokine	Cytokine effect	In vivo	In vitro	Reference
Hypothalamic neurons	Interleukin 1	Decreased activity in warm sensitive neurons	X		111
		Increased activity in cold sensitive neurons	X		111
		Induction CRF production	X		112,113
	Interleukin 2	Decreased discharge frequency in ventro-medial nucleus	X		114
		Increased discharge frequency in supraoptic and paraventricular nuclei	X		114
Pituitary cells	Interleukin 1	Up-regulation POMC mRNA		X	115
		Induction ACTH production		X	116
	Interleukin 2	Up-regulation POMC mRNA		X	115
PC12 cells	Interferon γ	Facilitation of NGF-induced differentiation		X	117
	Interleukin 6	Differentiation		X	118
Neuroblastomas	Interferon γ	Neurite outgrowth		X	119
Sciatic nerve	Interleukin 1	Up-regulation NGF mRNA		X	120

[a]CRF, corticotropin releasing factor; POMC, prospiomelanocortin; ACTH, adrenocorticotropic hormone; NGF, nerve growth factor.

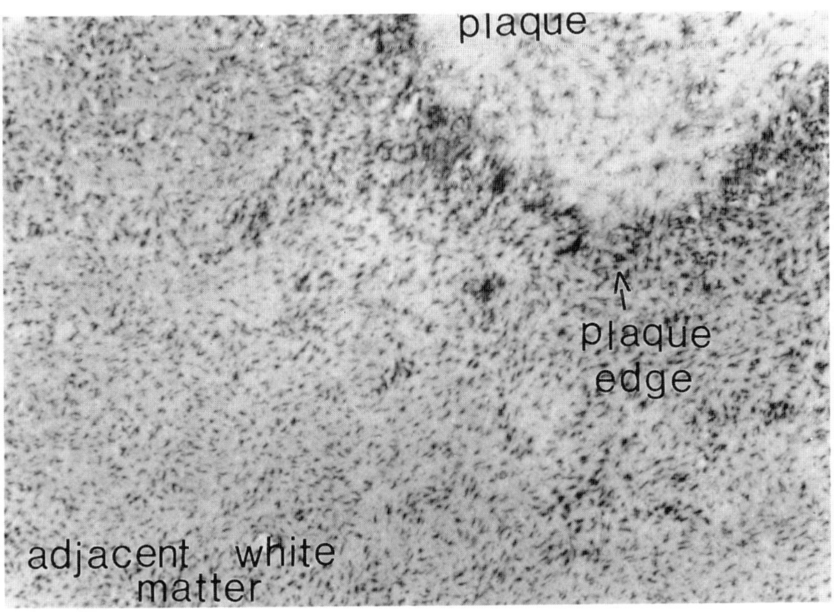

FIG. 1. HLA-DR-positive cells at edge of lesion in MS brain. Frozen sections were air dried and fixed with acetone for 10 min at room temperature. Primary monoclonal antibody, anti-HLA-DR (Becton Dickinson, 1:100), was added for 30 min at room temperature in a humidified chamber. After washing with PBS, peroxidase-labeled secondary antibody goat anti-mouse (Tago, Inc, 1:50) was applied to section for 30 min. Slide was rinsed with phosphate buffered saline (PBS) for 10 min. Substrate used was aminoethyl carbazole, applied for 10 min, after which the slide was rinsed 10 min.

DR+ reactive astrocytes at the lesion edge in MS plaques (Fig. 2) (22,33). Endogenous inhibitors of induction of class II antigens are present in the CNS: prostaglandin E (PGE) and transforming growth factor-beta (TGFβ) as well as norepinephrine (34–36).

Astrocytes produce and release several factors to which they also respond in an autocrine fashion by proliferation or differentiation. Both primary astrocytes and astrocytoma cell lines produce Interleukin-1 (IL1) and proliferate to IL1 *in vivo* (37–40) and *in vitro* (41–43). Proliferation and differentiation are astrocytic changes that are hallmarks of disease and trauma in animal models (44,45) as well as human disease like MS, AIDS, and viral infections of the CNS (46,47). IL1 can, in fact, be seen in MS plaques (14). With respect to tumors *in vivo,* IL1 may lead to enhanced tumor cell growth and indirectly stimulate tumor expansion by inducing neovascularization (38, 41,43). Part of this effect may be due to IL1 induction of TNF, which in its own right enhances angiogenesis and astrocytoma proliferation (38,43). In the context of normal brain homeostasis, IL1 may be involved in fetal and neonatal brain development, and in adults it regulates fever, slow-wave sleep, and appetite as well as participating in the adrenal-hypothalamic pituitary axis (48,49).

IL1 and TNF can coinduce each other (48,50,51) and can induce PGE, thereby allowing them to regulate temperature (48,51). In addition, endogenous mechanisms such as neurotransmitters, protein kinase C stimulators, or infectious events in the absence of an overt immune inflammation may acutely induce PGE production in astrocytes and febrile episodes (52–54). Both PGE and LTB_4 production is elevated in MS *in vitro, in vivo,* and *in situ* in the lesions (14,33,55,56). Ultimately, the long-term consequences of IL1 and PGE/LTB_4 production involve endothelial cells, alterations in vascular permeability, blood-brain barrier changes, and an increase in inflammatory infiltrates leading to the perpetuation of the autoimmune response (57,58). Furthermore, TNF stimulates respiratory burst of macrophages and neutrophils (59, 60). Release of toxic products of macrophages are implicated in the fatal hemorrhagic necrosis seen in cerebral malaria (61) and in membrane lipid peroxidation and the damage to myelin and the oligodendrocyte in MS (62,63).

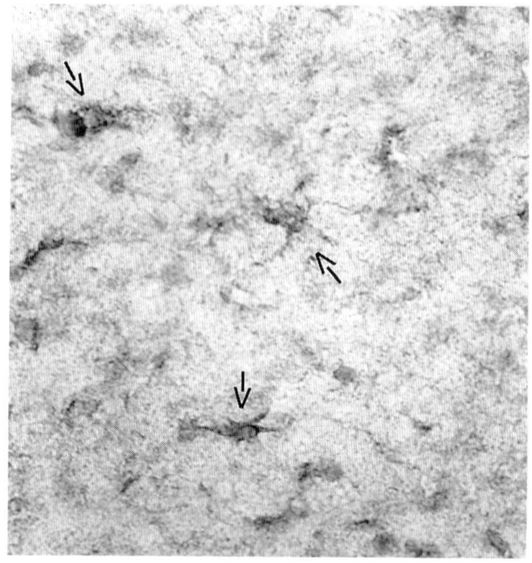

FIG. 2. TNF alpha-staining of astrocytes in MS plaques. Staining procedures and most reagents are identical to those in Fig. 1. Primary antibody used here was monoclonal anti-TNF α (supplied by Cetus Corp., 1:100). *Arrows* point to astrocytes.

Astrocytes produce Interleukin 3 (IL3), Interleukin 6 (IL6), and granulocyte-macrophage colony stimulating factor (GMCSF), all of which stimulate microglia cells (64–69). The role of these two cells in diseases like CNS AIDS and MS is thus of significant interest.

The autologous negative regulation of astrocytes function appears to be mediated through PGE production and possibly by a transforming growth factor beta-like protein (GT_SF) produced by astrocytes (9,41, 70–72). TGFβ has been shown to be a natural antagonist of TNFα in that it reduces

DR, down-regulates transferrin and IL2 receptor expression, and inhibits IFNγ production (60,70,71,73). The TGFβ_2-like protein produced by astrocytes inhibits T-cell proliferation to the same degree as TGFβ (70,71). TGFβ suppresses oxygen burst by macrophages (60). TGFβ may thus facilitate a coordinated endogenous wound healing by reduction of inflammation and lipid peroxidation and promotion of tissue repair.

Much investigation has focused on the ability of the myelin-producing cell, the oligodendrocyte, to proliferate *in vitro* in response to a variety of factors including lymphokines and astrocyte- and neuron-derived factors (74–78). Nevertheless, the questions of whether mature myelinating oligodendrocytes can divide and what *in vivo* factors are involved still exist (79). Certainly there is now agreement that immature oligodendrocyte (nonmyelin attached) or their precursors can proliferate *in vivo*, and regeneration of myelin does occur in both animal models of demyelination and in MS (80–86).

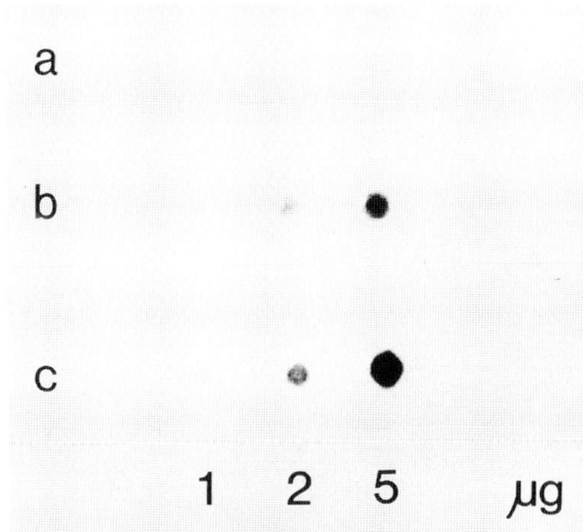

FIG. 3. Up-regulation of MBP mRNA in primary rat oligodendrocytes in response to IL2. Total RNA was prepared from primary rat oligodendrocytes. Row *a* is embryonic kidney control RNA. Row *b* is uninduced rat oligodendrocytes. Row *c* is IL2-induced rat oligodendrocytes. The 1.5-kb EcoR1 fragments of the rat MBP complementary deoxyribonucleic acid (cDNA) clone was labeled by random priming and hybridized at high stringency. Micrograms of total RNA blotted are indicated.

Many groups have investigated the induction of proliferation and differentiation of oligodendrocytes or their precursors *in vitro* and *in vivo*. Crude T-cell supernatants as well as purified recombinant proteins induce cell division and production of myelin basic protein or myelination (8,12,37,63,87–93). Although precursors do not respond to IL2, mature oligodendrocytes proliferate and differentiate in response to IL2 (Fig. 3) (87,88,94). The low-affinity IL2 receptors on human oligodendroglioma cell lines and primary rat oligodendrocytes suggest that recombinant IL2 may be cross-reacting with a receptor for an endogenous brain protein, a kind of CNS IL2 analogue. A preliminary assessment of the human brain IL2R alpha chain (p50) using T-cell reagents [a complementary deoxyribonucleic acid (cDNA) probe and monoclonal antibodies] suggests that this receptor chain may be similar but not identical to that on T cells (reviewed in ref. 95). Availability of the cDNA probe and antibodies to the recently cloned, signal-transducing IL2 receptor chain (p70–75) will allow further assessment of the similarities of this receptor complex in these two systems (96,97). If the IL2 binding receptor complex in the CNS is entirely novel, its purification and cloning will provide the tools for extracting the putative endogenous ligand that behaves like IL2. In this regard, Nieto-Sampedro and Chandry (98) have extracted an IL2-like protein from injured rat brain that behaves like IL2 functionally but is of a higher molecular weight (98). It should be noted that although IL2 and IL2R are regularly seen during inflammation of the CNS (Fig. 4), they may also occur in normal healthy adult white matter (14,20,21,55,98). The identification of a brain IL2-like protein if one exists is important because IL2 seems to be oligodendroglial specific, having no direct-effect astrocytes (88,90,98).

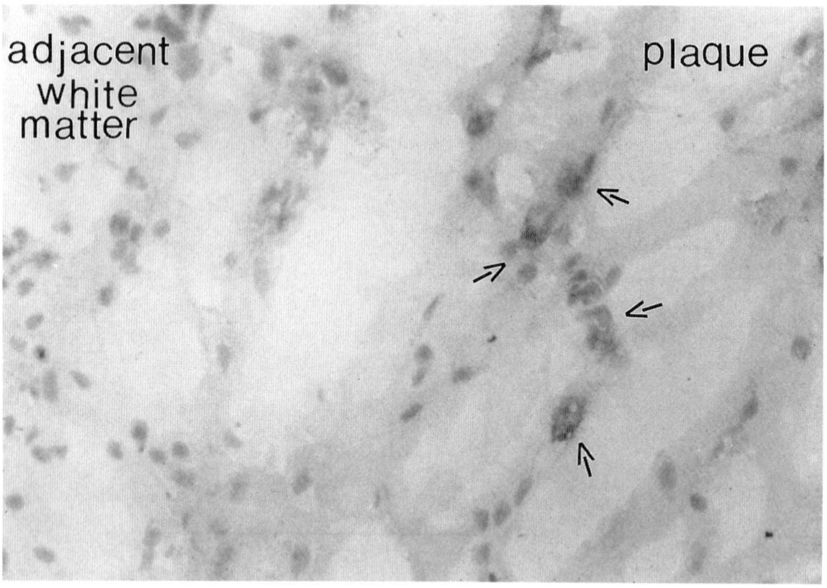

FIG. 4. IL2 at lesion edge of MS plaque. Staining was performed as in Fig. 1. Primary monoclonal antihuman IL2 (gift of Dr. Steven Gillis, 1:100). *Arrows* mark IL2-positive cells.

Although primary rat oligodendrocytes appear to be targets for a variety of growth factors, they do not secrete any factors that act in a autocrine or paracrine fashion, distinguishing them from microglia and astrocytes. However, malignant oligodendroglioma cells do produce IL1 and respond to it by proliferation (92). Interestingly, these human cell clones produce and release IL1α and not IL1β. IL1β is the major species produced by human macrophages and the only species for which messenger ribonucleic acid (mRNA) has been identified in the CNS (W. Farrar, *personal communication*). Further, work on primary normal human oligodendrocytes and malignant çell lines will establish the exact circumstances under which these cells produce and respond to factors. The abortive attempt at remyelination and proliferation by oligodendrocytes at the lesion edge in MS may well be due to the antiproliferative effects of PGE and TGFβ_2 produced by reactive astrocytes (see above). Therapeutic modalities using oligodendrocyte growth-inducing factors such as glial growth promoting factor (89,91) or IL2 must then also include inhibitors of these negative cytokines to allow for breakthrough of the stimulatory growth effects.

MICROGLIAL CELLS

Microglia migrate into the CNS as monocyte/macrophages from the bone marrow and periphery (99,100). These are highly plastic cells that change morphologically as they differentiate in the CNS (100,101). Interesting phenotypic changes occur as microglia differentiate or migrate away from the blood-brain barrier. A loss of the CD4, leukocyte common antigen, and transferrin receptor occurs (101). That is to say, there is an increased density of these markers in newly immigrated microglia or microglia in the choroid plexus or median eminens, areas possibly related to a first line defense role. When stimulated by trauma or disease, microglia deep in the parenchyma will be reactivated to express CD4 (101).

Microglia can be induced to express Ia by

IFNγ and can present antigen in an MHC-restricted way *in vitro* and *in vivo* (13,67). Like their blood counterparts they bear C_3R and FcR and are phagocytic; they also mediate tumor cell cytotoxicity and can be stimulated to generate reactive oxygen. In this regard they are capable of immune-mediated defense in the CNS (67,102). Microglia also produce IL1, PGE, and TNF *in vivo* and *in vitro* (103–105). Although the ability of microglia to proliferate *in vivo* is still open (101), they have been shown to proliferate *in vitro* to IL3 and GMCSF (67, 68,103). Because of their ability to migrate within the brain, to be regulated by astrocytic growth factors, and to produce oxygen radicals and putatively toxic cytokines such as TNF, microglia have been hypothesized to be major endogenous effectors in CNS diseases like EAE and EAN in animals and MS in humans (106–109). They certainly might play a role very early or quite late in the autoimmune process when infiltrating macrophages from the blood are at a minimum.

CONCLUSIONS

We now know that immune cells cross the blood-brain barrier on a regular basis and this infiltration is exaggerated during trauma and disease. Immune factors activate glia to produce cytokines that can perpetuate an autocrine response or maintain an inflammatory leukocyte response. Factors that are normally produced in lower concentrations for day-to-day homeostasis of neurophysiological and behavioral states, when produced in increased concentrations, can dramatically alter blood-brain barrier permeability and glial or neuronal cell function. Some of the interesting questions for future work are (1) How do the cell surface receptors and molecules such as IL2R, IFNγ R, and Ia function in the developing and normal healthy adult brain? (2) Are there endogenous ligands for these molecules in the brains (e.g., IL2 and IFNγ analogues)? (3) what synergies exist among factors to maintain a normal glial cell response in the CNS? (4) Are two signals required for glial cell response under abnormal conditions like trauma or diseases: one involving an endogenous neurotransmitter, the other an exogenous infectious agent or lymphokine? Is this a safety feature of the CNS? (5) What endogenous feedback mechanisms exist to shut down glial cell responses? (6) At the molecular level, what are the bases for different forms of a growth factor or its receptor in different tissues: gene duplication, alternative polyadenylation sites, alternative exon usage, or alternative tissue-specific promoters?

Sorting out the factors and receptors that are CNS-specific or shared with the immune system will help in our understanding of gliosis, demyelination, neural degeneration, and remyelination.

REFERENCES

1. Babinski J. Recherches sur l'anatomie pathologique de la sclerose en plaques et etude comparative des diverses varietes de scleroses de la moelle. *Arch Physiol* 1885;5–6:186–207.
2. Williams AF. A year in the life of the immunoglobulin super family. *Immunol Today* 1987;8: 298–303.
3. Bartlett PF. Pluripotential hemopoietic stem cells in adult mouse brain. *Proc Natl Acad Sci* 1982;79:2722–2725.
4. Choi BH. Hematogenous cells in the central nervous system of developing human embryo and fetuses. *J Comp Neurol* 1981;196:683–694.
5. Lent R, Linden R, Cavalcante LA. Transient populations of presumptive macrophages in the brain of the developing hamster as indicated by endocytosis of blood-brain horseradish peroxidase. *Neuroscience* 1985;15:1203–1215.
6. Ting JP-Y, Nixon DF, Weiner LP, Frelinger JA. Brain Ia antigens have a bone marrow origin. *Immunogenetics* 1982;92:1–7.
7. Wekerle H, Linington C, Lassmann H, Meyermann R. Cellular immune reactivity within the CNS. *TINS* 1986;9:271–277.
8. Barna BP, Chou SM, Jacobs B, Ransohoff RM, Hahn JF, Bay JW. Enhanced DNA synthesis of human glial cells exposed to human leukocyte products. *J Neuroimmunol* 1985;10:151–158.
9. Fontana A, Hengartner H, deTribolet N, Weber E. Gliobastoma cells release interleukin 1 and factors inhibiting interleukin 2 mediated effects. *J Immunol* 1984;132:1837–1844.

10. Fontana A, Dubs R, Merchant R, Balsiger S, Grob PJ. Glia cell stimulatory factor (GSF): a new lymphokine. Part 1. Cellular sources and partial purification of murine GSF, role of cytoskeleton and protein synthesis in its production. *J Neuroimmunol* 1981;2:55–71.
11. Fontana A, Jost R, Balsiger S, Grob P, Grieder A. Involvement of cyclic AMP in the regulation of lymphokine induced glia cell stimulation. *Dev Brain Res* 1982;2:505–511.
12. Merchant RE, Fontana A, Nelson AT, Young HF. Mitogen activated lymphocytes of normals and glioma patients produce factors with anti glioblastoma activity *in vitro*. *J Neuroimmunol* 1986;11:1–14.
13. Hayes GM, Woodroofe MN, Cuzner ML. Microglia are the major cell type expressing MHC class II in human white matter. *J Neurol Sci* 1987;80:25–37.
14. Hofman F, Von Hanwher RI, Dinarello CA, Mizel SB, Hinton D, Merrill JE. Immunoregulatory molecules and IL2 receptors identified in multiple sclerosis brain. *J Immunol* 1986;136: 3239–3245.
15. Lisak RP, Hirayama M, Kuchmy D, et al. Cultured human and rat oligodendrocytes and rat Schwann cells do not have immune response gene associated antigen (Ia) on their surface. *Brain Res* 1983;289:285–292.
16. Traugott U. Multiple sclerosis: relevance of class I and class II MHC-expressing cells to lesion development. *J Neuroimmunol* 1987;16: 283–302.
17. Traugott U, Scheinberg LC, Raine CS. On the presence of Ia-positive endothelial cells and astrocytes in multiple sclerosis lesions and its relevance to antigen presentation. *J Neuroimmunol* 1985;8:1–14.
18. Frank E, Pulver M, deTribolet N. Expression of class II major histocompatibility antigens on reactive astrocytes and endothelial cells within the gliosis surrounding metastases and abscesses. *J Neuroimmunol* 1986;12:29–36.
19. Hickey WF, Osborn JP, Kirby WM. Expression of Ia molecules by astrocytes during acute experimental allergic encephalomyelitis in Lewis rats. *Cell Immunol* 1985;91:528–535.
20. Luber-Narod J, Rogers J. Immune system associated antigens expressed by cells of the human central nervous system. *Neurosci Lett* 1988;94:17–22.
21. Rogers J, Luber-Narod J, Styren SD, Civin WH. Expression of immune system associated antigens by cells of the human central nervous system: relationship to the pathology of Alzheimer's disease. *Neurobiol Aging* 1988;9:339–349.
22. Traugott U, Lebon P. Interferonγ and Ia antigen present on astrocytes in active chronic multiple sclerosis lesions. *J Neurol Sci* 1988; 84:257–264.
23. Liu Y, King N, Kesson A, Blanden RV, Mullbacher A. Flavivirus infection upregulates the expression of class I and class II major histocompatibility antigens on and enhances T cell recognition of astrocytes *in vitro*. *J Neuroimmunol* 1989;21:157–168.
24. Massa PT, Dorries R, terMeulen V. Viral particles induce Ia antigen expression on astrocytes. *Nature* 1986;320:543–546.
25. Olsson T, Maehlen J, Love A, Klareskog L, Norrby E, Kristensson K. Induction of class I and class II transplantation antigens in rat brain during fatal and non-fatal measles virus infection. *J Neuroimmunol* 1987;16:215–224.
26. Rodriquez M, Pierce ML, Howie EA. Immune response gene products (Ia antigens) on glial and endothelial cells in virus-induced demyelination. *J Immunol* 1987;138:3438–3442.
27. Hirsch MR, Wetzerbin J, Pierres M, Goridis C. Expression of Ia antigens by cultures astrocytes treated with gamma interferon. *Neurosci Lett* 1983;41:199–204.
28. Mauerhoff T, Pujol-Borell R, Mirakian R, Bottazzo GF. Differential expression and regulation of major histocompatibility products in neural and glial cells of the human fetal brain. *J Neuroimmunol* 1988;8:271–289.
29. Wong GHW, Bartlett PF, Clark-Lewis I, Battye F, Schrader JW. Inducible expression of H-2 and Ia on brain cells. *Nature* 1984;310:688–691.
30. Massa PT, terMeulen V. Analysis of Ia induction on Lewis rat astrocytes *in vitro* by virus particles and bacterial adjuvants. *J Neuroimmunol* 1987;13:259–271.
31. Lavi EA, Suzumura EA, Murasko DM, Murray EM, Silberberg DH, Weiss SR. Tumor necrosis factor induces expression of MHC class I antigens on mouse astrocytes. *J Neuroimmunol* 1988;18:245–253.
32. Massa PT, Schimpl A, Wecker E, terMeulen V. Tumor necrosis factor amplifies measles virus-mediated Ia induction astrocytes. *Proc Natl Acad Sci* 1987;84:7242–7245.
33. Hofman FM, Hinton DR, Johnson K, Merrill JE. Tumor necrosis factor identified in multiple sclerosis brain. *J Exp Med* 1989;170:607–612.
34. Frohman EM, Vayuvegula B, van den Noort S, Gupta S. Norepinephrine inhibits gamma-interferon-induced MHC class II (Ia) antigen expression on cultured brain astrocytes. *J Neuroimmunol* 1988;17:89–101.
35. Snyder DS, Beller DI, Unanue ER. Prostaglandins modulate macrophage Ia expression. *Nature* 1982;299:163–165.
36. Zuber P, Kuppner MC, deTribolet N. Transforming growth factor-beta 2 downregulates HLA-DR antigen expression on human malignant glioma cells. *Eur J Immunol* 1988;18: 1623–1626.
37. Giulian D, Lachman LB. Interleukin-1 stimulation of astroglia proliferation after brain injury. *Science* 1985;228:497–499.
38. Giulian D, Woodward J, Young DG, Krebs JF, Lachman LB. Interleukin-1 injected into mammalian brain stimulates astrogliosis and neovascularization. *J Neurosci* 1988;8:2485–2490.
39. Nieto-Sampedro M, Berman MA. Interleukin-1

like activity in rat brain: sources, targets, and effects of injury. *J Neurosci Res* 1987;17:214–219.

40. Politis MJ, Lindsay JK. Exogenous interleukin regulates growth of C6 tumors *in vivo*. *Acta Neuropathol* 1988;77:182–185.
41. Fontana A, Kristensen F, Dubs R, Gemsa D, Weber E. Production of prostaglandin E and an interleukin 1 like factor by cultured astrocytes and C6 glioma cells. *J Immunol* 1982;29:2413–2419.
42. Ito J-I, Kato T, Wakabayashi S, Hara F, Tanaka R, Kato K. Autocrine regulation of glial proliferation and differentiation: the induction of cytodifferentiation of postmitotic normal glioblast by growth promoting factor from astrocytoma cell. *Brain Res* 1986;374:335–341.
43. Lachman LB, Brown DC, Dinarello CA. Growth-promoting effect of recombinant interleukin 1 and tumor necrosis factor for a human astocytoma cell line. *J Immunol* 1987;138:2913–2916.
44. Raine CS, Traugott U. Immunopathology of the lesion in multiple sclerosis and chronic relapsing experimental allergic encephalomyelitis In: Vandenbark AA, Raus JCM, eds. *Immunoregulatory processes in experimental allergic encephalomyelitis and multiple sclerosis*, vol 7. Amsterdam: Elsevier, 1984;151–212.
45. Sakai K, Tabira T, Endoh M, Steinman L. Ia expression in chronic relapsing experimental allergic encephalomyelitis induced by long term cultured T cell lines in mice. *Lab Invest* 1986;54:345–352.
46. Adams CWM. The general pathology of multiple sclerosis: morphological and chemical aspects of the lesions. In: Hallpike JF, Adams CWM, Tourtellotte WW, eds. *Multiple sclerosis: pathology, diagnosis, and management*. Baltimore: Williams and Wilkins, 1983;203–240.
47. Gabuzda DH, Hirsch MS. Neurologic manifestations of infection with human immunodeficiency virus. Clinical feature and pathogenesis. *Ann Int Med* 1987;107:383–391.
48. Dinarello CA. Biology of interleukin 1. *FASEB J* 1988;2:108–115.
49. Giulian D, Young DG, Woodward J, Brown DC, Lachman LB. Interleukin-1 is an astroglial growth factor in the developing brain. *J Neurosci* 1988;8:709–714.
50. Chezzi P, Dinarello CA. IL1 induces IL1. *J Immunol* 1988;140:4238–4244.
51. Dinarello CA, Cannon JG, Wolff SM, et al. Turmor necrosis factor (cachectin) is an endogenous pyrogen and induces production of interleukin 1. *J Exp Med* 1986;163:1433–1435.
52. Hartung H-P, Heininger K, Schafer B, Toyka KV. Substance P and astrocytes: stimulation of the cyclooxygenase pathway of arachidonic acid metabolism. *FASEB J* 1987;2:48–51.
53. Hartung H-P, Schafer B, Heininger K, Toyka KV. Recombinant interleukin 1 stimulates eicosanoid production in rat primary culture astrocytes. *Brain Res* 1989;489:113–119.
54. Hartung H-P, Klaus V, Toyka VKV. Phorbol diester TPA elicits prostaglandin E release from cultured rat astrocytes. *Brain Res* 1987;417:347–349.
55. Merrill JE, Hofman FM. Regulatory molecules in the CSF and CNS. In: Lowenthal H, Raus J, eds. *Cellular and humoral immunological components of cerebrospinal fluid in multiple sclerosis*. New York: Plenum Publishing, 1987;381–390.
56. Merrill JE, Strom SR, Ellison GW, Myers LW. *In vitro* study of mediators of inflammation in multiple sclerosis. *J Clin Immunol* 1989;9:84–96.
57. Janzer RC, Raff MC. Astrocytes induce blood-brain barrier properties in endothelial cells. *Nature* 1987;325:253–257.
58. Martin S, Maruta K, Burkart V, Gillis S, Kolb H. IL1 and IFN-γ increase vascular permeability. *Immunology* 1988;64:301–305.
59. Ferrante A, Nandeskor M, Bates EJ, Goh DHB, Beard J. Tumor necrosis factor beta (lymphotoxin) inhibits locomotion and stimulates the respiratory burst and degranulation of neutrophils. *Immunology* 1988;63:507–512.
60. Tsunawaki S, Sporn M, Ding A, Nathan C. Deactivation of macrophages by transforming growth factorβ. *Nature* 1988;334:260–262.
61. Grau GE, Fajardo LF, Piguet PF, Allet B, Lambert P-H, Vassalli P. Tumor necrosis factor (cachectin) is an essential mediator in murine cerebral malaria. *Science* 1987;237:1210–1212.
62. Robbins DS, Shirazi Y, Drysdale B-E, Lieberman A, Shin HS, Shin ML. Production of cytotoxic factor for oligodendrocytes by stimulated astrocytes. *J Immunol* 1987;139:2593–2597.
63. Selmaj KW, Raine CS. Tumor necrosis factor mediates myelin and oligodendrocyte damage *in vitro*. *Ann Neurol* 1988;23:339–346.
64. Farrar WL, Vinocour M, Hill JM. *In situ* hybridization histochemistry localization of interleukin-3 mRNA in mouse brain. *Blood* 1989;73:137–140.
65. Frei K, Bodmer S, Schweidel C, Fontana A. Astrocytes of the brain synthesize interleukin 3–like factors. *J Immunol* 1985;135:4044–4047.
66. Frei K, Bodmer S, Schwerdel C, Fontana A. Astrocyte-derived interleukin 3 as a growth factor for microglia cells and peritoneal macrophages. *J Immunol* 1986;137:3521–3527.
67. Frei K, Siepl C, Groscurth P, Bodmer S, Schwerdel C, Fontana A. Antigen presentation and tumor cytotoxicity by interferon-γ-treated microglial cell. *Eur J Immunol* 1987;17:1271–1278.
68. Giulian D, Ingeman JE. Colony stimulating factors as promoters of ameboid microglia. *J Neurosci* 1988;8:4707–4717.
69. Frei K, Malipiero UV, Leist TP, Zinkernagel RM, Schwab ME, Fontana A. On the cellular source and function of interleukin 6 produced in the central nervous system in viral diseases. *Eur J Immunol* 1989;19:689–694.

70. Siepl C, Bodmer S, Frei K, et al. The glioblastoma-derived T cell suppressor factor/transforming growth factor-β_2 inhibits T cell growth without affecting the interaction of interleukin 2 with its receptor. *Eur J Immunol* 1988;18:593–600.
71. Wahl SM, Hunt DA, Wong HL, et al. Transforming growth factor β is a potent immunosuppressive agent that inhibits IL1-dependent lymphocyte proliferation. *J Immunol* 1988; 140:3026–3032.
72. Wrann M, Bodmer S, deMartin R, et al. T cell suppressor factor from human gliobastomas cells is a 12.5kd protein closely related to transforming growth factorβ. *EMBO J* 1987;6:1633–1636.
73. Ding AH, Nathan CF. Trace levels of bacterial lipopolysaccharide prevent interferonγ or tumor necrosis factorα from enhancing mouse peritoneal macrophage respiratory burst capacity. *J Immunol* 1987;139:1971–1977.
74. Dubois-Dalcq M. Characterization of a slowly proliferative cell along the oligodendrocyte differentiation pathway. *EMBO J* 1987;6:2587–2595.
75. Edgar AD, Pfeiffer SE. Extracts from neuron-enriched cultures of chick telencephalon stimulate the proliferation of rat oligodendrocytes. *Dev Neurosci* 1985;7:206–215.
76. Mellstrom K, Hoglund AS, Nister M, Heldin CH, Westermark B, Lindberg U. The effect of platelet-derived growth factor on morphology and motility of human glial cells. *J Muscle Res Cell Motil* 1983;4:589–609.
77. Sensenbrenner M, Delaunoy J-P, Labourdette G, Pettman B. Effects of brain extracts on the proliferation and the maturation of astroglial and oligodendroglial cells in culture. *Biochem Soc Trans* 1981;10:424–426.
78. Wood PM, Bunge RP. Evidence that axons are mitogenic for oligodendrocytes isolated from adult animals. *Nature* 1986;320:756–758.
79. Ludwin SK, Bakkar DA. Can oligodendrocytes attached to myelin proliferate? *J Neurosci Res* 1988;8:1239–1244.
80. Arenella LS, Herndon RM. Mature oligodendrocytes, division following experimental demyelination in adult animals. *Arch Neurol* 1984;41:1162–1165.
81. Herndon RM, Price DL, Weiner LP. Regeneration of oligodendroglia during recovery from demyelinating disease. *Science* 1977;195:693–694.
82. Ludwin SK. An autoradiographic study of cellular proliferation in remyelination of the central nervous system. *Am J Pathol* 1979;95:683–696.
83. Ludwin SK. Proliferation of mature oligodendrocytes after trauma to the central nervous system. *Nature* 1984;308:274–275.
84. Prineas JW, Connell F. Remyelination in multiple sclerosis. *Ann Neurol* 1979;5:22–31.
85. Prineas JW, Kwon EE, Cho E-S, Sharer LR. Continual breakdown and regeneration of myelin in progressive multiple sclerosis plaques in multiple sclerosis. *Ann NY Acad Sci* 1984; 436:11–32.
86. Raine CS, Scheinberg L, Waltz JM. Multiple sclerosis oligodendrocyte survival and proliferation in an active established lesion. *Lab Invest* 1981;45:534–546.
87. Benveniste EN, Herman PK, Whitaker JN. Myelin basic protein specific RNA levels in interleukin-2 stimulated oligodendrocytes. *J Neurochem* 1987;79:1274–1279.
88. Benveniste EN, Merrill JE. Stimulation of oligodendroglial proliferation and maturation by interleukin-2 *Nature* 1986;321:610–613.
89. Benveniste EN, Merrill JE, Kaufman SE, Golde DW, Gasson JC. Purification and characterization of a human T-lymphocyte derived glial growth promoting factor. *Proc Natl Acad Sci* 1985;82:3930–3934.
90. Benveniste EN, Tozawa H, Gasson JC, Quan S, Golde DW, Merrill JE. Response of human glioblastoma cells to recombinant interleukin-2. *J Neuroimmunol* 1988;17:301–314.
91. Merrill JE, Kutsunai S, Mohlstrom C, Hofman F, Groopman J, Golde DW. Proliferation of astroglia and oligodendroglia in response to human T cell-derived factors. *Science* 1984; 224:1428–1430.
92. Merrill JE, Matsushima K. Production of and response to interleukin 1 by cloned human oligodendroglioma cell lines. *J Biol Reg Homeo Agents* 1988;2:4–13.
93. Yong VW, Kim SU, Kim MW, Shin DH. Growth factors for human glial cells in culture. *Glia* 1988;1:113–123.
94. Saneto RP, Altman A, Knobler RL, Johnson HM, DeVellis J. Interleukin 2 mediates the inhibition of oligodendrocyte progenitor cell proliferation *in vitro*. *Proc Natl Acad Sci* 1986; 83:9221–9225.
95. Merrill JE. Effect of lymphokines and monokines on glial cells *in vitro* and *in vivo*. In: Goetzl E, Spector NH, eds. *Neuroimmune networks: physiology and disease*. New York: Alan R Liss, 1989;12–20.
96. Hatakeyama M, Tsudo M, Minamoto S, et al. Interleukin 2 receptor β chain gene: generation of three receptor forms by cloned human α and β chain cDNAs. *Science* 1989;244:551–556.
97. Takeshita T, Goto Y, Tada K, Nagata K, Asao H, Sugamura K. Monoclonal antibody defining a molecule possibly identical to the p75 subunit of the interleukin 2 receptor. *J Exp Med* 1989;169:1323–1332.
98. Nieto-Sampedro M, Chandry KG. Interleukin 2–like activity in injured rat brain. *Neurochem Res* 1987;12:723–727.
99. Hickey WF, Kimura H. Perivascular microglia cells of the CNS are bone-marrow derived and present antigen *in vivo*. *Science* 1988;239:290–292.
100. Perry VH, Gordon S. Macrophages and microglia in the nervous system. TINS 1988; 11:273–277.

101. Perry VH, Gordon S. Modulation of CD4 antigen on macrophages and microglia in rat brain. *J Exp Med* 1987;166:1138–1143.
102. Sonderer B, Wild P, Wyler R, Fontana A, Peterhaus E, Schwyzer M. Murine glia cells in culture can be stimulated to generate reactive oxygen. *J Leukocyte Biol* 1987;42:463–473.
103. Gebicke-Haerter PJ, Bauer J, Schobert A, Northoff H. Lipopolysaccharide-free conditions in primary astrocyte cultures allow growth and isolation of microglial cells. *J Neurosci* 1989;9:183–194.
104. Giulian D, Baker TJ, Shih L-CN, Lachman LB. Interleukin 1 of the central nervous system is produced by ameboid microglia. *J Exp Med* 1986;164:594–604.
105. Hetier E, Ayala J, Denefle P, et al. Brain macrophages synthesize interleukin 1 and interleukin-1 mRNAs *in vitro*. *J Neurosci Res* 1988; 21:391–397.
106. Brosnan CF, Selmaj K, Raine CS. Hypothesis: a role for tumor necrosis factor in immune-mediated demyelination and its relevance to multiple sclerosis. *J Neuroimmunol* 1988;18:87–94.
107. Hartung H-P, Schafer B, Heininger K, Toyka KV. Suppression of experimental autoimmune neuritis by the oxygen radical scavengers superoxide dismutase and catalase. *Ann Neurol* 1988;23:453–460.
108. Symons TA, Bundick RV, Suckling AJ, Rumsby MG. Cerebrospinal fluid interleukin-1 like activity during chronic relapsing experimental allergic encephalomyelitis. *Clin Exp Immunol* 1987;68:648–654.
109. Woodroofe MN, Bellamy AS, Feldmann M, Davison AN, Cuzner ML. Immunocytochemical characterization of the immune reaction in the central nervous system in multiple sclerosis. Possible role for microglia in lesion growth. *J Neurol Sci* 1986;74:135–152.
110. Nishida T, Nakai S, Kawakami T, Aihara K, Nishino N, Hirai Y. Dexamethasone regulation of the expression of cytokine mRNAs induced by interleukin 1 in the astrocytoma cell line 4373MG. *FEBS Lett* 1989;243:25–29.
111. Hori T, Shibata M, Nakashima T, et al. Effects of interleukin 1 and arachidonate on the preoptic and anterior hypothalamic neurons. *Brain Res Bull* 1988;20:75–82.
112. Berkenbosch F, VanOers J, del Rey A, Tilders F, Besedovsky H. Corticotropin releasing factor-producing neurons in the rat activated by interleukin 1. *Science* 1987;238:524–526.
113. Sapolsky R, Rivier C, Yamamoto G, Plotsky P, Vale W. Interleukin 1 stimulates the secretion of hypothalamic corticotropin-releasing factor. *Science* 1987;238:522–524.
114. Bindoni M, Perciavalle V, Berretta S, Belluardo N, Diamantstein T. Interleukin 2 modifies the bioelectric activity of some neurosecretory nuclei in the rat hypothalamus. *Brain Res* 1988;462:10–14.
115. Brown SL, Smith LR, Blalock JE. Interleukin 1 and interleukin 2 enhance proopiomelanocortin gene expression in pituitary cells. *J Immunol* 1987;139:3181–3183.
116. Bernton EW, Beach JE, Holaday JW, Smallridge RC, Fein HG. Release of multiple hormones by a direct action of interleukin 1 on pituitary cells. *Science* 1987;238:519–521.
117. Improta T, Salvatore AM, Di Luzio A, Romeo G, Coccia EM, Calissano P. INF-gamma facilitates NGF-induced neuronal differentiation in PC12 cells. *Exp Cell Res* 1988;179:1–9.
118. Satoh T, Nakamura S, Taga T, et al. Induction of neuronal differentiation in PC12 cells by B cell stimulatory factor 2/interleukin 6. *Mol Cell Biol* 1988:3546–3549.
119. Sugimoto T, Horii Y, Hino T, et al. Differential susceptibility of HLA class II antigens induced by gamma-interferon in human neuroblastoma cell lines. *Cancer Res* 1989;49:1824–1828.
120. Lindholm D, Heumann R, Meyer M, Thoenen H. Interleukin 1 regulates synthesis of nerve growth factor in non-neuronal cells at rat sciatic nerve. *Nature* 1987;380:658–659.

Volume Transmission in the Brain: Novel Mechanisms for Neural Transmission,
edited by K. Fuxe and L.F. Agnati.
Raven Press, Ltd., New York © 1991.

21

Diffusion of Ions and Transmitters in the Brain Cell Microenvironment

Charles Nicholson and Margaret E. Rice

Department of Physiology and Biophysics, New York University Medical Center, New York, New York 10016

Diffusion is an inescapable process that influences and constrains numerous facets of brain function. Understanding diffusion provides a window onto many vital processes. This chapter will show both theoretically and experimentally that a suitably modified diffusion equation can accurately describe the migration of several ions and neurotransmitters, with emphasis on dopamine (DA), in the brain cell microenvironment. These results provide a basis for further generalization and discussion of extracellular volume transmission of signals in the brain.

Diffusing substances may be classified by their potential roles as "informational" or "energetic" (1,2). In this chapter we will be mainly concerned with informational substances. These can be defined (1,3) by the characteristic that they act on cells at low concentration to change system and cellular behavior. Informational substances must have specific receptors and must be presented to the receptor in sufficient concentration to activate it. One set of examples is the numerous experiments of Feldberg (4), which show that the addition of neuroactive substances to the ventricles brings about many behavioral changes. The ongoing controversy about the mismatches in localization of transmitters and receptors (5; see also this volume) raises important issues about informational substances and diffusion. In contrast, energetic substances serve the metabolic needs of cells in general, and neurons in particular. Energetic substances are characterized not only by their role in the life of a cell but also by their relatively high concentration (1). The movements of oxygen and glucose from blood vessel to cell are classic examples that have engendered much theoretical work (e.g., ref. 6).

The domain through which substances move in brain tissue is primarily the extracellular space. This is a major component of the brain cell microenvironment and comprises the narrow spaces between cell membranes. Its geometry may be likened to the soap phase of an aggregate of soap bubbles, and this complex of sheet-like spaces and three-dimensional intersections has a distinct and quantifiable effect on the transport properties of brain tissue.

QUANTITATIVE BASIS FOR DIFFUSION IN THE BRAIN

Substances move through the brain interstices almost entirely by diffusion. Bulk flow does occur, but it is extremely slow (7). We now know (see ref. 8 for review) that diffusion in complex media, such as porous aggregates or brain tissue, obeys

Fick's law, modified to incorporate two new parameters: the tortuosity and volume fraction.

In a complex medium, the diffusion coefficient, D, is reduced by the square of the tortuosity, λ, so that we obtain an apparent diffusion coefficient $D^* = D/\lambda^2$. Tortuosity is related to the fact that a diffusing particle encounters membranous obstructions as it executes a random walk in the intercellular clefts. The free and apparent diffusion coefficients for some substances are presented in Table 1. The data in this table show that the values of D differ by a factor of less than 3 among the chosen substances and also that the diffusion coefficient is quite sensitive to temperature [as shown by the tetramethylammonium (TMA^+) data]. The apparent diffusion coefficients are also shown for a value of $\lambda = 1.6$.

There are theoretical grounds for proposing a relation between molecular weight (M) and the diffusion coefficient; for molecules of low M the preferred relation is that D is proportional to the reciprocal of the square root of M, for a given temperature (9,10). This relation is shown in the last column of Table 1, where the product $DM^{1/2}$ is displayed. Davson and Danielli (ref. 14, Table IV, Chapter V) present data for several molecules ($32 < M < 180$; temperature not specified), and their average value for $DM^{1/2}$ is 7.1, which fits some of the substances in Table 1. TMA^+, α-naphthalene sulfonate (ANS^-), K^+, and Ca^{2+}, however, all show noticeable deviations. Indeed, the fact that K^+ and Ca^{2+} have similar molecular weights but different diffusion coefficients implies that a simple relation between D and M cannot work for both these ions.

The second factor that modifies the classic diffusion equation is the volume fraction, α, of the extracellular space. If a given quantity of substance is released into the extracellular space, it becomes more concentrated than it would in a free medium. In the past there has been considerable discussion about the size of the extracellular space but through the pioneering work of Van Harreveld (11) and others it is now accepted that about 20% of brain tissue is extracellular space (i.e., $\alpha = 0.2$). This value is confirmed by the studies to be reported here.

The two parameters, λ and α, are geo-

TABLE 1. *Diffusion coefficients for some substances*[a,b]

Ion	M	Temp (°C)	D (cm²/sec × 10⁶)	Ref	D^* (cm²/sec × 10⁶)	$DM^{1/2}$ (× 10⁵)
TMA^+	74.1	10	7.6	c	3.0	—
TMA^+	74.1	25	10.0	c	4.0	8.6
TMA^+	74.1	37	13.0	c	5.1	—
ANS^-	174.1	25	6.6	c,f	2.7	8.7
K^+	39.1	25	18.4	d	7.2	11.5
Ca^{2+}	40.1	25	7.7	e	3.0	4.9
Ascorbate	175.1	25	5.3	f	2.2	7.0
Epinephrine	183.2	25	5.1	f	2.1	6.9
Norepinephrine	169.2	25	5.5	f	2.2	7.2
Dopamine	153.1	25	6.0	f	2.4	7.4
Serotonin	176.2	25	5.4	f	2.2	7.2

[a] M, molecular weight; Temp, temperature. D^*, D/λ^2, in which the tortuosity, λ, is assumed to be 1.6; TMA^+, tetramethylammonium; ANS^-, α-naphthalene sulfonate.

[b] Note that diffusion coefficients were measured in various electrolyte solutions; consult original sources for details.

[c] C. Nicholson and M. Rice (*unpublished data*), measured in isotonic NaCl.

[d] Ref. 68.

[e] Ref. 69.

[f] Ref 70.

metric in nature and dimensionless in form. In brain tissue, a third factor, uptake, enters the diffusion equation. In its simplest form, this appears as a term, k, (s^{-1}) for nonspecific loss of material across the cell membranes caused by the concentration gradient and leakiness (that could be mediated through endocytosis for large molecules). For many biologically active compounds, especially transmitters and neuromodulators, more complex forms of nonlinear uptake are applicable (12) but do not allow analytical solutions to the diffusion equation.

Starting with the basic concepts of Fick's Law (13) and applying the method of volume averaging developed for describing convection in complex media (8,14), the roles of α, λ, and k may be expressed by the equation (8)

$$(D/\lambda^2)\nabla^2 C + Q/\alpha - kC = \partial C/\partial t \quad [1]$$

where C is concentration, t is time, Δ is the gradient operator, and Q is a source term. This is a member of the class of so-called parabolic partial differential equations and is similar to that found in the theory of heat conduction(15) and in cable theory as typically applied to neurons (16). Thus, in addition to specific solutions for diffusion problems (e.g., ref. 13), a considerable body of analytical and numerical solutions is available in the literature.

Some Solutions to the Diffusion Equation for Complex Media

Equation 1 can be solved for a variety of geometries, initial conditions, and source terms. For example, if the diffusing substance is released as a brief puff, as may be the case with natural transmitter release or pressure ejection from a micropipette (Fig. 1A), assuming spherical symmetry (ref. 17; unpublished data),

$$C = \{UC_f\lambda^3/[8(D\pi t)^{3/2}\,\alpha]\} \quad [2]$$
$$\times \exp[(-r^2\lambda^2/4Dt) - kt]$$

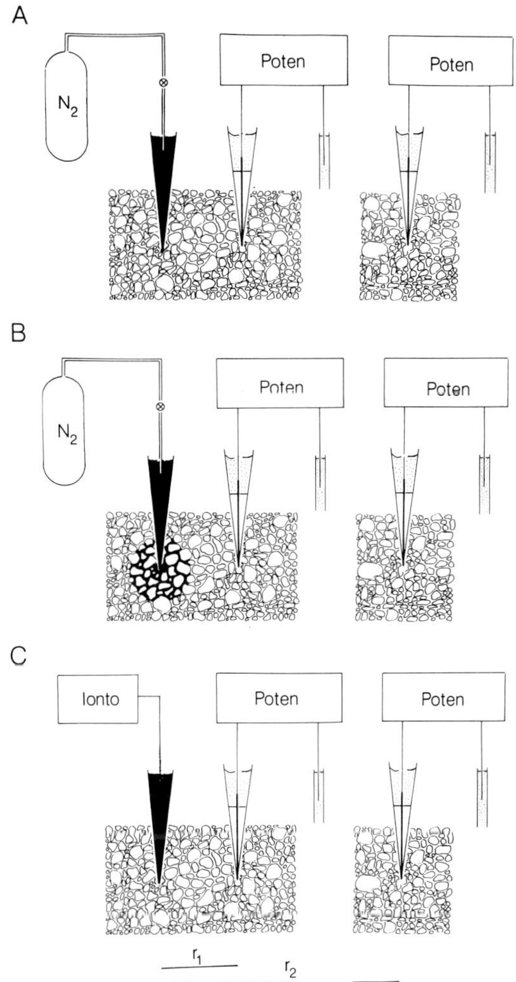

FIG. 1. Three experimental paradigms for measuring diffusion. In each case a source micropipette ejects the neuroactive substance (assumed to be dopamine here), and two carbon fiber voltammetric microelectrodes (CFVMs) connected to potentiostats (Poten) and other equipment for high-speed cyclic voltammetry (61) record the resulting tissue concentrations as continuous functions of time. One CFVM is close to the source (e.g., $r_1 = 20$ μm, as in Fig. 2) and one is far away, as shown by the break in the figure (e.g., $r_2 = 200$ μm). **A:** The neuroactive substance is ejected by a brief pulse of nitrogen (N_2) and initially forms a tiny bolus at the tip of the ejecting electrode. This corresponds to the case described by Equation 2. **B:** Pressure ejection releases a larger amount that infiltrates the extracellular space for significant distance. This case corresponds to Equation 3. **C:** Neuroactive substance (assumed to be charged) is released by iontophoresis. This corresponds to Equation 4.

where a volume U at concentration C_f is released. This solution assumes that all the material is released in a tiny volume (or that the observation point is so far away that the release site "looks" like a point source). Calculated concentration versus time profiles for a brief localized release of DA in typical brain tissue is shown in Fig. 2A–C. Particularly striking is the effect of distance on both the shape and amplitude of the DA signal. Figure 2C shows the effect of simple uptake on the concentration profile; the effect is to both attenuate and sharpen it.

If the released material is not confined to a point but instead infiltrates the interstices of the extracellular space and occupies a finite volume (Fig. 1B), defined by a sphere of radius b within the tissue (including both the intracellular and extracellular phases) so that $b = (3U/4\pi\alpha)^{13}$, the solution to Equation 1 is (17)

$$\begin{aligned} C = {} & (C_f/2)(\mathrm{erf}[\lambda(r + b)/2(Dt)^{1/2}] \\ & - \mathrm{erf}[\lambda(r - b)/2(Dt)^{1/2}] \\ & - [2/(\lambda r)](Dt/\pi)^{1/2} \qquad [3] \\ & \mathrm{x}\{\exp[-\lambda^2(r - b)^2/(4Dt)] \\ & - \exp[-\lambda^2(r + b)^2(4Dt)]\}) \end{aligned}$$

In this expression "erf" represents the error function. No uptake term is included in equation 3 because we do not have an analytic solution for that case at present. The effect of the finite source volume is illustrated in Fig. 2A (inset), where it is seen, for the values employed here, that there is little difference in the DA distribution even though the DA initially spreads out to a radius of 15 μm and measurements are at 20 μm. Note, though, that the times of the peaks are different. At greater distances the difference is quite negligible. The reason for the small effect of finite volume here lies in the small amount ejected, which was chosen to be typical of real experiments. If a bolus of, say, 100 μm diameter were ejected, the two solutions (Equations 2 and 3) would considerably differ over a much greater distance. Eventually, however, at sufficient distance, they would again be indistinguishable.

Another possibility is that the infiltrating solution forms a solution-filled cavity in the brain (so $\lambda = 1 = \alpha$ for this cavity). A numerical solution has been illustrated for this situation (17), but our experiments suggest that this case is not common.

In many experimental situations, an ion or transmitter is introduced into the brain cell microenvironment by iontophoresis (Fig. 1C). This means that the substance is released at a point with a constant flux for a finite period of time, S. The theoretical curves for iontophoresis are given by the solution (8) to Equation 1 with $Q = In/zF$ (I is iontophoresis current applied to the iontophoresis electrode, n is the transport number, z is the number of charges associated with substance iontophoresed, and F is Faraday's electrochemical equivalent). Assuming spherical symmetry, the solution is

$$\begin{aligned} C &= G(t) \quad t<S \\ C &= G(t) - G(t - S) \; t>S \\ G(w) &= (Q\lambda^2/8\pi D\alpha r)\{\exp[r\lambda(k/D)^{1/2}] \\ & \quad \mathrm{x}\ \mathrm{erfc}[r\lambda/2(Dw^{1/2} \qquad [4] \\ & \quad + (kw)^{1/2}] \\ & \quad + \exp[-r\lambda(k/D)^{1/2}]\mathrm{erfc}[r\lambda/2(Dw)^{1/2} \\ & \quad - (kw)^{1/2}]\} \end{aligned}$$

In this expression "erfc" represents the complementary error function. Figure 2A, B, and D shows the concentrations generated in the tissue for equal amounts of substance released by iontophoresis and by pressure ejection (i.e., $InS/zF = UC_f$). It is striking how much the concentration profiles differ close to the source (Fig. 2A) but become more similar with distance (Fig. 2B). Note the differences in both time and amplitude scales between A and B. Again, Fig. 2D shows the effect of uptake on the

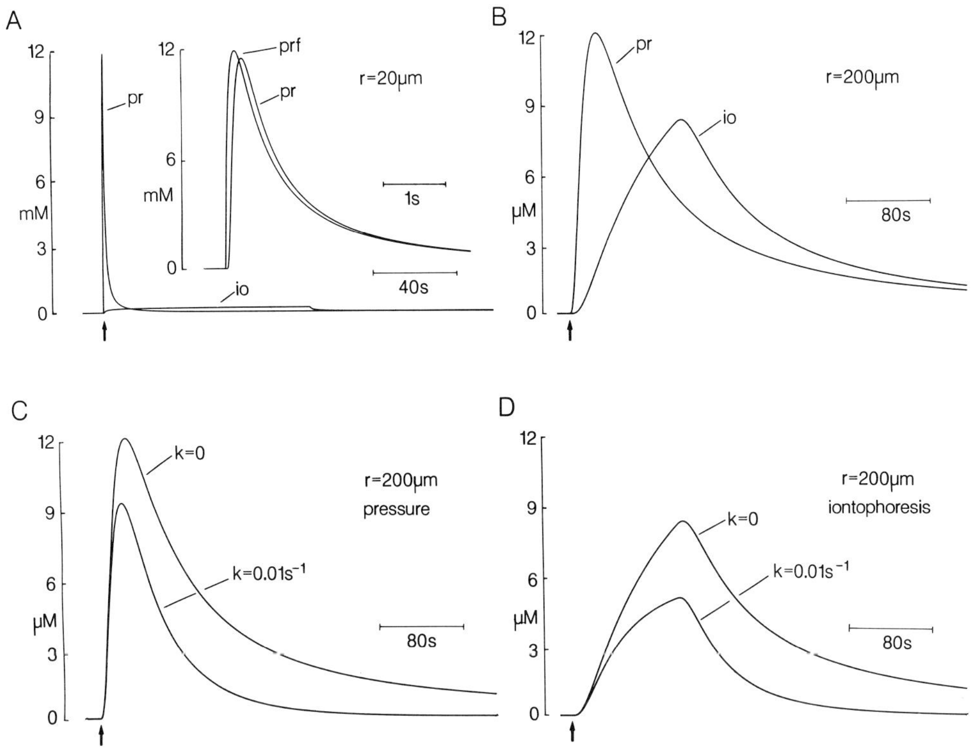

FIG. 2. Theoretical comparision of concentration profiles generated by pressure ejection and iontophoresis. In this example it was assumed that diffusion properties of the medium were characterized by $\lambda = 1.6$ and $\alpha = 0.2$, and $Q = 2.6 \times 10^{-13}$ equivalents of DA (or g.mol, because DA is singly charged) were released by each of the two methods. The diffusion coefficient for DA was taken as 6.0×10^{-6} cm²/sec (25°C), and the uptake term k was zero for A and B. **A:** Profiles labeled "pr" and "prf" represent instantaneous release (*arrow* in figure) of a volume (U) of 2.6 pγL of DA at a concentration (C_f) of 100 mM (i.e., $UC_f = Q$). For the pr profiles, it was assumed that all released material was actually localized at a point (curve given by Equation 2), whereas prf curves (Equation 3) were based on the more realistic assumption that released DA occupied a finite spherical volume of tissue of radius b defined by $b = (3U/4\pi\alpha)^{1/3}$. For the given parameters, $b = 15$ μm. The measuring CFVM was assumed to be 20 μm away from the ejecting electrode (see Fig. 1A and B). Only by displaying the calculations on a fast time scale (*inset,* A) can the difference in the pr and prf curves be seen. The prf curve rises earlier and to a slightly greater height than the pr curve because the outer boundary of the injected material is closer to the measuring point than when the entire mass is considered as concentrated at a point. Because of this similarity, all subsequent pressure-ejection calculations were only done for the point source case, i.e., using Equation 2. The curve "io" was generated by an iontophoresis pulse of duration $S = 100$ sec, beginning at the *arrow,* at a current strength $I = 5$ nA, and assuming a transport number for the iontophoretic electrode of $n = 0.05$. Thus $Q = InS/zF$ and $z = 1$ for DA. This curve was generated from Equation 4. The radical difference in diffusion curves for pr and io cases is evident in A; note that the peak amplitude of the pr curve is just over 12 mM at this measuring position. **B:** Diffusion profiles generated with the two paradigms are displayed for a sensing CVFM at a distance of 200 μm from the release electrode. Note first the 1,000-fold decrease in amplitude scale for 10-fold increase in distance. Second, the two profiles are now much more commensurate. This indicates that as distance from the source increases the source characteristics are smoothed out. **C** and **D:** The effects of nonspecific uptake on the diffusion profiles are shown at a distance of 200 μm for pressure and iontophoresis, respectively. A relatively strong uptake of $k = 0.01$/sec, at the upper end of our measured values, was used and shows that it has the effect of sharpening the curves. In particular, uptake reduces the amplitude and the duration of the falling phases.

concentration profile generated by iontophoresis.

Are α and λ related?

In the derivation of the parameters α and λ (8,14), they are treated as independent quantities. In reality, however, it is not clear that a three-dimensional *homogeneous* medium (one that has the same local structure at all points) with arbitrary α and λ could exist. Rather it seems, intuitively, that some constraints would occur. Interestingly enough, evidence has been found for this in many porous media in which an empirical relation, known as Archie's Law (see ref. 1), seems to be valid. This law has the form

$$\lambda^2 = \alpha^{-\beta} \quad [5]$$

where β is a constant between 1/2 and 2/3. The meagre data on the brain (1) provide some support for this relation, and our recent preliminary data using hypernatremia to alter the volume of the extracellular space (18) are consistent with this relation (Nicholson, Rice, and Cserr, *unpublished data*), but other preliminary studies are not (19). If it were possible to establish a relation between these two parameters, it would imply that the construction of brain tissue enforced certain constraints on its architecture and would enable one parameter to be calculated from the other.

TECHNIQUES

The ideas expressed by Equation 1 have developed over many decades in response to improvements in the ability to quantify diffusion in nervous tissue. Most of the remainder of this chapter will be devoted to describing results obtained with microelectrodes, but first it is relevant to briefly review some of the results obtained with the older radiotracer method and comment on the modern microdialysis method.

Radiotracers

The first measurements of diffusion in nervous tissue seem to go back to radiotracer studies by Harris and Burn (20) and others (see ref. 8). This technique was the basis for the extensive investigations of Fenstermacher and co-workers (summarized in refs. 21–23). The latter studies relied on the counting of radioactive tracer in sectioned tissue at a certain time after the tracer had been introduced into the tissue, thus providing a static picture of the spatial concentration profile at the instant of fixation. Much of the published work with radiotracers focused on the spinal cord and caudate nucleus in cats, dogs, rabbits, and monkeys. Substances used for diffusion included sucrose, inulin, and ethylenediaminetetraacetic acid (EDTA) (24,25) and a variety of metabolites and amino acids (26–28). These techniques were also applied in preliminary studies to the diffusion of sucrose, inulin, and EDTA in the dogfish (22). The wealth of data in these investigations is not easily summarized, but briefly, "well-behaved" extracellular compounds like sucrose and EDTA gave volume fractions of about 0.13 for white matter and 0.18 for gray matter together with tortuosities of 1.5 to 2 while the other compounds had variable distribution spaces and parameters depending on the extent to which they were actively transported across membranes.

Microdialysis

Data on diffusion can also be obtained by the use of *in vivo* microdialysis (29). An understanding of diffusion is actually crucial to the interpretation of recovery in terms of tissue concentration. This fact has been neglected in many studies in which it has been assumed that calibration in a free solution (beaker) would directly yield recovery factors applicable to brain tissue. Recently, however, new theoretical and experimental treatments of microdialysis that incorporate α and λ have appeared (30–32).

Ion Selective Microelectrodes

The introduction and development (see refs. 33,34) of the liquid-membrane ion-selective microelectrode (ISM) provided a means for the dynamic analysis of diffusion. The ISM consists of a double-barreled micropipette, one side of which is a conventional saline-filled microelectrode while the other barrel contains a liquid ion-exchanger in its tip. By a suitable choice of exchanger, different ions can be sensed by means of the logarithmic Nernst potential that develops across the liquid membrane. Much recent impetus for this method has come from the neutral carrier ligands developed in Prof. Simon's laboratory at the ETH, Zurich (33). During the past decade our laboratory has pursued this approach extensively (1,8,17,34–37) basing our technique on the original method of Lux and Neher (38). We have used the results to justify and refine the theory of diffusion in complex aggregates of brain cells (8,17). A typical experiment consists of releasing an extracellular marker, such as TMA^+ (see Table 1), from a micropipette by either iontophoresis or pressure ejection and recording its arrival 100 to 200 μm away with a suitable ISM. By appropriate calibration procedures and curve-fitting, α, λ, and *k* can be determined.

The work from our laboratory using ISM methods has primarily concentrated on the cerebellum and cerebral cortex of the rat (8,36) together with studies in the cerebellum of the turtle and skate (1,17; *unpublished data*) and some measurements on the guinea pig slice (39,40). Again the major finding was that volume fraction was about 0.2 and tortuosity about 1.6 for all these regions, species, and ions.

One limitation of the ISM approach to diffusion is that it is restricted to a small group of charged compounds for which ISMs exist. In practice, this includes (8) the cations TMA^+ and tetraethylammonium (TEA^+), the anion probes α-naphthalene sulfonate (ANS^-) and AsF_6^- together with Ca^{2+}, K^+, and certain other substances such as penicillin (37).

Carbon Fiber Voltammetric Microelectrodes

To extend microelectrode methods to transmitters, another microsensor can be used, known as the carbon fiber voltammetric microelectrode (CFVM). By cycling this electrode through potentials where an electroactive compound oxidizes and reduces, the corresponding electrode currents can be used to measure the concentration of the compound. The original neurobiological application of this tool was the result of work of Prof. Adams at the University of Kansas (see ref. 34). The contemporary CFVM consists of an 8- to 12-μm-diameter carbon fiber sealed in the tip of a glass micropipette. It is frequently treated and coated to improve sensitivity and selectivity. Among analytical techniques, chronoamperometry, differential pulse voltammetry, and high-speed cyclic voltammetry are all in use (34). The advantage of these electrochemical methods is that electroactive compounds can be detected, and these include serotonin and DA, as well as several of their metabolites.

A few studies have used CFVMs or the earlier carbon paste microelectrodes (41–43) to measure the diffusion of electroactive compounds in the brain following the paradigm established with the ISMs. Particularly striking were the data of Rice et al. (42) in the rat caudate nucleus, showing that while anionic molecules diffused in accord with the tortuosity and volume fractions established for other tissues, DA and other biological cations moved much more slowly than predicted. A possible explanation, supported by other data (41) and some preliminary studies of our own (see later), suggests that the results of Rice et al.

may have been influenced by incomplete suppression of uptake mechanisms for DA.

EXPERIMENTAL RESULTS FOR ION DIFFUSION

Diffusion of TMA^+

TMA^+ has proved to be a good extracellular ion for probing the extracellular space. It is comparatively small (molecular weight 74.1, see Table 1), is not readily taken up by cells, and is nontoxic (8).

During the past decade we have made iontophoretic studies using TMA^+ in the gray matter of several animals. Figure 3 illustrates results from a mammal, reptile, fish, and an invertebrate. We find that there is some variation from trial to trial, but the mean values over about 50 trials distributed over five to 10 similar preparations are consistent. The data in Fig. 3 represent single samples from the data set, not the mean values. A complete analysis of the full sets of data will be published elsewhere in the future.

The urethane-anesthetized rat preparation has been extensively used in our diffusion experiments (8,35,36). Figure 3A shows a diffusion profile from the cerebral cortex. As in other graphs from this figure, Equation 4 is fitted to the experimental data using a simplex algorithm with α, λ, and k as the parameters to be varied to obtain best fit. Values of D and n are determined by separate experiments in dilute agar gel (34). In each graph of Fig. 3, the experimental and fitted theoretical curves are superimposed. Noise is seen on the experimental curve because of small movements of the animal caused by the pulsatile blood flow and respiration.

Another mammalian preparation that we have studied is the cerebellar slice from the guinea pig (39,40). In that preparation the mean values for tortuosity and volume fraction were slightly different from other preparations (uptake was not measured at that time). Whether these differences were due to conditions in those slices or other factors we do not know. Recent studies (44) in the guinea pig hippocampal slice using similar methods also found values that differed somewhat from other mammalian intact preparations. One possible explanation for the latter variations is provided by the recent finding of inhomogeneity in the value of α in the hippocampal slice (45).

Much of our recent work has centered around the isolated turtle cerebellum. This preparation (46,47) has a remarkable ability to withstand anoxia, which makes it an ideal isolated preparation because, in contrast to the slice, it retains all the three-dimensional neuronal circuitry. A typical record, illustrated in Fig. 3B, shows that diffusion in the cold-blooded reptile is similar to that in the mammal. In making this comparison and the ones shown later in the figure, it should be noted that the temperature dependence of the free diffusion coefficient has been taken into account. Because the turtle cerebellum is an isolated preparation, there is an absence of noise.

Another series of our studies, done at the Mount Desert Island Biological Laboratory, Maine, (48,49), has focused on the cerebellum of the skate. In this marine elasmobranch fish, the extracellular fluid has an unusual ionic composition. $[Na^+]_o$ elevated to almost 300 mM (in contrast to that of other vertebrates, including teleost fishes, is about 150 mM), and in addition, there is 350 mM of urea. These modifications bring the total osmolality of the extracellular fluid to about 1,000 mOsmol, which is isotonic with sea water. In addition the elasmobranch is the only known species with a glial, rather than endothelial, blood-brain barrier (50). These anomalies make the elasmobranch an interesting animal for study. Nevertheless, extracellular diffusion (Fig. 3C) appears similar to that in the other two species illustrated in Fig. 3. Again noise is seen due to the use of an intact anesthetized animal.

Figure 3D depicts diffusion measurements from an invertebrate, *Sepia officinal-*

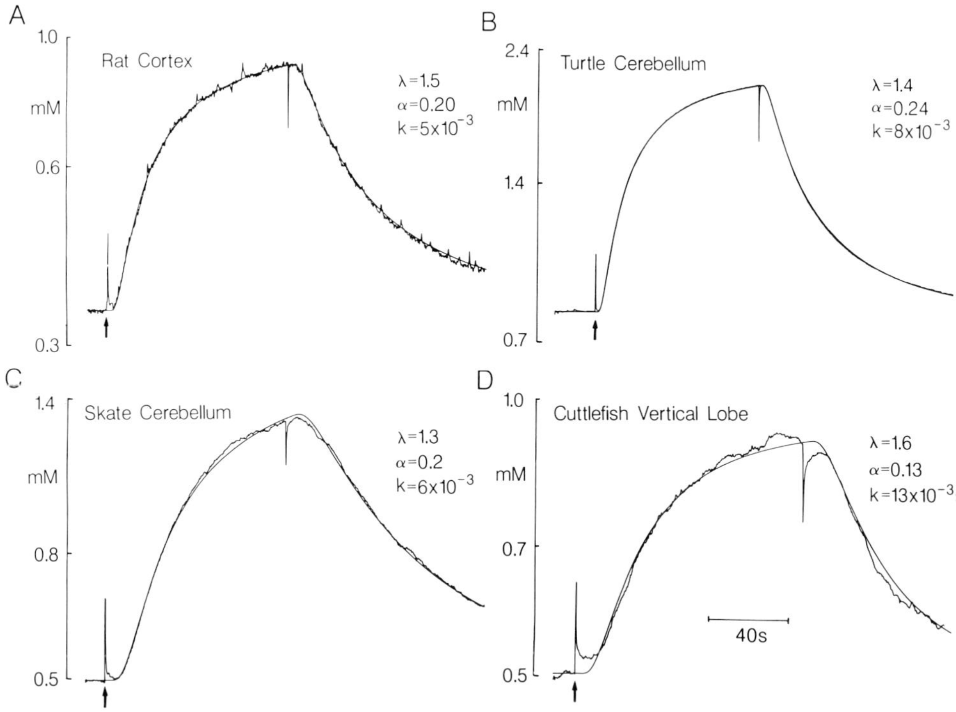

FIG. 3. Experimental diffusion profiles for TMA^+ in four species using iontophoresis. In each case iontophoresis commenced at the time indicated by the *arrow* and continued for the duration specified later in this legend. The time scales of all the figures are the same (see scale in D), but the amplitude calibrations, which are logarithmic (with allowance for ISM interference), differ between panels. In each case a theoretical curve defined by Equation 4 was fitted to the experimental data by varying λ, α, and *k* until the best fit was obtained. The values of these parameters that optimized the curve fit are given in each panel; other data are listed later in this legend. A, C, and D have more experimental noise because they were recorded in intact anesthetized preparations. B was recorded in an isolated cerebellum. In all experiments a continuous forward bias current, I_b, was applied to the iontophoresis electrode to maintain tip concentration (see ref. 48) and provide a stable baseline; this was then stepped to a new value I_m to generate the diffusion curve. **A:** record taken at about 400 μm depth in rat cerebral cortex. Temperature 35°C, free diffusion coefficient of TMA^+ at this temperature $D = 1.3 \times 10^{-5}$ cm²/sec, distance from source $r = 186$ μm, iontophoretic current step 69 nA, iontophoresis duration 90 sec, transport number of source electrode 0.40. **B:** record taken in molecular layer of isolated turtle cerebellum. Temperature 22°C, free diffusion coefficient of TMA^+ at this temperature $D = 0.98 \times 10^{-5}$ cm²/sec, distance from source $r = 144$ μm, iontophoretic current step 100 nA, iontophoresis duration 80 sec, transport number of source electrode 0.50. **C:** record taken in molecular layer of skate. Temperature 15°C, free diffusion coefficient of TMA^+ at this temperature $D = 0.85 \times 10^{-5}$ cm²/sec, distance from source $r = 247$ μm, iontophoretic current step 129 nA, iontophoresis duration 90 sec, transport number of source electrode 0.55. **D:** Record taken about 600 μm deep in the vertical lobe of the cuttlefish. Temperature 15°C, free diffusion coefficient of TMA^+ at this temperature $D = 0.85 \times 10^{-5}$cm²/sec, distance from source $r = 202$ μm, iontophoretic current step 53 nA, iontophoresis duration 112 secs, transport number of source electrode 0.31.

is, commonly called the cuttlefish. This animal belongs to the family of cephalopods that includes squid and octopus. These are the most highly developed invertebrates and exhibit considerable intelligence. The nervous system consists of a collection of lobes, some large (51). It is the mass of the neural tissue in certain of these lobes that has permitted preliminary diffusion measurements (Nicholson, Miyan, and Abbott, *unpublished data*) in the Laboratory of the Marine Biological Association, Plymouth, England. Figure 3D shows a diffusion curve that is similar to those for the three vertebrates, again with noise caused by respiratory movement in this intact anesthetized preparation. Our data indicate that the tortuosity in this preparation is quite similar to that in the vertebrates, but the volume fraction appears less. We do not presently know if the latter result is due to ischemic conditions in the tissue of this preparation or is representative of the normal tissue. A study with radiotracers (52) has indicated a volume fraction of 20%. We plan further studies to resolve the issue.

In summary, our findings with TMA^+ are that λ and α are similar from species to species in the regions that we have looked at, although we are presently uncertain about the true value of α in the cuttlefish. Studies with other ions from our own laboratory (8) and studies from other laboratories using this method (19,45,53) also support these data. Finally, we note that there is general agreement with the earlier radiotracer data. The value for k, the nonspecific uptake, is also fairly consistent at about 3.0 to 9.0×10^{-3}/sec, except in the cuttlefish, in which it seems to be about twice this value.

Diffusion of Some Endogenous Ions

It may be asked why experiments are conducted with foreign ions when several endogenous species can be sensed with available ISMs. This question can be answered by considering each of those endogenous ions in turn.

Na^+ and Cl^- are present in concentrations exceeding 100 mM in the extracellular space. This means that for a sensor, such as an ISM, with a logarithmic sensitivity, to detect a change in these ions over the ambient levels, the change must be several mM. This lack of sensitivity is further aggravated by the relatively poor selectivity of the available liquid membranes for these two ions (34). As a consequence of these factors, no measurements of diffusion of these ions with ISM methods have been made to date.

K^+ is particularly interesting and has been the most studied physiologically of all extracellular ions (see ref. 54 for review). The extracellular baseline value is low at 3 mM and excellent ISMs based on valinomycin (33,34) now exist for sensing K^+. Among all the ions considered, this exhibits the most anomalous migration behavior, however. When K^+ is iontophoresed in the extracellular space the diffusion curves appear normal (35,38), but the volume fraction routinely exceeds 100%. This can be explained by the ability of K^+ to cross membranes. This occurs both by transport mechanisms and by the so-called spatial buffer (55,56). In the spatial buffer mechanism, $[K^+]_0$ accumulation in the vicinity of one location on an extended cell [i.e., one that constitutes an electrotonic "core conductor" (16)] leads to local depolarization, which then sets up a circulating current that carries K^+ into the cell at that point and out elsewhere (55). Spatial buffering is particularly suited to glial cells because of the high potassium permeability of their membranes. This mechanism also generates low-frequency extracellular potentials and is thought to be important in the retina (56) among other tissues. Finally the existence of a "ceiling" (57) for $[K^+]_0$ in most brain regions [i.e., a level that is not exceeded even during strong stimulation (and is usually in the vicinity of 12 mM)] implies nonlinear control mechanisms that are not fully understood at this time.

The final extracellular ion of major interest is Ca^{2+}. In common with K^+, it has a low baseline, in this case about 1.2 mM, and excellent ISMs can be fabricated to detect it. Its diffusion behavior was also thought to be anomalous (58), in this case caused by interaction within negative surface charges on cells, but later studies by Nicholson and Rice (36) did not substantiate this finding. In the latter experiments, a solution of TMA^+ and Ca^{2+} was pressure ejected into the rat cortex and the known diffusion characteristics of TMA^+ used to estimate the amount of ions injected and so determine the diffusion characteristics of Ca^{2+} according to Equation 2. This was necessary because the iontophoresis of Ca^{2+} is unreliable. Similar studies in both turtle (Nicholson and Rice, *unpublished data*) and skate (49; *unpublished data*) have indicated the generality of this result.

EXPERIMENTAL RESULTS FOR DA DIFFUSION

To test adequately ideas about transmitter migration in the extracellular space, we must be able to measure these neuroactive substances. The CFVM has proved to be a powerful tool for measuring certain catecholamines and indolamines. Among these the most research effort has been expended on DA. Apart from being an intrinsically interesting transmitter, the devastating effects of its absence in Parkinson's disease have made this substance a topic of intense study.

In the normal striatum, studies with CFVMs have revealed that DA migration is strongly curtailed by the action of uptake that obeys Michaelis-Menten kinetics (e.g., ref. 12). The evidence for this is seen mainly in experiments in which uptake blockers, (e.g., nomifensine) are applied. In the presence of the blocker, the stimulus-evoked DA release is greatly enhanced (12). It is thought that such uptake is mediated by synaptic terminals (re-uptake) and possibly glia. It has also been suggested that charged sites on membranes hinder DA migration (42).

To begin to examine DA diffusion more quantitatively, we have refined the technology of CFVMs and high-speed cyclic voltammetry and begun a series of experiments to test systematically the issues involved.

The first question that we have examined is whether DA can diffuse normally in essentially nondopaminergic brain regions. Such areas would be expected to possess glia with nonspecific uptake systems and have fixed charges on cell surfaces but to lack the re-uptake sites associated with dopaminergic terminals. We have pursued our initial experiments in three regions of the turtle brain (Fig. 4), the olfactory bulb, cortex, and cerebellum (Rice and Nicholson, *unpublished data*). We have found that in all these regions, DA diffuses in a similar manner to TMA^+. Thus both ions are governed by the same tortuosity and volume fraction. Nonspecific uptake is also similar to that found with TMA^+ (demonstrating its nonspecificity), except in the olfactory bulb, where it may be slightly enhanced.

Looking at Fig. 4, it is apparent that the detection limit for DA concentration in these experiments is in the micromolar range. Although this is three orders of magnitude better than that typical for many ISMs, it is still not at the level required to work in the physiological range in which DA may be active. The work of Church et al. (59) and Gonon (60) demonstrates that basal levels in the striatum are about 20 to 50 nM. It is important to have sensors that work at this level because uptake mechanisms will tend to saturate at much greater concentrations. These facts have prompted us to develop CFVMs with nanomolar sensitivity (61). The performance of these electrodes is demonstrated in Fig. 5. This shows both the results for an agarose gel and for the isolated putative striatum of the turtle. In these preliminary experiments, we also obtained similar diffusion parameters for this possibly dopaminergic region of the turtle brain. It is unwise to over-interpret these data at this time, because they are so preliminary, but they may indicate

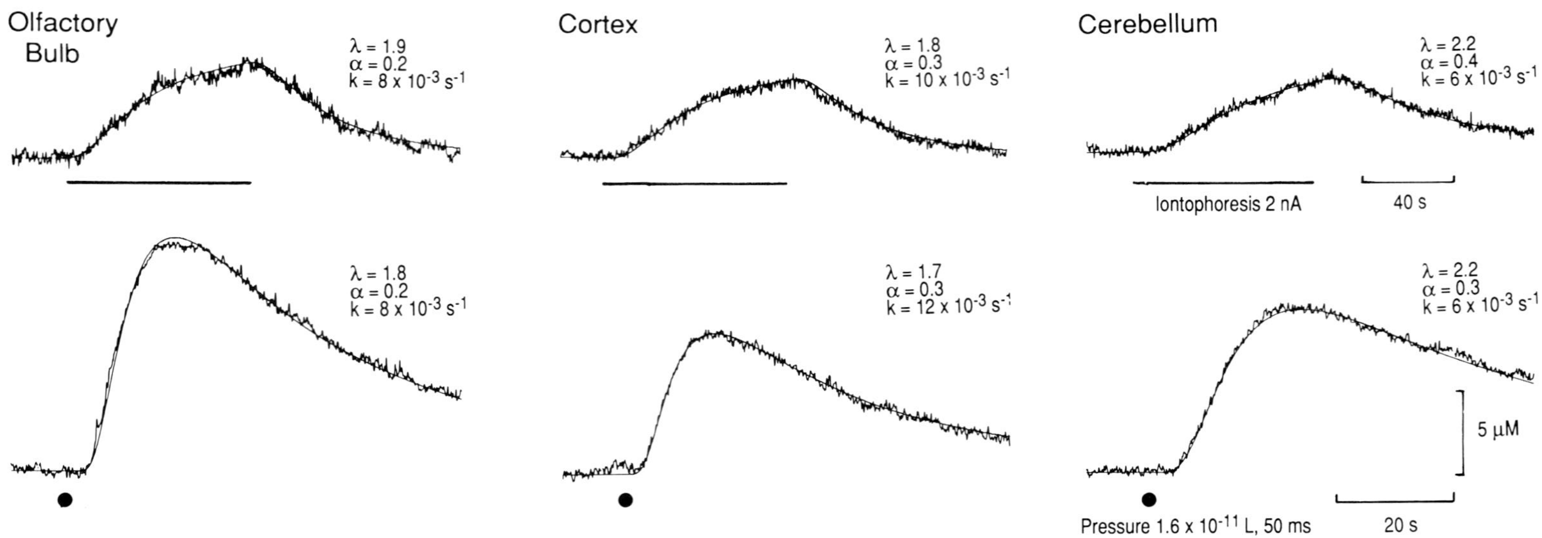

FIG. 4. Experimental diffusion profiles for DA in three different regions of the turtle brain using both iontophoresis and pressure ejection. In all experiments the isolated brain region was used, and data were recorded using a CFVM and high-speed cyclic voltammetry. All records were taken with the same electrode array that consisted of an iontophoresis micropipette, a pressure ejection micropipette, and a CFVM. The spacing between the two source electrodes and the sensor was 150 μm, the temperature was 25°C (free diffusion coefficient for DA was 6.0×10^{-6} cm²/sec), and the transport number of the iontophoresis electrode was $n = 0.09$. The top row of figures indicates the iontophoretic curves (with theoretical curves fitted as in Fig. 3 using Equation 4). The increment in iontophoretic current and its duration is indicated in the figure. The lower row of records shows the pressure data, time of ejection indicated by *dot,* and volume ejected indicated in last panel, curves fitted with Equation 2. Note that the amplitude scale is linear for this technique and the calibration (last record) is in μM.

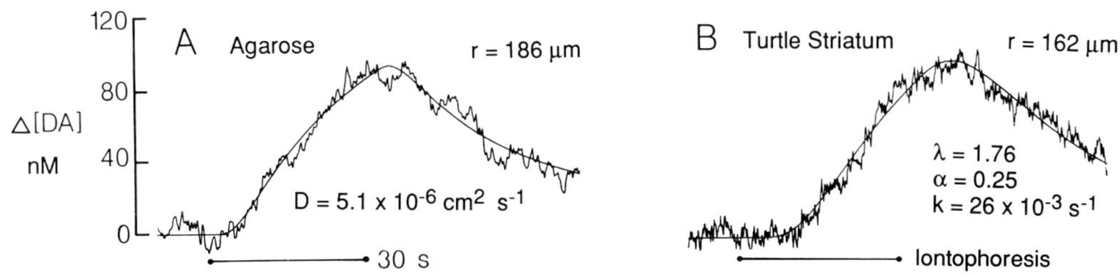

FIG. 5. Demonstration of diffusion measurement of DA with nanomole detection limit. Using improvements in CFVM techniques, including averaging 16 records (61), it was possible to increase detection capabilities shown in Fig. 4 by a factor of 100. **A:** measurement of free diffusion coefficient in dilute agarose gel for an iontophoresis current step of 10 nA with a transport number $n = 0.003$. **B:** diffusion measurements in putative isolated striatum of the turtle. Ionotophoresis current step was 1 nA, and the transport number was $n = 0.01$. Temperature 25°C. Practical resolution limit with the time resolution shown is presently about 20 nM.

that this region of the turtle brain differs from that in mammals in regard to uptake. The importance of these data, however, lies in the fact that we have a working tool that is both sensitive and fast enough to explore diffusion of DA in the brain.

DISCUSSION

The idea that transmitters or neuromodulators not only act within the strict confines of conventional synapses but may also pervade the extracellular space and act on populations of receptors has been a persuasive concept, but elusive to prove. Several recent papers (2,62–66) presented at a symposium on synaptic and nonsynaptic communication in the brain as well as the papers in this volume marshal the available evidence.

Looking at the phylogenetic origins of the nervous system, there is some evidence that it originated around a central fluid-filled canal. Indeed, as Nieuwenhuys (63, 67) has pointed out, even in advanced vertebrates a "paracore" exists around the canals that is unusually rich in receptor diversity, implying that an abundance of transmitters is active. If one imagines that some of these molecules may enter from the canals or be released from processes that do not make specific postsynaptic contacts, the diversity of receptors is an essential prerequisite for selectivity. If we accept this possibility, the question of how neuroactive compounds move from their release site to possibly remote targets becomes important.

When first confronted with an electron micrograph, it is tempting to deny that substances could move through the tiny spaces between the cellular membranes. Our data and that from many other laboratories in the recent past, however, have shown that the channels between cells are sufficiently wide to allow the migration of neuroactive compounds. Of equal significance, we can construct accurate theoretical models for the extracellular diffusion that enable us to describe the concentration profiles as functions of time and space. Of particular importance are the two dimensionless geometrical parameters α and λ. For many substances, however, we will need more elaborate models of uptake or degradation in the future.

These results enable an often-asked question to be discussed. The question is: How far does a substance diffuse in the brain? It can be immediately recognized from the nature of the solution to the diffusion equation that a substance released at a point in the brain will, given sufficient time, spread throughout the entire tissue. This implies that the final concentration at any point will be small compared with the original concentration in the immediate vicinity of the release site at the time of release. In practice, molecules are also lost from the system because of uptake, degradation, and loss at brain boundaries. Such factors may be collectively described as the sink properties of the tissue for that molecule; in contrast to α and λ, most of the sink properties will be substance-specific. The presence of these nonideal factors means that all molecules can be purged from the brain after a finite time. The real question, however, needs to be restated as: Given a medium with defined diffusion parameters and sink properties and given a receptor with defined sensitivity for a specified neuroactive molecule, what is the maximum distance from the release site at which a receptor can be activated after the release of a defined number of molecules?

Thus the elucidation of the mechanics of volume transmission in the extracellular space depends on three types of information, the diffusion characteristics of the brain, the sink properties of the brain for the molecule under discussion, and the receptor activation requirements for that molecule. As this chapter shows, we have a reasonable understanding of the first topic. We are developing an understanding of sink properties on a molecule-by-molecule basis, and the receptor characteristics are being quantified in many laboratories at this time.

ACKNOWLEDGMENT

This work was supported by U.S. Public Health Service grants NS-13742 and NS-07745 (MER) and a grant from the American Parkinson Disease Association.

REFERENCES

1. Nicholson C, Rice ME. The migration of substances in the neuronal microenvironment. *Ann NY Acad Sci* 1986;481:55–71.
2. Nicholson C. Issues involved in the transmission of chemical signals through the brain extracellular space. *Acta Morphol Neerl-Scand* 1988/89;26:69–80.
3. Schmitt FO. Molecular regulators of brain function: a new view. *Neuroscience* 1984;13:991–1001.
4. Feldberg W. History of *in vivo* analysis of brain functions elicited from its inner and outer surfaces. *Ann NY Acad Sci* 1986;473:1–20.
5. Herkenham M. Mismatches between neurotransmitters and receptor localizations in brain: observations and implications. *Neuroscience* 1987;23:1–38.
6. Hill AV. Diffusion of oxygen through tissues. In: Hill AV, ed. *Trials and trails in physiology*. Baltimore: Williams & Wilkins, 1965;208–241.
7. Cserr HF. Role of secretion and bulk flow of brain interstitial fluid in brain volume regulation. *Ann NY Acad Sci* 1988;529:9–20.
8. Nicholson C, Phillips JM. Ion diffusion modified by tortuosity and volume fraction in the extracellular microenvironment of the rat cerebellum. *J Physiol* 1981;321:225–257.
9. Davson H, Danielli JF. *The permeability of natural membranes*, (reprint of 1952 edition). Darien, Connecticut: Hafner, 1970;48–56.
10. Rubinow SI. *Introduction to mathematical biology*. New York: Wiley, 1975;196–277.
11. Van Harreveld A. The extracellular space in the vertebrate central nervous system. In: Bourne GH, ed. *The structure and function of nervous tissue*, vol IV. New York: Academic, 1972;447–511.
12. Wighman RM, Amatore C, Engstrom RC, Hale PD, Kristensen EW, Kuhr WG, May LJ. Real-time characterization of dopamine overflow and uptake in the rat striatum. *Neuroscience* 1988;25:513–523.
13. Crank J. *The mathematics of diffusion*, 2nd ed. Oxford: Clarendon Press, 1975.
14. Lehner FK. On the validity of Fick's law for transient diffusion through a porous medium. *Chem Eng Sci* 1979;43:821–825.
15. Carslaw HS, Jaeger JC. *Conduction of heat in solids*, 2nd ed. Oxford: Clarendon Press, 1959.
16. Jack JJB, Noble D, Tsien RW. *Electric current flow in excitable cells*. Oxford: Clarendon Press, 1975.
17. Nicholson C. Diffusion from an injected volume of a substance in brain tissue with arbitrary volume fraction and tortuosity. *Brain Res* 1985; 333:325–329.
18. Nicholson C, Cserr HF, Rice ME, DePasquale M, Patlak CS. Volume and ion regulation in both intracellular and extracellular compartments determined by diffusion, gravimetric and ion analysis. *Acta Physiol Scand* 1989;136(Suppl 582):24.
19. Lundbaek JA, Hansen AJ. Relationship between the extracellular space and tortuosity in brain and red cell suspensions. *Acta Physiol Scand* 1989;136(Suppl 582):74.
20. Harris EJ, Burn GP. The transfer of sodium and potassium ions between muscle and the sur-

rounding medium. *Trans Faraday Soc* 1949;45:508–528.
21. Fenstermacher JD, Patlak CS, Blasberg RG. Transport of material between brain extracellular fluid, brain cells and blood. *Fed Proc* 1974;33:2070–2074.
22. Fenstermacher JD, Patlak CS. The exchange of material between cerebrospinal fluid and brain. In: Cserr HF, Fenstermacher JD, Fencl V, eds. *Fluid environment of the brain*. New York: Academic Press, 1975;201–214.
23. Fenstermacher JD, Kaye T. Drug "diffusion" within the brain. *Ann NY Acad Sci* 1988;531:29–39.
24. Levin VA, Fenstermacher JD, Patlak CS. Sucrose and inulin space measurements of cerebral cortex in four mammalian species. *Am J Physiol* 1970;219:1528–1533.
25. Kessler JA, Fenstermacher JD, Owens ES. Spinal subarachnoid perfusion of rhesus monkeys. *Am J Physiol* 1976;230:614–618.
26. Kessler JA, Patlak CS, Fenstermacher JD. 3-Methoxy-4-hydroxyphenyleneglycol (MHPG) transport from the spinal cord during spinal subarachnoid perfusion. *Brain Res* 1976;102:131–141.
27. Kessler JA, Patlak CS, Fenstermacher JD. Transport of 5-hydroxy-3-indoleacetic acid by spinal cord during subarachnoid perfusion. *Brain Res* 1976;116:471–483.
28. Fenstermacher JD, Davson H. Distribution of two model amino acids from cerebrospinal fluid to brain and blood. *Am J Physiol* 1982;242:F171–F180.
29. Ungerstedt U. Measurement of neurotransmitter release by intracranial dialysis. In: Marsden CA, ed. *Measurement of neurotransmitter release* in vivo. London: Wiley, 1984;81–105.
30. Benveniste H, Hansen AJ, Ottosen NS. Determination of brain interstitial concentrations by microdialysis. *J Neurochem* 1989;52:1741–1750.
31. Lindefors N, Amberg G, Ungerstedt U. Intracerebral microdialysis: I. Experimental studies of diffusion kinetics. *J Pharmacol Meth* 1989; 22:141–156.
32. Amberg G, Lindefors N. Intracerebral microdialysis: II. Mathematical studies of diffusion kinetics. *J Pharmacol Meth* 1989;22:157–183.
33. Ammann D. *Ion–selective microelectrodes*. Berlin: Springer-Verlag, 1986.
34. Nicholson C, Rice ME. Use of ion-selective microelectrodes and voltammetric microsensors to study brain cell microenvironment. In: Boulton AA, Baker GB, Walz W, eds. *Neuromethods: the neuronal microenvironment*. Clifton, New Jersey: Humana, 1988;247–361.
35. Nicholson C, Phillips JM, Gardner-Medwin AR. Diffusion from an iontophoretic point source in the brain: role of tortuosity and volume fraction. *Brain Res* 1979;169:580–584.
36. Nicholson C, Rice ME. Calcium diffusion in the brain cell microenvironment. *Can J Physiol Pharmacol* 1987;65:1086–1091.
37. Lehmenkuhler A, Kersting U, Nicholson C. Diffusion of penicillin in agar and cerebral cortex of the rat. *Brain Res* 1988;444:181–183.
38. Lux HD, Neher E. The equilibration time course of $[K^+]_0$ in cat cortex. *Exp Brain Res* 1973; 17:190–205.
39. Hounsgaard J, Nicholson C. Potassium accumulation around individual Purkinje cells in cerebellar slices. *J Physiol* 1983;340:359–388.
40. Nicholson C, Hounsgaard J. Diffusion in the slice microenvironment and implications for physiological studies. *Fed Proc* 1983;42:2865–2868.
41. Dayton MA, Ewing AG, Wightman RM. Diffusion processes measured at microvoltammetric electrodes in brain tissue. *J Electroanal Chem* 1983;146:189–200.
42. Rice ME, Gerhardt GA, Hierl PM, Nagy G, Adams RN. Diffusion coefficients of neurotransmitters and their metabolites in brain extracellular fluid. *Neuroscience* 1985;15:891–902.
43. Rice ME, Nicholson C. Serotonin migration in the neuronal microenvironment. *Ann NY Acad Sci* 1986;481:381–382.
44. Lipinski H-D, Bingmann D. Diffusion in slice preparations bathed in unstirred solutions. *Brain Res* 1987;437:26–34.
45. Traynelis SF, McBain CJ, Dingledine R. Seizure susceptibility in hippocampal slices: relation to regional variation in extracellular space (*Abstr.*). *Soc Neurosci Abstr* 1989;15:702.
46. Chan CY, Nicholson C. Modulation by applied electric fields of Purkinje and stellate cell activity in the isolated turtle cerebellum. *J Physiol* 1986;371:89–114.
47. Hounsgaard J, Nicholson C. The isolated turtle brain and the physiology of Neuronal circuits. In: Jahnsen H, ed. *Preparations of vertebrate central nervous system* in vitro. Chichester: Wiley, 1990 (in press).
48. Nicholson C, Rice ME. Diffusion characteristics of skate (Raja erinacea) cerebellum measured with tetramethylammonium and ion-selective microelectrodes. *Mount Desert Island Biol Lab Bull* 1985;25:54–55.
49. Nicholson C, Rice ME. Calcium diffusion in the cerebellum of the skate (*Raja erinacea*). *Mount Desert Island Biol Lab Bull* 1986;26:149–150.
50. Cserr HF, Bundgaard M. Blood-brain interfaces in vertebrates: a comparative approach. *Am J Physiol* 1964;246:R277–R288.
51. Boycott BB. The functional organization of the brain of the cuttlefish *Sepia officinalis*. *Proc Soc (B)* 1961;153:503–534.
52. Abbott NJ, Bundgaard M, Cserr HF. Brain vascular volume, electrolytes and blood-brain interface in the cuttlefish *Sepia officinalis* (Cephalopoda). *J Physiol* 1985;368:197–212.
53. Svoboda J, Sykova E. Activity-related extracellular K^+ and volume changes in spinal cord of the rat. *Acta Physiol Scand* 1989;136(Suppl 582):86.
54. Sykova E. Extracellular K^+ accumulation in the central nervous system. *Prog Biophys Mol Biol* 1983;42:135–189.
55. Gardner-Medwin AR. A new framework for assessment of potassium-buffering mechanisms. *Ann NY Acad Sci* 1986;481:287–302.

56. Newman EA. Regional specialization of the membrane of retinal glial cells and its importance to K^+ spatial buffering. *Ann NY Acad Sci* 1986;481:273–286.
57. Heinemann U, Lux HD. Ceiling of stimulus induced rises in extracellular potassium concentration in the cerebral cortex of cat. *Brain Res* 1977;120:231–249.
58. Morris ME, Krnjevic K. Slow diffusion of Ca^{2+} in the rat's hippocampus. *Can J Physiol Pharmacol* 1981;59:1022–1025.
59. Church WH, Justice JB Jr, Neill DB. Detecting behaviorally relevant changes in extracellular dopamine with microdialysis. *Brain Res* 1987; 412:397–399.
60. Gonon FG. Nonlinear relationship between impulse flow and dopamine released by rat midbrain dopaminergic neurons as studied by *in vitro* electrochemistry. *Neuroscience* 1988;24:19–28.
61. Rice ME, Nicholson C. Measurement of nanomolar dopamine diffusion using low-noise perfluorinated ionomer coated carbon fiber microelectrodes and high-speed cyclic voltammetry. *Anal Chem* 1989;61:1805–1810.
62. Buma P. Synaptic and nonsynaptic release of neuromediators in the central nervous system. *Acta Morphol Neerl-Scand* 1988/89;26:81–113.
63. Nieuwenhuys R, Veening JG, Van Domburg P. Core and paracores; some new chemoarchitectural entities in the mammalin neuraxis. *Acta Morphol Neerl-Scand* 1988/89;26:131–163.
64. Stoof JC. Localization and pharmacology of some dopamine receptor complexes in the striatum and the pituitary gland: synaptic and nonsynaptic communication. *Acta Morphol Neerl-Scand* 1988/89;26:115–130.
65. Swanson LW. The neural basis of motivated behavior. *Acta Morphol Neerl-Scand* 1988/89;26: 165–176.
66. Korf J, Loopuijt LD. Synaptic and non-synaptic striatal dopamine D2 receptors: possible implications in normal and pathological behaviour. *Acta Morphol Neerl-Scand* 1988/89;26:177–190.
67. Nieuwenhuys R. *Chemoarchitecture of the brain*. Berlin: Springer-Verlag, 1985.
68. Stokes RH. The diffusion coefficient of eight uni-univalent electrolytes in aqueous solution at 25°. *J Am Chem Soc* 1950;72:2243–2247.
69. Wang JH. Tracer-diffusion in liquids. IV. Self-diffusion of calcium ion and chloride ion in aqueous calcium chloride solutions. *J Am Chem Soc* 1953;75:1769–1770.
70. Gerhardt G, Adams RN. Determination of diffusion coefficients by flow injection analysis. *Anal Chem* 1982;54:2618–2620.

Volume Transmission in the Brain: Novel Mechanisms for Neural Transmission, edited by K. Fuxe and L.F. Agnati. Raven Press, Ltd., New York © 1991.

22

Action at a Distance: The Extracellular Spread of Chemicals in the Nervous System

Aryeh Routtenberg

Northwestern University, Chicago, Illinois 60611

The present chapter reviews work published 15 years ago, although I shall not resist the temptation to discuss the Descarries/Bloom debate, which is still of current interest.

THIRST FOR KNOWLEDGE: VOLUME TRANSMISSION LOUD AND CLEAR

Volume transmission, which by itself could mean a sound check at a rock concert or problems of an Encyclopedia Brittanica warehouse shipping clerk, actually refers to the idea that chemical material can influence intercellular communication via nonsynaptic interactions. For this to occur, it needs to be shown that such material can, in fact, move through the extracellular space with some degree of freedom. Those whose tastes run to the more classical favoring a conventional model of synaptic transmission may eschew hubris because it has not been shown that a transmitter substance, initially present in the presynaptic terminal, actually moves across the synaptic cleft and binds to a receptor, altering ionic conductance; debate still occurs as to the vesicular versus nonvesicular origin of transmitter release.

Work carried out in my laboratory in the late 1960s suggested that transmitters could readily move in the extracellular space. A bit of reminiscence will set the stage for why this research was carried out (such activities are indulgences of both graybeard professors and flour beetles) (1). In the early 1960s, it had been suggested that the induction of drinking by the cholinergic agonist, carbachol, could be observed in a variety of brain locations. This suggested a thirst circuit to some. The extensive nature of the circuit raised serious doubts in my mind, and I suggested that the carbachol was accessing the ventricles and then acting on some distant central site (2). This would be volume transmission to be sure, except that such an event would not occur physiologically because acetylcholine, unlike carbachol, would be readily hydrolyzed. This suggestion led me in two directions: One, what, if any, was that site? Second, could chemicals injected into the brain at a particular site spread to other locations?

The first question led to the discovery of the subfornical organ, the site of action of compounds (i.e., carbachol, acetylcholine, angiotensin) that induce drinking (3–8). This work has been carried on by Simpson and others, and to the best of my knowledge, it remains the key site of action of dipsogenic compounds. For whimsical reasons, in part, I note that thirst is initiated by either volumetric or osmotic factors and the subfornical organ functions in the regulation of volume factors. Because of its enriched innervation by the vasculature, its impoverished blood-brain barrier, its central location in telencephalic-lateral ventricular area immediately below the fornix, and its sensitivity to a variety of transmitter and peptide first messengers, it would be a valuable model to study volume or extracellular signaling in the nervous system, where "volume" has two meanings. To summarize briefly, we demonstrated that a particular site within the central nervous system, the subfornical organ, could represent the access point for injections at distant targets. Indeed, in one study especially relevant to this point, it was shown that lesions of this organ prevented carbachol-induced drinking (7).

But could chemicals injected at one point actually be shown to move from that point to a rather distant point? To answer this question, John Sladek, then a graduate student, Bill Bondareff, then professor of anatomy at Northwestern, and I tracked transmitter movement after injecting biogenic amines into the brain. We used the Falck-Hillarp histochemical fluorescence method discovered in Sweden to visualize the location of the chemical.

To our surprise we found that injection of dopamine into the caudate/putamen complex led to considerable spread within the complex and into adjoining regions (9). Consistent with the ventricular hypothesis proposed earlier, we observed highly fluorescent ventricular walls and ependyma, which under control conditions showed no fluorescence whatsoever. Finally, in one case, we observed the rapid transport of dopamine to contralateral caudate/putamen. For part of its course the dopamine traveled in the corpus callosum, presumably between the axonal fibers. Ironically, this observation was relevant to the finding that unilateral implantation of medullary autografts was therapeutic in Parkinson's disease (10).

The spread of chemical through the extracellular matrix was of interest, too, because there was a considerable debate at that time (not necessarily resolved now) concerning the amount of extracellular space in the nervous system. Van Harreveld (12) and others, such as my colleague Bondareff, argued that the beautiful images of Dr. Palay [see Chap. 4 in this volume, but especially Peters, Palay, and de Webster (11)] representing cross sections of processes as "pieces of puzzle" interlocking with one another was not a reflection of the *in vivo* state. Rather, if one used physical rather than chemical fixation methods, one observed considerably more extracellular space and not the smooth surfaces and neat fit portrayed in aldehyde fixed material. This debate can be accessed by reference to van Harreveld and Fifkova (12) and studies of co-workers Bondareff and Pysh (13) and Bondareff and Narotzky (14).

If this increased extracellular space is even in part correct, the spread of chemical observed could occur through these eddys and rivulets. Joined by Dave McClone, then a resident in neurosurgery, and Bob Narotzky, then a medical student, we demonstrated that the movement of chemical in the caudate putamen was not by simple diffusion (15). To our surprise, we discovered that there was preferential flow of chemical from the site of injection caudad. Because much of it was within the bundles of fibers in the internal capsule, which conducts impulses caudally, we proposed that the chemical was carried along by some principle associated with orthodromic axonal conduction. Reaching for straws we suggested electro-osmosis. Surely alternative physical mechanisms exist, and in my view, would be worthwhile investigating, for it appears that a principle of operation in volume transmission remains rather interesting but clearly not understood.

For those skitterish about entering byzantine areas of research, I would note that we also observed an age-dependent alteration in the extent of movement of chemical.

Aged animals showed a major decrease in the extent of chemical spread, which we attributed to a reduced extracellular space described previously by Bondareff, Narotzky, and Routtenberg (16). This hypothesis was given further credence by our finding that agents typically used in neurosurgery to reduce swelling and which reduce extracellular space, also reduced the amount of spread of injected chemical. In sum, these results clearly indicated that chemicals in the brain have broader access than may have previously been thought.

THE DESCARRIES/BLOOM DEBATE

In this volume one will find excellent updated summaries of this controversy. Briefly, Descarries proposes that amine transmission, particularly norepinephrine and serotonin, is not by conventional synaptic communication, but is more hormonal in nature, there being many instances of presynaptic terminal profiles without postsynaptic specialization presence. Bloom, in a detailed analysis, indicates that these instances can be accounted for by sampling problems arising with single sections and an inadequate number of sections when serial sections were studied. To this outside observer, it would appear that adherent points ("puncta adhaerentia") could exist on all synapses (although they be small) and that the influence of the chemical could then be wider spread than thought. If our spread of chemical studies has any bearing on the normal physiology of the brain, it is this: Chemicals released at points in the nervous system where modest junctional specialization is evident can spread for distances well beyond the point of release. This will provide a local "milieu interieur," such that a local state will be established. This could be modular in nature, because there could be barriers to diffusion in a variety of brain locations, not only the hippocampus (17). Thus, diffusion of chemical does not necessarily mean a nonspecific blurring of anatomical boundaries. The only problem is that the anatomy that this proposal suggests has not been described.

CODA

This brief review illustrates well the intent of the editor: There are principles of communication in the nervous system that have been hinted at but that still remain outside the central dogma that tends to infiltrate much of current cellular and molecular neurobiology. In reviewing this earlier research endeavor, I have the sense that one could exploit technological advances made since this work was done some 20 years ago to determine the extent to which the extracellular movement of chemicals and action at a distance would regulate the physiology of the central nervous system.

REFERENCES

1. Alloway TA, Routtenberg A. "Reminiscence" in the cold flour beetle, *Tenebrio molitor*. *Science* 1967;158:1066–1067.
2. Routtenberg A. Drinking induced by carbachol: thirst circuit or ventricular modification? *Science* 1967;157:838–839.
3. Routtenberg A, Simpson J. Carbachol-induced drinking at ventricular and subfornical organ sites of application. *Life Sci* 1971;10:481–490.
4. Simpson JB, Routtenberg A. The subfornical organ and carbachol-induced drinking. *Brain Res* 1972;45:135–152.
5. Simpson JB, Routtenberg A. The subfornical organ: site of drinking elicitation by angiotensin-II. *Science* 1973;181:1172–1174.
6. Simpson JB, Routtenberg A. Subfornical organ: acetylcholine application elicits drinking. *Brain Res* 1974;70:157–164.
7. Simpson JB, Routtenberg A. Subfornical organ lesions reduce intravenous angiotension-induced drinking. *Brain Res* 1975;88:154–161.
8. Simpson J, Routtenberg A. Subfornical organ: a dipsogenic site of action of angiotensin-II. *Science* 1978;201:379–380.
9. Routtenberg A, Sladek J, Bondareff W. Histochemical fluorescence after application of neurochemicals to caudate nucleus and septal area *in vivo*. *Science* 1968;161:272–274.

10. Drucker-Colin R, Madrazo I, Ostrosky-Solis F, Shkurovich M, Franco R, Torres C. Adrenal medullary tissue transplants in the caudate nucleus of Parkinson's patients. *Prog Brain Res* 1988;78:567–575.
11. Peters A, Palay SL, de Webster HF. *The fine structure of the nervous system*. 1970.
12. Van Harrevald A, Fifkova E. Rapid freezing of deep cerebral structures for electron microscopy. *Anat Rec* 1975;182:377–386.
13. Bondareff W, Pysh JJ. Distribution of the extracellular space during postnatal maturation of rat cerebral cortex. *Anat Rec* 1968;160:773–780.
14. Bondareff W, Narotzky R. Age changes in the neuronal microenvironment. *Science* 1972;176: 1135–113x.
15. Bondareff W, Routtenberg A, Narotzky R, McLone DG. Intrastriatal spreading of biogenic amines. *Exp Neurol* 1970;28:213–229.
16. Bondareff W, Narotzky R, Routtenberg A. Intrastriatal spread of catecholamines in senescent rats. *J Gerontol* 1971;26:163–167.
17. Ruth R, Routtenberg A. Possible diffusion barriers within rat dentate gyrus. *Fed Proc* 1980; 30(3):597.

Volume Transmission in the Brain: Novel Mechanisms for Neural Transmission, edited by K. Fuxe and L.F. Agnati. Raven Press, Ltd., New York © 1991.

23

Amino Acid Gradients in the Choroid Plexus-Cerebrospinal Fluid-Extracellular Fluid System

*Anders Hamberger, *Britta Nyström, **Carsten Wikkelsø, and †Herbert Silfvenius

**Institute of Neurobiology and Departments of **Neurology and †Neurosurgery, University of Göteborg, Göteborg, Sweden*

AMINO ACIDS IN THE CEREBROSPINAL FLUID

A recent review on the subject of volume transmission (1) gave the impression that it concerns mainly monoamines and, furthermore, that it is a misconception. According to the volume transmission concept, transmitters play modulatory roles as chemical signals, working fairly diffusely via the extracellular fluid (ECF). However, amino acids act, perhaps better than many other compounds, according to the volume transmission ideas because amino acids

- are present not only in nerve terminals, but in all glial as well as in neuronal cells
- are present at a 1,000-fold of the monoamine concentration in the brain
- are most likely the transmitters in at least 90% of the synapses
- are present in a metabolic pool in addition to in the transmitter pool; however, the former may easily leak neuroactive substances
- are being inactivated as transmitters exclusively by uptake, into terminals and glia
- induce receptor-mediated nerve cell death in an excitotoxic fashion, i.e., by flooding the extracellular space
- enter and leave the cells of the brain as osmoequivalents to avoid excessive cell and organ volume changes
- are influenced in the patient by systemic acute or chronic metabolic disturbances, which affect the neurological state

With amino acids, the question is not whether the transmitter in the ECF is from a terminal in contact with a postsynaptic cell; it is whether it comes from terminals at all.

Routinely, the access to the amino acid concentrations in the neuronal environment is via the cerebrospinal fluid (CSF), the internal fluid of the CNS, a protein-poor plasma product that fills the intercellular spaces as well as the ventricular system. The composition of the fluid appears to be fairly similar in the interstitium and in the cavities. The lumbar CSF gives at best an average of the situation in the periventricular CNS tissue, because the secretion from the choroid plexus is modified biochemically by exchange with the CNS on its way through the ventricular system (2). The approach to establish whether a compound enters the CSF over the blood-brain barrier (BBB) or over the blood-choroid plexus barrier involves blood infusions, CSF gradient studies, ventriculocisternal perfusions, etc. (3,4). The choroid plexus secretion has so far only been characterized with respect to ions (5). Most available evidence suggests that a substantial portion of not only solutes but fluid volume of the CSF comes from the periventricular brain (6). Undoubtedly, this dual origin of the CSF has a physiological relevance and the contribution from the choroid plexus would hardly have a composition identical to that coming from the brain tissue. This component can be subdivided further into what passes over from the blood and what is produced in the CNS. A reasonable assumption would be a nephron-like system in the brain, particularly as the CSF represents the only way of fluid disposal for the brain to the blood beside exchange at the level of the capillary endothelium.

SAMPLING OF EXTRACELLULAR FLUID

We have sampled the fluid exsudate from the pial surface of the cortex and from the chorid plexus in patients during neurosurgery to obtain amino acid "fingerprints" of the ECF. A suitable technique for the sampling purpose is microdialysis (7–9), but we have recently had good results with a type of filter paper. This is a 5 × 10-mm sample application piece (LKB 1850-901), which is left on the brain or plexus surface for 10 sec. The amount of fluid taken up was close to 10μl. Application pieces and dialysis tubings were applied side by side on the pial surface, and the amino acid concentrations in the perfusion fluid from the dialysis tubing compared well with the fluid extracted from the paper. Amino acids were measured with liquid chromatography after precolumn derivatization (10).

The patterns of soluble proteins were analyzed by isoelectric focusing (11) to characterize independently the origin of fluid samples. The protein patterns of serum, brain ECF, plexus ECF, and CSF had characteristic differences. The high concentration of transthyretin in the plexus ECF confirms the origin of the sample because transthyretin is synthetized within the choroid plexus cells and secreted into the CSF (12). The low concentration of the gamma-trace protein in the plexus sample gives further support because this protein has been demonstrated in cortical neurons (13) and appears to enter the CSF by diffusion. The tau-globulin, formed by desialization of transferrin in neural cells (14), was also of high concentration in the cortical ECF. In summary, the distribution of proteins gave good support for the distinction of the two ECF samples and basis for a discussion of amino acid concentrations in the CSF relative to that in the ECF of the tissue surrounding the ventricles. The ECF of the CNS has become available for biochemical characterization largely with the dialysis technique, but other ECF sampling techniques similar to the filter paper have also been used (15). The sampling from tissue surfaces when the skull is opened and the ventricles are exposed is clearly open to criticism: The fluid dynamics in the brain are altered as soon as the cranium is opened, and the pial surface was employed to sample ECF although the ependymal surface would have been more appropriate.

CONCENTRATION GRADIENTS FOR AMINO ACIDS

Although proteins in the CSF are as low as 0.5% of their concentration in blood plasma, amino acids average about 30% of the plasma concentration, ranging from 1% (glutamic acid) to 100% (glutamine) of the plasma concentration (Fig. 1). Glutamic acid is a good example to show the importance of BBB. The concentration of this excitatory transmitter was 0.5 μM in the CSF. The 120-fold reduction in concentration of glutamate contrasts to the concentration of glutamine, which was more or less the same in CSF and blood plasma.

In the comparison with the concentra-

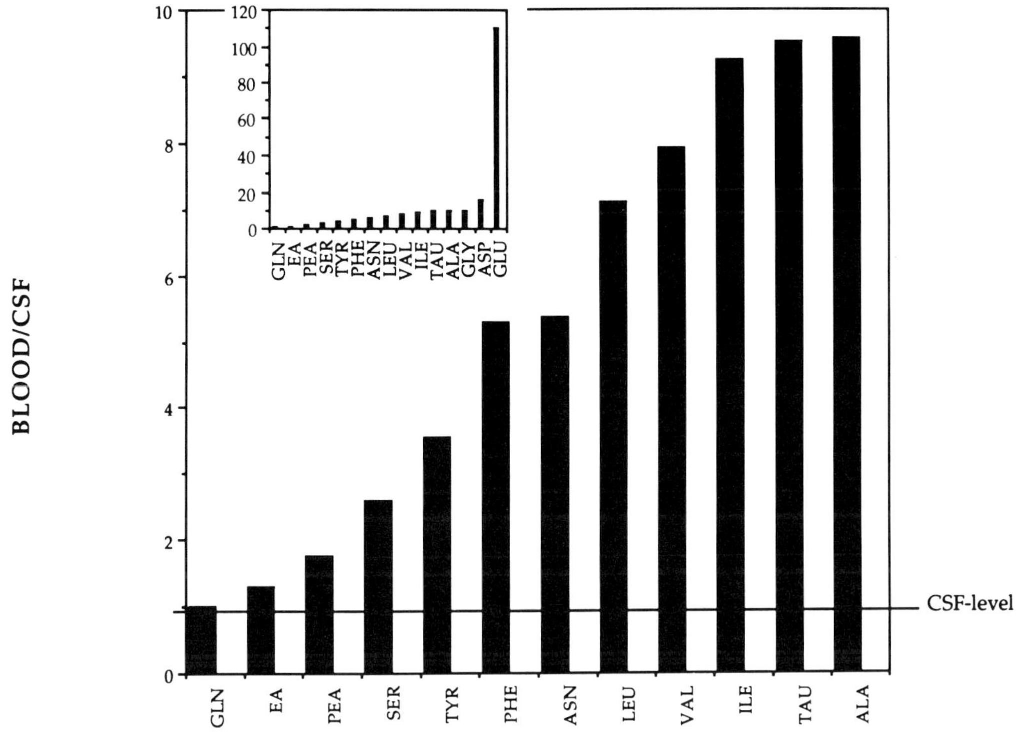

FIG. 1. Ratio of amino acid concentration in blood plasma and CSF (man). *Insert* shows high ratio for glutamate. CSF concentration of each amino acid = 1.

tions in the CSF, the ECF collected from the cortical surface was higher with respect to glutamine, ethanolamine, glycine, and taurine (Fig. 2). Alanine and serine showed similar concentrations in the brain ECF and the CSF, whereas other amino acids were lower in the ECF. In Fig. 3, the amino acid concentration in the freshly secreted choroid plexus ECF is compared with that in the cortical ECF and in the lumbar CSF. Strikingly, the choroid plexus ECF is poor in glutamine, the concentration being 25% of that in lumbar CSF as well as of that in blood plasma. The situation for ethanolamine was similar. However, the ECF from the choroid plexus was rich in alanine and phosphoethanolamine and the concentrations of serine, tyrosine, valine, isoleucine, and phenylalanine were also higher than in the CSF.

Two amino acids, taurine and asparagine, do not fit into the model that either the cortical ECF or the choroid ECF has a higher concentration than the CSF: Taurine, because both the plexus ECF and the cortical ECF have higher concentrations than the CSF, and asparagine, which is lower than in CSF in both ECF compartments.

GLUTAMINE CYCLE

The difference in concentration for amino acids between the CSF and the ECF of the choroid plexus and the brain, respectively, is interpreted to indicate that the newly secreted CSF rapidly equilibrates with the brain tissue. However, an active transport for amino acids against a concentration gradient from the CSF to the blood via the choroid plexus (16) has also been indicated. It is of particular interest that the

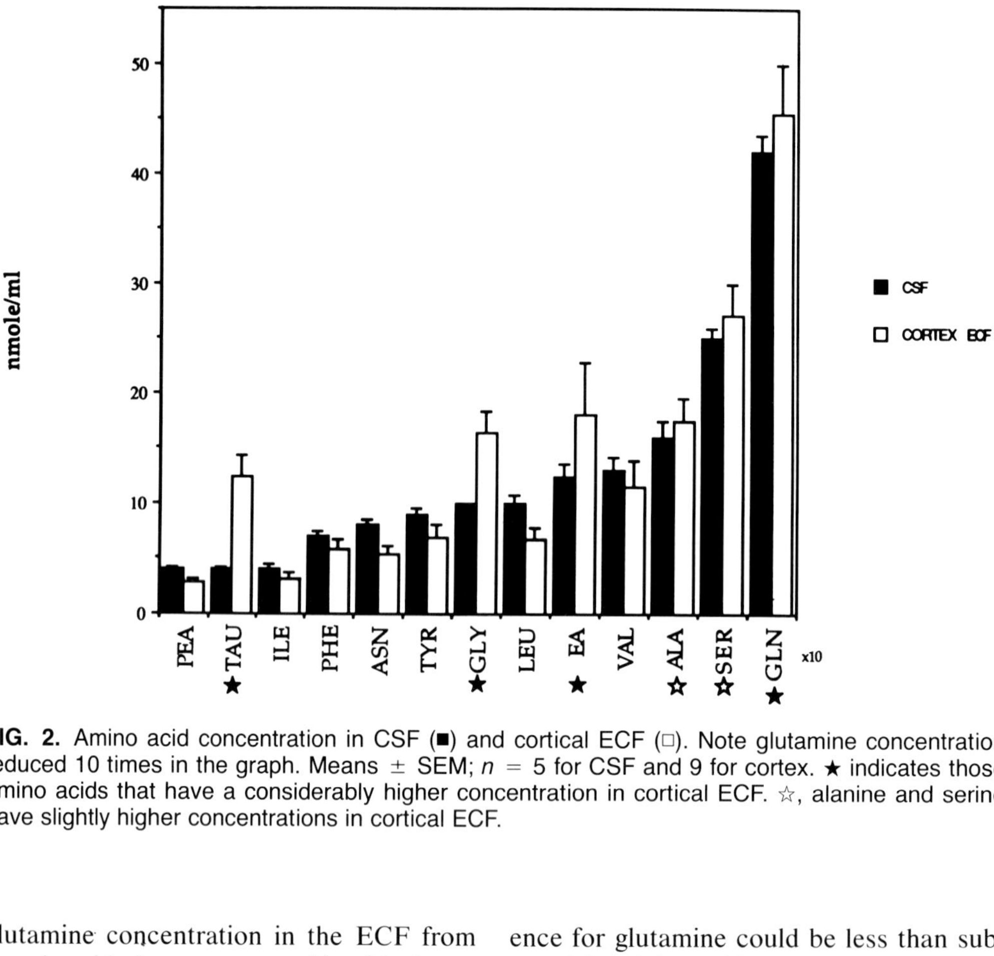

FIG. 2. Amino acid concentration in CSF (■) and cortical ECF (□). Note glutamine concentration reduced 10 times in the graph. Means ± SEM; n = 5 for CSF and 9 for cortex. ★ indicates those amino acids that have a considerably higher concentration in cortical ECF. ☆, alanine and serine have slightly higher concentrations in cortical ECF.

glutamine concentration in the ECF from the choroid plexus was considerably lower than in the lumbar CSF, which in turn is close to the concentration in the plasma and the brain ECF. Glutamine is quantitatively dominating in the CSF; as much as 85% of the amino acid nitrogen is in glutamine, and the corresponding values for the choroid plexus and blood plasma are 25% and 35%, respectively. The brain is well equipped for glutamine formation; immunohistochemically, glutamine synthetase is localized to the astrocytic glia (17). A transfer of glutamine from the brain to the CSF implies a net glutamine production in the brain. A difference in the concentration of such a quantitatively important compound could act as a driving force to move water out of the brain (18). The arterio-venous difference for glutamine could be less than substantial (19) but still accounts for an addition of glutamine over the ependymal border, because part of it could be derived from the blood. The increased net production of glutamine in hyperammonemia as measured by arteriovenous difference (20), in parallel with hypertrophic astrocytes rich in glutamine synthetase, is, however, supportive of the brain–CSF transfer of glutamine.

Another compound that may enter the CSF from the brain because of its relatively high concentration in the cortical ECF is ethanolamine, which also occurs at high intracellular concentrations in the brain and may be a precursor for membrane phospholipids. The release of ethanolamine into the hippocampal and striatal ECF has been re-

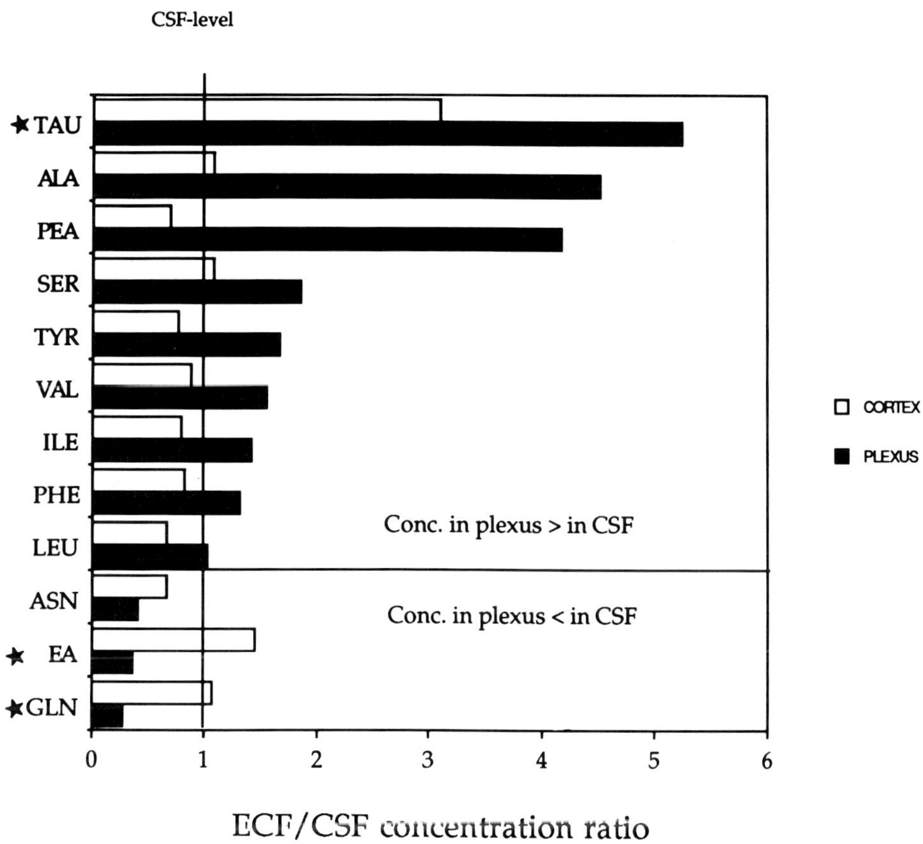

FIG. 3. Amino acid concentration in cortical ECF (□) and ECF of the choroid plexus (■) relative to that of CSF (= 1). ★ shows amino acids also indicated in Fig. 2, i.e., with a higher concentration in cortical ECF.

ported during reversible brain ischemia (21).

BLOOD–CSF PATHWAY

Another side of the difference between what passes over the BBB and the blood–CSF barrier is, i.e., the possibility that certain amino acids are supplied over the ependymal cells into the brain. Spector (22) reviewed the criteria for a substance that enters the brain cells mainly from the CSF and found these to be filled by some of the vitamins and nucleosides. A group of amino acids (i.e., taurine, alanine, and serine) had a higher concentration in the plexus ECF than in the CSF. This is suggestive of an uptake into the periventricular CNS. The concentration of taurine in the brain is one of the highest among the amino acids even though its rate of synthesis in brain is low (23). The brain needs exogenous taurine, but this need is counteracted by an efficient BBB for taurine (24). A transfer from the blood via the CSF may be of quantitative importance not only for taurine but also for some other amino acids (Fig. 4). However, until data on the transport characteristics are available, the direction of transport, if any, is inconclusive.

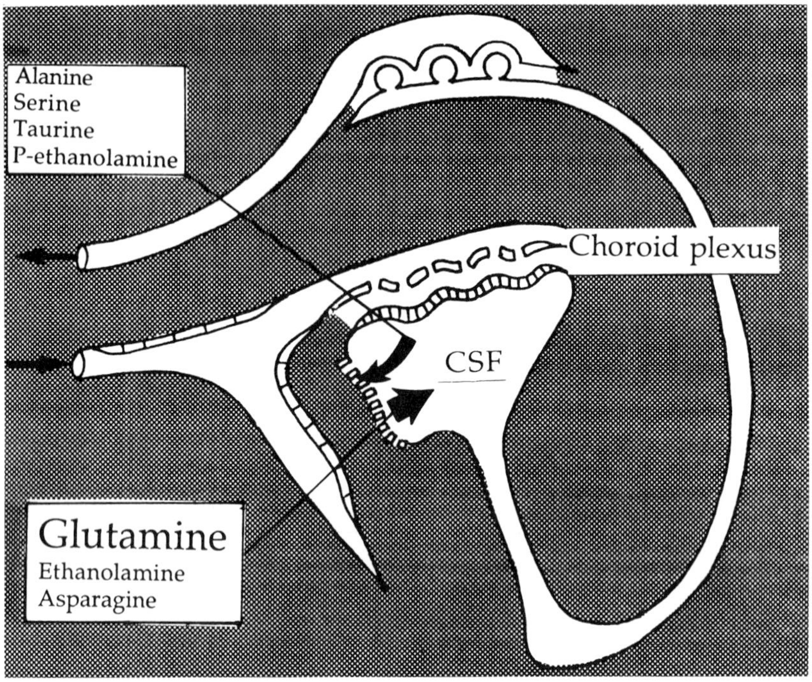

FIG. 4. Schematic representation of hypothetical flow of amino acids in CSF. *Arrows* on far left indicate venous and arterial systems.

ACKNOWLEDGMENTS

This study was supported by grants from the Swedish Medical Research Council (grant no. 12X-00164) and from Riksbankens Jubileumsfond.

REFERENCES

1. Parnavelas JG, Papadopoulos GC. The monoaminergic innervation of the cerebral cortex is not diffuse and nonspecific. *TINS* 1989;12:315–319.
2. Davson H, Welch K, Segal MB, eds. *The physiology and pathophysiology of the cerebrospinal fluid.* New York: Churchill Livingstone, 1987.
3. Wiechert P. Uber die Permeabilität der Blut-Liquor-Schranke fur einige Aminosäuren. *Acta Biol Med Germanica* 1963;10:305–310.
4. Franklin GM, Dudzinski DS, Cutler RWP. Amino acid transport into the cerebrospinal fluid of the rat. *J Neurochem* 1975;25:367–372.
5. Ames A, Sakanoue M, Endo S. Na, K, Ca, Mg and Cl concentrations in choroid plexus fluid and cisternal fluid compared with plasma ultrafiltrate. *J Neurophysiol* 1964;27:672–681.
6. Milhorat TH, Hammock MK. Cerebrospinal fluid as reflection of internal milieu of brain. In: Wood JH, ed. *Neurobiology of cerebrospinal fluid,* vol 2. New York: Plenum Press, 1983;1–23.
7. Ungerstedt U, Pycock C. Functional correlates of dopamine neurotransmission. *Bull Schweiz Akad Med Wiss* 1974;1287:1–5.
8. Hamberger A, Jacobson I, Lehmann A, Sandberg M. Dynamics of the cellular and extracellular compartments of brain amino acids—visions and reality of the dialysis approach. In: Hunter G, ed. *Amino acid availability in health and disease.* Berlin: Springer-Verlag, 1988;81–90.
9. Hamberger A, Nyström B, Silfvenius H. Extra- and intracellular amino acids in the CNS of patients with epilepsy and other neurological disorders. In: Avanzini G, ed. *Proc IV Woniep.* 1988.
10. Lindroth P, Mopper K. High performance liquid chromatographic determination of subpicomole amounts of amino acids by precolumn fluorescence derivatization with *o*-phtaldialdehyde. *Anal Chem* 1979;51:1667–1674.
11. Vesterberg O. Isoelectric focusing of proteins in polyacrylamide gels. *Biophys Acta* 1972;257:11–19.
12. Herbert J, Wilcox JN, Pham KTC, et al. Transthyretin: a choroid plexus-specific transport

protein in human brain. *Neurology* 1986;36:900–911.
13. Löfberg H, Grubb AO, Brun A. Human brain cortical neurons contain gamma-trace. Rapid isolation, immunohistochemical and physicochemical characterization of human gamma-trace. *Bio Med Res* 1981;3:298–306.
14. Gallo P, Bracco F, Morara S, Battistin L, Tavoloto B. The cerebrospinal fluid transferrin/tau proteins. *J Neurol Sci* 1985;70:81–92.
15. Go KG, Patberg WR, Teelken AW, Gazendam J. The starling hypothesis of capillary fluid exchange in relation to brain edema. In: Pappius Feindel, eds. *Dynamics of brain edema*. Berlin: Springer, 1976;63–67.
16. Bito L, Davson H, Levin E, Murray M, Snider N. The concentrations of free amino acids and other electrolytes in cerebrospinal fluid, *in vivo* dialysate of brain and blood plasma of the dog. *J Neurochem* 1966;13:1057–1076.
17. Norenberg MD, Martinez-Hernandez A. Fine structural localizations of glutamine synthetase in astrocytes of rat brain. *Brain Res* 1979; 161:303–310.
18. van Gelder NM. Metabolic interactions between neurons and astroglia: glutamine synthetase, carbonic anhydrase and water balance. In: Jasper HH, van Gelder NM, eds. *Basics mechanisms of neuronal excitability*. New York: Alan R Liss, 1983;5–29.
19. Cansiano C, Farber MO, Cardelli-Cansiano P, et al. Plasma amino acid levels across the brain in patients with respiratory failure. *Eur J Clin Invest* 1983;1:87–89.
20. Gjedde A, Lockwood AH, Duffy TE, Plum F. Cerebral blood flow and metabolism in chronically hyperammonemic rats: effect of an acute ammonia challenge. *Ann Neurol* 1978;3:325–330.
21. Hagberg H, Lehmann A, Sandberg M, Nyström B, Jacobson I, Hamberger A. Ischemia-induced shift of inhibitory and excitatory amino acids from intra- to extracellular compartments. *J Cereb Blood Flow Metab* 1985;5:413–419.
22. Spector R. Nucleoside and vitamin homeostasis in the mammalian central nervous system. In: Cserr H, ed. *The neuronal microenvironment*. New York: New York Academy of Science, 1986;221–230.
23. Huxtable RJ, Franconi F. Introduction: biochemistry, nutrition and development. In: Huxtable RJ, Franconi F, Giotti A, eds. *Advance in experimental medicine and biology (The biology of taurine methods and mechanisms)*. New York: Plenum Press, 1987;25–28.
24. Kuriyama K, Ida S, Ohkuma S, Tanaka Y. Alteration of cerebral bisynthesis of taurine in spontaneously hypertensive and 3-acetylpyridine intoxicated rats. In: Oja S, Ahtee L, Kontro P, Paasonen M, eds. *Taurine: biological actions and clinical perspectives*. New York: Alan Liss, 1985;91–103.

Volume Transmission in the Brain: Novel Mechanisms for Neural Transmission, edited by K. Fuxe and L.F. Agnati. Raven Press, Ltd., New York © 1991.

24

Role of the Cerebrospinal Fluid in Volume Transmission Involving the Choroid Plexus

Christer Nilsson, Maria Lindvall-Axelsson, and Christer Owman

Department of Medical Cell Research, Section of Neurobiology, University of Lund, Sweden

The cerebrospinal fluid (CSF) is a clear and colorless fluid that surrounds the inner and outer surfaces of the brain and fills the intracranial cavities that are not occupied by neuronal tissue or blood vessels. The total CSF volume is approximately 125 ml, with a four- to fivefold turnover during each 24-hr period (1). The direction of CSF flow is from the lateral, third and fourth ventricles, through the foramina of Luschkae and Magendie, to the cisterna magna and the outer surface of the brain. From cisterna magna, the CSF flows into the subarachnoid space surrounding the cortex and the spinal cord. The flow of CSF is driven partly by the production of CSF in the ventricles but especially from the pulsations of the brain in connection with systole and diastole (2). This ensures a rapid flow and mixing of CSF at the interfaces between CSF and brain. In addition, the blood vessels penetrating the cortex from the outer surface of the brain are surrounded by the subarachnoid space down to the arteriolar level, creating the Virchow-Robin spaces, thus increasing the contact area between CSF and brain tissue. There are no barriers between the CSF and the extracellular fluid of the brain parenchyma, neither in the Virchow-Robin spaces nor over the ependyma lining the brain ventricles. The close association of some brain structures (e.g., the cerebral cortex, choroid plexus, and the circumventricular organs) to the CSF indicates that these areas could be potential targets for extracellular signaling mediated by the CSF.

The CSF is mainly secreted by the choroid plexuses that are located in each of the four brain ventricles (3), the additional part probably originating from the cerebral capillaries (4,5) or arachnoid tissue (3). The choroid plexus is a villous structure, which is a specialized part of the ventricular ependyma invaginating into the ventricles and surrounded by CSF. It is composed of a single, secretory epithelium covering a highly vascularized, connective tissue stroma. Characteristically, the epithelial cells are joined together with tight junctions that form the barrier between blood and CSF, whereas the capillaries in the choroid plexus are fenestrated, thus differing from the capillaries in the brain parenchyma (6).

The functions of the CSF are closely related to the functions of the choroid plexus. Apart from giving mechanical support and protection to the brain, the CSF is in equilibrium with the extracellular fluid, contributing to the homeostasis of the brain, and also serves as a transport route for hormones and nutrients to the brain, as will be further elaborated below.

This chapter will discuss two aspects of possible volume transmission in the CSF. The first involves the choroid plexus as a target for chemical mediators in the CSF regulating choroid plexus function; the second deals with the choroid plexus as a route for endocrine signals targeted to the brain.

THE CHOROID PLEXUS AS A TARGET FOR CSF VOLUME TRANSMISSION

The 5-Hydroxytryptamine 1C Receptor Subtype in the Choroid Plexus

In 1985, the presence of a new 5-hydroxytryptamine (5-HT) receptor subtype was described in choroid plexus from the pig (7), designated the 5-HT_{1C} receptor subtype, and was soon reported to be present also in rat, rabbit, cow, mouse, and humans (8,9). The receptor was found to be present at a high density (as much as 3.1 pmol/mg protein in the rat) (8). It was first thought that the 5-HT_{1C} receptor was uniquely located in the choroid plexus, but recent receptor binding and messenger ribonucleic acid (mRNA) hybridization studies also located it to the substantia nigra, globus pallidus, ventromedial hypothalamus, and other areas of the brain, although at an at least 10-fold lower density (9–11), where it has been postulated to be involved in the initiation of migraine (12).

The activation of the 5-HT_{1C} receptor is strongly coupled to the stimulation of phosphoinositol hydrolysis (13) but does not affect the activity of adenylate cyclase (14). Despite the 5-HT_{1C} receptor having a relatively high affinity for 5-HT, thus placing it in the 5-HT_1 receptor class, the pharmacology of this receptor resembles the 5-HT_2 receptor more closely than the other 5-HT_1 receptors. For example, compounds like α-methyl-5-HT, LY 53857, ritanserin, ketanserin, and cyproheptadine, which have been believed to be selective for 5-HT_2 receptors, also interact with the 5-HT_{1C} receptor (15). In addition, the 5-HT_2 receptor is also coupled to the phosphoinositol second-messenger system (16). Some differences in pharmacology of the two receptors do exist, however, which will be elaborated on below.

The apparent similarities between the 5-HT_{1C} and 5-HT_2 receptors were confirmed by the cloning of both receptors (17,18). It turned out that both the 5-HT_{1C} and 5-HT_2 receptors are composed of seven transmembrane regions, in common with other G protein–coupled receptors, and that there is a 51% homology between the two receptors (17). In contrast, there is only a 35% homology between the 5-HT_{1A} receptor to the 5-HT_{1C} and 5-HT_2 receptors, respectively (17).

Possible Sources of 5-HT Acting on the Choroid Plexus

On the basis of chemical determinations of 5-HT in the choroid plexus, combined with lesions of the 5-HT-containing neurons in the raphe nuclei, it has been proposed that the choroid plexus receives 5-HT-containing nerve fibers from the brainstem (19). Immunocytochemical studies in the rat using antibodies toward 5-HT (20; H.W.M. Steinbusch, *personal communication*), as well as autoradiographic techniques visualizing the uptake of ^{3}H-tryptamine (21), were not able to visualize any 5-HT-containing nerve fibers in the choroid plexus tissue proper or on its surface. In both these studies (20,21), the remarkable finding was made, however, that there was a rich plexus of 5-HT-containing nerve fibers lying on the surface of the ependyma lining the cerebral ventricles, lacking any apparent synaptic contact with the ependymal cells themselves (21). This is in good agreement with our own immunocytochemical studies in the rat with antibodies toward 5-HT (Fig. 1). This leads to an alternative explanation of the lesion experiments described above (19), namely, that the nerve fibers in the so-called supra-

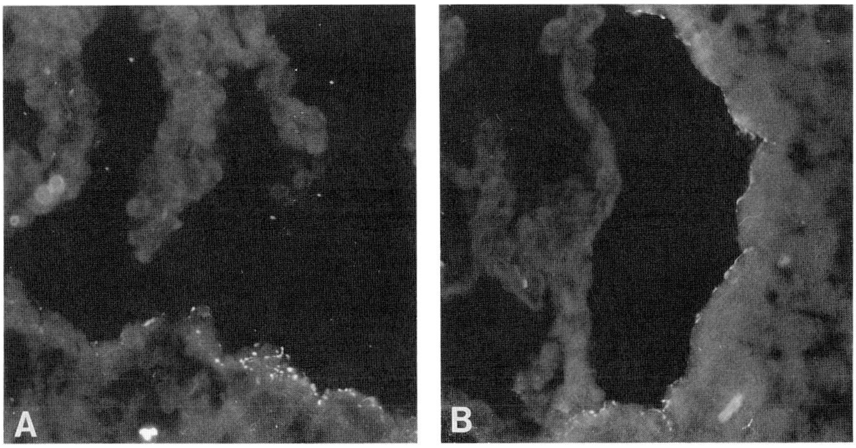

FIG. 1. Photomicrograph of the choroid plexus and ventricle wall of the lateral (**A**) and third (**B**) ventricle, in 10-μm sections of rat brain. Sections were stained by indirect immunohistochemistry using fluorescein isothiocyanate–coupled antibodies. A prominent plexus of nerve fibers can be seen overlying the ventricular ependyma, whereas the choroid plexus tissue itself (*upper left* part of both A and B) are without nerve fibers.

ependymal nerve plexus releases 5-HT into the CSF, which is partly taken up by the 5-HT uptake system described in the choroid plexus (22). After a lesion of the raphe system, the levels of 5-HT in the CSF would fall, and less 5-HT will be taken up in the choroid plexus, accounting for the lower levels recorded. The function of this uptake of 5-HT may probably be part of an elimination mechanism through breakdown by monoamine oxidase (23) as part of the choroid plexus barrier function. This situation resembles that seen in the cerebral arteries, where noradrenergic, sympathetic nerves take up 5-HT from the surrounding extracellular fluid and thus become immunoreactive in certain situations (24–26). Finally, the 5-HT-induced phosphoinositide hydrolysis described in the choroid plexus was not affected by tetrodotoxin and could not be mimicked by KC1 or veratrine, indicating that 5-HT is neither released from nerves nor acts indirectly through the release of another neurotransmitter (13). There is no reason, however, why 5-HT released into the CSF could not have an effect in itself on the choroid plexus epithelium, mediated by the functional 5-HT_{1C} receptors present in this tissue (see above). Supporting this notion is the study by Conn et al. (27), which showed that lesion of the raphe system in rats induced supersensitivity of the 5-HT_{1C} receptor in the choroid plexus, indicating a tonic serotonergic influence on the receptor. Although this does not tell us anything about the route by which 5-HT reaches the choroid plexus, it is notable that the 5-HT_{1C} receptor may be localized on the apical surface of the choroid plexus epithelial cells (28).

In conclusion, there are several indications that 5-HT, released from supraependymal nerve fibers in contact with the CSF, can reach the choroid plexus by way of the CSF, bind to the apically located 5-HT_{1C} receptors, and elicit a functional response mediated by the hydrolysis of phosphatidyl inositol. To our knowledge there is no convincing evidence for a direct innervation of the choroid plexus by 5-HT-containing nerves. This hypothesis is summarized in Fig. 2.

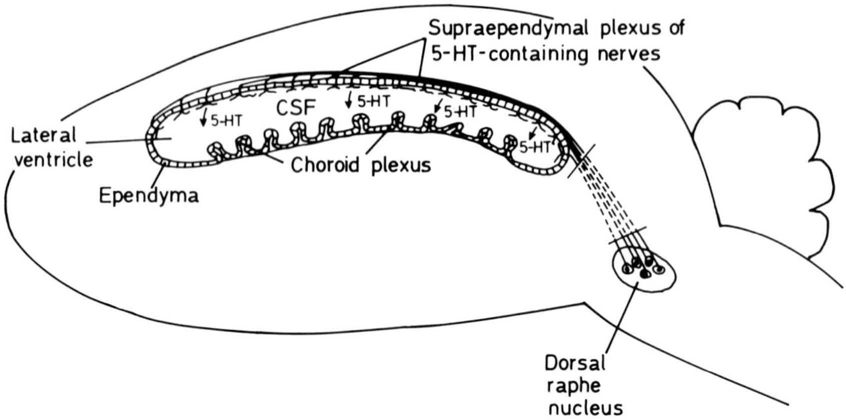

FIG. 2. Schematic picture of rodent brain, showing a possible pathway for 5-HT influencing the choroid plexus (only choroid plexus of a lateral ventricle is shown). Nerve fibers originating in serotonergic neurons of the dorsal raphe nucleus project to the ependyma of all four ventricles, terminating in a dense supraependymal plexus in direct contact with CSF. Released 5-HT might diffuse into the CSF and reach the choroid plexus by bulk flow.

Effects of 5-HT on CSF Production and Other Functions of the Choroid Plexus

In several series of studies we have used the ventriculo cisternal perfusion method developed by Pappenheimer et al. (29), modified for use on rabbits (30) to quantify the rate of bulk CSF production. Briefly, the animal is anesthetized, and both femoral arteries and one femoral vein are catheterized for blood pressure recording, blood sampling, and drug injection, respectively. The animal is tracheotomized and put on artificial ventilation for the control and regulation of blood Pco_2, Po_2, and pH, which are monitored throughout the experiment. After these initial operations, the animal is placed in a specially made hammock and the head fixed in a metal frame. A hole is drilled through the bone of the skull and an infusion needle placed in the underlying right lateral ventricle. The muscles covering the atlanto-occipital membrane in the neck are separated and a needle placed in the cisterna magna. The ventricular system is perfused at a rate of 30 μl/min with mock CSF containing ^{14}C-inulin. The perfusion pressure is registered continuously along with blood pressure. CSF production is calculated from the dilution of the ^{14}C-inulin radioactivity according to Heisey et al. (31):

$$V_f = V_p\,[(C_i - C_o)/C_o]$$

where V_f = rate of CSF production, V_p = rate of perfusion, C_i and C_o = concentration of radioactivity (in dpm) in the inflow and outflow of artificial CSF, respectively.

Using the method described above, we have previously shown that 5-HT lowers CSF production in the rabbit dose-dependently with up to 40% (32,33). This reduction was only partly blocked by the 5-HT_2-receptor antagonist ketanserin (32), whereas the commonly used 5-HT_{1C}-receptor antagonist mesulergine (7,9) completely blocked the 5-HT-induced reduction in CSF production (33). Thus, we believe that 5-HT decreases CSF production via a direct action on the 5-HT_{1C} receptors on the secretory epithelium of the choroid plexus in a manner similar to that described for the noradrenergic sympathetic nerves acting through epithelial β_1-adrenoceptors (30). The results are also in agreement with a study in dogs, in which intravenous infusion of 5-hydroxytryptophan decreased CSF production by 40% (34). However, as the cerebral vessels contain vasoconstrictory 5-HT_2 receptors, and ketanserin and mesulergine may interact with both the 5-HT_2 and 5-HT_{1C} receptors, although with different affinities, a reduction in CSF production induced by 5-HT could be partly associated with a reduced blood perfusion of the choroid plexus. However, this possibility is not supported by the recent findings that 5-HT administered into the left atrium in monkeys and dogs increases plexus blood flow by 100% and 200%, respectively (35), which indicates that changes in choroid plexus blood flow and CSF production induced by 5-HT are not immediately linked to each other. Because as much as 70% to 80% of the CSF is produced by the choroid plexus (3), it seems unlikely that the reported reduction in CSF production induced by 5-HT would be due to an effect on a source different from the choroid plexus. Regulation of bulk CSF production does not need to be the primary role of 5-HT and the 5-HT_{1C} receptor in the choroid plexus, however. In cultures of choroid plexus epithelial cells, 5-HT increases the synthesis of the plasma protein transferrin (36), thus raising the possibility that 5-HT can change the composition of the CSF secreted from the choroid plexus. When rat brain RNA is injected into *Xenopus* oocytes, 5-HT stimulates an electrophysiological response pharmacologically compatible with an effect on 5-HT_{1C} receptors (37), although it is not yet possible to interpret what the physiological significance of this finding would be for the function of choroid plexus epithelium.

Several lines of evidence indicate that the 5-HT_{1C} receptors present in the choroid plexus are a target for 5-HT released into and transported through the CSF (8,20, 21,27,32,36). The concentration of 5-HT in the rat CSF has been measured to approximately 2 nM (38), which is similar to the concentration found in humans (39). Although some ligands such as [^{125}I]LSD and [^{3}H]mesulergine bind with high affinity (K_d = 2 nM) to choroid plexus membranes (7,8,40), the binding of 5-HT is noticeably weaker, with a K_d approximately 20 nM (7,40). This suggests that the majority of the 5-HT_{1C} receptors are not occupied during normal conditions and that the serotonergic tone on the choroid plexus is low. It is also possible that 5-HT concentrations show circadian or diurnal variation so that activation of the 5-HT_{1C} receptors occurs only periodically (e.g., during night time). Studies investigating the effect of 5-HT on choroid plexus function (16,32,36) actually show an effect in concentration ranges of 10^{-8} to 10^{-5}M higher than would be expected from the binding studies and the concentration of 5-HT in the CSF. It should be noted, though, that several factors such as the pronounced uptake and breakdown of 5-HT in the choroid plexus epithelium might be a reason why high concentrations of 5-HT are needed when measuring CSF production rates *in vivo* (32). This is supported by the fact that the monoamine oxidase inhibitor nialamide potentiates the effect of 10^{-7}M 5-HT threefold (32).

Other Evidence for the Choroid Plexus Being a Target for CSF Volume Transmission

In addition to 5-HT, at least three other neural or neuroendocrine messengers seem to be involved in the regulation of choroid plexus function, without being present in choroid plexus nerve fibers, and thus probably reaching the choroid plexus epithelium by CSF volume transmission. Vasopressin and atrial natriuretic peptide (ANP) both have a high density of receptors on choroid plexus epithelium (41–43) and are present in central neurons (44,45) as well as in the CSF (46–50). Ultrastructural studies of choroid plexus tissue incubated with vasopressin indicates that the peptide might increase fluid secretion from the tissue (51), whereas physiological evidence using the ventriculocisternal perfusion technique points to the opposite (52,53). In Brattleboro rats, which are genetically deficient in vasopressin, the intracerebral fluid dynamics after intravenous infusion of hyperosmolar mannitol is severely disturbed (54). ANP has been shown to decrease CSF production at low concentrations (55), and the receptor density is increased in hydrocephalic mice (56). Both vasopressin and ANP could thus have an important regulatory role in CSF dynamics.

Ultrastructural studies have also shown that the monoamine melatonin, synthesized in the pineal gland, might influence fluid secretion from the choroid plexus (57). It is conceivable that there is a pathway through the CSF between the anatomically closely related pineal gland and the third ventricle choroid plexus and maybe also the more distant lateral and fourth choroid plexuses.

THE CHOROID PLEXUS AND CSF AS AN ENDOCRINE PATHWAY TO THE BRAIN

Plasma Protein Synthesis and Secretion From the Choroid Plexus

During the past few years, it has become increasingly clear that the choroid plexus synthesizes and secretes a number of different plasma proteins (58) that were previously thought to be produced only in the liver. Among these proteins are transthyretin (TTR; formerly called prealbumin), transferrin, ceruloplasmin (59–61), and insulin-like growth factor II (IGF II) (62,63); and the mRNA for other proteins has been identified as well (G. Schreiber, *personal communication*).

As the blood-brain and blood-CSF barriers are almost completely impermeable to proteins in the plasma, it is reasonable to assume that certain transport proteins (e.g., TTR and transferrin) might be required in greater concentrations in the intracerebral compartment to maintain brain homeostasis. Indeed, although the CSF protein concentration is low, certain proteins (e.g., TTR) have a much higher CSF to plasma ratio than would be expected from their molecular weight and size. The plasma protein that has been most extensively studied in this respect is TTR and its possible role in transport of thyroid hormone from the periphery to brain. Substantial evidence points to a route for thyroxine (tetraiodothyronin), but not triiodothyronin, in the direction: blood → choroid plexus → CSF → brain (64). In this manner, the CSF might provide a route for endocrine signals directed to the brain parenchyma. The plasma protein synthesis and secretion in the choroid plexus and its functional aspects have been reviewed more extensively elsewhere (58,65).

CONCLUDING REMARKS

There is strong reason to assume that the CSF constitutes an important pathway in volume transmission. This particular system actually represents one of the early known transport routes for neurosecretory material in regions close to such areas as the subcommisural organ and the median eminence. The present short overview provides recent data on the involvement of serotonergic mechanisms and the choroid plexus in a more modern aspect of volume transmission, not least involving the 5-HT_{1C} type of receptors, which are highly concentrated in the choroid plexus epithelium. Besides being a target in such volume transmission, the choroid plexus also represents an important link in the transport of hormones and other blood-borne substances into the brain through the expression of specific transport proteins in the plexus epithelium. These two types of functions of the CSF and the choroid plexus further emphasize the increasingly diffuse border between the classical information transmission taking place in synapses and the broadcasting of information that characterizes endocrine systems.

ACKNOWLEDGMENTS

This work was supported by the Swedish Medical Research Council, grant no. 14X-732/5680 and 21X-07517.

REFERENCES

1. Heimer L. *The human brain and spinal cord.* New York/Berlin/Heidelberg/Tokyo: Springer-Verlag, 1983;37–48.
2. Mark AS, Feinberg DA, Brant-Zawadzki MN. Changes in size and magnetic resonance signal intensity of the cerebral CSF spaces during the cardiac cycle as studied by gated, high-resolution magnetic resonance imaging. *Invest Radiol* 1987;22:290–297.
3. McComb JG. Recent research into the nature of cerebrospinal fluid formation and absorption. *J Neurosurg* 1983;59:369–383.
4. Cserr HF, DePasquale M, Pullen CS. Convection of cerebral interstitial fluid and its role in brain volume regulation. *Ann NY Acad Sci* 1986;481:123–136.
5. Rowland LP. Blood-brain barrier, cerebrospinal fluid, brain edema, and hydrocephalus. In: Kandal ER, Schwartz JH, eds. *Principles of neural science,* 2nd ed. New York/Amsterdam/Oxford: Elsevier, 1985;837–844.
6. Brightman MW. The anatomic basis of the blood-brain barrier. In: Neuwelt EA, ed. *Implications of the blood-brain barrier and its manipulation,* vol. 1. New York/London: Plenum, 1989;53–83.
7. Pazos A, Hoyer D, Palacios JM. The binding of serotonergic ligands to the porcine choroid plexus: characterization of a new type of serotonin recognition site. *Eur J Pharmacol* 1985; 106:539–546.
8. Yagaloff KA, Hartig PR. ^{125}I-Lysergic acid diethylamide binds to a novel serotonergic site on rat choroid plexus epithelial cells. *J Neurosci* 1985; 5:3178–3183.
9. Pazos A, Probst A, Palacios JM. Serotonin receptors in the human brain—III. Autoradiographic mapping of serotonin-1 receptors. *Neuroscience* 1987;21:97–122.
10. Hoffman BJ, Mezey E. Distribution of serotonin 5-HT_{1C} receptor mRNA in adult rat brain. *FEBS Lett* 1989;247:453–462.
11. Molineaux SM, Jessell TM, Axel R, Julius D. 5-HT_{1C} receptor is a prominent serotonin receptor subtype in the central nervous system. *Proc Natl Acad Sci USA* 1989;86:6793–6797.
12. Fozard JR, Gray JA. 5-HT_{1C} receptor activation: a key step in the initiation of migraine? *TIPS* 1989;10:307–309.
13. Conn PJ, Sanders-Bush E. Agonist-induced phosphoinositide hydrolysis in choroid plexus. *J Neurochem* 1986;47:1754–1760.
14. Palacios JM, Markstein R, Pazos A. Serotonin-1C sites in the choroid plexus are not linked in a stimulatory or inhibitory way to adenylate cyclase. *Brain Res* 1986;380:151–154.
15. Hoyer D, Waeber C, Schoeffter P, Palacios JM, Dravid A. 5-HT_{1C} receptor-mediated stimulation of inositol phosphate production in pig choroid plexus. *Naunyn Schmiedebergs Arch Pharmacol* 1989;339:252–258.
16. Conn PJ, Sanders-Bush E. Relative efficacies of piperazines at the phosphoinositide hydrolysis-linked serotonergic (5-HT_2 and 5-HT_{1C}) receptors. *J Pharmacol Exp Ther* 1987;242:552–557.
17. Pritchett DB, Bach AWJ, Wozny M, et al. Structure and functional expression of cloned rat serotonin 5-HT_2 receptor. *EMBO J* 1988;7:4135–4140.
18. Julius D, MacDermott AB, Axel R, Jessel TM. Molecular characterization of a functional cDNA encoding the serotonin 1C receptor. *Science* 1988;241:5558–5564.
19. Moskowitz MA, Liebmann JE, Reinhard JF Jr, Schlossberg A. Raphe origin of serotonin-containing neurones within choroid plexus of the rat. *Brain Res* 1979;169:590–595.
20. Steinbusch HWM. Serotonin-immunoreactive neurons and their projections in the CNS. In: Björklund A, Hökfelt T, Kuhar MJ, eds. *Handbook of chemical neuroanatomy, vol. 3. Classical transmitters and transmitter receptors in the CNS, part II.* Amsterdam/New York/Oxford: Elsevier, 1984;68–125.
21. Chan-Palay V. Serotonin axons in the supra- and subependymal plexuses in the leptomeninges; their roles in local alterations of cerebrospinal fluid and vasomotor activity. *Brain Res* 1976; 102:103–130.
22. Tochino Y, Schanker LS. Transport of serotonin and norepinephrine by the rabbit choroid plexus *in vitro. Biochem Pharmacol* 1965;14:1557–1566.
23. Lindvall M, Hardebo JE, Owman C. Barrier mechanisms for neurotransmitter monoamines in the choroid plexus. *Acta Physiol Scand* 1981;108:215–221.
24. Chang JY, Owman C, Steinbusch HWM. Evidence for coexistence of serotonin and noradrenaline in sympathetic nerves supplying brain vessels of guinea pig. *Brain Res* 1988;438: 237–246.
25. Chang JY, Ekblad E, Kannisto P, Owman C. Serotonin uptake into cerebrovascular nerve fibres of rat, visualization by immunohistochemistry, disappearance following sympathectomy, and release during electrical stimulation. *Brain Res* 1989;492:79–88.
26. Jackowski A, Crockard A, Burnstock G. Ultrastructure of serotonin-containing nerve fibres in the middle cerebral artery of the rat and evidence for its localisation within catecholamine-containing nerve fibres by immunoelectron microscopy. *Brain Res* 1988;443:159–165.
27. Conn PJ, Janowsky A, Sanders-Bush E. Denervation supersensitivity of 5-HT_{1c} receptors in rat choroid plexus. *Brain Res* 1987;400:396–398.
28. Giordano J, Hartig PR. *In vivo* labelling indi-

cates that CSF serotonin activates the 5-HT_{1C} receptor on the apical surface of the choroid plexus epithelium. *Neuroscience* 1987;22(Suppl.): 75.

29. Pappenheimer JR, Heisey SR, Jordan EF, Downer JD. Perfusion of the cerebral ventricular system in unanesthetized goats. *Am J Physiol* 1962; 203:763–774.
30. Lindvall M, Edvinsson L, Owman Ch. Effect of sympathomimetic drugs and corresponding receptor antagonists on the rate of cerebrospinal fluid production. *Exp Neurol* 1979;64:132–145.
31. Heisey SR, Held D, Pappenheimer JR. Bulk flow and diffusion in the cerebrospinal fluid system of the goat. *Am J Physiol* 1962;203:775–781.
32. Lindvall-Axelsson M, Mathew C, Nilsson C, Owman Ch. Effect of 5-hydroxytryptamine on the rate of cerebrospinal fluid production in rabbit. *Exp Neurol* 1988;99:362–368.
33. Lindvall-Axelsson M, Nilsson C, Owman Ch, Svensson P. Involvement of 5-HT_{1C}-receptors in the production of CSF from the choroid plexus. In: Seylaz J, ed. *Neurotransmission and cerebrovascular function*. New York/Amsterdam/ Oxford: Elsevier, 1989;237–240.
34. Maeda K. Monoaminergic effect on cerebrospinal fluid production. *Nihon Univ J Med* 1983; 25:155–174.
35. Faraci FM, Mayhan WG, Heistad DD. Effect of serotonin on blood flow to the choroid plexus. *Brain Res* 1989;478:121–126.
36. Tsutsumi M, Skinner MK, Sanders-Bush E. Transferrin gene expression and synthesis by cultured choroid plexus epithelial cells. Regulation by serotonin and cyclic adenosine 3′,5′-monophosphate. *J Biol Chem* 1989;264:9626–9631.
37. Lübbert J, Snutch TP, Dascal N, Lester HA, Davidson N. Rat brain 5-HT_{1C} receptors are encoded by a 5-6 kbase mRNA size class and are functionally expressed in injected *Xenopus* oocytes. *J Neurosci* 1987;7:1159–1165.
38. Anderson GM, Teff KL, Young SN. Serotonin in cisternal cerebrospinal fluid of the rat: measurement and use as an index of functionally active serotonin. *Life Sci* 1987;40:2253–2261.
39. Gjerris A, Stub Sørensen A, Rafaelsen OJ, Werdelin L, Alling C, Linnoila M. 5-HT and 5-HIAA in cerebrospinal fluid in depression. *J Affective Disord* 1987;12:13–22.
40. Hoyer D, Srivatsa S, Pazos A, Engel G, Palacios JM. [^{125}I]LSD labels 5-HT_{1C} recognition sites in pig choroid plexus membranes. Comparison with [^{3}H]mesulergine and [^{3}H]5-HT binding. *Neurosci Lett* 1986;69:269–274.
41. Phillips PA, Abrahams JM, Kelly J, et al. Localization of vasopressin binding sites in rat brain by *in vitro* autoradiography using a radioiodinated V_1 receptor antagonist. *Neuroscience* 1988;27:749–761.
42. Bianchi C, Gutkowska J, Ballak M, et al. Radioautographic localization of ^{125}I-atrial natriuretic factor binding sites in the brain. *Neuroendocrinology* 1986;44:365–372.
43. Mantyh CR, Kruger L, Brecha NC, Mantyh PW. Localization of specific binding sites for atrial natriuretic factor in the central nervous system of rat, guinea pig, cat and human. *Brain Res* 1987;412:329–342.
44. Baldino F Jr, Ruth JL, Davis LG. Nonradioactive detection of vasopressin mRNA with *in situ* hybridization histochemistry. *Exp Neurol* 1989;104:200–207.
45. Standaert DG, Needleman P, Saper CB. Organization of atriopeptin-like immunoreactive neurons in the central nervous system of the rat. *J Comp Neurol* 1986;253:315–341.
46. Schwartz WJ, Reppert SM. Neural regulation of the circadian vasopressin rhythm in cerebrospinal fluid: a pre-eminent role for the suprachiasmatic nuclei. *J Neurosci* 1985;5:2771–2778.
47. Barreca T, Franceschini R, Siani C, et al. Diurnal pattern of plasma and cerebrospinal fluid vasopressin levels in hydrocephalic patients: absence of a circadian rhythm and of a correlation between plasma and cerebrospinal fluid variations. *Horm Res* 1988;30:28–31.
48. Marumo F, Masuda T, Ando K. Presence of the atrial natriuretic peptide in human cerebrospinal fluid. *Biochem Biophys Res Commun* 1987;143: 813–818.
49. Masuda T, Ando K, Marumo F. The existence of low concentrations of atrial natriuretic peptide (ANP) in canine cerebrospinal fluid which does not correlate with plasma ANP levels. *Neurosci Lett* 1988;88:93–99.
50. Marumo F, Masuda T, Masaki Y, Ando K. The presence of atrial natriuretic peptide in canine cerebrospinal fluid and its possible origin in the brain. *J Endocr* 1988;119:127–131.
51. Liszczak TM, Black PM, Foley L. Arginin vasopressin causes morphological changes suggestive of fluid transport in rat choroid plexus epithelium. *Cell Tissue Res* 1986;246:379–385.
52. Szmydynger-Chodobska J, Chodobski A, Szczepanska-Sadowska E. Effect of vasopressin on cerebrospinal fluid formation. *Eur J Clin Invest* 1987;17(2 part 2):54.
53. Szczepanska-Sadowska E, Szmydynger-Chodobska J, Chodobski A. Effect of vasopressin on cerebrospinal fluid formation and composition under aniisoosmotic conditions. *Eur J Clin Invest* 1987;18(2 part 2):27.
54. DePasquale M, Cserr HF. Regulation of brain sodium content during hypernatremic stress in the Brattleboro rat. *Ann NY Acad Sci* 1987;529: 283–285.
55. Steardo L, Nathanson JA. Brain barrier tissues: end organs for atriopeptins. *Science* 1987; 235:470–473.
56. Tsutsumi K, Niwa M, Himeno A, et al. α-Atrial natriuretic peptide binding sites in the rat choroid plexus are increased in the presence of hydrocephalus. *Neurosci Lett* 1988;87:93–98.
57. Decker JF, Quay WB. Stimulatory effects of melatonin on ependymal epithelium of choroid plexuses in golden hamsters. *J Neural Transm* 1982;55:53–67.

58. Schreiber G, Aldred AR. Pathophysiological aspects of plasma protein formation in the choroid plexus. In: Johansson B, Owman C, Widner H, eds. *Pathophysiology of the blood-brain barrier: long term consequences of barrier dysfunction for the brain*. Amsterdam/New York/Oxford, Elsevier, 1990 (*in press*).
59. Dickson PW, Howlett GJ, Schreiber G. Rat transthyretin (prealbumin). Molecular cloning, nucleotide sequence, and gene expression in liver and brain. *J Biol Chem* 1985;260:8214–8219.
60. Aldred AR, Grimes, A, Schreiber G, Mercer JFB. Rat ceruloplasmin. Molecular cloning and gene expression in liver, choroid plexus, yolk sac, placenta, and testis. *J Biol Chem* 1987;262:2875–2878.
61. Aldred AR, Dickson PW, Marley PD, Schreiber G. Distribution of transferrin synthesis in brain and other tissues in the rat. *J Biol Chem* 1987;262:5293–5297.
62. Stylianopoulou F, Herbert J, Soares MB, Efstratiadis A. Expression of the insulin-like growth factor II gene in the choroid plexus and the leptomeninges of the adult rat central nervous system. *Proc Natl Acad Sci USA* 1988;85:141–145.
63. Ichimiya Y, Emson PC, Northrop AJ, Gilmour RS. Insulin-like growth factor II in the rat choroid plexus. *Mol Brain Res* 1988;4:167–170.
64. Dickson PW, Aldred AR, Menting JGT, Marley PD, Sawyer WH, Schreiber G. Thyroxine transport in choroid plexus. *J Biol Chem* 1987;262:13907–13915.
65. Schreiber G. Synthesis, processing, and secretion of plasma proteins by the liver and other organs and their regulation. In: *The plasma proteins, vol. 5*. Academic Press, 1987;293–363.

Volume Transmission in the Brain: Novel Mechanisms for Neural Transmission,
edited by K. Fuxe and L.F. Agnati.
Raven Press, Ltd., New York © 1991.

25

Activity-Related Ionic and Volume Changes in Neuronal Microenvironment

E. Syková

Laboratory of Neurohumoral Regulation, Institute of Physiological Regulations, Czechoslovak Academy of Sciences, Prague, Czechoslovakia

The conception of the neuronal and glial cells microenvironment is largely based on research results of the past 15 years. The contemporary view is that the microenvironment is a dynamic structure whose composition and volume can alter and thereby significantly influence functions of neurons and glia and the signal transmission in the nervous system. The transmembrane flux of ions that is produced by neuronal and fiber activity will modify the ionic composition of the fluid in the surrounding extracellular space (ECS). During action potential or during excitatory postsynaptic potential (EPSP), Na^+ and/or Ca^{2+} enters the intracellular compartment and K^+ leaves it. In addition to the typical Na^+ and K^+ exchange, there are dynamic changes in the concentrations of Ca^{2+}, Cl^-, and H^+.

The microenvironment of nerve cells is usually identified with the ECS. Actually, however, the question is not one of relation between these two compartments, but rather that the nervous system is divided into neurons, glial cells, and the ECS (i.e., three compartments with complex interrelations). The close proximity of neuronal elements in the CNS and the narrowness of the intercellular spaces provide a basis not only for interaction between the cells themselves, but also between the elements and their microenvironment. Chemical substances, ions, neurotransmitters, and enzymes that can cross the cell membranes can easily find their way through the ECS to adjacent cellular elements. In this way the microenvironment can ensure nonsynaptic communication between the relevant neurons. Signalization can be coded by modulation of the chemical composition of the ECS in the vicinity of the cell membrane and does not require classic connection by axons, dendrites, and synapses. This conception also includes the transport of chemical substances through the intracellular environment of cellular elements like glial cells, so that the spatial relation between nerve cells acquires new significance.

This review deals with ionic changes in the neuronal microenvironment, including changes in the extracellular pH (pH_e) and their mechanisms, and with the physiological significance of these changes. It contributes new findings on dynamic changes in the volume of the ECS and their possible physiological significance. This research was primarily made possible by a new method, allowing the measurement of the dynamic ionic changes in living tissue in the immediate vicinity of nerve cells [i.e., the method using ion-selective microelectrodes (ISMs)] (1–3). The questions have been analyzed in the brain, at spinal cord level, and also in the sensory organs in experiments on various animals *in vivo* as well as *in vitro*. The findings presented in this review support the suggestion of the role of the microenvironment as a communication and modulation channel whose ionic and chemical changes, volume changes, and anatomical structure significantly influence the complex function of the neurons and glial cells in nervous tissue and the sensory organs.

ION-SELECTIVE MICROELECTRODES

ISMs are miniature potentiometric sensors whose main component part is an electrochemical membrane. The first ISMs were made of ion-selective glass (1). Their disadvantage was their relatively small selectivity and the large size of the active surface. Today, ISMs based on liquid membranes are the most perfect sensors for measuring dynamic changes in the concentration of biologically important ions *in vivo*. Their theory and preparation have been described in several monographs (4–8).

When introduced into tissues or solutions in which the activity of the respective ion is to be measured, a Nernst potential develops across the ion-exchanger membrane (i.e., one measures a potential that changes with the activity of the ion for which the ion exchanger is selective). However, when the ISM is inserted into neural tissue, it records not only electrical potential differences generated mainly by free ion concentration but also all types of electrical activity (e.g., membrane and action potentials) and slow potentials.

Distortion of the microelectrode potential (which provides information about actual and transient changes of a given ion concentation) by electrical activity in the tissue can be avoided by placing the reference electrode as close to the tip of the ISM as possible. This requirement is best fulfilled by the double-barrel glass micropipette (Fig. 1), which has in one channel a liquid ion exchanger while the other channel serves as the reference electrode (usually filled with saline). Because the reference pipette also records electrical activity in the tissue, the signal from it is used to cancel out the undesired component. Suitable liquid-ion exchangers (natural and synthetic ionophores) exist for K^+, Na^+, Ca^{2+}, Cl^-, and a few ions not normally present in neural tissue [e.g., Mg^{2+}, Li^+, tetraethylammonium (TEA^+), and tetramethylammonium (TMA^+)] (7). Electrode impedance always exceeds 10^8 ohms. Therefore a low noise level differential amplifier is required, with an input resistance of about 10^{14} ohms, compensation of capacitance by negative feedback, and an arrangement for measurement of electrode impedance. Currently used ISMs can have a response time of only several milliseconds (depending on the exchanger used) and relatively high selectivity to only one ion. They are therefore ideal for measurements of fast changes of ion activity in small volumes of tissues.

Measurements of Dynamic Changes in ECS Volume

The ISMs are currently also successfully used to study how substances migrate in the ECS (see Chap. 21, *this volume*). These studies lead to the estimation of the ECS size [i.e., volume fraction (α) and the ECS tortuosity (λ)], a factor increasing the length of diffusion pathways (8–10). In other words, the geometry of the ECS affects ion movements in at least two ways: by ECS volume and by ECS tortuosity. Studies using ISMs enable us to investigate ion diffusion in specific regions of the CNS in relation to time.

Dynamic changes in the size of the ECS are studied by means of the intophoretic administration of ions that do not cross the cell membranes and therefore remain in the ECS. Their concentration in the ECS is in inverse proportion to the size of the ECS

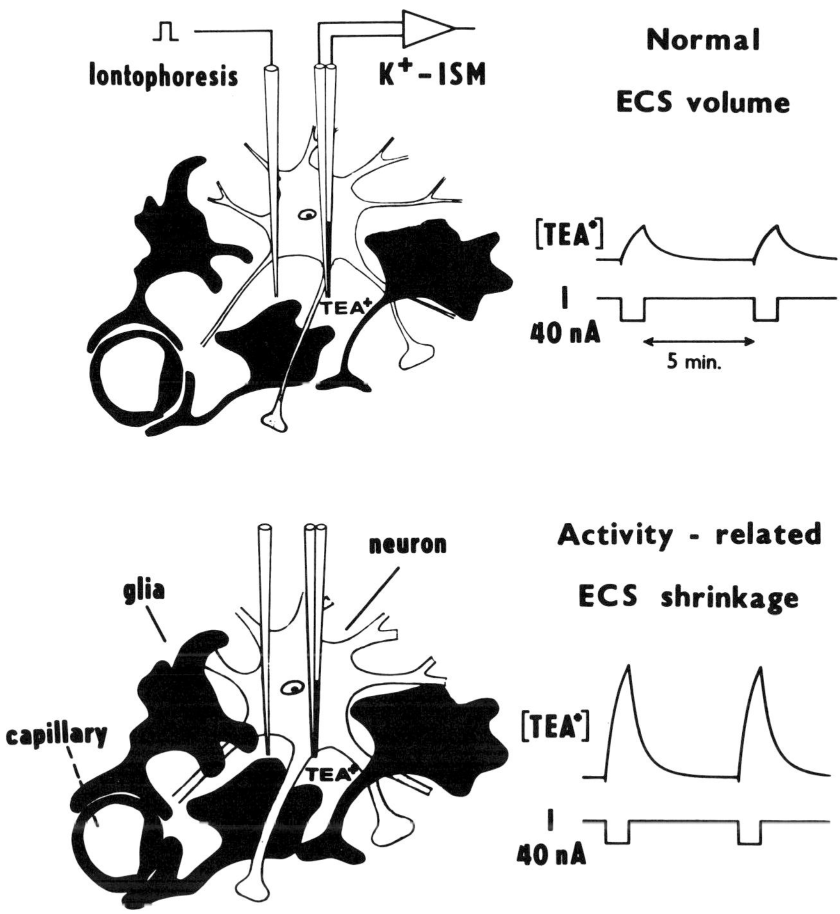

FIG. 1. Scheme of experimental set-up when measuring dynamic ionic and volume changes in extracellular space (ECS). Details are given in text. TEA^+, tetraethylammonium; K^+-ISM, K^+-selective microelectrode.

(9). The ion exchanger for K^+ (Corning 477317) is highly sensitive to tetraalkylammonium ions and to choline, which, in small concentrations, are not toxic and do not cross cell membranes (10,11). Choline, TEA^+, or TMA^+ are therefore used for testing changes in the size of the ECS. Figure 1 is a schematic representation of the principle of the measurement of dynamic changes in the size of the ECS in the CNS. Using a current of 10 to 100 nA, these ions are administered 5 to 60 sec into the ECS with a single- or multichannel iontophoretic pipette. The tip of the K^+-selective microelectrode (K^+-ISM) with which we record ion concentration changes during administration is 40 to 200 μm distant from the tip of the iontophoretic pipette. If this distance needs to be known exactly (for the determination of changes in the tortuosity (λ) or actual volume fraction (α) of the ECS), the two electrodes must be glued together with cement and the intertip distance measured exactly under the microscope. After they have been joined together, the electrodes are introduced into the nervous tissue. The ions are administered at regular intervals (1–10 min), always with the same current. The few values obtained before starting stimulation or the relevant experiment are regarded as the control if they do not differ from one another by more than 5%. The change in the size of the ECS can be computed from the changes in the concentration of the given ions, according to the formula

$$\text{Percentual decrease in ECS} = \left(1 - \frac{\Delta\,[\mathrm{TEA}^+]\ \text{before activity}}{\Delta\,[\mathrm{TEA}^+]\ \text{after activity}}\right) \times 100$$

The same method can be used for computing and determining changes in α and λ. For determination, however, it is best to administer the ions into the nervous tissue for 50 to 60 sec and to know the intertip distance of the two electrodes. When measurements in the nervous tissue have been completed, the procedure is repeated after withdrawing the microelectrodes into a thick layer of 0.3% agar poured over the surface of the nervous tissue. Measurement of the concentration change in the agar gives us the free-diffusion values for the relevant ion, in which $\alpha = 1$ and $\lambda = 1$. K^+-ISMs are calibrated in appropriate calibration solutions containing TEA^+, TMA^+, or choline. The diffusion coefficient values, α and λ, in the brain or spinal cord are determined by computation after Nicholson and Phillips (10,12), (see also Chap. 21, *this volume*). The results of these studies show that if we incorporate factors α and λ in Fick's law, diffusion in the CNS can be described fairly satisfactorily. The micropipette from which TEA^+ or TMA^+ is released iontophoretically acts as the source $Q = In/F$. The appropriate solution of the diffusion equation is

$$C\,(r,t) = \frac{Q\,\lambda^2}{4\pi D\alpha r}\,\mathrm{erfc}\left(\frac{r\lambda}{2\sqrt{Dt}}\right)$$

in which C is the concentration; I, the iontophoretic current increment; n, the transport number; F, Faraday's constant; r, the distance of ISM from the ion source (iontophoretic pipette); D, the diffusion coefficient; t, the time; and erfc, complementary error function.

EXTRACELLULAR VARIATIONS IN K⁺ ACTIVITY

In the first experiments using K^+-ISMs, it was already found that extracellular K^+ concentration, $[K^+]_e$, in the ECS of the gray matter of the dorsal spinal horns in rats increased by several mmol/l during stimulation of the sciatic nerve (13,14). Since then, increases in $[K^+]_e$ have been described in whole series of studies on the brain, spinal cord, peripheral nerves, and sensory organs of different animal species, *in vivo* and *in vitro,* after both electrical and adequate stimulation and during the pathological events (for review, see refs. 15–18). The results of all those studies can be summed up as follows:

- K^+ are released from active neurons (or receptor cells) and from primary afferent fibers (i.e., postsynaptically as well as presynaptically);
- K^+ accumulation in ECS depends on the frequency, intensity, and duration of stimulation;
- K^+ are redistributed and cleared from the ECS chiefly through activation of the membrane Na^+/K^+ pump and by the glial cells;
- K^+ that accumulates temporarily in ECS depolarizes neurons and nerve fibers in the vicinity, affects their excitability and transmitter release, and modulates the efficacy of impulse transmission; and
- K^+ released from active elements depolarizes glial cells and affects their function.

Activity-related Changes in $[K^+]_e$

The resting $[K^+]_e$ in CNS can significantly differ from that in cerebrospinal fluid. Moreover, the level of $[K^+]_e$ varies in different individuals in the same animal species as well as in different parts of the nervous system in the same animal. It has been demonstrated in the rat spinal cord

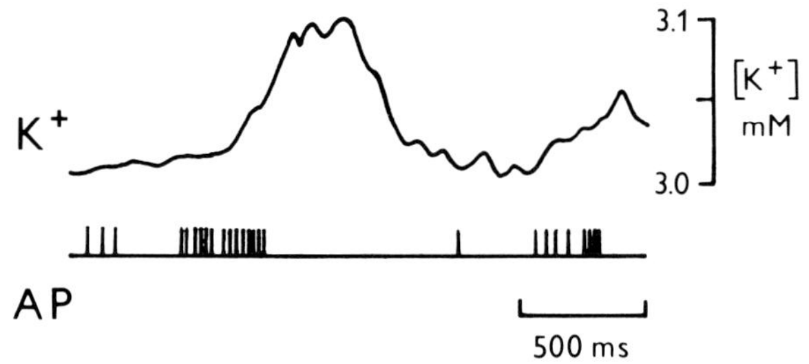

FIG. 2. *Top line:* elevation of $[K^+]_e$ in unstimulated reticular formation of rat associated with bursts of firing of cells. *Bottom line:* neuronal firing recorded by reference barrel. AP, action potentials. (Modified from ref. 21, with permission.)

(19), in the frog spinal cord (20), in the rat mesencephalic reticular formation (21,22), and in the rat cortex (23) that the level of $[K^+]_e$ varies between 2.6 and 5.0 mmol/liter in the unstimulated animal, apparently caused by differences in the spontaneous neuronal activity level.

The mesencephalic reticular formation (MRF) is known as a structure with a high spontaneous activity level. A great many neurons in the MRF display spontaneous activity characterized by an incidence of bursts of action potentials (i.e., of groups of a few action potentials following each another in shorter intervals than the intervals between the discharges). Dynamic increases in $[K^+]_e$ by 0.02 to 0.2 mmol/liter in the immediate vicinity of individual neurons of the MRF were found (Fig. 2). Their amplitude was dependent on the distance of K^+-ISM from the neuronal membrane and on the spontaneous activity frequency and duration (21). If the averaging technique was used, the increase in $[K^+]_e$ could be recorded after a single action potential. The maximum $[K^+]_e$ change recorded after a single action potential appeared with latency of 5 to 20 msec after termination of concomitantly recorded extracellular action potential and the return to resting level took about 60 msec. The observed changes amounted to only 0.01 to 0.02 mmol/liter; the smaller the distance of the electrode from the neuron, the greater the size of the change in $[K^+]_e$ and the sooner the maximum of the increase in $[K^+]_e$ is reached. The possibility of bringing the microelectrode right up to the neuron or even into the intercellular space about 15 nm width is limited by the size of the microelectrode. The $[K^+]_e$ changes in the narrow intercellular cleft after a single action potential are therefore in actual fact several times greater than the values we measured.

The interneurons in the dorsal spinal horns of mammals and amphibians also often have a high spontaneous activity level, and the $[K^+]_e$ is therefore not the same everywhere (see Fig. 3). We found that the $[K^+]_e$ level in the submeningeal fluid was close to 3.0 mmol/liter and the resting level in the superficial layers in the dorsal spinal horns and in ventral spinal horns was hardly any different. In deeper layers of the dorsal spinal horns and in the intermediate region (i.e., around the interneurons, however, the resting $[K^+]_e$ value is 0.5 to 1.5 mmol/liter higher (20,21). In experiments on the isolated frog spinal cord, it was demonstrated that these depth differences in the resting level were actually the outcome of spontaneous activity of the interneurons in the dorsal horns, because after a block of synaptic transmission or temperature changes (procedures that caused the spontaneous activity to disappear), these differences also disappeared (20).

Elevations in $[K^+]_e$ are associated with neural activity evoked by various kinds of stimulation. The main sources of K^+ are stimulated neurons, unmyelinated fibers, and unmyelinated terminals of myelinated axons. The reason for the increase is obvious, as K^+ is released with each impulse, and therefore ongoing activity results in

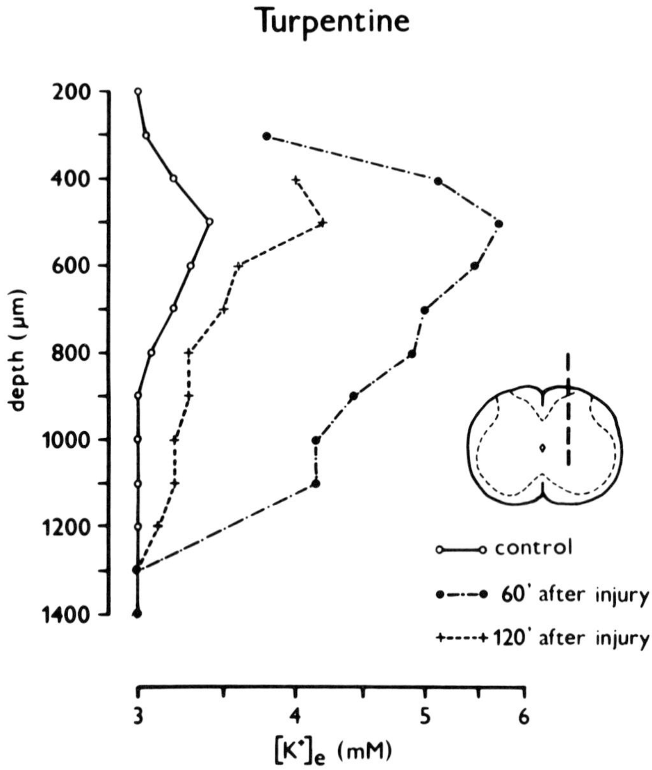

FIG. 3. Effect of subcutaneous injection of turpentine on resting $[K^+]_e$ level at different depths in segment L_4 of rat spinal cord. Control: $[K^+]_e$ depth profile before injury. $[K^+]_e$ was raised 60 min after turpentine by about 3 mmol/liter. A partial return to resting values was found after 120 min. (Modified from ref. 19, with permission.)

substantial K^+ accumulation in narrow intercellular clefts. For example, it was shown that K^+ accumulates preferentially in deeper layers of the dorsal horn of the mammalian as well as amphibian spinal cord after electrical stimulation of peripheral nerves or adequate stimulation of the skin. K^+ activity increase was also found in lumbar segments in response to electrical stimulation of somatosensory cortex or the caudal part of the medulla. The observed increase in $[K^+]_e$ varies between about 0.1 to 6.0 mmol/liter and depends on stimulation frequency, intensity, and duration. An increase of 0.1 to 0.5 mmol/liter has been found after a single electrical stimulus applied to the dorsal root, to peripheral nerve, or after a single adequate stimulus (light touch or pinch) applied to the skin on the hindlimb (Fig. 4). Repeated stimulation leads to summation of responses. An increase of 3 to 6 mmol/liter found in the rat and cat spinal cord after 15 sec of repetitive electrical stimulation (30–100 Hz) of peripheral nerves of the hindlimb. When stimulation is continued, no greater changes in K^+ are found because a steady state is established, which is the result of concurrent release and clearance of K^+ (Fig. 4). This so-called "ceiling level" can be exceeded only during epileptic seizures, anoxia, or spreading depression, when $[K^+]_e$ can reach levels as high as 20 to 50 mmol/liter (for review, see ref. 18).

A long-term increase in $[K^+]_e$ in the lower dorsal horn by 0.3 to 3.0 mmol/liter was evoked by chemical and thermal injury (i.e., by subcutaneous injection of formalin

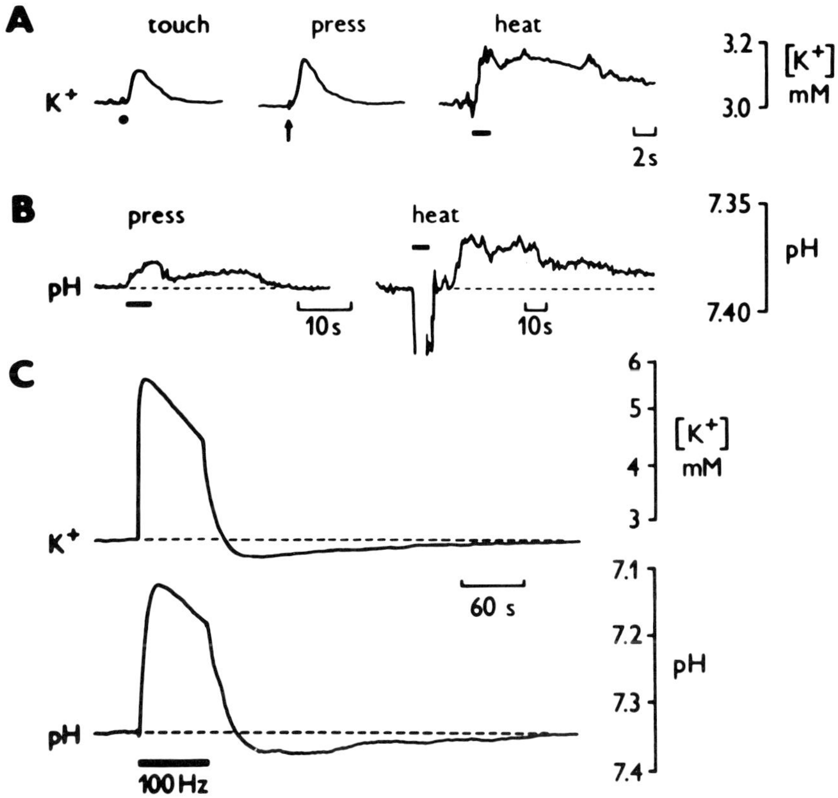

FIG. 4. (**A**) K^+ changes and (**B**) pH_e changes in rat spinal cord in response to adequate stimulation [i.e., to light touch, strong pressure of the toes, and application of about 1 ml of hot water to the skin of the paw (heat)]. **C:** $[K^+]_e$ and pH_e changes in dorsal horn of segment L_4 of rat spinal cord evoked by repetitive bipolar electrical stimulation with acupuncture needles (100 Hz, 60 sec) in plantar muscles of the ipsilateral hind paw. (Modified from refs. 19 and 60, with permission.)

or turpentine into the rat hind paw or by thermal injury of the hind paw) (19). The increase began 5 to 15 min after injury and persisted for more than 2 hr (Fig. 3). The K^+ increases evoked by injury as well as by acute nociceptive stimuli were blocked by preinjecting the hind paw by local anesthetic procaine. However, when procaine was applied 20 min or later after injury, the evoked long-term rise in $[K^+]_e$ was not blocked. This shows that the long-term increase in $[K^+]_e$ results from self-sustained abnormal neuronal firing induced in dorsal horns by injury.

Similar activity-related changes in $[K^+]_e$ have been found in brain and in receptor organs: in cerebral cortex during stimulation of various cortical pathways (24–29), in cerebellum after climbing fiber-related activation of Purkinje cells or when the parallel fiber afferents are directly stimulated (30–32), in hippocampus in response to electrical stimulation of fimbria (33–35), in the organ of Corti during acoustic stimulation (36), and in the retina during light stimulation (37,38).

Clearance of Accumulated K^+ from ECS

Active neurons that lose K^+ and thereby induce a change in the ionic composition of the extracellular fluid can significantly affect the function of neighboring neuronal elements. Although this may be an important way in which neurons can interact and integrate their activity for a time, stability of nervous tissue function requires prompt and exact renewal of the composition of the neuronal environment. In fact, there are two problems that need to be resolved simultaneously: (1) Excessive and prolonged changes in $[K^+]_e$ must be prevented to avoid disturbances of function and irreversible changes; and (2) the neurons must regain the K^+ released during their activity, otherwise they would no longer be able to function. Many observations support the conclusion that excessive changes in $[K^+]_e$ are cleared by active uptake across neuronal membranes. However, fluctuations in $[K^+]_e$ are also compensated by passive transport across the cell membrane, by voltage gradients within the tissue, by diffusion in the ECS, by transport across the blood-brain barrier, and by water movements between different compartments (18,39).

The conclusion that K^+ is cleared by means of active transport is well-supported (40–42). For example, with continuous stimulation $[K^+]_e$ falls from the ceiling level during stimulation; when stimulation stops, $[K^+]_e$ decreases below "resting" level (i.e., there is a poststimulation "undershoot" (Fig. 4). In particular, the lowering of $[K^+]_e$ below the baseline suggests the participation of active transport processes in the redistribution of accumulated K^+. Its metabolic character was demonstrated by high sensitivity to oxygen supply, blood pressure, body temperature, and anesthetic drugs. Furthermore, the decrease of $[K^+]_e$ during stimulation as well as the undershoot were blocked by ouabain, a known inhibitor of Na^+/K^+ pump activity.

Another mechanism that contributes to K^+ clearance is a temporary storage in other nonactivated cells (i.e., inactive neurons or glial cells). In glial cells two types of uptake processes must be considered: (1) uptake of K^+ driven by Na^+/K^+ pump, and (2) K^+ spatial buffering. The latter operates as follows. At a site of local buildup of $[K^+]_e$, glial cells are depolarized; therefore a current develops between these depolarized glial cells and more distant glial cells to which they are electrically coupled. This current flow drives K^+ into the glia in regions where K^+ activity is higher and out of the distant glia where K^+ activity is low (43,44). Two properties of glial cells make them suitable for K^+ clearance: (1) Glial cell membranes are almost exclusively permeable to K^+, and (2) low-resistance interconnections (gap junctions) exist between the glial cells, which permit K^+ to move through the extensive glial syncytium. The experiments on the retina (38), the brain (45–47) and the spinal cord (48) support the role of glial cells spatial buffer mechanism in the clearance of K^+.

The removal of K^+ from ECS by ouabain-sensitive uptake has been demonstrated on tissue cultures of astrocytes (49) as well as of oligodendrocytes (50). Another route for K^+ clearance is diffusion through ECS. Recent studies have shown that the movement of K^+ and of other ions is influenced by geometry of the ECS. As already stated above, the K^+ diffusion in ECS is slowed down by at least two factors, tortuosity and volume fraction (9,10) (see also Chap. 21, *this volume*). Finally, it seems unlikely that K^+ exchange between the extracellular fluid and the blood could influence the time course of dynamic K^+ changes occurring during neuronal activity, because this process is extremely slow (for review, see refs. 18 and 39).

EXTRACELLULAR VARIATIONS IN Ca^{2+}, Na^+, AND Cl^- ACTIVITY

In addition to $[K^+]_e$ changes, neuronal activity is accompanied by changes in the concentration of the other ions in the microenvironment of the CNS (for review, see refs. 18 and 39). In the case of Na^+ and Cl^-, it is difficult to measure the changes directly, as the sensitivity of the relevant ISMs is relatively small in a high resting extracellular concentration of these ions. Changes in Na^+ and Cl^- can be measured only if they exceed several mmol/liter; they are thus hard to demonstrate during electrical stimulation and are impossible to demonstrate during adequate stimulation. The dramatic changes in $[Na^+]_e$ and $[Cl^-]_e$ occur during anoxia, epilepsy, and spreading depression.

A decrease in $[Ca^{2+}]_e$ was found during excitation in different parts of the brain and spinal cord, during electrical stimulation of afferent input, and in pathological states. The decrease in $[Ca^{2+}]_e$ is caused by movements of Ca^{2+} into neurons and presynaptic terminals. The resting $[Ca^{2+}]_e$ level in the brain and spinal cord of amphibians and mammals varies from 1.2 to 1.6 mmol/liter. Repetitive electrical stimulation causes a 0.1 to 0.5 mmol/liter drop in the resting level (i.e., $[Ca^{2+}]_e$ falls to less than 1.0 mmol/liter. When stimulation is discontinued, recovery takes place quickly, like recovery of the increase in $[K^+]_e$, and an "overshoot" to concentrations higher than the original resting level appears. The depth profile of stimulation-induced $[Ca^{2+}]_e$ changes in the cortex, cerebellum, thalamus, and spinal cord is the same as the depth profile of $[K^+]_e$ changes (51–53). The decrease in $[Ca^{2+}]_e$ is proportional to the intensity, frequency, and duration of stimulation. If electrical stimulation lasts longer than 10 to 30 sec, the $[Ca^{2+}]_e$ changes do not increase any further (i.e., the "ceiling" value is reached). The time course of the return of $[Ca^{2+}]_e$ changes to the resting level is similar to the one for the return of $[K^+]_e$ and pH_e changes (53).

Functional Significance of Activity-Related K^+ and Ca^{2+} Changes

It has been proposed that transient ionic changes in the extracellular microenvironment might be a powerful mechanism in the regulation of CNS function. With the exception of pathological states such as spreading depression, anoxia, and epilepsy, a change in extracellular K^+ and Ca^{2+} activity alone might serve as a signal. These two ions can influence neuronal excitability, transmitter release, intercellular communication, and glial cell function (for review, see refs. 18 and 39).

The extent to which the increase of $[K^+]_e$ could affect neuronal function has been discussed for a long time, and several mechanisms for its action have been suggested. K^+ accumulation has been accepted as one of two causal factors (together with γ-aminobutyric acid) in primary afferent depolarization, which is the mechanism underlying presynaptic inhibition. It is assumed that the increase in $[K^+]_e$ that is associated with repetitive neuronal activity reduces transmitter release by curtailing the presynaptic spike amplitude by presynaptic depolarization. Moreover, this effect of raised $[K^+]_e$ could be enhanced by concomitantly decreasing $[Ca^{2+}]_e$. Both these ionic changes would therefore affect transmitter release in the same direction. Instead of its presynaptic action, the increase in $[K^+]_e$ could influence impulse transmission postsynaptically, by depolarization of neurons and glia. K^+-mediated neuronal depolarization enhances neuronal excitability, which under some circumstances may dominate over its presynaptic inhibitory effect. K^+ accumulation may therefore function as a negative-positive feedback control system. However, the causal role of K^+ activity in learning, memory, and initiating or terminating seizures is far from clear. Similarly, some pathological states such as hypoxia, ischemia, and spreading depression probably do not result from increased $[K^+]_e$, but rather lead to K^+ accumulation in the ECS.

We may consider activity-related ionic changes from a different perspective. As demonstrated with ISMs, these changes may last for many seconds even after adequate stimuli (i.e., nociceptive stimuli, injury, acoustic stimuli, and visual stimuli). Therefore it was suggested that changes in $[K^+]_e$ and $[Ca^{2+}]_e$ may modulate nervous system activity in a manner analogous to that proposed for local hormones (i.e., neuromodulators, including neuropeptides). For example, the idea has recently developed that K^+ accumulation in the spinal dorsal horn plays some role in poststimulation analgesia (54). The antinociceptive effect of transcutaneous electrical stimulation, used in clinical practice in humans for pain relief, is well known. Its long-lasting modulatory effects may result not only from the action of endogenous opioid peptides, but also from evoked changes in extracellular ionic composition in the spinal dorsal horn.

EXTRACELLULAR pH TRANSIENTS

pH, although considerably well-regulated both in intracellular and extracellular microenvironments in excitable tissues, has been shown to change transiently under a variety of physiological and pathological conditions. The use of pH-sensitive microelectrodes enabled us to follow the dynamic changes in pH in nervous tissue resulting from neuronal activity. A large number of studies indicate that the neuronal activity gives rise to changes in extracellular pH (pH_e).

Transient alkaline–acid changes in pH_e were found in cortex (55), hippocampus (56), cerebellum (57–59), spinal cord (53,60, 61), and peripheral nerves (62,63) during electrical stimulation of afferent input, seizures, spreading depression, and ischemia (56,64–68). It has been found that polarized neurons appear to change their surface pH (69). No simple monophasic changes occurred in all these studies during neuronal activity. Alkaline and acid transients, ranging from 0.01 to 0.4 pH units, have been found to occur in different sequences and in latencies ranging from 10 msec to more than 1 min. The mechanisms of activity-related transient pH_e changes under physiological conditions *in vivo* and the mechanisms by which pH_e is stabilized are far from clear. It has been suggested in several studies that stimulation-evoked alkaline–acid shifts result from the release of metabolic products after activation of the Na/K pump. Other mechanisms may involve membrane transport of H^+ or OH^- and HCO_3^- or CO_3^{-2} by specific membrane transport systems (for review, see refs. 61 and 70).

Activity-Related Changes in pH_e

In this chapter, the results of our studies about transient pH_e changes in the spinal cord of rats *in vivo* and in the isolated spinal cord of frogs are presented and compared with findings in the brain. The pH_e resting levels studied so far in the CNS vary between 7.10 to 7.35 pH units (i.e., is slightly alkaline) (55,57,60,66,68). Regional differences in pH_e were found in spinal cord of the rat. The pH_e baseline in unstimulated spinal cords varied between 7.15 to 7.35 pH units (60). The pH_e was most acid in the lower dorsal horn (laminae III–VII), about 7.15 (i.e., about 0.2 pH units lower than that measured in the cerebrospinal fluid). The variations in pH_e baseline had a similar depth profile as the variations in $[K^+]_e$ baseline. An alkaline shift in pH_e baseline by 0.05 to 0.10 pH units was found after blockade of synaptic activity. Our findings suggest that pH_e "resting" level in the lower dorsal horn is more acid than that in the upper dorsal horn, ventral horn, and cerebrospinal fluid, respectively, because of release of H^+ or equivalent ions during spontaneous activity of spinal interneurons.

Transient acid shifts in pH_e of 0.01 to 0.05 pH units were found when single electrical pulses were applied to the peripheral nerves or when acute nociceptive stimuli (i.e., pinch, press, heat) were applied to the hind paw (Fig. 4). Peripheral tissue injury evoked by subcutaneous injection of turpentine or by application of 1 ml of hot oil onto the plantar surface of the hind paw produced a long-term decrease in pH_e baseline in the lower dorsal horn of about 0.05 to 0.1 pH units (60). The decrease in pH_e began 2 to 10 min after injury and persisted for more than 2 hr. The onset of pH_e changes had similar latencies and time course as the increase in $[K^+]_e$ (19).

Electrical nerve stimulation at a frequency of 10 to 100 Hz elicited typical biphasic alkaline–acid shifts in cerebellum (57–59), triphasic alkaline–acid–alkaline changes in the dorsal horns of rats (Fig. 4) and alkaline–acid–acid changes in the dorsal horns of isolated spinal cords of frogs (53,71). The amplitude and duration of the pH_e changes increased with the intensity, duration, and frequency of stimulation. The dominating acid shift was preceded by a small initial alkaline shift of about 0.01 pH units, which lasted 2 to 5 sec. During stimulation, pH_e decreased by 0.15 to 0.25 pH units. When the stimulation was discontinued, pH_e transients returned to the baseline, and the former acid shift was followed by an alkaline shift (alkaline undershoot) of about 0.02 to 0.05 pH units. The alkaline undershoot recorded in the rat spinal cord *in vivo* was associated with the poststimulation K^+ undershoot (Fig. 4). Although the poststimulation K^+ undershoot and alkaline undershoot are present regularly in the mammalian spinal cords *in vivo*, they are present only rarely in the frog spinal cord. The recovery in the isolated frog spinal cord was frequently not smooth, but a second spontaneous acid shift occurred during poststimulation recovery to the original pH_e baseline, which was associated with a spontaneous increase in $[K^+]_e$.

The pH_e changes are closely related to the changes in $[K^+]_e$. Any spontaneous or stimulus-evoked increase in $[K^+]_e$ was accompanied by an acid shift in pH_e. However, the rise time of the stimulation-evoked acid shift was slower than the rise time of the increase in $[K^+]_e$. The $[K^+]_e$ ceiling level in rat spinal cord was attained in about 6 to 8 sec, whereas the ceiling level of the acid shift was reached in about 15 to 20 sec. When stimulation lasted longer, a gradual decrease of both transients, $[K^+]_e$ and pH_e, occurred immediately after the ceiling levels were reached. Recovery of the $[K^+]_e$ and pH_e changes to the original baselines took about 5 min and had similar time courses. The poststimulation alkaline undershoot below the original pH_e baseline had a similar time course to the poststimulation K^+ undershoot. These results suggest that the recovery of the activity-related K^+ and pH_e changes may have common underlying mechanisms, the most important being Na^+/K^+ pump activity and perhaps glial cell spatial buffering (48,72).

To examine further the relation between $[K^+]_e$ and pH_e we superfused the isolated spinal cords of frogs with Ringer solution containing elevated $[K^+]_e$. The elevation in $[K^+]_e$ depolarized the neuronal elements, fibers and glial cells and was accompanied by prompt and progressive acid shift of about 0.2 pH units. During the superfusion, the pH_e remained acid and returned to the baseline only when $[K^+]_e$ returned to the standard bath concentration of 3 mmol/liter K^+.

It can be assumed that dynamic pH_e changes are partly of postsynaptic (i.e., neuronal) origin and partly of presynaptic origin (i.e., caused by depolarization of primary afferent fibers). The effect of a block of synaptic transmission on pH_e changes was studied in the isolated frog spinal cord by superfusing the spinal cord with Ringer solution containing only 0.5 mmol/liter $CaCl_2$ and either 2 to 4 mmol/liter $MnCl_2$ or 20 mmol/liter $MgCl_2$. The pH_e changes evoked by repetitive stimulation were markedly reduced; spontaneous activity, the initial alkaline shift, and the second poststimulation acid shift always disappeared completely and the amplitude of the acid shifts during stimulation fell to 25% to 35% (61). These results demonstrate that depolarized postsynaptic elements are responsible for most of the changes in the pH_e. Like other dynamic extracellular ionic changes, however, some of the pH_e changes are the result of depolarization of the primary afferent fibers themselves.

To clarify the mechanisms of the observed changes in pH_e we studied the effects of various inhibitors of enzymatic activity and of ion transport mechanisms across cell membranes in spinal cord (53, 60,61,71). In summary, the initial alkaline shift of about 0.01 pH units as well as the dominating acid shift of about 0.1 to 0.2 pH units were accelerated and increased by acetazolamide, showing that the high buffering capacity of the extracellular fluid may hamper the resolution of acid–base perturbations. The initial alkaline shift was significantly decreased by La^{3+}, a H^+ channel blocker. The stimulation-evoked acid shifts were blocked by amiloride, stilbene derivates (SITS and DIDS), furosemide, and La^{3+} and, therefore, have a complex mechanism that includes Na^+/H^+ exchange, Cl^-/HCO_3^- exchange, and/or $Na^+/Cl^-/H^+/HCO_3^-$ antiport, and/or Na^+-HCO_3^- cotransport, K^+-Cl^- cotransport, and H^+ efflux through voltage sensitive H^+ channels. None of the inhibitors with exception of La^{3+} significantly depressed acid shifts evoked in the isolated spinal cord by 10 mmol/liter K^+. The poststimulation alkaline shift (alkaline undershoot) was blocked by ouabain and reflects coupled clearance of K^+ and H^+ by active transport processes.

The mechanisms of pH_e changes in spinal cord may differ to a certain extent from those in brain (for discussion see refs. 57, 59, and 61) and from those under pathological conditions (68) and remain to be fully elucidated. For example, the role of glial cells in pH_e homeostasis remains to be recognized, owing to their high permeability to HCO_3^- and the nature of membrane transport processes that regulate their pH_i (73–76).

Functional Significance of pH_e Changes

The results of studies carried out during the past few years have shown that pH_e changes accompanying neuronal activity, as well as changes in pathological states, can influence impulse transmission in the CNS. Changes in pH_e of only a few tenths of pH unit have been shown to evoke marked effects on neuronal Na^+ and Ca^{2+} channels (77,78). Rapid pH_e changes over a range of a few tenths of pH unit activate Na^+ currents, which may temporarily depolarize sensory neurons (78–81); proton-induced current is carried primarily by Na^+ through Ca^{2+} channels. An increase in the pH_e probably transforms Ca^{2+} channels from voltage-gated state to proton-gated state. This alters the permeability of the Ca^{2+} channel, so that Na^+ transport has priority. The proton-activated increase in membrane permeability for Na^+, as the response to a drop in the pH_e, could act as a signal of nociceptive stimulation. These changes have been demonstrated in the sensory neurons of spinal ganglia, and the same type of receptor can be localized in nerve endings, where it acts as a pH sensor. About half the sensory neurons in the trigeminal ganglia were found to be proton-sensitive; the majority were small neurons with a diameter of less than 26 μm. Proton receptors exhibited partial or complete desensitization. These findings indicate that proton receptors can play an important role in the transmission of nociceptive information.

It has been shown in several studies that pH_e changes influence excitability of neurons (82,83). The results indicate that H^+ in the extracellular fluid has an inhibitory effect on the excitability of neurons. Moreover, the pH_e changes influence cell metabolism, proliferative capacity, cell-to-cell coupling, and glial cell function (for review, see ref. 70). The role of pH_e changes in the activity-related cell volume changes is supported by experiments on the isolated frog spinal cord. An acid change of 0.2 to 0.4 pH units in the pH of the Ringer solution reduced the ECS volume, whereas the alkaline changes increased it (A. Chvátal and E. Syková, *unpublished results*). In pathological states like ischemia, anoxia, and nervous tissue injury, acidosis is considered to be an important factor influencing the functional recovery of nervous tissue (84–87).

It can thus be concluded that activity-related dynamic changes in the pH_e can be an important modulating and controlling factor in CNS. The functional significance of dynamic pH_e changes is, however, still an open question.

DYNAMIC CHANGES IN ECS VOLUME

There is growing evidence that the dynamic changes in the ECS volume occur locally, in concert with the ionic changes associated with neuronal activity. Reversible increase in the cell size accompanied by diminution of the ECS have been demonstrated in the cerebral and cerebellar cortex during spreading depression (88,89), during epileptic activity (90,91), and after repetitive direct electrical stimulation in the cat sensorimotor cortex (46,92) in the rat optic nerve (62,93) and in the isolated frog spinal cord (94,95). Under more physiological conditions the dynamic changes in the ECS volume were studied in the rat spinal cord (96,97). In the spinal dorsal horns the absolute and relative changes in ECS volume were computed after measuring the changes in the concentration of TEA^+, TMA^+, or choline administered into the ECS with an iontophoretic pipette before and after electrical and adequate stimulation of an afferent input. In the anesthetized animals that were in good condition, with blood pressure greater than 100 mm Hg, and that were not exposed to electrical or noxious stimulation, the ECS volume fraction was 0.23 ± 0.01 and tortuosity 1.42 ± 0.05 ($n = 27$). The values did not vary significantly with depth in the spinal dorsal horn.

Long-term decrease in the ECS volume has been found in the lower dorsal horn (laminae III–V) after repetitive electrical stimulation of hind limb nerves. If a nerve was repetitively stimulated through two needles introduced under the skin of the paw, we observed up to a 30% reduction of the ECS volume in the dorsal horns of segment L_4 (Fig. 5). The decrease in the ECS volume had a characteristic depth profile. The maximum changes were always observed at sites of the greatest neuronal activity (i.e., where the maximum $[K^+]_e$, $[Ca^{2+}]_e$, and pH_e changes were found). The higher the stimulation frequency and intensity and the greater the concomitant ionic changes, the greater was the decrease in ECS volume. The decrease persisted after stimulation had been discontinued, and it regularly lasted longer than the stimulation-evoked ionic changes. Comparison of the course of $[K^+]_e$ changes with the reduction in size of the ECS showed that the latter develops more slowly than the increase in $[K^+]_e$ but often lasts much longer. The longer the stimulation time, the longer the duration of dynamic changes in ECS volume, even when no further changes occurred in $[K^+]_e$. In the isolated frog spinal cord, the decrease in ECS volume persisted for 30 to 60 min after stimulation of the dorsal root. In the spinal cord of the rat, the volume changes after the stimulation of the peripheral nerve lasted for 15 to 25 min.

The decrease in the ECS volume was also found during chronic nociceptive stimulation and peripheral injury. The decrease was evoked in rat spinal dorsal horns by chemical injury of the ipsilateral hind paw induced by subcutaneous injection of 0.1 to 0.2 ml of turpentine or by thermal injury of a small skin area on the planta evoked by application of 2 ml of hot oil. The decrease in ECS volume amounted to as much as 30% to 50% (i.e., the ECS volume decreased from about 23% to about 12%) and was greatest in places of maximal increases in neuronal activity accompanied by maximum changes in $[K^+]_e$, $[Ca^{2+}]_e$, and pH_e. The volume changes after thermal injury persisted for more than 40 min and after the injection of turpentine for more than 1 hr. The ECS volume decrease evoked by injury had a similar time course to concomitantly recorded $[K^+]_e$ and pH_e changes in the dorsal horns and to the time course of nociceptive reaction as tested on naive and unrestrained rats by measuring the duration of the licking response (19,60).

These data show that repetitive electrical stimulation of afferent input and peripheral

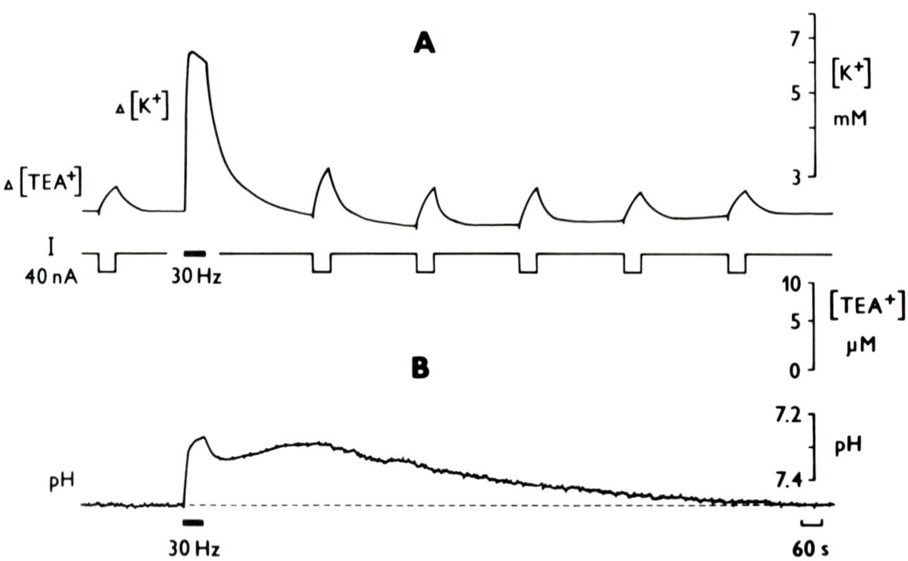

FIG. 5. A: Effect of repetitive electrical stimulation of dorsal root VIII (30 Hz, 60 sec) on ECS volume in segment VIII of the isolated frog spinal cord. Current (40 nA, the *bottom recordings*) was applied with a pipette filled with 5 mmol/liter TEA^+. *Top records* show changes in [TEA^+] during iontophoretic applications and change in $[K^+]_e$ induced by stimulation. **B:** pH_e changes induced by stimulation. Three microelectrodes (iontophoretic pipette, K^+-ISM, and pH-ISM) were inserted separately from the dorsal spinal cord surface. Recordings from a depth of about 300 μm.

tissue injury produce not only transmembrane ionic fluxes, which result in the dynamic ionic changes attributable to a long-term neuronal discharge of dorsal horn neurons, but also produce significant and long-term dynamic changes in ECS volume.

Mechanism of the ECS Volume Changes

This section deals only with the mechanisms that lead to enlargement of the intracellular compartment of the CNS accompanied by compensatory reduction of the size of the ECS. It thus does not deal with disturbances of the intravascular compartment or with the mechanism of volume changes involving an increase in total water volume in CNS. The mechanisms that have been so far suggested merely contribute basic information, and their participation in the observed dynamic activity-related volume changes require further research.

The mechanism of volume changes in CNS after repetitive stimulation can be deduced from changes in tissue osmolarity and from ion shifts across the cell membranes. Dietzel et al. (46) reported that intraneuronal osmolarity can rise during direct cortical stimulation by as much as 30 mOsmol/liter, because the number of intracellular particles rises owing to splitting of large molecules during increased metabolism of neurons. This leads to the intake of water and increase in cell volume, which, as estimated by above authors, would reduce the ECS size by less than 10%. The ion shifts across cell membranes accompanied by intake of water are, therefore, likely to be most significance in determination of the neuronal and glial cell volume. The three main types of membrane pathways for diffusible solutes may be considered (98): primary active transport, secondary active transport, and ion conductance pathways.

Primary active transport is defined as direct coupling of the energy produced by metabolism to the movement of ions across cell membranes. The Na^+/K^+ ATPase plays the important role in the regulation of cell volume. The injection of ouabain into brain tissue showed that a block of Na^+/K^+ ATPase leads to a drastic increase in cell volume (99). Impairment of Na^+ and K^+ transport in the neurons and glial cells can be, therefore, an important factor in the origin and regulation of the observed volume changes. In our experiments on the isolated frog spinal cord, the administration of ouabain led to a twofold larger decrease in the size of the ECS after stimulation compared with the situation in which Na^+/K^+ ATPase activity was not blocked; and the dynamic changes in the ECS volume lasted longer (A. Chvátal and E. Syková, *unpublished observations*). The activity of the Na^+/K^+ pump is thus probably an important factor in the mechanism of the activity-related changes and in the regulation of cell volume.

Secondary active transport, either electrogenic or electroneutral, involves various movements of ions across the membrane that are coupled to another ion. Between so-far described cotransports and antiports are the Na^+/H^+ exchanger, Cl^-/HCO_3^- exchanger, K^+-Cl^- cotransport, and others important for regulation of cell pH (for review see ref. 98). Alteration in cell pH may affect the volume of the cells by changing the membrane permeabilities or by affecting the cell metabolism. Finally, the regulation of cell volume is related to ion conductance pathways, mainly to K^+, Na^+, and Cl^-. The regulation may involve activation and inactivation of channels.

The Functional Significance of Dynamic Changes in the ECS Volume

Significant changes may occur in ECS volume, not only in pathological states, but also under physiological conditions (Fig. 6). It must be, however, admitted that we do not know the functional significance of these changes and can therefore only sum up hypotheses that have been put forward or emerge from the findings. Trubatch and colleagues (100) showed that the dendritic spines in the frog brain increased in size, and they grew shorter but thicker. Because the impedance of the elements is directly proportional to their length and in inverse proportion to their diameter, the impedance of the dendritic spines decreases and the effectiveness of synaptic transmission increases (101). The changes observed in the shape and dimension of postsynaptic elements may thus provide a morphological basis for the facilitation of transmission in nervous tissue.

In our experiments on preparations consisting of the isolated frog spinal cord connected with an isolated hind limb, we observed that decrease of the threshold for the flexor reflex after repetitive electrical transcutaneous stimulation of the hind limb had followed the same time course as the decrease in the ECS (94). These measurements indicate that poststimulation decrease of the ECS volume can be a cause of hyperexcitability and synaptic hyperactivity in the dorsal spinal horns.

According to Van Harreveld and Fifková (102), fluid uptake by dendritic spines can be the basis of long-term potentiation of transmission occurring after electrical stimulation. An increase in synaptic transmission efficacy may also be one of the mechanisms leading to plastic neuronal changes (101). The given volume changes can influence both the formation of short-term memory traces and long-term memory (e.g., through their effect on proteosynthesis) (103,104).

CONCLUSION

I have described many of the changes in the composition and volume of ECS in the CNS and have specified the possibilities of their functional significance. The basic question that still remains a subject for fur-

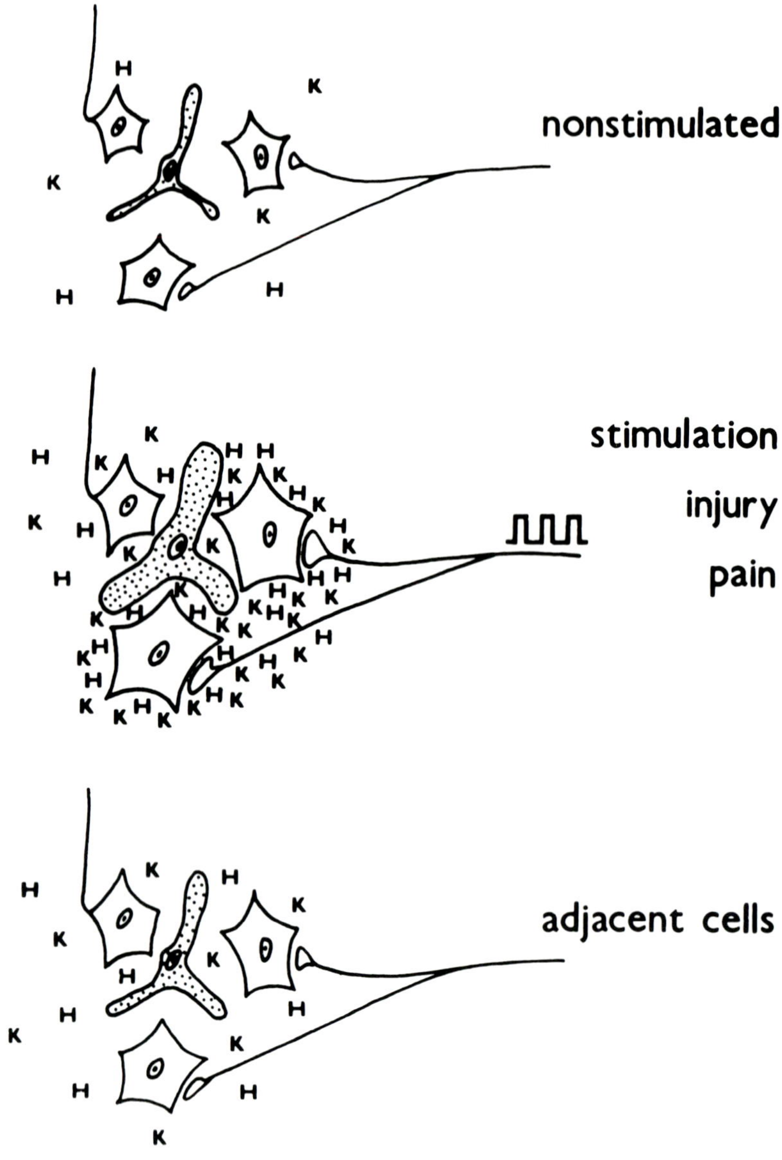

FIG. 6. Schematic illustration of dynamic changes in ECS volume and changes in extracellular K^+ and H^+ concentration in unstimulated CNS tissue, in the immediate vicinity of stimulated cells, and in the neighborhood of stimulated cells (adjacent cells).

ther research is to ascertain to what extent the ECS ensures intercellular interaction and integration. Does it provide a kind of "wireless" connection alongside the familiar and well-described connection via nerve fibers and synapses? The studies cited in this review strongly support the idea that the microenvironment of neurons and glia can serve as a fast communication channel, as well as a slow modulatory system that is capable of processing information in excitable tissues.

REFERENCES

1. Hinke JAM. Glass microelectrodes for measuring intracellular activities of sodium and potassium. *Nature* 1959;184:1257–1258.
2. Walker JL Jr. Ion specific liquid ion exchanger microelectrodes. *Anal Chem* 1971;43:89A–93A.
3. Vyskočil F, Kříž N, Bureš J. Potassium-selective microelectrodes used for measuring the extracellular brain potassium during spreading depression and anoxic depolarization in rats. *Brain Res* 1972;39:255–259.
4. Thomas RC. *Ion-sensitive intracellular microelectrodes*. London: Academia Press, 1978.
5. Koryta J. *Medical and biological applications of electrochemical devices*. Chichester: John Wiley & Sons, 1980.
6. Syková E, Hník P, Vyklický L. *Ion-selective microelectrodes and their use in excitable tissues*. New York: Plenum Press, 1981.
7. Ammann D. *Ion-selective microelectrodes*. Berlin: Springer-Verlag, 1986.
8. Nicholson C, Rice ME. Use of ion-selective microelectrodes and voltammetric microsensors to study brain cell microenvironment. In: Boulton AA, Baker GB, Walz W, eds. *Neuromethods: the neuronal microenvironment*. New York: The Humana Press, 1988;247–361.
9. Nicholson C, Phillips JM, Gardner-Medwin AR. Diffusion from an iontophoretic point source in the brain: role of tortuosity and volume fraction. *Brain Res* 1979;169:580–584.
10. Nicholson C, Phillips JM. Ion diffusion modified by tortuosity and volume fraction in the extracellular micro-environment of the rat cerebellum. *J Physiol* 1981;321:225–257.
11. Kříž N, Syková E. Sensitivity of K^+-selective microelectrodes to pH and some biologically active substances. In: Syková E, Hník P, Vyklický L, eds. *Ion-selective microelectrodes and their use in excitable tissues*. New York: Plenum Press, 1981;25–39.
12. Phillips J, Nicholson C. Anion permeability in spreading depression investigated with ion-selective microelectrodes. *Brain Res* 1979;173: 567–571.
13. Krnjevič K, Morris ME. Extracellular K^+ activity and slow potential changes in spinal cord and medulla. *Can J Physiol Pharmacol* 1972; 50:1214–1217.
14. Vyklický L, Syková E, Kříž N, Ujec E. Poststimulation changes of extracellular potassium concentration in the spinal cord of the rat. *Brain Res* 1972;45:608–611.
15. Somjen GG. Extracellular potassium in the mammalian central nervous system. *Annu Rev Physiol* 1979;41:159–177.
16. Syková E. K^+ changes in the extracellular space of the spinal cord and their physiological role. *J Exp Biol* 1981;95:93–109.
17. Syková E, Kříž N, Hájek I. Extracellular K^+ accumulation in the spinal cord and its role in primary afferent depolarization and poststimulation analgesia. In: Kesler M, Harrison DK, Hoper J, eds. *Ion measurements in physiology and medicine*. Berlin: Springer-Verlag, 1985; 214–220.
18. Syková E. Extracellular K^+ accumulation in the central nervous system. *Prog Biophys Mol Biol* 1983;42:135–189.
19. Svoboda J, Motin V, Hájek I, Syková E. Increase in extracellular potassium level in rat spinal dorsal horn induced by noxious stimulation and peripheral injury. *Brain Res* 1988;458: 97–105.
20. Syková E, Kříž N, Preis P. Elevated extracellular potassium concentration in unstimulated spinal dorsal horns of frogs. *Neurosci Lett* 1983;43:293–298.
21. Syková E, Rothenberg S, Krekule I. Changes of extracellular potassium concentration during spontaneous activity in the mesencephalic reticular formation of the rat. *Brain Res* 1974; 79:333–337.
22. Satoh T, Watanabe K, Eguchi K. Enhancement during REM sleep of extracellular potassium ion activity in the reticular formation. *Brain Res* 1979;174:180–183.
23. Moghaddam B, Adams RN. Regional differences in resting potassium levels of rat brain. *Brain Res* 1987;406:337–340.
24. Prince DA, Lux HD, Neher E. Measurement of extracellular potassium activity in cat cortex. *Brain Res* 1973;50:489–495.
25. Mutani R, Futamachi KJ, Prince DA. Potassium activity in immature cortex. *Brain Res* 1974;75:27–39.
26. Heinemann U, Lux HD. Undershoots following stimulus-induced rises of extracellular potassium concentration in the cerebral cortex of cat. *Brain Res* 1975;93:63–76.
27. Heinemann U, Lux HD. Ceiling of stimulus-induced rises in extracellular potassium concentration in the cerebral cortex of cat. *Brain Res* 1977;120:231–249.
28. Lothman E, LaManna JC, Cordingley G, Rosenthal M, Somjen G. Responses of electrical potential, potassium levels and oxidative metabolic activity of the cerebral neocortex of cats. *Brain Res* 1975;88:15–36.
29. Futamachi KJ, Mutani R, Prince DA. Potassium activity in rabbit cortex. *Brain Res* 1974;75:5–25.
30. Bruggencate ten G, Nicholson C, Stöckle H. Climbing fiber evoked potassium release in cat cerebellum. *Pflügers Arch* 1976;367:107–109.
31. Stöckle H, Bruggencate ten G. Fluctuation of extracellular potassium and calcium in the cerebellar cortex related to climbing fiber activity. *Neuroscience* 1980;5:893–901.
32. Nicholson C, Bruggencate ten G, Stöckle H, Steinberg R. Calcium and potassium changes in extracellular microenvironment of cat cerebellar cortex. *J Neurophysiol* 1978;41:1026–1039.
33. Krnjevič K, Morris ME, Reiffenstein RJ. Changes in extracellular Ca^{2+} and K^+ activity accompanying hippocampal discharges. *Can J Physiol Pharmacol* 1980;58:579–583.
34. Morris ME. Measurements of ion activity in

the CNS: extracellular K^+ and Ca^{2+} in the hippocampus. In: Syková E, Hnik P, Vyklický L, eds. *Ion-selective microelectrodes and their use in excitable tissues*. New York: Plenum Press, 1981;241–246.
35. Krnjevič K, Morris ME, Reiffenstein RJ. Stimulation-evoked changes in extracellular K^+ and Ca^{2+} in pyramidal layers of the rat's hippocampus. *Can J Physiol Pharmacol* 1982;60:1643–1657.
36. Johnstone BM, Patuzzi R, Syka J, Syková E. Stimulus-related potassium changes in the organ of Corti of guinea-pig. *J Physiol* 1989; 408:77–92.
37. Karwoski CJ, Proenza LM. Sources and sinks of light-evoked $\Delta[K^+]_o$ in the vertebrate retina. *Can J Physiol Pharmacol* 1987;65:1009–1017.
38. Karwoski CJ, Lu Hong-Kai, Newman EA. Spatial buffering of light-evoked potassium increases by retinal Müller (glial) cells. *Science* 1989;244:505–620.
39. Nicholson C. Dynamics of the brain cell microenvironment. *Neurosci Res Prog Bull* 1980; 18:177–322.
40. Kříž N, Syková E, Vylický L. Extracellular potassium changes in the spinal cord of the cat and their relation to slow potentials, active transport and impulse transmission. *J Physiol* 1975;249:167–182.
41. Krnjevic K, Morris ME. Factors determining the decay of K^+ potentials and focal potentials in the central nervous system. *Can J Physiol Pharmacol* 1975;53:923–934.
42. Syková E, Hájek I, Chvátal A, Kříž N, Diatchkova GI. Changes in extracellular potassium accumulation produced by opioids and naloxone in frog spinal cord: relation to changes of Na-K pump activity. *Neurosci Lett* 1985; 59:285–290.
43. Orkand RK, Nicholls JG, Kuffler SW. Effects of nerve impulses on the membrane potential of glial cells in the central nervous system of amphibia. *J Neurophysiol* 1966;29:788–806.
44. Coles JA, Orkand RK. Modification of potassium movement through the retina of the drone (*Apis mellifera* ♂) by glial uptake. *J Physiol* 1983;340:157–174.
45. Gardner-Medvin AR, Nicholson C. Changes of extracellular potassium activity produced by electric current through brain tissue in the rat. *J Physiol* 1983;335:375–392.
46. Dietzel I, Heinemann U, Hofmeier G, Lux HD. Stimulus-induced changes in extracellular Na^+ and Cl^- concentration in relation to changes in the size of the extracellular space. *Exp Brain Res* 1982;46:73–84.
47. Dietzel I, Heinemann U, Lux HD. Relations between slow extracellular potential changes, glial potassium buffering, and electrolyte and cellular volume changes during neuronal hyperactivity in cat brain. *Glia* 1989;2:25–44.
48. Syková E, Orkand RK, Chvátal A, Hájek I, Kříž N. Effects of carbon dioxide on extracellular potassium accumulation and volume in isolated frog spinal cord. *Pflügers Arch* 1988; 412:183–187.
49. Walz W, Hertz L. Ouabain-sensitive and ouabain-resistant net uptake of potassium into astrocytes and neurons in primary cultures. *J Neurochem* 1982;39:70–77.
50. Kettenmann H, Syková E, Orkand RK, Schachner M. Glial potassium uptake following depletion by intracellular ionophoresis. *Pflügers Arch* 1987;410:1–6.
51. Nicholson C, Brugencate ten G, Senekowitsch R. Large potassium signals and slow potentials evoked during aminopyridine or barium superfusion in cat cerebellum. *Brain Res* 1976;113: 606–610.
52. Heinemann U, Lux HD, Gutnick MJ. Extracellular free calcium and potassium during paroxysmal activity in the cerebral cortex of cat. *Exp Brain Res* 1977;27:237–243.
53. Chvátal A, Jendelová P, Kříž N, Syková E. Stimulation-evoked changes in extracellular pH, calcium and potassium activity in the frog spinal cord. *Physiol Bohemoslov* 1988;37:203–212.
54. Syková E, Kříž N, Hájek I. Role of endogenous opiates and extracellular K^+ accumulation in the inhibition of frog spinal reflexes by electrical skin stimulation. *Physiol Bohemoslov* 1975;34:548–561.
55. Urbanics R, Leniger-Follert E, Lübbers DW. Time course of changes of extracellular H^+ and K^+ activities during and after direct electrical stimulation of the brain cortex. *Pflügers Arch* 1978;378:47–53.
56. Krishtal OA, Osipchuk YuV, Schelest TN, Smirnoff SV. Rapid extracellular pH transients related to synaptic transmission in rat hippocampal slices. *Brain Res* 1987;436:352–356.
57. Kraig RP, Ferreira-Filho CR, Nicholson C. Alkaline and acid transients in cerebellar microenvironment. *J Neurophysiol* 1983;49:831–850.
58. Chesler M, Chan CY. Stimulus induced extracellular pH transients in the *in vitro* turtle cerebellum. *Neuroscience* 1988;27:941–948.
59. Rice ME, Nicholson C. Behavior of extracellular K^+ and pH in skate (*Raja erinacea*) cerebellum. *Brain Res* 1988;461:328–334.
60. Syková E, Svoboda J. Extracellular alkaline-acid-alkaline transients in the rat spinal cord evoked by peripheral stimulation. *Brain Res* 1990;512:181–189.
61. Syková E. Activity-related extracellular pH transients in spinal cord. *Verh Dtsch Zool Ges* 1989;82:153–163.
62. Ransom BR, Carlini WG, Connors B. Brain extracellular space: developmental studies in rat optic nerve. *Ann NY Acad Sci* 1985;481:87–105.
63. Endres W, Grafe P, Bostock H, Bruggencateten G. Changes in extracellular pH during electrical stimulation of isolated rat vagus nerve. *Neurosci Lett* 1986;64:201–205.
64. Lehmenkühler A, Zidek W, Staschen M, Caspers H. Cortical pH and pCa in relation to DC potential shifts during spreading depression and asphyxiation. In: Syková E, Hnik P, Vyklický L, eds. *Ion-selective microelectrodes and their use in excitable tissues*. New York: Plenum Press, 1981;225–229.

65. Harris RJ, Symon L. Extracellular pH, potassium, and calcium activities in progressive ischemia of rat cortex. *J Cereb Blood Flow Metab* 1984;4:178–186.
66. Mutch WAC, Hansen AJ. Extracellular pH changes during spreading depression and cerebral ischemia: mechanisms of brain pH regulation. *J Cereb Blood Flow Metab* 1984;4:17–27.
67. Somjen GG. Acidification of interstitial fluid in hippocampal formation caused by seizures and by spreading depression. *Brain Res* 1984;311: 186–188.
68. Seisjö BK, von Harewehr R, Nergelius G, Nevander G, Ingwar M. Extra- and intracellular pH in the brain during seizures and in the recovery period following the arrest of seizure activity. *J Cereb Blood Flow Metab* 1985;5:47–57.
69. Thomas RC. Changes in the surface pH of voltage-clamped snail neurones apparently caused by H^+ fluxes through a channel. *J Physiol* 1988;398:313–324.
70. Thomas RC. *Proton passage across cell membranes*. Chichester: John Wiley & Sons, 1988.
71. Syková E, Svoboda J, Chvátal A, Jendelová P. Extracellular pH and stimulated neurons. In: Thomas RC, ed. *Proton passage across cell membranes*. Chichester: John Wiley & Sons, 1988;220–232.
72. Astion ML, Orkand RK. Electrogenic Na^+/HCO_3^- cotransport in neuroglia. *Glia* 1988;1: 355–357.
73. Astion ML, Coles JA, Orkand RK. Effects of bicarbonate on glial cell membrane potential in *Necturus* optic nerve. *Neurosci Lett* 1987;76: 47–52.
74. Deitmer JW, Schlue WR. The regulation of intracellular pH by identified glial cells and neurones in the central nervous system of the leech. *J Physiol* 1987;388:261–283.
75. Kettenmann H, Schlue WR. Intracellular pH regulation in cultured mouse oligodendrocytes. *J Physiol* 1988;406:147–162.
76. Chesler M, Kraig RP. Intracellular pH transients of mammalian astrocytes. *J Neurosci* 1989;9:2011–2019.
77. Iljima T, Ciani S, Hagiwara S. Effects of the external pH on Ca channels: experimental studies and theoretical considerations using a two-site, two-ion model. *Proc Natl Acad Sci USA* 1986;83:654–658.
78. Konnerth A, Lux HD, Morad M. Proton induced transformation of calcium channel in chicks dorsal root ganglion cells. *J Physiol* 1987;386:603–633.
79. Krishtal OA, Pidoplichko VI. A receptor for protons in the nerve cell membrane. *Neuroscience* 1980;5:2325–2327.
80. Krishtal OA, Pidoplichko VI. A receptor for protons in the membrane of sensory neurons may participate in nociception. *Neuroscience* 1981;6:2599–2601.
81. Krishtal OA, Pidoplichko VI. A "receptor" for protons in small neurons of trigeminal ganglia: possible role in nociception. *Neurosci Lett* 1981;24:243–246.
82. Gruol DL, Barker JL, Huang LM, McDonald JF, Smith TG Jr. Hydrogen ions have multiple effects on the excitability of cultured mammalian neurons. *Brain Res* 1980;183:247–252.
83. Balestrino M, Somjen GG. Concentration of carbon dioxide, interstitial pH and synaptic transmission in hippocampal formation of the rat. *J Physiol* 1988;396:247–266.
84. Rehncrona S, Rosen I, Siesjö BK. Excessive cellular acidosis: an important mechanism of neuronal damage in the brain. *Acta Physiol Scand* 1980;110:435–437.
85. Siesjö BK. Cell damage in the brain: a speculative synthesis. *J Cereb Blood Flow Metab* 1981;1:155–185.
86. Siesjó BK. Cerebral circulation and metabolism. *J Neurosurg* 1984;60:883–908.
87. Kraig RP, Petito CK, Plum F, Pulsinelli WA. Hydrogen ions kill brain at concentrations reached in ischemia. *J Cereb Blood Flow Metab* 1987;7:379–386.
88. Phillips JM, Nicholson C. Anion permeability in spreading depression investigated with ion-sensitive microelectrodes. *Brain Res* 1979;173: 567–571.
89. Hansen AJ, Olsen CE. Brain extracellular space during spreading depression and ischemia. *Acta Physiol Scand* 1980;108:355–365.
90. Heinemann U, Dietzel I. Extracellular potassium concentration in chronic alumina cream foci of cats. *J Neurophysiol* 1984;3:421–434.
91. Dietzel I, Heinemann U. Dynamic variation of the brain cell microenvironment in relation to neuronal hyperactivity. *Ann NY Acad Sci* 1986;481:72–86.
92. Dietzel I, Heinemann U, Hofmeier G, Lux HD. Transient changes in the size of the extracellular space in the sensorimotor cortex of cats in relation to stimulus-induced changes in potassium concentration. *Exp Brain Res* 1980;40: 432–439.
93. Ransom BR, Yamate CL, Connors BW. Activity dependent shrinkage of extracellular space in rat optic nerve: a developmental study. *J Neurosci* 1985;5:532–535.
94. Syková E. Modulation of spinal cord transmission by changes in extracellular K^+ activity and extracellular volume. *Can J Physiol Pharmacol* 1987:65:1058–1066.
95. Syková E, Kříž N, Chvátal A, Motin V, Hájek I. Activity-related transient changes in the sizes of the extracellular space in the frog spinal cord. *Physiol Bohemoslov* 1987;36:379.
96. Svoboda J, Syková E. Activity-related extracellular K^+ and volume changes in spinal cord of rat. *Acta Physiol Scand* 1989;136 Suppl. 582:86.
97. Syková E, Svoboda J. Activity related volume and ionic changes in spinal cord of the rat. *Proc IUPS XVII* 1989;148.
98. Macknight ADC. Principles of cell volume regulation. In: Berlyne KM, Brooklyn NY, eds. *Renal physiology biochemistry*, vols 3–5. 1988; 114–141.
99. Cornog JL, Gonatas NK, Feierman JR. Effects

of intracerebral injection of ouabain on the fine structure of rat cerebral cortex. *Am J Pathol* 1967;51:573–590.

100. Trubatch J, Loud AU, Van Harreveld A. Quantitative stereological evaluation of KCl-induced ultrastructural changes in frog brain. *Neuroscience* 1977;2:963–974.
101. Rall W. Dendritic spines, synaptic potency and neuronal plasticity. In: Woody CD, Brown KA, Crow, Jr. TJ, Krispel JD, eds. *Cellular mechanisms subserving changes in neuronal activity,* vol 3. Los Angeles: Brain Information Service, 1974;13–22.
102. Van Harreveld A, Fifková E. Swelling of dendritic spines in the fascia dentata after stimulation of the perforant fibres as a mechanism of post-tetanic potentiation. *Exp Neurol* 1975;49: 736–749.
103. Van Harreveld A, Fifková E. Involvement of glutamate in memory formation. *Brain Res* 1974;81:455–467.
104. Cherkin A, Eckardt MJ, Gerbrant XX. Memory: proline induces retrograde amnesia in chicks. *Science* 1976;193:242–244.

Volume Transmission in the Brain: Novel Mechanisms for Neural Transmission, edited by K. Fuxe and L.F. Agnati. Raven Press, Ltd., New York © 1991.

26

Relation Between Impulse Flow and Extracellular Catecholamine Levels as Studied by *In Vivo* Electrochemistry in CNS

F.G. Gonon, M.F. Suaud-Chagny, C.C. Mermet, and M. Buda

INSERM U 171 and CNRS URA 1195, Centre Hospitalier Lyon-Sud, 69310, Pierre-Bénite, France

Fuxe and Agnati (1) postulated that there are two types of chemical transmission, namely, the "wiring" transmission (WT) and the "volume" transmission (VT). WT is defined as being "the classical type of electrochemical transmission which rests on the physical substrate of the neuronal chain," whereas "VT is a humoral type of transmission" (1). The speed of transmission is much higher for WT than for VT. According to the classical view, WT is triggered by a single action potential and occurs within 1 msec. Moreover, WT is highly localized (i.e., restricted to the synaptic cleft), whereas VT occurs far from the release site.

In the present chapter we will discuss some data regarding the chemical transmission mediated by dopamine (DA) and norepinephrine (NE) in the central nervous system (CNS) in an attempt to answer the following question: Are DA and NE involved in WT? Answering this question amounts to knowing when and where DA and NE are released. The features of DA release will be discussed first and separately from those of NE release. The discussion will be restricted to DA and NE release occurring in brain areas that are densely innervated either by DA and NE terminals [i.e., in striatum, nucleus accumbens, and olfactory tuberculum on the one hand and in the paraventricular hypothalamic nucleus (PVN) on the other hand].

WHEN IS DA RELEASED?

Firing Activity of DA Neurons in CNS

According to the classical view, neurotransmitter release is triggered by nerve action potentials. Thus, variations in the discharge rate of dopaminergic neurons should trigger parallel variations in DA release. Although the relation between impulse flow and DA release is complex (see below), the monitoring of the discharge rate represents an indirect but valid approach for investigating the transmission mediated by DA. In this field, the literature is extremely abundant. In this section we will briefly recall some features that dopaminergic neurons have in common.

In rats these neurons exhibit a mean discharge rate that ranges from 0 to 8 Hz. Two types of discharge pattern can be distinguished: either a single spike mode or a bursting mode (Fig. 1). Most bursts consist of three to 10 spikes whose interspike inter-

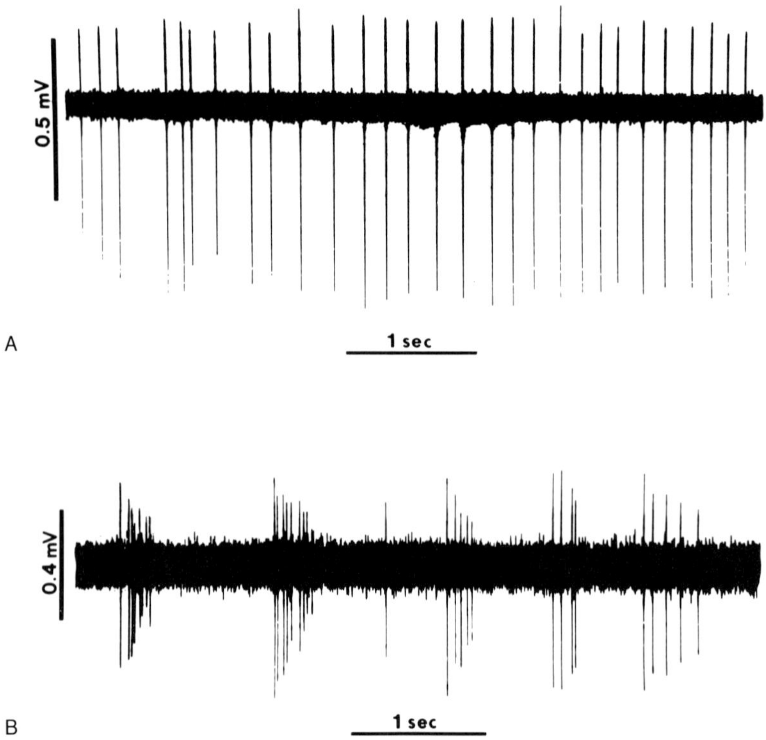

FIG. 1. Typical firing pattern of two dopaminergic neurons recorded in a freely moving rat. Single spike discharge mode is illustrated in **A.** A predominant bursting pattern is shown in **B.** Most dopaminergic neurons exhibit an intermediate degree of burst firing and oscillate between single spike pattern and bursting mode. (From ref. 2, with permission.)

val varies from 50 to 100 msec. Individual DA neurons can switch from the single spike mode to the bursting mode (and vice versa), and this switch is associated with a moderate (+50%) increase in their mean discharge rate. However, the mean frequency is not predictive of the degree of bursting (2–4).

In anesthetized animals, the spontaneous firing activity is mainly of the single spike type. However, some bursts are usually observed from DA neurons recorded in the ventral tegmental area (A 10 group). This bursting activity depends on a neuronal input originating in the frontal cortex and mediated by an excitatory amino acid because it is selectively abolished either by cooling the frontal cortex or by a kynuranate i.c.v. injection, whereas the single spike mode is unaffected (5,6). In unanesthetized freely moving animals, the firing activity of DA neurons depends on the state of vigilance. During sleep and quiet waking, it is close to that recorded from anesthetized animals, whereas during active waking, the occurrence of bursts is much more frequent. Moreover, brief sensorial stimuli trigger a burst in many DA neurons (7).

All these data suggest that the functional response of DA neurons to physiological stimuli is not a single spike but a burst of three to 10 spikes that lasts for 0.1 to 0.4 sec. The single spike mode might correspond to a resting activity without direct functional significance with regard to "wiring" DA transmission. This is further supported by *in vitro* studies showing that when the firing activity is recorded from brain slices, the bursting mode is no longer observed, whereas the single spike pattern is preserved (8).

Relation Between Impulse Flow and DA Release

In Vivo Monitoring of Extracellular DA: Technical Problems

First attempts to monitor *in vivo* the extracellular DA level were performed using perfusion with push-pull cannulae. Unexpected observations were reported with this technique. For example, DA release recorded from the striatum was not affected during a 10-min sensorial stimulation, but a delayed increase was observed after the end of the stimulation (9). However, this technique is too slow (one measurement every 10 min) to monitor rapid variations in DA release.

In the early 70s, Kissinger et al. (10) made the first attempt to monitor extracellular DA by a much faster technique: *in vivo* electrochemistry. During the course of its expansion, many difficulties, mainly related to its lack of selectivity, were encountered. In the present state of the art, two experimental approaches are considered to be reliable for monitoring DA release with *in vivo* electrochemistry (11). The first requires that the animals be pretreated with a monoamine oxidase inhibitor (usually pargyline) and allows the recording of a basal oxidation signal, which is entirely due to extracellular DA concentration (12). Its main drawback is that the pargyline treatment potentiates DA release (13). With the second approach, a pharmacological treatment is not required, but only the DA release, which is evoked by brief electrical stimulations, can be recorded (13,14). Up to now, no reliable electrochemical technique allows the basal extracellular DA level from untreated animals to be monitored (11).

Finally, *in vivo* dialysis is an appealing approach for monitoring DA release.

Is Basal Extracellular DA Level Related to the Impulse Flow?

To block locally the impulse flow, tetrodotoxin (TTX) was added to the perfusion medium of a dialysis probe. Thus, a decrease in DA release was observed, although its amplitude depended on the experimental conditions (15). Likewise, we monitored the extracellular DA level with *in vivo* electrochemistry from pargyline-treated rats, and we observed that an electrolytic lesion of the dopaminergic pathway induced an immediate 70% decrease in this level (12,13). Therefore, in basal conditions most of the extracellular DA level is related to the spontaneous impulse flow.

Variations in DA Release Induced by Electrical Stimulations

In vivo dialysis showed that electrical stimulations of the dopaminergic pathway induced an increase in DA release (16). However, the kinetics of this evoked release cannot be examined with accuracy with this slow technique. The *in vivo* electrochemical monitoring of DA release evoked by stimulations as short as 1 sec revealed that this release started immediately and lasted as long as the stimulation did, provided that the frequency of the stimulation was in a physiological range (i.e., less than 20 Hz) (12,13,17). Moreover, the amplitude of the evoked DA release depends on the stimulation frequency in a nonlinear manner: It is low when the frequency is less than 5 Hz, but it is strongly enhanced between 10 and 14 Hz (Fig. 2). Thus, DA terminals convert action potentials into a chemical signal as a high pass filter in the physiological range of frequency (13).

The observation that DA neurons exhibit two types of discharge pattern led Grace

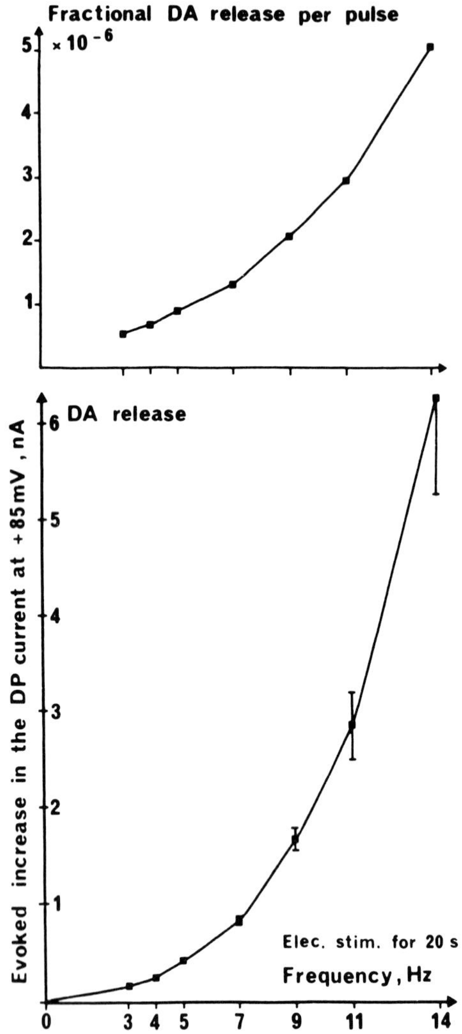

FIG. 2. Effect of stimulation frequency on the amplitude of evoked DA release recorded by *in vivo* electrochemistry from the olfactory tuberculum. *Lower part* of the figure shows original data reported in a previous study (13). Briefly, electrical stimulations of the dopaminergic pathway were applied for 20 sec at various frequencies. They induced an immediate increase in the oxidation current recorded by differential pulse amperometry at the oxidation potential of DA (+85 mV). This increase is entirely due to evoked DA release. Fractional DA release (i.e., DA release expressed as a fraction of the whole DA tissue content) was calculated from these original data following the same assumptions as previously published as regards NE release (30). Briefly, absolute increases in the oxidation current were expressed in terms of increases in extracellular DA concentration on the basis of electrode calibration *in vitro* after *in vivo* recording (e.g., this increase reached 0.57 μM with a 20-sec stimulation at 14 Hz). Then, the increase in extracellular DA concentration was compared with the whole DA content, which corresponds to a mean concentration of 80 μM, taking into account that DA is released in the extracellular space, which represents 20% of whole brain volume (for a 14-Hz stimulation, this fractional release was $0.57 \times 0.2/80 = 1.42 \times 10^{-3}$). Finally, for each frequency, fractional release was divided by the number of pulse in the corresponding stimulation (280 pulses for a 20-sec stimulation at 14 Hz).

and Bunney (3) to hypothesize that "bursting may be as important as (mean) firing frequency in affecting DA release." To test this assumption, we investigated the effect of electrical stimulations, which closely mimicked both types of discharge pattern. When the mean frequency was the same (5 Hz), the bursting stimulation was twice more potent than the regularly spaced one. Finally, we took into account the fact that the switch from the single spike discharge pattern to the bursting mode is associated with a 50% increase in the mean frequency. Therefore, we compared the effect of a regularly spaced stimulation at 4 Hz with that of a bursting stimulation at 6 Hz. The latter was found to be six times greater than the former in triggering DA release (13).

Variations in DA Release Induced by Physiological Stimuli

Because of technical limitations (see above), few studies have reported on variations in extracellular DA level induced by physiological stimuli. An *in vivo* dialysis study reported a 70% increase in DA release from the nucleus accumbens when the rat was eating as compared with a control period of quiet waking (18).

Electrophysiological Effects of DA on Target Cells

The role of DA on postsynaptic neurons in the striatum and the nucleus accumbens has been investigated by numerous electrophysiological studies. This role appears extremely complex. Nevertheless, two independent groups reported on an important finding that is relevant to the present discussion. When DA release was triggered by electrical stimulations of the dopaminergic pathway, it deeply altered the response of postsynaptic neurons to other stimulations provided that the DA release was triggered by one burst consisting of five to 10 pulses at 10 Hz. Single pulse stimulations of the DA pathway were ineffective (19,20).

Conclusion

When all the data discussed above are fitted together, the following picture emerges. Firstly, variations in extracellular DA level closely parallel those of the impulse flow (except, of course, when these variations are induced by certain pharmacological treatments). Thus, in physiological conditions, DA release is mainly triggered by action potentials. Secondly, DA neurons respond to a single stimulus by a burst of three to 10 spikes lasting from 0.1 to 0.4 sec, and this is highly consistent with the fact that this kind of burst is required to trigger a significant DA release. In summary, our answer to the first question (i.e., when is DA released) is the following: In physiological conditions, a significant DA release is triggered by one burst of action potential that lasts for no more than 1 sec.

WHERE IS DA RELEASED?

Quantitative Estimates of Extracellular DA Concentration

In brain areas that are densely innervated by DA terminals, the whole DA content is about 12 μg/g of tissue. This corresponds to a mean concentration of 80 μM. Because DA terminals occupy less than 1% of the whole tissue volume, the mean concentration inside the DA varicosities therefore reaches 8 mM. In 1980, the extracellular DA concentration was measured from the striatum by *in vivo* electrochemistry: It was less than 50 nM (21). Thus, for the first time in the living brain, it provided a quantitative estimate of the steep gradient of DA concentration between the inside of the terminals and the extrasynaptic extracellular space. This estimate was, afterward, confirmed by *in vivo* dialysis (16). Recently, we improved the sensitivity of our technique, and we showed that in physiological conditions, the extracellular DA concentration varies depending on the impulse flow, from 10 to 100 nM when recorded from brain area densely innervated by DA terminals (12,13).

Evidence Supporting That DA Transmission Occurs Inside a Synaptic Cleft

Recent studies show that most DA terminals make symmetric synaptic contacts in densely innervated areas (striatum, accumbens, olfactory tuberculum) (22,23). Moreover, most DA synaptic contacts occur on the neck of a dendritic spine (23). This provides an anatomical support to electrophysiological studies showing that DA can modulate the excitatory input to striatal spiny neurons (20). It is obvious that carbon fiber electrodes record the DA concentration in the extrasynaptic extracellular space, which represents about 20% of the whole brain volume (24). Therefore, we hypothesized that DA is released inside the synaptic cleft and that the DA concentra-

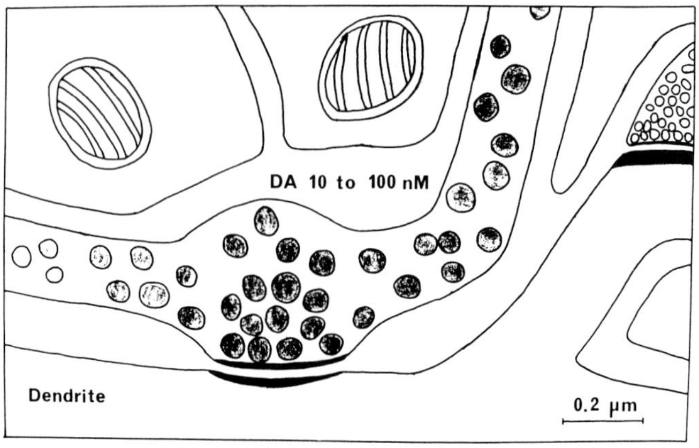

FIG. 3. Hypothetic scheme of a dopaminergic terminal fiber forming an "en passant" symmetric synaptic contact with a dendritic shaft. It was drawn according to data discussed in Fig. 4. We strove to depict the extrasynaptic extracellular space, which represents 20% of whole brain volume, and to preserve the scale. The volume of the synaptic cleft represents a minute fraction of remaining extracellular space. Therefore, if the DA content of one vesicle is released inside the cleft, DA concentration might reach a much higher level (10–100 μM).

tion that is actually measured reflects at a low level, through dilution and diffusion, the DA release at these sites (12,13). In fact, although the amount of quantal DA that is released inside the synaptic cleft seems extremely low because it might correspond to the DA content of one vesicle (see Functional Meaning of Bursts, Second Hypothesis section), the volume of the synaptic cleft is so small compared with the remaining extracellular space that inside the synaptic cleft, DA concentration might reach much higher levels (10–100 μM) (Fig. 3). This is further supported by *in vitro* studies investigating the role of DA in the retina, showing that DA modulates the response to light at a concentration of 10 to 50 μM (25). Finally, this estimate is consistent with the affinity of D_1 dopaminergic receptors (26).

This synaptic hypothesis might explain some unexpected observations concerning the effect of local injections of exogenous DA. A detailed discussion of these observations has been published elsewhere (13). Briefly, local injections of exogenous DA induce electrophysiological effects, which differ from that observed when the release of endogenous DA is triggered either by electrical stimulation of the DA pathway or by indirect DA agonists. Johnson et al. (27) concluded that "locally applied DA does not *normally* contact postsynaptic dopaminergic receptors."

WHEN IS NE RELEASED IN CNS?

Firing Activity of Noradrenergic Neurons in CNS

Although this phenomenon has not yet been systematically investigated, it is obvious from many studies that, as well as dopaminergic neurons, the noradrenergic ones exhibit two kinds of discharge pattern (28,29). On the one hand, in anesthetized animals as well as during sleep and quiet waking, the spontaneous activity is slow (0.5–3 Hz) and regular. On the other hand, many kinds of sensory stimuli trigger a burst of three to 20 action potentials and the frequency within the burst reaches 20 Hz (29).

Relation Between Impulse Flow and NE Release

Few studies have been devoted to the *in vivo* monitoring of NE release whether by dialysis or by electrochemistry. These studies have been discussed by detail in a recent report (30). We came to the following conclusions. Firstly, in the absence of drug treatment, variations in the impulse flow induce parallel changes in NE release. Secondly, the relation between NE release and the frequency of the impulse flow is not linear: Noradrenergic terminals act as a high pass filter in the range of the physiological frequency (0–20 Hz). Therefore, it is likely that the release, which is induced by bursts, is enhanced.

Conclusion

The effect of endogenous NE released by electrical stimulation on the discharge rate of target cells has been investigated in several brain areas including two hypothalamic nuclei, which are densely innervated by noradrenergic terminals: The supraoptic nucleus and the PVN. In both nuclei, electrical stimulations of the ascending noradrenergic pathway mainly induced an excitation of the target cells. In some studies this effect was triggered by single pulse stimulations (31,32), whereas trains of 25 pulses were used in another (33). However, to our knowledge, the comparison between the effects of these two kinds of stimulation has not yet been investigated.

Therefore, it is likely that, as for dopaminergic neurons, a significant NE release is triggered by a burst of action potential. However, the experimental evidence is weaker than as regards DA release.

WHERE IS NE RELEASED?

NE Release from Sympathetic Terminals

The features of NE release have been investigated in more detail in peripheral tissues than in the CNS (34). For example, the neurogenic contraction of the rat tail artery is almost entirely caused by NE released from sympathetic terminals (35). *In vitro,* the strength of the contraction increases with the NE concentration from 10 nM to 10 μM, and the half of the maximal response is induced by a 100-nM NE concentration. Electrical stimulations induce both contraction and NE release *in vitro*. In a recent study the local extracellular NE concentration was monitored by electrochemistry at the surface of the rat tail artery. We applied electrical stimulations, which induced the same contractile response as that obtained by NE at 100 nM, and we measured the local NE concentration, which was reached at the arterial surface. We found a value of 120 nM (*unpublished observations* by J. Bao, M. Msghina, L. Stjärne and F.G. Gonon).

Therefore, although the release sites are highly localized (i.e., restricted to varicosities) and although the NE release is intermittent (34), as regards the rat tail artery, the postsynaptic response to NE is triggered by a low concentration of NE spread over the whole arterial surface rather than by a high NE level restricted inside a synaptic space. This conclusion is consistent with the following observations. Sympathetic terminal fibers innervating arterioles form, every 8 to 10 μm, varicose enlargements that contain a high density of vesicles. Discrete close appositions between the muscle cells and most varicosities have a mean area of 1.2 μm^2. Local ionophoretic ejections of NE over the arterial surface produce at most sites a localized constriction of the arteriole. However, subsequent histochemistry demonstrates that these sites are not related to the location of the sympathetic fibers (36). In conclusion, the contraction of the rat tail artery seems to be triggered by a diffuse chemical signal, which has its source at some distance in a population of release sites.

NE Release from the CNS

In physiological conditions the extracellular NE concentration recorded from the PVN ranges from 10 nM to 100 nM

(30,37,38). Because the density of noradrenergic terminals in PVN is as high as for the dopaminergic ones in the striatum and because the above estimate is the same as for DA recorded from the striatum, this suggests that our synaptic hypothesis concerning DA transmission is also valid as regards NE in PVN (30,38). Morphological studies support this view because most noradrenergic terminals form synaptic contacts with target cells in PVN (39).

Contrary to DA, local injections of NE seem to mimic closely the excitatory effects of endogenous NE release on neurosecretory cells (40). Therefore, a NE transmission involving a population of varicosities and mediated by the low NE extracellular concentration cannot be ruled out. However, no firm conclusion can be drawn because the concentration of injected NE used in this study (50 μM–1 mM) was much higher than the physiological extracellular NE concentration (40).

However that may be, it is certain that in PVN, the NE transmission occurs inside this nucleus. In fact, when extracellular catechols are monitored by electrochemistry just outside the PVN (except from the zona incerta or from the preoptic area, where a catechol signal from another source can be recorded), no basal signal was detected and electrical stimulations of the noradrenergic pathway did not induce the appearance of a catechol signal (*unpublished observations* by M.F. Suaud-Chagny, C.C. Mermet and F.G. Gonon).

CONCLUSION AND OPEN QUESTIONS

DA Is Involved in a "Wiring" Transmission

From all the experimental evidence discussed in the sections *When Is DA Released?* and *Where Is DA Released?*, the following picture emerges: A phasic and functionally significant DA release is triggered by one burst of some action potentials and mediates a chemical transmisison, which occurs inside synaptic clefts. Therefore, according to the criteria of Fuxe and Agnati, this transmission, which is rapid and highly localized, is of the "wiring" type.

Is NE Involved in a "Wiring Transmission" in CNS?

The NE release in PVN has many features in common with DA release in striatum, nucleus accumbens, and olfactory tuberculum. Moreover, in PVN, NE has at least one definite function: It is involved in the response of the corticotropic axis to stress (38). Therefore, we are tempted to conclude that NE transmission in PVN is of the "wiring" type, too. However, it is not clear whether this transmission is localized inside a synaptic cleft. Firstly, tenuous but significant differences exist between DA and NE release. For example, the relation between DA release and frequency of the impulse flow is close but not identical to that regarding NE (13,30). Secondly, because the efficient NE release from sympathetic terminals does not seem to be localized inside a synaptic cleft, cautiousness is imperative as regards the CNS.

Beyond these conclusions, the experimental data reviewed above raise two open questions.

First Open Question: Functional Meaning of Bursts?

According to the classical view based on the study of the neuromuscular junction, the chemical neurotransmission can be triggered by a single action potential. In contrast to this classical view, a burst of three to 20 action potentials is required to trigger a phasic catecholaminergic transmission. In the next section, the functional meaning of this type of triggering will be discussed.

Second Open Question: Site of Autoregulation of Catecholamine Release?

The data discussed in the section *Where Is DA Released?* suggest that the phasic and functionally significant DA release occurs inside the synaptic cleft. This might be true, too, as regards NE transmission in PVN. Although the extrasynaptic extracellular catecholamine (CA) concentration ranges from 10 to 100 nM, this does not imply that this low chemical signal does not play a functional role. In fact, experimental evidence supports the view that this signal mediates the inhibitory feedback, which controls the phasic CA release. Such evidence and its functional implications will be reviewed in the last section of this chapter.

FUNCTIONAL MEANING OF BURSTS

Introduction

The fact that a burst of action potential is required to trigger a significant release has long been established as regards neurosecretory cells. For example, the milk ejection reflex is due to a massive release of ocytocin in blood circulation, which is triggered by a burst of action potentials lasting for 4 sec at 40 Hz (41). This suggests that the mechanism by which the impulse flow is converted to transmitter release is similar for neurosecretory cells and for CA neurons. But this does not imply that DA and NE are brain hormones. We rather assume that in CNS, DA and NE are involved in a wiring transmission.

The crudest answer to the question of the functional meaning of bursts is that a burst is required to trigger a significant DA release. This is true but tautological. At the neuromuscular junction, a single action potential is necessary. In CNS it is likely that this is true also as regards the chemical transmission mediated by excitatory amino acids. Because DA transmission, which is triggered by a burst, occurs within a period that is two order of magnitude longer, this important quantitative difference calls for an interpretation.

First Hypothesis: The Need for a Long-Lasting Signal

In an interesting theoretical paper (42), Shepherd and Brayton suggested that dendritic spines receiving two synaptic contacts "are well suited for generating logic operations," which combines these two inputs. This comparison with an elementary unit of a computer seems especially relevant as regards the medium-size spiny neurons of the striatum. In fact, morphological and electrophysiological studies have shown that the excitatory afferences originating in the cortex form an asymmetric synaptic contact with the head of a dendritic spine, whereas an inhibitory input mediated by DA is achieved by a symmetric synapse on the neck of the same spine (Fig. 4) (20,23).

However, a difficulty with this model is

> the need for the signals to arrive nearly at the same time in order to perform the AND operations. Since biological neural networks seem to be asynchronous, it would be difficult to imagine that the timing of the various signals is so exquisitely arranged as to meet all the required simultaneity. A solution to this is to postulate that a signal is not considered a single pulse but a train of pulses. This in effect extends the period of time for signals to overlap and thus to reinforce each other as required by the AND operation (42).

As regards DA transmission, this theoretical consideration seems especially relevant. In fact, as mentioned in the first section, the response of striatal neurons to a single pulse electrical stimulation in the cortex is modulated by stimulation of the DA pathway, provided that a train of five pulses is applied while single pulses are ineffective (20).

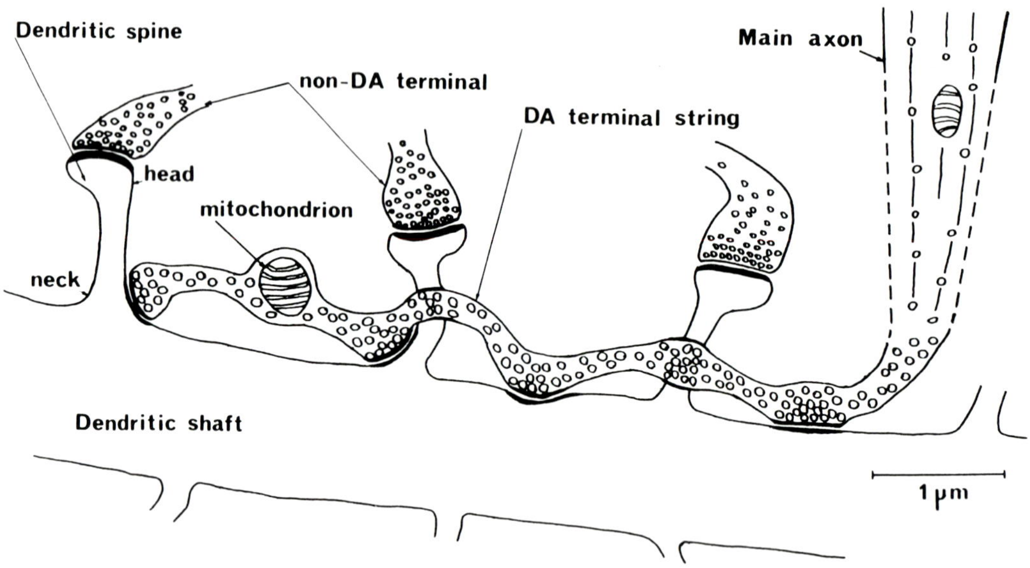

FIG. 4. Schematic drawing of a dopaminergic string of varicosities in close apposition with a dendrite of a medium-size spiny neuron in the striatum. Although it is purely imaginary, we strove to preserve the scale. This drawing has been reconstructed from data reported by Freund et al. (23). In their preterminal part, dopaminergic terminal fibers are unmyelinated and have a diameter between 0.2 and 0.4 μm, with a slight enlargement at the synaptic junction. Occasional larger varicosities are usually nonsynaptic and contain a mitochondrion. These DA terminal strings form many "en passant" symmetric synapses with dendritic elements, either at the neck of a dendritic spine (56% of all DA synapses) or with the dendritic shaft (36%). They contain large (40–80 nm), round vesicles throughout their entire length, and these vesicles are usually more concentrated near the synaptic contact. Main DA axons are thicker (0.5–1 μm in diameter), are never found to make synaptic contacts, and contain many microtubules and few vesicles. In addition, many dendritic spines receive an asymmetrical synapse on their head. The figure shows a DA terminal string forming five synaptic contacts. It is likely that the number of contacts is actually much higher (see text). Finally, the figure does not show that DA fibers are highly branched.

Second Hypothesis: The Elementary Secretory Unit Is Not a Single Varicosity but a String of Varicosities

The Catecholaminergic Innervation in CNS is Redundant

In the rat striatum, the number of DA varicosities per medium-size spiny neuron is 4,500 to 8,000 (43). Because each DA neuron innervates a limited part of the striatum and because the number of DA neurons innervating the rat striatum is about 8,000 (44), each striatal neuron is thus innervated by several varicosities originating in the same DA neuron. In other words, as already shown for many types of neurons in CNS (see, e.g., the innervation of Purkinje cells by climbing fibers), the DA innervation is redundant. More precisely, morphological studies show that one dopaminergic terminal fiber forms several successive "en passant" synaptic contacts with elements of the same dendritic segment (23) (Fig. 4). Does every individual varicosity along one fiber release DA every time the parent axon is invaded by a single action potential?

NE Release from Peripheral Terminals Is Quantal and Highly Intermittent

The fractional release of NE (i.e., the amount of NE released by a single action potential expressed as a fraction of the whole NE tissue content) from peripheral sympathetic terminals increases from 10^{-6} to 10^{-5} with increasing frequency of the stimulations. Thus, considering that varicosities contain less than 500 vesicles, the fractional NE release is much smaller than that which should be observed if every varicosity releases the content of one vesicle in response to each action potential. Recent studies show that this is not due to the failure of an action potential to invade all varicosities. Moreover, NE release appears to be quantal and corresponds to the exocytosis of the whole content of one vesicle. Therefore, from the low fractional release it was inferred that NE release is highly intermittent, and this is strongly supported by experimental data. The probability of quantal release increases with increasing frequency of action potential (34).

In CNS, the CA Release Might Be Quantal and Highly Intermittent

The increase in extracellular CA concentration, which is induced by electrical stimulation, is the net effect of the evoked overflow of CA diminished by the clearance. However, as discussed previously with regard to NE release, when short-lasting electrical stimulations were applied at physiological frequencies, this increase represents a fair estimate of the overflow (30). In fact, in these conditions, inhibition of the reuptake slowed down the clearance but moderately enhanced the slope and the amplitude of the increase in extracellular NE concentration.

As previously reported concerning NE, this estimate of CA overflow can be compared with the whole CA content and then divided by the number of pulses for each stimulation. Thus, the fractional NE release per pulse was estimated: It reached 5×10^{-6} for stimulation at 10 Hz (30). Figure 2 shows that the fractional DA release ranges from 5×10^{-7} to 5×10^{-6}, depending on the frequency. This low fractional release is consistent with other reports showing that much larger DA releases induced by high-frequency stimulations were not followed by a decrease in the whole DA content (14).

The fractional CA release in CNS appears as low as in peripheral tissues. Moreover, it is facilitated with increasing frequency of action potential to the same extent and in the same range of frequency. Finally, vesicular storage appears essential because inhibition of CA uptake in vesicles by reserpine treatment induces a rapid and pronounced decrease in the evoked release of DA and NE (30,45). Therefore, similarly to peripheral tissue, it is likely that the low fractional CA release in CNS corresponds to a highly intermittent quantal release.

The Elementary Secretory Unit Might Be a Terminal String

On the one hand, the probability of a quantal CA release triggered by a single spike from an individual varicosity in CNS might be as low as in peripheral tissue (i.e., ~0.002) (34). Because the fractional release per pulse is enhanced with increasing frequency, this probability is five- to 10-fold increased when an action potential occurs within a train of three to 10 spikes (as many as 20 for NE neurons). Therefore, the probability that an individual varicosity releases one quanta during one train can be up to 100 times higher than that triggered by a single spike. Nevertheless, the probability that an individual varicosity releases the CA content of one vesicle every time the parent fiber is invaded by a burst seems smaller than one.

On the other hand, the CA innervation is redundant. The elementary secretory unit might not be an individual varicosity, but a terminal string forming as many as 200 synaptic contacts with a dendrite, as illustrated by Fig. 4 (this number of 200 synapses is suggested by analogy with the end-plate at the neuromuscular junction, which consists of about 200 release sites). Therefore, the probability that the content of at least one vesicle is released somewhere along a string by a burst might be 1. In other words, facilitation of CA release by the bursting discharge mode combined with redundancy of the CA innervation might contribute, in a probabilistical manner, to overcome the intrinsically low probability of quantal release from individual varicosities.

Conclusion

The two hypotheses discussed above are not mutually exclusive. Both tend to explain how the temporal and spatial features of CA release might contribute to the safety of the CA transmission in CNS.

SITE OF AUTOREGULATION OF CA RELEASE

Introduction

As regards NE release from sympathetic neurons, it has been demonstrated that it is controlled by an inhibitory feedback, which is due to the action of extracellular NE on alpha$_2$ receptors (34). Concerning DA and NE, this autoregulation has been widely documented in CNS by *in vitro* studies from brain slices. *In vivo* dialysis and *in vivo* electrochemistry show that CA release is inhibited by CA agonists and is enhanced by antagonists acting at the terminal level of a D_2 dopaminergic receptor for DA release and on an alpha$_2$ adrenoreceptor as regards NE. However, evidence showing that endogenous extracellular DA and NE actually inhibit their own release in physiological conditions is indirect (45,46).

Extrasynaptic Extracellular CA Might Be the Signal for Autoregulation

In physiological conditions, the extrasynaptic extracellular CA concentration is 10 to 100 nM (see the first section). However, autoreceptors are more sensitive to CA than postsynaptic receptors. In fact, *in vitro* studies from brain slices show that CA release is inhibited by exogenous CA in the 10- to 100-nM range (47,48). Therefore, the chemical signal driving autoregulation might be the extrasynaptic extracellular level. Thus, although the functional CA release is a brief and localized event, autoregulation might involve a population of varicosities. This hypothesis is supported by a recent study concerning transmitter release from sympathetic nerve terminals: The release from some varicosities enclosed in an extracellular pipette was indirectly recorded by electrophysiology, and this release was inhibited by the addition of clonidine *outside* the pipette (49). Therefore, the intermittent and quantal release of CA from an individual varicosity within a string might be influenced at a distance by the mean extrasynaptic extracellular CA concentration bathing the string.

GENERAL CONCLUSION

The distinction made by Fuxe and Agnati between "wiring" and "volume" transmission provides a new conceptual frame for chemical transmission in CNS. In our opinion, in brain area that is densely innervated by catecholaminergic terminals, DA and NE are involved in a "wiring" rather than in a "volume" transmission. However, the features of this "wiring" transmission are not those of the neuromuscular junction. Firstly, it is not triggered by a single action potential but by a burst of three to 20 spikes. Secondly, the elementary secretion site might be a string of varicosities rather than a single bouton. These features suggest that the mechanisms of CA transmission are intrinsically probabilistic in nature but do not mean that the function of this transmission is soft and uncertain. On the contrary, the function of NE in the rat tail artery and in the PVN is definite and strong, and this is also true as regards the

DA transmission in densely innervated brain areas. In other words, these CA transmissions exhibit a high degree of safety.

The virtue of the concept of Fuxe and Agnati is that it points up some basic features of the chemical transmission in CNS. Thus, it should stimulate new experimental research. However, in our opinion, it should not be taken in a too-restrictive sense, as a strict alternative. The features of the "wiring" transmission in CNS might be in keeping with a continuum from the synaptic cleft to the local extracellular microenvironment of a dendritic segment and from 1 msec to a 1 sec.

ACKNOWLEDGMENTS

This work was supported by INSERM U 171 and CNRS UA 1195. We are indebted to Pr. L. Stjärne for exciting discussions, and we thank Dr. J. Carew for revising the English manuscript.

REFERENCES

1. Fuxe K, Agnati LF, Zoli M, Bjelke B, Zini I. Some aspects of the communicational and computational organization of the brain. *Acta Physiol Scand* 1989;135:203–216.
2. Freeman AS, Meltzer LT, Bunney BS. Firing properties of substantia nigra dopaminergic neurons in freely moving rats. *Life Sci* 1985; 36:1983–1994.
3. Grace AA, Bunney BS. Intracellular and extracellular electrophysiology of nigral dopaminergic neurons. I. Identification and characterization. *Neuroscience* 1983;10:301–315.
4. Grace AA, Bunney BS. The control of firing pattern in nigral dopamine neurons: burst firing. *J Neurosci* 1984;4:2877–2890.
5. Grenhoff J, Tung CS, Svensson TH. The excitatory amino acid antagonist kynurenate induces pacemaker-like firing of dopamine neurons in rat ventral tegmental area *in vivo*. *Acta Physiol Scand* 1988;134:567–568.
6. Svensson TH, Tung CS. Local cooling of prefrontal cortex induces pacemaker-like firing of dopamine neurons in rat ventral tegmental area *in vivo*. *Acta Physiol Scand* 1989;136:135–136.
7. Miller JD, Sanghera MK, German DC. Mesencephalic dopaminergic unit activity in the behaviorally conditioned rat. *Life Sci* 1981;29:1255–1263.
8. Sanghera MK, Trulson ME, German DC. Electrophysiological properties of mouse dopamine neurons: *in vivo* and *in vitro* studies. *Neuroscience* 1984;12:793–801.
9. Nieoullon A, Cheramy A, Glowinski J. Nigral and striatal dopamine release under sensory stimuli. *Nature* 1977;269:340–342.
10. Kissinger PT, Hart JB, Adam RN. Voltammetry in brain tissue—a new neurophysiological measurement. *Brain Res* 1973;55:209–213.
11. Marsden C, Joseph MH, Kruk ZL, et al. *In vivo* voltammetry—present electrodes and methods. *Neuroscience* 1988;25:389–400.
12. Gonon F, Buda M. Regulation of dopamine release by impulse flow and by autoreceptors as studied by *in vivo* voltammetry in the rat striatum. *Neuroscience* 1985;14:765–774.
13. Gonon FG. Nonlinear relationship between impulse flow and dopamine released by rat midbrain dopaminergic neurons as studied by *in vivo* electrochemistry. *Neuroscience* 1988;24:19–28.
14. Kuhr WG, Ewing AG, Near JA, Wightman RM. Amphetamine attenuates the stimulated release of dopamine *in vivo*. *J Pharmacol Exp Ther* 1985;232:388–394.
15. Westerink BHC, De Vries JB. Characterization of *in vivo* dopamine release as determined by brain microdialysis after acute and subchronic implantations: methodological aspects. *J Neurochem* 1988;51:683–687.
16. Imperato A, Di Chiara G. Trans-striatal dialysis coupled to reverse phase high performance liquid chromatography with electrochemical detection: a new method for the study of the *in vivo* release of endogenous dopamine and metabolites. *J Neurosci* 1984;4:966–977.
17. Kuhr WG, Wightman RM. Real-time measurement of dopamine release in rat brain. *Brain Res* 1986;381:168–171.
18. Radhakishun FS, Van Ree JM, Westerink BHC. Scheduled eating increases dopamine release in the nucleus accumbens of food-deprived rats as assessed with on-line brain dialysis. *Neurosci Lett* 1988;85:351–356.
19. Toan DL, Schultz W. Responses of rat pallidum cells to cortex stimulation and effects of altered dopaminergic activity. *Neuroscience* 1985;15: 683–694.
20. Vives F, Mogenson GJ. Electrophysiological study of the effects of D_1 and D_2 dopamine antagonists on the interaction of converging inputs from the sensory-motor cortex and substantia nigra neurons in the rat. *Neuroscience* 1986;17: 349–359.
21. Gonon F, Buda M, Cespuglio R, Jouvet M, Pujol JF. *In vivo* electrochemical detection of catechols in the neostriatum of anaesthetized rats: dopamine or DOPAC? *Nature* 1980;286:902–904.
22. Bouyer JJ, Joh TH, Pickel VM. Ultrastructural localization of tyrosine hydroxylase in rat nucleus accumbens. *J Comp Neurol* 1984;227:92–103.
23. Freund TF, Powell JF, Smith AD. Tyrosine hydroxylase-immunoreactive boutons in synaptic contact with identified striatonigral neurons, with particular reference to dendritic spines. *Neuroscience* 1984;13:1189–1215.

24. Nicholson C, Philips JM. Ion diffusion modified by tortuosity and volume fraction in the extracellular microenvironment of the rat cerebellum. *J Physiol Lond* 1981;321:225–257.
25. Teranishi T, Negishi K, Kato S. Dopamine modulates S-potential amplitude and dye-coupling between external horizontal cells in carp retina. *Nature* 1983;301:243–246.
26. Kebabian JW, Petzold GL, Greengard P. Dopamine-sensitive adenylate cyclase in caudate nucleus of rat brain, and its similarity to the "dopamine receptor." *Proc Natl Acad Sci* 1972; 69:2145–2149.
27. Johnson SW, Hoffer BJ, Freedman R. Investigation of the failure of parenterally administered haloperidol to antagonize dopamine released from micropipettes in the caudate. *J Neurosci* 1986;6:572–580.
28. Akaike T. Periodic bursting activities of locus coeruleus neurons in the rat. *Brain Res* 1982; 239:629–633.
29. Elam M, Svensson TH, Thorén P. Locus coeruleus neurons and sympathetic nerves activation by cutaneous sensory afferents. *Brain Res* 1986;366:254–261.
30. Suaud-Chagny MF, Mermet C, Gonon F. Electrically evoked noradrenaline release in the rat hypothalamic paraventricular nucleus studied by *in vivo* electrochemistry: characterization and facilitation by increasing the stimulation frequency. *Neuroscience* 1990;34:411–422.
31. Day TA, Ferguson AV, Renaud LP. Facilitatory influence of noradrenergic afferents on the excitability of rat paraventricular nucleus neurosecretory cells. *J Physiol* 1984;355:237–249.
32. Tanaka J, Kaba H, Saito H, Seto K. Inputs from the A1 noradrenergic region to hypothalamic paraventricular neurons in the rat. *Brain Res* 1985;335:368–371.
33. Day TA, Renaud LP. Electrophysiological evidence that noradrenergic afferents selectively facilitate the activity of supraoptic vasopressin neurons. *Brain Res* 1984;303:233–240.
34. Stärjne L. Basic mechanisms and local modulation of nerve impulse-induced secretion of neurotransmitters from individual sympathetic nerve varicosities. *Rev Physiol Biochem Pharmacol* 1989;112:1–137.
35. Cheung DW. Neural regulation of electrical and mechanical activities in the rat tail artery. *Pflugers Arch* 1984;400:335–337.
36. Luff SE, McLachlan EM, Hirst GDS. An ultrastructural analysis of the sympathetic neuromuscular junctions on arterioles of the submucosa of the guinea pig ileum. *J Comp Neurol* 1987;257:578–594.
37. Mermet C, Gonon F. *In vivo* voltammetric monitoring of noradrenaline release and catecholamine metabolism in the hypothalamic paraventricular nucleus. *Neuroscience* 1986;19:829–838.
38. Mermet C, Gonon F. Ether stress stimulates noradrenaline release in the hypothalamic paraventricular nucleus. *Neuroendocrinology* 1988; 47:75–82.
39. Liposits ZS, Phelix C, Paull WK. Electron microscopic analysis of tyrosine hydroxylase, dopamine-β-hydroxylase and phenylthanolamine-*N*-methyl-transferase immunoreactive innervation of the hypothalamic paraventricular nucleus in the rat. *Histochemistry* 1986;84:105–120.
40. Day TA, Randle JCR, Renaud LP. Opposing alpha- and beta-adrenergic mechanisms mediate dose-dependent actions of noradrenaline on supraoptic vasopressin neurones *in vivo*. *Brain Res* 1985;358:171–179.
41. Poulain DA, Wakerley JB. Electrophysiology of hypothalamic magnocellular neurones secreting oxytocin and vasopressin. *Neuroscience* 1982; 7:773–808.
42. Shepherd GM, Brayton RK. Logic operations are properties of computer-simulated interactions between excitable dendritic spines. *Neuroscience* 1987;21:151–165.
43. Doucet G, Descarries L, Garcia S. Quantification of the dopamine innervation in adult rat neostriatum. *Neuroscience* 1986;19:427–445.
44. Björklund A, Lindvall O. Dopamine-containing systems in the CNS. In: Björklund A, Hökfelt T, eds. *Handbook of chemical neuroanatomy*. Elsevier 1984;55–122.
45. Suaud-Chagny MF, Buda M, Gonon F. Pharmacology of electrically evoked dopamine release studied in the rat olfactory tubercle by *in vivo* electrochemistry. *Eur J Pharmacol* 1989;164: 273–283.
46. Mermet C, Suaud-Chagny MF, Gonon F. Electrically evoked noradrenaline release in the rat hypothalamic paraventricular nucleus studied by *in vivo* electrochemistry: autoregulation by alpha-2 receptors. *Neuroscience* 1990;34:423–432.
47. Reimmann W, Zumstein A, Jackisch R, Starke K, Hertting G. Effect of extracellular dopamine on the release of dopamine in the rabbit caudate nucleus: evidence for a dopaminergic feedback inhibition. *Naunyn Schmiedebergs Arch Pharmacol* 1979;306:53–60.
48. Taube HD, Starke K, Borowski E. Presynaptic receptor systems on the noradrenergic neurones of rat brain. *Naunyn Schmiedebergs Arch Pharmacol* 1977;299:123–141.
49. Stjärne L, Stjärne E. Evidence that alpha 2-agonists may inhibit sympathetic transmitter secretion by an effect "upstream" of varicosities. *Acta Physiol Scand* 1988;133:433–434.

Volume Transmission in the Brain: Novel Mechanisms for Neural Transmission, edited by K. Fuxe and L.F. Agnati. Raven Press, Ltd., New York © 1991.

27
Biochemistry of Peptides

Viktor Mutt

Department of Biochemistry II, Karolinska Institute, S-104 01 Stockholm, Sweden

SOME ASPECTS OF BIOSYNTHESIS

In 1953, du Vigneaud and co-workers (1) described the synthesis of a nonapeptide with the amino acid sequence that they (2) and Tuppy (independent of the others) (3) had determined for ox oxytocin. The natural and synthetic peptides were found to have identical activities in several bioassay systems. This proved that a peptide could be a hormone and not, as had sometimes been suggested, admittedly with decreasing conviction, only a carrier for some ephemeral hormonal principle (4). Since then, many peptide hormones have been isolated, had their amino acid sequences determined, and had replicates synthesized. Indeed it may seem as if all interesting naturally occurring peptides were of a hormonal character. This is, of course, not so. Long before the amino acid sequences had been established for oxytocin and vasopressin (5,6), this had been done for carnosine (7,8) and for glutathione (9,10, and references therein), both of which had been isolated from mammalian tissues, and for glutathione, which was isolated from yeast. Also the amino acid sequence of the antibiotic decapeptide gramicidin S from *Bacillus brevis* had been determined by 1947 (11). Carnosine, glutathione, and gramicidine S are not formed by fragmentation of precursor proteins, but from amino acids by the action of peptide-specific synthetases. Many peptides with antibiotic properties are formed in this manner, and much work is currently in progress in many laboratories on such synthetases [e.g., on the synthetase involved in the formation of delta-(L-alfa-aminoadipyl)L-cysteinyl-D-valine, the starting tripeptide for the formation of penicillin and cephalosporin] (12).

Glutathione is mainly an intracellular peptide with a variety of different functions. Because it regulates the activities of several enzymes (13,14), it may be called a regulatory peptide, acting in its cell of origin. This is distinctive to hormonal peptides, which act as messengers between different cells. Glutathione may indeed also act as a hormonal peptide, although not in free form, but as a component of a leukotriene (15). Typical hormonal peptides are, however, formed by proteolytic cleavage of precursor proteins. This was first recognized for the extracellular formation of angiotensin from angiotensinogen (16) and of bradykinin from kininogen (17). The possibility that proteolytic cleavage might be involved in the intracellular formation of peptide hormones was implicit in the observation by Sachs and Takabatake (18) that the biosynthesis of vasopressin in dog hypothalamus took place "via the intermediate formation of a precursor molecule which contains the hormone in a bound biologically inactive form." It was, however, the investigations of Steiner and Oyer (19) on the biosynthesis of insulin and Chrétien and Li (20) on the formation of gamma-lipotropin from beta-lipotropin, together with the vast amount of work (but in different laboratories) on the biosynthesis of different hormonal peptides that followed these observations, that led to the realization that hormonal peptides are, as a rule, formed from prohormones. Proteolytic cleavage of the latter C-terminally to pairs of basic amino acid residues flanking the hormonal sequence releases the peptide, and this is followed by removal by carboxypeptidase activity of the pair that extends it C-terminally (21). It was further found that the biosynthesis of the prohormone also involved a proteolytic step: *In vitro* translation of the messenger ribonucleic acid (mRNA) in the cell-free system that did not contain cell membranes did not result in the prohormone itself, but in a preprohormone in which the prohormone was extended N-terminally by a "pre"-sequence. This was shown early for the rat insulins (22) and ox parathyroid hormone (23). The pre-sequence is a synonym for the "signal sequence" of other secreted proteins (24,25). In some cases [e.g., rat (26) and ox (27) pituitary prolactin, rat (28) and ox (27) growth hormone, and human placental lactogen (29)], it was found that the pre-sequence was not attached to the N-terminus of a prohormone, but to that of the hormone itself. This was confirmed for rat growth hormone by the use of another technique, one that was soon to dominate the field of peptide hormone precursor research: preparation, using recombinant deoxyribonucleic acid (DNA) methodology, of DNA complementary to the mRNA of the hormone precursor, determination of its nucleotide sequence, and deduction from this of the amino acid sequence of the protein (30). It was thereby evident that there is no absolute requirement for a hormonal sequence to be flanked by pairs of basic amino acid residues in its precursor, because signal peptides are known to have neutral C-terminal amino acid residues for attachment to the hormonal or prohormonal sequence (31). Renin affects the release of angiotensin from angiotensinogen by catalyzing the hydrolytic cleavage of a bond between neutral amino acid residues (32).

It is now well known that during the biosynthesis of a peptide hormone precursor protein, the pre-sequence is removed at the ribosomal level during passage of the precursor through the membrane of the rough endoplasmatic reticulum. The prohormone or hormone is then transported to the Golgi apparatus, where proteolytic cleavage(s) of the prohormone and, in the case of some hormones, various kinds of posttranslational processing takes place, followed by packaging of the mature hormone into secretory vesicles, ready for secretion out of the cell in response to some appropriate stimulus. The time course for the different biosynthetic steps has been determined in several cases, especially clearly for the parathyroid hormone (33).

Immunochemical (34) investigations and investigations of biosynthesis (35,36) showed that corticotropin, with its alpha-melanotropin component, and beta-lipotropin, with its beta-melanotropin and beta-endorphin components, are biosynthesized as parts of the same preprohormone. This was confirmed when the amino acid sequence of prepro-opiomelanocortin (37) was deduced from the complementary DNA (cDNA) nucleotide sequence by Nakanishi et al. (38). These workers found that the preprohormone sequence N-terminally to the lipotropin and corticotropin sequences contained a dodecapeptide sequence with structural similarity to alpha- and beta-melanotropin and was flanked by pairs of residues of basic amino acids. They suggested that this sequence was that of a previously unknown hormonal peptide, gamma-melanotropin. This way of scanning amino acid sequences of proteins for pairs of basic amino acid sequences flanking possible hormonal sequences has found wide application (e.g., refs. 39–42).

In prepro-opiomelanocortin, single copies of different hormonal peptides are found. This is the case also in, for example, prepro-vasoactive intestinal peptide (VIP)/(PHI) (peptide with N-terminal histidine and C-terminal isoleucine amide)/(PHM) (the human form of (PHI) with methionine instead of isoleucine) (43,44), preproneurotensin/neuromedin N (45), and mammalian preproglucagon/glucagon-like peptide I/glucagon-like peptide II (46). In other cases, several copies of an identical hormonal sequence may be found in the same preprohormone: Proenkephalin A has six copies of the methionine–enkephalin sequence, in addition to a single copy of that of leucine enkephalin (47,48); prepro-thyrotropin-releasing hormone (TRH) has five TRH sequences (49); and in the skin of *Xenopus laevis,* there are three different preprocerulеins containing, respectively, one, three, and four copies of the decapeptide sequence (50).

Rosenfeld et al. (51) made the interesting observation that the primary RNA transcript of the rat gene for the dotriacontapeptide calcitonin was processed alternatively to mRNA for preprocalcitonin, primarily in the thyroid "C" cells, or to that for the preprohormone of an earlier unknown hexatriacontapeptide, the "calcitonin gene-related peptide" (CGRP), primarily in the hypothalamus. Later a second calcitonin/CGRP gene, giving rise to a slightly different form of CGRP, was shown to be present in the human (52) as well as in the rat (53) genome.

Other cases have since been described in which the primary transcript of the gene for a hormonal peptide is spliced to mRNAs for nonidentical preprohormones. Thus the endecapeptide tachykinin substance P (SP) has been shown to derive from three different preprotachykinins. It was first found (54) that in ox brain, there is an alpha-preprotachykinin, which contains a single SP sequence as the only tachykinin sequence, but also a beta-preprotachykinin, in which there is (C-terminally to the SP sequence and separated from it by a nonacosapeptide) the sequence of the decapeptide tachykinin neurokinin alpha (NKA) (55), also known as neuromedin L (56) or substance K (54). Later a gamma-preprotachykinin was discovered that also contained the SP and NKA sequences, with the NKA sequence C-terminal to the SP sequence but separated by a tetradecapeptide identical in the bovine form to the nonacosapeptide of beta-preprotachykinin, except for deletion of the 6–21 sequence of the latter (57). The three different preprotachykinins have been found to be present also in rat brain (57,58). Interestingly, the relative amounts in which they are formed has been found to be tissue-specific in the ox (54) but not in the rat (59).

The primary RNA transcript of the gene for the gastrin-releasing polypeptide (GRP) (60), a mammalian peptide related structurally to amphibian skin bombesin (61), has likewise been found to be spliced into three different mRNA forms. However, the preprohormones into which they are translated differ only outside of the GRP sequence (62).

A typical peptide has an amino group at one end and a carboxyl group at the other. GRP and the tachykinins, however, like many other peptides, lack free C-terminal carboxyl groups but have alpha-carboxamide groups instead. The occurrence of this type of structure in peptides was originally observed in oxytocin and vasopressin, both of which have residues of glycine amide C-terminally. The way in which such amides are formed, however, long remained unknown. In 1972, Gross (63) suggested that they were not formed by the amidation of free carboxyl groups, but by cleavage of peptide chains (63). This has proved to be the case, although the actual mechanism suggested by Gross has not been found to be operative in nature.

Unraveling of what the mechanism—or at least the only mechanism hitherto observed—was, started with the observation (64) that the alpha-melanotropin, which has C-terminal valine amide, has the same amino acid sequence as the N-terminal tridecapeptide of corticotropin, in which Val-13 is followed by Gly-14, and that the bee venom peptide melittin has C-terminal glutaminylamide, whereas promelittin has C-terminal glutaminylglycine (65). It is now clearly established (66,67) that if processing of a prohormone leads to a hormonal peptide with C-terminal glycine, an additional reaction takes place, catalyzed by a Cu^{2+}- and ascorbic acid–dependent peptidylglycine hydroxylase, leading to hydroxylation of the glycine residue on its alpha-carbon, followed by dismutation to glyoxalic acid and a hormonal peptide with a C-terminal alpha-amide.

Hitherto, no naturally occurring peptide with a C-terminal alpha-amide structure has been found to lack some kind of hormonal or alternatively toxic activity. Because of this, it has been reasonable to look in tissue extracts—or rather various fractions of peptides from such extracts—for yet unknown peptides with C-terminal alpha-amide structures, with the expectation that they might be of hormonal nature. Such work has indeed led to the discovery of a number of peptides that first after their isolation have been found to have hormonal peptide (68–73).

SOME PEPTIDE GROUPS

In the early 1950s it was found that oxytocin and vasopressin, despite differences in physiological activities, belonged to one structural group because they had highly similar amino acid sequences. Since then, a number of additional groups of structurally clearly related hormonal peptides have been discovered [e.g., a glucagon-secretin group (74,75), a gastrin-cholecystokinin group (76,77), or an opioid peptide group (78)]. These groups refer to peptides in the same species. The various peptides in them may form subgroups of their own on the basis of species variability (e.g., when it was found that pig vasopressin differs from ox vasopressin by having Lys-8 instead of Arg-8 (5) or human gastrin from pig gastrin by having Leu-5 instead of Met-5 (79). Earlier, Sanger (80) had found that insulins from different species had different amino acid compositions. At times, when comparing peptides from distantly related species, it may not be clear whether two structurally similar peptides belong to a species-determined subgroup of the same peptide or different members of the group. For instance, the amphibian skin decapeptide cerulein is structurally similar to cholecystokinin and gastrin. In addition to other similarities, the three peptides have identical C-terminal pentapeptide sequences. Therefore they are usually referred to as representing three members of the same group. However, as long as neither amphibian gastrin nor cholecystokinin has been isolated, the possibility remains that cerulein in part belongs to one or another of these subgroups. The recent finding of material in the brain that reacts with antisera specific to cerulein rather than to gastrin or cholecystokinin, however, suggests that there may exist a yet unknown "mammalian cerulein" and (81) that cerulein consequently may form its own subgroup. Another long-known group has expanded to a remarkable degree. Originally defined as the glucagon-secretin group it was shown to include also VIP (82), the gastric inhibitory (or glucose-dependent insulinotropic) polypeptide (GIP) (83,84), PHI (70), the growth hormone–releasing hormone (GHRH) (85,86), as well as the two extended forms of glucagon glicentin (87) and oxyntomodulin (88). Recently it has expanded still more. Elucidation, by way of the corresponding cDNA nucleotide sequences, the amino acid sequences of mammalian preproglucagons, first for the hamster (46) and ox (89) forms, revealed that they contain, in addition to the glucagon sequence and C-terminally to it, two other sequences showing similarity to it. It seems probable that these sequences were released in the glucagon-like peptides (GLP) I (closest to the glucagon sequence) and II. Fish had been found to have the sequence of only one GLP, in addition to the glucagon sequence in their preproglucagons, but, however, have two distinct preproglucagons (90). It was evident that if GLP-I were to be released from mammalian preproglucagon by cleavage C-terminally to pairs of basic amino acid residues flanking its sequence, followed by removal of the two residues extending the released peptide, this would result in a heptatriacontapeptide GLP-I with a C-terminal residue of glycine. This heptatriacontapeptide would, however, not be glucagon-like in one important respect: A typical feature of the glucagon, secretin, VIP, GIP, PHI, GHRH group of peptides is that they have a residue of phenylalanine in position 6. As pointed out earlier (91), this would not be the case for the heptatriacontapeptide. However, if an additional cleavage took place between its Arg-6 and His-7 or, alternatively, if the prohormone was initially cleaved at this position, the resulting entriacontapeptide would have the typical His-1 Phe-6 relation. It would, however, also have C-terminal glycine, a typical amidation signal. Until recently, in all known mammalian peptides with C-terminal alpha-amide structures, the amidated residue was one of a neutral amino acid, and indeed the only vertebrate peptide with a charged amidated residue was the GLP, a nonacosapeptide with C-terminal arginine amide from the preproglucagon II of the anglerfish (92). An immunochemical investigation of the proteolytic fragmentation of preproglucagon in the rat pancreas and intestine (93) indicated distinct differences and also suggested that GLP-I may be released in the intestine in different forms, one of which was a triacontapeptide with C-terminal arginine amide. Strong although not conclusive evidence for the occurrence of such a C-terminally amidated triacontapeptide in extracts of pig intestine was provided by Holst et al. (94), and soon thereafter conclusive evidence was provided for its presence in extracts from rat intestine (95).

More recently, its presence in extracts of human as well as pig intestine has been clearly established (96). It has been found to have a hormonal activity (i.e., stimulation of the release of insulin) (94,95, 97), which applies also to its C-terminally glycine-extended form (98). It is often referred to as GLP-I-7-36 amide instead of simply GLP-I. This is presumably based on the unfounded assumption that because many hormonal peptides are released by cleavages C-terminally to pairs of basic amino acid residues in their precursors, all should be released in this fashion.

In the tritriacontapeptide GLP-II, which is indeed released from its precursor according to the usual pattern, there is the expected positional relation between N-terminal His and Phe-6. Little is, however, yet known about possible hormonal activities for GLP-II (99).

It has been established for most members of the glucagon-secretin, etc., group of peptides, first for glucagon itself by Rall and Sutherland (100), that they stimulate adenylate cyclase activity in their target cells.

Recently, from sheep hypothalamic tissue, Miyata et al. (101) isolated an octatriacontapeptide with C-terminal lysine amide that has a His-1 Phe-6 positional relation typical for the group and an overall sequence identity to sheep VIP of 68% in its N-terminal octacosapeptide sequence. It is, however, some 1,000 times stronger than VIP in stimulating adenylate cyclase activity in rat anterior pituitary cells in culture. The octatriacontapeptide has been named pituitary adenylate cyclase activating polypeptide with 38 residues (PACAP 38).

The secretin-glucagon, etc., group is long known, as is the gastrin-cholecystokinin (cerulein?) group. Several others have been discovered more recently [e.g., the pancreatic polypeptide–PYY-NPY group (102,103) or the heart atrial natriuretic peptide (cardiodilatin) (104,105)–brain natriuretic peptide (106) group]. A recent one is the endothelin (107)–vasoactive intestinal contractor (VIC) (108) group. The encosapeptides of this group, each with two disulfide bridges, show a remarkable structural similarity (109) to the sarafotoxins (i.e., toxic encosapeptides from the venom of the snake *Antractaspis engaddensis*) (110).

In addition to the encosapeptide endothelin, a C-terminally extended form of it, a nonatriacontapeptide, has been found to be present in small amounts relative to those of the encosapeptide in supernatants of cultures of pig aortic endothelial cells (mentioned in ref. 111). The existence of this endothelin-39 had been predicted earlier from the cDNA-nucleotide sequence–deduced amino acid sequence of pig preproendothelin (107). The contractile activity of endothelin-39 on isolated rat aorta was found to be weak as compared with endothelin-21 (112). The formation of endothelin-21 from endothelin-39 does not take place by cleavage at a pair of basic amino acid residues, but by a cleavage of a tryptophanyl–valine bond (107).

In the case of the parathyroid hormone, however, the longest form, a tetraoctacontapeptide, appears to be an important, hormonally active circulating form, although N-terminal fragments of it, at least down to a tetratriacontapeptide, are also still active (113,114). Much interest is currently focused on a parathyroid hormone–like peptide (PTH-LP) that interacts with receptors for the parathyroid hormone and appears to be responsible for the long-known association of certain nonparathyroid tumors with hypercalcemia (115,116, and references therein). Other hormonal peptides for which C-terminally extended forms have been shown to be hormonally active are glucagon-37 or oxyntomodulin, so named because of its stimulating activity on adenylate cyclase in the oxyntic area of the rat stomach (117,118), and PHI (PHM).

It was shown (119) that the processing of pro-VIP/PHI (PHM for the human form) in adrenal pheochromocytoma resulted in the release of not only PHM and VIP, but alternatively also of a dotetracontapeptide with C-terminal valine, histidine valine 42 (PHV-42), in which the PHM sequence is extended C-terminally by a pentadecapeptide. It is obvious that for its release, cleavage of the prohormone C-terminally to the Lys-Arg sequence flanking the PHM sequence at its N-terminus had taken place as usual but that the Gly-Lys-Arg sequence flanking it from its C-terminus had not been processed, and the extension had been continued for an additional dodecapeptide until and including the valine residue preceding the Lys-Arg sequence that flanks the VIP sequence from its N-terminus in the prohormone. PHV-42 was found to be considerably more potent than PHM in relaxing rat isolated uterus and smooth muscle preparations of rat stomach and guinea pig trachea.

PHV-42 and PHI were found to be present in extracts of rat intestine but in smaller amounts than PHI-Gly (i.e., PHI extended C-terminally by a residue of glycine) (120). Earlier it had been shown that secretin-Gly (121), secretin-Gly-Lys-Arg (122), and VIP-Gly-Lys-Arg (123) were present in extracts of pig intestine, although apparently in much lower amounts than the mature amidated forms of secretin and VIP. The hormonal activities of the extended forms were roughly the same as for the mature forms.

Immunochemical evidence has been provided for the presence of C-terminally extended forms of cholecystokinin in extracts of pig cerebral cortex (124) and of similarly extended forms of gastrin in extracts of pig stomach (125). Recently it was shown that both immunoreactive gastrin-Gly and cholecystokinin-Gly are present also in human plasma, where their concentrations increase in connection with meals (126).

Such forms should, however, have low cholecystokinin-like, respectively gastrin-like bioactivity because it is well known that the C-terminal phenylalanine amide residue is important for gastrin-like activity (127).

However, fully active forms of gastrin as well as cholecystokinin are known that are longer than the one originally isolated, because of N-terminal extensions over pairs of basic amino acid residues (128–130). The same applies to somatostatin (131–133).

Recently, in extracts of pig brain, a hexatriacontapeptide, a product of beta-preprotachykinin, has been found in which the NKA (alternatively named substance K) sequence is extended N-terminally by a hexacosapeptide with C-terminal Lys-Arg (134). An analogously formed product of gamma-preprotachykinin, neuropeptide gamma, or gamma-preprotachykinin-(72-92)-peptide amide (135) has been isolated from extracts of rabbit intestine (136) and also shown to be produced, together with NKA and SP, by a mouse pituitary cell line (AtT-20) after infection with a vaccinia virus recombinant expressing gamma-preprotachykinin (59).

Not only hormonal peptides come in groups; many others do so, too. Two recently discovered peptides are "pig cecropin" and PEC-60, so named because of its terminal amino acid residues in one-letter notation, and the total number of residues was discovered through its ability to inhibit the release of insulin from the isolated perfused rat pancreas, admittedly a hormonal type of action. However, when its amino acid sequence was determined, it became evident that PEC-60 belonged to a peptide group that also included the secretory (or Kazal) type of pancreatic trypsin inhibitor, PSTI, (137–139) and the "monitor peptide," which had been isolated from rat bile–pancreatic juice and shown to stimulate the release into the blood stream of cholecystokinin on introduction into the lumen of the intestine in rats (140). The true physiological role of PEC-60 is not yet clear. It does not inhibit trypsin (141).

As to pig cecropin, Zasloff (142) recently described the isolation from extracts of *Xenopus laevis* skin of two tricosapeptides with antibacterial activities and closely similar amino acid sequences. He named these peptides *magainins* (142). Because it is well known that many peptides that occur in amphibian skin are also in mammalian intestine and brain (143), we, in collaboration with Boman and co-workers at the Department of Microbiology, Stockholm University, have undertaken a search for antibacterial peptides in extracts of pig small intestine. We have obtained evidence for the occurrence there of several such peptides and have hitherto isolated one, an entriacontapeptide exhibiting antibacterial activity, particularly against certain types of gram-negative bacteria (144). Structurally the entriacontapeptide clearly belongs to a group of peptides named *cecropins* and was described earlier only in insects, in which they obviously participate in defense against bacteria (145). Whether antibacterial activity is the main function of pig cecropin or whether it has some other yet unknown important function remains to be elucidated.

UNEXPECTED ACTIVITIES OF FRAGMENT PEPTIDES

A remarkable development of recent years has been the finding that in several cases, fragments of long-known peptide hormones have hormonal activities of their own, different from those of the unfragmented peptides. Thus it was found that the 1-7 fragment of the endecapeptide SP was antinociceptive in mice on intracerebroventricular injection, just like SP itself, but lacked the typical effects of the latter on smooth muscle contraction and blood pressure (146). The dipeptide glycylglutamine corresponding to the C-terminal sequence of both beta-endorphin and motilin was found to inhibit the firing of rat brainstem neurons (147). Later evidence was presented for the possibility that glycylglutamine was metabolized to glycyl-glutamic acid, which acted as a neurotrophic factor for the maintenance of acetylcholinesterase and butyrylcholinesterase in the preganglionically denervated superior cervical ganglion of the cat (148).

When the C-terminal phenylalanine residue was removed from angiotensin-8, the residual heptapeptide stimulated the release of vasopressin from the rat hypothalamo-neurohypophysial system *in vitro* as efficiently as angiotensin-8 but lacked the vasopressor activity typical of the latter (149). It was subsequently found that angiotensin 1-7 occurred naturally in extracts of rat brain (150).

A peptide with the sequence of the N-terminal encosapeptide of glucagon was found to lack the typical metabolic effects of glucagon but to inhibit electrically evoked contractions of the isolated guinea pig ileum and pentagastrin-stimulated gastric acid secretion in cats as efficiently as glucagon (151). In their actions on motor activity of the human sphincter of Oddi, subtle differences were found between intravenously administered glucagon and the glucagon (1-21) peptide (152). The same applies to their inhibitory activity on secretin-stimulated pancreatic secretion in dogs (153). Another fragment of glucagon, glucagon(22-29) was found to have 1,000 times higher inhibitory activity on the Ca^{2+} pump in rat liver cell membranes than glucagon itself (154). It has often been stated that all the activities of cholecystokinin are exerted by its C-terminal part. Recently, however, it was shown that the 1-21 fragment of cholecystokinin-33 had a lipolytic effect on human adipose tissue *in vitro* (155), whereas the C-terminal octapeptide of cholecystokinin had no such effect.

It may also be mentioned that the 4-9 fragment of vasopressin, which does not exhibit the typical vasopressor effect of vasopressin, has been found to improve the memory of rats on intracerebral administration (156).

PEPTIDES WITH HORMONE-LIKE ACTIVITIES FROM NONPROHORMONAL PROTEINS

Another surprising development has been the finding that several proteins with well-known functions (other than prohormonal) of their own, nevertheless give rise to fragments with, at least pharmacologically, hormone-like activities.

Thus Schally and co-workers (157) isolated a tetradecapeptide with adrenocorticotropic hormone (ACTH)-releasing activity on rat pituitary fragments from extracts of pig hypothalami and showed that it corresponded to the 33-46 sequence of the alpha chain of hemoglobin.

Takagi and co-workers (158) isolated the dipeptide Tyr-Arg from extracts of ox brain, which corresponds to the C-terminal dipeptide of the alpha chain of ox hemoglobin, and showed that it had analgetic activity on intracerebroventricular administration to mice, presumably caused by an enkephalin-releasing action. This dipeptide has been named *kyotorphin*. Subsequently, a pentapeptide corresponding to the C-terminal pentapeptide of the alpha chain of ox hemoglobin was isolated from ox brain extracts, found to have analgetic activity, and named *neo-kyotorphin* (159).

We too have described the isolation of a hemoglobin fragment, a dotriacontapeptide corresponding to the C-terminal sequence of the beta chain of hemoglobin from pig intestine (160). We pointed out that, in size, this peptide resembles a number of peptides with hormonal properties. We have, however, not found any such activity for it, but neither have we looked extensively.

Much more interest has been attracted to the certain peptides with direct opiate-like activities that have been found to be released on proteolytic degradation of casein, casomorphins (161,162), or in more general denomination exorphins because peptides with similar activities are released not only from casein, but also from other food proteins (e.g., wheat gluten) (163–165) and from mitochondrial cytochrome b (166). The casomorphins have been found to be opiate-like also in their ability to stimulate electrolyte absorption in the rabbit ileum *in vitro* (167). Recently they have also been shown to influence—evidently acting from the luminal side—the uptake of leucine in everted sacs of the rat jejunum (168). A hexapeptide fragment of casein, albeit, in

C-terminally methylated form, was found to act as an opioid antagonist (169). These opioid agonists and the antagonist are by far not the only nonprohormonal protein fragments that have been found to have some kind of hormonal activity. For instance, a fragment of bovine serum albumin was shown to enhance the stimulating action of insulin on fatty acid synthesis from glucose in rat adipose tissue (170).

A nonapeptide showing structural and functional similarities to neurotensin has been isolated from pepsin-treated fractions of plasma from several mammalian species (171). It evidently derives from a plasma protein that is similar to albumin, presumably a variant of the quantitatively predominant form of it.

Extraction of mammalian, as well as avian, gastric tissue with dilute aqueous HCl was found to result in the enzymatically catalyzed release from certain tissue proteins of peptides with immunoreactivities and bioactivities similar to those of the amphibian skin peptide xenopsin (172).

Extracts of rabbit stomach were found to contain material with immunoreactivity similar to that of the endecapeptide physalaemin, another amphibian skin peptide. This led to the isolation from such extracts of an octapeptide in which the amino acid residues were in four positions identical to those in the 1-8 peptide of physalaemin. The peptide did not, however, have any physalaemin-like bioactivity and possibly represented a fragment of a longer peptide with such activity (173).

Certain peptides derived from casein stimulate the activity of the mammalian immune system and possibly also its development (174,175). Similar observations have recently been reported for synthetic peptides corresponding to short sequences of a number of mammalian proteins (176).

In this connection, there is evidence suggesting that in the production of antibodies, protein antigens are not recognized as such but are broken down to short peptides, which are then bound to major histocompatibility complex proteins for presentation to T cells (177,178, and references therein).

INFLUENCE ON THE SUBJECT BY GENERAL DEVELOPMENTS IN THE FIELD

With few exceptions such as insulin (179), glucagon (180), the pancreatic polypeptide (181), the human epidermal growth factor/urogastrone (182), (Leu)enkephalin (183,184), and (Met)enkephalin (185), small(ish) peptides isolated from natural sources have not been crystallized, and their three-dimensional (3-D) structures consequently have not been determined by X-ray crystallographic methods. Because of this, in many discussions, structure of a peptide has meant only amino acid sequence. Nevertheless, in other cases, attempts have been made to gain insight into higher-order structures of peptides in solution. Thus, for instance, measurements of optical rotatory dispersion (ORD) and circular dichroism (CD) spectra showed that secretin in water had a low but distinct degree of helical structure (186), and a long helical stretch was identified in beta-endorphin (187). For the C-terminal tetrapeptide of gastrin [and of cholecystokinin], Trp-Met-Asp-Phe-NH_2 in trifloroethanol CD spectra suggested a folded confirmation, bringing the aromatic rings closer together. This conformation was seen also in aqueous solution but to a lesser extent (188).

Also for the desulfated C-terminal heptapeptide of cholecystokinin, Tyr-Met-Gly-Trp-Met-Asp-Phe-NH_2, measurements of fluorescence enhancement suggested that a folded conformation had brought the aromatic rings closer together than they would have been in an extended conformation (189). From the 3-D structures of peptides for which they had been determined by X-ray crystallography, plausible suggestions have been made concerning 3-D structures of structurally related peptides [i.e., for secretin and VIP from the glucagon structure (190) and for neuropeptide Y and peptide YY from that of the "pancreatic polypeptide"] (181). Attempts to use nuclear magnetic resonance (NMR) spectroscopy for the study of peptide conformations began long ago. Such studies on oxytocin and vasopressin were reported in 1969 (191). An ever-increasing number of peptides have been studied by this technique, and recent important advances in methodology seem to be making it the method of choice for such studies (192,193). Recently, NMR studies have, for instance, been reported for gastrin (194,195), cholecystokinin (196–198), the epidermal growth factor/urogastrone (199), secretin (200), bombesin (201,202), (Leu)enkephalin (203), and a peptide corresponding to the 132-153 residues of sperm whale myoglobulin, of which it constitutes an immunodominant region (178). A recent comparison of the structure of the small protein thioredoxin as determined by X-ray crystallography and NMR spectroscopy, respectively, adds support to the reliability of both these methodologies (204).

Another technique increasingly used for the determination of peptide higher-order structures is computer-assisted calculations of molecular dynamics, leading to energy-minimized structures (205). Such investigations are being carried out either in conjunction with NMR studies [e.g., in the case of cholecystokinin (197,198), (Leu)enkephalin (203), and secretin (206)] or alone [e.g., in the case of secretin (207) or of the C-terminal octapeptide of cholecystokinin (208)]. The availability of a crystal structure for avian neuropeptide Y facilitated the computer-assisted construction of a structural model for its mammalian homologue (209). Perhaps not unexpectedly, calculations have shown that a peptide in solution may have more than one energetically reasonable conformation (203,208). Molecular dynamics simulations and molecular dynamics calculations are being extensively used in the design of analogues to hormonal peptides [e.g., in the recent description of the design of superpotent cyclic alpha-melanotropins (210)].

The amino acid sequences are known—mainly by deduction from experimentally determined cDNA nucleotide sequences—for specific receptors to a number of hormonal peptides and hormonal proteins on the surfaces of their target cells. They are known for two different subunits of the insulin receptor (211,212) and for the receptors for the epidermal growth factor/urogastrone (213), the low-density lipoproteins (214), Interleukin 2 (215), the platelet-derived growth factor (216–218), angiotensin (219), neurokinin alpha/substance K (220), substance P (221), prolactin (222), lutropin-choriogonadotropin (223,224), the nerve growth factor (225), the pituitary growth hormone (226), the yeast *Saccharomyces cerevisiae* alpha factor (227), erythropoetin (228), and the atrial natriuretic peptide/cardiodilatin (229).

This is a short list, because for several of the ligands, receptor sequences have been determined in more than one species. Undoubtedly, receptor amino acid sequences will soon be known for many additional hormonal peptides. Amino acid sequence alone is not enough to describe the structure of a protein. Higher-order structures must also be determined, and functionally important posttranslational modifications may exist, which are not apparent from the primary sequence, especially if this has not been determined directly, but by deduction from a nucleotide sequence. However, knowledge of amino acid sequences forms a firm basis for investigations of these problems.

Recent reviews of work on different second messengers involved in the intracellular actions that follow on the interactions of various peptidal (or other) ligands with their specific receptors are by Garbers (230) for cyclic guanosine 3′,5′-monophosphate (cyclicGMP), by de Gunzburg (231), and Levitzki (232) for cyclic adenosine 3′,5′-

monophosphate (cyclicAMP), by Berridge and Irvine (233) and by Nahorski and Potter (234) for the diacylglycerol–inositol polyphosphate combinations, by Pelech and Vance (235) for the diacylglycerol eicosanoid combinations, and by Low and Saltiel (236) for glycosyl-phosphatidyl-inositols. The intracellular guanine-nucleotide-binding proteins (237) that in various ways take part in the cellular activities that follow on hormone receptor interaction (i.e., in some cases by influencing the activities of the enzymes producing second messengers, in others more directly) have been recently reviewed by Gilman (238), Neer and Clapham (239), and Spiegel (240). A recent, short, general review of second-messenger systems for peptide hormones is by Hanley and Steiner (241).

Recent advances that may be of great importance for developments in the hormonal peptide field are:

1. elucidation of the amino acid sequences of several membrane-bound guanylate cyclases and of a subunit of a cytosolic, heterodimeric form of the enzyme by different groups of workers, and the recognition that the membrane-bound enzyme can act directly as a receptor for certain hormonal peptides (references in ref. 230);
2. elucidation of the amino acid sequences of the adenylate cyclases from two species of yeast (242,243) and from a mammal (244);
3. elucidation of the amino acid sequence of an intracellular membrane-bound receptor protein for inositol-1,4,5-trisphosphate (245,246). This protein is most probably involved in the mechanism by which increases in intracellular inositol-1,4,5-trisphosphate concentrations lead to increases in cytosolic Ca^{2+} concentrations by release of Ca^{2+} from intracellular stores (247).

The discovery that receptors known to react with morphine and other poppy alkaloids of the phenantrene series (248,249) react also with endogenous substances of peptide nature (250–252) seems to be the first clear demonstration of the possibility of peptidal and nonpeptidal substances reacting with common receptors. Several other cases have been discovered since then. Thus dibutyryl cyclic GMP was found to antagonize the action of cholecystokinin on the pancreas, gallbladder, and ileum, presumably by competing with the hormone for receptor sites in these organs (253,254). Later certain 3-substituted 5-phenyl-1,4-benzodiazepines were found to be efficient antagonists—active on peroral administration—of one of the two known types of cholecystokinin receptors, the A receptors, with mainly peripheral localization (255). Recently, certain other benzodiazepines have been found to be selective antagonists to the mainly centrally located cholecystokinin B receptors (256), and a pseudopeptide has been described, which is an agonist of the A type but an antagonist of the B type of cholecystokinin receptors (257). An interesting structural and conformational analogy between cholecystokinin and certain ergot alkaloids has been observed by Hughes and Andrews (258). Another nonpeptide that has been found to interact with a receptor for a peptide is erythromycin, which has been found to be an agonist of the motilin receptor (259,260). A few additional cases are known and have been discussed in a recent review (256).

ACKNOWLEDGMENTS

The investigations from the author's laboratory referred to in this chapter have been supported by grants from the Swedish Medical Research Council (Project 13X-01010), KI fonder, KabiGen AB, and Skandigen AB.

REFERENCES

1. du Vigneaud V, Ressler C, Swan JM, Roberts CW, Katsoyannis PG, Gordon S. The synthesis

of an octapeptide amide with the hormonal activity of oxytocin. *J Am Chem Soc* 1953; 75:4879–4880.
2. du Vigneaud V, Ressler C, Trippett S. The sequence of amino acids in oxytocin, with a proposal for the structure of oxytocin. *J Biol Chem* 1953;205:949–957.
3. Tuppy H. The amino-acid sequence in oxytocin. *Biochim Biophys Acta* 1953;11:449–450.
4. Pierce JG, du Vigneaud V. Studies on high potency oxytocic material from beef posterior pituitary lobes. *J Biol Chem* 1950;186:77–84.
5. du Vigneaud V, Lawler HC, Popenoe EA. Enzymatic cleavage of glycinamide from vasopressin and a proposed structure for this pressor-antidiuretic hormone of the posterior pituitary. *J Am Chem Soc* 1953;75:4880–4881.
6. Acher R, Chauvet J. La structure de la vasopressine de boeuf. *Biochim Biophys Acta* 1953;12:487–488.
7. Barger G, Tutin F. Carnosine, constitution and synthesis. *Biochem J* 1918;12:402–407.
8. Baumann L, Ingvaldsen T. Concerning histidine and carnosine. The synthesis of carnosine. *J Biol Chem* 1918;35:263–276.
9. Harington CR, Mead TH. Synthesis of glutathione. *Biochem J* 1935;29:1602–1611.
10. Meister A. On the discovery of glutathione. *TIBS* 1988;13:185–188.
11. Consden R, Gordon AH, Martin AJP, Synge RLM. Gramicidin S: the sequence of the amino-acid residues. *Biochem J* 1947;41:596–602.
12. White RL, DeMarco AC, Shapiro S, Vining LC, Wolfe S. Measurement of gamma-(L-alpha-aminoadipyl)-L-cysteinyl-D-valine synthetase activity in *Streptomyces clavuligerus* by high-performance liquid chromatography after precolumn derivatization with *o*-phthaldialdehyde. *Anal Biochem* 1989;178:399–403.
13. Cappel RE, Gilbert HF. Thiol/disulfide exchange between 3-hydroxy-3-methylglutaryl-CoA reductase and glutathione. *J Biol Chem* 1988; 263:12204–12212.
14. Ondarza RN. Enzyme regulation by biological disulfides. *Biosci Rep* 1989;9:593–604.
15. Hammarström S, Murphy RC, Samuelsson B, Clark DA, Mioskowski C, Corey EJ. Structure of leukotriene C. Identification of the amino acid part. *Biochem Biophys Res Commun* 1979;91:1266–1272.
16. Skeggs LT Jr, Lentz KE, Kahn JR, Shumway NP, Woods KR. The amino acid sequence of hypertensin II. *J Exp Med* 1956;104:193–197.
17. Elliott DF, Lewis GP, Horton EW. The structure of bradykinin—a plasma kinin from ox blood. *Biochem Biophys Res Commun* 1960; 3:87–91.
18. Sachs H, Takabatake Y. Evidence for a precursor in vasopressin biosynthesis. *Endocrinology* 1964;75:943–948.
19. Steiner DF, Oyer PE. The biosynthesis of insulin and a probable precursor of insulin by a human islet cell adenoma. *Proc Natl Acad Sci USA* 1967;57:473–480.
20. Chrétien M, Li CH. Isolation, purification, and characterization of gamma-lipotropic hormones from sheep pituitary glands. *Can J Biochem* 1967;45:1163–1174.
21. Zimmerman M, Mumford RA, Steiner DF. Precursor processing in the biosynthesis of proteins. *Ann NY Acad Sci* 1980;343.
22. Chan SJ, Keim P, Steiner DF. Cell-free synthesis of rat preproinsulins: characterization and partial amino acid sequence determination. *Proc Natl Acad Sci USA* 1976;73:1964–1968.
23. Habener JF, Potts JT Jr, Rich A. Pre-proparathyroid hormone. *J Biol Chem* 1976;251:3893–3899.
24. Blobel G, Sabatini DD. Ribosome-membrane interaction in eukaryotic cells. In: L.A. Manson, ed. *Biomembranes,* vol 2. New York, London: Plenum Press, 1971;193–195.
25. Milstein C, Brownlee CG, Harrison TM, Mathews MB. A possible precursor of immunoglobulin light chains. *Nature (New Biol)* 1972; 239:117–120.
26. Maurer RA, Gorski J, McKean DJ. Partial amino acid sequence of rat pre-prolactin. *Biochem J* 1977;161:189–192.
27. Lingappa VR, Devillers-Thiery A, Blobel G. Nascent prehormones are intermediates in the biosynthesis of authentic bovine pituitary growth hormone and prolactin. *Proc Natl Acad Sci USA* 1977;74:2432–2436.
28. Sussman PM, Tushinski RJ, Bancroft FC. Pregrowth hormone: product of the translation *in vitro* of messenger RNA coding for growth hormone. *Proc Natl Acad Sci USA* 1976;73:29–33.
29. Birken S, Smith DL, Canfield RE, Boime I. Partial amino acid sequence of human placental lactogen precursor and its mature hormone form produced by membrane-associated enzyme activity. *Biochem Biophys Res Commun* 1977;74:106–112.
30. Seeburg PH, Shine J, Martial JA, Baxter JD, Goodman HM. Nucleotide sequence and amplification in bacteria of structural gene for rat growth hormone. *Nature* 1977;270:486–494.
31. von Heijne G. Signal sequences. The limits of variation. *J Mol Biol* 1985;184:99–105.
32. Skeggs LT Jr, Kahn JR, Lentz K, Shumway NP. The preparation, purification, and amino acid sequence of a polypeptide renin substrate. *J Exp Med* 1957;106:439–453.
33. Potts JT Jr, Kronenberg HM, Habener JF, Rich A. Biosynthesis of parathyroid hormone. *Ann NY Acad Sci* 1988;343:38–55.
34. Lowry PJ, Hope J, Silman RE. The evolution of corticotrophin, melanotrophin and lipotrophin. In: James VHT, ed. *Endocrinology,* vol 2. Amsterdam, Oxford: Excerpta Medica, 1977; 71–76.
35. Mains RE, Eipper BA, Ling N. Common precursor to corticotropins and endorphins. *Proc Natl Acad Sci USA* 1977;74:3014–3018.
36. Roberts JL, Herbert E. Characterization of a common precursor to corticotropin and β-lipotropin: Cell-free synthesis of the precursor and identification of corticotropin peptides in the

molecule. *Proc Natl Acad Sci USA* 1977;74:4826–4830.

37. Chrétien M, Benjannet S, Gossard F, et al. From β-lipotropin to β-endorphin and "pro-opio-melanocortin." *Can J Biochem* 1979;57:1111–1121.
38. Nakanishi S, Inoue A, Kita T, et al. Nucleotide sequence of cloned cDNA for bovine corticotropin-β-lipotropin precursor. *Nature* 1979;278:423–427.
39. Amara SG, Jonas V, O'Neil JA, et al. Calcitonin COOH-terminal cleavage peptide as a model for identification of novel neuropeptides predicted by recombinant DNA analysis. *J Biol Chem* 1982;257:2129–2132.
40. MacIntyre I, Hillyard CJ, Murphy PK, Reynolds JJ, Gaines Das RE, Craig RK. A second plasma calcium-lowering peptide from the human calcitonin precursor. *Nature* 1982;300:460–462.
41. Hoffmann W, Richter K, Kreil G. A novel peptide designated PYL[a] and its precursor as predicted from cloned mRNA of *Xenopus laevis* skin. *EMBO J* 1983;2:711–4.
42. Newcomb R, Fisher JM, Scheller RH. Processing of the egg-laying hormone (ELH) precursor in the bag cell neurons of *Aplysia*. *J Biol Chem* 1988;263:12514–12521.
43. Itoh N, Obata K-i, Yanaihara N, Okamoto H. Human preprovasoactive intestinal polypeptide contains a novel PHI-27-like peptide, PHM-27. *Nature* 1983;304:547–549.
44. Bloom SR, Christofides ND, Delamarter J, Buell G, Kawashima E, Polak JM. Diarrhoea in VIPoma patients associated with cosecretion of a second active peptide (peptide histidine isoleucine) explained by single coding gene. *Lancet* 1983,II:1163–1165.
45. Dobner PR, Barber DL, Villa-Komaroff L, McKiernan C. Cloning and sequence analysis of cDNA for the canine neurotensin/neuromedin N precursor. *Proc Natl Acad Sci USA* 1987;84:3516–3520.
46. Bell GI, Santerre RF, Mullenbach GT. Hamster preproglucagon contains the sequence of glucagon and two related peptides. *Nature* 1983;302:716–718.
47. Gubler U, Seeburg P, Hoffman BJ, Gage LP, Udenfriend S. Molecular cloning establishes proenkephalin as precursor of enkephalin-containing peptides. *Nature* 1982;295:206–208.
48. Noda M, Furutani Y, Takahashi H, et al. Cloning and sequence analysis of cDNA for bovine adrenal preproenkephalin. *Nature* 1982;295:202–206.
49. Lechan RM, Wu P, Jackson IMD, et al. Thyrotropin-releasing hormone precursor: characterization in rat brain. *Science* 1986;231:159–161.
50. Richter K, Egger R, Kreil G. Sequence of preprocaerulein cDNAs cloned from skin of *Xenopus laevis*. *J Biol Chem* 1986;261:3676–3680.
51. Rosenfeld MG, Mermod J-J, Amara SG, et al. Production of a novel neuropeptide encoded by the calcitonin gene via tissue-specific RNA processing. *Nature* 1983;304:129–135.
52. Amara SG, Arriza JL, Leff SE, Swanson LW, Evans RM, Rosenfeld MG. Expression in brain of a messenger RNA encoding a novel neuropeptide homologous to calcitonin gene-related peptide. *Science* 1985;229:1094–1097.
53. Steenbergh PH, Höppener JWM, Zandberg J, Lips CJM, Jansz HS. A second human calcitonin/CGRP gene. *FEBS Lett* 1985;183:403–407.
54. Nawa H, Hirose T, Takashima H, Inayama S, Nakanishi S. Nucleotide sequences of cloned cDNAs for two types of bovine brain substance P precursor. *Nature* 1983;306:32–36.
55. Kimura S, Okada M, Sugita Y, Kanazawa I, Munekata E. Novel neuropeptides, neurokinin α and β, isolated from porcine spinal cord. *Proc Japan Acad* 1983;59(B):101–104.
56. Minamino N, Kangawa K, Fukuda A, Matsuo H. Neuromedin L: a novel mammalian tachykinin identified in porcine spinal cord. *Neuropeptides* 1984;4:157–166.
57. Kawaguchi Y, Hoshimaru M, Nawa H, Nakanishi S. Sequence analysis of cloned cDNA for rat substance P precursor: existence of a third substance P precursor. *Biochem Biophys Res Commun* 1986;139:1040–1046.
58. Krause JE, Chirgwin JM, Carter MS, Xu ZS, Hershey AD. Three rat preprotachykinin mRNAs encode the neuropeptides substance P and neurokinin A. *Proc Natl Acad Sci USA* 1987;84:881–885.
59. Krause JE, MacDonald MR, Takeda Y. The polyprotein nature of substance P precursors. *BioEssays* 1989;10:62–69.
60. McDonald TJ, Jörnvall H, Nilsson G, et al. Characterization of a gastrin releasing peptide from porcine non-antral gastric tissue. *Biochem Biophys Res Commun* 1979;90:227–233.
61. Erspamer V, Falconieri Erspamer G, Inselvini M. Some pharmacological actions of alytesin and bombesin. *J Pharm Pharmacol* 1970;22:875–876.
62. Sausville EA, Lebacq-Verheyden A-M, Spindel ER, Cuttitta F, Gazdar AF, Battey JF. Expression of the gastrin-releasing peptide gene in human small cell lung cancer. *J Biol Chem* 1986;261:2451–2457.
63. Gross E. Structural relationships in and between peptides with α,β-unsaturated amino acids. In: Meienhofer J, ed. *Chemistry and biology of peptides. Proc 3rd Am. Peptide Symp.* Ann Arbor, MI: Ann Arbor Scientific Publishers, 1972;671–678.
64. Smyth DG. The peptide hormones: molecular and cellular aspects. *Nature* 1975;257:89–90.
65. Suchanek G, Kreil G, Hermodson MA. Amino acid sequence of honeybee prepromelittin synthesized *in vitro*. *Proc Natl Acad Sci USA* 1978;75:701–704.
66. Eipper BA, Mains RE. Peptide α-amidation. *Annu Rev Physiol* 1988;50:333–344.
67. Bradbury AF, Smyth DG. Enzyme-catalysed

peptide amidation. *Eur J Biochem* 1987;169: 579–584.

68. Tatemoto K, Mutt V. Chemical determination of polypeptide hormones. *Proc Natl Acad Sci USA* 1978;75:4115–4119.
69. Tatemoto K, Mutt V. Isolation of two novel candidate hormones using a chemical method for finding naturally occurring polypeptides. *Nature* 1980;285:417–418.
70. Tatemoto K, Mutt V. Isolation and characterization of the intestinal peptide porcine PHI (PHI-27), a new member of the glucagon-secretin family. *Proc Natl Acad Sci USA* 1981;78: 6603–6607.
71. Tatemoto K, Carlquist M, Mutt V. Neuropeptide Y—a novel brain peptide with structural similarities to peptide YY and pancreatic polypeptide. *Nature* 1982;296:659–660.
72. Tatemoto K, Rökaeus Å, Jörnvall H, McDonald TJ, Mutt V. Galanin—a novel biologically active peptide from porcine intestine. *FEBS Lett* 1983;164:124–128.
73. Tatemoto K, Efendić S, Mutt V, Makk G, Feistner GJ, Barchas JD. Pancreastatin, a novel pancreatic peptide that inhibits insulin secretion. *Nature* 1986;324:476–478.
74. Bromer WW, Sinn LG, Staub A, Behrens OK. The amino acid sequence of glucagon. *J Am Chem Soc* 1956;78:3858–3860.
75. Jorpes JE, Mutt V, Magnusson S, Steele BB. Amino acid composition and N-terminal amino acid sequence of porcine secretin. *Biochem Biophys Res Commun* 1962;9:275–279.
76. Gregory H, Hardy PM, Jones DS, Kenner GW, Sheppard RC. The antral hormone gastrin. Structure of gastrin. *Nature* 1964;204:931–933.
77. Mutt V, Jorpes JE. Structure of porcine cholecystokinin-pancreozymin. *Eur J Biochem* 1968; 6:156–162.
78. Morley JS. Chemistry of opioid peptides. *Br Med Bull* 1983;39:5–10.
79. Bentley PH, Kenner GW, Sheppard RC. Structures of human gastrins I and II. *Nature* 1966;209:583–585.
80. Sanger F. Species differences in insulins. *Nature* 1949;164:529.
81. Varro A. Identification and characterization of a novel CCK-like immunoreactive peptide in pig CNS. *Neurochem Int* 1989;14:505–510.
82. Said SI, Mutt V. Isolation from porcine-intestinal wall of a vasoactive octacosapeptide related to secretin and to glucagon. *Eur J Biochem* 1972;28:199–204.
83. Brown JC, Dryburgh JR. A gastric inhibitory polypeptide II: the complete amino acid sequence. *Can J Biochem* 1971;49:867–872.
84. Jörnvall H, Carlquist M, Kwauk S, et al. Amino acid sequence and heterogeneity of gastric inhibitory polypeptide (GIP). *FEBS Lett* 1981;123:205–210.
85. Guillemin R, Brazeau P, Böhlen P, Esch F, Ling N, Wehrenberg WB. Growth hormone-releasing factor from a human pancreatic tumor that caused acromegaly. *Science* 1982;218:586–587.
86. Rivier J, Spiess J, Thorner M, Vale W. Characterization of a growth hormone-releasing factor from a human pancreatic islet tumour. *Nature* 1982;300:276–278.
87. Thim L, Moody AJ. The primary structure of porcine glicentin (proglucagon). *Regul Pept* 1981;2:139–150.
88. Bataille D, Tatemoto K, Gespach C, Jörnvall H, Rosselin G, Mutt V. Isolation of glucagon-37 (bioactive enteroglucagon/oxyntomodulin) from porcine jejuno-ileum. *FEBS Lett* 1982; 146:79–86.
89. Lopez LC, Frazier ML, Su C-J, Kumar A, Saunders GF. Mammalian pancreatic preproglucagon contains three glucagon-related peptides. *Proc Natl Acad Sci USA* 1983;80:5485–5489.
90. Lopez LC, Li W-H, Frazier ML, Luo C-C, Saunders GF. Evolution of glucagon genes. *Mol Biol Evol* 1984;1:335–344.
91. Mutt V. Questions answered and raised by work on the chemistry of gastrointestinal and cerebrogastrointestinal hormonal polypeptides. *Chemica Scripta* 1986;26B:191–207.
92. Andrews PC, Hawke DH, Lee TD, Legesse K, Noe BD, Shively JE. Isolation and structure of the principal products of preproglucagon processing, including an amidated glucagon-like peptide. *J Biol Chem* 1986;261:8128–8133.
93. Mojsov S, Heinrich G, Wilson IB, Ravazzola M, Orci L, Habener JF. Preproglucagon gene expression in pancreas and intestine diversifies at the level of post-translational processing. *J Biol Chem* 1986;261:11880–11889.
94. Holst JJ, Ørskov C, Vagn Nielsen O, Schwartz TW. Truncated glucagon-like peptide I, an insulin-releasing hormone from the distal gut. *FEBS Lett* 1987;211:169–174.
95. Kreymann B, Yiangou Y, Kanse S, Williams G, Ghatei MA, Bloom SR. Isolation and characterisation of GLP-1 7-36 amide from rat intestine. *FEBS Lett* 1988;242:167–170.
96. Ørskov C, Bersani M, Johnsen AH, Højrup P, Holst JJ. Complete sequences of glucagon-like peptide-1 from human and pig small intestine. *J Biol Chem* 1989;264:12826–12829.
97. Matsuyama T, Komatsu R, Namba M, Watanabe N, Itoh H, Tarui S. Glucagon-like peptide-1 (7-36 amide): a potent glucagonostatic and insulinotropic hormone. *Diabetes Res Clin Pract* 1988;5:281–284.
98. Mojsov S, Weir GC, Habener JF. Glucagonlike peptide I (7-37) is a potent stimulator of insulin secretion. In: Marshall GR, ed. *Peptides: chemistry and biology: Proc. 10th Am. Peptide Symp.* Leiden: Escom, 1988;644–646.
99. Ørskov C, Buhl T, Rabenhøj L, Kofod H, Holst JJ. Carboxypeptidase-B-like processing of the C-terminus of glucagon-like peptide-2 in pig and human small intestine. *FEBS Lett* 1989; 247:193–196.
100. Rall TW, Sutherland EW. Formation of a cyclic adenine ribonucleotide by tissue particles. *J Biol Chem* 1958;232:1065–1076.
101. Miyata A, Arimura A, Dahl RR, et al. Isolation

of a novel 38 residue-hypothalamic polypeptide which stimulates adenylate cyclase in pituitary cells. *Biochem Biophys Res Commun* 1989; 164:567—574.

102. Kimmel JR, Hayden LJ, Pollock HG. Isolation and characterization of a new pancreatic polypeptide hormone. *J Biol Chem* 1975;250:9369–9376.
103. Chance RE, Johnson MG, Hoffmann JA, Lin T-M. Pancreatic polypeptide: a newly recognized hormone. In: Baba S, Kaneko T, Yanaihara N, eds. *Proinsulin, insulin, C-peptide*. Amsterdam, Oxford: Excerpta Medica, 1979;419–425.
104. Flynn TG, de Bold ML, de Bold AJ. The amino acid sequence of an atrial peptide with potent diuretic and natriuretic properties. *Biochem Biophys Res Commun* 1983;117:859–865.
105. Forssmann WG, Hock D, Lottspeich F, et al. The right auricle of the heart is an endocrine organ. *Anat Embryol* 1983;168:307–313.
106. Sudoh T, Kangawa K, Minamino N, Matsuo H. A new natriuretic peptide in porcine brain. *Nature* 1988;332:78–81.
107. Yanagisawa M, Kurihara H, Kimura S, et al. A novel potent vasoconstrictor peptide produced by vascular endothelial cells. *Nature* 1988; 332:411–415.
108. Saida K, Mitsui Y, Ishida N. A novel peptide, vasoactive intestinal contractor, of a new (endothelin) peptide family. *J Biol Chem* 1989; 264:14613–14616.
109. Lee CY, Chiappinelli VA. Similarity of endothelin to snake venom toxin. *Nature* 1988; 335:303
110. Takasaki C, Tamiya N, Bdolah A, Wollberg Z, Kochva E. Sarafotoxins S6: several isotoxins from *Atractaspis engaddensis* (burrowing asp) venom that affect the heart. *Toxicon* 1988;26: 543–548.
111. Shinmi O, Kimura S, Sawamura T, et al. Endothelin-3 is a novel neuropeptide: isolation and sequence determination of endothelin-1 and endothelin-3 in porcine brain. *Biochem Biophys Res Commun* 1989;164:587–593.
112. McMahon EG, Fok KF, Moore WM, Smith CE, Siegel NR, Trapani AJ. *In vitro* and *in vivo* activity of chymotrypsin-activated big endothelin (porcine 1-40). *Biochem Biophys Res Commun* 1989;161:406–413.
113. Potts JT Jr, Tregear GW, Keutmann HT, et al. Synthesis of a biologically active N-terminal tetratriacontapeptide of parathyroid hormone. *Proc Natl Acad Sci USA* 1971;68:63–67.
114. Habener JF, Kemper BW, Rich A, Potts JT Jr. Biosynthesis of parathyroid hormone. *Rec Prog Horm Res* 1977;33:249–308.
115. Mangin M, Ikeda K, Dreyer BE, Broadus AE. Isolation and characterization of the human parathyroid hormone-like peptide gene. *Proc Natl Acad Sci USA* 1989;86:2408–2412.
116. Shigeno C, Yamamoto I, Kitamura N, et al. Interaction of human parathyroid hormone-related peptide with parathyroid hormone receptors in clonal rat osteosarcoma cells. *J Biol Chem* 1988;263:18369–18377.
117. Tager HS, Steiner DF. Isolation of a glucagon-containing peptide: primary structure of a possible fragment of proglucagon. *Proc Natl Acad Sci USA* 1973;70:2321–2325.
118. Bataille D, Tatemoto K, Gespach C, Jörnvall H, Rosselin G, Mutt V. Isolation of glucagon-37 (bioactive enteroglucagon/oxyntomodulin) from porcine jejuno-ileum. *FEBS Lett* 1982; 146:79–86.
119. Yiangou Y, Di Marzo V, Spokes RA, Panico M, Morris HR, Bloom SR. Isolation, characterization, and pharmacological actions of peptide histidine valine 42, a novel prepro-vasoactive intestinal peptide-derived peptide. *J Biol Chem* 1987;262:14010–14013.
120. Cauvin A, Vandermeers A, Vandermeers-Piret M-C, Rathe J, Robberecht P, Christophe J. Peptide histidine isoleucinamide (PHI)-(1-27)-Gly as a new major form of PHI in the rat small intestine. *Endocrinology* 1989;125:1296–1302.
121. Carlquist M, Rökaeus Å. Isolation of a proform of porcine secretin by ion-exchange and reversed-phase high-performance liquid chromatography. *J Chromatogr* 1984;296:143–151.
122. Gafvelin G, Carlquist M, Mutt V, A proform of secretin with high secretin-like bioactivity. *FEBS Lett* 1985;184:347–352.
123. Gafvelin G, Andersson M, Dimaline R, Jörnvall H, Mutt V. Isolation and characterization of a variant form of vasoactive intestinal polypeptide. *Peptides* 1988;9:469–474.
124. Rehfeld JF, Hansen HF. Characterization of preprocholecystokinin products in the porcine cerebral cortex. *J Biol Chem* 1986;261:5832–5840.
125. Sugano K, Aponte GW, Yamada T. Identification and characterization of glycine-extended post-translational processing intermediates of progastrin in porcine stomach. *J Biol Chem* 1985;260:11724–11729.
126. Del Valle J, Sugano K, Yamada T. Glycine-extended processing intermediates of gastrin and cholecystokinin in human plasma. *Gastroenterology* 1989;97:1159–1163.
127. Morley JS. Structure–function relationships in gastrin-like peptides. *Proc R Soc* 1968;B170: 97–111.
128. Gregory RA. A review of some recent developments in the chemistry of the gastrins. *Bioorg Chem* 1979;8:497–511.
129. Mutt V. Further investigations on intestinal hormonal polypeptides. *Clin Endocrinol* 1976; 5(Suppl):175s–183s.
130. Eysselein VE, Reeve JR Jr, Shively JE, Hawke D, Walsh JH. Partial structure of a large canine cholecystokinin (CCK_{58}): amino acid sequence. *Peptides* 1982;3:687–691.
131. Pradayrol L, Jörnvall H, Mutt V, Ribet A. N-terminally extended somatostatin: the primary structure of somatostatin-28. *FEBS Lett* 1980; 109:55–58.
132. Schally AV, Huang W-Y, Chang RCC, et al. Isolation and structure of pro-somatostatin: a putative somatostatin precursor from pig hy-

pothalamus. *Proc Natl Acad Sci USA* 1980;77: 4489–4493.

133. Böhlen P, Brazeau P, Benoit R, Ling N, Esch F, Guillemin R. Isolation and amino acid composition of two somatostatin-like peptides from ovine hypothalamus: somatostatin-28 and somatostatin-25. *Biochem Biophys Res Commun* 1980;96:725–734.
134. Tatemoto K, Lundberg JM, Jörnvall H, Mutt V. Neuropeptide K: isolation, structure and biological activities of a novel brain tachykinin. *Biochem Biophys Res Commun* 1985;128:947–953.
135. Takeda Y, Krause JE. Gamma-preprotachykinin-(72-92)-peptide amide potentiates substance P–induced salivation. *Eur J Pharmacol* 1989;161:267–271.
136. Kage R, McGregor GP, Thim L, Conlon JM. Neuropeptide-gamma: a peptide isolated from rabbit intestine that is derived from gamma-preprotachykinin. *J Neurochem* 1988;50:1412–1417.
137. Kazal LA, Spicer DS, Brahinsky RA. Isolation of a crystalline trypsin inhibitor-anticoagulant protein from pancreas. *J Am Chem Soc* 1948; 70:3034–3040.
138. Tschesche H, Wachter E. Die Primärstruktur des spezifischen Trypsininhibitors II (Kazal-Typ) aus Schweinepankreas. *Hoppe Seylers Z Physiol Chem* 1970;351:1449–1459.
139. Bartelt DC, Greene LJ. The primary structure of the porcine pancreatic secretory trypsin inhibitor I. *J Biol Chem* 1971;216:2218–2229.
140. Iwai K, Fushiki T, Fukuoka S-i. Pancreatic enzyme secretion mediated by novel peptide: monitor peptide hypothesis. *Pancreas* 1988;3: 720–728.
141. Agerberth B, Söderling-Barros J, Jörnvall H, et al. Isolation and characterization of a 60-residue intestinal peptide structurally related to the pancreatic secretory type of trypsin inhibitor: influence on insulin secretion. *Proc Natl Acad Sci USA* 1989;86:8590–8594.
142. Zasloff M. Magainins, a class of antimicrobial peptides from *Xenopus* skin: isolation, characterization of two active forms, and partial cDNA sequence of a precursor. *Proc Natl Acad Sci USA* 1987;84:5449–5453.
143. Erspamer V, Melchiorri P. Active polypeptides: from amphibian skin to gastrointestinal tract and brain of mammals. *TIPS* 1980;1:391–395.
144. Lee J-Y, Boman A, Chuanxin S, et al. Antibacterial peptides from pig intestine: isolation of a mammalian cecropin. *Proc Natl Acad Sci USA* 1989;86:9159–9162.
145. Boman HG, Hultmark D. Cell-free immunity in insects. *Annu Rev Microbiol* 1987;41:103–126.
146. Stewart JM, Hall ME, Harkins J, et al. A fragment of substance P with specific central activity: SP(1-7). *Peptides* 1982;3:851–857.
147. Parish DC, Smyth DG, Normanton JR, Wolstencroft JH. Glycyl glutamine, an inhibitory neuropeptide derived from β-endorphin. *Nature* 1983;306:267–270.
148. Koelle GB, Sanville UJ, Wall SJ. Glycyl-L-glutamine, a precursor, and glycyl-L-glutamic acid, a neurotrophic factor for maintenance of acetylcholinesterase and butyrylcholinesterase in the preganglionically denervated superior cervical ganglion of the cat *in vivo*. *Proc Natl Acad Sci USA* 1985;82:5213–5217.
149. Schiavone MT, Santos RAS, Brosnihan KB, Khosla MC, Ferrario CM. Release of vasopressin from the rat hypothalamo-neurohypophysial system by angiotensin-(1-7) heptapeptide. *Proc Natl Acad Sci USA* 1988;85:4095–4098.
150. Chappell MC, Brosnihan KB, Diz DI, Ferrario CM. Identification of angiotensin-(1-7) in rat brain. *J Biol Chem* 1989;264:16518–16523.
151. Jørgensen KD, Weis JU, Diamant B. Dissociation of the spasmolytic and metabolic effects of glucagon. *Eur J Pharmacol* 1983;90:315–323.
152. Ponce J, Garrigues V, Pertejo V, et al. Effect of intravenous glucagon and glucagon-(1-21)-peptide on motor activity of sphincter of Oddi in humans. *Dig Dis Sci* 1989;34:61–64.
153. Itoh H, Matsuyama T, Namba M, et al. Effect of glucagon-(1-21)-peptide on secretin-stimulated pancreatic exocrine secretion in anesthetized dogs. *Life Sci* 1989;44:819–825.
154. Mallat A, Pavoine C, Dufour M, Lotersztajn S, Bataille D, Pecker F. A glucagon fragment is responsible for the inhibition of the liver Ca^{2+} pump by glucagon. *Nature* 1987;325:620–622.
155. Richter WO, Schwandt P. Cholecystokinin 1-21 stimulates lipolysis in human adipose tissue. *Horm Metab Res* 1989;21:216–217.
156. Strupp BJ. Improvement of memory by a vasopressin fragment: importance of individual differences in mnemonic function. *Behav Neurosci* 1989;103:743–754.
157. Schally AV, Huang WY, Redding TW, et al. Isolation, structural elucidation and synthesis of a tetradecapeptide with *in vitro* ACTH-releasing activity corresponding to residues 33-46 of the α-chain of porcine hemoglobin. *Biochem Biophys Res Commun* 1978;82:582–588.
158. Takagi H, Shiomi H, Ueda H, Amano H. A novel analgesic dipeptide from bovine brain is a possible Met-enkephalin releaser. *Nature* 1979;282:410–412.
159. Kiso Y, Kitagawa K, Kawai N, et al. Neo-kyotorphin (Thr-Ser-Lys-Tyr-Arg), a new analgesic peptide. *FEBS Lett* 1983;155:281–284.
160. Carlquist M, Jörnvall H, Mutt V. A C-terminal fragment of hemoglobin β-chains in extracts of porcine upper intestine. *FEBS Lett* 1984;170: 28–32.
161. Brantl V, Teschemacher H, Henschen A, Lottspeich F. Novel opioid peptides derived from casein (β-casomorphins). *Hoppe Seylers Z Physiol Chem* 1979;360:1211–1216.
162. Brantl V. Novel opioid peptides derived from human β-casein: human β-casomorphins. *Eur J Pharmacol* 1984;106:213–214.
163. Zioudrou C, Streaty RA, Klee WA. Opioid peptides derived from food proteins. *J Biol Chem* 1979;254:2446–2449.
164. Loukas S, Varoucha D, Zioudrou C, Streaty

RA, Klee WA. Opioid activities and structures of α-casein-derived exorphins. *Biochemistry* 1983;22:4567–4573.

165. Huebner FR, Lieberman KW, Rubino RP, Wall JS. Demonstration of high opioid-like activity in isolated peptides from wheat gluten hydrolysates. *Peptides* 1984;5:1139–1147.
166. Brantl V, Gramsch C, Lottspeich F, Henschen A, Jaeger K-H, Herz A. Novel opioid peptides derived from mitochondrial cytochrome b: cytochrophins. *Eur J Pharmacol* 1985;111:293–294.
167. Hautefeuille M, Brantl V, Dumontier A-M, Desjeux J-F. *In vitro* effects of β-casomorphins on ion transport in rabbit ileum. *Am J Physiol* 1986;250:G92–97.
168. Ermisch A, Brust P, Brandsch M. β-Casomorphins alter the intestinal accumulation of L-leucine. *Biochim Biophys Acta* 1989;982;79–84.
169. Yoshikawa M, Tani F, Ashikaga T, Yoshimura T, Chiba H. Purification and characterization of an opioid antagonist from a peptic digest of bovine *K*-casein. *Agric Biol Chem* 1986;50: 2951–2954.
170. Ueno A, Hong Y-M, Arakaki N, Takeda Y. Insulin-stimulating peptide from a tryptic digest of bovine serum albumin: purification and characterization. *J Biochem* 1985;98:269–278.
171. Carraway RE, Mitra SP, Cochrane DE. Structure of a biologically active neurotensin-related peptide obtained from pepsin-treated albumin(s). *J Biol Chem* 1987;262:5968–5973.
172. Carraway RE, Feurle GE. Generation of xenopsin-related peptides during acid extraction of gastric tissues. *J Biol Chem* 1985;260:10921–10925.
173. Wilson WE, Harvan DJ, Hamm C, et al. Physalaemin-like immunoreactive peptides from rabbit stomach. *Int J Pept Protein Res* 1986; 28:58–66.
174. Parker F, Migliore-Samour D, Floc'h F, et al. Immunostimulating hexapeptide from human casein: amino acid sequence, synthesis and biological properties. *Eur J Biochem* 1984;145: 677–682.
175. Migliore-Samour D, Jollès P. Casein, a prohormone with an immunomodulating role for the newborn? *Experientia* 1988;44:188–193.
176. Ivanova VP, Sorochinskaya EI, Lozhkina TK, Anokhina VV. Immunomodulating and analgesic activity of synthetic fragments of proteins and immunopeptides. *Ukr J Biochem* 1988; 60:3–9.
177. Buus S, Colon S, Smith C, Freed JH, Miles C, Grey HM. Interaction between a "processed" ovalbumin peptide and Ia molecules. *Proc Natl Acad Sci USA* 1986;83:3968–3971.
178. Waltho JP, Feher VA, Lerner RA, Wright PE. Conformation of a T cell stimulating peptide in aqueous solution. *FEBS Lett* 1989;250:400–404.
179. Blundell TL, Cutfield JF, Cutfield SM, et al. Atomic positions in rhombohedral 2-zinc insulin crystals. *Nature* 1971;231:506–511.
180. Sasaki K, Dockerill S, Adamiak DA, Tickle IJ, Blundell T. X-ray analysis of glucagon and its relationship to receptor binding. *Nature* 1975; 257:751–757.
181. Glover ID, Barlow DJ, Pitts JE, et al. Conformational studies on the pancreatic polypeptide hormone family. *Eur J Biochem* 1984;142:379–385.
182. Higuchi Y, Morimoto Y, Horinaka A, Yasuoka N. Crystallization and preliminary X-ray studies on human epidermal growth factor. *J Biochem* 1988;103:905–906.
183. Smith GD, Griffin JF. Conformation of (Leu^5) enkephalin from X-ray diffraction: features important for recognition at opiate receptor. *Science* 1978;199:1214–1216.
184. Camerman A, Mastropaolo D, Karle I, Karle J, Camerman N. Crystal structure of leucine-enkephalin. *Nature* 1983;306:447–450.
185. Mastropaolo D, Camerman A, Camerman N. Crystal structure of methionine-enkephalin. *Biochem Biophys Res Commun* 1986;134:698–703.
186. Bodanszky A, Ondetti MA, Mutt V, Bodanszky M. Synthesis of secretin. IV. Secondary structure in a miniature protein. *J Am Chem Soc* 1969;91:944–949.
187. Hollósi M, Kajtár M, Gráf L. Studies on the conformation of β-endorphin and its constituent fragments in water and trifluoroethanol by CD spectroscopy. *FEBS Lett* 1977;74:185–189.
188. Pham Van Chuong P, Penke B, De Castiglione R, Fromageot P. Conformational study of gastrin and C-terminal peptides by circular dichroism. In: Rosselin G, Fromageot P, Bonfils S, eds. *Hormone receptors in digestion and nutrition.* Amsterdam: Elsevier/North Holland Biomedical Press, 1979,33–44.
189. Schiller PW, Natarajan S, Bodanszky M. Determination of the intramolecular tyrosine-tryptophan distance in a 7-peptide related to the C-terminal sequence of cholecystokinin. *Int J Pept Protein Res* 1978;12:139–142.
190. Blundell TL, Dockerill S, Sasaki K, Tickle IJ, Wood SP. The relation of structure to storage and receptor binding of glucagon. *Metabolism* 1976;25(suppl 1):1331–1339.
191. Johnson LF, Schwartz IL, Walter R. Oxytocin and neurohypophyseal peptides: spectral assignment and conformational analysis by 220 MHz nuclear magnetic resonance. *Proc Natl Acad Sci USA* 1969;64:1269–1275.
192. Wüthrich K. Protein structure determination in solution by nuclear magnetic resonance spectroscopy. *Science* 1989;243:45–50.
193. Bax A. Two-dimensional NMR and protein structure. *Annu Rev Biochem* 1989;58:223–256.
194. Mammi S, Goodman M, Peggion E, Foffani MT, Moroder L, Wuensch E. Conformational studies on gastrin related peptides by high resolution ^{1}H-n.m.r. *Int J Pept Protein Res* 1986; 27:145–152.
195. Mammi S, Mammi NJ, Peggion E. Conformational studies of human Des-Trp^1, Nle^{12}-minigastrin in water-trifluoroethanol mixtures by

^{1}H NMR and circular dichroism. *Biochemistry* 1988;27:1374–1379.

196. Koizuka I, Watari H, Yanaihara N, Nishina Y, Shiga K, Nagayama K. Proton nuclear magnetic resonance study of CCK octapeptide. *Biomed Res* 1984;5(suppl):161–168.
197. Loomis RE, Lee P-C, Tseng C-C. Conformational analysis of the cholecystokinin C-terminal octapeptide: a nuclear magnetic resonance and computer-simulation approach. *Biochim Biophys Acta* 1987;911:168–179.
198. Fournié-Zaluski MC, Belleney J, Lux B, et al. Conformational analysis of cholecystokinin CCK_{26-33} and related fragments by ^{1}H NMR spectroscopy, fluorescence-transfer measurements, and calculations. *Biochemistry* 1986; 25:3778–3787.
199. Carver JA, Cooke RM, Esposito G, Campbell ID, Gregory H, Sheard B. A high resolution ^{1}H NMR study of the solution structure of human epidermal growth factor. *FEBS Lett* 1986;205: 77–81.
200. Gronenborn AM, Bovermann G, Clore GM. A ^{1}H-NMR study of the solution conformation of secretin. *FEBS Lett* 1987;215:88–94.
201. Di Bello C, Gozzini L, Tonellato M, et al. Mono and two-dimensional 500-MHz characterization of synthetic bombesin and related peptides in DMSO and DMSO-water. *Biopolymers* 1989;28:421–440.
202. Cavatorta P, Spisni A, Szabo AG, Farruggia G, Franzoni L, Masotti L. Conformation of bombesin in buffer and in the presence of lysolecithin micelles: NMR, CD, and fluoresence studies. *Biopolymers* 1989;28:441–463.
203. Vesterman B, Saulitis J, Betins J, Liepins E, Nikiforovich GV. Dynamic space structure of the Leu-enkephalin molecule in DMSO solution. *Biochim Biophys Acta* 1989;998:204–209.
204. Dyson HJ, Holmgren A, Wright PE. Assignment of the proton NMR spectrum of reduced and oxidized thioredoxin: sequence-specific assignments, secondary structure, and global fold. *Biochemistry* 1989;28:7074–7087.
205. McCammon JA, Gelin BR, Karplus M. Dynamics of folded proteins. *Nature* 1977;267:585–590.
206. Clore GM, Nilges M, Brünger A, Gronenborn AM. Determination of the backbone conformation of secretin by restrained molecular dynamics on the basis of interproton distance data. *Eur J Biochem* 1988;171:479–484.
207. Akhmedov NA, Agaeva GA, Popov EM. Structure-function organization of the secretin molecule. II. Direct conformational problem. *Mol Biol (Mosk)* 1987;21:174–181.
208. Kreissler M, Pesquer M, Maigret B, Fournié-Zaluski MC, Roques BP. Computer simulation of the conformational behavior of cholecystokinin fragments: conformational families of sulfated CCK8. *J Comput Aided Mol Design* 1989;3:85–94.
209. Allen J, Novotný J, Martin J, Heinrich G. Molecular structure of mammalian neuropeptide Y: analysis by molecular cloning and computer-aided comparison with crystal structure of avian homologue. *Proc Natl Acad Sci USA* 1987;84:2532–2536.
210. Al-Obeidi F, Hadley ME, Pettitt BM, Hruby VJ. Design of a new class of superpotent cyclic α-melanotropins based on quenched dynamic simulations. *J Am Chem Soc* 1989;111:3413–3416.
211. Ullrich A, Bell JR, Chen EY, et al. Human insulin receptor and its relationship to the tyrosine kinase family of oncogenes. *Nature* 1985; 313:756–761.
212. Ebina Y, Ellis L, Jarnagin K, et al. The human insulin receptor cDNA: the structural basis for hormone-activated transmembrane signalling. *Cell* 1985;40:747–758.
213. Ullrich A, Coussens L, Hayflick JS, et al. Human epidermal growth factor receptor cDNA sequence and aberrant expression of the amplified gene in A431 epidermoid carcinoma cells. *Nature* 1984;309:418–425.
214. Russell DW, Schneider WJ, Yamamoto T, et al. Domain map of the LDL receptor: sequence homology with the epidermal growth factor precursor. *Cell* 1984;37:577–585.
215. Hatakeyama M, Tsudo M, Minamoto S, et al. Interleukin-2 receptor beta chain gene: generation of three receptor forms by cloned human α and β chain cDNA's. *Science* 1989;244: 551–556.
216. Yarden Y, Escobedo JA, Kuang W-J, et al. Structure of the receptor for platelet-derived growth factor helps define a family of closely related growth factor receptors. *Nature* 1986; 323:226–232.
217. Claesson-Welsh L, Eriksson A, Morén A, et al. cDNA cloning and expression of a human platelet-derived growth factor (PDGF) receptor specific for β-chain-containing PDGF molecules. *Mol Cell Biol* 1988;8:3476–3486.
218. Gronwald RGK, Grant FJ, Haldeman BA, et al. Cloning and expression of a cDNA coding for the human platelet-derived growth factor receptor: evidence for more than one receptor class. *Proc Natl Acad Sci USA* 1988;85:3435–3439.
219. Jackson TR, Blair LAC, Marshall J, Goedert M, Hanley MR. The *mas* oncogene encodes an angiotensin receptor. *Nature* 1988;335:437–440.
220. Masu Y, Nakayama K, Tamaki H, et al. cDNA cloning of bovine substance-K receptor through oocyte expression system. *Nature* 1987;329:836–838.
221. Yokota Y, Sasai Y, Tanaka K, et al. Molecular characterization of a functional cDNA for rat substance P receptor. *J Biol Chem* 1989;264: 17649–17652.
222. Kelly PA, Boutin J-M, Jolicoeur C, et al. Purification, cloning, and expression of the prolactin receptor. *Biol Reprod* 1989;40:27–32.
223. McFarland KC, Sprengel R, Phillips HS, et al. Lutropin-choriogonadotropin receptor: an un-

usual member of the G protein–coupled receptor family. *Science* 1989;245:494–499.
224. Loosfelt H, Misrahi M, Atger M, et al. Cloning and sequencing of porcine LH-hCG receptor cDNA: variants lacking transmembrane domain. *Science* 1989;245:525–528.
225. Radeke MJ, Misko TP, Hsu C, Herzenberg LA, Shooter EM. Gene transfer and molecular cloning of the rat nerve growth factor receptor. *Nature* 1987;325:593–597.
226. Leung DW, Spencer SA, Cachianes G, et al. Growth hormone receptor and serum binding protein: purification, cloning and expression. *Nature* 1987;330:537–543.
227. Blumer KJ, Reneke JE, Courchesne WE, Thorner J. Functional domains of a peptide hormone receptor: the α-factor receptor (STE2 gene product) of the yeast *Saccharomyces cerevisiae*. *Cold Spring Harbor Symp Quant Biol* 1988;LIII:591–603.
228. D'Andrea AD, Lodish HF, Wong GG. Expression cloning of the murine erythropoietin receptor. *Cell* 1989;57:277–285.
229. Chinkers M, Garbers DL, Chang M-S, et al. A membrane form of guanylate cyclase is an atrial natriuretic peptide receptor. *Nature* 1989; 338:78–83.
230. Garbers DL. Cyclic GMP and the second messenger hypothesis. *Trends Endocrin Metab* 1989;1:64–67.
231. de Gunzburg J. Le mode d'action de l'AMP cyclique chez les procaryotes et les eucaryotes, CAP et protéine kinases AMPc dépendantes. *Biochimie* 1985;67:563–582.
232. Levitzki A. Regulation of hormone-sensitive adenylate cyclase. *TIPS* 1987;8:299–303.
233. Berridge MJ, Irvine RF. Inositol phosphates and cell signalling. *Nature* 1989;341:197 205.
234. Nahorski SR, Potter BVL. Molecular recognition of inositol polyphosphates by intracellular receptors and metabolic enzymes. *TIPS* 1989; 10:139–144.
235. Pelech SL, Vance DE. Signal transduction via phosphatidylcholine cycles. *TIBS* 1989;14:28–30.
236. Low MG, Saltiel AR. Structural and functional roles of glycosyl-phosphatidylinositol in membranes. *Science* 1988;239:268–275.
237. Rodbell M, Birnbaumer L, Pohl SL, Krans HMJ. The glucagon-sensitive adenyl cyclase system in plasma membranes of rat liver. *J Biol Chem* 1971;246:1877–1882.
238. Gilman AG. G proteins: transducers of receptor-generated signals. *Annu Rev Biochem* 1987; 56:615–649.
239. Neer EJ, Clapham DE. Roles of G protein subunits in transmembrane signalling. *Nature* 1988;333:129–134.
240. Spiegel AM. Receptor–effector coupling by G-proteins. Implications for endocrinology. *Trends Endocrin Metab* 1989;1:72–76.
241. Hanley RM, Steiner AL. The second-messenger system for peptide hormones. *Hosp Prac* 1989;24:59–70.
242. Kataoka T, Broek D, Wigler M. DNA sequence and characterization of the *S. cerevisiae* gene encoding adenylate cyclase. *Cell* 1985;43:493–505.
243. Yamawaki-Kataoka Y, Tamaoki T, Choe H-R, Tanaka H, Kataoka T. Adenylate cyclases in yeast: a comparison of the genes from *Schizosaccharomyces pombe* and *Saccharomyces cerevisiae*. *Proc Natl Acad Sci USA* 1989;86:5693–5697.
244. Krupinski J, Coussen F, Bakalyar HA, et al. Adenylyl cyclase amino acid sequence: possible channel- or transporter-like structure. *Science* 1989;244:1558–1564.
245. Ferris CD, Huganir RL, Supattapone S, Snyder SH. Purified inositol 1,4,5-trisphosphate receptor mediates calcium flux in reconstituted lipid vesicles. *Nature* 1989;342:87–89.
246. Furuichi T, Yoshikawa S, Miyawaki A, et al. Primary structure and functional expression of the inositol 1,4,5-trisphosphate–binding protein P_{400}. *Nature* 1989;342:32–38.
247. Streb H, Irvine RF, Berridge MJ, Schulz I. Release of Ca^{2+} from a nonmitochondrial intracellular store in pancreatic acinar cells by inositol-1,4,5-trisphosphate. *Nature* 1983;306:67–69.
248. Beckett AH, Casy AF. Synthetic analgesics: stereochemical considerations. *J Pharm Pharmacol* 1954;6:986–1001.
249. Goldstein A, Lowney LI, Pal BK. Stereospecific and nonspecific interactions of the morphine congener levorphanol in subcellular fractions of mouse brain. *Proc Natl Acad Sci USA* 1971;68:1742–1747.
250. Hughes J. Search for the endogenous ligand of the opiate receptor. *Neurosci Res Prog Bull* 1975;13:55–58.
251. Terenius L. Narcotic receptors in guinea pig ileum and rat brain. *Neurosci Res Prog Bull* 1975;13:39–42.
252. Hughes J, Smith TW, Kosterlitz HW, et al. Identification of two related pentapeptides from the brain with potent opiate agonist activity. *Nature* 1975;258:577–579.
253. Peikin SR, Costenbader CL, Gardner JD. Actions of derivatives of cyclic nucleotides on dispersed acini from guinea pig pancreas. *J Biol Chem* 1979;254:5321–5327.
254. Davison JS, Farashah AN, Pearson GT, Petersen OH; Dibutyryl cyclic GMP: a specific inhibitor of the action of CCK-like peptides on the pancreas, gallbladder and ileum. *J Physiol* 1980;305:77P–78P.
255. Evans BE, Rittle KE, Bock MG, et al. Design of nonpeptidal ligands for a peptide receptor: cholecystokinin antagonists. *J Med Chem* 1987;30:1229–1239.
256. Freidinger RM. Nonpeptide ligands for peptide receptors. *TIPS* 1989;10:270–274.
257. Mendre C, Rodriguez M, Gueudet C, et al. A pseudopeptide that is a potent cholecystokinin agonist in the peripheral system is able to inhibit the dopamine-like effects of cholecysto-

kinin in the striatum. *J Biol Chem* 1988;263: 10641–10645.

258. Hughes RA, Andrews PR. Structural and conformational analogy between cholecystokinin and ergopeptines. *J Pharm Pharmacol* 1987; 39:339–343.

259. Kondo Y, Torii K, Omura S, Itoh Z. Erythromycin and its derivatives with motilin-like biological activities inhibit the specific binding of ^{125}I-motilin to duodenal muscle. *Biochem Biophys Res Commun* 1988;150:877–882.

260. Peeters T, Matthijs G, Depoortere I, et al. Erythromycin is a motilin receptor agonist. *Am J Physiol* 1989;257:G470–G474.

Volume Transmission in the Brain: Novel Mechanisms for Neural Transmission,
edited by K. Fuxe and L.F. Agnati.
Raven Press, Ltd., New York © 1991.

28

Peptides in Neurotransmission

W. Zieglgänsberger, H.-U. Dodt, R. Deisz, and H. Pawelzik

Clinical Neuropharmacology, Max Planck Institute for Psychiatry, Munich, Federal Republic of Germany

Current research suggests that neuropeptides play a key role in synaptic signaling and allow communication within specific subsystems of the brain. In addition to the classical synaptic mechanisms, information transfer by neuropeptides can also occur via the extracellular space (1,2). Neuropeptides are considered to be major candidates for integrative actions and long-term changes in neuronal excitability, which alter behavior, mood, and mental processes, as well as endocrine and autoregulatory functions (3). The mechanisms underlying the physiological role of neuropeptides in the mammalian peripheral (4) and central nervous system (CNS) (5,6; see Chap. 2, *this volume*) remain to be elucidated.

NEUROPEPTIDES AS NEURONAL SIGNALING MOLECULES

The mere presence of a neuropeptide in neuronal elements does not warrant any speculation about its role as an intercellular messenger. The specific distribution in the neuropil, the Ca^{2+}-dependent release, and the distinct actions that some neuropeptides exert on neurons of the mammalian CNS give credence to the belief that synaptic and nonsynaptic intercellular communication may, in part, be peptidergic in nature. The slow time course of their actions, the mismatch between their site of release, and the location of their receptors [i.e., on (a) the soma-dendritic membrane, (b) presynaptic terminals, (c) glia, or (d) blood vessels] suggest a more diffuse site of action, in addition to a clearly targeted action through conventional synaptic contacts. There is evidence that highly selective receptors for neuropeptides are not restricted to subsynaptic areas but are distributed over the entire soma-dendritic membrane of central neurons (7). Recent publications (8–10) report that type II astrocytes, a glia cell type almost exclusively found in the white matter, expresses L-glutamate receptors as well as receptors for neuropeptides (see Chap. 2, *this volume*). The molecular structure of a binding protein for kainate, a subtype of the L-glutamate receptor, has most recently been characterized on glial membranes (11). These neuron–glia interactions occur in a part of the CNS that was always considered to be involved only as a passive link in the information transfer. Their role in intercellular signaling remains to be elucidated.

MODE OF ACTION OF NEUROPEPTIDES IN THE CNS

Some of the hitherto characterized actions of neuropeptides on neuronal targets are assumed to be modulatory (i.e., peptides alter the response of a target neuron to a classical neurotransmitter without interfering with neuronal excitability by a shift in membrane permeability) (12,13). Intercellular communication by means of synaptic transmission triggered by the rapid opening and closing of ionic channels may transiently be altered by these signal molecules, which are co-released (see Chap. 33, *this volume*) from either the same terminals or adjacent neuronal structures (14). Pre- and postsynaptic interactions between neuropeptides and classical neurotransmitters for monoamines or amino acids have been demonstrated in various neuronal systems (6). The mechanism of differential release of coexisting neuropeptides and neurotransmitter(s) is still unknown. It has been suggested that high-frequency stimulation preferentially releases neuropeptides, most probably from dense-core vesicles (see Chap. 33, *this volume*). The findings reported in this contribution also tentatively suggest that low-frequency versus high-frequency stimulation of neocortical interneurons releases different neurotransmitters or different combinations of them. However, the physiological significance of the multiplicity of messengers involved in cotransmission is still unknown, and the multiplicity of messengers and receptors certainly complicates the interpretation of synaptic activity and receptor responses. A major breakthrough would be the development of more selective receptor antagonists, which might be able to probe the actions of the various components of the response.

Several neuropeptides influence transmembrane ion fluxes at pre- and postsynaptic sites as well as intracellular enzyme activity through G-protein-coupled mechanisms. It has been shown that neuropeptides can increase the intracellular Ca^{2+} concentration and trigger the Ca^{2+}-dependent metabolic cascades, as well as intracellular mechanisms involving cyclic adenosine monophosphate (cyclicAMP), inositol phosphate, or other second-messenger systems, through activation of voltage- and ligand-gated Ca^{2+} channels or mobilization of Ca^{2+} from intracellular stores (15,16). After they are released, neuropeptides are not taken up again but are degraded by relatively unspecific peptidases localized in the extracellular environment. Peptidase inhibitors, which prolong the actions of neuropeptides, may gain therapeutic potential (a) once neuropeptides have been tailored, which can cross the blood-brain barrier, or (b) by enhancing the action of tonically active peptide-releasing neurons (Chap. 29, *this volume*).

OPIOID PEPTIDES AND SYNAPTIC FUNCTION IN THE NEOCORTEX

The opioidergic innervation of neocortical neurons is mainly derived from proenkephalin A and proenkephalin B (-dynorphin) precursors. As with most other neuropeptides, proenkephalin A– and proenkephalin B–derived opioid peptides are only rarely contained in long projecting systems but function as neurotransmitter or neurmodulators of interneurons (see Chap. 2, *this volume*). Only the proopiomelanocortin (POMC) projection system, which originates mainly from the nucleus arcuatus and the mediobasal hypothalamus, gives rise to a large number of collaterals. Although no afferent fibers have been demonstrated, POMC-derived opioid peptides may contribute to opioid actions, for example, in neocortical neurons through mechanisms that do not involve synaptic contacts (1) (see below).

None of the three opiate receptor subtypes have yet been cloned. Purification and affinity labeling of the receptors support the concept that μ-,δ-, and κ-opiate receptors are discrete molecular entities. This concept was initially derived from binding and activity profiles, different sensitivity of the agonists to naloxone, and the protection and inactivation kinetics. At present, the structure of the opiate receptor is unknown but probably will resemble other inhibitory receptors coupled to G proteins characterized by their seven transmembrane helices and a large cytoplasmic loop. A coupling of the opiate receptors to Gi/Go proteins has been shown (17).

The fragments of POMC and proenkephalin A have affinities mainly to the μ and δ binding sites and those of proenkephalin B (-dynorphin) mainly to the κ binding site. Only the synthetic compounds display almost exclusive binding at one site. In contrast, all the endogenous opioid peptides bind to more than one type of binding site. The different levels of the various receptor subtypes in a brain area will also likely be a decisive factor in determining the final response of the target cell for the exogenously applied, as well as the endogenous, opioids (18–20). There is evidence that the μ and δ subtypes of the opiate receptor are colocalized on various neurons (see below).

The predominantly inhibitory effects of opioid peptides on spontaneous or chemically evoked neuronal discharge activity are qualitatively similar throughout the mammalian central and peripheral nervous systems (12,13,21). Some major exceptions to the usual depressions exist (e.g., the naloxone-reversible excitatory responses seen in the majority of hippocampal pyramidal cells). The bulk of the extracellular and intracellular data supports a disinhibitory mechanism of action for the single-unit excitatory effects of these opioid peptides in the hippocampus. In these neurons the excitation may actually be indirect, resulting from a primary inhibitory effect on neighboring inhibitory gamma-aminobutyric acid (GABA)ergic interneurons leading to excitation by disinhibition (22). Pyramidal cells of the hippocampus *in vitro* show little or no direct transmembrane effects of morphine or opioid peptides. Most studies on the hippocampal slice indicate that the enkephalins and β-endorphin primarily reduce the size of recurrent and feedforward inhibitory postsynaptic potentials (IPSPs) in both CA1 and CA3 cell groups, thus supporting a disinhibitory mechanism (23,24).

The inhibition by opiate agonists is generally stereospecific, blocked by naloxone, and is reduced during tolerance development after chronic administration of opiates. These findings suggest the involvement of the empirically defined opiate receptors. The last decade has seen an increase in intracellular studies, leading to a better understanding of the mechanisms of opioid function. As with other neuropeptides, several mechanisms can be triggered at different sites by opioid peptides. As many as four mechanisms involving pre- and postsynaptically located receptors have been postulated to account for the depressant effects of opioids on central and peripheral mammalian neurons (12,13).

A synopsis of histochemical, ultrastructural, and electrophysiological data suggests a preferential postsynaptic interaction between opioid peptides and transmitters released from adjacent terminals. The substantial decrease in opiate binding after deafferentation and the finding that opioids can reduce release of synaptic transmitters suggest an additional presynaptic site of action, which might involve the blockade of voltage-sensitive Ca^{2+} channels. The final outcome of release of endogenous opioids thus may depend on the local circuitry as well as the strategic location of the receptor subtypes. Unfortunately, the difficulties of recording from presynaptic terminals in mammalian CNS generally preclude direct examination of the actual mechanisms involved.

To circumvent this problem a study was performed in neuroblastoma x glioma hybrid cells (NG 108-15), in which the tight-seal whole-cell variant of the patch-clamp technique was used to analyze the functional coupling of opiate receptors to voltage-dependent Ca^{2+} channels. As previously shown for the relatively unselective agonist D-Ala^{2}-D-Leu^{5}-enkephalin (25), both μ- (D-Ala^2-NMe-Phe^4-Gly^{5}ol) (DAGO) and δ-selective (D-Pen^2-D-Pen^5-enkephalin) (DPDPE) agonists reversibly reduced the Ca^{2+} currents and the increase in intracellular Ca^{2+} following depolarizing pulses measured with an intracellular fluoresence technique (FURA II). Because the action of the δ-preferring agonist DPDPE could be antagonized in some cells with a receptor-selective opioid antagonist (ICI) without altering the effect of the μ-preferring agonist DAGO, it is suggested that both μ and δ receptors are colocalized and are probably linked to the same effector system (W. Zieglgänsberger et al., *unpublished data*). These results are in accord with previous investigations reporting the activation of a pertussis toxin–sensitive G protein (Go) as an essential step in mediating the inhibition of voltage-activated Ca^{2+} channels by opioids (25).

Also studies performed in cultured dorsal root ganglia neurons, which could provide insights into the presynaptic effects of opioid peptides, revealed that opioid agonists reduce Ca^{2+} influx. In these neurons (derived from either chick or rat neonates), none of the opioid peptides tested altered membrane potentials, but they did reduce the duration of the action potential, presumably through an enhancement of some K^+ conductance, e.g., Ca^{2+}-dependent K^+ or the delayed rectifier conductance. Dorsal root ganglion cells acutely isolated from adult rats do not show this opiate effect on action potential duration. It seems possible that either the relevant (somatic) opiate receptors disappear entirely with maturity or they migrate down the axons to the terminals in the spinal cord (12,13,26).

A series of findings obtained in various sites in the CNS have been reported that showed that opioid peptides induce hyperpolarization and that the increased conductance involves K^+ channels (27). More recent data suggest that the K^+ channels involved in the opioid-induced hyperpolarization are the same voltage-dependent (inwardly rectifying) channels activated by several other inhibitory transmitters (e.g., muscarinic receptors, $GABA_B$ receptors, and α_2 adrenoreceptors) and that such activation (28) involves a pertussis toxin–sensitive guanosine triphosphate (GTP) binding protein (29,30). Because the effects of the inhibitory agonists are nonadditive, it is suggested that the agonists open the same population of potassium channels and inhibit cell firing through an increase in potassium conductance and consequent hyperpolarization. In newborn rats, currents through single channels were activated by DAGO (31). The unitary inward currents showed conductances of approximately 45 pS. An intimate coupling between the receptor and the potassium channel was suggested by these studies. It was postulated that μ-opiate receptors, α_2 adrenoreceptors, and probably also somatostatin receptors are directly coupled to potassium channels through GTP binding proteins in the membrane.

Interestingly, in the zona compacta of the substantia nigra of the rat, the principal cells, which are by far the majority of cells encountered in that area, are hyperpolarized by dopamine and the $GABA_B$ agonist baclofen, whereas μ- and δ-opiate receptor agonists had no effect on the membrane potential (32). The coupling of the κ receptors to voltage-dependent Ca^{2+} channels is suggested from several investigations in different preparations (33). The molecular events that take place subsequent to the activation of the opiate receptor subtypes have not yet been elucidated in details.

Potassium channels are widely distributed in the nervous tissue and are ubiquitous membrane proteins. However, in various sites in the CNS, opioid peptides do not change the membrane potential or membrane resistance (32), but decrease the magnitude and rate of rise of the excitatory postsynaptic potentials (EPSPs) (21,24). Postsynaptic modulation of the actions of excitatory amino acid transmitters, which mediate these EPSPs (34), was one of the first described actions of opioid peptides and opiate alkaloids on central neurons (13).

In a recent study, we investigated the actions of the μ- and δ-receptor-specific opioid peptides DAGO and DPDP in neocortical neurons of the rat, employing intracellular recording including single-electrode voltage-clamp techniques. Despite the fact that neocortical neurons have potassium conductances (e.g., the inwardly rectifying), these neurons are not hyperpolarized by opiates. Applications of either agonist caused small invariable changes in membrane potential, which were difficult to separate from slow spontaneous fluctuations (± 2 mV) and current/voltage relation measurements revealed little if any changes. At the same time, the excitatory postsynaptic potentials evoked by low stimulus intensities were reduced on average by 50%. At higher stimulus intensities, the attenuation of the postsynaptic transients became less pronounced. Both DAGO and DPDP, applied either iontophoretically or added to the perfusion medium, attenuated the depolarizing responses to iontophoretically applied L-glutamate. Single-electrode voltage-clamp measurements showed that opioids reduced the inward currents (EPSC) triggered by synaptic stimulation and by iontophoretic application of L-glutamate (or quisqualate). The blocking effect of opioids on L-glutamate-induced depolarizations was still present when the synaptic transmission was blocked by high Mg^{2+} and low Ca^{2+} concentrations in the perfusion medium, suggesting a postsynaptic site of action. It was postulated that the opiates interfere with the chemically excitable cationic channels permeable to Na^+ and K^+ also opened by synaptically released excitatory transmitters.

It is concluded from these results that the predominant effect of opioid peptides on neocortical neurons is a modulation of non-*N*-methyl-D-aspartate (NMDA) responses because in contrast to L-glutamate (and quisqualate) evoked depolarizations, the NMDA-induced depolarizations and inward currents were either not influenced or reduced only slightly by opioids. The depolarizations, elicited by NMDA were blocked by D-2-amino-5-phosphono-valeric acid (APV), which, at the concentrations employed (5–50 μM), did not affect L-glutamate-induced depolarizations.

In accord with previous studies, we showed that in these neurons NMDA and L-glutamate-induced inward currents are (a) triggered through the activation of different receptors, (b) carried by a different ionic composition, and (c) display different ion and voltage sensitivities. NMDA-receptor activation leads to an inward current, which increases with more depolarized holding potentials and is carried to a great extent by Ca^{2+} ions entering along with Na^+ ions. This current is markedly potentiated by removal of Mg^{2+} ions (34,35).

In contrast to the repetitive firing induced by L-glutamate caused by the voltage sensitivity of the NMDA response, regenerative currents are induced, which increase with depolarization and induce burst-like firing patterns. This latter effect is exploited to differentially activate interneurons (see Fig. 1). In addition to the four main L-glutamate-receptor subtypes (NMDA, quisqualate, kainate, and 2-amino-4-phosphone-butyrate), a fifth receptor has been characterized that is coupled to a G protein. This receptor subtype may gain particular interest because it might provide the trigger for long-term changes through intracellular metabolic actions (see below), which are also affected by other signaling molecules like neuropeptides (34). Most recently, a cDNA clone from a rat forebrain cDNA library was isolated that encodes a functional L-glutamate receptor (36). It has been shown that this protein forms an ion channel but bears only slight resemblance to other members of the ligand-gated-channels superfamily characterized by four trans-

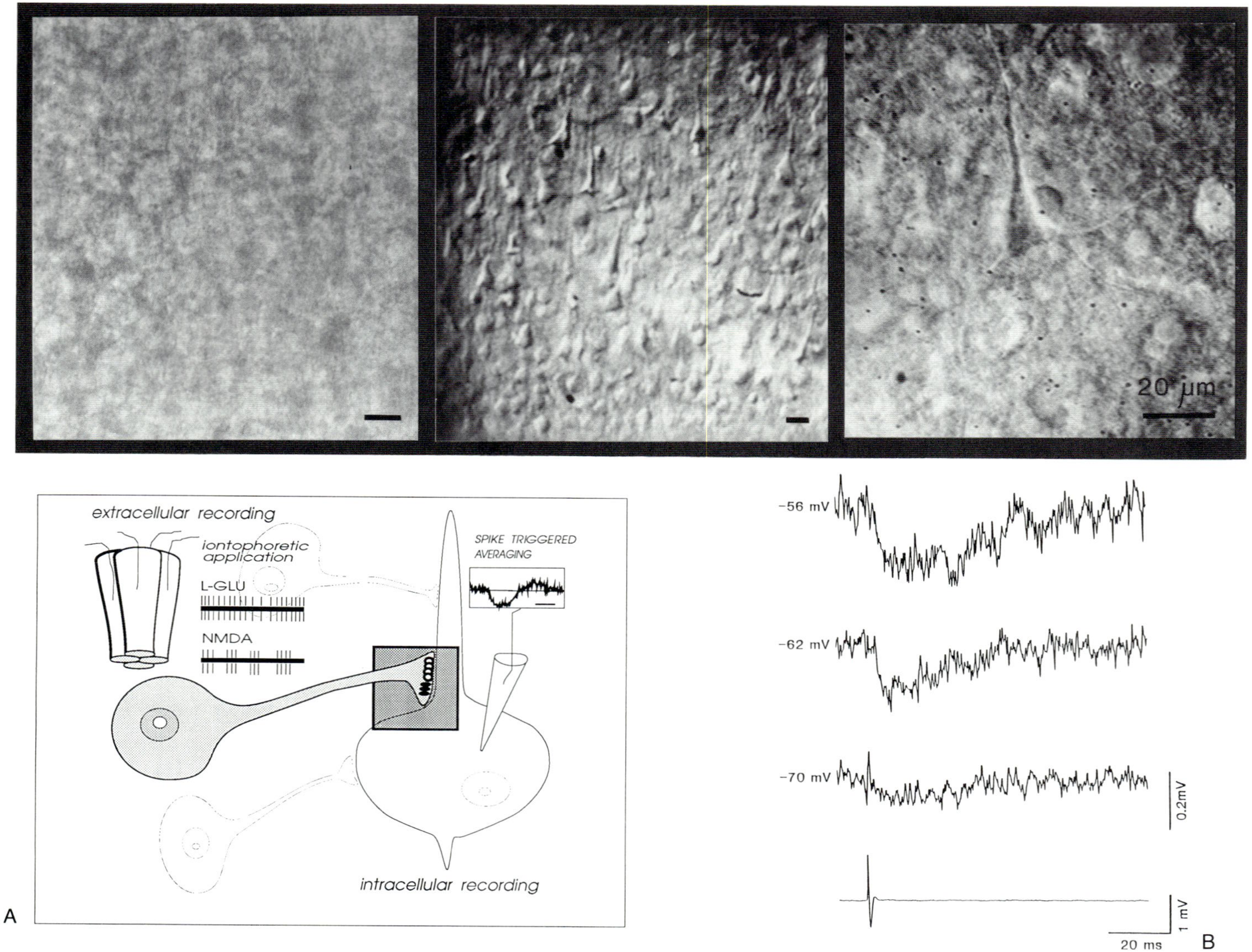
20 µm
extracellular recording
iontophoretic application
L-GLU
NMDA
SPIKE TRIGGERED AVERAGING
intracellular recording
-56 mV
-62 mV
-70 mV
0.2mV
1 mV
20 ms

FIG. 1. A B

A: Scheme is illustrating simultaneous intra- and extracellular recording from a pyramidal neuron and an interneuron (spike-triggered averaging). Most of the information that has been gathered so far about actions of neuropeptides in CNS derives from extracellular recording of single-unit activity in combination with iontophoretic or pneumatic administration from multibarreled micropipettes in regions containing a high density of specific binding sites (13). Introduction of *in vitro* preparations of mammalian CNS has introduced interesting possibilities in analysis of neuropeptide actions. In the present study, coronal slices (nominally 300–500 μm thick) were cut from frontal poles of each cerebral cortical hemisphere of Sprague-Dawley rats (120–160 g). In experiments in which spike-triggered averaging or videomicroscopy (see below) was employed, a brain slice chamber with a coverslip bottom mounted on a rigid stage was used. Slices were continuously superfused and were kept submerged under a thin film of liquid (medium in mM: NaCl, 122.75; KCl, 5.00; NaH_2PO_4, 1.25; $CaCl_2$, 2.50; $MgSO_4$, 1.30; $NaHCO_3$, 26; glucose, 10; gassed continuously with 95% O_2/5% CO_2; final pH, 7.4 at 36–37°C). Intracellular recordings were made from neurons in layers II and III of dorsomedial frontal neocortex. Glass microelectrodes were usually filled with potassium acetate (4 M, pH 7.0). Postsynaptic potentials were evoked by bipolar electrical stimulation (0.05-msec duration) of superficial cortical layers (0.5–1.0 mm from the recording electrode; for details see ref. 44). With a novel combination of infrared illumination, differential interference contrast, and image intensification by video, unstained living neurons have been visualized up to a depth of 50 to 100 μm in 300-μm thick slices (45). Local neuronal networks and bundling of apical dendrites of pyramidal cells were visible. Slices displayed normal field and intracellular potentials when their viability was controlled during videomicroscopy. After a pyramidal neuron was impaled intracellularly, the discharge activity of an extracellularly recorded nearby neuron (*shaded neuron* in schematic illustration) was evoked or enhanced by iontophoretic application of either L-glutamate (L-Glu) or NMDA from a multibarreled micropipette and used to trigger the averaging. L-Glutamate (30–80 nA/5–20 sec) most commonly evoked repetitive discharge, whereas NMDA (5–15 nA/5–10 sec) evoked high–frequency burst discharges. Responses of the target neuron to two different discharge patterns evoked in the neighboring neuron were compared. The present sample includes 62 intracellular recordings (membrane potential: -80.9 mV ± 9.7 SD; input resistance: 29.6 MΩ ± 4.1 SD; action potential amplitude: 100.7 mV ± 9.0 SD). Drug testing was started 30 to 40 min after impalement. Two-tailed Student's *t*-tests were used for statistical comparisons of two sample means. Drugs were applied iontophoretically from five-barreled micropipettes (overall tip size. 3–5 μm resistance. 10–50 MΩ) placed in the vicinity of the recording electrode. For placement of the multibarreled micropipette, the depolarizing response to iontophoretically applied L-glutamate (phoretic currents: 50–150 nA) was used as a search stimulus. Backing currents of 10 to 20 nA and automatic current neutralization were used routinely. DPDPE, DAGO, VIP, adrenergic agonists (isoprenalin, phenylephrine, clonidine), and the opiate antagonist naloxone were added to the perfusion medium or were applied by pressure (10–100 kPa) from micropipettes filled with diluted solutions (10^{-8}–10^{-5} M). The action of neuropeptides persisted in Ca^{2+}-free/high-Mg^{2+} medium, suggesting that they were mediated through postsynaptically located receptors. High-pressure liquid chromatography techniques were employed to measure the iontophoretic release of the neuropeptides from multibarreled micropipettes and to control for their integrity. In the current range employed, the release of neuropeptides was linearly related to phoretic current. Neuropeptides appeared intact after they were removed from micropipettes previously used in the experiment (for 1–2 hr) or when they were added (10^{-6} and 10^{-7} M) to the superfusion medium and were collected after contact with the slice. In 68 intracellularly and 166 extracellularly recorded neurons, 19 single-axonsynaptic connections were revealed. In 17 cases, action potentials of the presynaptic cell resulted in depolarization of the target cell with an average amplitude of 162 $\pm$ 111 μV and a latency to peak of 15 to 50 msec. Two purely inhibitory connections were found. Excitatory postsynaptic potentials were followed in nine of 17 neurons by hyperpolarization potentials with a time to peak of about 60 msec and amplitudes of 27 $\pm$ 16.2 μV. The time-to-peak of these potentials is about 100 $\pm$ 20 msec and lies between the time-to-peak of fast ($GABA_A$) and slow (150–250 msec) inhibitory postsynaptic potentials in neocortical pyramidal cells. Thus this potential may represent a novel synaptic component, which can only be revealed by spike-triggered averaging. Activation of interneurons with different frequencies lacked clearly distinct responses. Attempts to separate peptidergic components by blockade of the $GABA_A$ component by bicuculline have been made. The neuropeptide(s) involved in remaining response is currently being investigated (H. Pawelzik et al., *unpublished observations*). **B:** Illustration of voltage dependence of an averaged single-axon inhibitory postsynaptic potential. Current injection changes the conventional IPSP evoked by orthodromic stimulation and this postsynaptic transient in the same manner. Reversal potential for both transients was in the same range (-72–-75 mV).

membrane regions. It remains to be shown whether these receptor subtypes belong to a novel class of ligand-gated channels.

VASOACTIVE INTESTINAL POLYPEPTIDE INCREASES NMDA-MEDIATED SYNAPTIC COMPONENT

In the same sample of neurons in which the opioid actions were studied, strong synaptic stimuli evoked APV-reversible synaptic components. These NMDA-mediated synaptic components were increased in a number of neurons by exogenously applied vasoactive intestinal polypeptide (VIP), as well as isoprenaline, the β-adrenergic agonist of norepinephrine (NE). VIP is contained in neocortical interneurons in laminae II and III (37; see also Chap. 2, *this volume*), with radially oriented bipolar processes branching in cylinders of about 30 μm diameter (38). In addition to afferent terminals from VIP containing interneurons, the pyramidal neurons in these laminae receive noradrenergic synaptic afferents, which are composed of fine axons and organized predominantly in a plane parallel to the pial surface (39). The cell bodies of this highly divergent trajectory are located in the locus coeruleus. Ample evidence suggests that their activation produces potent and specific forms of conditional actions on their target neurons.

When tested on basal discharge activity, NE and VIP appears to be mainly inhibitory on neuronal discharge activity (40). However, NE may function as an "enabler" (i.e., it enhances the responsiveness of target neurons to other excitatory and inhibitory synaptic inputs, which may be active simultaneously) (6). Current research suggests that the convergence of NE and VIP on neocortical neurons might create metabolic hot spots on neocortical neurons by acting synergistically to increase cyclic AMP levels (41–43; see also Chap. 2, *this volume*). As both VIP and NE were shown to increase the excitability of pyramidal cells to synaptic inputs, the response of a well-defined cortical domain could selectively be altered. Long-lasting VIP responses and event-related NE effects could provide the basis for the induction of neuronal plasticity in exactly addressable cortical columns.

ACKNOWLEDGMENTS

We wish to thank Drs. Harvey Karten, Floyd Bloom, and George Siggins for their invaluable criticisms. Also Mrs. C. H. Hilf for technical assistance and Mrs. Nancy Callahan for her manuscript assistance.

REFERENCES

1. Agnati LF, Fuxe K, Zoli M, Ozini I, Toffano G, Ferraguti F. A correlation analysis of the regional distribution of central enkephalin and β-endorphin immunoreactive terminals and of opiate receptors in adult and old male rats. Evidence for the existence of two main types of communication in the central nervous system: the volume transmission and the wiring transmission. *Acta Physiol Scand* 1986a;128:201–207.
2. Agnati LF, Fuxe K, Merlo Pich E, et al. Aspects of the integrative capacities of the central nervous system: evidence for "volume transmission" and its possible relevance for receptor–receptor interactions. In: Fuxe K, Agnati LF, eds. *Receptor–receptor interactions: a new intramembrane integrative mechanism*. London: Macmillan. 1986b;236–249.
3. Fuxe K, Agnati LF, Hartstrand A, et al. Principles for the hormone regulation of wiring transmission and volume transmission in the central nervous system. *Acta Physiol Scand* 1986;128: 1–53.
4. North RA. Mechanisms of autonomic integration. In: Bloom FE, ed. *Handbook of physiology, volume on intrinsic regulatory systems of the brain*. Bethesda, Maryland: The American Physiological Society, 1986;115–153.
5. Bloom FE. General features of chemically identified neurons. In: Björklund A, Hökfelt T, eds. *Handbook of chemical neuroanatomy,* vol 2. The Netherlands: Elsevier Biomedical Press, 1984;1–22.
6. Bloom FE. Neurotransmitters: past, present, and future directions. *FASEB J* 1988;2:32–41.
7. Herkenham H. Mismatches between neurotransmitter and receptor localization in brain: observations and implications. *Neuroscience* 1987;23:1–38.

8. Sontheimer H, Kettenmann H, Backus KH, Schachner H. Glutamate opens Na^+/K^+ channels in cultured astrocytes. *Glia* 1988;1:328–336.
9. Marrero H, Astion ML, Coles JA, Orkand RK. Multiple conductance channels in type-2 cerebellar astrocytes activated by excitatory amino acids. *Nature* 1989;339:378–380.
10. Usowicz MM, Gallo V, Cull-Candy SG. Multiple conductance channels in type-2 cerebellar astrocytes activated by excitatory amino acids. *Nature* 1989;339:380–383.
11. Gregor P, Mano I, Maoz I, McKoewn M, Teichberg VI. Molecular structure of the chick cerebellar kainate-binding subunit of a putative glutamate receptor. *Nature* 1989;342:689–691.
12. Zieglgänsberger W. Central control of nociception. In: Bloom FE, ed. *Handbook of physiology, volume on intrinsic regulatory systems of the brain*. Bethesda, Maryland: The American Physiological Society, 1986;581–645.
13. Siggins GR, Gruol DL. Synaptic mechanisms in the vertebrate central nervous system. In: Bloom FE, ed. *Handbook of physiology, volume on intrinsic regulatory systems of the brain*. Bethesda, Maryland: The American Physiological Society, 1986;1–114.
14. Nicoll RA. Neurotransmitter regulated ion channels. *Science* 1988;241:545–555.
15. Berridge MJ. Inositol triphosphate and diacylglycerol: two interacting second messengers. *Annu Rev Biochem* 1987;56:159–193.
16. Chuang De-Maw. Neurotransmitter receptors and phosphoinositide turnover. *Annu Rev Pharmacol Toxicol* 1989;29:71–110.
17. Simonds WF. The molecular basis of opioid receptor function. *Endocr Rev* 1988;9:200–212.
18. Mansour A, Khachaturian H, Lewis ME, Akil H, Watson SJ. Autoradiographic differentiation of mu, delta, and kappa opioid receptors in the rat forebrain and midbrain. *J Neurosci* 1987;7:2445–2464.
19. Tempel A, Zukin RS. Neuroanatomical patterns of the mu, delta, and kappa opioid receptors of rat brain as determined by quantitative *in vitro* autoradiography. *Proc Natl Acad Sci USA* 1987;84:4308–4412.
20. Harlan RE, Shivers BD, Romano QJ, Howells RD, Pfaff DW. Localization of preproenkephalin mRNA in the rat brain and spinal cord by *in situ* hybridization. *J Comp Neurol* 1987;258:159–184.
21. Deisz RA, Madamba S, Moore S, Siggins GR, Sutor B, Zieglgänsberger W. Regulation of neuronal excitability by opioid peptides: intracellular analysis in several brain regions. In: Illes P, Farsang C, eds. *Regulatory role of opioid peptides*. Weinheim, New York: VCH Press, 1988;147–164.
22. Zieglgänsberger W, French ED, Siggins GR, Bloom FE. Opioid peptides may excite hippocampal pyramidal neurons by inhibiting adjacent inhibitory interneurons. *Science* 1979;205:415–417.
23. Siggins GR, Zieglgänsberger W. Morphine and opioid peptides reduce inhibitory synaptic potentials in hippocampal pyramidal cells *in vitro* without alteration of membrane potential. *Proc Natl Acad Sci USA* 1981;78:5235–5239.
24. Siggins GR, Zieglgänsberger W. How neuropeptides alter neuronal excitability. *Proc Collegium Internationale Neuro-Psychopharmacologicum*, Munich (*in press*).
25. Hescheler T, Rosenthal W, Trautwein G, Schultz G. The GTP-binding protein, Go, regulates neuronal calcium channels. *Nature* 1987;325:445–447.
26. Williams JT, Zieglgänsberger W. Mature spinal ganglion cells are not sensitive to opiate receptor mediated actions. *Neurosci Lett* 1981;21:211–216.
27. North RA, Williams JT, Surprenant AM, Christie MJ. μ and δ receptors belong to a family of receptors that are coupled to potassium channels. *Proc Natl Acad Sci USA* 1987;84:5487–5491.
28. Christie MJ, North RA. Agonists at μ-opioid, M_2-muscarinic and $GABA_B$-receptors increase the same potassium conductance in rat lateral parabrachial neurones. *Br J Pharmacol* 1988; 95:896–902.
29. McFadzean I. The ionic mechanisms underlying opioid actions. *Neuropeptides* 1988;11:173–180.
30. Williams JT, North RA, Tokimasa T. Inward rectification of resting and opiate-activated potassium currents in rat locus coeruleus neurons. *J Neurosci* 1988;8:4299–4306.
31. Miyake M, Christie MJ, North RA. Single potassium channels opened by opioid in rat locus coeruleus neurons. *Proc Natl Acad Sci USA* 1989;86:3419–3422.
32. Lacy MQ, Mercuri NB, North RA. Two cell types in rat substantia nigra zona compacta distinguished by membrane properties and the actions of dopamine and opioids. *J Neurosci* 1989;9:1233–1241.
33. Macdonald RL, Werz MA. Dynorpin A decreases voltage-dependent calcium conductance of mouse dorsal root ganglion neurones. *J Physiol (Lond)* 1986;377:237–249.
34. Monaghan DT, Bridges RJ, Cotman CW. The excitatory amino acid receptors: their classes, pharmacology, and distinct properties in the function of the central nervous system. *Annu Rev Pharmacol Toxicol* 1989;29:365–402.
35. Cotman CW, Monaghan DT, Ganong AT. Excitatory amino acid neurotransmission: NMDA receptors and Hebb-type synaptic plasticity. *Annu Rev Neurosci* 1988;11:61–80.
36. Hollmann M, O'Shea-Greenfield A, Rogers SW, Heinemann S. Cloning by functional expression of a member of the glutamate receptor family. *Nature* 1989;342:643–648.
37. Magistretti PJ, Morrison JH. VIP neurons in the neocortex. *Trends Neurosci* 1985;8:7–8.
38. Morrison JH, Magistretti PJ, Benoit R, Bloom FE. The distribution and morphological characteristics of the intracortical VIP-positive cell: an immunohistochemical analysis. *Brain Res* 1984; 292:269–282.
39. Parnavelas JG, Papadopoulos GC. The mono-

aminergic innervation of the cerebral cortex is not diffuse and nonspecific. *Trends Neurosci* 1989;12:315–320.

40. Ferron A, Siggins GR, Bloom FE. Vasoactive intestinal polypeptide acts synergistically with norephinephrine to depress spontaneous discharge rate in cerebral cortical neurons. *Proc Natl Acad Sci USA* 1985;82:8810–8812.
41. Magistretti PJ, Schorderet M. VIP and noradrenaline act synergistically to increase cyclic AMP in cerebral cortex. *Nature* 1984;308:280–284.
42. Magistretti PJ, Schorderet M. Norepinephrine and histamine potentiate the increases in cyclic adenosine 3′:5′-monophosphate elicited by vasoactive intestinal polypeptide in mouse cerebral cortical slices: mediation by a_1-adrenergic and H_1-histaminergic receptors. *J Neurosci* 1985;5:363–368.
43. Magistretti PJ, Morrison JH. Noradrenaline and vasoactive intestinal peptide-containing neuronal systems in neocortex: functional convergence with contrasting morphology. *Neuroscience* 1988;24:367–378.
44. Howe JR, Sutor B, Zieglgänsberger W. Baclofen reduces postsynaptic potentials of rat neocortical neurones by an action separate from its hyperpolarizing action. *J Physiol* 1987;384:539–569.
45. Dodt HU, Zieglgänsberger W. Infrared DIC-videomicroscopy of living brain slices (*Abstr*) *Soc Neurosci Abstr* 1989;15:280.

Volume Transmission in the Brain: Novel Mechanisms for Neural Transmission, edited by K. Fuxe and L.F. Agnati. Raven Press, Ltd., New York © 1991.

29

Neuropeptidases and Neuropeptide Inactivation in the Brain

Jean-Charles Schwartz, Marie-Louise Bouthenet, Bruno Giros, Claude Gros, Catherine Llorens-Cortes, and Hélène Pollard

Unité de Neurobiologie et Pharmacologie (U. 109) de l'INSERM, Centre Paul Broca, 75014 Paris, France

It is often assumed that peptidergic neurotransmission in the central nervous system (CNS) differs from transmission by amino acids and even monoamines by several characteristic features: It is of slower onset and longer duration and occurs over larger distances. More than on experimental evidence, this assumption seems largely based on analogies with the actions exerted by peptides in the peripheral nervous system or as hormones.

Because these spatio-temporal characteristics of peptidergic signals may be largely governed by the distance at which the neuropeptides may diffuse before being inactivated, it is of obvious relevance to define the processes responsible for their inactivation. During the past decade, these processes have been the most extensively studied in the case of enkephalins, these opioid pentapeptides being small molecules for which identification of metabolic pathways has represented a relatively simple task in comparison with larger neuropeptides (reviewed in refs. 1–4).

Endogenous enkephalins [i.e., Tyr-Gly-Gly-Phe-Met (YGGFM) and Tyr-Gly-Gly-Phe-Leu (YGGFL)] are substrates for several brain peptidases, but the participation of some of them in their physiologic inactivation such as angiotension-converting enzyme, cytosolic aminopeptidases, or dipeptidyl aminopeptidases could be progressively excluded despite some controversies. However, it now rather clearly appears that two ectopeptidases (i.e., integral membrane peptidases with their active sites oriented toward the extracellular space) are mainly if not solely responsible for the physiological inactivation of the opioid pentapeptides. Cleavage of the Tyr^1-Gly^2 bond seems to occur mainly if not solely under the action of aminopeptidase M (EC 3.4.11.2), an enzyme abundant in microvillar membranes of the kidney and intestine (5), but whose presence in brain was only recently detected (6).

In addition, cleavage of the Gly^3-Phe^4 bond selectively occurs under the action of enkephalinase (membrane metallo-endopeptidase EC 3.4.24.11), an enzyme abundant in kidney and intestine but also present in cerebral membranes (7). Both peptidases display a rather large substrate specificity, and it appears that the critical factor in their function is that they are strategically located to ensure an efficient hydrolysis of the released enkephalins. For both enzymes, careful analysis of their specificity has been the first step in a rational design of potent inhibitors, which were thereafter shown to protect endogenous enkephalins and to mimic in a naloxone-reversible manner some (but not all) biological effects of exogenous opiates.

Finally, using a sensitive radioimmunoassay (RIA), the tripeptide Tyr-Gly-Gly (YGG) was characterized as an endogenous constituent of rodent brain tissues and shown to originate from the hydrolysis of the Gly^3-Phe^4 bond of enkephalins by enkephalinase (8–10). Measurement of the endogenous YGG level progressively appears to provide a new means of evaluating the dynamic state of enkephalins under a variety of conditions and assessing the *in vivo* effects of inhibitors of the two enkephalin-hydrolyzing peptidases.

MOLECULAR AND BIOCHEMICAL PROPERTIES OF ENKEPHALINASE (EC 3.4.24.11)

Enkephalinase is a zinc-containing metallopeptidase inserted into the plasma membrane of a variety of cells. Its recent molecular cloning (11,12) has shown it to be constituted of a 742-amino-acid sequence comprising six potential *N*-linked glycosylation sites and a 23-amino-acid long hydrophobic domain that is likely to span the lipid bilayer of membranes. This molecular structure is consistent with the conclusion that enkephalinase in rat brain is an "ectopeptidase" (i.e., an enzyme with its active site directed toward the extracellular space), a topography that enables it to attack neuropeptides released into this space (13,14).

When the enzyme was first isolated from kidney and designated "neutral proteinase from kidney brush borders," only the (slow) hydrolysis of large substrates such as insulin or glucagon was characterized (5). After its identification in cerebral membranes and the demonstration that it could efficiently cleave the Gly^3-Phe^4 amide bond of enkephalins (7), the hydrolysis of numerous other neuropeptides was demonstrated (reviewed in refs. 2–4,15). Hydrolysis of these peptides occurs more or less efficiently as judged, in each case, from the specificity constant (i.e., the ratio of catalytic constant to Michaelis constant) (Table 1). This specificity constant is a biochemical parameter that allows us to predict to some extent the relative rate of hydrolysis of different peptides at low concentrations (relative to the K_M, which is generally in the micromolar range). From this it appears that the substrate specificity of enkephalinase is relatively large; it mainly acts on small peptides, cleaving the amide bonds in which the amino group is donated by aromatic or hydrophobic amino acid residues, particularly when the latter are situated at the C-terminal part of the molecule.

These features explain that the enzyme is essentially inactive on opioid peptides of the dynorphin and β-endorphin families, in contrast to those derived from the preproenkephalin A genes (except the octapeptide), which are excellent substrates. Substance P, atrial natriuretic factor, and to a lesser extent, bradykinin and neurotensin appear to be good substrates, whereas cholecystokinin and angiotensin also appear to be cleaved at relatively low rates. It should be underlined, however, that it is impossible to predict from this sole parameter whether the peptidase is involved in the physiological inactivation of a given neuropeptide. Other parameters have to be taken into consideration, such as the relative localizations of the enzyme and the peptide-releasing nerve endings or the presence of other, possibly more efficient peptidases in the vicinity. Hence, not only the substrate specificity but also the localization of peptidases appears to represent important factors in determining their "physiological specificity."

TABLE 1. *Hydrolysis of various neuropeptides by purified enkephalinase*

Peptides	Sites of hydrolysis	Specificity constant[ab] (k_{cat}/k_M)	References
1°/Opioid peptides			
(Leu^5) enkephalin	YGGFL	100	59
(Met^5) enkephalin	YGGFM	300	59
Enkephalin heptapeptide	YGGFMRF	21	60
Enkephalin octopeptide	YGGFM . .	NH	
Dynorphin (1–9)	YGGFLRRIR	<3	60
Dynorphin (1–13)	YGGFLRRIRPKLK	NH	61
α-Neoendorphin	YFFFLRYPK	NH	61
β-Neoendorphin	YGGFLRKYP	NH	61
β-Endorphin (31 amino acids)	YGGFM . .	NH	60
2°/Other neuropeptides			
Cholecystokinin	DYMGWMDF-NH_2	10	34
Substance P	RPKQQFFGLM-NH_2	362	62
Neurokinin A	HPKTDSFVGLM-NH_2	ND	62
Neurokinin B	DMWDFFVGLM-NH_2	ND	62
Bradykinin	RPPGFSPFR	63	63
Angiotensin II	DRVYIHPF	24	63
Atrial natriuretic factor	S . . . CF^{8} . . SFRY	100	C. Gros et al. (*unpublished data*)
Neurotensin	pELYENLPRRPYIL	37	64

[a]NH, not hydrolyzed; ND, not determined.

[b]Specificity constants relative to that of (Leu^5) enkephalin (= 100). The value of the latter, determined with rat kidney enkephalinase purified to homogeneity, was 90/μM/min (59).

LOCALIZATION OF ENKEPHALINASE IN THE CENTRAL NERVOUS SYSTEM

Gross dissections, subcellular fractionation, and lesion studies initially indicated that enkephalinase was expressed in a markedly heterogeneous manner in the mammalian brain and that its activity was high in areas or neuronal fractions enriched in enkephalin nerve endings or receptors (reviewed in refs. 1,2,16).

More recently, the distribution of the enzyme was established with more details using ^{3}H-labeled inhibitors (17,18) or a peroxidase-labeled antibody (19).

With the use of a [^{125}I]iodinated monoclonal antibody, a detailed mapping of the enzyme in rat brain and spinal cord could be recently established and compared with enkephalin and substance P immunoreactivity (18,20). Moreover, the same monoclonal antibody was used to localize the enzyme at the ultrastructural level (A. Beaudet et al., *unpublished data*).

Immunoreactivity is undetectable in white matter areas (e.g., the corpus callosum or fornix) and has a laminar pattern in, for example, the cerebral cortex or hippocampal formation. Hence, even without performing the immunodetection at the cellular level, a major neuronal localization of the peptidase is suggested. In addition, on sagittal sections the enzyme appears to be localized all along a pathway linking the striatum to the globus pallidum, the entopeduncular nucleus, and the substantia nigra, indicating that enkephalinase is expressed on the whole neuronal surface including axons and is not restricted to synaptic areas (Fig. 1). That this massive lo-

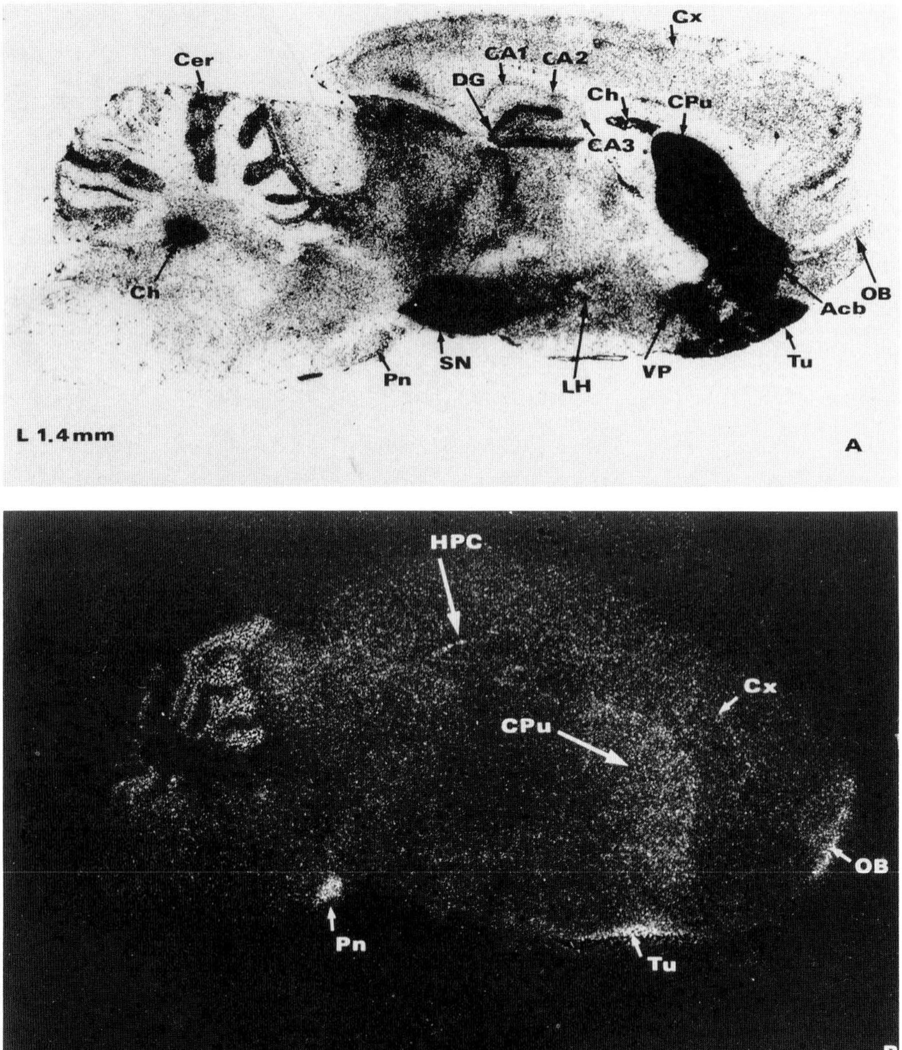

FIG. 1. Localization of enkephalinase immunoreactivity (**A**) and enkephalinase mRNA (**B**) in sagittal sections of rat brain. Peptidase was detected autoradiographically, using a ^{125}I-labeled monoclonal antibody, whereas mRNA was detected by *in situ* hybridization, using a P_{32}-labeled complementary deoxyribonucleic acid (cDNA) probe (21,22). Note absence of hybridization signal at the level of substantia nigra, where an abundant immunoreactivity is present. Acb, accumbens nucleus; Cev, cerebellum; Ch, choroid plexus; Cx, cerebral cortex; Ca1, Ca2, CA3, fields of Ammon's horn; CPu, caudate putamen; DG, dentate gyrus; LH, lateral hypothalamic area; OB, olfactory bulb; Pn, pontine nucleus; SN, substantia nigra; Tu, olfactory tubercle; VP, ventral pallidum.

calization corresponds to the presence of the enzyme on striatopallidal and striatonigral pathways was shown by the effects of intrastriatal kainate lesions as well as by the absence of any enkephalinase messenger ribonucleic acid (mRNA) detectable by *in situ* hybridization in the substantia nigra (20–22). The presence of a high enkephalinase immunoreactivity in a striatopallidal pathway known to contain a large set of enkephalinergic fibers (23) would be consistent with the idea that enkephalingeric neurons express the peptidase all along their plasma membranes, in the same way as cholinergic neurons express acetylcholinesterase. At the ultrastructural level, enkephalinase immunoreactivity appears to be present in membranes of a large percentage of striatal neurons, lining most extracellular spaces in some areas (A. Beaudet, *personal communication*). It is clear that such a disposition is well-adapted to prevent the diffusion of peptidergic signals at any long distance of the terminals from which they emanate.

The massive expression of the peptidase in a striatonigral pathway known to contain many substance P fibers may suggest that this localization corresponds to the inactivation of tachykinins at the level of the substantia nigra, particularly pars reticulata. This view would be consistent with the observations that enkephalinase inhibitors accelerate dopamine turnover in caudate putamen and nucleus accumbens (24–27) because these drugs protect endogenous substance P released from nigral slices (28) and substance P excites nigral dopaminergic neurons. Nevertheless, naloxone blocks the actions of enkephalinase inhibitors on dopamine turnover and modifies substance P release from nigral slices; this suggests that endogenous enkephalins are involved in the two effects.

In other areas (e.g., the dorsal horn of the spinal cord or the spinal nucleus of the trigeminal nerve, the superficial layers containing both substance P and enkephalin (ENK) terminals), a dense enkephalinase (ENKase) immunoreactivity is present (Fig. 2). The latter may largely account for the naloxone-reversible analgesic activity of enkephalinase inhibitors (29), assuming that these inhibitors mainly act by delaying the local inactivation of endogenous enkephalins, rather than that of substance P. Hence, in this area and several others in which enkephalin and substance P neurons interact, the pharmacological actions of inhibitors seem more consistent with a role of the peptidase in enkephalin than in substance P inactivation.

In addition to its major expression in neuronal cells, the peptidase is also present in nonneuronal cells even in the CNS (e.g., a strong enkephalinase immunoreactivity is found in epithelial cells of the choroid plexuses, the ependymal cells, in scattered cells of the pituitary). It seems likely that some of these localizations reflect a protective function of the peptidase, the entrance of circulating peptides in intact form into the CNS being thereby prevented. Also the low but still easily measurable level of a soluble form of enkephalinase in cerebrospinal fluid, possibly formed in the choroid plexuses, may contribute in preventing any distance transmission of susceptible neuropeptides through the subarachonoidal spaces and extracellular fluids of the brain parenchyma (Table 2). Furthermore, the peptidase is present in a large number of peripheral tissues, sometimes with a much higher abundance than in the CNS (30), particularly in the kidney brush borders (5) and lung (31). It is clear that these localizations presumably correspond to functions other than neuropeptide inactivation. For instance, the participation of enkephalinase in the inactivation of circulating atrial natriuretic factor was demonstrated (32), and its participation, together with other peptidases, in the processes of hydrolysis of short peptides into amino acids directly absorbable from the intestine or kidney tubular lumen seems likely.

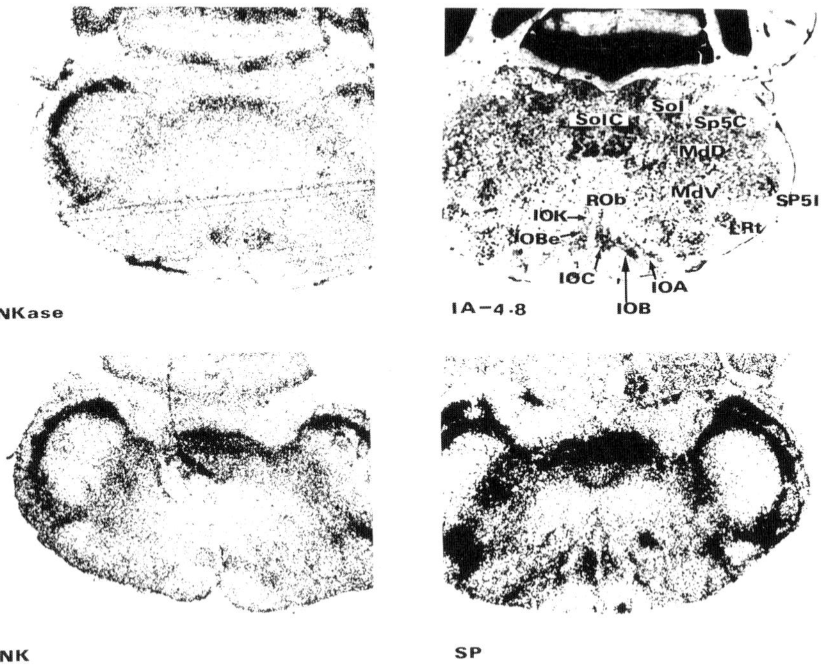

FIG. 2. Distributions of ENKase, ENK, and substance P (SP) immunoreactivities at the level of the rat spinal nucleus of the trigeminal nerve. Immunoreactivities were detected autoradiographically, using ^{125}I-labeled antibodies. A corresponding Nissl-stained section is also shown. Note good matches of three immunoreactivities at the level of external layers of the nucleus and that of enkephalinase and enkephalins in the cerebellum. MdD, MdV, medullary reticular nucleus, dorsal and ventral parts; ROb, raphe obscurus nucleus; Sol, nucleus of the solitary tract; SolC, nucleus of the solitary tract, commusural part; Sp5C, Sp5I, triderminal nucleus, caudal and interpolar parts; IOA, IOB, IOC, inferior olive, subnuclei of medial nucleus; IOBe, inferior olive, beta subnucleus.

TABLE 2. *Characterization of a soluble form of enkephalinase in human lumbar cerebrospinal fluid (CSF) and comparison with the enzyme activity in cerebral membranes*[a]

	Mouse brain homogenate	Human CSF
V_{max} (nmol/mg protein/min)	26 ± 1	0.014 ± 0.003
K_M (μM)	73 ± 11	54 ± 8
K_i values (μM)		
Thiorphan	0.002	0.002
Phosphoramidon	0.002	0.004
Captopril	100	100
(Met^5) enkephalin	7	19

[a]Enzyme activity measured using succinyl-Ala-Ala-Phe (7-amido, 4-methyl coumarin) as a substrate (60).

ROLE OF ENKEPHALINASE IN NEUROPEPTIDE INACTIVATION IN THE CNS AS DEDUCED FROM THE EFFECTS OF ITS INHIBITORS

The design and use of potent and selective enkephalinase inhibitors has progressively allowed us to define the role of the peptidase (1,2,15). Its participation in endogenous enkephalin inactivation *in vitro* was demonstrated in studies with depolarized striatal slices, a model in which enkephalinase inhibitors enhance the recovery of the intact peptide (33). A similar model was more recently used to demonstrate the role of enkephalinase in endogenous substance P hydrolysis in nigral slices, but a participation of endogenous enkephalins in the enhanced recovery was evoked (28). Similar studies with other cerebral neuropeptides, also known as substrates for the peptidase, are still lacking. Moreover, in the case of endogenous cholecystokinin, another substrate, similar studies could rule out its participation in this neuropeptide inactivation (34). In the latter case the absence of any significant protective effect of enkephalinase inhibitors seems to be due to the fact that another peptidase, a serine peptidase more active on cholecystokinins than enkephalinase and/or more adequately localized, plays this role (35).

Hence it appears that a combination of biochemical and topographic factors ensures the "functional specificity" of CNS peptidases toward the various neuropeptides.

In the case of enkephalins the participation of enkephalinase in their inactivation was confirmed by the demonstration that inhibitors elicit numerous opioid-like, naloxone-reversible responses (reviewed in refs. 2,15).

These include antinociception in rodents (29) and analgesia in humans (36,37). Also modifications in the turnover of cerebral monoamines, similar to those elicited by opiates, which consist in acceleration of striatal dopamine or cortical serotonin turnover and decrease in norepinephrine turnover, occur in a naloxone-reversible manner after administration of enkephalinase inhibitors (25–27,38,39). In agreement with these observations, enkephalins and enkephalinase are present at the level of the groups of perikarya from which the corresponding monoaminergic systems emanate (40). By activating mesolimbic dopamine neurons, enkephalinase inhibitors stimulate in a naloxone-reversible manner spontaneous locomotor activity in rodents (25,41).

Other central opioid-like, naloxone-reversible effects include inhibition of reflex contractions of the urinary bladder (42) and "antidepressant" activity in a forced swimming test (39,43,44), or changes in susceptibility to experimental seizures (45).

The fact that all *in vivo* central effects of enkephalinase inhibitors reported so far are attributable to protection of endogenous enkephalins may suggest that *in vivo,* these neuropeptides are unique substrates of the peptidase. However, we should also consider the fact that its participation in the inactivation of other neuropeptides has been less actively investigated, because pharmacological tools such as peptide antagonists are generally not available in their case.

MOLECULAR PROPERTIES AND LOCALIZATION OF AMINOPEPTIDASE M (EC 3.4.11.2)

The participation of peculiar, membrane-bound aminopeptidase activity in the inactivation of endogenous enkephalins in the CNS was suggested by the observations that (Met^5) enkephalin released by depolarization of striatal slices could be only partially protected by enkephalinase inhibitors. Total protection required the copresence of the aminopeptidase inhibitor bestatin, whereas puromycin, another inhibitor of cytosolic aminopeptidases, was largely ineffective in this respect (14). The search for a membrane-bound aminopeptidase sensitive to bestatin but poorly sensitive to puromycin resulted in the isolation and identification of aminopeptidase M from rat cerebral membranes (6).

The primary structure of aminopeptidase M of human (46) and rat origin (47) was determined by cloning of its complementary deoxyribonucleic acid (cDNA) (Fig. 3). The enzyme is a glycoprotein comprising 966 amino acids. A highly hydrophobic segment, 24 amino acids in length, located close to the amino terminus of the protein is likely to serve as the membrane-spanning domain, confirming that the enzyme is membrane-bound. Hence, like enkephalinase, it consists of a short intracellular tail and a long extracellular domain, which contains the active site. The sequence includes nine potential *N*-linked glycosylation sites and one potential sulfation site. It also contains an eight-amino-acid sequence, showing high homology with sequences in other metallopeptidases known to serve as the zinc-binding domain. This defines a family of enzymes that, otherwise show, rather limited homology.

From a biochemical point of view, aminopeptidase M is characterized by its ability to hydrolyze a large variety of short peptides at the level of the N-terminal amino acid, particularly when the latter is a hydrophobic or aromatic amino acid residue.

In the brain, microvessels represent a major localization of aminopeptidase M, as shown by their intense immunostaining by antibodies raised against the renal enzyme as well as by characterization of the enzyme activity in a microvascular fraction (48). Rather high immunoreactivity also seems present in choroid plexuses (Fig. 4).

In addition, aminopeptidase-M activity is enriched (by five- to sixfold) in fractions containing synaptic membranes from rat cerebral cortex. Although contamination of this fraction by microvascular elements cannot be entirely excluded, it is likely that the presence of the ectopeptidase in synaptic elements accounts for its role in the inactivation of endogenous enkephalins, as demonstrated in pallidal slices (49).

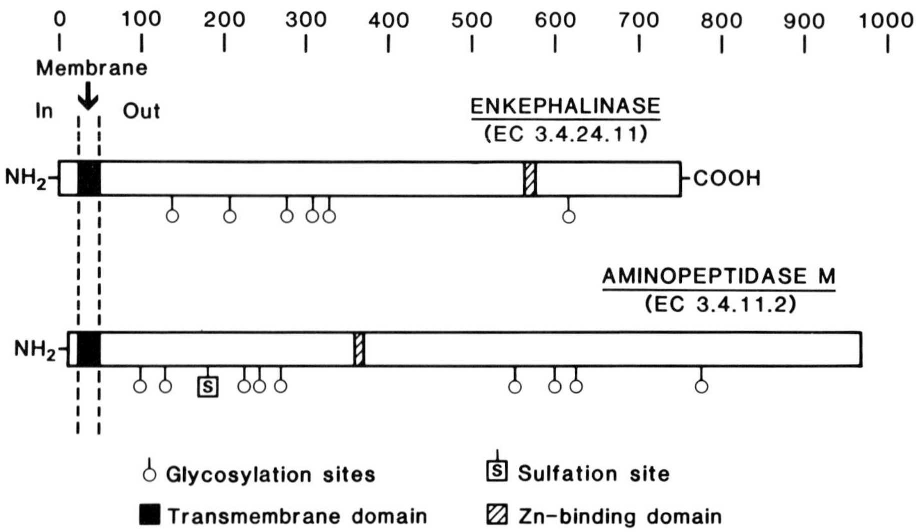

FIG. 3. Schematic structures of enkephalinase and aminopeptidase M as deduced from cloning of their cDNAS.

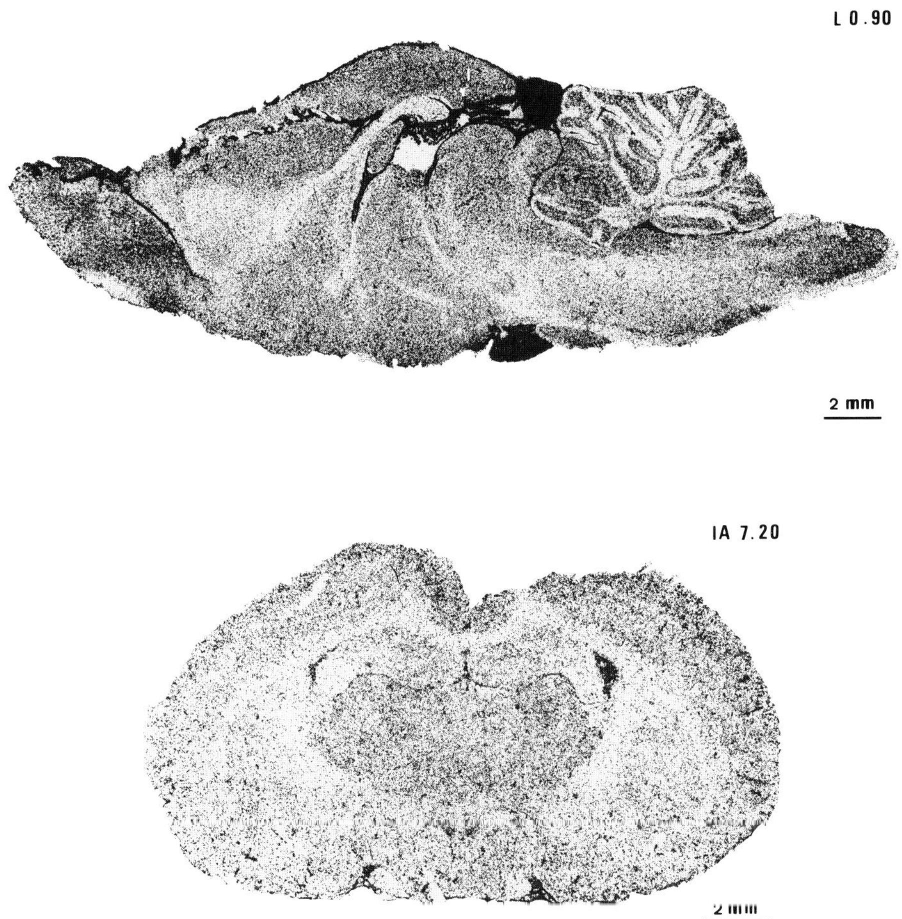

FIG. 4. Localization of aminopeptidase M immunoreactivity in rat brain. Peptidase was detected autoradiographically, using a polyclonal antibody directed against aminopeptidase M raised in rabbit (6) and a secondary ^{125}I-labeled anti-rabbit immunoglobulin antibody. Note intense immunoreactivity in choroid plexuses and meninges as well as the dotted distribution of immunoreactivity in most cerebral areas, mainly corresponding to a major vascular localization of the antigen. Note also differential distribution of the immunoreactivity in the brain parenchyma, which may reflect the localization of peptidase in neuronal membranes, as suggested by subcellular fractionation and lesion studies.

The role of aminopeptidase M in enkephalin inactivation *in vivo* is supported by the opioid-like effects of aminopeptidase inhibitors such as bestatin or phethiol, particularly when these compounds are associated with enkephalinase inhibitors. Hence a robust naloxone-reversible analgesia results from such treatments in rodents (14,50–52) and even in humans (37) and were also found with a single compound inhibiting both enzymes (53).

Whereas aminopeptidase M seems to play a role as important as that of enkephalinase in the inactivation of endogenous enkephalins, it seems to differ from it by at least two aspects: Its substrate specificity is larger, and its distribution is more homogeneous. Both features are consistent with the idea that aminopeptidase M might represent a "neuropeptidase" of wide functional implication in the central nervous system.

THE TRIPEPTIDE TYR-GLY-GLY, AN EXTRACELLULAR METABOLITE OF ENKEPHALINS

The tripeptide Tyr-Gly-Gly (YGG) appears as a metabolic product of endogenous enkephalins (8,10). The parallel regional distributions of YGG and enkephalins in rodent brain is consistent with this idea, YGG levels representing about 10% of enkephalin levels. Ablation of proenkephalin A–containing neurons in the striatopallidal pathway elicited by intrastriatal kainate is accompanied by significant decreases in YGG in both the projection area and the injected area, which contains both perikarya and terminals of short enkephalin neurons (49).

Another important indication is provided by the identification of enkephalinase as the

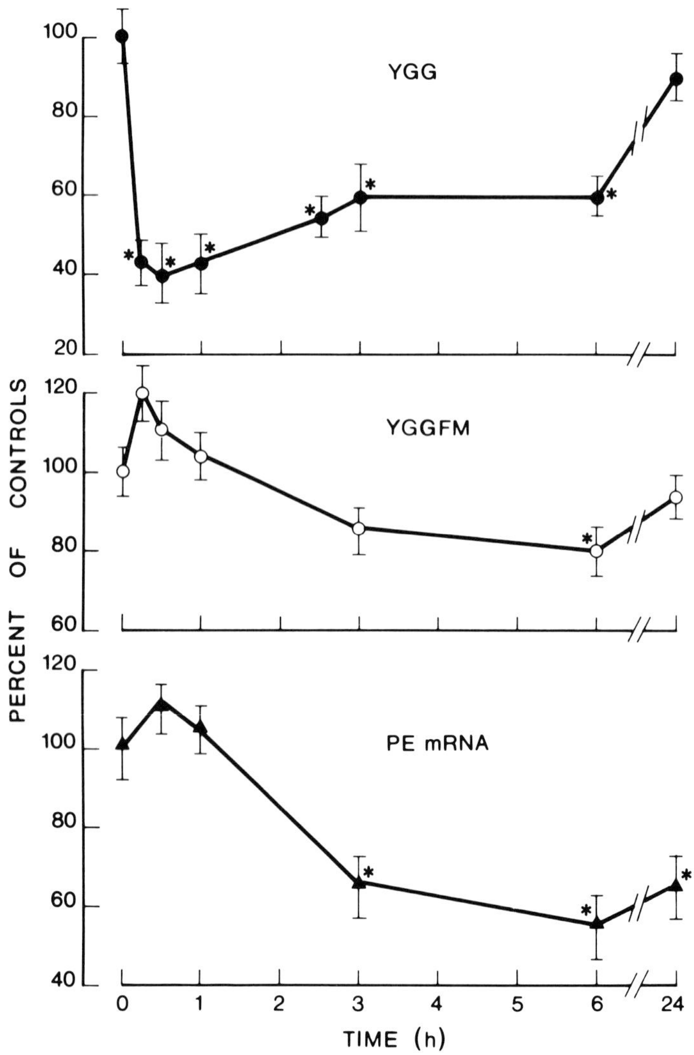

FIG. 5. Effects of muscimol and diazepam on levels of YGG, YGGFM, and PE A mRNA in mouse striatum. Animals were killed after administration of 0.75 mg/kg muscimol and 2.5 mg/kg diazepam. Peptide levels were evaluated by RIA and proenkephalin A mRNA abundance by Northern blot analysis.*, $p<0.05$ as compared to controls.

YGG-forming enzyme *in vitro* and *in vivo* (9,54). The substrate specificity of the enzyme is consistent with the idea that endogenous YGG is formed by hydrolysis of peptides with a YGGF . . . sequence. Taking into account the orientation of the active site of the enzyme, which faces the extracellular space, this strongly suggests that YGG is formed extracellularly, an idea consistent with both its subcellular distribution in mouse brain and its predominant localization in the incubation medium of cerebral slices (9,52,54). The effects of aminopeptidase inhibitors, which block not only one of the two degradative pathways of enkephalins but also the hydrolysis of YGG itself, leads to a progressive increase in YGG level both *in vitro* and *in vivo,* and its rate reflects that of enkephalin release in brain and spinal cord (55).

All these features are consistent with the idea that the tripeptide YGG is in a dynamic state in extracellular fluids of the CNS and that its steady-state level reflects the rate of enkephalin release. In agreement with this view, stimulation of $GABA_A$ receptors, which is known to reduce enkephalin release *in vitro* and *in vivo* (56–58), elicits a rapid and long-lasting reduction of striatal YGG levels in mice treated with a combination of muscimol and diazepam (Fig. 5). Interestingly, this change is accompanied by a decrease of proenkephalin (PE) A mRNA, but the latter occurs only after several hours and outlasts that of YGG level. Meanwhile, (Met^5)enkephalin (YGGFM) levels are only slightly and transiently affected. This suggests that inhibition of enkephalin release is the initial change after inhibition of enkephalin neuron activity by GABA and that a compensatory decrease in enkephalin biosynthesis occurs with some delay, helping to maintain constant (or nearly constant) levels of the neuropeptide.

Measurements of the change in YGG levels in brain or spinal cord perfusates have also shown (or confirmed) that enkephalin release is markedly decreased by barbiturates (10) and selectively increased by some nociceptive stimuli (55). Hence measurements of the level of this extracellular metabolite constitutes a useful index of enkephalin neuron activity.

CONCLUSIONS

The studies performed during the past decade on the metabolism of endogenous enkephalins in brain indicate that two well-defined metallopeptidases take part in the inactivation of these neuropeptides. Both are ectoenzymes, their localization within the plasma membrane, with their active site oriented toward the extracellular space, being presumably essential for local inactivation of released neuropeptides. Both seem expressed by neuronal as well as nonneuronal cell populations. The localization of enkephalinase to neuronal populations in the vicinity of enkephalin neurons seems largely predominant in brain, but it is not restricted to synaptic areas, the peptidase being expressed on the entire neuronal surface. However, the nonneuronal (vascular) localization of aminopeptidase M is predominant in the brain, but the detailed distribution of this enzyme has not been established so far. The cooperative action of these two "neuropeptidases" is likely to ensure an effective local inactivation of enkephalins, as evidenced by the characteristic opioid-like responses that result from their inhibition. This also suggests that the local inactivation of released neuropeptides should prevent their diffusion at large distances, but this hypothesis was never tested directly in the mammalian brain. It also remains to be established whether similar mechanisms apply to other neuropeptides, and if so, the relevant peptidases must be identified.

REFERENCES

1. Schwartz JC, Malfroy B, De la Baume S. Biological inactivation of enkephalins and the role

of enkephalin-dipeptidylcarboxypeptidase ("enkephalinase") as neuropeptidase. *Life Sci* 1981; 29:1715–1740.
2. Schwartz JC. Enkephalinase inhibitors as drugs. In: Sandler M, Smith HJ, eds. *Design of enzyme inhibitors as drugs*. New York: Oxford University Press, 1989;206–220.
3. McKelvy JF, Blumberg S. Inactivation and metabolism of neuropeptides. *Annu Rev Neurosci* 1986;9:415–434.
4. Lynch DR, Snyder SH. Neuropeptides: multiple molecular forms, metabolic pathways and receptors. *Annu Rev Biochem* 1986;55:773–799.
5. Kenny AJ. Proteinases associated with cell membranes. In: Barrett AJ, ed. *Proteinases in mammalian cells and tissues*. Amsterdam: North Holland Publishers, 1977;393–444.
6. Gros C, Giros B, Schwartz JC. Identification of aminopeptidase M as an enkephalin-inactivating enzyme in rat cerebral membranes. *Biochemistry* 1985;24:2179–2185.
7. Malfroy B, Swerts JP, Guyon A, Rogues BP, Schwartz J-C. High affinity enkephalin-degrading peptide in brain is increased after morphine. *Nature (Lond.)* 1978;276:523–526.
8. Llorens-Cortes C, Schwartz JC, Gros C. Detection of the tripeptide Tyr-Gly-Gly, a putative enkephalin metabolite in brain, using a sensitive radioimmunoassay. *FEBS Lett* 1985;189:325–328.
9. Llorens-Cortes C, Gros C, Schwartz JC. Study of endogenous Tyr-Gly-Gly, a putative enkephalin metabolite, in mouse brain: validation of a radioimmunoassay localization, and effects of peptidase inhibitors. *Eur J Pharmacol* 1985; 119:183–191.
10. Llorens-Cortes C, Gros C, Schwartz JC. Steady-state level and turnover rate of the tripeptide Tyr-Gly-Gly as indexes of striatal enkephalin release *in vivo* and their reduction during pentobarbital anesthesia. *Proc Natl Acad Sci USA* 1986;83:6226–6230.
11. Malfroy B, Schofield PR, Kuang W, Seeburg P, Mason AJ, Menzel WJ. Molecular cloning and aminoacid sequence of rat enkephalinase. *Biochem Biophys Res Commun* 1987;144:59–66.
12. Devault A, Lazure C, Nault C, et al. Aminoacid sequence of rabbit kidney neutral endopeptidase 24.11 (enkephalinase) deduced from a complementary DNA. *EMBO J* 1987;6:1317–1322.
13. De la Baume S, Patey G, Schwartz JC. Subcellular distribution of enkephalin dipeptidylcarboxypeptidase ("enkephalinase") in rat brain. *Neuroscience* 1981;6:315–321.
14. De la Baume S, Yi CC, Schwartz JC, Chaillet P, Marcais-Collado H, Costentin J. Participation of both "enkephalinase" and aminopeptidase activities in the metabolism of endogenous enkephalins. *Neuroscience* 1983;8:143–151.
15. Schwartz JC, Costentin J, Lecomte JM. Pharmacology of enkephalinase inhibitors. *Trends Pharmacol Sci* 1985;6:472–476.
16. Schwartz JC. Metabolism of enkephalins and the concept of neuropeptidase. *Trends Neurosci* 1983;6:15–18.
17. Waksman G, Hamel E, Fournie-Zaluski MC, Roques BP. Autoradiographic comparison of the distribution of the neutral endopeptidase "enkephalinase" of μ and δ opioid receptors in rat brain. *Proc Natl Acad Sci USA* 1986;83:1523–1527.
18. Pollard H, De la Baume S, Bouthenet ML, Schwartz JC, Ronco P, Verroust P. Characterisation of two probes for the localisation of enkephalinase in rat brain: [^{3}H]thiorphan and a ^{125}I-labeled monoclonal antibody. *Eur J Pharmacol* 1987;133:155–164.
19. Matsas R, Kenny AJ, Turner AJ. An immunohistochemical study of endopeptide 24.11 (enkephalinase) in the pig nervous systems. *Neuroscience* 1986;18:991–1012.
20. Pollard H, Llorens-Cortes C, Couraud JY, Ronco P, Verroust P, Schwartz JC. Enkephalinase (EC 3.4.24.11) is highly localized to a striatonigral pathway in rat brain. *Neurosci Lett* 1987;77:267–271.
21. Pollard HY, Wilcox JN, Malfroy B, Ronco P, Verroust P, Schwartz JC. Localization of enkephalinase (EC 3.4.24.11) in rat brain: radioimmunohistochemical and mRNA hybridization data. In: Cros J, Meunier JC, Hamon M, eds. *Advances in the Biosciences*, vol 75: *Progress in opioid research*. New York: Pergamon Press, Oxford, 1989;317–320.
22. Wilcox JN, Pollard H, Moreau J, Schwartz JC, Malfroy B. Localization of enkephalinase mRNA in rat brain by *in situ* hybridization: comparison with immunohistochemical localization of the protein. *Neuropeptides* 1989;14:77–83.
23. Cuello AC, Paxinos G. Evidence of a long Leu-enkephalin striato-pallidal pathway in rat brain. *Nature* 1978;271:178–180.
24. Algeri S, Alstein M, De Simone M, Guarda-Basso GM. *In vivo* potentiation of [D-Ala^2]Met enkephalin amide central effects after administration of an enkephalinase inhibitor. *Eur J Pharmacol* 1981;74:207–209.
25. Kalivas PW, Richardson-Carlson R. Endogenous enkephalin modulation of dopamine neurons in ventral tegmental area. *Am J Physiol* 1986;R243–R249.
26. Llorens-Cortes C, Schwartz JC. Changes in turnover of cerebral monoamines following inhibition of enkephalin metabolism by thiorphan and bestatin. *Eur J Pharmacol* 1984;104:369–374.
27. Wood PL. Phasic enkephalinergic modulation of nigrostriatal dopamine metabolism: potentiation with enkephalinase inhibitors. *Eur J Pharmacol* 1982;82:119–120.
28. Mauborgne A, Bourgoin S, Benoliel JJ, et al. Enkephalinase is involved in the degradation of endogenous substance P released from slices of rat substance nigra. *J Pharmacol Exp Ther* 1987;243:674–680.
29. Roques BP, Fournie-Zaluski MC, Soraco E, et al. The enkephalinase inhibitor thiorphan shows antinociceptive activity in mice. *Nature (Lond)* 1980;288:286–288.
30. Llorens-Cortes C, Schwartz JC. Enkephalinase activity in rat peripheral organs. *Eur J Pharmacol* 1981;69:113–116.
31. Swerts JP, Perdrisot R, Patey G, De la Baume S,

Schwartz JC. Enkephalinase is distinct from brain "angiotensin-converting enzyme." *Eur J Pharmacol* 1979;57:279–281.

32. Gros C, Souque A, Schwartz JC. Protection of atrial natriuretic factor against degradation, diuretic and natriuretic responses after *in vivo* inhibition of enkephalinase (EC 3.4.24.11) by acetorphan. *Proc Natl Acad Sci USA* 1989;86: 7580–7584.
33. Patey G, De la Baume S, Schwartz JC, et al. Selective protection of methionine enkephalin released from brain slices by enkephalinase inhibition. *Science* 1981;212;1153–1155.
34. Zuzel KA, Rose C, Schwartz JC. Assessment of the role of "enkephalinase" in cholecystokinin inactivation. *Neuroscience* 1985;15:149 158.
35. Rose C, Camus A, Schwartz JC. A serine peptidase responsible for the inactivation of endogenous cholecystokinin in brain. *Proc Natl Acad Sci* 1988;85:8326–8330.
36. Floras P, Bidabe AM, Caille JM, Simonnet G, Lecomte JM, Sabathie M. Double blind study of effects of enkephalinase inhibitor on adversive reactions to myelography. *Am J Neuroradiol* 1983;4:653–655.
37. Meynardier J, Dalmas S, Lecomte JM, Gros C, Schwartz JC. Potent analgesic effects of inhibitors of enkephalin metabolism administered intrathecally to cancer patients. *Pain Clin* 1988; 2:201–205.
38. Petit F, Hamon M, Fournie-Zaluski MC, Roques BP, Glowinski J. Further evidence for a role of γ-opiate receptors in the presynaptic regulation of newly synthesized dopamine release. *Eur J Pharmacol* 1986;126:1–9.
39. Lecomte JM, Constentin J, Vlaiculescu A, et al. Pharmacological properties of acetorphan: a parenterally active "enkephalinase" inhibitor. *J Pharmacol Exp Ther* 1986;237:937–944.
40. Pollard H, Bouthenet ML, Moreau J, et al. Detailed immunoautoradiographic mapping of enkephalinase (EC 3.4.24.11) in rat central nervous system: comparison with enkephalins and substance P. *Neuroscience* 1989;30:339–376.
41. Michael-Titus A, Constentin J. Analgesic effects of metapramine and evidence against the involvement of endogenous enkephalins in the analgesia induced by tricyclic antidepressants. *Pain* 1987;31:391–400.
42. Hisamitsu T, De Groat WC. The inhibitory effect of opioid peptides and morphine applied intrathecally and intracerebroventricularly on the micturation reflex in the cat. *Brain Res* 1984; 298:51–65.
43. Ben Natan L, Chaillet P, Lecomte JM, Marcais-Collado H, Uchida G, Costentin J. Involvement of endogenous enkephalins in the mouse "behavioral despair" test. *Eur J Pharmacol* 1984; 97:301–304.
44. De Felipe MC, Jimenez I, Castro A, Fuentes JA. Antidepressant action of imipramine and inpridole in mice is enhanced by inhibitors of enkephalin-degrading peptidases. *Eur J Pharmacol* 1989;159:175–180.
45. Ukponmwan DE, Dzoljic MR. Enkephalinase inhibition antagonizes the increased susceptibility to seizure induced by REM sleep deprivation. *Psychopharmacology* 1984;83:229–232.
46. Olsen J, Cowell GM, Konigshofer E, et al. Complete aminoacid sequence of human intestinal aminopeptidase N as deduced from cloned DNA. *FEBS Lett* 1988;238:307–314.
47. Malfroy B, Kado-Fong M, Gros C, Giros B, Schwartz JC, Helmiss R. Molecular cloning and aminoacid sequence of rat kidney aminopeptidase M: a member of a superfamily of zinc-metallopeptidases. *Biochem Biophys Res Commun* 1989;161:236–241.
48. Solhonne B, Gros C, Pollard H, Schwartz JC. Major localization of aminopeptidase M in rat brain microvessels. *Neuroscience* 1987;22:225–232.
49. Giros B, Gros C, Solhonne B, Schwartz JC. Characterization of aminopeptidases responsible for the inactivation of endogenous (Met^5)enkephalin in brain slices using peptidase inhibitors and aminopeptidase M antibodies. *Mol Pharmacol* 1986;29:281–287.
50. Zhang AZ, Yang HYT, Costa E. Nociception, enkephalin content and dipeptidyl carboxypeptidase activity in brain of mice treated with exopeptidase inhibitors. Neuropharmacology 1982; 21:625–630.
51. Chaillet P, Marcais-Collado H, Costentin J, Yi CC, De la Baume S, Schwartz JC. Inhibition of enkephalin metabolism by, and antinociceptive activity of bestatin, an aminopeptidase inhibitor. *Eur J Pharmacol* 1983;86:329–336.
52. Gros C, Giros B, Schwartz JC, Vlaiculescu A, Costentin J, Lecomte JM. Potent inhibition of cerebral aminopeptidases by carbaphethiol, a parenterally active compound. *Neuropeptides* 1988;12:111–118.
53. Fournie-Zaluski MC, Chaillet P, Bouboutou R, et al. Analgesic effects of kelatorphan, a new highly potent inhibitor of multiple enkephalin degrading enzymes. *Eur J Pharmacol* 1984; 102:525–528.
54. Giros B, Llorens-Cortes C, Gros C, Schwartz JC. The endogenous tripeptide Try-Gly-Gly as a possible metabolite of opioid peptides in rat brain: identification, regional distribution, effects of lesions and formation in depolarized slices. *Peptides* 1986;7:669–677.
55. Llorens-Cortes C, Gros C, Schwartz JC, Clot AM, Le Bars D. Changes in levels of the tripeptide Tyr-Gly-Gly as an index of enkephalin release in the spinal cord: effects of noxious stimuli and parenterally active peptidase inhibitors. *Peptides* 1989;10:609–614.
56. Osborne H, Herz A. K^+-evoked release of Met-enkephalin from rat striatum *in vitro:* effect of putative neurotransmitters and morphine. *Naunyn Schmiedebergs Arch Pharmacol* 1980;310: 203–209.
57. Harsing G, Yang HYT, Costa E. Evidence for γ-aminobutyric acid (GABA) mediation in the benzodiazepine inhibition of the release of Met^5-enkephalin elicited by depolarization. *J Pharmacol Exp Ther* 1982;220:616–620.
58. Bourgoin S, Artaud F, Cesselin F, Glowinski J, Hamon M. Local and remote effect of intracau-

date administration of GABA-related drugs on Met-enkephalin release in the basal ganglia. *Brain Res* 1985;361:361–372.
59. Malfroy B, Schwartz JC. Enkephalinase from rat kidney. Purification, characterization and study of substrate specificity. *J Biol Chem* 1984;259:14365–14370.
60. Schwartz JC, Gros C, Giros B, et al. Metabolic inactivation of enkephalius. In: Vizi ES, Magyark, eds. *Regulation of transmitter function: basic and clinical aspects*. Amsterdam: Elsevier, 1984;217–228.
61. Hersh LB. Reaction of opioid peptides with neutral endopeptidase ("enkephalinase"). *J Neurochem* 1984;43:487–493.
62. Matsas R, Fulcher IS, Kenny JA, Turner, AJ. Substance P and leu enkephalin are hydrolyzed by an enzyme in pig caudate synaptic membranes that is identical with the endopeptidase of kidney microvilli. *Proc Natl Acad Sci USA* 1983;80:3111–3115.
63. Gafford JT, Skidgel RA, Erdos EG, Hersh LB. Human kidney "enkephalinase," a neutral metalloendopeptidase that cleaves active peptides. *Biochemistry* 1983;22(13):3265–3271.
64. Checler F, Emson PC, Vincent JP, Kitabgi P. Inactivation of neurotentin by rat brain synaptic membranes cleavage of the Pro^{-10}–Tyr^{11} bond by endopeptidase 24.11 (enkephalinase) and a peptidase different from proline-endopeptidase. *J Neurochem* 1984;43:1295–1301.

Volume Transmission in the Brain: Novel Mechanisms for Neural Transmission,
edited by K. Fuxe and L.F. Agnati.
Raven Press, Ltd., New York © 1991.

30

Role of Calcium-Dependent Proteases in Volume Regulation

Michel Baudry

Department of Biology, University of Southern California, Los Angeles, California

One of the striking features of neurons is the extreme diversity of the shape of dendritic spines and of synaptic contacts (1). It has been theorized that this morphological diversity is matched by a corresponding functional diversity that might reflect the plastic property of synaptic transmission (2,3). In other words, the variable strength of a given synaptic contact is generally assumed to be a function of the shape of the dendritic spine (4–6). Although much is known about the nature of the proteins constituting the postsynaptic densities, the spine apparatus, and the dendritic cytoskeleton, little is known concerning the dynamics of the assembly of these elements and of the mechanisms underlying the regulation of spine shape. What we do know is that synaptic contacts, dendritic spines, and even dendritic branches are modifiable as a result of certain patterns of synaptic activity as well as under pathological conditions. Thus, it has been shown by several laboratories that high-frequency electrical stimulation produces structural modifications of the dendritic spines of the pyramidal cells in the CA1 field (7–9) and of the granule cells in the dentate gyrus (10,11). On the basis of quantitative electron microscopy studies, it has been concluded that both the shape and the number of synaptic contacts were modified. The change in shape was interpreted as a transformation of spines from an ellipsoidal to a spherical configuration, whereas the increase in synaptic contacts was due to an increase in the number of shaft contacts and especially of the short sessile spine contacts. The increase in contact number was present early after the stimulation (<15 min) and persisted as long as 8 hr after the stimulation (8). It has been recently proposed that this effect is due to the transformation of short spine synapses to stubby synapses as a result of the collapse of the spine cytoskeleton (12). By eliminating the high-resistance spine neck, such a transformation could result in a much more efficient synaptic contact and could provide the structural basis for the long-term increase in synaptic efficacy [long-term potentiation (LTP)] that accompanies high-frequency electrical stimulation (13–16).

Deafferentation is another instance that is followed by a profound reorganization of dendritic structures. Most of the deafferented spines in the molecular layer of the dentate gyrus collapse as a result of a lesion of the perforant path, the major input pathway to the dentate gyrus (17,18). In addition, the dendritic trees of the granule cells shrink and exhibit an abnormal morphology (19,20). These modifications are similar to those observed in aging or in certain neurodegenerative diseases. These neuronal alterations are also associated with profound modifications of glial cells. In particular, astrocytes become first hypertrophied and at later stages atrophied, whereas microglia proliferate enormously and migrate toward the denervated zones (21,22). The signals responsible for these glial reactions are not clearly identified, although a number of cytokinins have been proposed to play critical roles in the initiation of glial proliferation and migration (23).

In all these preceding examples, the cell cytoskeleton exhibits various degrees of reorganization that are assumed to be responsible for the changes in cell morphology. In this chapter, we will review the current evidence supporting the view that calcium-dependent proteases (calpains) play a major role in regulating the dynamics of cell cytoskeleton in a large number of physiological and pathological conditions. We will first review the biochemical properties of this class of proteases before discussing the data pertinent to the functions of these enzymes. The final section will attempt to present a general model of cytoskeleton regulation and to relate the dynamics of the cell cytoskeleton to the main theme of this volume (i.e., to the question of how changes in cell processes might produce long-lasting alterations in synaptic transmission).

PROPERTIES OF Ca^{2+}-DEPENDENT PROTEASES

A number of recent reviews (24–27) have been devoted to the subject of Ca^{2+}-dependent proteases, and we will only summarize here the pertinent information.

Biochemical Properties

Two forms of neutral calcium-dependent proteases have been shown to be present in a variety of tissues and species, one form with a high sensitivity (low μM) and one with a low sensitivity (mM) to Ca^{2+} (28,29). These two enzymes have been cloned and sequenced in both human and rabbit tissues (30–32). The sequences indicate that these two proteins are distinct, although some homologies are present. Both enzymes consist in a large (80 kda) catalytic and a light (30 kda) regulatory subunit. The large subunits are different for calpains I and II, and they both comprise four different domains, two of which have homologies with other thiol proteases and with calmodulin (33,34). The light subunit comprises two domains and is the same for calpain I and II. It is now generally accepted that native calpain is inactive in the absence of calcium and therefore represents a proenzyme. In the presence of calcium, the protein changes conformation and undergoes a limited autoproteolysis (probably an intramolecular process) that results in the formation of an active protease. The calcium sensitivity of both enzymes thus reflects the sensitivity for the autolysis and not for the actual proteolysis of other substrates.

Both endogenous activator (35,36) and inhibitor (37) of calpains have been identified. The endogenous inhibitor, calpastatin, is a large molecule that differs in molecular weight depending on the tissue of origin (38). It has also been cloned and sequenced, and the sequence revealed the existence of several repeat units, which are thought to represent the inhibitory domain (39,40). The activator is a small protein (15–20 kDa) and stimulates both calpain I and II 10–25-fold, although it does not change the calcium sensitivity of either protease.

In addition to these endogenous regulators, it has been proposed that phospholipids play an important role in activating the proenzyme, by reducing the concentration of calcium required for calpain autolysis (41). Based on this observation, several authors have suggested that translocation of calpain to the cell membrane followed by interactions with phospholipids represents an initial step in generating an active form of the protease (26,34,42).

Cellular and Subcellular Localization

In crude homogenates from brain tissues, only calpain II activity can be measured (43). Calpain II is found mainly in soluble fractions, with some activity present in the crude mitochondrial/synaptosomal fraction. However, the specific activity of calpain I is highest in this fraction, and with further subfractionation, calpain I is recovered in the fractions containing myelin and synaptosomes. These results indicate that calpain I is predominantly localized in soluble form in small cellular processes (e.g., dendritic branches, spines, boutons, and glial processes), whereas calpain II is mainly found in cytosolic compartments. A somewhat similar localization of calpains has been obtained with immunohistochemistry (44–46). Calpain II immunoreactivity is observed mainly in axons and glia, whereas calpain I immunoreactivity is detected in cell bodies and proximal dendrites of neurons. Using immersion fixation procedures and electron microscopic immunocytochemistry, calpain I–like immunoreactivity is also found in spines, postsynaptic densities, and dendrites (46).

Levels of calpain II activity in soluble fractions of adult rat brain vary about 10-fold across brain regions, with telencephalic structures exhibiting the lowest levels of activity, and pons-medulla and cerebellum exhibiting the highest (47). In contrast, calpain I activity is relatively similar throughout different brain structures. These data should, however, be considered only approximate because they might reflect different distributions of endogenous activator and inhibitor, as well as different degrees of subcellular localization.

Phylogeny and Ontogeny

The comparative studies of the levels of calpain activity in soluble fractions of brains from several classes of vertebrates provided important information concerning the phylogeny of calpains (48).

Fish brain have undetectable levels of activity, and amphibians and reptilians have low levels. Birds and mammals have the largest levels of activity. Across several orders of mammals, brain size is inversely correlated with level of calpain activity in cortex. As brain size is closely correlated with maximal life span, it follows that calpain activity in cortex is inversely and almost linearly correlated with maximal life span (49). This is even true for the order Chiroptera, as bats exhibit both an unusually long life span for a small-brained mammal but also low levels of calpain activity (50). When complementary deoxyribonucleic acid (cDNA) probes for calpains become available, it will be interesting to determine whether this relation is due to a parallel variation in the levels of messenger ribonucleic acid (mRNA) coding for calpains or to some other factors.

Similarly, and in good agreement with the principle that ontogeny recapitulates phylogeny, levels of calpain activity are high immediately at birth in the forebrain and then decline during the first week postnatally (for both calpain I and II) (47). Similar changes are observed for calpain II in the forebrain but not in the hindbrain. In this case, too, it will be interesting to compare these changes with those of the mRNA coding for calpain I and II.

SUBSTRATES OF CALCIUM-DEPENDENT PROTEASES

Before reviewing the data concerning the substrates of calpains, it is important to stress that there is an inherent problem in determining which proteins are natural substrates of proteases in general. Given enough incubation time and enough protease, it is likely that every protein could be degraded by calpains. Few studies have determined kinetic parameters that could discriminate between preferred versus nonpreferred substrates. In general, two different classes of proteins have been shown to be good candidates for physiological substrates of calpains: cytoskeletal proteins and calcium-dependent enzymes (Table 1).

Cytoskeletal Proteins

Most cytoskeletal proteins (e.g., spectrin, neurofilament proteins, microtubule-associated proteins, erythrocyte band 3, actin, and tubulin) have been shown to be potential substrates of calpains in test tube experiments (51–54). Similarly, myelin basic protein and glial fibrillary acidic protein (GFAP) are degraded by calpains (55–58).

Only in few cases has it been shown that the calpain-mediated degradation of these substrates takes place *in situ*. In the case of brain spectrin, calpain-mediated degradation generates a major breakdown product with an apparent M_r of 150 kDa, which is resistant to further degradation and can be quantified on Western blots stained with spectrin antibodies (59). This antigen is found in purified postsynaptic densities (60), suggesting that it could be formed in synaptic contacts during normal physiological activity or that it is generated during the preparation of the subcellular fractions.

TABLE 1. *Substrates of calcium-dependent neutral proteases*

Cytoskeletal proteins
Microtubule-associated proteins (MAPs)
Fodrin or spectrin
Myosin
Ankyrin
Neurofilament proteins
Tubulin
Myofibril proteins
Glial fibrillary acidic protein
Amyloid precursor protein
Enzymes
Kinase C
Ca^{2+}/calmodulin kinase
CyclicAMP–dependent phosphodiesterase
Myosin light-chain kinase
Calcineurin
Calcium-dependent transglutaminase

Furthermore, the amount of this antigen is dramatically increased under a number of experimental conditions (see below), indicating that calpain activation and spectrin degradation occur *in situ*. Similarly, the initial observations that calcium-induced neurofilament and microtubule disruptions were prevented by protease inhibitors were taken as evidence supporting a role for calpain in the regulation of neurofilament and microtubule assembly/disassembly/turnover (51,61).

Another recently identified substrate of calpains is the amyloid precursor protein (APP) (62). Although the function of the APP is not known, its structure indicates that it has a transmembrane domain with the majority of the molecule facing the extracellular space (63). In this case, too, there are no convincing data to indicate that the APP is degraded by calpains *in situ*.

Calcium-Dependent Regulatory Enzymes

In addition to structural proteins, calpains have been shown to produce limited proteolysis of a number of Ca-dependent enzymes (Table 1). In particular, several protein kinases, including protein kinase C, cyclic adenosine monophosphate (cyclic AMP)-dependent protein kinase, Ca-calmodulin-dependent protein kinase, and myosin light-chain kinase, are partially degraded by calpains, and the resulting products have in general lost their regulatory components (64–66). Thus, protein kinase C and the Ca-calmodulin kinase are transformed into constitutive kinases, and this conversion has been proposed to play a critical role in long-term adaptive changes in synaptic transmission (67).

Several protein phosphatases including calcineurin as well as the calcium-calmodulin-dependent cyclic nucleotide phosphodiesterase are partially degraded by calpains (68,69). Finally, calpain also partially degrades and activates calcium-dependent transglutaminases (70).

As mentioned above, little evidence is available to indicate that these conversions take place under normal or pathological conditions, but these data suggest the possibility that calpains may play a critical role in triggering cascades of biochemical reactions with a wide range of intracellular targets.

FUNCTIONS OF CALCIUM-DEPENDENT PROTEASES

Until recently, the precise functions of calpains have remained largely speculative. This was due not only to the difficulty of measuring the activity of these enzymes *in situ* but also to the lack of specific inhibitors that could penetrate rapidly and easily inside cells. Thus, the possible functions of calpains were inferred mainly from the nature of their potential substrates (Table 2) and the limited use of various protease inhibitors. Another approach to identify the functions of these enzymes is to determine under which conditions they are activated and the consequences of their activation.

Calpain Activation After Lesions or Neurotoxin Injections

Because a role for calpains had been proposed in degenerative conditions in muscles and peripheral nerves (71,72), we were interested in determining whether calpain could be activated as a result of various types of insults to the CNS. For this, we took advantage of the ability to quantify the calpain-mediated spectrin breakdown products (bdps) using Western blots stained with antibodies against spectrin. Lesions of the entorhinal cortex eliminate the majority of the inputs to the outer molecular layer of the dentate gyrus. This model of deafferentation has been extensively used to study the anatomical, physiological, and biochemical consequences of deafferentation in the CNS (73–76). An increase in bdps is evident as early as 4 hr after the lesion (77). The increase is maximal 2 days after the lesion, and a small increase is still detected 2 and 4 weeks after the lesion. Constant intraventricular infusion of the protease inhibitor leupeptin, started 3 days before the lesion, produces a 60% to 70% reduction in the amount of bdps observed 1 day after the lesion. The effect of the lesion on the amount of bdps is biphasic with a massive increase occurring between 1 and 4 days and a second period between 96 hr and 27 days. It was postulated that the first period corresponds to the massive structural rearrangement that takes place postsynaptically (i.e., elimination of postsynaptic densities, collapsing of dendritic spines, shrinkage of dendritic trees), whereas the second phase might be related to a long-lasting modification of the turnover of spectrin and possibly other structural proteins. These results were complemented by studies of the changes in the pattern of spectrin-like immunoreactivity in the dentate gyrus after the lesion (78). Thus, 2 days after the lesion, the outer molecular layer displays a large increase in immunoreactivity, which is markedly reduced after the intraventricular infusion of leupeptin. At later times, the increased immunoreactivity shifts toward the inner border of the outer molecular layer and is mainly due to the staining of reactive astrocytes.

Similarly, injection of the neurotoxin colchicine has been widely used to produce selective neuronal death (78–80). Intrahippocampal colchicine injection markedly

TABLE 2. *Functions of calcium-dependent neutral proteases*

By degrading cytoskeletal proteins, functions in:
cell morphology
cell locomotion
intracellular transport
By limited proteolysis of enzymes, involvement in:
intracellular regulation of various functions
amplification of second-messenger systems
By limited degradation of specific proteins, participation in:
targeting proteins for further degradation
activation of soluble proteins

increases the level of bdps between 1 and 2 days after injection, and the increase in bdps is reduced by pretreatment with leupeptin but elevated by treatment with the lysosomal inhibitor chloroquine (81). In this case, too, increased staining of the molecular layer is observed immunohistochemically with spectrin antibodies as well as abnormal staining of reactive astrocytes.

Other experimental conditions resulting in neurodegenerative abnormalities have been shown to be accompanied with increased levels of bdps and thus are likely to be associated with calpain activation. Thus, injection of the excitotoxin kainic acid directly intracerebroventricularly (82) or peripherally (81) produces massive increase in the levels of bdps. Brief periods of ischemia also produce selective neuronal death particularly in field CA1 of the hippocampus, and such treatment results in large increases in bdp levels (83).

Calpain Activation Under Physiological Conditions

If widespread calpain activation and spectrin degradation occur under pathological conditions, it is legitimate to ask whether similar events take place in a more spatially and temporally restricted fashion under physiological conditions. In particular, the physiological activation of one of the subtypes of the receptor for the excitatory amino acid transmitter glutamate [the *N*-methyl-D-aspartate (NMDA) receptor] has been postulated to accomplish precisely this function. By requiring the simultaneous presence of the transmitter and enough depolarization of the membrane to release the ionic channel from a voltage-dependent magnesium blockade (84,85), this receptor is ideally suited to produce spatially localized and temporally brief increases in calcium levels in postsynaptic structures. In agreement with this idea, brief exposure of hippocampal slices to the agonist of these receptors NMDA results in a rapid increase in bdp levels (86). NMDA-receptor activation is also produced by brief periods of high-frequency electrical stimulation and is a necessary condition to elicit LTP of synaptic transmission (87–89). We have discussed in length elsewhere the hypothesis that calpain activation and spectrin degradation were key steps in the sequence of events leading from high-frequency stimulation to long-term changes in synaptic efficacy (90). Suffice it to mention here that the protease inhibitor leupeptin prevents the formation of LTP both *in vivo* when it is infused for several days intraventricularly (91) and *in vitro* when hippocampal slices are incubated for several hours in its presence (92). More recently, two other calpain inhibitors have been shown to prevent the induction of LTP in the hippocampal slice preparation (S. Del Cerro et al., *unpublished data*). Finally, it should also be noted here that intraventricular administration of leupeptin produces impairment in the learning of both spatial and olfactory memory tasks (93,94).

CALPAIN AND DYNAMIC REGULATION OF CELL STRUCTURE

The data reviewed in the preceding section indicate that calpain activation and spectrin degradation take place under conditions associated with structural reorganization of one type or another, spatially restricted in the case of physiological events and widespread under pathological conditions. These data have been taken as evidence that the main function of calpains in a variety of cells is to participate in the dynamic regulation of the cell cytoskeleton and therefore of cell structure. There is an ample literature reviewing the role of the cytoskeleton in the regulation of cell structure and shape, and probably the best examples of the potential roles of calpain in shape regulation are provided by studies of simple cellular systems such as the erythrocytes and the blood platelets. In both cases, it has been clearly demonstrated that morphological alterations are accompanied with calpain activation and degradation of various cytoskeletal proteins including spectrin (95,96).

Although the precise mechanisms linking structural protein degradation and morphological alterations are not entirely known, a number of general rules have emerged from the studies of these simple systems (97):

1. Partial proteolysis of the protein meshwork underlying the plasma membrane results in changes in cell surface receptor distribution/properties and releases the membrane from mechanical constraints;
2. Proteolysis of structural proteins produces a transient conformational state that forms the basis for the formation of a new long-lasting conformation; and
3. Cytoskeletal proteins are synthesized in excess and newly synthesized proteins can be rapidly assembled to provide for new configurations.

Several additional factors are likely to contribute in some ways to the dynamics of the disassembly–reassembly process. The nature and extent of proteolysis is regulated not only by the duration and amplitude of the calcium signal that triggers it but also by other modulators. In particular calmodulin has been shown to influence both the nature of the spectrin breakdown products and the rate of spectrin proteolysis (98). In this respect, it is worth mentioning that a number of calmodulin antagonists have been shown to block LTP induction in rat hippocampal slices, and although it has been proposed that this effect is due to the inhibition of phosphorylation reactions (99,100), it could as well be due to an interaction with the degradation of structural proteins. The assembly–disassembly process is also modulated by phosphorylation reactions. These reactions participate in microtubule formation (101) and also modify the affinity of cytoskeletal proteins for their attachment sites on the membrane. These interactions provide another link between a variety of second messengers (cAMP, phospholipids) and structural reorganizations. Finally, phospholipids and phospholipases are also good candidates to provide a link between membrane events and cytoskeleton reorganization. Recent data suggest that phospholipase A_2 (PLA_2) activity is regulated in part by the state of myosin–actin interaction in certain cell types (102). In turn, PLA_2 activity, by modifying the structure/distribution of phospholipids in the membrane, regulates the conformation of a variety of membrane proteins, including ion channels and transmitter receptors (103,104).

To summarize, it is our hypothesis that the main function of calcium-dependent proteases is to provide a rapid mechanism for producing structural modifications as a result of physiological or pathological conditions. The variety of factors that modulate and participate in the final reorganization process is likely to be responsible for the variability of shapes that was discussed earlier. Moreover, the ubiquitous nature of this mechanism indicates that morphological alterations can take place in neuronal as well as in glial structures. The role of glial cells has remained elusive, but recently an intense effort has been devoted to obtain a better understanding of their functions in normal and pathological conditions. The existence in glial cells of a mechanism producing rapid structural modifications provides a potential explanation for the hypertrophic or atrophic reactions that these cells exhibit under both physiological or pathological conditions (22,23,105).

Finally, the properties of this mechanism reinforce the idea that structures themselves contain information about the past and present events and that structural modification provides a way of modifying the informational content of the brain.

ACKNOWLEDGMENTS

This work is supported by NIH grants NS-18427 and NSF grant BNS 12156. We wish to acknowledge Claudine Jutras for her excellent secretarial assistance. This work is also the result of a long and friendly collaboration with Dr. Gary Lynch and the many people who worked with us at UCI.

REFERENCES

1. Jones EG, Powell TPS. Morphological variations in the dendritic spines of the neocortex. *J Cell Sci* 1969;5:509–519.
2. Rall W. Dendritic spines, synaptic potency and neuronal plasticity. In: Woody CD, Brown KA, Crow TJ, Knispel JD, eds. *Cellular mechanisms subserving changes in neuronal activity*, vol 3. Los Angeles: UCLA Brain Info Service Research Report, 1974;13–26.
3. Cross RG, Perkel DH. The function of dendritic spines: a review of theoretical issues. *Behav Neural Biol* 1985;44:151–185.
4. Rall W, Segev I. Functional possibilities of synapses on dendrites and on dendritic spines. In: Edelman GM, Gall WE, Cowan WM, eds. *Synaptic function*. New York: John Wiley, 1987; 625–636.
5. Wilson CJ. Passive cable properties of dendritic spines and spiny neurons. *J Neurosci* 1984;4:281–297.
6. Harris KM, Stevens JK. Dendritic spines of CA1 pyramidal cells in the rat hippocampus: serial electron microscopy with reference to their biophysical characteristics. *J Neurosci* 1989;9:2982–2997.
7. Lee K, Schottler F, Oliver M, Lynch G. Brief bursts of high-frequency stimulation produce two types of structural change in rat hippocampus. *J Neurophysiol* 1980;44:247–258.
8. Chang FLF, Greenough WT. Transient and enduring morphological correlates of synaptic activity and efficacy change in the rat hippocampal slice. *Brain Res* 1984;309:35–46.
9. Wenzel J, Matthies H. Morphological changes in the hippocampal formation accompanying memory formation and long-term potentiation. In: Weinberger N, McGaugh J, Lynch G, eds. *Memory systems of the brain*. New York: The Guilford Press, 1985;150–170.
10. Desmond NL, Levy WB. Synaptic correlates of associative potentiation/depression: an ultrastructural study in the hippocampus. *Brain Res* 1983;265:21–30.
11. Andersen P, Blackstad T, Hulleberg G, Line Vaaland J, Trommald M. Changes in spine morphology associated with long-term potentiation in rat dentate granule cells. *Proc Physiol Soc* 1988.
12. Lynch G, Kessler M, Arai A, Larson J. The nature and causes of hippocampal long-term potentiation. In: Zimmer J, Ottersen OP, eds. *Understanding the brain through the hippocampus: the hippocampal region as a model for studying structure and function*. 1989 (*in press*).
13. Bliss TVP, Lomo T. Long-lasting potentiation of synaptic transmission in the dentate area of the anaesthetized rabbit following stimulation of the perforant path. *J Physiol (Lond)* 1973; 232:331–356.
14. Baudry M, Larson J, Lynch G. Long-term changes in synaptic efficacy: potential mechanisms and implications. In: Landfield P, Deadwyler S, eds. *Long-term potentiation: mechanisms and key issues*. 1987;107–135.
15. Brown TH, Chang VC, Ganong AH, Keenan CL, Kelso SR. Biophysical properties of dendrites and spines that may control the induction and expression of long-term synaptic potentiation. *Neurol Neurobiol* 1988;35:201–264.
16. Desmond NL, Levy WB. Anatomy of associative long-term synaptic modification. *Neurol Neurobiol* 1988;35:201–264.
17. Matthews DA, Cotman C, Lynch G. An electron microscopic study of lesion-induced synaptogenesis in the dentate gyrus of the adult rat. I. Magnitude and time-course of degeneration. *Brain Res* 1976;115:1–21.
18. Steward O, Vinsant SL. The process of reinnervation in the dentate gyrus of the adult rat: a quantitative electron microscopic analysis of terminal proliferation and reactive synaptogenesis. *J Comp Neurol* 1983;214:370–386.
19. Caceres A, Steward O. Dendritic reorganization in the denervated dentate gyrus of the rat following entorhinal cortical lesions: a Golgi and electron microscopic analysis. *J Comp Neurol* 1983;214:387–403.
20. Parnavelas JG, Lynch G, Brecha N, Cotman CW, Globus A. Spine loss and regrowth in hippocampus following deafferentation. *Nature* 1974;248:71–73.
21. Rose G, Cotman CW, Lynch GS. Hypertrophy and redistribution of astrocytes in the deafferented hippocampus. *Brain Res Bull* 1976;1:87–92.
22. Gall C, Rose G, Lynch G. Proliferative and migratory activity of glial cells in the deafferented hippocampus. *J Comp Neurol* 1978;183:539–550.
23. Gage FH, Olegniczak P, Armstrong DM. Astrocytes are important for sprouting in the septohippocampal circuit. *Exp Neurol* 1988;102:2–13.
24. Zimmerman UJP, Schlaepfer WW. Calcium-activated neutral protease (CANP) in brain and other tissues. *Prog Neurobiol* 1984;23:63–78.
25. Pontremoli S, Melloni E. Extralysosomal protein degradation. *Annu Rev Biochem* 1986; 55:455–481.
26. Mellgren RL. Calcium-dependent proteases: an

enzyme system active at cellular membranes? *FASEB J* 1987;1:110–115.

27. Bond JS, Butler PE. Intracellular proteases. *Annu Rev Biochem* 1987;56:333–364.
28. Murachi T, Hatanaka M, Yasumoto Y, Nakayama N, Tanaka K. A quantitative distribution study on calpain and calpastatin in rat tissues and cells. *Biochem Int* 1981;2:651–656.
29. Siman R, Baudry M, Lynch G. Purification from synaptosomal plasma membranes of calpain I, a thiol-protease activated by micromolar calcium concentrations. *J Neurochem* 1983;41: 950–956.
30. Emori Y, Kawasaki H, Sugihara H, Imajoh S, Kawashima S, Suzuki K. Isolation and sequence analyses of cDNA clones for the large subunits of two isozymes of rabbit calcium-dependent protease. *J Biol Chem* 1986;261: 9465–9471.
31. Aoki K, Imajoh S, Ohno S, et al. Complete amino acid sequence of the large subunit of the low Ca^{++}-requiring form of human Ca^{++}-activated neutral protease (CANP) deduced from its cDNA sequence. *FEBS Lett* 1986; 205:313–317.
32. Imajoh S, Aoki K, Ohno S, et al. Molecular cloning of the cDNA for the large subunit of the high-Ca^{++}-requiring form of human Ca^{++}-dependent neutral protease. *Biochemistry* 1988; 27:8127–8128.
33. Ohno S, Emori Y, Imajoh S, Kawasaki H, Kisaragi M, Suzuki U. Evolutionary origin of a calcium-dependent protease by fusion of genes for a thiol protease and a calcium binding protein? *Nature* 1984;312:566–570.
34. Suzuki K. Calcium-activated neutral protease, domain structure and activity regulation. *Trends Biochem Sci* 1987;12:103–105.
35. De Martino GN, Blumenthal DK. Identification and partial purification of a factor that stimulates calcium-dependent proteases. *Biochem J* 1982;21:4303–4310.
36. Takeyama Y, Nakanishi H, Uratsuji Y, Kishimoto A, Nishizuka Y. A calcium-protease activator associated with brain microsomal elements. *FEBS Lett* 1986;194:110–116.
37. Murachi T. Calpain and calpastatin. *Trends Biochem Sci* 1983;8:167–171.
38. Murachi T, Tanaka K, Hatanaka M, Murakami T. Intracellular Ca^{++}-dependent protease (calpain) and its high-molecular weight endogenous inhibitor (calpastatin). *Adv Enzyme Regul* 1981; 19:407–424.
39. Takano E, Maki M, Mori H, et al. Pig heart calpastatin: identification of repetitive domain structures and anomalous behavior in polyacrylamide gel electrophoresis. *Biochemistry* 1988;27:1964–1972.
40. Emori Y, Kawasaki H, Imajoh S, Imahori K, Suzuki K. Endogenous inhibitor for calcium-dependent cysteine protease contains four internal repeats that could be responsible for its multiple reactive sites. *Proc Natl Acad Sci USA* 1987;84:3590–3594.
41. Coolican SA, Hathaway DR. Effect of L-α-phosphatidylinositol on a vascular smooth muscle Ca^{2+}-dependent protease. *J Biol Chem* 1984;259:11627–11630.
42. Imajoh S, Kawasaki H, Suzuki K. The amino-terminal hydrophobic region of the small subunit of calcium-activated neutral protease (cANP) is essential for its activation by phosphatidylinositol. *J Biochem* 1986;99:1281–1284.
43. Baudry M, Dubrin R, Lynch G. Subcellular compartmentalization of calcium-dependent and calcium-independent neutral proteases in brain. *Synapse* 1987;1:506–511.
44. Siman R, Gall C, Perlmutter L, Christian C, Baudry M, Lynch G. Distribution of calpain I, an enzyme associated with degenerative activity, in rat brain. *Brain Res* 1985;347:399–403.
45. Hamakubo T, Kannagi R, Murachi T, Matus A. Distribution of calpains I and II in rat brain. *J Neurosci* 1986;6:3103–3111.
46. Perlmutter L, Siman R, Gall C, Baudry M, Lynch G. The ultrastructural localization of calcium-activated protease "calpain" in rat brain. *Synapse* 1988;2:79–86.
47. Simonson L, Baudry M, Siman R, Lynch G. Regional distribution of soluble calcium-activated proteinase activity in neonatal and adult rat brain. *Brain Res* 1985;327:153–159.
48. Baudry M, Simonson L, DuBrin R, Lynch G. A comparative study of soluble calcium-dependent proteolytic activity in vertebrate brain. *J Neurobiol* 1986;17:15–28.
49. Lynch G, Larson J, Baudry M. Proteases, neuronal stability and brain aging. An hypothesis. In: Crook T, Bartus R, Ferris S, Gershen S, eds. *Treatment development strategies for Alzheimer's disease*. Mark Powley Associates, 1986;119–149.
50. Baudry M, DuBrin R, Beasley L, Leon M, Lynch G. Bat's brain contains low levels of calcium-dependent protease activity: implication for the cellular mechanisms of aging. *Neurobiol Aging* 1985;7:255–258.
51. Sandoval IV, Weber K. Calcium-induced inactivation of microtubule formation in brain extracts. *Eur J Biochem* 1978;92:463–470.
52. Siman R, Baudry M, Lynch G. Calcium-activated proteases as possible mediators of synaptic plasticity. In: Edelman G, Gall W, Cowan WM, eds. *Synaptic function*. New York: John Wiley, 1987;519–548.
53. Sato C, Nishizawa K, Nakayama T, et al. Rapid proteolysis of brain MAP-1 related cytoskeleton-associated 350 kd protein by purified calpain. *Cell Struct Func* 1986;11:253–261.
54. Malik MN, Meyers LA, Iqbal K, Sheikh AM, Scotto L, Wisniewski HM. Calcium-activated proteolysis of fibrous proteins in central nervous system. *Life Sci* 1981;29:795–802.
55. Schlaepfer WW, Zimmerman UJP. Calcium-mediated breakdown of glial filaments and neurofilaments in rat optic nerve and spinal cord. *Neurochem Res* 1981;6:243–251.
56. De Armond SJ, Fajardo M, Naughton SA, Eng L. Degradation of glial fibrillary acidic protein

by a calcium-dependent proteinase: an electroblot study. *Brain Res* 1983;262:275–280.

57. Sato S, Miyatake T. Degradation of myelin basic protein by calcium-activated neutral protease (CANP)-like enzyme in myelin and inhibition by E-64 analogue. *Biomed Res* 1982; 3:461–467.
58. Singh I, Singh AK. Degradation of myelin proteins by brain endogenous neutral protease. *Neurosci Lett* 1983;39:77–82.
59. Siman R, Baudry M, Lynch G. Brain fodrin: substrate for the endogenous calcium-activated protease calpain I. *Proc Natl Acad Sci USA* 1984;81:3276–3280.
60. Carlin RK, Bartelt DC, Siekevitz P. Identification of fodrin as a major calmodulin-binding protein in postsynaptic density preparations. *J Cell Biol* 1983;96:443–448.
61. Billger M, Wallin K, Karlsson JO. Proteolysis of tubulin and microtubule-associated proteins 1 and 2 by calpain I and II. Difference in sensitivity of assembled and disassembled microtubules. *Cell Calcium* 1988;9:33–44.
62. Davis LG, Card JP, Meade RP, Siman R. Proteolytic processing of β-amyloid precursor protein by calpain I (*Abstr*). *Abstr Soc Neurosci* 1989;15:1377.
63. Kang J, Lemaire HG, Unterbeck A, et al. The precursor of Alzheimer's disease amyloid A4 protein resembles a cell-surface receptor. *Nature* 1987;325:733–736.
64. Inoue M, Kishimoto A, Takai Y, Nishizuka Y. Studies on a cyclic-nucleotide independent protein kinase and its proenzyme in mammalian tissues. II. Proenzyme and its activation by calcium-dependent protease from rat brain. *J Biol Chem* 1977;252:7610–7616.
65. Suzuki K, Imajoh S, Emori Y, Kawasaki H, Mirami Y, Ohno S. Calcium-activated neutral protease and its endogenous inhibitor. *FEBS Lett* 1987;220:271–276.
66. Ito M, Tanaka TI, Nunoki K, Hidaka N, Suzuki K. The Ca^{++}-activated protease (calpain) modulates Ca^{++} calmodulin dependent activity of smooth muscle myosin light chain kinase. *Biochem Biophys Res Commun* 1987;145:1321–1326.
67. Kennedy M. Synaptic memory molecules. *Nature* 1988;335:770–772.
68. Tallant EA, Brumley LM, Wallace RW. Activation of a calmodulin-dependent phosphatase by a Ca^{++}-dependent protease. *Biochemistry* 1988;27:2205–2211.
69. Kincaid RL, Stith-Coleman IE, Vaughan M. Proteolytic activation of calmodulin-dependent cyclic nucleotide phosphodiesterase. *J Biol Chem* 1985;260:9009–9014.
70. Ando Y, Imamura S, Murachi T, Kannagi R. Calpain activates two transglutaminases from porcine skin. *Arch Dermatol Res* 1988;280:380–384.
71. Leonard JP, Salpeter MM. Agonist-induced myopathy at the neuromuscular junction is mediated by calcium. *J Cell Biol* 1979;82:811–819.
72. Schlaepfer WW, Hasler MB. Characterization of the calcium-induced disruption of neurofilaments in rat peripheral nerve. *Brain Res* 1979;168:299–309.
73. Lynch G, Rose G, Gall C, Cotman CW. The response of the dentate gyrus to partial deafferentation. In: Santini M, ed. *Golgi centennial symposium proceedings*. New York: Raven Press, 1975;305–317.
74. Cotman CW, Nieto-Sampedro M, Harris EW. Synapse replacement in the nervous system of adult vertebrates. *Physiol Rev* 1981;61:684–784.
75. Steward O, Cotman CW, Lynch G. A quantitative autoradiographic and electrophysiological study of the reinnervation of the dentate gyrus by the contralateral entorhinal cortex following ipsilateral entorhinal lesions. *Brain Res* 1976;114:181–200.
76. Baudry M, Lynch G. Regulation of calcium transport in rat hippocampal mitochondria during development and following denervation. *Prog Brain Res* 1985;63:107–119.
77. Seubert P, Ivy G, Larson J, et al. Lesions of entorhinal cortex produce a calpain-mediated degradation of brain spectrin in dentate gyrus. I. Biochemical studies. *Brain Res* 1988;459: 226–232.
78. Ivy G, Seubert P, Lynch G, Baudry M. Lesions of entorhinal cortex produce a calpain-mediated degradation of brain spectrin in dentate gyrus. II. Anatomical studies. *Brain Res* 1988; 459:233–240.
79. Goldschmidt RB, Steward O. Neurotoxic effects of colchicine: differential susceptibility of CNS neuronal populations. *Neuroscience* 1982; 7:695–714.
80. Nakagawa Y, Nakamura S, Kase Y, Noguchi T, Ishihara T. Colchicine lesions in the rat hippocampus mimic the alterations of several markers in Alzheimer's disease. *Brain Res* 1987; 408:57–64.
81. Seubert P, Nakagawa Y, Ivy G, Vanderklish P, Baudry M, Lynch G. Intrahippocampal colchicine injection results in spectrin proteolysis. *Neuroscience* 1989;31:195–202.
82. Siman R, Noszek JC, Kegerise C. Calpain I activation is specifically related to excitatory amino acid induction of hippocampal damage. *J Neurosci* 1989;9:1579–1590.
83. Seubert P, Lee K, Lynch G. Ischemia triggers NMDA receptor-linked cytoskeletal proteolysis in hippocampus. *Brain Res* 1989;492:366–370.
84. Mayer ML, Westbrook GL, Guthrie PB. Voltage-dependent block by Mg^{2+} of NMDA responses in spinal cord neurones. *Nature* 1984; 309:261–263.
85. Nowak L, Bregestovski P, Ascher P, Herbert A, Prochiantz A. Magnesium gates glutamate-activated channels in mouse central neurones. *Nature* 1984;307:462–465.
86. Seubert P, Larson J, Oliver M, Baudry M, Lynch G. Stimulation of NMDA receptors induce proteolysis of spectrin in hippocampus. *Brain Res* 1988;460:189–194.

87. Collingridge GL, Kehl SJ, McLennan H. The antagonism of amino-acid-induced excitations of rat hippocampal CA1 neurones *in vitro*. *J Physiol* 1983;334:19–31.
88. Harris EW, Ganong H, Cotman CW. Long-term potentiation in the hippocampus involves activation of *N*-methyl-D-aspartate receptors. *Brain Res* 1984;323:132–137.
89. Larson J, Lynch G. Role of *N*-methyl-D-apartate receptors in the induction of synaptic potentiation by burst stimulation patterned after the hippocampal θ-rhythm. *Brain Res* 1988; 441:111–118.
90. Lynch G, Baudry M. Brain spectrin, calpain and long-term changes in synaptic efficacy. *Brain Res Bull* 1987;18:809–815.
91. Staubli U, Larson J, Baudry M, Thibault O, Lynch G. Chronic administration of a thiol-proteinase inhibitor blocks long-term potentiation of synaptic responses. *Brain Res* 1988;444:153–158.
92. Oliver M, Baudry M, Lynch G. Exposure to leupeptin interferes with the development of LTP in hippocampal slices. *Brain Res* 1989 (*in press*).
93. Staubli U, Baudry M, Lynch G. Leupeptin, a thiol-proteinase inhibitor, causes a selective impairment of spatial maze performance in rats. *Behav Neurol Biol* 1984;40:58–69.
94. Staubli U, Baudry M, Lynch G. Olfactory discrimination learning is blocked by leupeptin, a thiol-proteinase inhibitor. *Brain Res* 1985;337: 333–336.
95. Bennett V. The membrane skeleton of human erythrocytes and its implications for more complex cells. *Annu Rev Biochem* 1985;54:273–304.
96. Fox JEB, Phillips DR. Stimulus-induced activation of the calcium-dependent protease within platelets. *Cell Mobility* 1983;3:579–588.
97. Lynch G, Bodsch W, Baudry M. Cytoskeletal proteins and the regulation of synaptic structure. *Neurol Neurobiol* 1988;37:217–243.
98. Seubert P, Baudry M, Dudek S, Lynch G. Calmodulin stimulates the degradation of brain spectrin by calpain. *Synapse* 1986;1:20–24.
99. Malenka RC, Kauer JA, Perkel DJ, et al. An essential role for postsynaptic calmodulin and protein kinase activity in long-term potentiation. *Nature* 1989;340:554–557.
100. Malinow R, Schulman H, Tsien RW. Inhibition of postsynaptic PKC or CaMKII blocks induction but not expression of LTP. *Science* 1989; 245:862–866.
101. Marchesi VT. Stabilizing infrastructure of cell membranes. *Annu Rev Cell Biol* 1985;1:531–561.
102. Dubose DA, Shepro D, Hechtman HB. Modulation of phospholipase A_2 lytic activity by action and myosin. *Inflammation* 1989;13:15–29.
103. Loh HH, Law LY. The role of membrane lipids in receptor mechanisms. *Annu Rev Pharmacol Toxicol* 1980;20:201–234.
104. Massicotte G, Kessler M, Lynch G, Baudry M. *N*-methyl-D-aspartate and quisqualate/AMPA receptors: differential regulation by phospholipase C. *Mol Pharmacol* 1990;37:278–285.
105. Hatton GI, Perlmutter LS, Salm AK, Tweedle CP. Dynamic neuronal-glial interactions in hypothalamus and pituitary: Implication for control of hormone synthesis and release. *Peptides* 1984;5(suppl 1):121–138.

Volume Transmission in the Brain: Novel Mechanisms for Neural Transmission, edited by K. Fuxe and L.F. Agnati. Raven Press, Ltd., New York © 1991.

31

Intracellular and Extracellular Processing of Neuropeptides

J. del Rio-Garcia and D. G. Smyth

National Institute for Medical Research, London NW7 1AA, England

The formation of neuroactive peptides in the central nervous system is known to take place by the same or similar mechanisms to those involved in the elaboration of peptide hormones in the periphery. Within the neurosecretory granule, where the processing events take place in an intracellular environment, peptide trasmitters are generated from polypeptide precursors by the action of specific endopeptidases, which cleave the propeptide at sites containing paired basic residues (Fig. 1) (1,2). The resulting fragments may on occasion undergo further proteolytic cleavage at sites other than paired basic residues, such as at a single arginine or leucine residue, but these cleavages appear to be secondary and the products may not retain the biological activity of their immediate precursors.

The primary fragments formed from the propeptides undergo trimming by the action of exopeptidases, which remove lysine and arginine (3) and, in some cases, histidine (4) from their C-termini, and a similar enzymic reaction has been proposed for removal of basic residues from the N-termini (5). Peptides terminating in glycine undergo a hydroxylation reaction, which leads to the formation of the C-terminal amide group, a characteristic feature of many biologically active peptides (6,7). In addition N-terminal acetylation may take place, and the side chains of the hydroxyamino acids tyrosine, serine, and threonine may undergo phosphorylation or sulfation. In some cases glycosylation may take place, but this reaction occurs at an early stage in the biosynthetic pathway. The main posttranslational processing events occur within the sequestered environment of the neuron, the precursor chain of the propeptide being biosynthesized in the cell body and the subsequent processing reactions taking place in the neurosecretory granules during transport to the nerve terminal.

It is not known whether secretion from terminals involves release from a homogeneous group of granules or whether different patterns of peptides are released in response to different signals. Certainly tissue-specific processing is a well-understood phenomenon, in which the degree and nature of the processing reactions differ markedly from one tissue to another. Thus the anterior pituitary processes pro-opiomelanocortin (POMC) [the prohormone of adrenocorticotrophic hormone (ACTH) and β-endorphin] to produce ACTH together with lipotropin and β-endorphin, whereas the pars intermedia processes the same prohormone to generate α-melanotropin ($ACTH_{1-13}$), CLIP (corticotropin like intermedian peptide) and *N*-acetylated forms of β-endorphin (Fig. 2) (8,9). Recent experiments, however, have shown that the processing reactions in anterior pituitary cells are sensitive to environmental conditions. Monolayer cultures of anterior pituitary

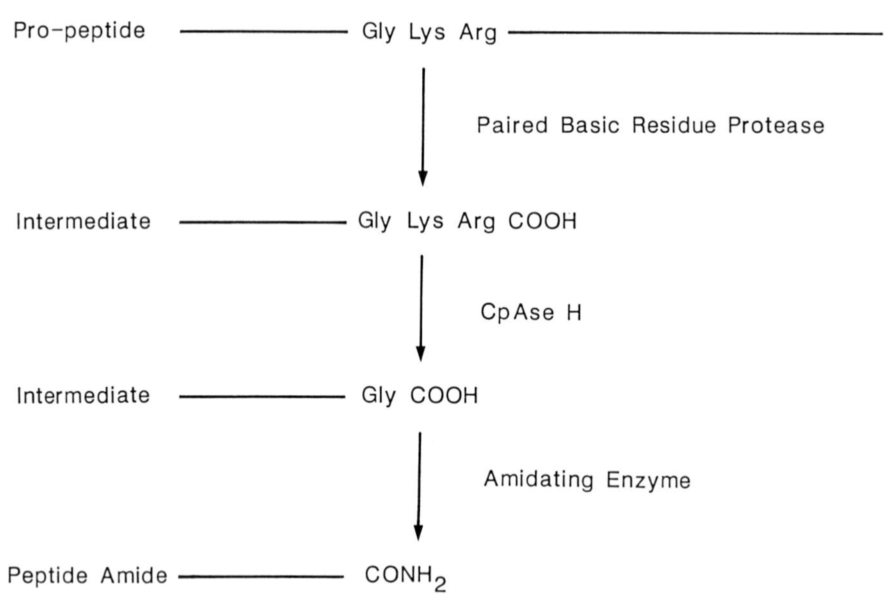

FIG. 1. General mechanism for intracellular processing of neuropeptides.

cells release predominantly β-endorphin when the secretion is stimulated acutely by corticotrophin-releasing factor (CRF); in contrast, during chronic stimulation the secreted β-endorphin is seen to be accompanied by its precursor lipotropin, and ultimately lipotropin is essentially the only peptide released (Fig. 3) (10). It is important to note that the cells at this point still contain β-endorphin in excess over lipotropin, but this β-endorphin is not released by the protracted exposure to CRF. These observations seem to point to the existence of different kinds of β-endorphin-producing cells (or granules), which differ in their response to stimulation by CRF.

Further evidence that processing reactions are sensitive to environmental signals has been reported in a study of pars intermedia cells transplanted on to the kidney where they no longer receive the dopaminergic innervation that normally inhibits secretion. At the new site, the cells secrete freely, but the processing patterns of the peptides they produce differ from the pattern that is characteristic of the pituitary pars intermedia. As in the experiments with the anterior pituitary cells grown in culture, the peptides produced by the cells in the new environment proved to be less processed than the products of normal pars intermedia cells, and the main product was ACTH, not its proteolytic fragment α-melanocyte-stimulating hormone (MSH) (11). From these experiments, there are reasonable grounds for considering that the nature of peptides released from a nerve terminal may be determined by secretion-promoting or secretion-inhibiting signals.

After release from a nerve terminal, a peptide transmitter passes into the extracellular space. It may react at a specific receptor on the postsynaptic membrane, leading to transmission of a signal in the postsynaptic neuron, or it may react at a receptor remote from the synapse by diffusion in the extracellular fluid. As a neurohormone it may be released directly into the blood. An example of the latter is offered by the releasing factors, which are generated neuronally and pass into the portal blood to act on the pituitary. Another is that of the neurohypophysial hormones,

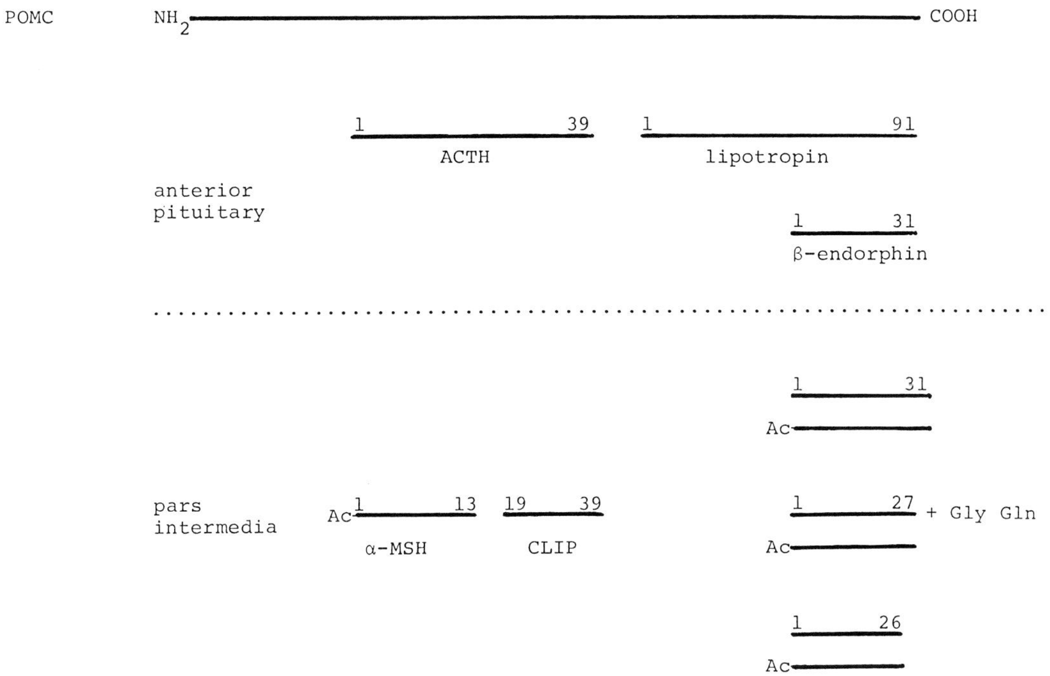

FIG. 2. Tissue specific processing of ACTH and β-endorphin in regions of rat pituitary. α-MSH, melanocyte-stimulating hormone, CLIP, corticotrophin-like intermediary peptide.

which are elaborated in specific nuclei in the hypothalamus, transported in vesicles along axons that terminate in the neural lobe of the pituitary, and ultimately released into the circulation.

Some peptide hormones appear to undergo processing at the moment of release from their cell of origin. For example, it has been reported that in *Xenopus* pituitary, α-MSH is elaborated and stored without the *N*-acetyl group normally present on the peptide in mammalian pituitary. The secreted form of *Xenopus* α-MSH, however, is an *N*-acetylated molecule, which implies that the acetylation reaction takes place coincidently with release (12). In another example, atrial natriuretic hormone in the heart appears to be biosynthesized principally as the prohormone, and the formation of active hormone from it takes place either at the surface of the secretory cell or in the circulation in response to change in extracellular fluid volume (13,14). Other examples of hormone activation taking place in extracellular fluid, rather than in the confined environment of a cell, are seen in the elaboration of tissue hormones such as angiotensin and bradykinin. In this case the activating enzymes are located at specific sites: the angiotensin-converting enzyme (ACE) is present in the lung, and the bradykinin-forming enzyme (Kallikrein) in the circulation, the production of bradykinin being stimulated during inflammation. Whether similar extracellular processing reactions occur in the brain is not known.

Recent work in this laboratory has provided evidence suggesting that extracellular processing may occur in the elaboration of a new class of peptide amides with putative neurotransmitter or neurohormone function. Previous studies have shown that the processing of thyrotropin-releasing hormone (TRH) in rat and ox hypothalamus, and in *Xenopus* skin, is essentially complete: the major product is the tripeptide

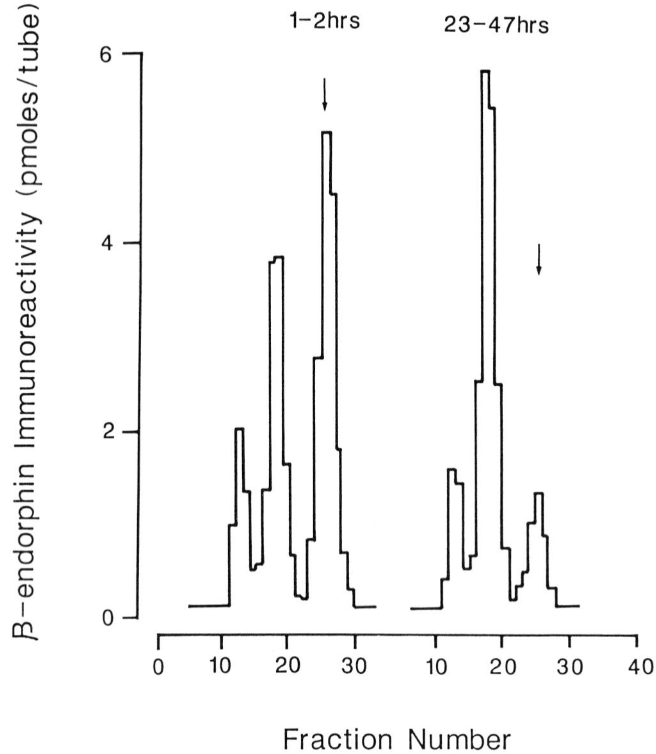

FIG. 3. Differential release of lipotropin and β-endorphin during stimulation by CRF of rat anterior pituitary cells *in vitro. Left hand panel:* acute stimulation; *right hand panel:* chronic stimulation, showing β-endorphin (third chromatographic component), the major peptide released at 1 hr, and lipotropin (second chromatographic component) at 24 hr. Anterior pituitary cells were grown in monolayer culture, and CRF concentration was maintained at 10^{-7} M by replacement of medium at intervals. Released peptides were resolved by gel exclusion chromatography and determined by radioimmunoassay using a β-endorphin antibody.

pyroglutamylhistidylproline amide, and only small amounts of partially processed extended forms of TRH are present in these tissues (15,16). In the longer forms, the extensions were found to occur exclusively from the C-terminus of the TRH sequence. When the processing of TRH was examined in the prostate complex of the rabbit, the processing patterns observed were completely different. In particular the extended forms of TRH predominated over the tripeptide, and the extensions to the TRH sequence were on the N-terminal side (17). After purification, the TRH-immunoreactive peptide in the prostate was isolated, and its structure was determined by sequencing and mass spectrometry. It proved to be a new form of TRH, pyroglutamylglutamylproline amide (Fig. 4), and subsequently the same peptide was identified in the semen (18,19).

The existence of a TRH-related peptide in an endocrine tissue led us to investigate the possibility that the same or similar peptides might occur in the brain, especially because TRH-immunoreactive substances different from TRH have been reported to

pGlu. Glu. Pro-amide

FIG. 4. Amino acid sequence of the TRH-like peptide isolated from rabbit prostate and human semen.

be present in portal blood (20). A new chromatographic method was therefore developed, which allowed TRH to be distinguished unequivocally from TRH-related peptides, and this method was applied to screen a wide range of tissues from the central nervous system (Fig. 5) (21). When weakly acidic conditions (1 M acetic acid) were used for extraction of the tissues, both TRH and TRH-related peptides were found

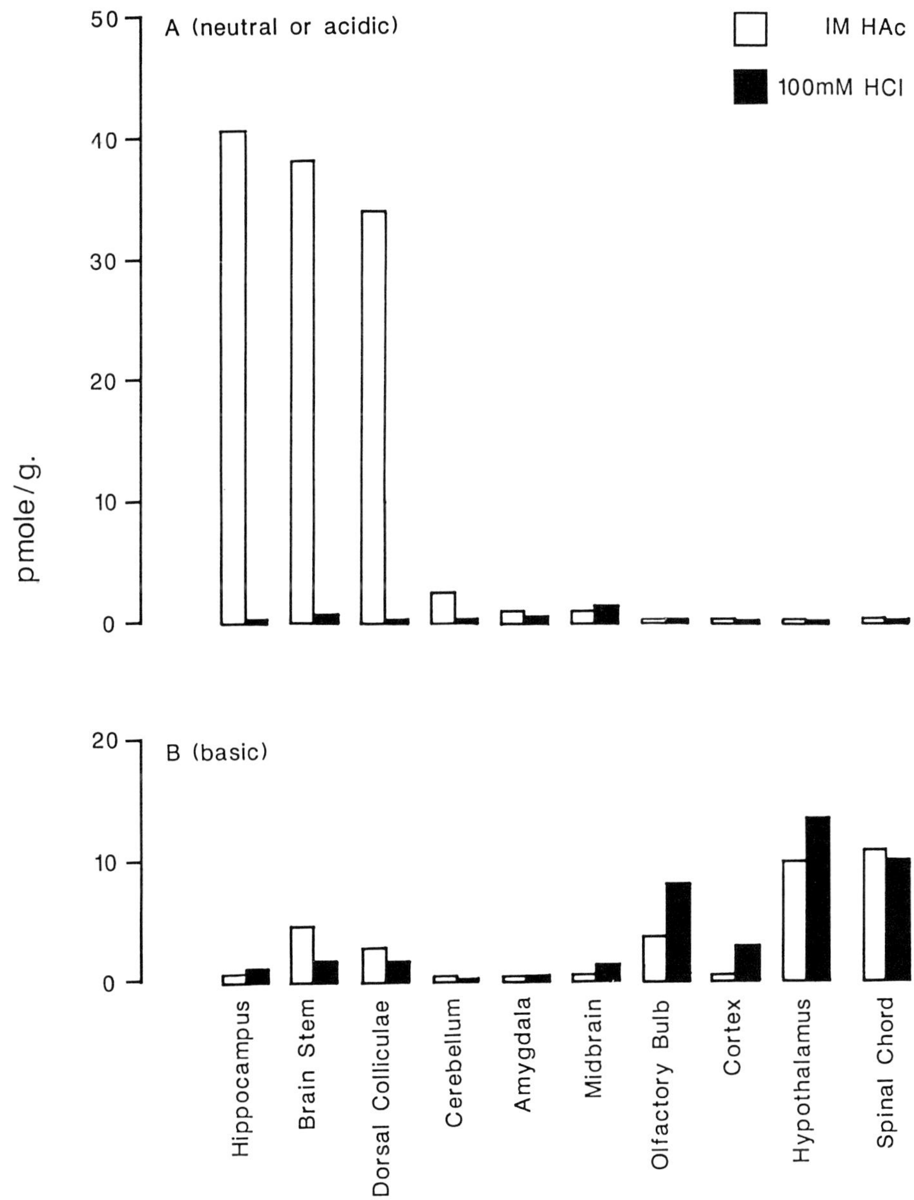

FIG. 5. TRH-like peptides (pyroglutamylpeptide amides) and TRH in regions of rat central nervous system. After extraction from dissected regions of the brain, peptides were separated from TRH on mini-columns of SP-Sephadex C25 in 50% acetic acid and were determined by radioimmunoassay with TRH antibody.

in the hippocampus, brainstem, dorsal colliculae, and cerebellum, and it is notable that in these tissues the amounts of the TRH-related peptides exceeded the amounts of TRH. In contrast, when 100 mM hydrochloric acid, acidified acetone, 1 M formic acid, 1 M acetic acid at 4°C or methanol was used for extraction, no TRH-related peptides were observed: the TRH immunoreactivity was identified solely with the known tripeptide hormone. These results suggest that the TRH-related peptides in the weakly acidic extracts of brain tissue are elaborated in precursor forms from which TRH immunoreactivity can be released during extraction.

Confirmation that the TRH-related peptides occur intracellularly in precursor forms was obtained by investigating the time course of release of TRH-like immunoreactivity during extraction of tissues in 1 M acetic acid. Studies with peptides in *Xenopus* skin or rat testis showed that the release of TRH immunoreactivity distinct from TRH increased up to a maximum at approximately 15 min during homogenization at weakly acid pH (3–3.5) and subsequently declined (Fig. 6). The decrease was probably due to the action of degrading enzymes. At the present time the identity of the TRH-like peptides in the central nervous system is not known.

It seems likely that the N-terminally extended forms of the pyroglutamyl peptide amides in brain may be the source of the TRH immunoreactive substances in portal blood. In this context, we have observed that no more than 10% of the TRH immu-

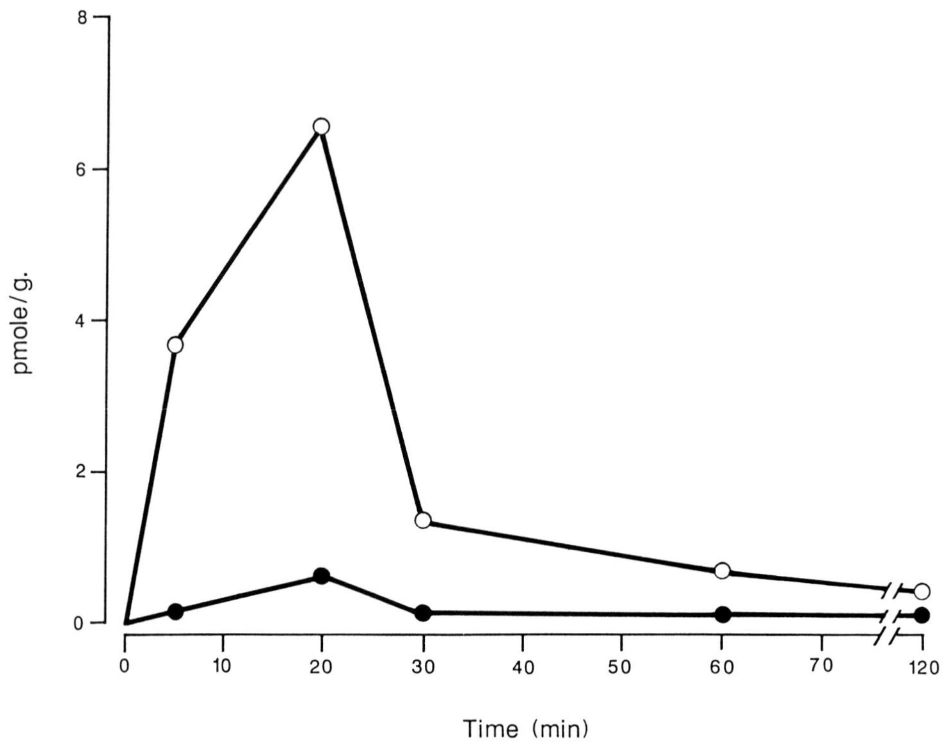

FIG. 6. Time course release of TRH immunoreactivity from rat testis. Tissues were homogenized in 1 M acetic acid at 4°C, and portions of homogenates were incubated either at 4°C (closed circles) or 37°C (open circles) for different times. TRH-related peptides were resolved from TRH on mini-columns of SP-Sephadex C25, and peptides were determined by radioimmunoassay, as in Fig. 5.

noreactivity in portal blood can be attributed to authentic TRH. In fact, when the TRH-like peptides are extracted from whole brain and resolved by gel filtration followed by kation exchange chromatography and high-pressure liquid chromatography (HPLC), seven components can be distinguished (Fig. 7) (22). The least retained of these peptides cochromatographed with pGlu-Glu-Pro amide (Glu_2-pEP amide). The remaining six peptides, which have a similar molecular size to TRH, also appear to be pyroglutamylpeptide amides, which lack a basic residue. This new series of peptides seems to be elaborated intracellularly but undergoes proteolytic conversion to generate TRH-immunoreactive forms in the extracellular fluid.

The possible functions of the new TRH-related peptides need to be explored. Because the C-terminal amide group is a common structural feature of many hormones and transmitters, it is likely that this new class of peptide amides will be found to possess biological activity and fulfil a physiological role, either as a neurotransmitter or a neurohormone.

SUMMARY

Seven new pyroglutamylpeptide amides (TRH-like peptides) have been shown to be present in rat brain. The peptides have a similar molecular size to TRH but possess neutral or acidic residues in place of histidine at position 2. The pyroglutamyl peptides appear to be elaborated in precursor forms that have a TRH-like sequence at their C-terminus; the TRH-like immunoreactivity is released during extraction at weakly acidic pH. The new peptides are lo-

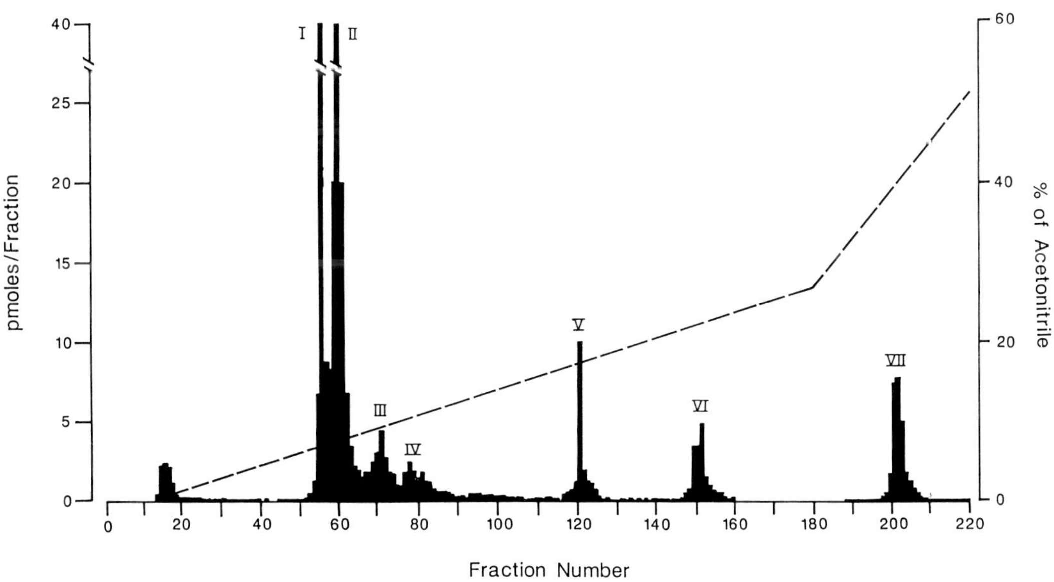

FIG. 7. Resolutions of TRH-like peptides, distinct from TRH, in rat brain by reverse phase HPLC (HPLC). Peptides were extracted from four rat brains in 1 M HAc, and purification was carried out by gel exclusion chromatography on Sephadex G50 in 50% acetic acid. To remove TRH, TRH-immunoreactive fraction was resolved on an SP-25 column (H^+ form) eluted in H_2O. Unretained peptides were fractionated by reverse phase HPLC using a C_{18} Microbondapak column in 10 mM HCl operated at a flow rate of 1.5 ml/min. Fractions were collected at 0.3-min intervals, and an acetonitrile gradient of 0.5%/min was applied up to fraction 180, with a 2% acetonitrile gradient thereafter. Peptides were determined by RIA with TRH antibody.

cated in specific regions of rat brain, where their concentrations exceed those of TRH. Conversion to low molecular weight TRH-like peptides may take place in the extracellular fluid.

ACKNOWLEDGMENTS

We are grateful to Dr. Hamish Fraser for generously donating TRH-antibody, and we thank Mrs. Rita Findon for typing the manuscript. This work was supported in part by a Wellcome Trust Fellowship to J.R-G.

REFERENCES

1. Docherty K, Steiner DF. Posttranslational proteolysis in polypeptide hormone biosynthesis. *Annu Rev Physiol* 1982;44:625–638.
2. Darby NJ, Smyth DG. Endopeptidases and prohormone processing. *Biosci Rep* 1990;10:1–13.
3. Fricker LD, Snyder SH. Purification and characterization of enkephalin convertase, an enkephalin-synthesizing carboxypeptidase. *J Biol Chem* 1983;258:10950–10955.
4. Smyth DG, Maruthainar K, Darby NJ, Fricker LD. Catalysis of slow C-terminal processing reactions by carboxypeptidase H. *J Neurochem* 1989;53:489–493.
5. Gainer H, Russell JT, Loh YP. An aminopeptidase activity in bovine pituitary secretory vesicles that cleaves the N-terminal arginine from β-lipotropin $_{60-65}$. *FEBS Lett* 1984;175:135–139.
6. Bradbury AF, Finnie MDA, Smyth DG. Mechanism of C-terminal amide formation by pituitary enzymes. *Nature* 1982;298:686–689.
7. Bradbury AF, Smyth DG. Biosynthesis of the C-terminal amide group in peptide hormones. *Biosc Rep* 1988;7:907–917.
8. Zakarian S, Smyth DG. Distribution of β-endorphin related peptides in rat pituitary and brain. *Biochem J* 1982;202:561–571.
9. Smyth DG, Zakarian S. β-Endorphin in brain. In: Buijs RM, Pevet P, Swaab DF, eds. *Chemical transmissions in the brain, Progress in brain research,* vol 55. Amsterdam, New York: Elsevier Biomedical Press, 1982;123–132.
10. Ham J, Smyth DG. Chronic stimulation of rat anterior pituitary cultures with CRF leads to the secretion of lipotropin. *Neuroendocrinology* 1985;44:433–438.
11. Iturriza FC, Eberle AN. Secretion of melanocyte stimulating hormone and adrenocorticotropin from transplanted pituitary pars intermedia in stressed and non stressed rats. *Neuroendocrinology* 1989;49:610–616.
12. Martens GJM, Jenks BG, van Overbecke AP. *N*α-acetylation is linked to α-MSH release from pars intermedia of the amphibian pituitary gland. *Nature* 1981;294:558–560.
13. Thibault G, Lazure C, Schiffrin EL, et al. Identification of a biologically active circulating form of rat atrial natriuretic factor. *Biochem Biophys Res Commun* 1985;130:981–986.
14. Thibault G, Garcia R, Gutkowska J, et al. The propeptide Asn-1-Tyr_{126} is the storage form of rat atrial natriuretic factor. *Biochem J* 1987;24: 265–272.
15. Cockle SM, Smyth DG. Processing of the thyrotropin releasing hormone (TRH) precursor in *Xenopus* skin and bovine hypothalamus: evidence for the existence of longer forms of TRH. *Regul Pept* 1986;14:217–227.
16. Cockle SM, Smyth DG. Specific processing of the TRH prohormone in rat brain and spinal cord. *Eur J Biochem* 1987;167:161–165.
17. Cockle SM, Morrell JM, Smyth DG. Thyrotropin-releasing hormone-related polypeptides in rabbit prostate and semen are different from those in rabbit hypothalamus. *J Endocrinol* 1989;120:31–36.
18. Cockle SM, Aitken A, Beg F, Smyth DG. A novel peptide, pyroglutamylglutamylproline amide, in the rabbit prostate complex, structurally related to thyrotrophin releasing hormone. *J Biol Chem* 1989;264:7788–7791.
19. Cockle SM, Aitken A, Beg F, Smyth DG. The TRH-related peptide pyroglutamylglutamylprolineamide is present in human semen. *FEBS Lett* 1989;252:113–117.
20. Sheward WJ, Harmar AJ, Fraser HM, Fink G. Thyrotropin-releasing hormone in rat pituitary stalk blood and hypothalamus: studies with high performance liquid chromatography. *Endocrinology* 1983;113:1865–1869.
21. del Rio-Garcia J, Smyth DG. Distribution of pyroglutamyl peptide amides in the central nervous system and periphery of the rat. *Neuroendocrinology* (*in press*).
22. del Rio-Garcia J, Smyth DG. Pyroglutamyl peptide amide related to TRH in rat brain: possible new class of neurotransmitters. *Biochem Soc Trans* 1990;18:434–435.

Volume Transmission in the Brain: Novel Mechanisms for Neural Transmission,
edited by K. Fuxe and L.F. Agnati.
Raven Press, Ltd., New York © 1991.

32

Peptidases and Proteases in Cerebrospinal Fluid, Significance in Volume Transmission

*Lars Terenius and **Fred Nyberg

**Department of Drug Dependence Research, Karolinska Institute, S-10401 Stockholm, Sweden; and **Department of Pharmacology, Uppsala University, 751 24 Uppsala, Sweden*

Neuropeptides constitute a large and still expanding group of putative neurotransmitters/neuromodulators in the nervous system. Frequently they exist in the same nerves, together with a conventional transmitter. The conventional transmitter is synthesized by a few enzymatic steps from an amino acid precursor (unless the amino acid itself assumes the transmitter function). Generation of neuropeptides also occurs by enzymatic (proteolytic) transformation. Here, however, the precursor is a protein of 10 to 30 kD molecular weight, which embeds the active peptide. To release this peptide, usually several enzymes are needed that act in a stepwise fashion. Whereas the biosynthesis of conventional transmitters occurs locally in the nerve terminals, the peptide precursors are formed in the cell soma and packed into vesicles in the Golgi apparatus (1). It is assumed that the proteolytic transformations occur during the axonal transport to the nerve terminals, where the mature peptide is stored before being released. The enzymatic steps are apparently highly specific, indicating that the enzymes involved are sequence-specific.

Vesicles storing neurotransmitters differ in size and are commonly divided into so-called small dense-cored vesicles (SDVs) and large dense-cored vesicles (LDVs). Neuropeptides are mainly stored in LDVs, whereas the conventional transmitters distribute almost equally between SDVs and LDVs. Because neuropeptides cross biological membranes poorly because of their size and hydrophilicity, release of the neuropeptide from the LDVs is likely to occur via exocytosis. This model of neuropeptide generation emphasizes a limited compartment, LDVs, which constitutes a small volume of the total nerve tissue where the metabolic transformation of neuropeptide precursors occurs. Thus, in contrast to conventional transmitters, there is probably no significant cytosolic pool of peptide transmitter. Therefore, most metabolic conversions will take place in the vesicular compartment, and the enzymes involved are probably stored with the neuropeptide precursor. To isolate and characterize such enzymes, several groups have therefore isolated LDVs by differential centrifugation of nerve tissue homogenates. However, the rather limited enrichment that is usually obtained and artifactual loss of enzyme into the nonparticulate fractions are limiting factors for this approach. We have turned to another approach, attempting to identify secreted enzymes. According to our hypothesis, the whole vesicular content including neuropeptide, eventually untransformed peptide precursor or smaller precursors, and the transforming enzymes are released by exocytosis on nerve activation.

Material that is released will enter the synaptic cleft and eventually reach the cerebrospinal fluid (CSF). We have found access to rather large volumes of human CSF, which can be processed chromatographically to purify different enzyme activities. During this work, it has become obvious that not only do proteolytic enzymes occur in CSF, but they also occur in substantial quantities. Moreover, the enzymes we are identifying are not just general proteases but have definite stringency and act on neuropeptides with potential functional relevance. Several of the enzymes have previously been identified as potentially significant for neuropeptide turnover (2).

NEUROPEPTIDE SYSTEMS, GENERAL

Enzymatic transformations of the neuropeptide precursor into the mature neuropeptide involves several steps. Because the different precursors have a similar general structure, every known precursor will go through the same principal steps (Fig. 1). The neuropeptide sequences are usually flanked by pairs of basic amino acids, which serve as targets for processing enzymes. At the first step, cleavage may or may not be trypsin-like. Different enzymes have different sequence specificities (e.g., ref. 3), strongly suggesting that in fact, there is a whole family of processing enzymes. The next step may be "trimming" of any remaining basic amino acid at the C-terminus or/and at the N-terminus. One trimming enzyme, carboxipeptidase E (also designated enkephalin convertase), has rather broad specificity and removes a C-terminal Arg, His, or Lys from a variety of peptide precursors and not just, as its trivial name suggests, enkephalyl peptides (4). Several aminopeptidases are known to be present in the brain. Aminopeptidase M seems to be exclusively expressed in cerebral blood vessels (5) and is therefore probably not a trimming enzyme. Other aminopeptidases have not been particularly well-characterized.

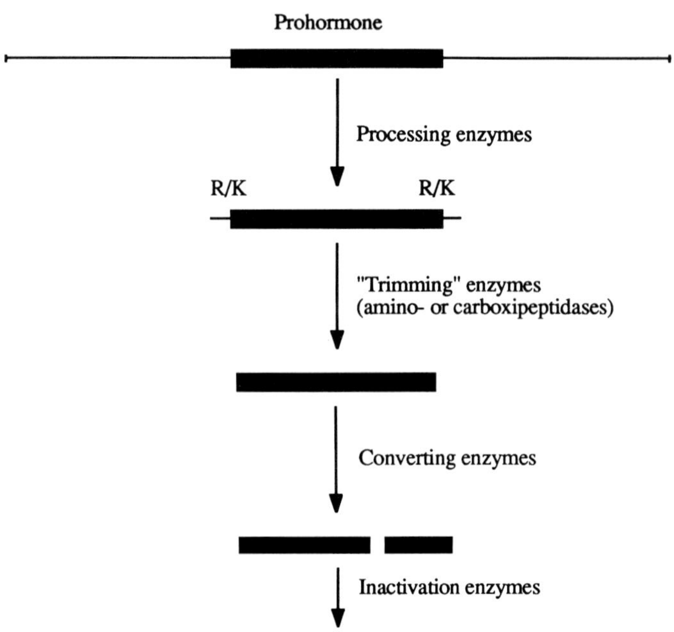

FIG. 1. Outline of the generation of bioactive neuropeptides by sequential enzymatic transformations. R, arginine; K, lysine.

TABLE 1. *Representative enzymes in human cerebrospinal fluid (CSF)[a]*

Characteristic	Amidating enzyme	ACE	DCE	E.C.24.11	SPE
Molecular weight (kD)	25	30,60,150	45	90	45
pH optimum	7–8	7–8	6.5–8	7	7–8
Class	Mono-oxygenase	Metallo-protease	Serine-protease	Metallo-protease	Thiol-protease
Substrates	SP-Gly[12]	ANGI ENK ENK-Arg[6]	DYN A DYN B	ENK SP α-NEO	SP
K_M (μM)			10[b]		20
Enzyme protein/ total CSF protein (%)	0.9		1.7		0.14

[a]ACE, angiotensin-converting enzyme; DCE, dynorphin-converting enzyme; SPE, substance P endopeptidase; ANGI, angiotensin I; DYN, dynorphin; ENK, enkephalin; NEO, neoendorphin; SP, substance P.
[b]K_M determined for DYN A.

An important converting reaction is the alpha-amidation of peptide precursors with a C-terminal Gly (6,7), a reaction that is a requirement for biologic activity of a number of neuropeptides [e.g., tachykinins, vasoactive intestinal polypeptide (VIP), and neuropeptide tyrosine (NPY)]. Also several endopeptidases act on the primary products of the proteolytic transformations. They may transform these products into secondary products that have biologic activity, albeit, sometimes of a different kind. These enzymes may therefore be called *converting enzymes*. These enzymes also act with considerable stringency and are therefore interesting targets for experimental work. At least four such enzymes have been found to occur at relatively high concentrations in human CSF: angiotensin-converting enzyme (ACE), dynorphin-converting enzyme (DCE), neutral endopeptidase EC 3.4.24.11 (enkephalinase), and a substance P endopeptidase. As will be discussed later, none of these enzymes has the single specificities indicated by their trivial names.

Chemical properties of some major CSF enzymes are summarized in Table 1. They clearly represent a heterogeneous group, both mechanistically and structurally. Concentrations of several enzymes in the CSF are remarkably high; if related to total protein, they occur in much higher concentrations than in the total brain. This is the predicted finding if they are actively secreted. It is also notable that these enzymes have a relatively low molecular weight, substantially lower than the molecular weight recorded in different organs. For instance, in the pituitary, alpha-amidating monooxygenase occurs in two forms with molecular weights 38 and 54 kD, respectively; in the brain, it is 100 kD (8). The low molecular weight may either be explained by different biosynthesis (altered splicing) or posttranslational changes including loss of carbohydrate, fragmentation, or dissociation. It is an attractive hypothesis that with lower molecular weight these enzymes diffuse more readily and globally throughout the CNS (9).

INDIVIDUAL ENZYMES IN CSF

Angiotensin-Converting Enzyme

Angiotensin-converting enzyme (E.C. 3.4.15.1) occurs in particularly high concentrations in endothelial cells of the lung, but it is also present in the brain. In the brain it is colocalized with the rather discrete neuronal system that stains with antisera directed against angiotensin. Moreover, there is a major pathway with high levels of ACE running from the striatum to the substantia nigra (10), a pathway otherwise known to contain substance P and dynorphin A immunoreactive neurons.

Mechanistically, ACE is defined as a carboxydipeptidase by its conversion of angiotensin I to angiotensin II, which has strong pressor effects. ACE inhibitors like captopril or enalapril are clinically used in hypertension. ACE is also able to act as a carboxydipeptidase against various enkephalyl peptides such as Met-enkephalin-Arg^6-Phe^7 (11), which derives from proenkephalin, and Leu-enkephalin-Arg^6 (12), which derives from prodynorphin. The latter reaction generates peptides with negligible opioid activity, and this reaction therefore represents an inactivation. The enkephalin pentapeptides, however, are relatively poor substrates for ACE. ACE also acts as an endopeptidase and cleaves substance P at the 8–9 bond besides the 9–10 bond. The 8–9 bond is favored by a ratio of 4:1 (13). It has also been reported that brain and lung ACE act differently on neurokinin A; the brain enzyme cleaves the peptide as rapidly as substance P, whereas the lung enzyme is relatively inactive on neurokinin A. This difference was not confirmed by Hooper et al. (14), who found no or negligible hydrolysis with ACE from either source. The ACE in CSF including the smaller fragments is similar to the lung enzyme (i.e., has low activity against neurokinin A) (I. Lantz et al., *unpublished observations*).

The functional significance of ACE in the brain is poorly understood. Obviously it is of importance for the angiotensin pathways. In the striatonigral pathway, ACE may act on substance P or prodynorphin-derived peptides. Preliminary experiments in which ACE inhibitors are given into the substantia nigra have indicated a suppression of dopamine release in the striatum, suggesting that ACE normally is important for peptide degradation in this area (M. Herrera-Marschitz et al., *unpublished observations*). ACE is present in the CSF in several catalytically active molecular sizes, all of which are smaller than that found in brain, molecular weight about 100 kD. In CSF, the smallest form is about 30 kD. This form maintains a high potency and cannot be distinguished mechanistically from forms of higher molecular weight. A form of intermediate size (~60 kD) is also present in CSF (I. Lantz et al., *unpublished observations*) and in rabbit lung (15). The cloning of ACE (16) and other recent studies have identified the active site in the C-terminal part of the molecule; it is therefore likely that the smaller forms in CSF represent fragments containing this domain.

Mainly because of readily available methodology, ACE activity has been measured in the CSF in different clinical states including basal ganglia disorders. Thus, levels are decreased in demented patients with Parkinson's disease or supranuclear palsy. The levels correlated with those of 3,4-dihydroxyphenylacetic acid (DOPAC) and homovanillic acid (HVA), suggesting an association to the defective dopamine function, and also with reduced acetylcholinesterase levels (17). These data are indirectly suggesting an origin of the enzyme in neurons rather than glia cells or epithelial linings in blood vessels or secretory cells.

Dynorphin-Converting Enzyme

This enzyme, originally described by Nyberg et al. (18), was defined by its capacity to convert peptides from prodynorphin, dynorphin A and B, and alpha-neoendorphin to Leu-enkephalin-Arg^6 and the corresponding C-terminal fragments. This enzyme is highly abundant in human CSF (Table 1). Later work has defined other enzymes with more stringent capacity to preferentially cleave dynorphin B (19). The dynorphin B–specific enzyme seems to be the dominating species in spinal cord and brain extracts (20; J. Silberring et al., *unpublished data*). An enzyme isolated from bovine and rat brain generates dynorphin A (1–8) from dynorphin A (21). However, this pathway of fragmentation cannot be observed with enzymes isolated from human CSF.

Cleavage of the prodynorphin-derived peptides generating Leu-enkephalin-Arg^6 causes a marked change in the biological activity of the peptides. Whereas the dynorphin peptides have selectivity against the kappa-opioid receptor, enkephalyl hexapeptides have specificity for delta-opioid receptors. The enkephalyl peptides are also more susceptible to further enzymatic degradation.

It was previously thought that enkephalins were almost uniquely produced in the proenkephalin pathways. However, more recent work has shown that conversion of prodynorphin peptides to enkephalins is considerable in several pathways, including spinal interneurons and a striatonigral pathway. In the substantia nigra, there are high concentrations of Leu-enkephalin-Arg^6. The concentrations of both the substrate, dynorphin A, and the product, Leu-enkephalin-Arg^6, are affected by long-term treatment of rats with dopamine D1–receptor or D2–receptor agonists or antagonists. The changes seem to occur independently (Table 2), suggesting that they can be partly explained by regulation at the enzymatic conversion step (22).

Dynorphin-converting enzyme activity has been measured in various clinical CSF specimens. Levels of activity have been found to be lower in women in late pregnancy than in nonpregnant women (23). There also seems to be a sex difference at least in patients with chronic pain, males having higher levels of enzyme (M. Hyppää et al., *unpublished observations*). Because dynorphins have strong analgesic activity if given intrathecally and at least some of the CSF enzyme may have a spinal origin, the higher enzyme levels probably indicate lower dynorphin activity and therefore tolerance to pain.

Neutral Endopeptidase EC 3.4.24.11 (Enkephalinase)

Despite its name, this enzyme is not very discriminative. The enkephalins are cleaved at the Gly^3-Phe^4 bond, which leads to complete loss of opioid activity. Many other peptides including substance P, cholecystokinin octapeptide, and neurotensin are also cleaved, usually on the N-terminal side of a hydrophobic amino acid (24). Endopeptidase-24.11 has a broad distribution outside the nervous system, with particularly high concentrations in the renal microvillar membranes, and is probably the major enzyme responsible for the rapid clearance of these and other peptides from plasma.

The significance of endopeptidase-24.11 has been amply illustrated by the introduction of specific inhibitors. If such inhibitors are microinjected into the brain of rats, they become less responsive to noxious stimuli (25). There is, however, little information

TABLE 2. *Effects of subchronic treatment (5 days) of rats with dopamine receptor agonists or antagonists on levels of dynorphin A (DYN A) or Leu-enkephalin-Arg^6 (ENK-Arg^6) in substantia nigra (22)^a*

	D_1 receptor		D_2 receptor	
Peptide	Agonist (SKF 38393)	Antagonist (SCH 23390)	Agonist (LY 171555)	Antagonist (Sulpirid)
DYN A	+32%*	−32%*	−10%	+44%**
ENK-Arg^6	+89%***	+20*	+81%***	−8%

$^a p<0.05$; $^{**}p<0.01$; $^{***}p<0.001$.

on endopeptidase-24.11 in the CSF. However, Spillantini et al. (26) and Lantz and Terenius (12) noticed that enkephalin degradation in human CSF was slowed down by addition of endopeptidase-24.11 inhibitors. The latter authors found that this endopeptidase was much less important for enkephalin degradation than aminopeptidases and less important for enkephalyl hexapeptide degradation than ACE. Substance P degradation to substance P (1–7) in CSF is also slower in the presence of endopeptidase-24.11 inhibitors. However, another enzyme described below seems more important. There is endopeptidase-24.11 in choroid plexus membranes, but levels are much lower than those of several aminopeptidases or dipeptidases (27). Few attempts have been made to use measurement of endopeptidase-24.11 in CSF as a clinical marker.

Substance P Endopeptidase

This enzyme has been named from its capacity to cleave substance P at the Phe^7-Phe^8 and, to a lesser extent, at the Phe^8-Gly^9 bonds (28). It bears resemblance to a previously described endopeptidase isolated from human brain, which, however, also cleaves at the Gln^6-Phe^7 bond (29). The reason for the discrepancy might be a difference in purity. This enzymatic reaction may be considered conversion because substance P (1–7) and (1–8) are biologically active. Substance P (1–7) has strong behavioral effects if given intracerebrally (30). It is a powerful antagonist against the irritant effects induced by substance P given intrathecally to mice (31) or to the dopamine release in striatum of rats induced by the injection of substance P into their substantia nigra (M. Herrera-Marschitz et al., *unpublished data*). Thus, the endopeptidase may have an autoregulatory function, modulating the actions of substance P.

Substance P endopeptidase does not act on two other tachykinins: neurokinin A or neuromedin B. It does not act on a variety of other neuropeptides with the exception of calcitonin gene-related peptide (CGRP) and somatostatin (32). CGRP also acts as a potent inhibitor of substance P degradation. This metabolic interaction may be a mechanism for the potentiating effects of CGRP on substance P action on rat spinal cord. CGRP can potentiate at doses in which it has no intrinsic activity of its own (33). The interaction between substance P and CGRP is interesting because these two peptides coexist in the nonmyelinated afferent fibers, known to be activated by noxious stimuli (33,34).

SOURCES OF ENZYMES THAT OCCUR IN THE CSF

In principle, enzymes that occur in the CSF can have several origins, and it is likely that a particular enzyme may derive from several sources (Fig. 2). Because the levels of the enzymes are so much higher in CSF than would be expected from just leakage, we propose that they have been actively secreted. Certain enzymes that occur in plasma at high concentrations, such as ACE, could theoretically also derive from plasma. However, given their size and the slow blood-brain barrier passage of molecules of a comparable size, this is an unlikely possibility.

Neuronal Projection in the CNS

Distribution of ACE in the rat CNS has been studied by several groups (35–37). There is one system overlapping the angiotensin-positive cells in the hypothalamic-hypophyseal tract that probably transforms intraneuronal angiotensin I to II. However, the main systems are in the basal ganglia, with no overlap to angiotensin fibers, including a pathway from the substantia nigra (38). The peptide substrate for this projection is not known, and it is not clear whether ACE coexists with a particular

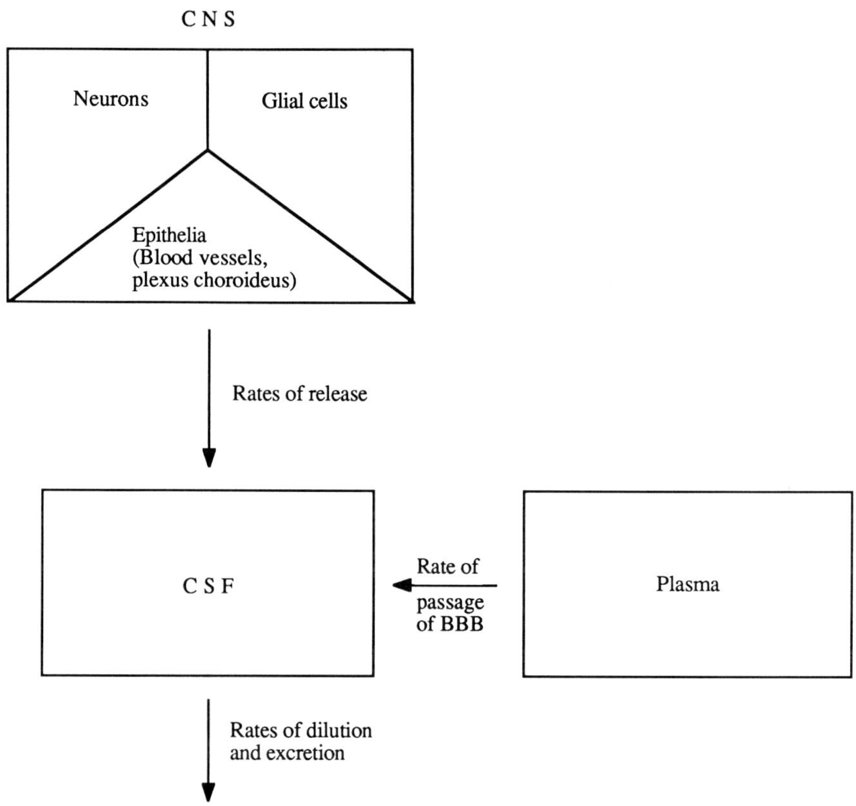

FIG. 2. Origins of proteolytic enzymes in cerebrospinal fluid (CSF). BBB, blood-brain barrier; CNS, central nervous system.

neuropeptide system. There are at least two main candidates: preprotachykinin including substance P, and dynorphins. Because ACE has a high affinity for dynorphin-related peptides and because microinjection of captopril into the projection area of the major pathway, the substantia nigra, causes behavioral effects that are naloxone-reversible (M. Herrera-Marschitz, *unpublished data*; L. Terenius et al., *unpublished data*), effects on the dynorphin system may be particularly relevant.

Distribution of endopeptidase-24.11 has been studied in some detail. It is found all over the CNS, with considerable overlap with enkephalin terminal fields. It is clearly associated with neurons, although few discrete pathways with endopeptidase-24.11 have been identified (39).

Mapping of both ACE and endopeptidase-24.11 has been facilitated by the existence of specific inhibitors that can be labeled to high specific radioactivity. Such inhibitors are not yet available for the other enzymes mentioned in this communication, and therefore mapping has so far not been possible.

Glial Cells

Many studies have shown that glial cells are capable of generating neuropeptides. Consequently they should express the enzymes critically involved in the biosynthesis of neuropeptides. This has also been shown to be the case. Several groups have used glial cells in primary culture. Thus,

Bauer and co-workers (40) have identified several proteolytic enzymes in astrocytes, including enkephalinase in high concentration, and aminopeptidases. Carboxypeptidase E is also present in cultured astrocytes. Interestingly, this enzyme is easily secreted from astrocytes into the cultivating medium, a process not facilitated by K^+ ion (41). Given the fact that glial cells outnumber the neurons, glial cells may be an important origin for CSF enzymes.

Choroid Plexus

The choroid plexus is a much-folded organ lined with epithelial cells on the ventricular side. Several proteolytic enzymes are richly represented in this organ, including ACE, several aminopeptidases, and carboxypeptidase M (27). The choroid plexus may serve an important role in the degradation and elimination of peptides from the CSF. Endopeptidase-24.11 and ACE are also present on pial membranes in the brain and spinal cord (27). A DCE has been recently isolated from human choroid plexus and found to be only partly similar to the enzyme in CSF. For instance, the CSF endopeptidase cleaves dynorphin A only at the Arg^6-Arg^7 bond, whereas the choroid plexus enzyme also cleaves at the Arg^7-Ile^8 bond. Another difference is the larger size of the enzyme isolated from the choroid plexus (F. Nyberg et al., *unpublished data*). Thus, at this stage of knowledge, the contribution of choroid plexuses to CSF enzyme levels seems rather insignificant.

GENERAL CHARACTERISTICS OF ENZYMES IN CSF

Most of the enzymes described here can also be isolated from brain tissue or spinal cord. Generally, the CSF enzymes have a smaller molecular weight than the corresponding enzymes extracted from tissue homogenates. A striking example is ACE, which exists in several forms (i.e., a 140-kD species but also two minor species of 60 kD and 30 kD that are fully active). The smaller forms cannot be distinguished functionally from the major big component (I. Lantz et al., *unpublished observations*).

In CNS tissue, DCE seems to be cytosolic in vesicles, whereas substance P endopeptidase is bound to vesicular membranes (42). Concentrations of the former in CSF are higher than those of the latter (Table 1). One explanation for the difference might be the difference in localization, DCE being more readily available for release. Both enzymes can be released by K^+-induced depolarization (42).

FUNCTIONAL SIGNIFICANCE OF CSF ENZYMES

It is proposed in this chapter that enzymes that appear in the CSF are primarily secreted from neurons during synaptic transmission. This forms the basis for the use of enzymes in CSF as markers. Such enzymes would consequently have a presynaptic localization of activity in peptidergic pathways. Enzymes that are abundant in CSF are candidates for isolation efforts and for development of inhibitors. Such potential inhibitors would be pharmacologic tools and potential therapeutic agents. Indirectly, they can also serve as probes for the neuronal pathways that express the enzymes. In fact, the existence of ACE inhibitors like captopril and endopeptidase-24.11 inhibitors have been instrumental for the histologic mapping of these enzymes and have confirmed their presynaptic localization in particular pathways. An additional possibility that deserves attention is that these enzymes are not just markers of synaptic activity but that they also act in the CSF as scavengers or modifiers of peptide transmission. If so, they may play an important role in volume transmission.

Practically nothing is known about how the biosynthesis of these enzymes is regulated. It is therefore interesting that chronic opiate treatment of rats increases the levels of DCE and substance P endopeptidase (43). It can be foreseen that the regulation of peptide-transforming enzyme biosynthesis and secretion will be important topics for future research.

CONCLUDING REMARKS

The high levels of certain enzymes in CSF strongly suggest that they are actively secreted, most probably from neurons. Besides low molecular weight neurotransmitters and peptides and various protein growth factors, certain enzymes could be important in signal transduction. The enzymes would be acting as scavengers for neuropeptides, some of which are extraordinarily potent. In this proposed role, enzymes would be acting globally, perhaps synchronizing widely separated neuropeptide systems. Most evidence for such a role is yet indirect. It will be of interest for future work to search for pathologic deviances in enzyme levels and their consequences.

It should also be emphasized that the present review is focused on opioid peptides and substance P, peptides of particular interest to the present reviewers. If other peptides would be investigated as substrates, it is likely that other enzymes would be identified. This seems an important extension of the present work, because it would potentially generate knowledge about peptide transformations, which presently are incompletely known.

ACKNOWLEDGMENTS

This work was supported by the Swedish Medical Research Council and the Magnus Bergvall Foundation.

REFERENCES

1. Gainer H. Precursor processing and the secretory vesicle. In: Schmitt FO, Bird S, Bloom FE, eds. *Molecular genetic neuroscience*. New York: Academic Press, 1982;171–187.
2. McKelvy JF, Blumberg S. Inactivation and metabolism of neuropeptides. *Annu Rev Neurosci* 1986;9:415–434.
3. Gluschankof P, Gomez S, Lepage A, et al. Role of peptide structure in the selective processing of peptide prohormones at basic amino acid pairs by endoproteases. *FEBS Lett* 1988;234: 149–152.
4. Fricker LD, Snyder SH. Enkephalin convertase: purification and characterization of a specific enkephalin-synthesizing carboxypeptidase localized to adrenal chromaffin granules. *Proc Natl Acad Sci* 1982;79:3886–3890.
5. Hersh LB, Aboukhair N, Watson S. Immunohistochemical localization of aminopeptidase M in rat brain and periphery: relationship of enzyme localization and enkephalin metabolism. *Peptides* 1987;8:523–532.
6. Bradbury AF, Finnie MDA, Smyth DG. Mechanism of C-terminal amide formation by pituitary enzymes. *Nature* 1982;298:686–688.
7. Mains RE, Eipper BA, Glembotski CC, Dores RM. Strategies for the biosynthesis of bioactive peptides. *Trends Neurosci* 1983;6:229–235.
8. Kizer JS, Busby WH Jr, Cottle C, Youngblood WW. Glycine-directed peptide amidation: presence in rat brain of two enzymes that convert *p*-Glu-His-Pro-Gly-OH into *p*-Glu-His-Pro-NH_2 (thyrotropin-releasing hormone). *Proc Natl Acad Sci* 1984;81:3228–3232.
9. Terenius L, Nyberg F. Neuropeptide-processing, -converting, and -inactivating enzymes in human cerebrospinal fluid. *Int Rev Neurobiol* 1988;30:101–121.
10. Arregui A, Emson PC, Spokes EG. Angiotensin-converting enzyme in substantia nigra: reduction of activity in Huntington's disease and after intrastriatal kainic acid in rats. *Eur J Pharmacol* 1978;52:121–124.
11. Yang HYT, Majane E, Costa E. Conversion of [Met^5]-enkephalin-Arg^6-Phe^7 to [Met^5]-enkephalin by dipeptidyl carboxypeptidase. *Neuropharmacology* 1984;20:891–894.
12. Lantz I, Terenius L. High enkephalyl peptide degradation, due to angiotensin-converting enzyme-like activity in human CSF. *FEBS Lett* 1985;193:31–34.
13. Skidgel RA, Defendini R, Erdös EG. Angiotensin I converting enzyme and its role in neuropeptide metabolism. In: Turner AJ, ed. *Neuropeptides and their peptidases*. New York: Weinheim, 1987;165–182.
14. Hooper NM, Kenny AJ, Turner AJ. The metabolism of neuropeptides. Neurokinin A (substance K) is a substrate for endopeptidase-24.11 but not for peptidyl dipeptidase A (angiotensin-

converting enzyme). *Biochem J* 1985;231:357–361.
15. Iwata K, Blacher R, Soffer RL, Lai C-Y. Rabbit pulmonary angiotensin-converting enzyme: the NH_2-terminal fragment with enzymatic activity and its formation from the native enzyme by NH_4OH treatment. *Arch Biochem Biophys* 1983; 227:188–201.
16. Soubrier F, Alhenc-Gelas F, Hubert C, et al. Two putative active centers in human angiotensin I-converting enzyme revealed by molecular cloning. *Proc Natl Acad Sci* 1988;85:9386–9390.
17. Zubenko GS, Volicer L, Direnfeld LK, Freeman M, Langlais PJ, Nixon RA. Cerebrospinal fluid levels of angiotensin-converting enzyme in Alzheimer's disease, Parkinson's disease and progressive supranuclear palsy. *Brain Res* 1985; 328:215–221.
18. Nyberg F, Nordström K, Terenius L. Endopeptidase in human cerebrospinal fluid which cleaves proenkephalin B opioid peptides at consecutive basic amino acids. *Biochem Biophys Res Commun* 1985;131:1069–1074.
19. Silberring J, Nyberg F. A novel bovine spinal cord endoprotease with high specificity for dynorphin B. *J Biol Chem* 1989;264:11082–11086.
20. Nyberg F, Silberring J. Conversion of the dynorphins to Leu-enkephalin in human spinal cord. *Prog Clin Biol Res* 1989;328:261–265.
21. Devi L, Goldstein A. A rat brain endopeptidase with unusual specificity towards dynorphin A. *Proc Natl Acad Sci* 1984;81:1842–1856.
22. Nylander I, Terenius L. Dopamine receptors mediate alterations in striato-nigral dynorphin and substance P pathways. *Neuropharmacology* 1987;26:1295–1302.
23. Lyrenäs S, Nyberg F, Lindberg B, Terenius L. Cerebrospinal fluid activity of dynorphin-converting enzyme at term pregnancy. *Obstet Gynecol* 1988;72:54–58.
24. Matsas R, Kenney AJ, Turner AJ. The metabolism of neuropeptides. The hydrolysis of peptides, including enkephalins, tachykinins and their analogues, by endopeptidase-24.11. *Biochem J* 1984;223:433–440.
25. Roques BP, Fournié-Zaluski MC, Soroca E, et al. The enkephalinase inhibitor thiorphan shows antinociceptive activity in mice. *Nature* 1980; 288:286–288.
26. Spillantini MG, Geppetti P, Fanciullacci M, Michelacci S, Lecomte JM, Sicuteri F. *In vivo* enkephalinase inhibition by acetorphan in human plasma and CSF. *Eur J Pharmacol* 1986; 125:147–150.
27. Bourne A, Barnes K, Taylor BA, Turner AJ, Kenny AJ. Membrane peptidases in the pig choroid plexus and on other cell surfaces in contact with the cerebrospinal fluid. *Biochem J* 1989; 259:69–80.
28. Nyberg F, Le Grevès P, Sundqvist C, Terenius L. Characterization of substance P (1–7) and (1–8) generating enzyme in human cerebrospinal fluid. *Biochem Biophys Res Commun* 1984; 125:244–250.
29. Lee C-M, Sandberg BEB, Hanley MR, Iversen LL. Purification and characterization of a membrane-bound substance P-degrading enzyme from human brain. *Eur J Biochem* 1981; 114:315–327.
30. Hall ME, Stewart JM. Substance P and behavior: opposite effects of N-terminal and C-terminal fragments. *Peptides* 1983;4:763–768.
31. Sakurada T, Kuwahara H, Takahashi K, Sakurada S, Kisara K, Terenius L. Substance P (1–7) antagonizes substance P-induced aversive behaviour in mice. *Neurosci Lett* 1988;95:281–285.
32. Le Grevès P, Nyberg F, Terenius L, Hökfelt T. Calcitonin gene-related peptide is a potent inhibitor of substance P degradation. *Eur J Pharmacol* 1985;115:309–311.
33. Wiesenfeld-Hallin Z, Hökfelt T, Lundberg JM, et al. Immunoreactive calcitonin gene-related peptide and substance P coexist in sensory neurons to the spinal cord and interact in spinal behavioral responses of the rat. *Neurosci Lett* 1984;52:199–204.
34. Skofitsch G, Jacobowitz DM. Calcitonin gene-related peptide coexists with substance P in capsaicin sensitive neurons and sensory ganglia of the cat. *Peptides* 1985;6:747–754.
35. Chevillard C, Saavedra JM. Distribution of angiotensin-converting enzyme activity in specific areas of the rat brain stem. *J Neurochem* 1982;38:281–284.
36. Snyder SH. Brain enzymes as receptors: angiotensin-converting enzyme and enkephalin convertase. *Ann NY Acad Sci* 1985;463:21–30.
37. Chai SY, Mendelsohn FAO, Paxinos G. Angiotensin converting enzyme in rat brain visualized by quantitative *in vitro* autoradiography. *Neuroscience* 1987;20:615–627.
38. Chai SY, Christie MJ, Beart PM, Mendelsohn FAO. Effects of nigral dopaminergic lesions and striatal excitotoxin lesions on brain converting enzyme. *Neurochem Int* 1987;10:101–107.
39. Matsas R, Kenny AJ, Turner AJ. An immunohistochemical study of endopeptidase-24.11 ("enkephalinase") in the pig nervous system. *Neuroscience* 1986;4:991–1012.
40. Horsthemke B, Hamprecht B, Bauer K. Heterogeneous distribution of enkephalin-degrading peptidases between neuronal and glial cells. *Biochem Biophys Res Commun* 1983;115:423–429.
41. Vilijn M-H, Das B, Kessler JA, Fricker LD. Cultured astrocytes and neurons synthesize and secrete carboxypeptidase E, a neuropeptide-processing enzyme. *J Neurochem* 1989;53:1487–1493.
42. Fried G, Franck J, Terenius L. Subcellular distribution of two enzymes with neuropeptide converting activity in rat spinal cord (*submitted*).
43. Persson S, Post C, Alari L, Nyberg F, Terenius L. Increased neuropeptide-converting enzyme activities in cerebrospinal fluid of opiate-tolerant rats. *Neurosci Lett* 1989;107:318–322.

Volume Transmission in the Brain: Novel Mechanisms for Neural Transmission,
edited by K. Fuxe and L.F. Agnati.
Raven Press, Ltd., New York © 1991.

33

Volume Transmission by Coreleased Peptides in the Autonomic Nervous System

J. M. Lundberg

Department of Pharmacology, Karolinska Institutet, S-104 01 Stockholm, Sweden

In the peripheral nervous system, tissue components are innervated by a diffuse plexus of autonomic motor and sensory nerves. These varicose fibers mostly do not form classical synapses but have neuroeffector gaps with ranges up to 1 μm. This forms the anatomical basis for diffusion of released mediators acting at distant sites, providing that they escape local degradation and/or reuptake. Acetylcholine (ACh) and norepinephrine (NA) have an efficient degradation and reuptake system, respectively, and are the established transmitters in the motor division of the autonomic nervous system. It seems clear, however, that certain effector responses to nerve stimulation remain in a variety of tissues in spite of blockade of classical transmitter mechanisms. These functional effects have been named non-adrenergic, non-cholinergic (NANC).

Increasing evidence suggests that a variety of peptides that have slow and long-lasting effects are involved as neuronal NANC messengers that act together with classical transmitters in many peripheral tissues (1). Because of high metabolic stability, neuropeptides may be especially suitable for the volume type of transmission in analogy with paracrine and hormonal secretion of similar substances from endocrine cells. Vasoactive intestinal polypeptide (VIP) and peptide histidine isoleucine (PHI) are present together with ACh in postganglionic parasympathetic nerves regulating blood flow and exocrine gland secretion (2). Neuropeptide tyrosine (NPY) occurs together with NA in postganglionic sympathetic nerves regulating, for example, vascular tone (3,4). Calcitonin gene-related peptide (CGRP) and multiple tachykinins such as substance P (SP) and neurokinin A (NKA) are present in peripheral branches of sensory nerves, which are characterized by specific sensitivity of capsaicin, the pungent agent in hot peppers. Local peptide release from these sensory nerves on various stimuli is likely to elicit slowly developing and long-lasting reactions such as neurogenic inflammation (vasodilatation and protein extravasation) (5), tachycardia (6), and visceral smooth muscle contraction (5) or relaxation (7). The concept that adenosine triphosphate (ATP) may mediate some rapid, short-lasting NANC effects in certain models (3,8) will not be discussed here because ATP is rapidly broken down and not likely to act at any longer distance from release sites.

ACh, VIP, AND PHI IN PARASYMPATHETIC TRANSMISSION

Parasympathetic mechanisms are of major importance for the autonomic control of exocrine gland function. In most glands, such as the salivary glands, the secretory response to parasympathetic stimulation is completely blocked by atropine, suggesting a muscarinic cholinergic receptor–mediated effect. The concomitant large and long-lasting increase in glandular blood flow is, however, a NANC response, which is reduced by muscarinic antagonists, only at low-frequency stimulation (2) (Fig. 1). VIP and PHI, which have potent vasodilatory activity, are coreleased with ACh on activation of parasympathetic postganglionic nerves. Whereas the ACh release dominated at low-frequency stimulation, major overflow of VIP/PHI mainly occurred on high frequencies when also atropine-resistant vasodilatation was observed (9) (Fig. 1). This difference in release may be related to a separate subcellular storage whereby ACh mainly is present in small clear vesicles and VIP in the large dense-cored vesicles (2). Atropine pretreatment was associated with parallel enhancement of nerve stimulation evoked VIP and PHI overflow, suggesting that the release of the peptides was regulated by the local biophase concentrations of ACh acting on prejunctional muscarinic receptors. Although VIP by itself did not trigger salivary secretion in the cat, VIP enhanced both the protein content and volume of ACh-induced secretion (9). The effects of VIP are mediated by specific receptors coupled to adenylate cyclase activation, and supersensitivity to VIP mechanisms develops in salivary glands after denervation, suggesting a physiological regulation of receptor function (9). Receptor binding data indicate that only one main type of VIP receptor is present, which also binds PHI with lower affinity. Therefore, VIP is, at present, a main candidate for parasympathetic NANC vasodilatation, although a contribution by PHI cannot be ruled out.

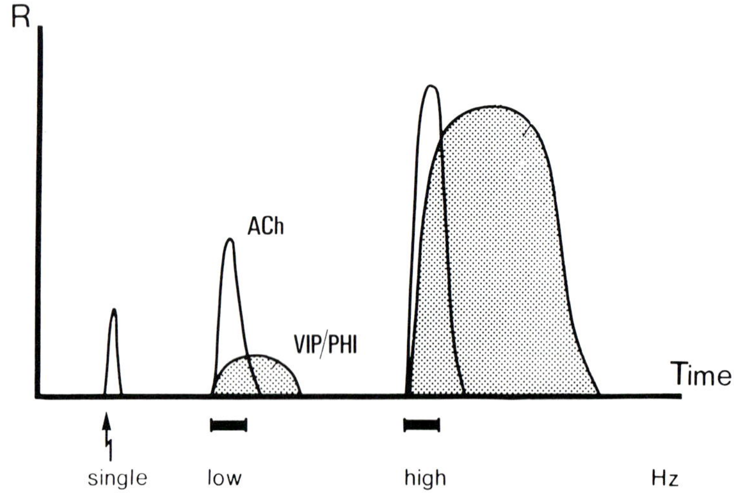

FIG. 1. Schematic illustration of multiple mediators involved in the vasodilatation response (R) evoked by parasympathetic nerve stimulation at different frequencies (Hz) in the cat submandibular salivary gland. Single impulse stimulation evokes a rapid and short-lasting vasodilatation, which is atropine sensitive (i.e., cholinergic). At low-frequency stimulation, most of the response is due to ACh with a small atropine-resistant component (shaded area), which is likely to be due to corelease of vasodilatory peptides like VIP and PHI. At high-frequency stimulation, most of the response is atropine-resistant (*shaded area*), long-lasting, and probably mediated by the coreleased peptides.

NA AND NPY IN SYMPATHETIC TRANSMISSION

Neuropeptide tyrosine is present together with NA in perivascular sympathetic fibers as well as in noradrenergic nerves to heart muscle and vas deferens (3,10). NPY is a potent vasoconstrictor agent *in vivo* and mimics the characteristic long-lasting appearance of the sympathetic responses, which remain on high-frequency sympathetic stimulation in the presence of adrenoceptor antagonists (Fig. 2). (3,9,11,12). Reserpine pretreatment, which interferes with granular storage of NA, is also associated with a progressive and marked depletion of NPY from perivascular sympathetic nerves. The reserpine-evoked depletion of NPY depends on a mechanism, which, in contrast to the effect on granular storage of NA, is entirely dependent on neuronal activation and excessive release in relation to resupply by axonal transport (12). Reserpine treatment combined with preganglionic sympathetic denervation thus creates a situation in which tissue levels of NPY remain intact while NA still is depleted by 98% to 99%. Under such conditions, there is an excellent correlation between the remaining long-lasting vasoconstriction and NPY release in the spleen, kidney, skeletal muscle, and nasal mucosa (8). The release of NPY on nerve stimulation is increased after reserpine, however, thus overestimating the contribution by NPY to the vasoconstrictor response under physiological conditions (12) (see below). NPY synthesis in ganglion cells seems to be dependent on neuronal activity, and both the expression of specific messenger ribonucleic acid (RNA) for NPY (13) and axonal transport of NPY are increased after reserpine (11).

Neuropeptide tyrosine binds to specific receptors on vascular smooth muscle cells whereby a large, amidated C-terminal portion of the peptide is a prerequisite for receptor binding, inhibition of forskolin-stimulated cyclic adenosine monophosphate (AMP) formation and vasoconstrictor effects *in vivo* (14).

The release of NPY compared with NA increases with frequency of stimulation (12), although even single nerve impulses

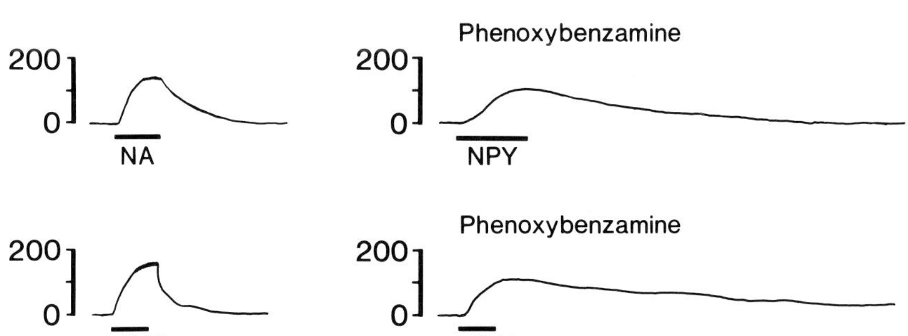

FIG. 2. *Top panel:* Perfusion pressure changes (mmHg) as indication of vasoconstriction in skeletal muscle (gracilis) of the dog evoked by intraarterial infusion of norepinephrine (NA) 3 nmol/min for two minor intraarterial infusions of NPY (2 nmol/min) after pretreatment with the alpha-adrenoceptor antagonist phenoxybenzamine (0.5 mg/kg intraarterially). *Lower panel:* Perfusion pressure changes on postganglionic sympathetic nerve stimulation (10 Hz for 2 min) before (*left curve*) and after pretreatment with phenoxybenzamine and beta-adrenoceptor antagonists (*right panel*). Note the long-lasting vasoconstriction evoked by NPY, which mimics the effects of sympathetic nerve stimulation in the presence of adrenoceptor antagonists.

may release NPY and cause vasoconstriction in the nasal mucosa of reserpinized animals (4). The mechanisms underlying the frequency-dependent differential release of NPY compared with NA may be related to the partly separate vesicular storage of these two agents in sympathetic nerve terminals. Thus, whereas NA is present both in small and large dense-cored vesicles, NPY preferentially occurs in the large vesicles (12). On prolonged stimulation, NPY release and tissue stores of NPY are depleted compared with NA in agreement with the view that unlike the efficient local synthesis of NA in nerve endings, NPY has to be resupplied by axonal transport, which is insufficient to maintain release for longer periods.

The secretion of both NA and NPY is regulated via the local biophase concentrations of released NA acting on prejunctional α_2 adrenoceptors both in experimental animals (15) and humans (16). Thus, α_2 agonists such as clonidine and UK14304 inhibit and the α_2 antagonists yohimbine and idazoxan increase stimulation-evoked NA and NPY release (Fig. 3). The enhanced NPY release after reserpine may thus be related to the absence of an inhibitory noradrenergic α_2-receptor-mediated feedback control (11). In accordance, agents such as desipramine, which increases local biophase concentrations of NA because of inhibition of reuptake, reduce NPY overflow (12). In addition, there is evidence that prejunctional angiotensin II mechanisms facilitate both NA and NPY release from sympathetic nerves in the kidney (17).

Neuropeptide tyrosine also exerts prejunctional inhibitory actions on NA release from perivascular nerves, heart, and vas deferens (3,10). Furthermore, activation of prejunctional NPY receptors by the structurally related peptide PYY inhibits stimu-

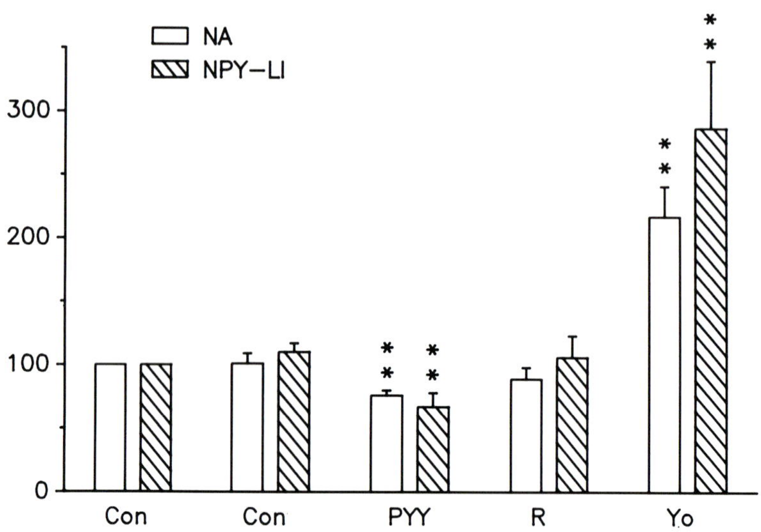

FIG. 3. Illustration of parallel prejunctional regulation of NA (*open bars*) and NPY-LI (*hatched bars*) overflow evoked by sympathetic nerve stimulation (5 Hz for 1 min) of the pig kidney *in vivo.* The release of NA and NPY is expressed as % control (con). First the control stimulation was repeated. Then, prejunctional NPY receptors were activated by the structurally related peptide PYY (1 μg/kg/min for 5 min), which does not interfere in the NPY radioimmunoassay. PYY depressed both NPY and NA overflow. After a recovery stimulation (R), yohimbine (Yo) (0.2 mg/kg) was given intravenously. Yohimbine increased both stimulation-evoked NPY and NA overflow, suggesting that the release of these two agents was inhibited by the local biophase concentration of NA acting on prejunctional alpha-2 adrenoceptors. $^{**}p<0.01$. Quade test for several related samples.

lation-evoked release of NPY itself (17) (Fig. 3). Because NPY seems to be mainly released at high frequencies, the relative importance of NPY mechanisms at the prejunctional level may then play a role while α_2-receptor-mediated inhibition of transmitter secretion is dominating at low frequencies.

Finally, there is evidence that NPY can facilitate the contractile effects of NA on smooth muscle both in blood vessels and vas deferens, suggesting also an indirect action of NPY via modulation of the postjunctional response of the coreleased classical transmitter (3,11).

The release of NPY, as revealed by increased plasma levels on sympathetic activation by bicycle exercise in healthy volunteers (3,16) and patients with cardiovascular disorders such as left heart failure (18), together with potent vasoconstrictor actions in humans *in vivo* (19), raises the possibility that NPY may be involved both in the physiological adaptation to stress and in the pathophysiology of human cardiovascular disorders. This remains to be established, however, with the use of specific NPY antagonists, which are not yet available. In addition to reserpine and the drugs that interfere with the noradrenergic α_2 feedback control of release, the classical sympatholytic agent guanethidine inhibits nerve stimulation–evoked NPY overflow (20). This illustrates some of the problems arising when trying to define the relative roles of the classical transmitter and the coreleased peptide in the transmission process.

CGRP AND TACHYKININS IN CAPSAICIN-SENSITIVE AFFERENTS

Peripheral branches of sensory nerves are present both in the skin and internal organs. Capsaicin causes acute excitation of a population of C-fiber afferents with subsequent central autonomic (both vasodilatation and secretion by parasympathetic mechanisms in certain organs as well as sympathetic vasoconstrictor responses in other regions) and protective reflexes such as cough and sneezing. Furthermore, there is a local release of multiple peptides via direct actions at the site of irritation and more distant via axon reflex mechanisms. CGRP and tachykinins (SP, NKA) seem to be present, not only in the same nerves but also in the same large dense-cored storage vesicles (5,7). The capsaicin-evoked release of tachykinins (SP, NKA) and CGRP is Ca^{2+}-dependent (7), suggesting exocytosis of material stored in vesicles. The capsaicin-induced release of peptides from sensory nerves is tetrodotoxin-sensitive in low concentrations, which illustrates the importance of nerve impulse propagation along axons (21), whereas at high concentration, capsaicin effects are tetrodotoxin-resistant and most likely caused by direct actions on the varicosities (7). The peptide release from sensory nerves evoked by electrical field stimulation is inhibited by μ-opiate, α_2-, and NPY-receptor activation (22). Ruthenium red is an agent that interferes with Ca^{2+} mechanisms in sensory nerves, which are triggered by capsaicin but not by bradykinin (23) or nicotine (24). Thus, local release of peptide mediators as well as desensitization development on capsaicin exposure can be blocked by Ruthenium red (24,25). After a high dose of capsaicin, the sensory nerves are damaged and tachykinins and CGRP depleted from peripheral organs. The functional effects linked to peptide release from sensory nerves (see below) are then absent (5,7). Therefore capsaicin pretreatment has been extensively used to evaluate the functional importance of sensory mechanisms.

At present distinct functional responses seem to be coupled to the respective sensory neuropeptides because of receptor specificity on target cells and differences in degradation rather than selective release or localization of varicosities (7).

Vasodilatation evoked by activation of capsaicin-sensitive sensory nerves is present in the skin (26), airways (27), and heart (28) and may be induced by chemical irritation, allergic reactions, or local ischemia. Histamine acting on H^{-1} receptors on sensory nerves may explain the flare reaction observed on mast cell degranulation evoked by antigen challenge both in the human skin (Fig. 4) (26) and the pig airways (27,29). The CGRP- and capsaicin-induced vasodilatation (*in vitro*) is independent of the endothelium, whereas the SP effect like that of ACh is mediated via release of an endothelium-derived relaxant factor (EDRF) (28). Tachyphylaxis to the SP effects does not change the capsaicin or CGRP relaxation of isolated blood vessels. CGRP is much more efficient as vasodilator compared with SP when injected into the extravascular space or given on to the airway mucosa (i.e., where the mediator release occurs from the sensory fibers located at the adventitio-medial junction of the blood vessels) (5). Therefore at present CGRP rather than tachykinins may mediate the vasodilation evoked by activation of sensory nerves.

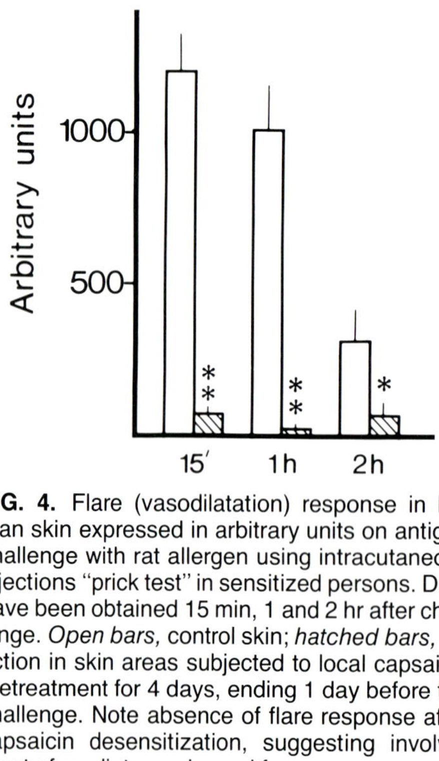

FIG. 4. Flare (vasodilatation) response in human skin expressed in arbitrary units on antigen challenge with rat allergen using intracutaneous injections "prick test" in sensitized persons. Data have been obtained 15 min, 1 and 2 hr after challenge. *Open bars,* control skin; *hatched bars,* reaction in skin areas subjected to local capsaicin pretreatment for 4 days, ending 1 day before the challenge. Note absence of flare response after capsaicin desensitization, suggesting involvement of mediators released from sensory nerves in this reaction. Presumably histamine released from mast cells activates sensory nerves, which in turn secrete multiple peptides such as CGRP with vasodilatory activity. $^{*}p<0.05$, $^{**}p<0.01$ using Mann-Whitney U-test.

The protein extravasation evoked by excitation of sensory nerves is pronounced in rodents such as rats. Thus, inhalation of irritant agents including capsaicin and cigarette smoke evoke protein extravasation in the airways via release of tachykinins such as SP from sensory nerves acting on neurokinin-1 receptors (7,30). These sensory mechanisms for protein extravasation in the airway mucosa are also activated by agents such as bradykinin, histamine (30), and hydrochloric acid (31).

The tachycardia and positive inotropic effect evoked by capsaicin on atria of rats and guinea pigs are most likely caused by CGRP released from sensory nerves acting on specific receptors leading to cyclic AMP elevation in myocytes (6).

The variable effects on contractility of visceral smooth muscle by local activation of capsaicin-sensitive nerves may be related to dominance of receptor mechanisms for certain peptides in specific tissues. Thus, the potent bronchoconstrictor action of capsaicin on guinea pig bronchi is caused presumably by tachykinins such as NKA acting on neurokinin-2 receptors, whereas CGRP is without effect (5,7). The long-lasting NANC contraction of guinea pig bronchi evoked by field stimulation *in vitro* and the atropine-resistant bronchoconstriction on vagal nerve stimulation *in vivo* also seem to be mediated by tachykinins released from peripheral branches of capsaicin-sensitive sensory nerves (5,30). In contrast, the inhibition of ureteral motility at low capsaicin concentrations could be related to CGRP actions, whereas the contractile response at high capsaicin doses suggests involvement of NKA (7). Peptides with opposing functional effects may thus be released from the same storage vesicles. The final effector response will, however, be determined by receptor sensitivity and distribution as well as diffusion barriers and degradation of the peptides. Inhibitors of neutral endopeptidase such as phosphoramidon and thiorphan have been shown to enhance the functional effects of tachykinins (32). Peptide inactivation mechanisms are slow compared with the classical transmitter and do not seem to be specific because both the capsaicin-evoked release of NKA and CGRP from the lung is enhanced by phosphoramidon (21).

The observation that 90% of the SP produced in the nodose ganglion (i.e., vagal sensory ganglion cells) is transported out into the peripheral rather than the central branches, suggests that the peptides exert their major actions in the periphery (33). The chemical that transmits the rapid signaling of sensory information from the periphery via the central branch synapsing in the CNS is less likely to be tachykinins or CGRP but rather a small molecular weight substance like glutamate, which is colocalized with the peptides (34).

It is concluded that autonomic motor and sensory nerves release a variety of chemical signals, which differ in many aspects. The autonomic control of peripheral organs is likely to depend on volume transmission to a major extent, especially with regard to peptidergic mechanisms.

ACKNOWLEDGMENTS

The present summary is based on data obtained with support from the Swedish Medical Research Council (14X-6554). For expert secretarial help I am grateful to Mrs. Evy Carlsson.

REFERENCES

1. Lundberg JM, Hökfelt T. Multiple co-existence of peptides and classical transmitters in peripheral autonomic and sensory neurons—functional and pharmacological implications. *Prog Brain Res* 1986;68:241–262.
2. Lundberg JM. Evidence for coexistence of vasoactive intestinal polypeptide (VIP) and acetylcholine in neurons of cat exocrine glands. Morphological, biochemical and functional studies. *Acta Physiol Scand* 1981;112(suppl 496):1–57.
3. Pernow J. Co-release and functional interactions of neuropeptide Y and noradrenaline in peripheral sympathetic vascular control. *Acta Physiol Scand* 1988;133:1–56.
4. Lacroix J-S. Adrenergic and non-adrenergic mechanisms in sympathetic vascular control of the nasal mucosa. *Acta Physiol Scand* 1989; 136:1–63.
5. Martling C-R. Sensory nerves containing tachykinins and CGRP in the lower airways. Functional implications for bronchoconstriction, vasodilatation and protein extravasation. *Acta Physiol Scand* 1987;130:1–57.
6. Franco-Cereceda A. Calcitonin gene-related peptide and tachykinins in relation to local sensory control of cardiac contractility and coronary vascular tone. *Acta Physiol Scand* 1988; 133:1–63.
7. Hua X-Y. Tachykinins and calcitonin gene-related peptide in relation to peripheral functions of capsaicin-sensitive sensory neurons. *Acta Physiol Scand* 1986;127:1–45.
8. Lundberg JM, Pernow J, Lacroix JS. Neuropeptide Y: sympathetic cotransmitter and modulator? *NIPS* 1989;4:13–17.
9. Lundberg JM, Martling C-R, Hökfelt T. Airways, oral cavity and salivary glands: classical transmitters and peptides in sensory and autonomic motor neurons. In: Björklund A, Hökfelt T, Owman C, eds. *Handbook of chemical neuroanatomy*. Elsevier Science Publishers B. V. 1988;391–444.
10. Lundberg JM, Terenius L, Hökfelt T, et al. Neuropeptide Y (NPY)-like immunoreactivity in peripheral noradrenergic neurons and effects of NPY on sympathetic function. *Acta Physiol Scand* 1982;116:477–480.
11. Lundberg JM, Pernow J, Franco-Cereceda A, Rudehill A. Effects of antihypertensive drugs on sympathetic vascular control in relation to neuropeptide Y. *J Cardiovasc Pharmacol* 1987; 10(suppl 12):551–568.
12. Lundberg JM, Rudehill A, Sollevi A, Fried G, Wallin G. Co-release of neuropeptide Y and noradrenaline from pig spleen *in vivo:* importance of subcellular storage, nerve impulse frequency and pattern, feedback regulation and resupply by axonal transport. *Neuroscience* 1989;28:475–486.
13. Schalling M, Franco-Cereceda A, Hökfelt T, Persson H, Lundberg JM. Increased neuropeptide Y messenger RNA and peptide in sympathetic ganglia after reserpine pretreatment. *Eur J Pharmacol* 1988;156:419–420.
14. Lundberg JM, Hemsén A, Larsson O, Rudehill A, Saria A, Fredholm BB. Neuropeptide Y receptor in pig spleen: binding characteristics, re-

duction of cyclic AMP formation and calcium antagonist inhibition of vasoconstriction. *Eur J Pharmacol* 1988;145:21–29.

15. Lundberg JM, Rudehill A, Sollevi A. Pharmacological characterization of neuropeptide Y and noradrenaline mechanisms in sympathetic control of pig spleen. *Eur J Pharmacol* 1989; 163:103–113.
16. Pernow J, Lundberg JM, Kaijser L. α-Adrenoceptor influence on plasma levels of neuropeptide Y-like immunoreactivity and catecholamines during rest and sympathoadrenal activation in humans. *J Cardiovasc Pharmacol* 1988;12:593–599.
17. Pernow J, Lundberg JM. Modulation of noradrenaline and neuropeptide Y (NPY) release in the pig kidney *in vivo:* involvement of alpha-2, NPY and angiotensin II receptors. *Naunyn Schmiedebergs Arch Pharmacol* 1989;340:379–385.
18. Hulting J, Ullman B, Franco-Cereceda A, Baerentz S, Lundberg JM, Sollevi A. Plasma NPY levels on admittance to a coronary care unit: elevated NPY levels in a subset of patients with left heart failure. *Cardiovasc Res* (*in press*).
19. Pernow J, Lundberg JM, Kaiser L. Vasoconstrictor effects in vivo and plasma disappearance rate of neuropeptide Y in man. *Life Sci* 1987;40:47–54.
20. Lundberg JM, Änggård A, Theodorsson-Norheim E, Pernow J. Guanethidine-sensitive release of neuropeptide Y-like immunoreactivity in the cat spleen by sympathetic nerve stimulation. *Neurosci Lett* 1984;52:175–180.
21. Kröll F, Karlsson J-A, Lundberg JM, Persson CGA. Capsaicin-induced bronchoconstriction and neuropeptide release in guinea pig perfused lungs. *J Appl Physiol* (*in press*).
22. Matran R, Martling C-R, Lundberg JM. Inhibition of cholinergic and non-adrenergic, non-cholinergic bronchoconstriction in the guinea pig mediated by neuropeptide Y and α_2-adrenoceptors and opiate receptors. *Eur J Pharmacol* 1988;163:15–23.
23. Amann R, Lembeck F. Ruthenium red selective prevents capsaicin-induced nociceptor stimulation. *Eur J Pharmacol* 1989;161:227–230.
24. Franco-Cereceda A, Lou Y-P, Lundberg JM. Ruthenium red differentiates between capsaicin and nicotine effects on cardiac sensory nerves. *Acta Physiol Scand* 1989;137:457–458.
25. Maggi CA, Santicioli P, Gepetti P, et al. The antagonism induced by ruthenium red of the actions of capsaicin on the peripheral terminals of sensory neurons: further studies. *Eur J Pharmacol* 1988;154:1–9.
26. Lundblad L, Lundberg JM, Änggård A, Zetterström O. Capsaicin-sensitive nerves and the cutaneous allergy reaction in man. *Allergy* 1987;42:20–25.
27. Alving K, Matran R, Lacroix JS, Lundberg JM. Allergen challenge induces vasodilatation in pig bronchial circulation via a capsaicin-sensitive mechanism. *Acta Physiol Scand* 1988;132:571–572.
28. Franco-Cereceda A, Lundberg JM, Rudehill A. Calcitonin gene-related peptide but not substance P mimics capsaicin-induced coronary vasodilatation in the pig. *Eur J Pharmacol* 1987; 142:235–243.
29. Saria A, Martling C-R, Yan Z, Theodorsson-Norheim E, Gamse R, Lundberg JM. Release of multiple tachykinins from capsaicin-sensitive sensory nerves in the lung by bradykinin, histamine, dimethylphenyl piperazinium, and vagal nerve stimulation. *Am Rev Resp Dis* 1988; 137:1330–1335.
30. Lundberg JM, Saria A. Polypeptide-containing neurons in airway smooth muscle. *Annu Rev Physiol* 1987;49:557–572.
31. Martling C-R, Lundberg JM. Capsaicin sensitive afferents contribute to acute airway edema following tracheal instillation of hydrochloric acid or gastric juice in the rat. *Anesthesiology* 1988;68:350–356.
32. Thompson JE, Sheppard D. Phosphoramidon potentiates the increase in lung resistance mediated by tachykinins in guinea-pig. *Am Rev Respir Dis* 1988;137:337–340.
33. Brimijoin S, Lundberg JM, Brodin E, Hökfelt T, Nilsson G. Axonal transport of substance P in the vagus and sciatic nerves of the guinea-pig. *Brain Res* 1980;191:443–457.
34. De Biasi S, Rustioni A. Glutamate and substance P coexist in primary afferent terminals in the superficial laminae of spinal cord. *Proc Natl Acad Sci USA* 1988;85:7820–7824.

Volume Transmission in the Brain: Novel Mechanisms for Neural Transmission, edited by K. Fuxe and L.F. Agnati. Raven Press, Ltd., New York © 1991.

34

Nonexocytotic Release of γ-Aminobutyric Acid: Implications for Volume Transmission

D.K. Meyer

Department of Pharmacology, University of Freiburg, 7800 Freiburg, Federal Republic of Germany

The following definition of "volume transmission" shall be used here: It is a special type of neurotransmission in which the transmitter and the activated receptors are not within the same synapse. There are two possible origins for the transmitter: (a) It is released within the area of a synapse but diffuses along a gradient to a distant receptor; and (b) the transmitter is released from a specialized area of the axon or terminal, which is extrasynaptic.

One of the morphological criteria of a synapse is the presence of granules that can store transmitter after its synthesis or reuptake. Release of transmitter from these storage vesicles is generally accepted to be the predominant form of release. This exocytotic release is strictly Ca^{2+}-dependent. Lately, however, several authors have reported that veratridine also enhances the release of neurotransmitters if there is no Ca^{2+} in the incubation medium. In the case of γ-aminobutyric acid (GABA), there is good evidence that release is not only Ca^{2+}-independent but that it is even independent of storage vesicles (i.e., nonexocytotic). If GABA can indeed be released from cytoplasmic pools, it becomes possible that this process does not only occur in synaptic areas, but also at other sites along the axons or terminals of the respective neurons. In conclusion, GABA could be a transmitter candidate for volume transmission.

In this chapter, the concept of nonexocytotic release of GABA will be developed in three parts:

1. Evidence for the Ca^{2+}-independent release of neurotransmitters,
2. Evidence that Ca^{2+}-independent GABA release is nonexocytotic, and
3. Modulation of Ca^{2+}-independent GABA release.

EVIDENCE FOR Ca^{2+}-INDEPENDENT RELEASE OF NEUROTRANSMITTERS

Veratridine binds to site 2 of the neuronal Na^+ channel and causes its persistent activation (1). The resulting influx of Na^+ depolarizes the membrane, which leads to an enhanced uptake of Ca^{2+} from the extracellular space. The ensuing increase in free intracellular Ca^{2+} levels, which plateau within 2 min, causes the release of neurotransmitters (e.g., of acetylcholine) (2,3). However, also in the absence of extracellular Ca^{2+}, veratridine is able to enhance the levels of free intracellular Ca^{2+} and to cause release of acetylcholine or norepinephrine (4,5). Mobilization of Ca^{2+} from intracellular stores by the increased intracellular concentration of Na^+ has been made responsible for these veratridine-induced phenomena (4,5). This explanation is based on the observations that Na^+ can

can regulate the release of Ca^{2+} sequestered in mitochondria (6,7). Reports that connected intracellular release of Ca^{2+} caused by inositol-1,4,5-triphosphate with peptide release from pituitary cells (8–11) have lately provided support to the hypothesis that release of Ca^{2+} from intracellular stores may be important for the regulation of cellular secretion of transmitters or hormones. This hypothesis may help to explain not only transmitter release, which is independent of extracellular Ca^{2+}, but also changes in transmitter release caused by agents that use phosphatidylinositol-4,5-bisphosphate metabolism as a second-messenger system.

EVIDENCE THAT Ca^{2+}-INDEPENDENT GABA RELEASE IS NONEXOCYTOTIC

The release of GABA from synaptosomes or brain slices, which can be induced by veratridine in Ca^{2+}-free medium, has been sometimes reported to be even higher than release found in Ca^{2+}-containing incubation medium (12–15). In analogy to other transmitters, veratridine-induced release of GABA in Ca^{2+}-free medium has been postulated to be an exocytotic process, which is initiated by secretion of Ca^{2+} from intracellular stores (15,16). However, there is good evidence for a different mechanism.

GABA released from nerve terminals is taken up into the cytosol of the surrounding GABA terminals by a carrier mechanism. Its driving force is an electrochemical gradient, which consists of the membrane electrical potential as well as the Na^+ gradient across the membrane (17). Veratridine-induced depolarization, which is due to an increase in intracellular Na^+ levels and the resulting reduction of the electrical potential, is enhanced in Ca^{2+}-free medium (14). In consequence, the driving force for GABA uptake into the neuron can be so much reduced (18) that another gradient comes into play, which may reverse the direction of GABA transport across the membrane (18–20). This chemical gradient is the difference between the cytosolic concentration of GABA, which is in the millimolar range, and its micromolar extracellular concentration (17,18,21).

Table 1 shows direct evidence for the crucial role of the GABA carrier in veratridine-induced release in Ca^{2+}-free medium. Two different inhibitors of GABA uptake were used: (a) SKF 89976-A inhibits the uptake of GABA by binding to the carrier without being transported (22). In this way it blocks the movement of the carrier. (b) In contrast, nipecotic acid competitively blocks the carrier without freezing its movements. It is transported by the carrier and thus may cause GABA release via heteroexchange (23). When the incubation medium contained Ca^{2+}, SKF 89976-A enhanced the veratridine-induced release of (^{3}H)GABA (Table 1). It probably did so by blocking the carrier, which caused re-uptake of exocytotically released (^{3}H)GABA. In contrast, SKF 89976-A prevented the stimulated (^{3}H)GABA release in Ca^{2+}-free incubation medium, which indicates that the GABA carrier is essential for this type of release. Nipecotic acid did not reduce the Ca^{2+}-independent release caused by veratridine, but even enhanced it (Table 1), probably by inducing heteroexchange. SKF 89976-A also blocked the veratridine-induced release of (^{3}H)GABA in the presence of nipecotic acid (Table 1), in this way confirming that a functionally intact carrier was essential for (^{3}H)GABA release under these circumstances.

GABA release caused by pump reversal may not only occur in the artifactual situation of veratridine-induced depolarization in Ca^{2+}-free medium, although it is most evident under these conditions. Instead, Na^+ influx during the normal action potential may also cause pump reversal and thus contribute to the release process, as has been suggested recently (17). Indeed, the electrically evoked release of (^{3}H)GABA from slices of rat striatum has been re-

TABLE 1. *Effects of SKF 89976-A (10 μM) and nipecotic acid (NIP) (0.5 μM) on release of (^{3}H)GABA as induced by veratridine (2.5 μM) from slices of rat neostriatum*[a]

Ca^{2+} (1.3 mM)-containing incubation medium:	
Ver	2.1 ± 0.4 (5)
Ver + SKF 89976-A	5.9 ± 1.2 (5)*
Ca^{2+}-free incubation medium:	
Ver	1.6 ± 0.4 (7)
Ver + NIP	5.2 ± 0.6 (4)*
Ver + SKF 89976-A	0.3 ± 0.1 (5)*
Ver + NIP + SKF 89976-A	0.6 ± 0.2 (4)

[a]After dissection, slices were incubated in medium composed of (mM): NaCl 118, KCl 4.8, $CaCl_2$ 1.3, $NaHCO_3$ 25, $MgSO_4$ 1.2, KH_2PO_4 1.2, glucose 11 (bubbled with 5% O_2 and 95% CO_2). Slices were loaded with ^{3}H-glutamine (34 nM) and distributed into plastic tubes. After 12-min prestimulation period (to measure basal release), they were stimulated for 12 min with veratridine. NIP and SKF 89976-A [which was a gift from Dr. W.E. Bondinell (Smith, Kline and French, Philadelphia, PA)] were present during prestimulation and stimulation periods. Incubation media of release experiments (i.e., basal and stimulated release), as well as homogenates of the slices, were extracted with the use of DOWEX 50 WX-4 column chromatography, using modification of the procedure described by Kemel et al. (53). To 0.95-ml aliquots of incubation media, 8 ml Triton X-100 (0.1% in distilled water) was added, and pH was adjusted to 4 with acetic acid. After centrifugation of homogenates, supernatants were adjusted to pH 4 with Na acetate after 8 ml Triton X-100 (0.1%) had been added. DOWEX 50 WX-4 resin was extensively washed, and pH was adjusted to 4; then columns were poured with approximately 2 ml of slurry and washed with Triton X-100 (0.1%). Columns were loaded with samples, washed with 10 ml of Triton X-100 (0.1%), and eluted with 2 ml of tris(hydroxymethyl)aminomethan (pH 7.4; 0.4 M). To the eluted radioactivity, 12 ml of Lumagel was added, and samples were measured in a β-scintillation counter. Eluates have been previously shown by high-pressure liquid chromatography analysis to contain more than 93% (^{3}H)GABA (34). Basal and stimulated release of (^{3}H)GABA are expressed as fractional rate of release with use of the following calculation: dpm of prestimulation period = A; dpm of stimulation period = B; dpm in slices after stimulation period = C. Fractional rate of basal release = $100 \times A : (A + B + C)$. Fractional rate of stimulated release = $100 \times B : (B + C)$. Student's t-test was used for statistical evaluation. *, significantly different from the respective controls ($p<0.05$). For explicit description of experimental procedures, see ref. 34.

ported not to be abolished by the lack of Ca^{2+} in the medium (24). In horizontal cells of catfish retina, GABA release via reversal of the transport direction of the carrier has been postulated to play a physiological role (25).

The hypothesis that the degree of membrane polarization determines the direction of net transport (17,18) would end the discussion on the predominant function of the GABA carrier, which was started by Martin in 1976 (26). The carrier would be neither restricted to the task of taking up GABA into the terminal to terminate the latter's action at postsynaptic sites nor would the carrier mainly release GABA from the cytosolic compartment of the neurons. Instead, the pump would have both functions. It would contribute to release of GABA during the depolarization of the membrane (i.e., during action potential). In addition, it would also take up GABA into the cell after repolarization of the membrane, thus terminating the action of the transmitter.

MODULATION OF Ca^{2+}-INDEPENDENT GABA RELEASE

The demonstration that Ca^{2+}-independent release of (^{3}H)GABA can be regulated by other neurotransmitters or neuromodulators would represent further, albeit, indirect evidence for a physiological relevance of nonexocytotic GABA release. Therefore, the following experiments were performed.

Dopamine is a transmitter of generally acknowledged relevance in the striatum. It exerts its effects via D_1 and D_2 receptors (27,28). To test whether dopamine-receptor stimulation affects the carrier-mediated release of GABA from slices of rat neostriatum, the D_2 receptor agonist RU 24926 (29) and the D_1 receptor agonist SKF 38393 (30) were used.

Although 0.01 μM RU 24926 did not affect the Ca^{2+}-independent release of (^{3}H) GABA, a concentration of 0.1 μM significantly increased the release by 33% (Fig. 1). The D_2 receptor antagonist domperidone (0.1 μM) (31) prevented this effect of RU 24926 (Fig. 1). Surprisingly, the effect of RU 24926 was also blocked by SCH 23390 (1 μM) (Fig. 1), which at this concentration blocks D_1 but not D_2 receptors (32,33).

Also the D_1 receptor agonist SKF 38393 (0.5–5 μM) enhanced the Ca^{2+}-independent release of (^{3}H)GABA from rat neostriatum slices (Fig. 2). As was to be expected, this effect was blocked by the D_1 receptor antagonist SCH 23390 (1 μM) (Fig. 2). However, also the D_2 receptor antagonist domperidone (0.1 μM) abolished the effect of SKF 38393 (Fig. 2). When used alone, neither domperidone (0.1 μM) nor SCH 23390 (1 μM) significantly affected the release of (^{3}H)GABA (data not shown).

In conclusion, in Ca^{2+}-free medium, the veratridine-induced release of (^{3}H)GABA from neostriatum slices, which is mainly from axoncollaterals of striatonigral projection neurons (34), was modulated by dopaminergic agents. Stimulation of D_1 as well as D_2 receptors by selective agonists increased the release of (^{3}H)GABA. Because the agonist at one receptor type was no longer effective, if the other receptor type was blocked by a selective antagonist, the release of (^{3}H)GABA was apparently only enhanced if striatal D_1 and D_2 receptors were simultaneously stimulated. The receptor type, which was not directly stimulated by the applied selective agonists, was probably activated by endogenous dopamine, which itself was released by the veratridine

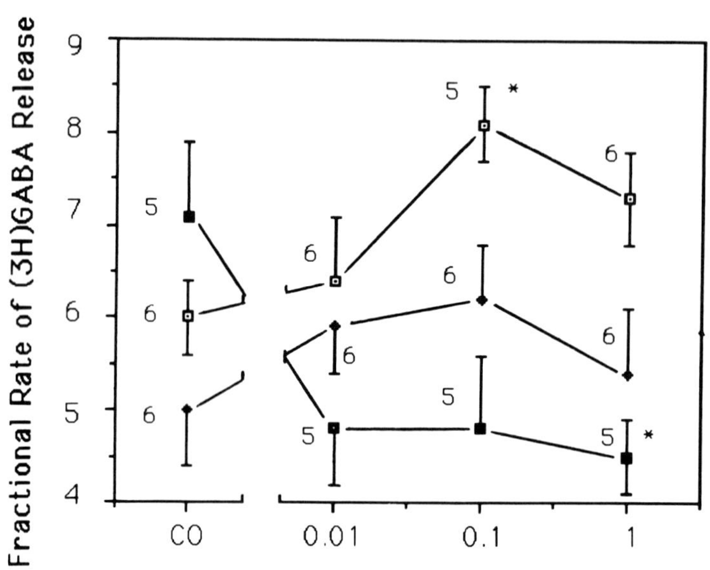

FIG. 1. Effects of RU 24926 (0.01–1 μM) on release of (^{3}H) GABA from slices of rat neostriatum as stimulated by veratridine (3 μM). The incubation medium did not contain Ca^{2+}. RU 24926 was used alone (□——□) or in the presence of domperidone (0.1 μM) (◆——◆) or SCH 23390 (1 μM) (■——■). Release is shown as fractional rate (for further explanation, see legend of Table 1). $\bar{x} \pm$ SEM, statistical evaluation was performed with Student's *t*-test and shows difference to respective control ($p<0.05$). Numbers show *n*.

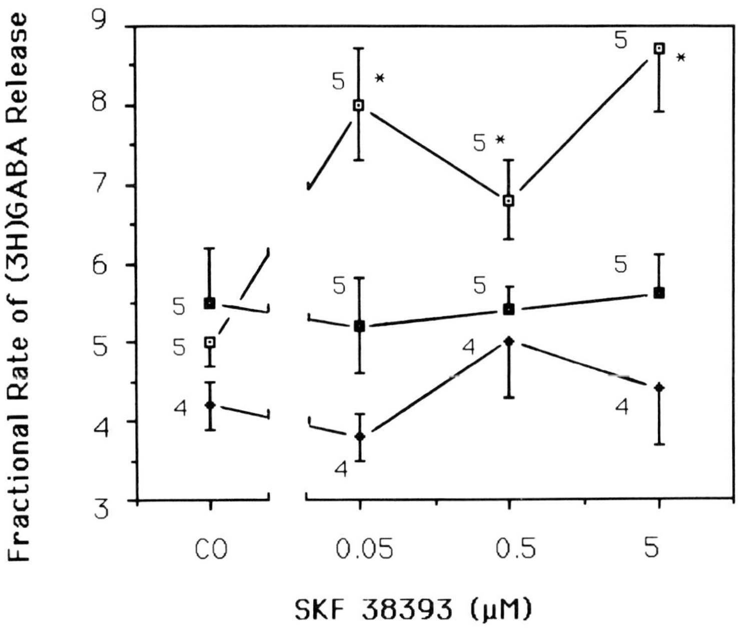

FIG. 2. Effects of SKF 38393 (0.05–5 μM) on release of (^{3}H) GABA from slices of rat neostriatum as stimulated by veratridine (3 μM). The incubation medium did not contain Ca^{2+}. SKF 38393 was used alone (□ ——— □) or in the presence of domperidone (0.1 μM) (◆ ——— ◆) or SCH 23390 (1 μM) (■ ——— ■). Release is shown as fractional rate (for further explanation, see legend of Table 1). × ± SEM, statistical evaluation was performed with Student's *t*-test and shows difference to respective control ($p<0.05$). Numbers show *n*.

stimulus. Thus, these data show that dopamine receptors of the D$_1$ and D$_2$ type synergistically affect the release of (^{3}H)GABA from axoncollaterals of striatonigral projection neurons.

For three reasons, the results obtained in this striatum preparation with dopamine cannot be directly compared with those of *in vivo* studies (35–38): (a) The neurons within the slice are deafferentated to a large extent; (b) secretion of all exocytotically released agents, such as neuropeptides, which seem to have an important role in the striatum, is nearly absent in Ca^{2+}-free incubation medium; and (c) those neurons that are still able to release their transmitter (e.g., amino acids or biogenic amines) are simultaneously stimulated in this activity by veratridine. In view of these restrictions, it is surprising that our results are in agreement with a recent hypothesis that is based on *in vivo* studies. It postulates that simultaneous stimulation of D$_1$ and D$_2$ receptors is necessary for certain types of motor behavior (39,40).

How do the present data compare with previous *in vitro* studies? These studies on incubated striatal slices have unequivocally shown that activation of dopamine D$_2$ receptors decreases the release of (^{3}H)GABA (41,42). A decrease in release of (^{3}H)GABA was also observed in the present experiments, if D$_2$ receptors were selectively stimulated (Fig. 1).

There is evidence that stimulation of D$_1$ receptors can enhance the release of (^{3}H)GABA from neostriatum slices (42). However, in contrast to our data, D$_1$ receptor activation also increased the release when D$_2$ receptors were simultaneously blocked by sulpiride (42). This difference is probably due to the different incubation media used.

In conclusion, our data do not only show

a clear-cut synergism of both receptor types but also that dopamine can decrease as well as increase the release of (^{3}H)GABA, depending on the absence or presence of this synergism. It is tempting to speculate that also *in vivo* dopamine has similar multiple effects in the striatum. In the past years, our knowledge of the intracellular effects of dopamine has been considerably improved. Thus, dopamine can activate the adenylate cyclase via stimulation of D_1 receptors and also inhibit it via D_2 receptor activation (43). In addition, dopamine D_2 receptors can also activate a subtype of K^+ channels and thus cause hyperpolarization (44) as well as reduce intracellular Ca^{2+} levels (45). An indirect connection with the phosphatidyl-4,5-bisphosphate system has been postulated to be present in pituitary cells (11). However, the yet known intracellular consequences of stimulation of both dopamine receptor types cannot be stringently connected with the mechanisms that bring about carrier-mediated release of (^{3}H)GABA.

CONCLUDING REMARKS

For the evaluation of the significance of nonexocytotic release of (^{3}H)GABA, the size of the cytosolic pool of (^{3}H)GABA is of obvious importance. If it has been artifactually enlarged, the amount of (^{3}H)GABA that is available for release by the carrier will be enhanced, and the importance of the latter for release will be overestimated. This mistake may occur if labeling of nerve terminals is performed by their incubation with (^{3}H)GABA, because it has been reported that even after a 120-min equilibration period (^{3}H)GABA had not been properly taken up by vesicles, but had been mainly accumulated in the cytosol (46). However, the storage vesicle, which have been reported only recently to have a considerable capacity for endogenous GABA (47,48), can be assumed to be responsible for exocytotic release (46). For this reason, in the present experiments, the slices were preloaded with (^{3}H)glutamine. It is taken up by various cells (49,50) but converted to (^{3}H)GABA only in GABA neurons, because they alone contain the enzyme glutamate decarboxylase (51). Newly synthetized (^{3}H)GABA is distributed into the vesicular and cytoplasmic compartment in the same way as is the respective unlabeled GABA. Thus it is not surprising that both types of GABA do not differ in the characteristics of their release (52). Therefore, the release of (^{3}H)GABA observed in these experiments should reflect that of endogenous GABA, and dopamine should also modulate the release of endogenous GABA from the cytosolic pool by depolarization-induced pump reversal.

By definition, nonexocytotic release of GABA is independent of storage vesicles. However, as has been shown, it is strictly dependent on the functional activity of the membrane carrier. Because the carrier sites may not only be in the area of synapses but along the axon and terminals, this type of GABA release may occur also in extrasynaptic areas. Consequently, GABA could qualify as a transmitter in "volume transmission."

REFERENCES

1. Catterall WA. Inhibition of voltage-sensitive sodium channels in neuroblastoma cells by antiarrhythmic drugs. *Mol Pharmacol* 1981;20:356–362.
2. Ashley RH. External calcium, intrasynaptosomal free calcium and neurotransmitter release. *Biochem Biophys Acta* 1986;854:207–212.
3. Adam-Vizi V, Ashley RH. Relation of acetylcholine release to Ca^{2+} uptake and intraterminal Ca^{2+} concentration in guinea-pig cortex synaptosomes. *J Neurochem* 1987;49:1013–1021.
4. Adam-Vizi V, Ligeti E. Release of acetylcholine from rat brain synaptosomes by various agents in the absence of external calcium ions. *J Physiol* 1984;353:505–521.
5. Schoffelmeer ANM, Mulder AH. ^{3}H-noradrenaline release from brain slices induced by an increase in the intracellular sodium concentration: role of intracellular calcium stores. *J Neurochem* 1983;40:615–621.
6. Crompton M, Moser R, Ludi M, Carafoli E. The

interrelations between the transport of sodium and calcium in mitochondria of various mammalian tissues. *Eur J Biochem* 1978;82:25–31.
7. Silbergeld EK. Na^+ regulates release of Ca^{2+} sequestered in synaptosomal mitochondria. *Biochem Biophys Res Commun* 1977;77:464–469.
8. Gershengorn MC, Thaw C. Calcium influx is not required for TRH to elevate free cytoplasmic calcium in GH_3 cells. *Endocrinology* 1983; 113:1522–1524.
9. Rebecci MJ, Gershengorn MC. Thyroliberin stimulates rapid hydrolysis of phosphatidylinositol 4,5-bisphosphate by a phosphodiesterase in rat mammotropic pituitary cells. *Biochem J* 1983; 216:287–291.
10. Martin TFJ. Thyrotropin-releasing hormone rapidly activates the phosphodiester hydrolysis of polyphosphoinositides in GH_3 pituitary cells. *J Biol Chem* 1983;258:14816–14822.
11. Vallar L, Meldolesi J. Mechanisms of signal transduction at the dopamine D_2 receptor. *TIPS* 1989;10:74–77.
12. Benjamin AM, Quastel JH. Locations of amino acids in brain slices from the rat. *Biochem J* 1972;128:631–646.
13. Szerb JC. Relationship between Ca^{2+}-dependent and independent release of ^{3}H-GABA evoked by high K^+, veratridine, or electrical stimulation from rat cortical slices. *Neurochemistry* 1979; 32:1565–1573.
14. Levi G, Gallo V, Raiteri M. A re-evaluation of veratridine as a tool for studying the depolarization-induced release of neurotransmitters from nerve endings. *Neurochem Res* 1980; 5:281–295.
15. Cunningham J, Neal MJ. On the mechanism by which veratridine causes a Ca^{2+}-independent release of GABA from brain slices. *Br J Pharmacol* 1981;73:655–667.
16. Sandoval ME. Sodium-dependent efflux of (^{3}H) GABA from synaptosomes is probably related to mitochondrial calcium mobilization. *J Neurochem* 1980;35:915–921.
17. Erecinska M. The neurotransmitter amino acid transport systems: a fresh outlook on an old problem. *Biochem Pharmacol* 1987;36:3547–3555.
18. Sihra T, Scott IG, Nicholls DG. Ionophore A23187, verapamil, protonophores, and veratridine influence the release of γ-aminobutyric acid from synaptosomes by modulation of the plasma membrane potential rather than the cytosolic calcium. *J Neurochem* 1984;43:1624–1630.
19. Hammerstad JP, Cutler RWP. Sodium ion movements and the spontaneous and electrically stimulated release of (^{3}H)GABA and (^{14}C)glutamic acid from rat cortical slices. *Brain Res* 1972; 47:401–413.
20. Haycock JW, Levy WB, Denner LA, Cotman CW. Effects of elevated $(K^+)_o$ on the release of neurotransmitters from cortical synaptosomes: efflux or secretion? *J Neurochem* 1978;30:1113–1125.
21. Berger SJ, Carter JG, Lowry OH. The distribution of glycine, GABA, glutamate and aspartate in rabbit spinal cord, cerebellum and hippocampus. *J Neurochem* 1977;28:149–158.
22. Larsson OM, Falch E, Krogsgaard-Larsson P, Schousboe A. Kinetic characterization of inhibition of γ-aminobutyric acid uptake into cultured neurons and astrocytes by 4,4-diphenyl-3-butenyl derivatives of nipecotic acid and guvacine. *J Neurochem* 1988;50:818–823.
23. Szerb JC. Effect of nipecotic acid, a γ-aminobutyric acid transport inhibitor, on the turnover and release of γ-aminobutyric acid in rat cortical slices. *J Neurochem* 1982;39:850–858.
24. Bernath S, Zigmond MJ. Characterization of (^{3}H)GABA release from striatal slices: evidence for a calcium-independent process via the GABA uptake system. *Neuroscience* 1988;27: 563–570.
25. Schwartz EA. Depolarization without calcium can release γ-aminobutyric acid from a retinal neuron. *Science* 1987;238:350–355.
26. Martin DL. Carrier-mediated transport and removal of GABA from synaptic regions. In: Roberts E, Chase TN, Tower DB, eds. *GABA in nervous system function*. New York: Raven Press, 1976;347–386.
27. Kebabian JW, Calne DB. Multiple receptors for dopamine. *Nature* 1979;277:93–96.
28. Leff S, Creese I. Dopamine receptors reexplained. *TIPS* 1983;463–467.
29. Euvrard C, Ferland L, Diapolo T, et al. Activity of two new potent dopaminergic agonists at the striatal and anterior pituitary levels. *Neuropharmacology* 1980;19:379–386.
30. O'Boyle KM, Waddington JL. Effects of the R- and S-enantiomers of the SK&F 38393 on ^{3}H-piflutixol and ^{3}H-spiperone binding to D_1 and D_2 dopaminergic sites. *Br J Pharmacol* 1983;80: 655P.
31. Baudry M, Martres MP, Schwartz JC. ^{3}H-Domperidone: a selective ligand for dopamine receptors. *Naunyn Schmiedebergs Arch Pharmacol* 1979;308:231–237.
32. Hyttel J. SCH 23390—the first selective dopamine D-1 antagonist. *Eur J Pharmacol* 1983; 91:153–154.
33. Hyttel J. Functional evidence for selective dopamine D-1 receptor blockade by SCH 23390. *Neuropharmacology* 1984;23:1395–1401.
34. Meyer DK, Conzelmann U, Schultheiss K. Effects of somatostatin-14 on the *in vitro* release of (^{3}H)GABA from slices of rat caudatoputamen. *Neuroscience* 1989;28:61–68.
35. Van der Heyden JAM, Venema K, Korf J. *In vivo* release of endogenous GABA from rat striatum: inhibition by dopamine. *J Neurochem* 1980;34:1338–1341.
36. Tossman U, Ungerstedt U. The effect of apomorphine and pergolide on the potassium-evoked overflow of GABA in rat striatum studied by microdialysis. *Eur J Pharmacol* 1986; 123:295–298.
37. Girault JA, Spampinato U, Savaki HE, Glowinski J, Besson MJ. *In vivo* release of (^{3}H) γ-aminobutyric acid in the rat neostriatum-I.

Characterization and topographical heterogeneity of the effects of dopaminergic and cholinergic agents. *Neuroscience* 1986;19:1101–1108.
38. Girault JA, Spampinato U, Glowinski J, Besson MJ. *In vivo* release of (^{3}H) γ-aminobutyric acid in the rat neostriatum-II. Opposing effects of D1 and D2 dopamine receptor stimulation in the dorsal caudate putamen. *Neuroscience* 1986; 19:1109–1117.
39. Robertson GS, Robertson HA. D_1 and D_2 dopamine agonist synergism: separate sites of action? *TIPS* 1987;295–299.
40. Walters JR, Carlson JH, Weick BG, Bergstrom DA. Neurophysiological examination of the role of D1-dopamine receptors in the regulation of neuronal activity in the basal ganglia. In: Goldstein M, Fuxe K, Tabachnick I, eds. *Central D_1 dopamine receptors*. New York: Plenum Publishing, 1988;145–158.
41. Brase DA. Pre- and postsynaptic striatal dopamine receptors: differential sensitivity to apomorphine inhibition of (^{3}H)dopamine and (^{14}C) GABA release *in vitro*. *J Pharm Pharmacol* 1980;32:432–433.
42. Bernath S, Zigmond MJ. Dopamine may influence striatal GABA release via three separate mechanisms. *Brain Res* 1989;476:373–376.
43. Stoof JC, Kebabian JW. Opposing roles for D-1 and D-2 dopamine receptors in efflux of cyclic AMP from rat neostriatum. *Nature* 1981;294: 366–368.
44. Lacey MG, Mercuri NB, North RA. On the potassium conductance increase activated by $GABA_B$ and dopamine D_2 receptors in rat substantia nigra neurones. *J Physiol* 1988;401:437–453.
45. Fujiwara H, Kato N, Shuntoh H, Tanaka C. D_2-dopamine receptor-mediated inhibition of intracellular Ca^{2+} mobilization and release of acetylcholine from guinea-pig neostriatal slices. *Br J Pharmacol* 1987;91:287–297.
46. Sihra TS, Nicolls DG. 4-Aminobutyrate can be released exocytotically from guinea-pig cerebral cortical synaptosomes. *J Neurochem* 1987;49: 261–267.
47. Fykse EM, Fonnum F. Uptake of γ-aminobutyric acid by a synaptic vesicle fraction isolated from rat brain. *J Neurochem* 1988;50:1237–1242.
48. Hell JW, Maycox PR, Stadler H, Jahn R. Uptake of GABA by rat brain synaptic vesicles isolated by a new procedure. *EMBO J* 1988;7:3023–3029.
49. Bradford HF, Ward HK, Thomas AJ. Glutamine—a major substrate for nerve endings. *J Neurochem* 1978;30:1453–1459.
50. Shank RP, Aprison MH. Glutamine uptake and metabolism by the isolated toad brain: evidence pertaining to its proposed role as a transmitter precursor. *J Neurochem* 1977;28:1189–1196.
51. McLaughlin BJ, Wood JG, Saito X, et al. The fine structural localization of glutamate decarboxylase in synaptic terminals of rodent cerebellum. *Brain Res* 1974;76:377–391.
52. Szerb JC. The release of (^{3}H)GABA formed from (^{3}H)glutamate in rat hippocampal slices. *Neurochem Res* 1983;8:341–351.
53. Kemel ML, Gauchy C, Glowinsky J, Besson MJ. Spontaneous and potassium-evoked release of (^{3}H)GABA newly synthetized from (^{3}H)glutamine in slices of rat substantia nigra. *Life Sci* 1979;24:2139–2150.

Volume Transmission in the Brain: Novel Mechanisms for Neural Transmission, edited by K. Fuxe and L.F. Agnati. Raven Press, Ltd., New York © 1991.

35

Release Properties of Intracerebrally Grafted Cholinergic and Monoaminergic Neurons as Studied by the Microdialysis Technique

Peter Kalén, Ola G. Nilsson, Olle Lindvall, and Anders Björklund

Section of Neurobiology, Department of Medical Cell Research, University of Lund, S-223 62 Lund, Sweden

The intracerebral microdialysis technique (1) has been successfully applied during recent years to monitor extracellular levels of a large number of compounds, including many transmitter substances and their metabolites (for review, see ref. 2). This method involves collection of perfusates from a dialysis tube implanted in a discrete region of the brain and subsequent analysis of the perfusates, using highly sensitive detection procedures such as high performance liquid chromatography (HPLC) with electrochemical or fluorimetric detection or radioenzymatic assays (see, e.g., refs. 1,2). Dopamine (DA) was the first transmitter to be studied with this technique (3), and today basal extracellular levels of virtually all the "classical" transmitter substances have been studied in this way (2). The relation between extracellular levels of DA and dopaminergic synaptic activity has been explored in several previous studies (see, e.g., refs. 4,5), whereas considerably less has been reported regarding the extracellular levels of acetylcholine (ACh) norepinephrine (NA), and serotonin (5-hydroxytryptamine; 5-HT).

In a series of experiments in intact and denervated rats, we have investigated the relation between extracellular levels of ACh, NA, and 5-HT and synaptic activity (6–11). The results from these and other studies (12–15) point toward a close correlation between extracellular levels of ACh, NA, and 5-HT and cholinergic, noradrenergic, and serotonergic synaptic activity. These observations and similar data from studies on striatal DA release (for review, see ref. 2) have provided the basis for the present series of experiments on ACh and monoamine release from transplanted neurons.

Previous studies have shown that cholinergic and monoaminergic neurons can survive grafting to adult host recipients (16). The implanted neurons grow to form extensive terminal networks in large areas of the surrounding host brain, and under some circumstances, the newly formed graft efferents have been shown to establish ultrastructurally normal synaptic contacts with elements in the host. Electrophysiological studies indicate that the cholinergic and monoaminergic graft–host interactions are functional, and there are data to suggest that such intracerebral transplants can correct for lesion-induced behavioral deficits (17,18).

In the present series of experiments, we have used the microdialysis technique to monitor ACh, NA, 5-HT, and DA release from grafts implanted in a previously denervated hippocampus or striatum. The experiments were aimed at investigating the degree of functional integration of the graft in the host brain by addressing the following questions: To what extent is the graft able to restore normal extracellular transmitter levels in the graft-reinnervated region? Is the transmitter overflow from the graft action potential dependent? Do the grafted neurons respond in a way similar to the innate ACh or monoamine innervation to transmitter selective drugs? To what extent can the graft-dependent transmitter overflow be modulated by influence from the host brain? At the present stage, the last question has been addressed only for cholinergic and noradrenergic grafts.

IMPLANTATION OF FETAL CHOLINERGIC AND MONOAMINERGIC NEURONS IN THE DENERVATED HIPPOCAMPUS AND STRIATUM

The cholinergic innervation of the hippocampal formation is known to originate in the septal-diagonal band (SD) area and reach the hippocampus primarily via the fimbria-fornix pathways (19). By disconnecting the hippocampus from the SD area through an aspirative lesion of the fimbria-fornix, the cholinergic innervation is reduced by approximately 90% (20,21). This denervation procedure has provided a highly useful model for the study of functional aspects of cholinergic implants (16,17). The cholinergic neurons to be implanted were obtained from the SD area of 15- to 16-day-old rat fetuses of the same strain. They were either injected stereotactically as a cell suspension into the denervated hippocampal formation 1 to 2 weeks postlesion or implanted as a solid piece of tissue into the cavity produced by the lesion.

The noradrenergic and serotonergic innervation of the hippocampus arises from the locus ceruleus (LC) (22) and mesencephalic raphe nuclei (MR) (23), respectively. By intraventricular injection of neurotoxins (6-hydroxydopamine for NA and 5,7-dihydroxytryptamine for 5-HT), the hippocampus can be selectively deprived of more than 90% of its noradrenergic and serotonergic innervation (24,25). Two to 4 weeks after neurotoxin treatment, grafts prepared from the LC or MR region of 13- to 14-day-old rat fetuses were injected as cell suspensions into the hippocampal formation.

The striatum was deprived of its dopaminergic innervation by a unilateral injection of 6-hydroxydopamine in the nigrostriatal bundle. The degree of DA depletion was checked for by amphetamine-induced rotation test (26). Cell suspension grafts were prepared from the substantia nigra (SN) of 15- to 16-day-old fetuses and injected into the denervated striatum. Only rats in which the dopaminergic grafts induced a significant improvement in the rotation test were used in the dialysis experiments.

DIALYSIS PROCEDURE

Dialysis loops were made from flexible cellulose tubing with an outer diameter of 0.3 mm (in the wet state) and a molecular weight cutoff of about 10,000. Either 8 or 4 mm of semipermeable membrane (covering a dorsoventral distance of 4 and 2 mm, respectively) were exposed to the brain tissue (Fig. 1). The *in vitro* recovery across the dialysis membrane for the 4-mm loop is typically 18% to 19%, and for the 2 mm loop, approximately 10%.

Dialysis probes were implanted 1 to 2 days before the experiment and secured to the scull using jeweler screws and dental acrylic. The probe was placed in an area innervated by the graft but was usually not in immediate apposition to the cell bodies of the graft. Intact and lesion-only animals

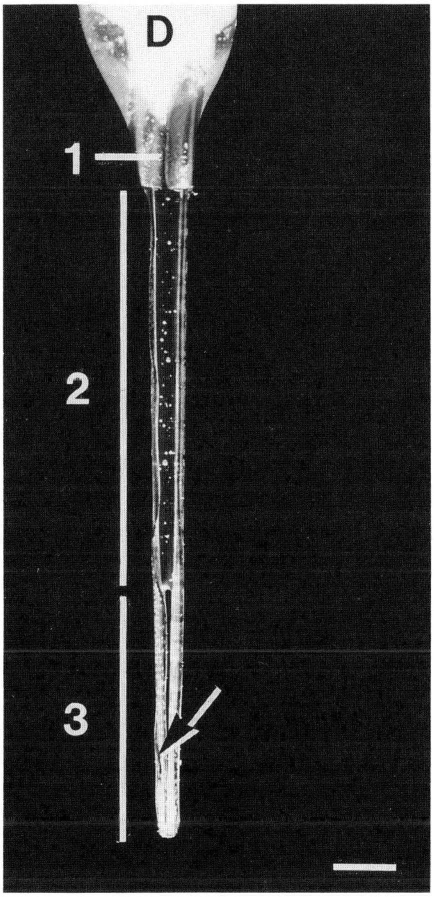

FIG. 1. Single-loop dialysis probe with 8 mm of semipermeable membrane (covering a dorsoventral distance of 4 mm). (1) Inlet and outlet stainless steel cannula attached parallel to each other with a drop of dental acrylic (D); (2) dialysis membrane covered by epoxy glue and supported by tungsten wires inserted into tubing. After implantation into the brain, the tungsten wires, which pass all the way through stainless steel cannulae, were pulled up; (3) semipermeable segment of dialysis membrane. Note suture thread (*arrow*) inserted to keep the loop open at the hairpin bend. Bar = 1 mm.

served as controls. For dialysis experiments under awake conditions (all ACh experiments; NA release monitored during handling), the rats were placed individually in plastic cages, and the probe inlet tubing connected to a liquid swivel, which was attached to a counterbalancing arm. Experiments in unconscious animals were performed under halothane anesthesia with the body temperature maintained at 37°C using incandescent light. Probes were perfused with a physiological Ringer solution at a constant flow rate of 2 μl/min, and samples were collected in 15- to 60-min fractions. ACh, NA, and DA samples were frozen in liquid nitrogen; indole samples were protected from air oxidation by L-cysteine and stored on ice before assay.

Perfusate was assayed for NA or DA by a catechol-*O*-methyltransferase (COMT)-based radioenzymatic method (27,28) that combined incubation and extraction steps from ref. 29 and separation steps from ref. 30. For a detailed description, see ref. 6. ACh, choline, 5-HT, and 5-hydroxyindoleacetic acid (5-HIAA) were assayed using HPLC with electrochemical or fluorimetric detection, respectively. Although extracellular ACh was possible to detect in most animals without acetylcholine esterase (AChE) inhibition, they were variable and near the detection limit. Therefore, 5 μM of neostigmine was added to the perfusion fluid during all manipulations. For details on the 5-HT and ACh assays, see refs. 7 and 11. Transmitter levels were expressed as moles per volume perfusate without correcting for recovery across the dialysis membrane.

EXTRACELLULAR TRANSMITTER LEVELS DURING STEADY-STATE CONDITIONS

Acetylcholine esterase staining and immunohistochemical analysis revealed that the density of graft-derived cholinergic, noradrenergic, dopaminergic, and serotonergic fibers 3 to 12 months postgrafting varied from clearly subnormal values in some of the MS, LC, and SN grafted animals to a marked hyperinnervation pattern in most of the MR grafted animals. In the hippocampus the outgrowing fibers appeared to establish a terminal pattern within the different subfields, which was similar to that of normal MS, LC, and MR innervation. Despite the variations in innervation density, the steady-state extracellular levels of ACh and monoamines recorded from the graft-reinnervated regions were similar to those obtained in intact control animals (Fig. 2; ACh levels monitored under continuous AChE inhibition). In fimbria-fornix and 6-hydroxydopamine lesioned control animals, the extracellular ACh, NA, and DA levels were close to or below the limit of detection, which suggests that the transmitters recovered in the perfusate of the transplanted rats was almost entirely graft-derived. When the septo-hippocampal pathway was transected in intact control animals during ongoing dialysis, the extracellular ACh levels were immediately and dramatically reduced, indicating that the ACh release is dependent on tonic activity of the cell bodies in the SD area. A similar instant reduction in extracellular levels of ACh was seen when a transection was made between a solid septal graft and the host hippocampus containing the dialysis probe. In the 5,7-DHT denervated hippocampi, where the tissue levels were reduced by 90% to 95%, the extracellular 5-HT levels had dropped to below the limit of detection at 2 weeks postlesion, but it had recovered to near-normal levels at 3 months. This suggests that the residual 5-HT terminals develop a compensatory hyperactivity in the chronically denervated target, making it possible to maintain a near-normal extracellular 5-HT level in the partially dener-

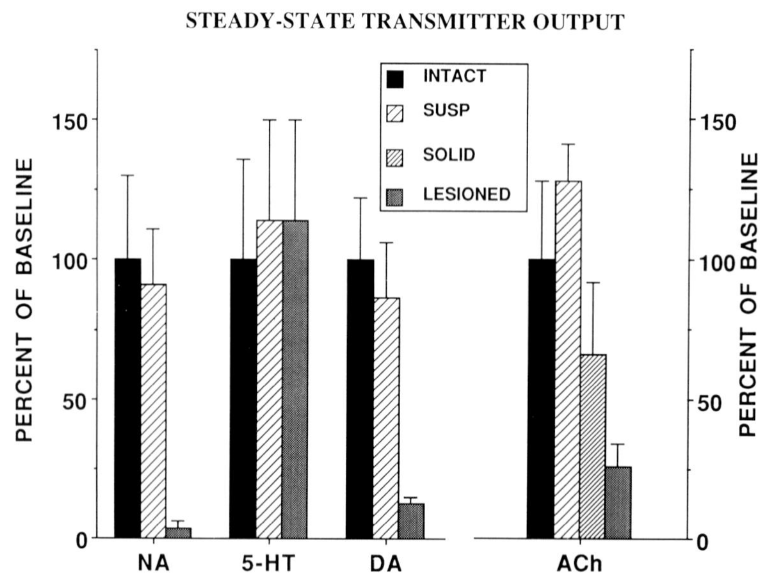

FIG. 2. Steady-state levels of monoamines and ACh in hippocampus (NA, 5-HT, ACh) and caudate-putamen (DA) of intact, grafted, and lesioned animals. Levels are expressed as percentage of levels obtained in the intact animals. Monoamines were monitored during halothane anesthesia and ACh under awake conditions. Noradrenergic, serotonergic, and dopaminergic grafts were implanted as cell suspensions. Cholinergic grafts were implanted either as solid pieces of tissue or as cell suspensions. In intact animals, NA concentration in the perfusate was 22.5 fmol/30 μl, using a 2-mm dialysis loop, and 5-HT and DA concentrations were 7.0 and 372 fmol/30 μl, respectively, using a 4-mm loop. ACh was recovered through a 2-mm loop under continuous choline esterase inhibition (neostigmine 5μM in the perfusion fluid) and amounted to 1,600 fmol/30 μl in intact animals. Values give means ± SEM of five to eight animals. [Data compiled from refs. 36 (NA), 37 (ACh), 47 (5-HT), and (Cenci MA, Kalén P, Björklund A, *unpublished observations*). (DA).]

vated area. The spared serotonin terminals were, however, unable to respond normally to pharmacological challenges (see below).

The ability of the MS, LC, MR, and SN grafts to restore steady-state extracellular ACh and monoamines to near-normal levels despite variations in intrinsic or graft-derived innervation density are in agreement with previous microdialysis studies on striatal dopaminergic grafts (31,32), and provide further evidence for the view that grafted neurons possess important autoregulatory features, despite their ectopic location, which makes them capable of compensating for abnormalities in innervation pattern. The mechanisms behind this phenomenon are only partially understood, but they probably involve autoreceptor-mediated regulation of neuronal firing rate, transmitter synthesis and release, and changes in transmitter elimination by terminal uptake (21,33).

TRANSMITTER RELEASE DURING PHARMACOLOGICAL CHALLENGES

Administration of transmitter selective uptake blockers to the perfusion fluid (desmethylimipramine 5 μM for NA, indalpine 1 μM for 5-HT, and nomifensine 10 μM for DA) induced substantial increases in extracellular monoamine levels both in the grafted and intact animals (Fig. 3). In the lesion-only animals, however, the extracellular monoamine levels during uptake blockade were still close to or below the limit of detection (Fig. 3). In the animals with hippocampal grafts of solid or cell suspension SD tissue, administration of the

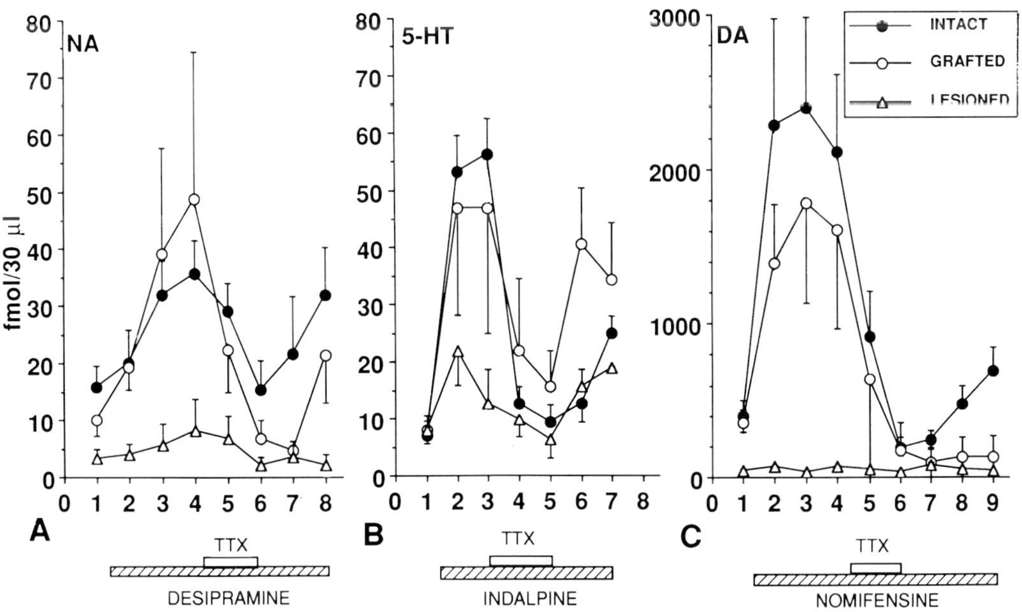

FIG. 3. Effect of transmitter selective uptake blockers and tetrodotoxin on monoamine output in hippocampus (NA, 5-HT) and caudate-putamen (DA) of intact (*filled circles*), grafted (*open circles*), and lesioned (*open triangles*) animals. Animals were anesthetized by halothane, and drugs were applied during periods indicated by *bars*. Values give means ± SEM of five to eight animals. [Data from refs. 36 (NA), 47 (5-HT), and (Cenci MA, Kalén P, Björklund A, *unpublished observations*). (DA).]

AChE inhibitor neostigmine (5 μM) through the dialysis probe increased extracellular ACh from barely detectable levels to stable steady-state levels. The increase was in the same order of magnitude as in the intact control animals, whereas in the animals with a fimbria-fornix lesion, the ACh output remained unaffected.

When the sodium channel blocker tetrodotoxin was added to the perfusion fluid at 1 μM under maintained uptake blockade or choline esterase inhibition, the same marked reduction in monoamine and ACh output was observed in the grafted as in the intact animals. The effect on monoamine output is illustrated in Fig. 3. Moreover, addition of a depolarizing concentration of KCl (100 mM) to the perfusion fluid induced dramatic increases in extracellular levels of NA and 5-HT and ACh in both intact and grafted hippocampi (Fig. 4). In contrast, the monoamine and ACh output in the lesioned control hippocampi was still low (Fig. 4).

The present data show that cholinergic and monoaminergic neurons, grafted to a previously lesioned target, can restore extracellular ACh, NA, 5-HT, and DA to near-normal levels despite variations in innervation density, not only under basal conditions, but also in the presence of an AChE inhibitor and selective uptake blockers. The dramatic increase in transmitter output obtained by depolarizing concentrations of KCl and the ability of tetrodotoxin to reduce transmitter overflow down to the levels seen in the denervated control animals indicate that the release was impulse-dependent. Taken together these findings suggest that the SD, LC, MR, and SN grafts are functionally active at the synaptic

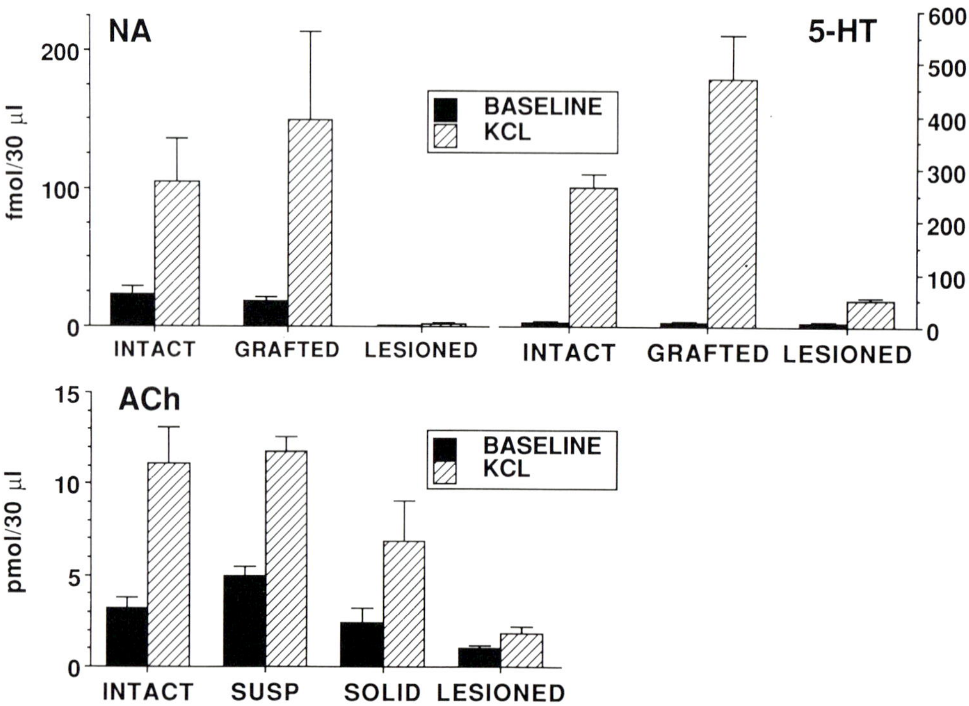

FIG. 4. Effect of membrane depolarization by KCl on hippocampal NA, 5-HT, and ACh output in intact, grafted, and lesioned animals. KCl was added to perfusion fluid at 100 mM during one dialysis sample (15 min for NA and ACh, 30 min for 5-HT). Solid, solid tissue grafts; susp, cell suspension grafts. Values give means ± SEM of five to eight animals. (Data from refs. 36,37,47.)

level and that they share at least some regulatory properties with the innate cholinergic and monoaminergic innervations.

EXTRACELLULAR LEVELS OF CHOLINE AND 5-HYDROXYINDOLEACETIC ACID

Hippocampal extracellular levels of choline, the precursor and breakdown product of ACh, were roughly 100 times higher than the extracellular ACh levels, and they were not significantly different in the normal, fimbria-fornix lesioned and SD grafted rats. Partial blockade of AChE by neostigmine reduced the choline levels by about 25% in the normal and grafted rats, whereas no effect was seen in the lesion-only rats. KCl induced an approximately 50% increase in choline levels in all three groups, whereas tetrodotoxin administration induced no clear changes in any of the groups. These observations suggest that only a small fraction of the extracellular choline was derived from intraneuronal ACh turnover and that most of the choline detected in the dialysis perfusate emanates from other sources such as the blood circulation (34).

In intact animals the hippocampal extracellular levels of 5-HIAA, the principal metabolite of serotonin, were approximately 1,000 times higher than the extracellular levels of serotonin. In the neurotoxin-treated animals, the 5-HIAA levels were reduced by more than 90% compared with normal, whereas in the MR grafted rats, the extracellular 5-HIAA levels were two to four times higher than in the intact control animals. These supranormal levels are compatible with the higher than normal density of serotonergic terminals formed by the grafted neurons in the host hippocampal formation as observed in tissue sections prepared for serotonin immunohistochemistry. Interestingly, the concentration of 5-HIAA in the perfusate represented a similar fraction of the surrounding tissue levels of 5-HIAA in both normal, lesioned, and grafted hippocampi (i.e., 10–15% under baseline conditions). Administration of an uptake blocker or tetrodotoxin induced no detectable changes in extracellular 5-HIAA in any of the groups, whereas KCl reduced the levels in the intact and grafted hippocampi. These observations are compatible with the view that extracellular levels of 5-HIAA are a poor indicator of synaptic activity and primarily reflect passive, unregulated diffusion of an intraneuronally formed metabolite (7–9).

HOST BRAIN INFLUENCE OVER HIPPOCAMPAL SD AND LC GRAFTS

Two different stimuli, electrical stimulation of the lateral habenula (LHb) at 15 Hz (0.3–0.5 mA) and handling of the rats (i.e., gentle stroking of the fur and tail) were applied to study the degree of functional integration of the SD and LC grafts in the host hippocampus.

The habenular complex, which projects through the fasciculus retroflexus to various regions in the upper brainstem including the interpeduncular nucleus, ventral tegmental area, substantia nigra, dorsal and median raphe nuclei, LC, and to a minor extent to the diagonal band area in the basal forebrain (for review, see ref. 35), has previously been identified as a powerful activator of hippocampal NA and ACh release (10,11). In normal animals kept under halothane anesthesia, the effect of LHb stimulation on hippocampal NA and ACh release seems to be attributed to a distinct activation of the habenular complex and its descending projections, because it is abolished by selective transection of the fasciculus retroflexus (36,37). Moreover, the effect seems to be relayed through the LC, because it is completely eliminated by a knife-cut transecting the ascending noradrenergic projections (i.e., the dorsal NA bundle) immediately rostral to the LC (10). The effect of habenular stimulation on ACh release has been studied under awake conditions. The route by which the stimulatory effect is mediated in the normal animal is not known. It is possible that the increase in ACh release during habenular stimulation is the result of a direct activation of cholinergic cell bodies in the diagonal band area. Alternatively, the effect may be mediated via the brainstem reticular formation, possibly via the ascending serotonergic and noradrenergic projections (38,39). It is, however, not mediated through a local spread of electricity to the cholinergic terminals in the hippocampal formation, because it is abolished by selective transection of the fasciculus retroflexus.

Handling of the animal under awake conditions may be regarded a complex environmental stimulus, which comprises a mixture of somatosensory stimulation, stress (40), and behavioral activation. Previous studies have shown that neurons of the SD and LC are activated by nonnoxious environmental stimuli (41–44). In normal animals, handling for 15 min induce a consistent increase in hippocampal NA and ACh output. For NA the response is rather small (50%) and transient (6,9,45), whereas the increase in extracellular ACh is more pronounced (100%), and the levels remain elevated also during the first 15 min posthandling (11,37).

In the LC grafted hippocampi, LHb stimulation induced an increase in NA release, which was virtually identical to the one observed in the intact rats (Fig. 5A). In contrast, the extracellular NA in the 6-hydroxydopamine lesioned control hippocampi remained at a level close to or below the limit of detection. Handling, however, induced somewhat variable responses in the grafted hippocampi, and the mean NA output did not differ significantly from baseline levels (Fig. 5B). It is likely that handling of awake animals represents a far more complex stimulus than electrical stimulation of a single structure under anesthe-

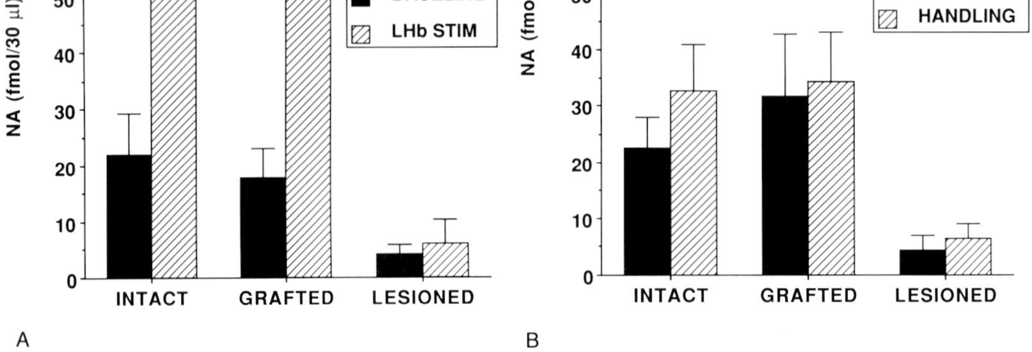

FIG. 5. Effects of electrical stimulation of the lateral habenular nucleus (15 Hz, 0.5 mA) **(A)** and handling (i.e., gentle stroking of the fur and tail) **(B)** on hippocampal NA output in intact, grafted, and lesioned animals. During habenular stimulation, animals were anesthetized with halothane, and during handling they were awake. The stimuli were applied continuously for 15 min. *Solid bars* represent steady-state NA levels, and *hatched bars* the NA levels obtained during stimulation. Values give means ± SEM of six to eight animals. (Data from ref. 36.)

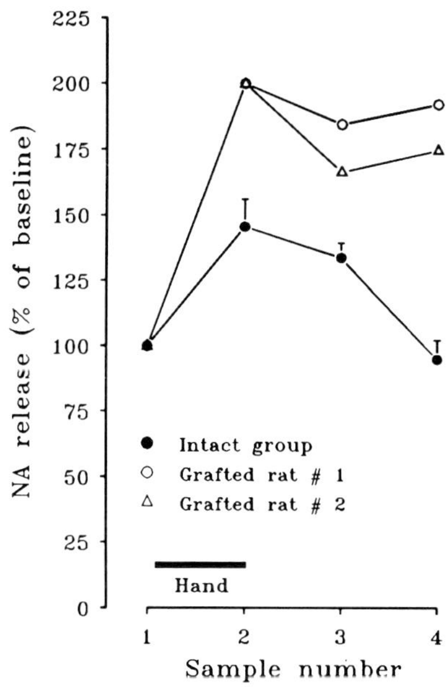

FIG. 6. Effect of handling on hippocampal NA release in two grafted animals (*open circles* and *open triangles*). The animals were awake, and handling was applied during a 15-min sample. The means ± SEM of eight intact control animals are given as a reference (*filled circles*). (Data from ref. 36.)

tized conditions. The present data may therefore indicate that the hippocampal LC grafts are under a regulatory influence by host brain afferents but that the grafts do not share the ability of the original hippocampal NA innervation to respond to complex environmental stimuli. Yet, in two of six grafted animals, which both responded also to LHb stimulation, a prompt twofold increase in NA output was obtained (Fig. 6). The response was, however, not identical to that obtained in the normal hippocampi, because the NA levels did not return to baseline levels immediately after cessation of the stimulus.

In the SD grafted animals, both habenular and handling stimulations were applied under awake conditions. Electrical stimulation of the LHb in awake animals has previously been found to induce a behavioral activation characterized by pronounced exploratory behavior, which continues as long as the stimulation is maintained (P. Kalén, *unpublished observations*). Therefore, both stimuli must be regarded as activators of multiple neuronal circuitries and transmitter systems in the brain. Similar to normal animals, both LHb stimulation and handling induced an increase in hippocampal ACh release in the SD grafted animals (Fig. 7). The effects were present both in the animals with solid grafts and in those with cell suspension grafts, and the effect was abolished after transection of the fasciculus retroflexus. These findings point toward a high level of functional integration of the cholinergic grafts in the host. The host-derived control of the septal grafts could be accomplished within the graft tissue itself and/or at the terminal level in the hippocampus. The activity of the normal cholinergic septohippocampal system is partly regulated from the brainstem reticular formation via its ascending serotonergic projections (39). These fibers terminate both in the SD area and in the hippocampal formation (23,46), which makes an interaction possible both at the cell body level and within the innervated target. Interestingly, we have found that intrahippocampal septal suspension grafts and, in particular, solid septal grafts receive a substantial ingrowth of host serotonergic fibers (37). Electrical stimulation of the LHb is known to induce an activation of ascending serotonergic projections (8), and it is possible that these projections may play a role in the activation of hippocampal ACh release in the normal rats and, perhaps also, in the SD grafted rats.

CONCLUSIONS

The present data show that the microdialysis technique provides a useful tool for the study of functional regulation of intracerebrally grafted cholinergic and monoam-

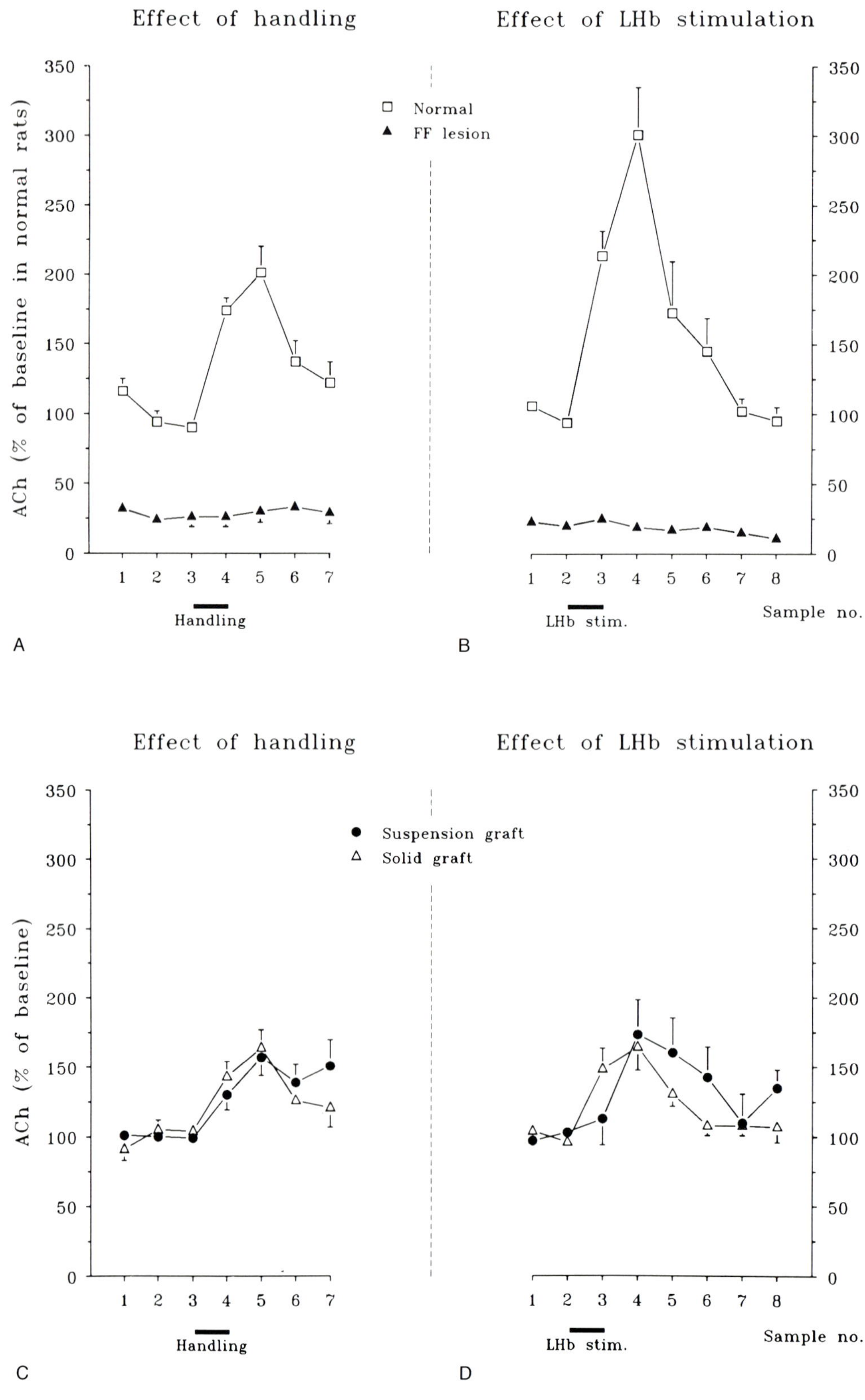
Effect of handling
Effect of LHb stimulation
ACh (% of baseline in normal rats)
Normal
FF lesion
Handling
LHb stim.
Sample no.
A
B
Effect of handling
Effect of LHb stimulation
ACh (% of baseline)
Suspension graft
Solid graft
Handling
LHb stim.
Sample no.
C
D

inergic neurons as well as for the analysis of graft–host interactions. Although the graft-derived fiber and terminal density varied substantially from one graft to another, the transmitters in the extracellular space were maintained at near-normal levels, not only under baseline conditions but also during K^+-induced depolarization, transmitter-selective uptake blockade, and tetrodotoxin. This suggests that the grafted neurons possess important autoregulatory properties despite their ectopic location. The results also show that the monoamine release in the graft-reinnervated host target is impulse-dependent and that the neurons are spontaneously functionally active at the synaptic level. Electrical stimulation of the lateral habenula (which has previously been identified as a powerful activator of the innate hippocampal cholinergic and noradrenergic innervations) produced a virtually identical increase in the release of these transmitters in the intact and grafted hippocampi. A complex environmental stimulus, such as handling, induced a consistent increase in ACh but a more variable response in NA release. These findings suggest that grafted cholinergic and noradrenergic neurons can be functionally activated by host brain inputs.

ACKNOWLEDGMENTS

We warmly thank Kerstin Fogelström, Karin Fyge, Birgit Haraldsson, and Gertrud Stridsberg for excellent technical assistance. The valuable support and technical advice of Anders Carlsson (Carnegie Medicin AB, Stockholm, Sweden) in setting up the analytical procedure for ACh is gratefully acknowledged. The present work was supported by grants from the Swedish Medical Research Council, the National Institutes of Health, the Thorsten and Elsa Segerfalk Foundation, the Greta and Johan Kock Foundation, and the Crafoord Foundation.

REFERENCES

1. Ungerstedt U. Measurement of neurotransmitter release by intracranial dialysis. In: Marsden CA, ed. *Measurement of neurotransmitter release* in vivo. *IBRO handbook series: methods in neurosciences*, vol 6. New York: John Wiley, 1984;81–107.
2. Westerink BHC, Damsma G, Rollema H, De Vries JB, Horn AS. Minireview: scope and limitations of *in vivo* brain dialysis: a comparison of its application to various transmitter systems. *Life Sci* 1987;41:1763–1776.
3. Ungerstedt U, Pycock C. Functional correlates of dopamine neurotransmission. *Bull Schweiz Acad Med Wiss* 1974;1278:1–13.
4. Zetterström T, Sharp T, Marsden CA, Ungerstedt U. *In vivo* measurement of dopamine and its metabolites by intracerebral dialysis: changes after *d*-amphetamine. *J Neurochem* 1983;41:1769–1773.
5. Imperato A, Di Chiara G. Trans-striatial dialysis coupled to reverse phase high performance liquid chromatography with electrochemical detection: a new method for the study of the *in vivo* dopamine release of endogenous dopamine and metabolites. *J Neurosci* 1984;4:966–977.
6. Kalén P, Kokaia M, Lindvall O, Björklund A. Basic characteristics of noradrenaline release in the hippocampus of intact and 6-hydroxydopamine lesioned rats as studied by *in vivo* microdialysis. *Brain Res* 1988;474:374–379.
7. Kalén P, Strecker RE, Rosengren E, Björklund A. Endogenous release of neuronal serotonin and 5-hydroxyindoleacetic acid in the caudate-putamen of the rat as revealed by intracerebral dialysis coupled to high-performance liquid chromatography with fluorimetric detection. *J Neurochem* 1988;51:1422–1435.
8. Kalén P, Strecker RE, Rosengren E, Björklund A. Regulation of striatal serotonin release by the lateral habenula-dorsal raphe pathway in the rat as demonstrated by *in vivo* microdialysis: role of excitatory amino acids and GABA. *Brain Res* 1989;492:187–202.

FIG. 7. Effects of handling (*A* and *C*) and electrical stimulation of the lateral habenular (LHb) nucleus (15 Hz, 0.3 mA) (*B* and *D*) on hippocampal ACh output in awake intact, lesioned, and grafted animals. Stimuli were applied continuously for 15 min. Values in lesioned rats are expressed in percentage of baseline in normal rats (A and B). Values give means ± SEM of five to six animals. (Data from ref. 37.)

9. Kalén P, Lindvall O, Björklund A. Electrical stimulation of the lateral habenula increases hippocampal noradrenaline release as monitored by *in vivo* microdialysis. *Exp Brain Res* 1989;76: 239–245.
10. Kalén P, Rosengren E, Lindvall O, Björklund A. Hippocampal noradrenaline and serotonin release over 24 hours as measured by the dialysis technique in freely moving rats: correlation to behavioral activity state, effect of handling and tail-pinch. *Eur J Neurosci* 1989;1:181–188.
11. Nilsson OG, Kalén P, Rosengren E, Björklund A. Acetylcholine release in the rat hippocampus as studied by microdialysis is dependent on axonal impulse flow and increases during behavioral activation. *Neuroscience,* 1990;36:325–338.
12. L'Heureux R, Dennis T, Curet O, Scatton B. Measurement of endogenous noradrenaline release in the rat cerebral cortex *in vivo* by transcortical dialysis: effects of drugs affecting noradrenergic transmission. *J Neurochem* 1986;46: 1794–1801.
13. Damsma G, Westerink BHC, de Vries JB, Van den Berg CJ, Horn AS. Measurement of acetylcholine release in freely moving rats by means of automated intracerebral dialysis. *J Neurochem* 1987;48:1523–1528.
14. Routledge C, Marsden CA. Comparison of the effects of selected drugs on the release of hypothalamic adrenaline and noradrenaline measured *in vivo*. *Brain Res* 1987;426:103–111.
15. Abercrombie ED, Jacobs BL. Systemic naloxone administration potentiates locus coeruleus noradrenergic neuronal activity under stressful but not non-stressful conditions. *Brain Res* 1988;441:362–366.
16. Björklund A, Stenevi U. Intracerebral neural implants: neuronal replacement and reconstruction of damaged circuitries. In: Cowan MW, Shooter EM, Stevens CF, Thompson RF, eds. *Annual Review of Neuroscience,* vol 7. Palo Alto, California: Annual Reviews, 1984;279–308.
17. Buzsáki G, Gage FH. Pathophysiology of the subcortically deafferented hippocampus: improvement and deterioration of function by fetal grafts. In: Gage F, Privat A, Christensen Y, eds. *Neuronal grafting and Alzheimer's disease*. Berlin: Springer-Verlag, 1989;103–119.
18. Dunnett SB. Short-term memory in rodent models of ageing: effects of cortical cholinergic grafts. In: Gage F, Privat A, Christensen Y, eds. *Neuronal grafting and Alzheimer's disease*. Berlin: Springer-Verlag, 1989;85–102.
19. Swanson LW, Mogenson GJ, Simerly RB, Wu M. Anatomical and electrophysiological evidence for a projection from the medial preoptic area to the "mesencephalic and subthalamic locomotor regions" in the rat. *Brain Res* 1987; 405:108–122.
20. Björklund A, Stenevi U. Reformation of the severed septohippocampal cholinergic pathway in the adult rat by transplanted septal neurons. *Cell Tiss Res* 1977;185:289–302.
21. Gage FH, Björklund A, Stenevi U. Local regulation of compensatory noradrenergic hyperactivity in the partially denervated hippocampus. *Nature* 1983;303:819-821.
22. Loy R, Koziell DA, Lindsey JD, Moore RY. Noradrenergic innervation of the adult rat hippocampal formation. *J Comp Neurol* 1980;189: 699–710.
23. Steinbusch HWM. Serotonin-immunoreactive neurons and their projections in the CNS. In: Björklund A, Hökfelt T, Kuhar MJ, eds. *Handbook of chemical neuroanatomy,* vol 3. Amsterdam: Elsevier Science Publishers B.V., 1984;68–125.
24. Baumgarten HG, Björklund A, Lachenmayer L, Nobin A. Evaluation of the effects of 5,7-dihydroxytryptamine on serotonin and catecholamine neurons in the rat CNS. *Acta Physiol Scand (Suppl)* 1973;391:1–19.
25. Björklund A, Nornes H, Gage FH. Cell suspension grafts of noradrenergic locus coeruleus neurons in rat hippocampus and spinal cord: reinnervation and transmitter turnover. *Neuroscience* 1986;18:685–698.
26. Ungerstedt U, Arbuthnott GW. Quantitative recording of rotational behavior in rats after 6-hydroxy-dopamine lesions of the nigrostriatal system. *Brain Res* 1970;24:485–493.
27. Da Prada M, Zürcher G. Simultaneous radioenzymatic determination of plasma and tissue adrenaline, noradrenaline and dopamine within the femtomolar range. *Life Sci* 1976;19:1161–1174.
28. Saller CF, Zigmond MJ. A radioenzymatic assay for catecholamines and dihydroxyphenylacetic acid. *Life Sci* 1978;23:1117–1150.
29. Peuler JD, Johnson GA. Simultaneous single isotope radioenzymatic assay of plasma norepinephrine, epinephrine and dopamine. *Life Sci* 1977;21:625–636.
30. Schmidt RH, Ingvar M, Lindvall O, Stenevi U, Björklund A. Functional activity of substantia nigra grafts reinnervating the striatum; neurotransmitter metabolism and [14C]-2-deoxy-D-glucose autoradiography. *J Neurochem* 1982; 38:737–748.
31. Zetterström T, Brundin P, Gage FH, et al. *In vivo* measurement of spontaneous release and metabolism of dopamine from intrastriatal nigral grafts using intracerebral dialysis. *Brain Res* 1986;362:344–349.
32. Strecker RE, Sharp T, Brundin B, Zetterström T, Ungerstedt U, Björklund A. Autoregulation of dopamine release and metabolism by intrastriatal nigra grafts as revealed by intracerebral dialysis. *Neuroscience* 1987;22:169–178.
33. Stricker EM, Zigmond MJ. Brain monoamines, homeostasis and adaptive behavior. In: Mountcastle VB, Bloom FE, Geiger SK, eds. *Handbook of physiology,* section I, vol IV: *Intrinsic regulatory systems of the brain*. Bethesda, Maryland: American Physiological Society, 1986;677–700.
34. Blusztajn JK, Wurtman RJ. Choline and cholinergic neurons. *Science* 1983;221:614–620.

35. Sutherland RJ. The dorsal diencephalic conduction system: a review of the anatomy and functions of the habenular complex. *Neurosci Behav Rev* 1982;6:1–13.
36. Kalén P, Cenci MA, Lindvall O, Björklund A. Host brain regulation of fetal locus coeruleus neurons grafted to the hippocampus of 6-hydroxydopamine treated rats as studied by intracerebral microdialysis. *Exp Neurol* 1991 (*submitted*).
37. Nilsson OG, Kalén P, Rosengren E, Björklund A. Acetylcholine release from intrahippocampal septal grafts is under control by the host brain. *Proc Natl Acad Sci USA* 1990;87:2647–2651.
38. Assaf SY, Miller JJ. The role of a raphe serotonin system in the control of septal unit activity and hippocampal desynchronization. *Neuroscience* 1978;3:539–550.
39. Vanderwolf CH. Cerebral activity and behavior: control by central cholinergic and serotonergic systems. *Int Rev Neurobiol* 1988;30:225–240.
40. Kvetnansky R, Sun CL, Lake CR, Thoa N, Torda T, Kopin IJ. Effect of handling and forced immobilization on rat plasma levels of epinephrine, norepinephrine, and dopamine-β-hydroxylase. *Endocrinology* 1978;103:1868–1874.
41. Dudar JD, Whishaw IQ, Szerb JC. Release of acetylcholine from the hippocampus of freely moving rats during sensory stimulation and running. *Neuropharmacology* 1979;18:673–678.
42. Aston-Jones G. Behavioral functions of locus coeruleus derived from cellular attributes. *Physiol Psychol* 1985;13:118–126.
43. Foote SL, Bloom FE, Aston-Jones G. The nucleus locus coeruleus: new evidence of anatomical and physiological specificity. *Physiol Rev* 1983;63:844–914.
44. Gilad GM, Mahon BD, Finkelstein Y, Koffler B, Gilad VH. Stress-induced activation of the hippocampal cholinergic system and the pituitary-adrenocortical axis. *Brain Res* 1985;347:404–408.
45. Kokaia M, Kalén P, Bengzon J, Lindvall O. Noradrenaline and 5-hydroxytryptamine release in the hippocampus during seizures induced by hippocampal kindling stimulation: an *in vivo* microdialysis study. *Neuroscience* 1989;32:647–656.
46. Azmitia EC, Segal M. An autoradiographic analysis of the differential ascending projections of the dorsal and median raphe nuclei in the rat. *J Comp Neurol* 1978;179:641–668.
47. Daszuta A, Kalén P, Strecker RE, Brundin P, Björklund A. Serotonin neurons grafted to the adult rat hippocampus. II. 5-HT release as studied by intracerebral microdialysis. *Brain Res* 1989;498:323–332.

Volume Transmission in the Brain: Novel Mechanisms for Neural Transmission, edited by K. Fuxe and L.F. Agnati. Raven Press, Ltd., New York © 1991.

36

Endogenous and Exogenous Nerve Growth Factor in the Central Nervous System

*Lars Olson, *Cynthia Wetmore, *Ingrid Strömberg, and **Ted Ebendal

**Department of Histology and Neurobiology, Karolinska Institutet, S-10401 Stockholm, Sweden; and **Department of Developmental Biology, Uppsala University, S-751 23 Uppsala, Sweden*

Nerve growth factor (NGF) is synthesized in many areas of the central nervous system (CNS), acting as a trophic signal for populations of cholinergic neurons and perhaps other neuronal systems. The reader is referred to recent reviews of NGF in CNS (1–4). Although certain cholinergic systems such as the projections from basal forebrain to hippocampus and cortex appear to be sensitive to and dependent on NGF throughout life, other systems such as cholinergic interneurons in striatum and α-motoneurons in the spinal cord (5) express the NGF receptor only during development and after injury, suggesting that they use NGF only during certain periods of development and/or after damage (5,6).

There are several reasons why it might be desirable to increase the central NGF levels. For instance, in aging (7) and Alzheimer's senile dementia (8–15), one major neuropathological finding is degeneration of the cholinergic projections from nucleus basalis to cortex cerebri. Whether or not this is due to a diminished NGF production in cortex, it might be possible to rescue cells in critical condition by addition of exogenous NGF to target areas because the remaining cholinergic neurons do express the NGF-receptor message and protein (16). Interestingly, NGF and its messenger ribonucleic acid (mRNA) both decrease in the aged rat brain (17). One novel approach for treatment of Parkinson's disease includes autografts of chromaffin tissue to the basal ganglia. We know from rat experiments that such grafted chromaffin cells survive in much greater numbers and transform to neuronal phenotypes if the grafts are supported by infusion of NGF.

In the following, we will briefly review the evidence for the presence and role of NGF in CNS and discuss strategies that may be employed to increase intracerebral NGF levels. In particular, and in the context of "volume transmission," we will describe the time course of diffusion and disappearance of NGF after local intraparenchymal injections.

NGF IS MADE BY NEURONS IN THE BRAIN

The highest concentration of NGF and NGF mRNA in the CNS is found in the hippocampal formation, but significant amounts are present also in cortex cerebri. Low levels are found, e.g., in striatum (18–20). When the regional levels of NGF, as measured by enzyme-linked immunoassays, are compared with the levels of NGF mRNA as measured by Northern blots, there is generally good agreement, suggesting that NGF has local paracrine actions. *In situ* hybridization (21) has clarified the nature of such interaction by revealing that neurons, rather than glial cells, appear to be the major (sole?) sites of synthesis of NGF in the adult unperturbed brain. Thus, strong NGF mRNA hybridization signals are seen over scattered cells in and around the pyramid cell layer and over hilar cells in the hippocampal formation (21). Similar to the interaction between neurons and their targets in the peripheral nervous system, it thus appears that the target cells of the cholinergic projections produce and secrete NGF to attract and maintain an appropriate cholinergic input. This interpretation is strengthened by studies showing that lesions that kill neurons while causing glial proliferation, decrease NGF-mRNA levels in hippocampus (21). Additional proof that NGF is indeed transferred from one neuron to another is provided by fimbria-fornix lesions. In such lesioned animals, hippocampal NGF levels increase when the cholinergic innervation is removed (22,23), presumably caused by, in part, a "pileup" of NGF in hippocampus in the absence of the normal uptake and retrograde transport by cholinergic nerve fibers.

The distribution of NGF receptor (NGF-R) has been studied extensively at both the mRNA and the protein level. High levels of NGF-R mRNA is found in cholinergic cell bodies in the basal forebrain. Their terminal networks express NGF-R-like immunoreactivity and bind labeled NGF. Although outside the scope of this review, it should, however, be mentioned that the distribution of the NGF-R appears more widespread than might have been anticipated by the regional distribution of NGF in the unperturbed adult rodent CNS (see, e.g., refs. 24,25). There are several possible explanations to this discrepancy including effects of low levels of endogenous NGF in certain areas, an increased production of NGF during development and/or after disturbances, and the possible detection limits of current techniques or cross-talk between NGF-R and receptors for other possible members of the NGF-brain-derived neurotropic factor (BDNF) family of growth factors (26).

STRATEGIES TO INCREASE CENTRAL NGF LEVELS

There are principally three different ways in which intracerebral NGF levels might be increased. The first approach, which has not yet been used experimentally, would be to attempt to increase endogenous levels of NGF. Although knowledge about factors that regulate NGF synthesis and turnover is scarce, it has recently been shown that inflammatory mediators such as lipopolysaccharide, interleukin-1-β, and prostaglandin E2 all increase NGF mRNA and protein in dissociated hippocampal structures, whereas agents that block inflammatory responses also block up-regulation of NGF mRNA and protein (27,28). It is conceivable that local treatment with inflammatory mediators might cause similar responses also *in vivo*.

The second approach to increase central NGF levels is to administer the purified growth factor. Using chronic indwelling cannulas or dialysis fibers attached to osmotic minipumps, this approach has been used to show increased survival and nerve fiber formation by grafts of chromaffin tissue (29) and to stimulate cholinergic neurons in a variety of experimental designs (see ref. 30). Although intraparenchymal infusions might be preferable to obtain localized effects, simple infusions into the lateral ventricle (31) have proven effective in promoting recovery of cholinergic biochemistry and function after excitotoxic lesions of nucleus basalis magnocellularis (32) and in rescuing axotomized septal hippocampal cholinergic neurons (33). It has indeed been shown that intraventricular infusion of NGF leads to increased levels of NGF in brain tissue such as hippocampus, as measured by two-site enzyme-linked immunoassay (34).

The relative degree of specificity that can be reached also after intraventricular infusion of NGF is attributed to the need for high-affinity NGF-R for neurons to respond. However, several other noncholinergic neuron systems have recently been suggested to be NGF-sensitive. In addition, the NGF-R is present in several circumventricular organs and areas, including area postrema (25). Thus, the intraventricular route of administration may cause a multitude of other, perhaps unwanted, effects in addition to the desired cholinotrophic activity.

The third approach to the problem of achieving long-term increases of central NGF levels is implantation of NGF-producing cell systems. There are at least two such systems of normal cells that have proven experimentally useful. Perhaps most straightforward is to use male mouse submandibular gland implants, which have been shown to stimulate cholinergic parameters such as cholinergic cell body size (35). One problem is the need for immunosuppression when the host animal is a rat; the second problem is the fact that exocrine salivary gland tissue does not sustain grafting well and may not even survive in an active state for longer periods of time. Schwann cells are also known to be a relatively rich source of NGF when deprived of nerve fibers (36). Attempts have therefore been made to use cografted peripheral nerve minces to provide support for grafts of chromaffin cells to striatum.

An interesting alternative to the use of normal cells is to transplant cell lines that have been genetically modified to secrete NGF (37,38). We have transfected mouse 3T3 fibroblasts with a large number of copies of the rat NGF gene and have shown that such transfected cells, designated 3E cells, produce large amounts of NGF mRNA and secrete NGF to the culture medium (38). Provided that the hosts are immunopressed, such cells grafted to the rat brain can induce sprouting of host cholinergic terminals in striatum and cortex, support grafts of fetal cholinergic neurons (38), and rescue septal cholinergic neurons that have been axotomized by a fimbria-fornix lesion (39). Similar rescuing effects after a fimbria-fornix lesion using a different (retroviral) transfection system and a rat cell line have been demonstrated by Rosenberg et al. (37).

The use of genetically modified cells is a promising new technique, permitting long-term local NGF support in the CNS. Several problems remain to be resolved, however, including immunological mismatch, which might be overcome by transfecting fibroblasts of host origin in primary culture. An additional problem with the established cell lines is the risk of tumor formation and, if such cells are transfected with other growth factors such as fibroblast growth factor (FGF), the possibility of direct autocrine effects of an unwanted nature on the transfected cells.

THE FATE OF INTRASTRIATAL NGF INJECTIONS

Nerve growth factor is a relatively large peptide dimer with negative charge, known to have "sticky" properties under various experimental conditions. These properties and the chances of being broken down in brain tissue prompted an immunohistochemical study of the time course of the spread of NGF after an acute injection directly into brain parenchyma (40). These studies were carried out using affinity-purified antibodies directed against mouse salivary NGF as well as against peptide sequences within mature NGF and proNGF. All these antibodies have been previously characterized (41,42). Four microliters of NGF-containing saline was slowly injected stereotactically into the center of striatum of normal adult rats. Total amounts varied from 2 ng to 1.1 μg.

Nerve growth factor protein levels in the

unperturbed adult rat striatum appear to be so low that they cannot be detected by antibodies directed against mouse NGF. Against this negative background, it became easy to visualize injected NGF in quantities at least down to 20 ng. Whether smaller amounts can be detected requires further study. The fluorescence detected in striatal parenchyma after NGF injections appears to be caused by mouse NGF, because (1) it is not seen after sham injections of cytochrome C, (2) it is not seen using antibodies directed against a peptide sequence in proNGF, and (3) using antibodies directed against NGF, the size and intensity of fluoresence correlates to the amount of NGF injected and the time after injection.

When 560 ng of NGF is injected into striatum, it rapidly diffuses throughout striatal neuropil. Already 15 min after injection, it has reached both the medial surface facing the lateral ventricle, particularly at its most dorsal aspect, as well as lateral and ventral areas of striatum, so that the fluorescent halo covers approximately two-thirds of a coronal section of striatum at the site of injection. NGF at this time is also found along the injection tract through cortex, and it appears as if there is facilitated lateral spreading within and in the dorsal and medial surface of corpus callosum. Close to the injection site there is both strong diffuse fluoresence in the neuropil and an accumulation of NGF-like immunoreactivity in nerve cell bodies and processes above the level of diffuse NGF-induced fluorescence. Immunoreactive cells appear to be medium-sized neurons and represent only a portion of all striatal neurons. The appearance of the fluorescent halo is shown in Fig. 1; Fig. 2A illustrates the presence of striatal neurons and nerve fibers with NGF-like immunoreactivity, and Fig. 2B shows the diffuse fluorescence of the neuropil seen at some distance from an injection.

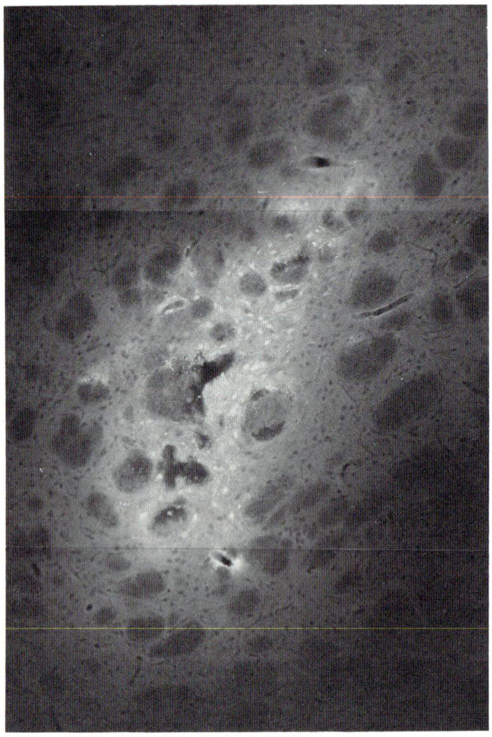

FIG. 1. NGF-like immunofluorescence 15 min after intrastriatal injection of 560 ng of purified mouse salivary NGF into rat brain. A halo of fluorescence surrounds injection site. Fluorescence intensity decreases in a gradient fashion with increasing distance from the site of injection. No fluorescence is found in myelinated fiber bundles of the internal capsule. Close to the injection site many neurons are strongly positive. Fluorescence microphotograph magnification ×120.

Nerve growth factor appears to spread even further during the next few hours while at the same time the fluorescence intensity rapidly decreses at the site of injection. A moderately fluorescent halo is seen 2 hr after injection, and weak fluorescence after 4 hr. Seven hours after injection of 560 ng NGF, only a faint small halo is seen. However, at this time, small fluorescent cells, probably glial in nature, are found. Twenty-four hours after injection of 560 ng NGF, the specific NGF-induced immunofluorescence has essentially disappeared and can also not be found 4 and 8 days after injection.

We conclude from these immunohistochemical studies, which are being reported

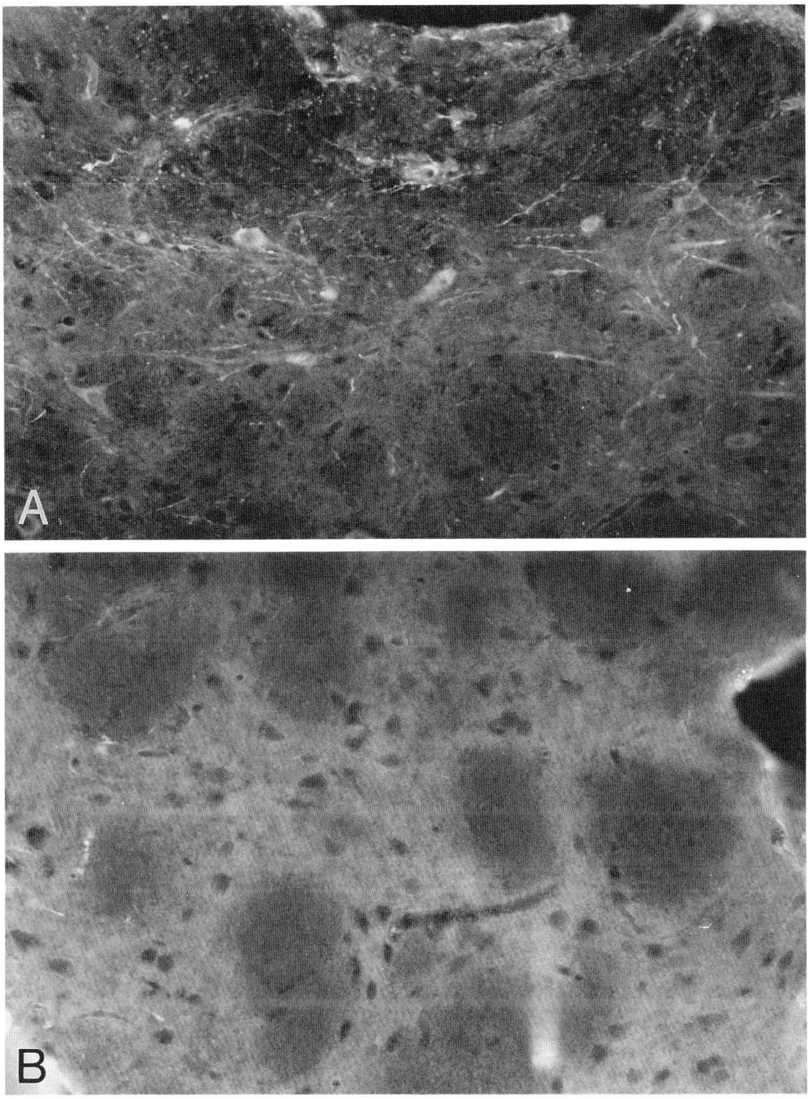

FIG. 2A and B. Close-up of various features of NGF-like immunofluorescence observed in striatal neuropil 15 min after injection of 560 ng of NGF. **A:** Appearance of small- and medium-sized NGF-positive neurons and processes close to the injection site. Not all neurons have accumulated NGF. **B:** Striatal neuropil at some distance from injection. Fluorescence of neuropil is clearly stronger than that of myelinated fiber bundles. Scattered nerve fibers are also positive, whereas neurons are negative. Fluorescence microphotographs. **A** and **B** magnification ×300.

in detail elsewhere (39,40), that immunohistochemistry is a sensitive way of localizing exogenous NGF injected into brain tissue and that NGF can spread relatively rapidly through intact brain parenchyma. These findings have implications for design of NGF delivery systems and, indirectly, bear on the possible modes of action of endogenous NGF in the CNS. Interestingly, in this respect, we have recently found that brain areas known to synthesize NGF (e.g., cortex cerebri and hippocampus) are influenced by NGF during development. Thus, when these areas taken from the fetal rat brain are grafted to the anterior chamber of adult hosts and treated with exogenous NGF, they stop growing at an earlier time, as if NGF acts as a maturation signal (43). It may be speculated that endogenous NGF produced in these areas during development not only acts to attract and maintain a cholinergic innervation, but also by local diffusion acts as a paracrine differentiation signal.

CONCLUDING REMARKS

The role for NGF as a trophic substance for cholinergic neurons in the CNS is firmly established. This role and the known effects of NGF in the periphery already suggest several possible clinical applications (e.g., supporting chromaffin autografts in parkinsonian patients by NGF and counteracting cholinergic degenerative changes by NGF in senile dementia of the Alzheimer type). Moreover, the recent discovery that injured α-motoneurons express NGF-R suggests that NGF treatment might be beneficial also, e.g., in amyotrophic lateral sclerosis. In addition to NGF, it has recently become evident that several other growth factors are present in and/or have effects in the CNS. These include the newly cloned second member of the NGF family, BDNF, which shows a surprising 60% amino acid homology to NGF (26); ciliary neurotrophic factor (CNTF), which has also recently been cloned (44) and shown to be completely different from the NGF-BDNF family; and factors with known effects outside the CNS (e.g., FGF).

We have discussed techniques of increasing central NGF levels using three approaches: either influencing endogenous NGF production, injecting purified growth factor, or finally, transplanting cells synthesizing growth factor. The fact that NGF is able to diffuse with relative ease through intact brain tissue as shown by our immunohistochemical studies of local NGF injections suggests that transplants of genetically modified cells designed to secrete growth factors might be a valid approach in designing novel treatments for neurodegenerative diseases.

ACKNOWLEDGMENTS

This work was supported by the Swedish Medical Research Council, the Swedish Natural Science Research Council, Magnus Bergvalls Stiftelse, Karolinska Institutets Fonder, Loo and Hans Ostermans Fond, and US PHS grants NS09199 and AG04418. C.W. was supported by Life & Health Insurance Medical Research Fund. We thank Eva Lindqvist, Susie Marczinko, and Monica Nyman for technical assistance, and Annika Kylberg and Stine Söderström for help with preparing the fibroblast-containing collagen gels. Ida Engqvist is gratefully acknowledged for editing and secretarial support.

REFERENCES

1. Thoenen H, Bandtlow C, Heumann R. The physiological function of nerve growth factor in the central nervous system: comparison with the periphery. *Rev Physiol Biochem Pharmacol* 1987;109:145–178.
2. Ebendal T. NGF in CNS: experimental data and clinical implications. *Prog Growth Factor Res* 1989:1:143–159.
3. Whittemore SR, Seiger Å. The expression, localization and functional significance of β-nerve growth factor in the central nervous system. *Brain Res* 1987;434:439–464.
4. Barde Y-A. Trophic factors and neuronal survival. *Neuron* 1989;2:1525–1534.
5. Ernfors P, Henschen A, Olson L, Persson H. Expression of nerve growth factor receptor mRNA is developmentally regulated and increased after axotomy in rat spinal cord motoneurons. *Neuron* 1989;2:1605–1613.
6. Eckenstein F. Transient expression of NGF-receptor-like immunoreactivity in postnatal rat brain and spinal cord. *Brain Res* 1988;446: 149–154.
7. Bartus RT, Dean RL, Beer B, Lippa AS. The cholinergic hypothesis of geriatric memory dysfunction. *Science* 1982;217:408–417.
8. Davies P, Maloney AJ. Selective loss of central

cholinergic neurons in Alzheimer's disease. *Lancet* 1976;8000:1403.

9. Perry EK, Perry RH, Blessed G, Tomlinson BE. Necropsy evidence of central cholinergic deficits in senile dementia. *Lancet* 1977;8004:189.
10. Perry RH, Candy JM, Perry EK, et al. Extensive loss of choline acetyltransferase activity is not reflected by neuronal loss in the nucleus of Meynert in Alzheimer's disease. *Neurosci Lett* 1982;33:311–315.
11. White P, Hiley CR, Goodhardt MJ, et al. Neocortical cholinergic neurons in elderly people. *Lancet* 1977;8013:668–671.
12. Arendt T, Bigl V, Arendt A, Tennstedt A. Loss of neurons in the nucleus basalis of Meynert in Alzheimer's disease, paralysis agitans and Korsakoff's disease. *Acta Neuropathol (Berl)* 1985; 43:388–393.
13. Bowen DM, Spillane JA, Curzon G, et al. Accelerated ageing or selective neuronal loss as an important cause of dementia? *Lancet* 1979; 8106:11–14.
14. Pearson RCA, Sofroniew MV, Cuello AC, et al. Persistence of cholinergic neurons in the basal nucleus in a brain with senile dementia of the Alzheimer's type demonstrated by immunohistochemical staining for choline acetyltransferase. *Brain Res* 1983;289:375–379.
15. Whitehouse P, Price DL, Struble RG, Clark AW, Coyle JT, De Long MR. Alzheimer's disease and senile dementia: loss of neurons in the basal forebrain. *Science* 1982;215:1237–1239.
16. Goedert M, Fine A, Dawbarn D, Wilcock GK, Chao MV. Nerve growth factor receptor mRNA distribution in human brain: normal levels in basal forebrain in Alzheimer's disease. *Brain Res Mol Brain Res* 1989;5:1–7.
17. Lärkfors L, Ebendal T, Whittemore SR, Persson H, Hoffer B, Olson L. Decreased level of nerve growth factor (NGF) and its messenger RNA in the aged rat brain. *Mol Brain Res* 1987;3:55–60.
18. Whittemore SR, Ebendal T, Lärkfors L, et al. Developmental and regional expression of β-nerve growth factor messenger RNA and protein in the rat central nervous system. *Proc Natl Acad Sci USA* 1986;83:817–821.
19. Korsching S, Auburger G, Heumann R, Scott J, Thoenen H. Levels of nerve growth factor and its mRNA in the central nervous system of the rat correlate with cholinergic innervation. *EMBO J* 1985;4:1389–1393.
20. Shelton DL, Reichardt LF. Studies on the expression of the β-nerve growth factor (NGF) gene in the central nervous system: level and regional distribution of NGF mRNA suggest that NGF functions as a trophic factor for several distinct populations of neurons. *Proc Natl Acad Sci USA* 1986;83:2714–2718.
21. Ayer-LeLievre C, Olson L, Ebendal T, Seiger Å, Persson H. Expression of the β-nerve growth factor gene in hippocampal neurons. *Science* 1988;240:1339–1341.
22. Korsching S, Heumann R, Thoenen H, Hefti F. Cholinergic denervation or the rat hippocampus by fimbrial transection leads to a transient accumulation of nerve growth factor (NGF) without change in mRNA NGF content. *Neurosci Lett* 1986;66:175–180.
23. Lärkfors L, Strömberg I, Ebendal T, Olson L. Nerve growth factor protein level increases in the adult rat hippocampus after a specific cholinergic lesion. *J Neurosci Res* 1987;18:525–531.
24. Koh S, Oyler GA, Higgins GA. Localization of nerve growth factor receptor messenger mRNA and protein in the adult rat brain. *Exp Neurol* 1989;106:209–221.
25. Johnson EM Jr, Yan Q. Immuohistochemical localization and biochemical characterization of nerve growth factor receptor in adult rat brain. *J Comp Neurol* 1989;290:585–598.
26. Leibrock J, Lottspeich F, Hohn A, et al. Molecular cloning and expression of brain-derived neurotrophic factor. *Nature* 1989;341:149–151.
27. Friedman WJ, Lärkfors L, Ayer-LeLièvre C, Ebendal T, Olson L, Persson H. Regulation of β-nerve growth factor expression by inflammatory mediators in hippocampal cultures. *J Neurosci Res* 1990 (*in press*).
28. Olson L, Ayer-LeLièvre C, Ebendal T, et al. Grafts, growth factors and grafts that make growth factors. *Progr Brain Res* 1990;82:55–65.
29. Strömberg I, Herrera-Marschitz M, Ungerstedt U, Ebendal T, Olson L. Chronic implants of chromaffin tissue in to the dopamine-denervated striatum. Effects of NGF on graft survival, fiber growth and rotational behavior. *Exp Brain Res* 1985;60:335–349.
30. Hefti F, Hartikka J, Knusel B. Function of neurotrophic factors in the adult and aging brain and their possible use in the treatment of neurodegenerative diseases. *Neurobiol Aging* 1989;10: 515–533.
31. Vahlsing HL, Varon S, Hagg T, et al. An improved device for continuous intraventricular infusions prevents the introduction of pump-derived toxins and increases the effectiveness of NGF treatments. *Exp Neurol* 1989;105:233–243.
32. Mandel RJ, Gage FH, Thal LJ. Spatial learning in rats: correlation with cortical choline acetyltransferase and improvement with NGF following NBM damage. *Exp Neurol* 1989;104:208–217.
33. Williams LR, Jodelis KS, Donald MR. Axotomy-dependent stimulation of choline acetyltransferase activity by exogenous mouse nerve growth factor in adult rat basal forebrain. *Brain Res* 1989;498:243–256.
34. Saffran BN, Woo JE, Mobley WC, Crutcher KA. Intraventricular NGF infusion in the mature rat brain enhances sympathetic innervation of cerebrovascular targets but fails to elicit sympathetic ingrowth. *Brain Res* 1989;492:245–254.
35. Springer JE, Collier TJ, Notter MFD, Loy R, Sladek JR Jr. Central nervous system grafts of nerve growth factor-rich tissue as an alternative source of trophic support for axotomized cholinergic neurons. *Prog Brain Res* 1988;78:401–407.
36. Heumann R, Lindholm D, Bandtlow C, et al. Differential regulation of mRNA encoding nerve growth factor and its receptor in rat sciatic nerve

durign development, degeneration, and regeneration: role of macrophages. *Proc Natl Acad Sci USA* 1987;84:8735–8739.

37. Rosenberg MB, Friedmann T, Robertson RC, et al. Grafting genetically modified cells to the damaged brain: restorative effects of NGF expression. *Science* 1988;242:1575–1578.
38. Ernfors P, Ebendal T, Olson L, Mouton P, Strömberg I, Persson H. A cell line producing recombinant NGF evokes growth responses in intrinsic and grafted central cholinergic neurons. *Proc Natl Acad Sci USA* 1989;86:4756–4760.
39. Strömberg I, Wetmore CJ, Ebendal T, Ernfors P, Persson H, Olson L. Rescue of forebrain cholinergic neurons after implantation of genetically modified cells producing recombinant NGF. *J Neurosci Res* 1990;25:405–411.
40. Wetmore C, Strömberg I, Ebendal T, Olson L. Immunohistochemical studies of NGF injections into brain tissue. 1990 (*in press*).
41. Olson L, Ayer-LeLièvre C, Ebendal T, Seiger Å. Nerve growth factor–like immunoreactivities in rodent salivary glands and testis. *Cell Tissue Res* 1987;248:275–286.
42. Ebendal T, Persson H, Larhammar D, Lundströmer K, Olson L. Characterization of antibodies to synthetic nerve growth factor (NGF) and proNGF peptides. *J Neurosci Res* 1989;22:223–240.
43. Eriksdotter-Nilsson M, Skirboll S, Ebendal T, Hersh L, Maussoulie J, Olson L. NGF treatment promotes development of basal forebrain tissue grafts in the anterior chamber of the eye. *Exp Brain Res* 1989;74:89–98.
44. Collins F. Purification, cloning, and expression of ciliary neurotrophic factor (CNTF). *Science* 1989;246:1023–1025.

Volume Transmission in the Brain: Novel Mechanisms for Neural Transmission,
edited by K. Fuxe and L.F. Agnati.
Raven Press, Ltd., New York © 1991.

37

Experimental Evidence for Volume Transmission by Analysis of Host–Graft Interaction Using Intrastriatal Adenohypophyseal Transplants in the Rat in Combination with Three-Dimensional Reconstruction

*Börje Bjelke, **Luigi F. Agnati, and *Kjell Fuxe

**Department of Histology and Neurobiology, Karolinska Institutet, S-104 01 Stockholm, Sweden; and **Department of Human Physiology, University of Modena, Modena, Italy*

Recently, the concept has been introduced that there exists, within the central nervous system, two principal types of communication: the classical wiring transmission involving synapses, and the volume transmission, which involves chemical and electrical communication via the extracellular fluid pathways in the central nervous system (CNS) (1–3), which includes the related concept of nonsynaptic transmission (4) and paracrine secretion (5,6). The early demonstration of extraneuronal monoamine fluoresence surrounding monoamine neurons after pharmacological treatments (7–10) may have been the first indication of the existence in the brain of volume transmission (fig. 1).

The concept of volume transmission appears capable of explaining a number of observations in neurotransmission that are difficult to explain solely on the basis of synaptic release of transmitters or of hormonal influences. Thus, long-lasting biological effects induced by electrical stimulation of pathways involved in, e.g., pain regulation can in this way be explained (11). The volume transmission also explains the early biological effects of transplants on various brain functions that occur at a time when no or little outgrowth of nerve fibers has taken place (12,13).

Finally, the transmitter-receptor mismatches (6,11,14–16) in the central nervous system, especially in the peptidergic neuronal systems, can be explained at least in part by the existence of chemical communication in the extracellular fluid pathways in the brain (volume transmission) (1,2,15–18).

In the present paper we have used an experimental model consisting of intrastriatally transplanted adenopituitaries to obtain experimental evidence for the existence of volume transmission. Thus, in this model we have studied the prolactin secretion from such transplants and analyzed how endogenous dopamine in the striatal networks may modulate the secretion of prolactin into the striatal neuropil from

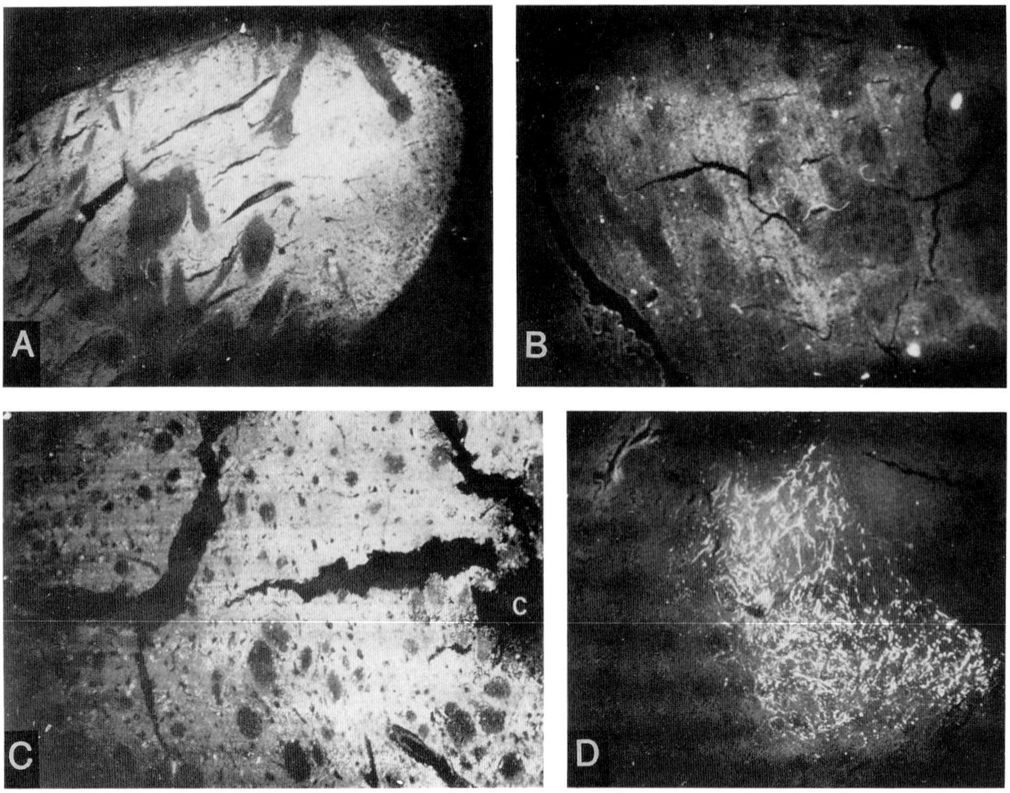

FIG. 1. A: Left nucleus caudatus-putamen of a reserpine–nialamide-pretreated rat after intrastriatal injection of dopamine (5 μg/5 μl, 1 hr before killing) at the level of commissura anterior. Section was taken from the cauda of nucleus caudatus-putamen posterior to site of injection. A diffuse green fluorescence was in the nucleus caudatus-putamen because the entire plexus of dopamine nerve terminals present in the figure accumulated the injected dopamine. ×53. **B:** Right nucleus caudatus-putamen (uninjected side) of the same rat shown in **A**. Area shown corresponds to that seen in **A**. Fluorescence is faint or nonexistent in the dense plexus of dopamine nerve terminals present in this section. ×53. **C:** Left nucleus caudatus-putamen of a reserpine–nialamide-pretreated rat after intrastriatal injection of dopamine (5 μg/5 μl, 1 hr before killing). Site (c) where the cannula tip was localized is illustrated to the *right. Black fissures* in the picture are due to freezing artifacts. To the *far left* there is a less marked accumulation of fluorescence in the dopamine nerve terminals. Strong fluorescence intensities were mainly observed in dopamine terminals lying within a 1-mm radius from the site of injection. ×73. **D:** Capsula interna and globus pallidus on the left side of the brain 1 hr after intrastriatal injection of dopamine (5 μg/5 μl) in reserpine–nialamide-pretreated rat. Diffuse green fluorescence is seen in the nucleus caudatus-putamen shown at the *upper left corner* of the picture. There is a marked accumulation of fluorescence in ascending nigro-neostriatal dopamine fibers, most of which lie within the capsula interna. These fibers accumulated fluorescence as far down as the rostral part of the substantia nigra. ×30. (From ref. 76, with permission.)

the transplanted adenopituitaries. The relevance of volume transmission for the biological effects of released dopamine from striatal adrenal medulla and nigral transplants will also be discussed.

ADENOPITUITARY TRANSPLANTS AND VOLUME TRANSMISSION

Peripheral endocrine transplants

The early work of adenopituitary transplantation mainly involved transplantation of the adenopituitary gland beneath the renal capsule or to the anterior chamber of the eye (19). The transplanted pituitary glands are mainly characterized by large numbers of prolactin-immunoreactive (IR) gland cells, which produce and release prolactin into the general blood circulation, sometimes resulting in hyperprolactinemia in nonhypophysectomized animals. The prolactin-containing gland cells are morphologically characterized by well-developed Golgi complexes, granular endoplasmatic reticulum, and exocytosis of the prolactin-containing granules (20,21). All such signs of secretion of prolactin in grafted pituitaries have been found to be counteracted by administration of D2-receptor agonists such as 2-Br-*a*-ergokryptine (21). Furthermore, injection of dopamine into the common carotide artery has also lead to a suppression of prolactin release as well as to a reduction in the size of the prolactin-containing cells. Despite the fact that these grafted pituitaries hypersecrete prolactin, biochemical determinations have demonstrated increased dopamine concentrations in the ectopic pituitaries of grafted rats in comparison with dopamine concentration present in *in situ* pituitaries. It cannot be excluded that the increased dopamine concentrations found can originate from an uptake from the local circulation. However, the ectopic pituitaries are also capable of synthesizing dopamine from L-tyrosine (22). The cellular localization of dopamine within the ectopic adenopituitary gland still remains to be demonstrated (23,24). The increased concentrations of dopamine found may, nevertheless, exert an inhibitory effect on prolactin secretion from the grafted pituitaries in view of the demonstration that antidopaminergic drugs can markedly increase prolactin levels in pituitary-grafted animals (25).

It has been demonstrated that the prolactin originating from transplanted pituitary glands or from prolactin-secreting tumors gives rise to elevated prolactin levels in the cerebrospinal fluid (26). Peripheral prolactin may thus influence central nervous system functions. In fact, hyperprolactinemia leads to marked alterations in dopamine use, not only within the hypothalamus and especially the median eminence, but also within extrahypothalamic dopaminergic pathways such as the nigrostriatal and mesolimbic dopamine pathways (27–30). These effects on the dopaminergic neurons may to a large extent mediate the behavioral effects of prolactin (31,32).

Hyperprolactinemia has also been demonstrated to increase striatal dopamine receptors (33–35). Such increases in striatal dopamine D2-receptor concentrations produced by prolactin may help to explain the reported reductions in motor behavioral deficits in senescent animals after chronic prolactin administration (36). All these biological effects of prolactin on brain, including also effects on dopamine-dependent behaviors (e.g., grooming and male copulatory behavior) may be mediated via prolactin binding sites, which have been reported to exist within the arcuate nucleus, the median eminence, the subcommissural organ, and the subfornical organ (37,38) Specific prolactin binding sites may thus exist inside the blood brain barrier in the CNS (38). It must, however, be considered that prolactin-like proteins can also be formed within the CNS (39). Thus, prolactin-like IR has been demonstrated in nerve cells of the arcuate nucleus, which probably give rise to prolactin-IR terminal networks found within the hypothalamus and the preoptic area (29,39–41). Thus, prolactin and prolactin-like peptides may also act as neurotransmitters and modulators in the CNS. Prolactin may be fragmented into biologically active peptides, which act via subpopulations of prolactin receptors or via their own receptors sensitive only to discrete prolactin fragments. All such observations on the existence of prolactin-IR neuronal populations give further support for the existence of volume transmission, because it seems likely that prolactin-like peptides can act on distant targets in the brain and reach distant receptors via the extracellular fluid pathways of the brain to exert their biological effects. The same is probably true for the adrenocorticotrophic hormone (ACTH) and *b*-endorphin peptides in the arcuate nerve cells (1,2,11,42–44).

Central endocrine transplants—Third ventricle transplants

A large number of experiments have been performed placing pituitary transplants within the third ventricle close to the median eminence. Such grafts have been found to sustain gonadal, adrenal, and thyroid weights probably because of successful reconstitution of pituitary-dependent hormone levels (19). The pars distalis of the pituitary glands in the third ventricle have been demonstrated to contain large numbers of gland cells containing luteinizing hormone (LH), growth hormone (GH), ACTH, and prolactin (45). The relation of the intracerebral pituitary grafts to the central neuropeptide systems has also been studied. It was found that the luteinizing hormone–releasing hormone (LHRH)-, somatostatin-, and ACTH-IR nerve fibers did not enter the graft, indicating that these neuropeptides influence the endocrine function of the pituitary graft via diffusion in the extracellular fluid pathways and the cerebrospinal fluid (45). These results thus give support for the existence of volume transmission in the hypothalamic control of pituitary graft function.

It should also be mentioned that the pituitary cell transplants into the cerebral ventricles promote growth of hypophysectomized rats and also suppress the pituitary function of the recipient host (46,47).

Adenohypophyseal homologous transplants in the rat neostriatum

In two recent papers, Bjelke and collaborators (48,49) demonstrated that adenohypophyseal transplants survive in the rat neostriatum and secrete prolactin into the surrounding extracellular space (see Figs. 2 and 3). In these experiments, young adult (6–8 weeks) specific pathogen-free female Sprague-Dawley rats were used both as donors and recipients. Immunocytochemical studies demonstrated that LH, ACTH, and prolactin-IR gland cells within the transplants were capable of surviving for many weeks in the rat neostriatum (see Fig. 3). It must be emphasized that no catecholamine (CA) terminals nor CA-containing gland cells could be demonstrated within the transplant itself. Furthermore, no tyrosine hydroxylase (TH)-IR could be demonstrated in the adenohypophyseal transplant (48). Evidence was obtained that prolactin is released into the neuropil of the neostriatum, because a diffusion zone of strong prolactin-like IR could be demonstrated in the neuropil surrounding the transplant (see Fig. 2). This zone of prolactin-IR disappeared on treatment with the dopamine D2-receptor agonist bromocriptine (50,51). Thus, these observations support the existence of volume transmission in the CNS because it shows a physiologically active neuronal interaction between host brain and graft, working via the extracellular fluid pathways (Figs. 4 and 5).

By analysis of the three-dimensional morphology of the zone of prolactin-like IR, using confocal laser scanning micros-

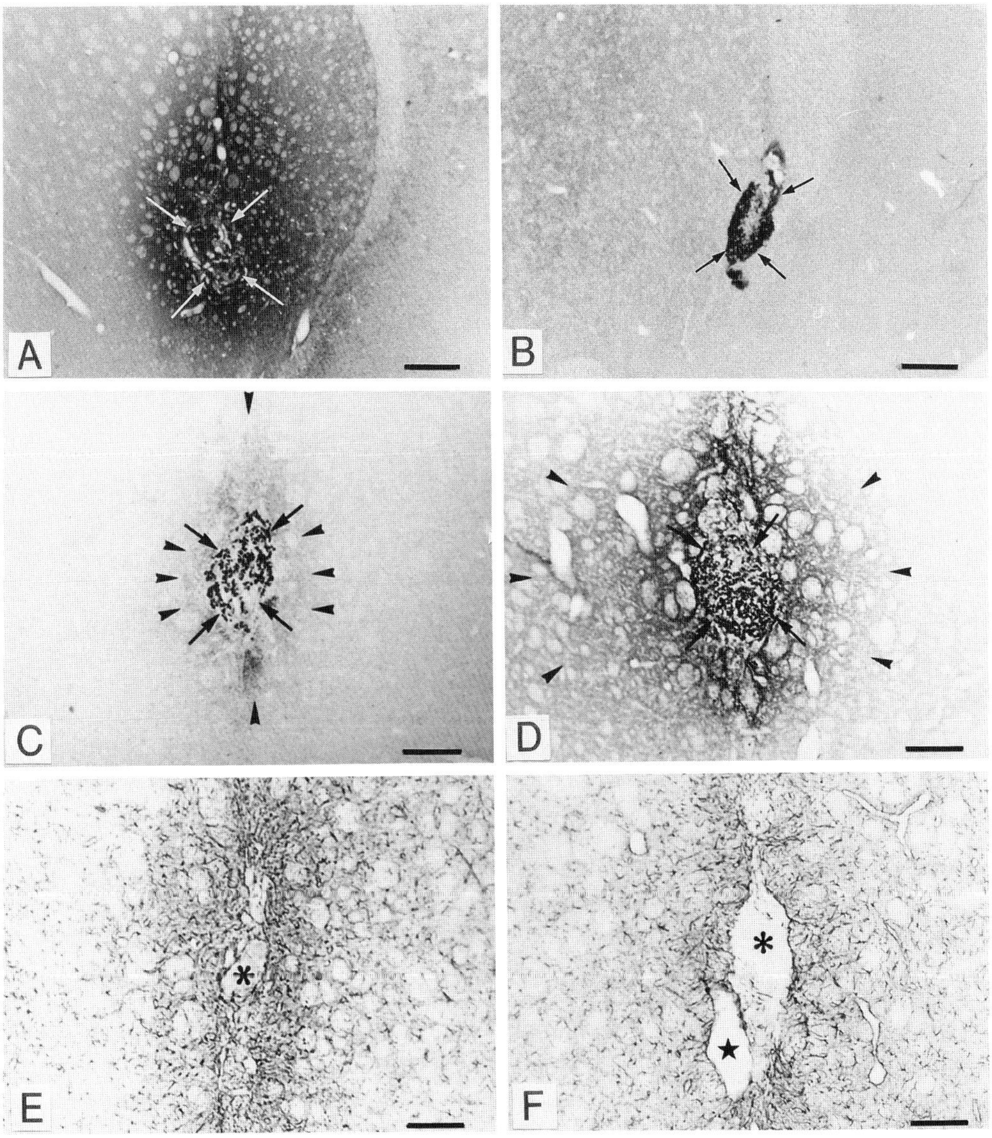

FIG. 2. A: Adenohypophyseal transplant in neostriatum of a female rat demonstrating prolactin-like immunoreactivity (IR) (control to B). The graft area border is marked (*white arrows*). Transplant area is 0.37 + 0.04 mm^2 and diffusion area 1.39 + 0.07 mm^2 (49). Primary prolactin antisera diluted 1:2,000. **B:** Coronal section of brain with an adenohypophyseal striatal transplant. Prolactin-like IR is shown in cells of the transplant but without a diffusion zone of prolactin-like IR due to bromocriptine treatment (for details see ref. 49). Transplant area 0.14 + 0.03 mm^2. *Black arrows* indicate transplant position. **C:** Intrastriatal adenopituitary transplant in a nondenervated female rat (control to D). Transplant border is marked by *arrows.* The diffusion zone of prolactin-IR in the surrounding neuropil is indicated by *arrowheads.* Because of the transplantation technique the diffusion zone is enlarged in the dorsal direction. (The section represents rat no. 3 in Table 1.) **D:** Intrastriatal adenopituitary transplant in a dopamine (DA) denervated neostriatum, showing an enlarged zone of prolactin-IR. Transplant border is marked by *arrows,* and diffusion zone is indicated by *arrowheads.* (Rat no. 9 in Table 1.) **E:** Intrastriatal adenopituitary transplant (indicated by *) labeled with antisera against glial fibrillary acidic protein (GFA). (Same animal as in C.) **F:** Intrastriatal adenopituitary transplant (indicated by *) in a DA denervated neostriatum stained for GFA. (Same animal as in D.) (Artifact is marked by a *star.*) A slight increase in GFA-IR appears around the graft and along the insertion channel in both nondenervated and denervated rats.

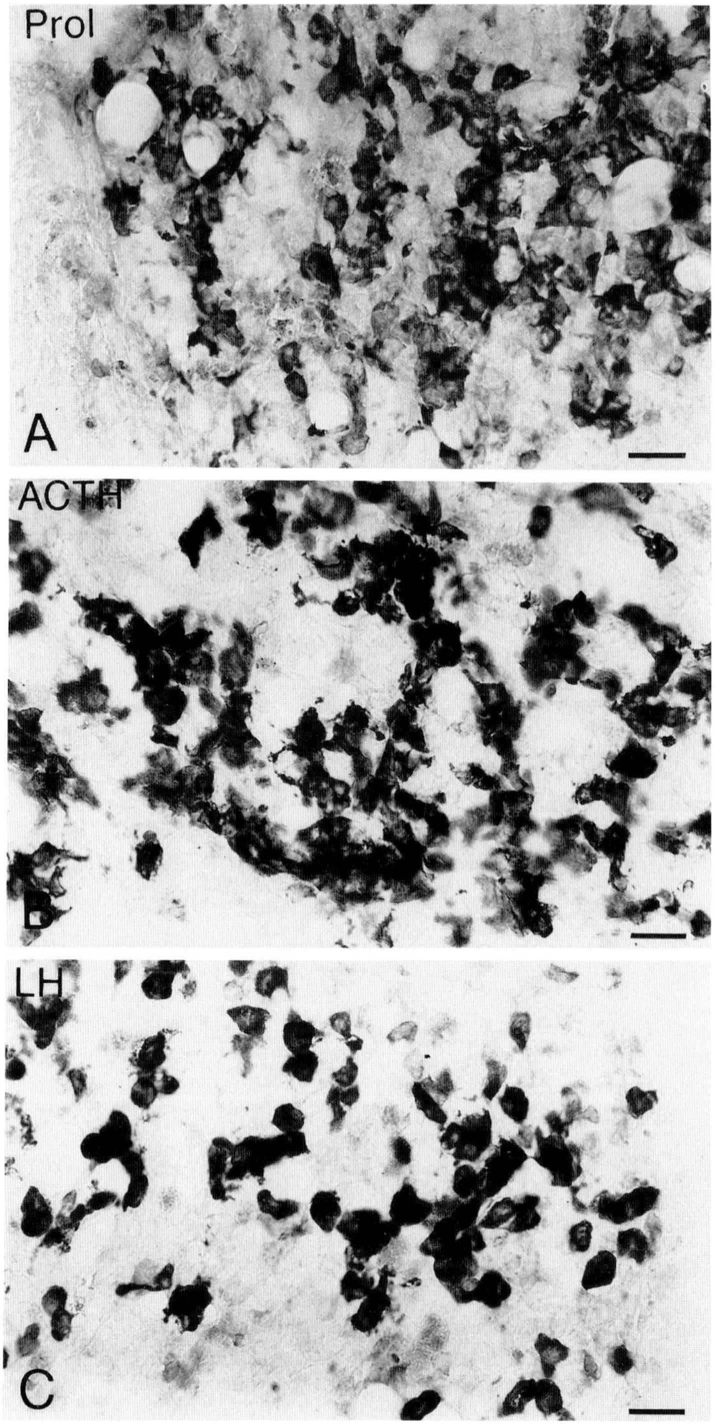

FIG. 3A–C. Adenohypophyseal striatal transplant with cell populations showing cytoplasmatic prolactin (Prol)-, ACTH-, and LH-like immunoreactivity. The ACTH-positive cells are polygonal, whereas prolactin- and LH-positive cells are more rounded, the latter being the largest of the gland cells studied. Bars = 20 mm.

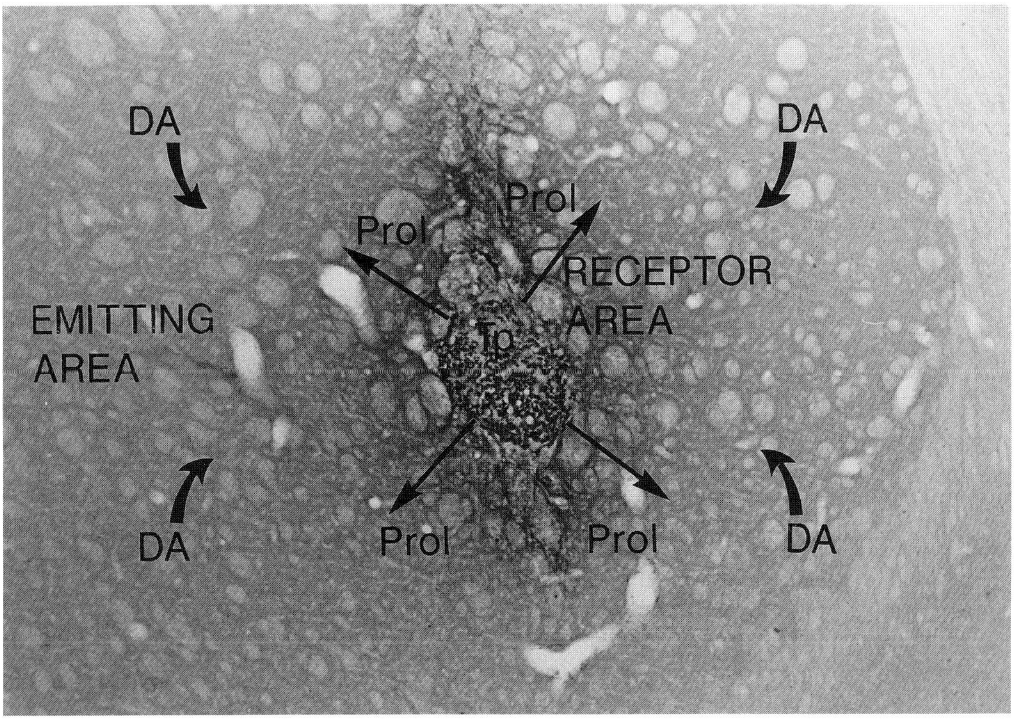

FIG. 4. Schematic presentation of the volume transmission hypothesis. Dopamine (DA) is released into the extracellular space from the terminals and synapses in the neostriatum, emitting area, and transported along the extracellular fluid pathways into the adenopituitary transplant (Tp) where it reaches its receptor site, D2 receptors, on the prolactin (Prol)-secreting cells, thus modulating the secretion of prolactin into the host extracellular space of neostriatum.

copy (SARASTRO, Sweden), it was possible to obtain support for the view that the released prolactin diffuses within the extracellular fluid pathways of the rat neostriatum (see Fig. 6; for methodological details, see refs. 52,53). Briefly, in a scanning laser microscope, which detects fluorescent light from a specimen, the depth discrimination property of confocal scanning can be used to carry out optical slicing of thick specimens. The recorded digital images constitute a three-dimensional raster covering a volume of the specimen. The images shown (see Fig. 6) have been obtained in stereo by making look-through projections of the digital data in different directions. Furthermore, for high-resolution reconstructions, the surface of the cells have been enhanced by a surface-shading procedure (53–55). The pictures demonstrate that the prolactin diffusing out from the transplant follows channels. It seems likely that these channels represent the extracellular fluid pathways. Thus, these images provide further evidence for the existence, at least under the present conditions, of volume transmission.

In another set of experiments we evaluated whether neostriatal dopamine may regulate the secretion of prolactin from the adenohypophyseal homologous transplants (49). Transplants were therefore stereotactically placed unilaterally into the intact and 6-hydroxydopamine (6-OHDA) denervated neostriatum. In the 6-OHDA denervated neostriatum, the zone of prolactin-like IR was increased in the neuropil surrounding the adenohypophyseal trans-

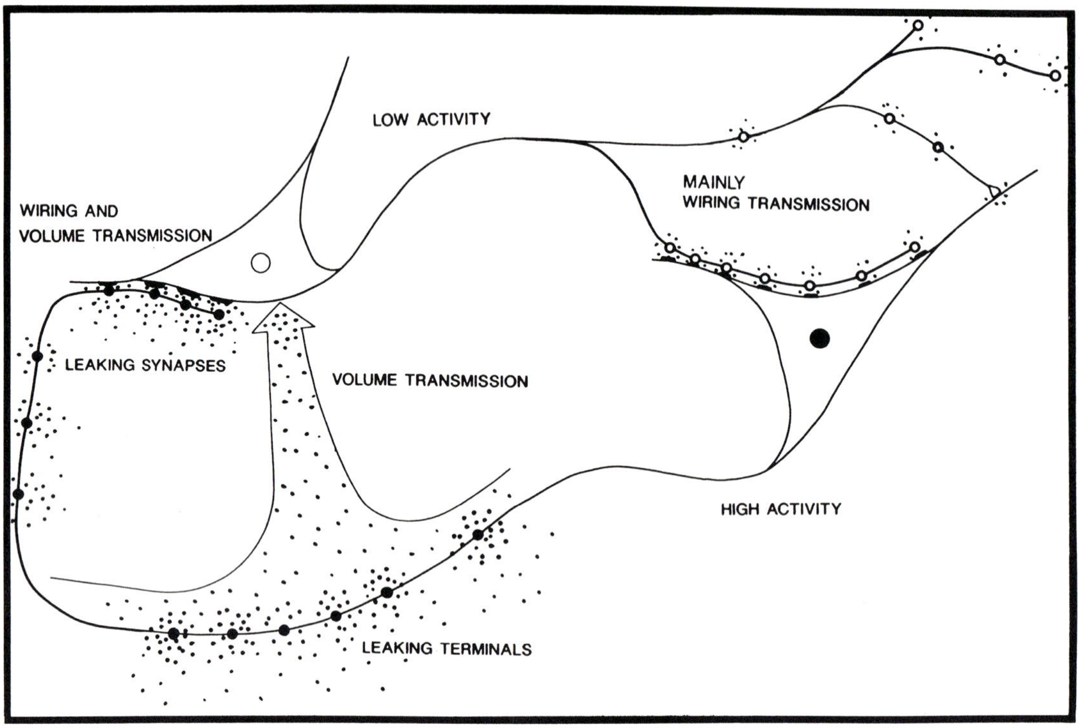

FIG. 5. Schematic illustration of wiring transmission (WT) and volume transmission (VT) in chemical communication in the CNS. It is illustrated that the chemical signals in VT originate both from leaking nerve terminals and from leaking synapses. It seems likely that the vast majority of synapses operate not only via WT but also via VT. When the nerve cell is in low activity, mainly WT mode is in operation, whereas in high-activity states, VT becomes more important in view of the available data in the literature (see Chap. 33, *this volume*). Thus, the results suggest that neuropeptides are mainly released when a high frequency of nerve impulses exists. The nucleus of the nerve cell with high activity is shown in *black,* and the nerve cell with low activity is shown as a *large open circle.* The *small black spots* along the axons represent varicosities operating via the VT or VT/WT (when synapses are formed) mode. The *open circles* along the axons of the neuron with low activity represent varicosities sometimes forming synapses mainly operating via the WT mode. The *very small black dots* surrounding the varicosities and synapses illustrate the variations in the concentrations of transmitters, especially neuropeptides, diffusing in the extracellular fluid. The synapses are indicated by postsynaptic thickenings.

plant compared with that of the transplant in the nondenervated neostriatum (see Fig. 2 and Table 1). The diffusion zones were compared by evaluating the ratio between the diffusion zone area and the transplant area. The prolactin-IR within the gland cells was not altered, but there was an increase in the mean gray values in the diffusion zone after the removal of the dopamine afferents to the neostriatum. The results obtained can be discussed along the following lines. It seems possible that the endogenous extracellular dopamine concentrations within the nondenervated neostriatum is sufficiently high to allow dopamine to diffuse into the transplant where it activates the high-affinity D2 receptors located on the prolactin-IR transplant cells, leading to decreased release of prolactin (56–58). It should be considered that the basal extracellular fluid dopamine concentrations in the intact animal within the neostriatum probably is in the range of 10 to 60 nM (59–61). Thus, these concentrations may be

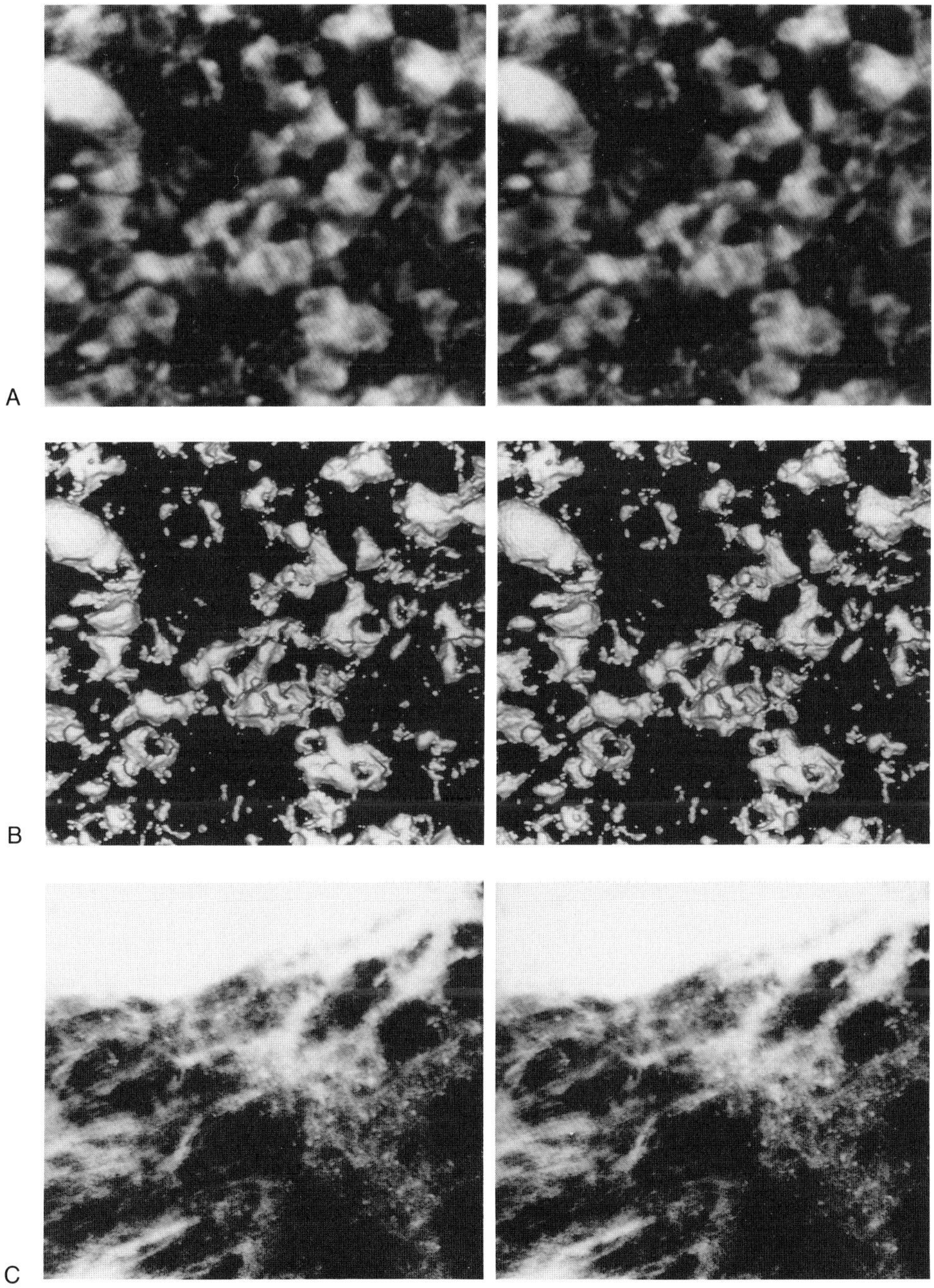

FIG. 6. Laser confocal microscopy showing a three-dimensional reconstruction of a part of an adenopituitary transplant in the neostriatum of the rat brain, stained for prolactin-like immunoreactivity (UCB, Belgium) and labeled with fluorescin-streptavidin-conjugated antirabbit antibody (Amersham, UK). **A:** Prolactin positive gland cells (bright cytoplasm) are shown in close proximity to one another. The *dark volume* represents prolactin-negative gland cells. **B:** Same image as in A but using the so-called surface shading factor, which is a mathematical operation for visualizing the surface of the cells. In this way an enhanced view of the topological morphology of prolactin-immunoreactive cells is obtained. **C:** The diffusion zone of prolactin adjacent to the transplant is visualized. The transplant is seen in the upper part of the picture (*bright area*), and diffusion pathways (*bright strands*) are radiating out from the transplant in putative extracellular fluid pathways. These images support the concept of volume transmission (48,49).

TABLE 1. *Morphometric and microdensitometric analysis of prolactin-like IR and area of transplants (T.P.) in the rat neostriatum with and without a 6-OHDA-induced degeneration of the dopamine terminals*[a]

	TP area (mm^2)	Prolactin-IR			
		TP MGV	Diffusion zone		Diffusion zone area/ TP area
			Area (mm^2)	MGV	
Nondenervated					
Rat 1	0.07	45	0.29	15	4.1
Rat 2	0.07	77	0.20	28	2.9
Rat 3	0.09	70	0.45	22	5.0
Rat 4	0.16	54	0.68	22	4.2
Rat 5	0.12	62	0.50	20	4.2
Rat 6	0.12	61	0.32	18	2.7
Means ± SEM	0.10 ± 0.01	62 ± 5	0.41 ± 0.07	21 ± 2	3.8 ± 0.4
Denervated					
Rat 7	0.08	63	0.99	25	12.4
Rat 8	0.05	95	0.93	32	18.6
Rat 9	0.10	67	1.36	30	13.6
Rat 10	0.21	59	1.19	25	5.7
Rat 11	0.12	68	0.81	27	6.8
Rat 12	0.14	62	1.51	28	10.8
Means ± SEM	0.12 ± 0.02	69 ± 5	1.13 ± 0.11**	28 ± 1*	11.3 ± 1.9**

[a]Prolactin-like IR has been demonstrated in coronal sections and stained according to the ABC technique and evaluated with the following parameters: Mean gray values (MGV), areas (mm^2) of TPs and diffusion zones, and the ratio between diffusion zone area and TP area are given for individual rats. Sections were analyzed where the TP area was of a maximum size. For details on measurements and treatment, see text. The significances refer to the corresponding parameters of the nondenervated group (**$p < 0.01$; *$p < 0.05$), Mann-Whitney U test.

high enough for activating the high-affinity dopamine D2 receptors of the prolactin-IR gland cells, and therefore an increased release of prolactin takes place from the transplants after removal of the dopaminergic afferents. A reduced degradation of prolactin, however, cannot be excluded as a possible explanation for these results. It should be pointed out that increased diffusion is an unlikely cause of the present observations, because such a phenomenon would lead to a reduced mean gray value of the diffusion zone. In contrast, the mean gray value was in fact increased after removal of the dopamine afferents.

Finally it should be considered that the blood-brain barrier probably is open within the area of the adenohypophyseal transplant as is the case for adrenal medulla transplants (62,63), although this remains to be demonstrated. If it is true that CA within the general circulation can exert a direct inhibitory effect on prolactin release from the transplant because of the failure of the blood-brain barrier, the following explanation may also be considered. The dopaminergic afferents within the neostriatum participate in the control of the blood-brain barrier, and on the disappearance of dopamine, circulating catecholamines may have increased difficulties in reaching the transplant and its D2 receptors, leading to increases in prolactin release. However, presently no evidence exists that dopamine in the neostriatum regulates blood-brain barrier function.

Taken together the present experimental results are compatible with the existence of an extracellular diffusion of dopamine from dopamine nerve terminals to reach high-affinity D1 and D2 receptors located in surrounding circuits. This type of chemical communication between neuronal networks in the brain has been defined as volume transmission, which uses extracellular fluid pathways instead of neuronal pathways (wiring transmission) (1,2,16,18). The existence of volume and not only wiring transmission in the central dopamine neurons may also help to explain the demonstration of mismatches in brain areas between high-affinity D1-dopamine receptors and dopamine nerve terminal systems (15).

In view of the fact that prolactin-IR nerve cells exist within the hypothalamus as do high-affinity ^{125}I-prolactin binding sites in many brain areas particularly the hypothalamus and limbic system (29,39–41,64), it would be of substantial interest to analyze the effects of adenohypophyseal transplants on the dopaminergic mechanisms (39) and on dopamine-dependent behaviors (65,66). In fact it has been shown that endogenous hyperprolactinemia enhances amphetamine and apomorphine-induced stereotypies. The first demonstration of high-affinity binding sites of ^{125}I-prolactin within the CNS was obtained in studies on homogenates or membrane fractions from the rabbit hypothalamus and substantia nigra by DiCarlo and colleagues (67–68). Finally, Schachter and colleagues have demonstrated the existence of prolactin messenger ribonucleic acid (mRNA) within the rat hypothalamus.

The fact that prolactin-like IR has been demonstrated within neurons of the CNS by itself supports the hypotheses on the existence of volume transmission within the brain. Thus, it may be assumed that prolactin and its fragments travel for long distances within the extracellular fluid of the brain as it does within the systemic circulation. In this context it is therefore relevant to mention the presence and widespread distribution of GH, thyroid-stimulating hormone (TSH), and LH throughout the brain of mammals as well as of extrahypothalamic brain prolactin (69–71). Evidence has been obtained that GH, TSH, and LH coeluate from Sephadex columns at the same place as their respective anterior pituitary hormones. These brain hypophyseal hormones are also active in bioassay systems. The brain GH, TSH, LH, and prolactin also appear to be independent of the anterior pituitary, because they exist for long periods after hypophysectomy.

These observations taken together would strongly favor the idea that brain hypophyseal hormones may be formed and released from nerve cells and/or glial cells to travel in the extracellular fluid to reach high-affinity receptors. Such events would represent a typical example of volume transmission in the brain. Efforts should, therefore, be focused on the analyses of the relation between the sites for the localization of hypophyseal hormone–IR and the sites for the localization of the high-affinity binding sites for the respective hypophyseal hormone in the CNS.

ADRENAL MEDULLARY IMPLANTS INTO THE RAT NEOSTRIATUM AND VOLUME TRANSMISSION

Many of these experiments have been performed on the dopaminergically denervated rat neostriatum (13,72–74). Using CA fluorescence histochemistry in combination with image analysis, CA was found to diffuse in all distances 1 to 1.5 mm within 2 min after implantation of the chromaffin tissue grafts. A fluorescent halo was also formed surrounding the implanted chromaffin cells, and this fluorescent halo remained also 100 min after implantation. These results are of substantial interest because they indicate that low molecular transmitters such as CA can diffuse in the extracellular fluid space rapidly, so that they can reach large parts of the entire neostriatum within 2 min (75,76). The biological relevance of the released and diffused CA was demonstrated by the elicitation of contralateral rotational behavior in these animals during a period of up to 200 min (74). These results may lead to the speculation that in volume transmission, diffusion of a chemical signal within the extracellular fluid during relatively short time periods (e.g., minutes) reaches a large volume and large numbers of neuronal networks. Based on these observations it also becomes easy to understand the mechanism for the therapeutic action of L-DOPA treatment in Parkinson's disease. Thus, L-DOPA can be decarboxylated in remaining dopamine nerve terminals and in other decarboxylating compartments of the neostriatum and then can diffuse into large parts of the neuropil of the neostriatum to exert effects on supersensitive dopamine receptors, leading to a therapeutic response (8,77). Thus, maintainance of dopamine function in Parkinson's disease is obtained by markedly increasing volume transmission in the few remaining dopamine terminals and increasing release and diffusion of dopamine into the extracellular fluid, and not by compensatory collateral sprouting.

Experiments also indicate that the diffuse release of CA into the extracellular space plays a role also for the biological effects of chronic implants of chromaffin tissue into the dopaminergically denervated neostriatum (72). Time periods of breakdown of the blood-brain barrier can exist also in transplantation in adult brain areas, so that dopamine from the transplant may be released into the microcirculation and taken up again via leaking capillaries into adjacent brain areas. Blood-brain barrier function must therefore be analyzed in these experiments to evaluate the role of microcirculation in this diffusion phenomenon. After treatment with nerve growth factor (NGF), however, there is an increased outgrowth of adrenergic fibers, which probably contributes to the release of CA by increasing the surface area of release and also by reducing the distance between release sites and high-affinity receptors in some parts of the neostriatum. The images obtained are compatible with the idea that the adrenergic fibers grown from the chromaffin cells may release CA and peptides at multiple sites along the fibers into the extracellular space to reach high-affinity dopamine and peptide receptors. Thus, it seems likely that the putative dopaminergic actions of grafted chromaffin tissue on a dopamine denervated striatum is obtained mainly via the use of volume transmission, involving leaking neuronal fibers and the extracellular fluid. In addition, however, it should be considered that adrenal medullary grafts can exert neurotrophic actions in those brains via, e.g., secreted peptides leading to enhanced recovery of dopaminergic neurons (78).

GRAFTED DOPAMINERGIC NEURONS AND VOLUME TRANSMISSION

A large number of studies indicate that nigral grafts give rise to a large plexus of CA nerve terminals in the denervated neostriatum (79–85). Ultrastructural analysis of graft and host interconnections have also demonstrated the development of synaptic contacts between TH-IR axon terminals and unlabeled dendrites. TH-IR dendrites have also been demonstrated, which are postsynaptic to unlabeled axon terminals. Based on such types of observations it has been implicated that discrete neuronal circuits are formed, which are responsible for a possible functional therapeutic action of the nigral grafts (86,87). Nevertheless, despite these observations, it has been found difficult in a number of studies to relate the outgrowth of nerve terminals from the graft to the dopaminergic actions of the nigral graft. It therefore seems likely that an important action of the nigral transplants and also other types of transplants in the brain is to release the transmitters from synaptic and nonsynaptic varicosities of the nerve terminal plexus to reach adjacent and distant high-affinity receptors via the extracellular fluid pathways (i.e., volume transmission). Thus, the leaking nerve terminal plexus is an important part of the substrate for volume transmission (7–10). Obviously, the synaptic contacts may contribute to the biological activity of the graft. In the case of dopaminergic transplants into the lateral ventricles, it is even more clear that such transplants operate via release and diffusion of CA from the transplant into the neostriatal neuropil to reach supersensitive dopamine receptors via the volume transmission in the denervated neostriatum in view of the discrepancy between outgrowth of dopamine fibers and behavior (88–92). Thus, we would like to propose the hypothesis that the precise wiring of the transplant, at least in the adult brain, may be of less importance compared with its ability to release CA and other transmitters from its plexus of terminals containing synaptic and nonsynaptic varicosities to reach adjacent and distant high-affinity receptors via the extracellular fluid (i.e., volume transmission).

CONCLUSIONS

Wiring transmission carries information with a high degree of speed and accuracy but requires more energy and space for operation than volume transmission. This mode of transmission is likely to be phylogenetically younger than volume transmission, which mediates information of a slow and long-lasting modality and with the benefit of being less demanding of energy and space. Thus, volume and wiring transmission should be looked on as common properties of communication in all neurons of the CNS and peripheral nervous system, but the balance between the two will vary with the system and the state. In the present study, analysis of prolactin diffusion from intrastriatal adenopituitary transplants has provided experimental support for diffusion of dopamine in the extracellular fluid from the neostriatal neuropil to the transplant, thereby controlling prolactin release in the absence of a dopamine innervation of the transplanted cells and thus providing evidence for volume transmission in the CNS.

ACKNOWLEDGMENTS

This work has been supported by a grant (04X-715) from the Swedish Medical Research Council. We want to express our warmest thanks to Mr. Johan Adler at the SARASTRO Company for spending his time and for being very enthusiastic in helping with the confocal microscopy and three-dimensional reconstructions. We also want to thank Mr. Lars Jonsson for technical support and encouragement, Mrs. Beth Andbjer, Miss Ulla Hasselrot for technical assistance, and Ms. Anne Edgren for excellent typing.

REFERENCES

1. Agnati LF, Fuxe K, Merlo Pich E, Benefenati F, Zini I, Goldstein M. Aspects on the information handling by the central nervous system: focus on cotransmission in the aged rat brain. *Prog Brain Res* 1986;68:291–301.
2. Agnati LF, Fuxe K, Zoli M, Zini I, Toffano G, Ferraguti F. A correlation analysis of the regional distribution of central enkephalin and β-endorphin immunoreactive terminals and of opiate receptors in adult and old male rats. Evidence for the existence of two main types of communication in the central nervous system: the volume transmission and the wiring transmission. *Acta Physiol Scand* 1986;128:201–207.
3. Fuxe K, Agnati LF, Härfstrand A, et al. Principles for the hormone regulation of wiring transmission and volume transmission in the central nervous system. In: Ganten D, Pfaff D, eds. *Current topics in neuroendocrinology*. Berlin: Springer-Verlag, 1988;1–53.
4. Vizi SE. *Non-synaptic interactions between neurons: modulation of neurochemical transmission. Pharmacological and clinical aspects*. Chichester: John Wiley, 1984.
5. Schmitt FO. Molecular regulators of brain function: a new view. *Neuroscience* 1984;13(4):991–1001.
6. Herkenham M. Mismatches between neurotransmitter and receptor localizations in brain: observations and implications. *Neuroscience* 1987;23(1):1–38.
7. Carlsson A, Jonason J, Lindqvist M, Fuxe K. Demonstration of extraneuronal 5-hydroxytryptamine accumulation in brain following membrane-pump blockade by chlorimipramine. *Brain Res* 1969;12:456–460.
8. Fuxe K, Ungerstedt U. Histochemical, biochemical and functional studies on central monoamine neurons after acute and chronic amphetamine administration. In: Costa E, Garattini S, eds. *Amphetamines and related compounds*. Milano: Proc Mario Negri Inst Pharmacol Res, 1970:257–288.
9. Butcher L, Engel J, Fuxe K. L-DOPA induced changes in central monoamine neurons after peripheral decarboxylase inhibition. *J Pharm Pharmacol* 1970;22:313–316.
10. Butcher LL, Engel J, Fuxe K. Behavioral, biochemical and histochemical analyses of the central effects of monoamine percursors after pe-

ripheral decarboxylase inhibition. *Brain Res* 1972;41:387–411.

11. Fuxe K, Agnati LF, Zoli M, et al. The opioid peptide systems: their organization and role in volume transmission and neuroendocrine regulation. In: Illes P, Farsang C, eds. *Regulatory roles of opioid peptides*. New York: VCH, Weinheim, 1988:33–68.
12. Brudin P, Björklund A. Survival, growth and function of dopaminergic neurons grafted to the brain. *Prog Brain Res* 1987;71:293–308.
13. Strömberg I, Herrera-Marschitz M, Hultgren L, Ungerstedt U, Olson L. Adrenal medullary implants in the dopamine-denervated rat striatum. I. Acute catecholamine levels in grafts and host caudate as determined by HPLC-electrochemistry and fluorescence histochemical image analysis. *Brain Res* 1984;297:41–51.
14. Kuhar MJ. The mismatch problem in receptor mapping studies. *TINS* 1985(May);190–191.
15. Fuxe K, Cintra A, Agnati LF, Härfstrand A, Goldstein M. Studies on the relationship of tyrosine hydroxylase, dopamine and cyclic AMP-regulated phosphoprotein-32 immunoreactive neuronal structures and D1 receptor antagonist binding sites in various brain regions of the male rat—mismatches indicate a role of D1 receptors in volume transmission. *Neurochem Int* 1988;13:179–197.
16. Fuxe K, Bunnemann B, Aronsson M, et al. Pre- and postsynaptic features of the central angiotensin systems. Indications for a role of angiotensin peptides in volume transmission and for interactions with central monoamine neurons. *Clin Exp Hypertensin* 1988;A10(suppl 1):143–168.
17. Zoli M, Agnati LF, Fuxe K, Bjelke B. Demonstration of NPY transmitter receptor mismatches in the central nervous system of the male rat. *Acta Physiol Scand* 1989;135:201–202.
18. Fuxe K, Agnati LF, Zoli M, Bjelke B, Zini I. Some aspects of the communicational and computational organization of the brain. *Acta Physiol Scand* 1989;135:203–216.
19. Tulipan NB, Zacur HA, Allen GS. Pituitary transplantation: part 1. Successful reconstitution of pituitary-dependent hormone levels. *Neurosurgery* 1985;16:331–335.
20. Ishibashi T, Shiino M. Ultrastructural and immunocytochemical changes of prolactin cells of grafted pituitary after the injection of dopamine in the albino rat. *Acta Anat* 1988;131:66–72.
21. Hiragaya T, Hoshino K. Immunohistochemical indentification of prolactin-producing cells in pituitary isografts in CBA mice. *Anat Anz* 1988;167:205–212.
22. Esquifino AI, Ramos JA, Tresguerres JAF. Possible role of dopamine in changes in LH and PRL concentrations after experimentally induced hyperprolactineamia in rats. *J Endocrinol* 1984;100:141–148.
23. Fernandez-Ruiz JJ, Cebeira M, Agrasal C, et al. Possible role of dopamine and noradrenaline in the regulation of prolactin secretion from an ectopic anterior pituitary gland in female rats. *J Endocrinol* 1987;113:45–49.
24. Fernandez-Ruiz JJ, Ubeda E, Cebeira M, et al. Modifications of plasma prolactin levels and catecholamine content in an ectopic anterior pituitary gland transplanted under the kidney capsule. *Horm Res* 1987;25:105–112.
25. Iturriza R, Rubio M, Gomez-Dumm L, Zieher L. Catecholamine metabolizing enzymes and synthesis of dopamine in normal and grafted pituitary partes distales. *Neuroendocrinology* 1983;37:371–377.
26. Login IS, MacLeod RM. Prolactin in human and rat serum and cerebrospinal fluid. *Brain Res* 1977;132:447–483.
27. Olson L, Fuxe K, Hökfelt T. The effect of pituitary transplants in the tubero-infundibular dopamine neurons in various endocrine states. *Acta Endocrinol* 1972;71:233–244.
28. Fuxe K, Eneroth P, Gustafsson J-Å, Löfström A, Skett P. Dopamine in the nucleus accumbens: preferential increase of DA turnover by rat prolactin. *Brain Res* 1977;122:177–182.
29. Fuxe K, Andersson K, Hökfelt T, et al. Prolactin–monoamine interactions in rat brain and their importance in regulation of LH and prolactin secretion. In: Robyn C, Harter M, eds. *Progress in prolactin physiology and pathology*. Amsterdam: Elsevier/North-Holland Biomedical Press, 1978;95–109.
30. Kovacs G, Drago F, Acsai L, Tihanyi A, Scapagnini U, Telegdy G. Catecholamine utilization in specific rat brain nuclei after short-term hyperprolactinaemia. *Brain Res* 1984;324:29–34.
31. Kovacs GL, Bohus B, Versteeg DHG. The interaction of posterior pituitary neuropeptides with monoaminergic neurotransmission: significance in learning and memory processes. *Prog Brain Res* 1980;53:123–140.
32. Drago F, Bohus B, Canonico PL, Scapagnini U. Prolactin induces grooming in the rat: possible involvement of nigro-striatal dopaminergic system. *Pharmacol Biochem Behav* 1981;15:61–63.
33. Hruska RE, Pitman KT, Silbergeld EK, Ludmer LM. Prolactin increases the density of striatal dopamine receptors in normal and hypophysectomized male rats. *Life Sci* 1982;30(6):547–553.
34. Eva C, Blengio M, Ferretti C, Muccioli G, Portaleone P. Ovine prolactin administration modifies 3H-spiperone binding to striatal membranes of rabbits. *Brain Res* 1982;251(2):388–390.
35. Fuxe K, Andersson K, Schwarcz R, et al. Studies on different types of dopamine nerve terminals in the forebrain and their possible interactions with hormones and with neurons containing GABA, glutamate and opioid peptides. In: Piorier LJ, Sourkes TL, Bedard PJ, eds. *Advances in Neurology*, vol 24. New York: Raven Press, 1979;199–214.
36. Joseph JA, Roth GS, Lippa AS. Reduction of motor behavioral deficits in senescent animals via chronic prolactin administration I. Rotational behavior. *Neurobiol Aging* 1986;7:31–35.
37. Walsh RJ, Posner BI, Kopriwa BM, Brauer JR. Prolactin binding sites in the rat brain. *Science* 1978;201:1041–1043.

38. Scapagnini U, Rizza V, Drago F, et al. Prolactin effects on the brain. In: MacLeod RM, Scapagnini U, eds. *Central and peripheral regulation of prolactin function*. New York: Raven Press, 1980;293–309.
39. Fuxe K, Hökfelt T, Eneroth P, Gustafsson J-Å, Skett P. Prolactin-like immunoreactivity: localization in nerve terminals of rat hypothalamus. *Science* 1977;196:899–900.
40. Toubeau G, Desclin J, Parmentier M, Pasteels JL. Cellular localization of a prolactin-like antigen in the rat brain. *J Endocrinol* 1979;83:261–266.
41. Toubeau G, Desclin J, Parmentier M, Pasteels JL. Compared localizations of prolactin-like and adrenocorticotropin immunoreactivities within the brain of the rat. *Neuroendocrinology* 1979;29:374–384.
42. Khachaturian H, Lewis ME, Kang T, Watson S. b-Endorphin, a-MSH, ACTH, and related peptides. In: Björklund A, Hökfelt T, eds. *Handbook of chemical neuroanatomy*, vol 4. Amsterdam: Elsevier, 1985;216–272.
43. Agnati LF, Fuxe K, Locatelli V, et al. Neuroanatomical methods for the quantitative evaluation of coexistence of transmitters in nerve cells. Analysis of the ACTH- and betaendorphin rat. *J Neurosci Methods* 1982;5:203–214.
44. Agnati LF, Fuxe K, Blake C, et al. Morphometrical characterization of the ACTH-endorphin and dopamine neuron systems of the mediobasal hypothalamus of the rat. Effects of neonatal monosodium glutamate treatment. *INSERM* 1983;110:87–102.
45. Knigge KM. Relationship of intracerebral pituitary grafts to central neuropeptide systems. *Am J Anat* 1980;158:549–563.
46. Weiss S, Bergland R, Page R, Turpen C, Hymer WC. Pituitary cell transplants to the cerebral venticles promote growth of hypophysectomized rats. *Proc Soc Exp Biol Med* 1978;159:409–413.
47. Kelsey RC, Hymer WC, Page R. Pituitary cell transplants to the cerebral ventricle suppress the pituitary of the recipient host. *Neuroendocrinology* 1981;33:312–316.
48. Bjelke B, Fuxe K, Agnati LF. Survival of adenohypophyseal homologous transplants in the rat striatum associated with prolactin-like immunoreactivity in the surrounding neuropil of the striatum. *Neurosci Lett* 1988;93:139–145.
49. Bjelke B, Fuxe K, Agnati LF. Increased diffusion of prolactin-like material into the brain neuropil from homologous adenohypophyseal transplants in the rat neostriatum after a 6-OH-dopamine induced degeneration of the mesostriatal dopamine neurons. *Neurosci Lett* 1989;107:33–38.
50. Fuxe K. Dopamine recepotor agonists in brain research and as therapeutic agents. *Trends Neurosci* 1979;2:1–4.
51. Flückiger E. Effects of bromocriptine on the hypothalamopituitary axis. *Acta Endocrinol* 1978;(suppl 216):111–117.
52. Carlsson K, Danielsson PE, Lenz R, Liljeborg A, Majlöf L, Åslund N. Three-dimensional microscopy using a confocal laser scanning microscope. *Optics Lett* 1985;10:53–55.
53. Brodin L, Ericsson M, Mossberg K, Hökfelt T, Ohta Y, Grillner S. Three-dimensional reconstruction of transmitter-identified central neurons by "en bloc" immunofluorescence histochemistry and confocal scanning microscopy. *Exp Brain Res* 1988;73:441–446.
54. Hamilton DK, Wilson T. Three-dimensional surface measurement using the confocal scanning microscope. *Appl Phys (B)* 1982;27:211.
55. Forsgren PO. Using vector representation to display 3-D data from a confocal laser scanning microscope. In: *Proceedings 5th Scandinavian Conference on Image Analysis*, Swedish Society for Automatic Image Analysis, Stockholm, 1987;367.
56. Caron MG, Beaulieu M, Raymond V, et al. Dopaminergic receptors in the anterior pituitary gland. *J Biol Chem* 1978;253:2244–2253.
57. Perkins NA, Westfall TC, Paul CV, Macleod R, Rogol AD. Effect of prolactin on dopamine synthesis in medial basal hypothalamus. Evidence for a short loop feedback. *Brain Res* 1979;160:431–444.
58. Calabro MA, MacLeod RM. Binding of dopamine to bovine anterior pituitary gland membranes. *Neuroendocrinology* 1978;25(1):32–46.
59. Gonon FG. Nonlinear relationship between impulse flow and dopamine released by rat midbrain dopaminergic neurons as studied by *in vivo* electrochemistry. *Neuroscience* 1988;24:19–28.
60. Imperto A, Di Chiara G. Trans-striatal dialysis coupled to reverse phase high performance liquid chromatography with electrochemical detection: a new method for the study of the *in vivo* release of endogenous dopamine and metabolites. *J Neurosci* 1984;4:966–977.
61. Zetterström T, Sharp T, Marsden CA, Ungersted U. *In vivo* measurement of dopamine and its metabolites by intracerebral dialysis: changes after d-amphetamine. *J Neurochem* 1983;41:1769–1773.
62. Rosenstein JM. Adrenal medulla grafts produce blood-brain barrier dysfunction. *Brain Res* 1987;414:192–196.
63. Brundin P, Widner H, Nilsson OG, Strecker RE, Björklund A. Intracerebral zenografts of dopamine neurons: the role of immunosuppression and the blood-brain barrier. *Exp Brain Res* 1989;75(1):195–207.
64. Fechner JH, Buntin JD. Localization of prolactin binding sites in ring dove brain by quantitative autoradiography. *Brain Res* 1989;487:245–254.
65. Drago F, Canonico P, Bitetti R, Scapagnini U. Systemic and intraventricular prolactin induces excessive grooming. *Eur J Pharmacol* 1980;65:457–458.
66. Drago F, Van Ree JM, Bohus B, DeWied D. Endogenous hyperprolactinaemia enhances amphetamine- and apomorphine-induced stereotype. *Eur J Pharmacol* 1981;72:249–253.
67. DiCarlo R, Muccioli G. Presence of specific pro-

lactin binding sites in the rabbit hypothalamus. *Life Sci* 1981;28:2229–2307.
68. DiCarlo R, Muccioli G, Lando D, Bellussi G. Further evidence for the presence of specific binding sites for prolactin in the rabbit brain. Preferential distribution in the hypothalamus and substantia nigra. *Life Sci* 1985;36:375–382.
69. Hojvat S, Emanuele N, Baker G, Connick E, Kirsteins L, Lawrence AM. Growth Hormone (GH), thyroid-stimulating hormone (TSH), and luteinizing hormone (LH)-like peptides in the rodent brain: non-parallel ontogenetic development with pituitary counterparts. *Dev Brain Res* 1982;4:427–434.
70. Emanuele N, Connick E, Howell T, et al. Hypothalamic luteinizing hormone (LH); characteristics and response to hypophysectomy. *Biol Reprod* 1981;25:321–326.
71. Emanuele NV, Baker G, McDonald D, Kirsteins L, Lawrence AM. The impact of aging on luteinizing hormone (LH) and thyroid-stimulating hormone (TSH) in the rat brain. *Dev Brain Res* 1985;20:179–183.
72. Strömberg I, Herrera-Marschitz M, Ungerstedt U, Ebendal T, Olson L. Chronic implants of chromaffin tissue into the dopamine-denervated striatum. Effects of NGF on graft survival, fiber growth and rotational behavior. *Exp Brain Res* 1985;60:335–349.
73. Strömberg I, Johnson S, Hoffer B, Olson L. Reinnervation of dopamine-denervated striatum by substantia nigra transplants: immunohistochemical and electrophysiological correlates. *Neuroscience* 1985;14(4):981–990.
74. Herrera-Marschitz M, Strömberg I, Olsson D, Ungerstedt U, Olson L. Adrenal medullary implants in the dopamine-denervated rat striatum. II. Acute behavior as a function of graft amount and location and its modulation by neuroleptics. *Brain Res* 1984;297:53–61.
75. Routtenberg A, Sladek J, Bondareff W. Histochemical fluorescence after application of neurochemicals to caudate nucleus and septal area *in vivo*. *Science* 1968;161:272–274.
76. Ungerstedt U, Butcher LL, Butcher SG, Anden NE, Fuxe K. Direct chemical stimulation of dopaminergic mechanisms in the neostriatum of the rat. *Brain Res* 1969;14:461–471.
77. Zigmond MJ, Berger TW, Grace AA, Stricker EM. Compensatory responses to nigrostriatal bundle injury. *Mol Chem Neuropathol* 1989;10: 185–200.
78. Bohn MC, Cupit L, Marciano F, Gash DM. Adrenal medulla grafts enhance recovery of striatal dopaminergic fibers. *Science* 1987;237:913–916.
79. Björklund A, Dunnett SB, Stenevi U, Lewis ME, Iversen SD. Reinnervation of the denervated striatum by substantia nigra transplants. Functional consequences as revealed by pharmacological and sensorimotor testing. *Brain Res* 1980;199:307–333.
80. Björklund A, Schmidt RH, Stenevi U. Functional reinnervation of the neostriatum in the adult rat by use of intraparenchymal grafting of dissociated cell suspensions from the substantia nigra. *Cell Tissue Res* 1980;212:39–45.
81. Björklund A, Stenevi U, Schmidt R, Dunnett SB, Gage FH. Intracerebral grafting of neuronal cell suspensions. II. Survival and growth of nigral cell suspensions implanted in different brain sites. *Acta Physiol Scand* 1983;(suppl 522):9–18.
82. Gage FH, Björklund A, Stenevi U, Dunnett SB. Intracerebral grafting of neuronal cell suspensions. VIII. Cell survival and axonal outgrowth of dopaminergic and cholinergic cells in the aged brain. *Acta Physiol Scand* 1983;(suppl 522):67–75.
83. Strömberg I, Almqvist P, Bygdeman M, et al. Human fetal mesencephalic tissues grafted to dopamine-denervated striatum of athymic rats: light- and electron-microscopical histochemistry and *in vivo* chromoamperometric studies. *J Neurosci* 1989;9:614–624.
84. Strömberg I, Almqvist P, Bygdeman M, et al. Intracerebral xenografts of human mesencephalic tissue into athymic rats: histochemical and *in vivo* electrochemical studies. *Proc NY Acad Sci* 1988;85:8331–8334.
85. Seiger Å, Bygdeman M, Goldstein M, Hoffer B, Strömberg I, Olson L. Human fetal catecholamine-containing tissues grafted intraocularly and intracranially to immuno-compromised rodent hosts. *Prog Brain Res* 1988;78:449–455.
86. Mahalik TJ, Finger TE, Strömberg I, Olson L. Substantia nigra transplants into denervated striatum of the rat: ultrastructure of graft and host interconnections. *J Comp Neurol* 1985; 240:60–70.
87. Freund TF, Bolam JP, Björklund A, et al. Efferent synaptic connections of grafted dopaminergic neurons reinnervating the host neostriatum: a tyrosine hydroxylase immunocytochemical study. *J Neurosci* 1985;5:603–616.
88. Carder RK, Snyder-Keller AM, Lund RM. Amphetamine- and stress-induced turning after nigral transplants in neonatally dopamine-depleted rats. *Dev Brain Res* 1987;33:315–318.
89. Carder RK, Snyder-Keller AM, Lund RM. Behavioral and anatomical correlates of immunologically induced rejection of nigral xenografts. *J Comp Neurol* 1988;277:391–402.
90. Snyder-Keller AM, Lund RD. Amphetamine sensitization of stress-induced turning in animals given unilateral dopamine transplants in infancy. *Brain Res* 1990;514(1):143–146.
91. Freed WJ, Perlow MJ, Karoum F, et al. Restoration of dopaminergic function by grafting of fetal substantia nigra to the caudate nucleus: long-term behavioural, biochemical and histochemical studies. *Ann Neurol* 1980;8:510–519.
92. Freed WJ, Olson L, Ko GN, et al. Intraventricular substantia nigra and adrenal medulla grafts: mechanisms of action and 3H-spiroperidol autoradiography. In: Björklund A, Stenevi U, eds. *Neural grafting in the mammalian CNS*. Amsterdam: Elsevier, 1985;471–490.

Volume Transmission in the Brain: Novel Mechanisms for Neural Transmission, edited by K. Fuxe and L.F. Agnati.
Raven Press, Ltd., New York © 1991.

38

Effects of Locus Ceruleus Transplants in a Seizure Model

O. Lindvall, J. Bengzon, P. Brundin, P. Kalén, and M. Kokaia

Department of Medical Cell Research, University of Lund, Lund, Sweden

Kindling is an animal model of epilepsy in which repeated administration of an initially subconvulsive electrical stimulus results in progressive intensification of stimulus-induced seizure activity, culminating in a generalized seizure (1,2). The initial stimulus evokes focal electrical seizure activity (so-called afterdischarge) without overt behavioral signs. Subsequent stimulations lead to the development of kindled seizures, which generally proceeds through five distinguishable grades (3): (a) facial clonus; (b) head nodding; (c) forelimb clonus; (d) rearing; and (e) rearing and falling. When the animal has exhibited grade 5 seizure, it is said to be kindled. This effect is permanent, and even if the animal is left unstimulated for as long as 12 months, it will respond to one of the first electrical stimuli with a grade 5 seizure (4). Kindling triggered by stimulation in the limbic system has been proposed to be analogous to complex partial epilepsy (also called *temporal lobe epilepsy*) (5,6), which is the most frequent type of epilepsy in adult humans (7).

A bulk of experimental evidence indicates that the noradrenergic locus ceruleus system exerts a powerful inhibitory action on the development of kindling. First, selective lesions of central noradrenergic neurons induced by intraventricular 6-hydroxydopamine (6-OHDA) (8,9) or systemic DSP-4 (10,11), as well as transections of the ascending fiber pathways from the locus ceruleus (12), lead to a marked facilitation of the rate of kindling. Second, electrical stimulation in the area of the locus ceruleus has been reported to retard the development of kindled seizures (13). Third, pharmacological depletion of brain norepinephrine (NA) (14) and blockade of alpha-2- but not alpha-1- or beta-adrenergic receptors (15) facilitate kindling, whereas systemic administration of an alpha-2 agonist has been found to suppress kindling development (15,16).

The series of studies reviewed here have been carried out with the following main objectives: first, to explore whether grafted locus ceruleus neurons can suppress seizure development in kindling similar to the intrinsic locus ceruleus system; and second, to characterize the mechanisms of action of locus ceruleus grafts in the kindling model of epilepsy.

INFLUENCE ON SEIZURE DEVELOPMENT

The capacity of grafted fetal locus ceruleus neurons to affect kindling epilepsy, induced by electrical stimulation in the hippocampus of NA-depleted rats, has been tested in two different experiments (for details, see refs. 17,18). In the first study we compared the development of kindled seizures in three groups of animals: Grafted animals first received intraventricular 6-OHDA and then bilateral cell suspension grafts from the locus ceruleus region of 13- to 14-day-old rat fetuses into the hippocampus; lesioned animals received intraventricular 6-OHDA and then glucose-saline injections into the hippocampus; control animals were given an intraventricular vehicle injection. After 6 to 11 months, kindling stimulation was performed in the hippocampus.

As shown in Fig. 1, the development of seizures was considerably faster in the NA-depleted rats than in the control animals. For example, the first grade 5 seizure was seen after eight stimulations in lesioned rats as opposed to a mean of 31 stimulations in controls. In the grafted rats, the first grade 5 seizure occurred after 22 stimulations, and the development of seizures was thus markedly retarded compared with the NA-depleted animals.

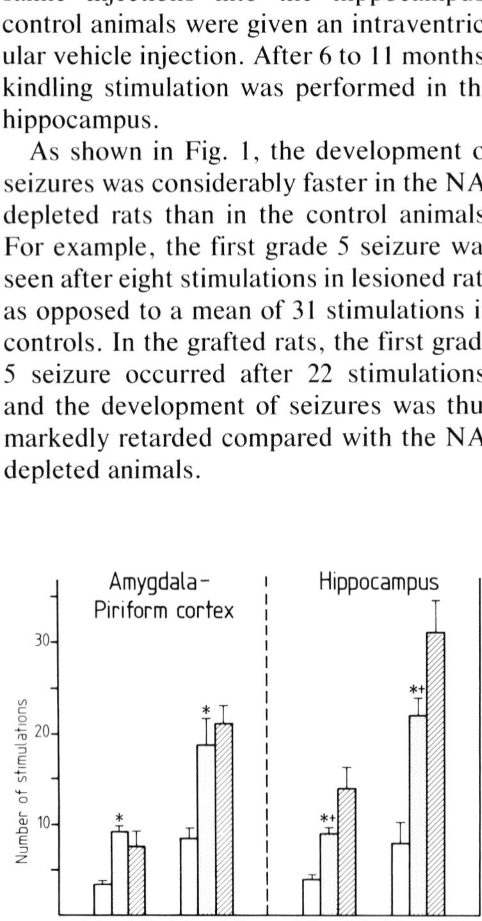

FIG. 1. Number of hippocampal kindling stimulations to reach the first grade 1 and 5 seizures, respectively, in normal rats ("controls" and in 6-OHDA-treated rats with either bilateral locus ceruleus grafts in the amygdala-pyriform cortex or hippocampus ("grafted"), or bilateral sham vehicle grafts ("lesioned"). *, Grafted rats significantly different from lesioned; †, grafted rats significantly different from controls. (Data from refs. 17,18.)

Fluorescence histochemical analysis revealed an extensive denervation of the forebrain NA system in the 6-OHDA-lesioned rats, with only few NA fibers remaining in the hippocampal formation. In the grafted rats there was a clear graft-derived NA fiber ingrowth in the dorsal two-thirds of the hippocampal formation, with a laminar distribution similar to that of normal locus ceruleus NA afferents. In contrast, only scattered NA fibers were observed in nongrafted forebrain areas. In the animals that kindled more slowly than the lesioned rats, 115 to 1,713 surviving NA neurons were found on the stimulated side. The number of stimulations needed to reach the first grade 5 seizure in the individual animals was significantly correlated to the degree of reinnervation of the stimulated hippocampus.

These data indicate that the increased susceptibility to hippocampal kindling seen in 6-OHDA-lesioned rats can be at least partly normalized by intrahippocampal locus ceruleus grafts. Although the intraventricular 6-OHDA lesion causes an extensive NA denervation throughout the forebrain, the noradrenergic reinnervation produced by the grafts was confined to parts of the hippocampal formation. This suggests that reinstatement of NA transmission locally in the hippocampus is sufficient to suppress the development of seizures in NA-depleted animals.

In the second study we explored the possibility that also when the locus ceruleus grafts are implanted distant to the stimulating electrode, in a region of central importance for the development and expression of kindled seizures, they can retard seizure development in NA-depleted animals. The amygdala-pyriform cortex was chosen as the implantation site, because lesions (19,20) as well as pharmacologically induced increase of gamma-aminobutyric acid (GABA)-ergic inhibition (21) in this area have been reported to influence hippocampal kindling. Rats were given 6-OHDA in the lateral ventricle and then divided into two groups: One group received locus ceruleus suspension grafts bilaterally into the amygdala-pyriform cortex, whereas the other was given sham vehicle transplants. A group of intact rats served as control. Two to three months later all animals were subjected to hippocampal kindling.

As seen in Fig. 1, the rats with bilateral locus ceruleus grafts in the amygdala-pyriform cortex showed a marked retardation of the development of kindled seizures as compared with sham grafted animals. Grade 5 was observed after a mean of 19 stimulations in the grafted animals, whereas only nine stimulations were required in the NA-depleted rats without grafts. The kindling rate in the grafted group was similar to that of normal controls.

Bilateral surviving grafts were observed in the amygdala-pyriform cortex of all animals implanted with locus ceruleus tissue. The mean number of NA neurons at each graft site (three on each side) was 125 (range 9–168). Because of sparing of part of the intrinsic catecholaminergic innervation and the inability of the histochemical procedure to differentiate between dopamine and NA-containing neurons, it was not possible to determine the degree of noradrenergic reinnervation by the grafts. Bundles of axons could, however, be seen to extend from the graft into the host brain and give rise to fiber plexa, e.g., in the pyriform cortex. The tip of the stimulating electrode was located in the caudal CA1–CA3 region and clearly outside the area of the graft-derived NA innervation. The dorsal hippocampus was practically devoid of catecholamine fibers, and only few were observed in the ventral part.

The findings show that implantation of fetal locus ceruleus tissue bilaterally into the amygdala-pyriform cortex retards the development of hippocampal kindling in hyperexcitable rats with prior lesions of their intrinsic forebrain catecholamine system. As illustrated in Fig. 1, the magnitude of this functional effect was comparable with that observed when the grafts were placed bilaterally in the hippocampus. Thus, implantation not only in the hippocampus itself but also at a site distant to the stimulating electrode can influence the rate of hippocampal kindling. However, the exact mechanisms by which the implantation of fetal locus ceruleus tissue into amygdala-pyriform cortex causes retardation of kindling rate remain to be elucidated. It may be suggested that growth of noncatecholaminergic neurons from the graft (which constitute the majority of implanted neurons) into the denervated amygdala-pyriform cortex could be responsible for the observed effects or that the fetal tissue could exert some trophic action and stimulate recovery mechanisms in the recipient's brain. In our opinion, the most likely explanation for the seizure-suppressant action of the graft is that the implanted locus ceruleus neurons release NA in the catecholamine-depleted amygdala-pyriform region. Reinstatement of NA neurotransmission in this area, remote from the stimulation site, thus seems to be sufficient to normalize the susceptibility of the NA-depleted animals to kindled seizures elicited by stimulation in the hippocampus.

NOREPINEPHRINE RELEASE

In a recent study (22), we used the intracerebral microdialysis technique of Ungerstedt (23) to monitor extracellular levels of NA in the hippocampus of intact rats during seizure activity induced by hippocampal kindling stimulation. There is good evidence that the extracellular NA levels recovered in the dialysis perfusate are neuronally derived and can be used as indices of noradrenergic synaptic activity (24). Our data showed that a generalized seizure in a kindled animal gave rise to a threefold increase of NA levels in the stimulated hippocampus (22). The maximal increase of NA levels occurred within the first minutes after onset of seizure activity, and the levels then tapered off, reaching baseline after 8 to 10 min. No significant increase of NA release was observed in response to a generalized seizure in 6-OHDA-lesioned rats, indicating that the high extracellular levels of NA measured in seizures were of neuronal origin.

We have monitored NA release in the hippocampus also from grafted locus ceruleus neurons during generalized seizures induced by hippocampal kindling stimulation (for details, see ref. 25). Rats were given 6-OHDA in the lateral ventricle and then either left untreated or implanted with locus ceruleus grafts bilaterally into the hippocampus. Two to three months later they were subjected to hippocampal kindling. One to three weeks after the last seizure, intracerebral microdialysis was performed according to the procedure described in detail by Kalén et al. (24) and Kokaia et al. (22).

The locus ceruleus grafts were able to normalize both basal and seizure-induced extracellular NA levels in the hippocampus of kindled animals (Fig. 2). The steady-state output of NA in the stimulated hippocampus of the previously denervated, grafted animals (58 ± 30 fmol/30 μl) was thus similar to the baseline level in intact, nonkindled animals in a parallel study

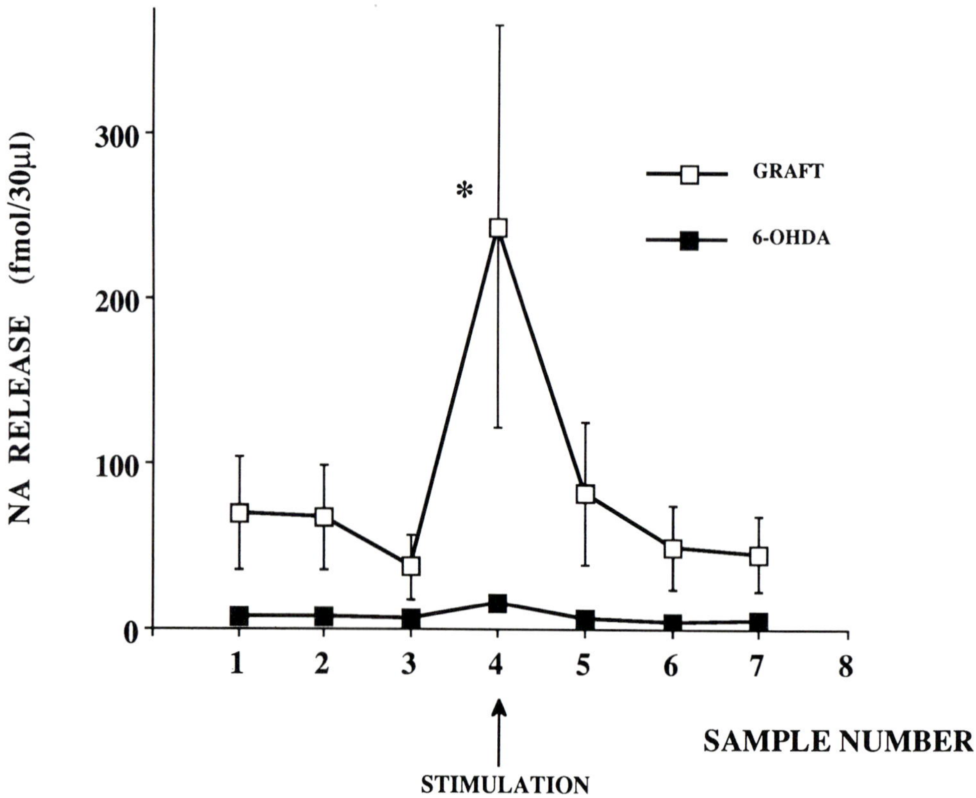

FIG. 2. Extracellular levels of hippocampal NA (as assessed in 15-min sample fractions) in 6-OHDA-treated animals with bilateral intrahippocampal locus ceruleus grafts and in rats with 6-OHDA lesion alone, respectively. Measurements were carried out during steady-state conditions and during generalized seizures evoked by electrical kindling stimulation of the hippocampus ipsilateral to the probe. Stimulation-evoked seizure lasted for about 2 to 3 min in both groups. *, Significantly different from 6-OHDA group and preceding steady-state level. (Data from ref. 25.)

(60±21 fmol/30 μl) (24) and higher than that found previously in intact, kindled animals (23±5 fmol/30 μl) (22). The NA output in the grafted rats increased about threefold after a generalized seizure (Fig. 2), which is similar to the increase seen in intact animals (22). Also, the peak increase occurred within the first minutes after the onset of seizure activity and then tapered off. The morphological analysis revealed medium-sized to large grafts in all animals with 272±90 noradrenergic cells in each hippocampus. The grafts gave rise to a noradrenergic reinnervation in the hippocampus of moderate to high density.

The grafted locus ceruleus neurons are thus able to respond by increased NA release to a kindled seizure similar to the intrinsic system. The conclusion that the high extracellular NA levels measured in seizures originate from the graft is strongly supported by the lack of any similar changes in the 6-OHDA-treated rats without grafts. However, the mechanisms underlying the increased release of NA from the grafts in response to seizures remain to be established. From our preliminary evidence obtained after transection of the ascending bundle from the locus ceruleus in intact rats (25), it seems that the effect of generalized kindled seizures on NA release is largely dependent on nerve impulse flow in the ceruleo-hippocampal NA neurons and only to a minor extent caused by local regulatory mechanisms in the hippocampus. This would suggest that the transmitter release from the grafted locus ceruleus neurons, despite their ectopic location in the hippocampus, can be regulated by the host brain through an influence at the cell body level.

ADRENERGIC RECEPTOR MECHANISMS

Recent experimental evidence has indicated that the dampening effect of NA on kindled seizures in the intact brain is mediated via postsynaptic alpha-2 receptors. In amygdala kindling, systemic administration of alpha-2-receptor antagonists has been reported to facilitate seizure development, whereas alpha-1- or beta-receptor antagonists were found to have no effect (15). Conversely, treatment with an alpha-2-receptor agonist reduced the rate of amygdala kindling (15,16). Also in hippocampal kindling with rapidly recurring seizures, pharmacological blockade of alpha-2 but not of beta receptors had a pronounced facilitatory effect on seizure development (26).

We have administered the alpha-2-adrenergic receptor antagonist idazoxan (IDX) to animals with locus ceruleus grafts in the hippocampus (for details, see ref. 27) for two main reasons: firstly, to provide further evidence that the seizure-suppressant action of such grafts in hippocampal kindling (17) is mediated via noradrenergic mechanisms; and secondly, to obtain some information about the subtype of postsynaptic adrenergic receptor involved in the graft-induced effect. Rats were given 6-OHDA in the lateral ventricle and then locus ceruleus suspension grafts bilaterally into the hippocampal formation. Two to three months later, the grafted animals and a group of normal rats were subjected to hippocampal kindling. The rats were divided into three groups: One group of grafted rats and the normal rats were given IDX before each kindling stimulation. The remaining grafted rats received vehicle injections.

Figure 3 shows the number of stimulations to reach the first grade 5 seizure in the three experimental groups. The kindling rate in grafted rats subjected to IDX treatment was significantly faster than in those receiving vehicle injections. Thus, the first grade 5 seizure was seen after 11 stimulations in the IDX-treated, grafted animals as opposed to 25 stimulations in the grafted rats not given IDX. The kindling rate in the IDX-treated, normal rats (8 stimulations to first grade 5 seizure) was similar to that in the corresponding grafted group (11 stimulations).

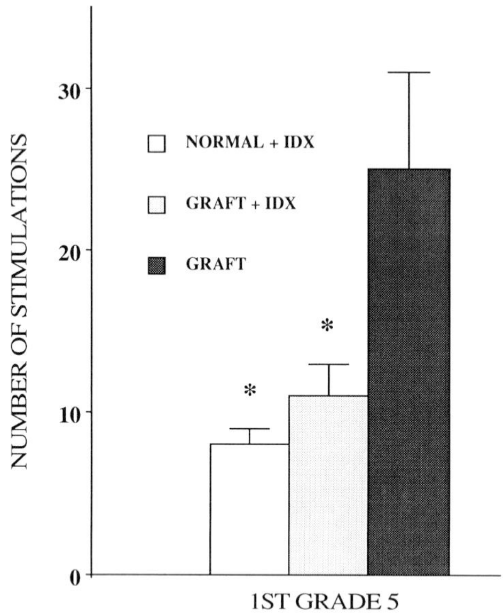

FIG. 3. Number of hippocampal kindling stimulations to reach the first grade 5 seizure in normal rats and in two groups of previously 6-OHDA-lesioned animals with bilateral intrahippocampal locus ceruleus grafts. Normal animals and one group of grafted rats received an injection of alpha-2-adrenergic receptor blocker idazoxan (IDX) before each kindling stimulation. For the statistical analysis, data were subjected to logarithmic transformation to reduce the heterogeneity of the variance between different groups. *, Significantly different from non-IDX-treated grafted rats. (Data from ref. 27.)

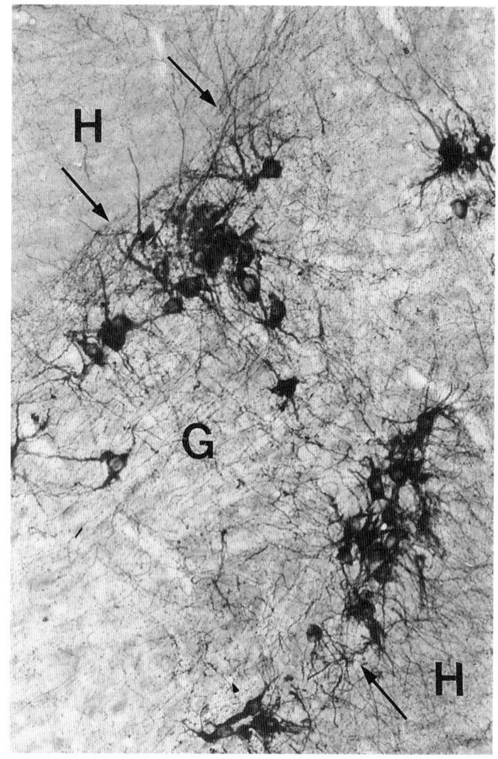

FIG. 4. Photomicrograph of dopamine-beta-hydroxylase-immunostained, most probably noradrenergic neurons, located within an intrahippocampal locus ceruleus graft. In this coronal section, numerous varicose fibers can be seen to extend from the graft (G) into the previously denervated host hippocampal CA1 region (H). *Arrows* point to the border between the graft and host tissue.

Bilateral locus ceruleus grafts were observed in the hippocampal formation in all brains subjected to microscopical analysis (Fig. 4). The mean number of noradrenergic neurons was 223 and 353 on the stimulated and nonstimulated side, respectively, of the IDX-treated, grafted rats and 295 and 304, respectively, on the two sides in the non-treated, grafted animals. The graft-derived noradrenergic reinnervation, which was of moderate to high density, was confined to the hippocampus and dentate gyrus.

The most likely explanation to the facilitation of the development of generalized seizures in the IDX-treated grafted animals (as well as in the normal rats) seems to be a blockade of postsynaptic alpha-2-adrenergic receptors (cf. refs. 15,26). Thus, there were no significant differences between the two groups of grafted animals in either afterdischarge threshold, degree of NA denervation outside the hippocampal formation, number of noradrenergic cells in the grafts, or distribution and density of the reinnervation in the host brain. In the study of Gellman et al. (15), no influence on kindling rate was observed after administration of either alpha-1- or beta-adrenergic receptor antagonists. The similar kindling rate in the IDX-treated normal and grafted rats described here therefore suggests the same specificity in the effect of NA released from intrinsic and grafted neurons (i.e., via stimulation of the alpha-2 subtype of adrenergic receptor). It should be remembered, though, that in the hippocampus of the intact brain, only low levels of alpha-2-receptor binding have been demonstrated except for a band of relatively high density localized in the stratum lacunosum-moleculare (28). Further studies are therefore needed to clarify in more detail the receptor mechanisms mediating the graft effect in the hippocampus, particularly to what extent and at which sites alpha-2 receptors are involved.

CONCLUSIONS

Grafts obtained from the pontine locus ceruleus region of rat fetuses have been reported to improve NA-dependent behavioral deficits, such as spinal reflexes and lo comotor activity in the lesioned spinal cord (29,30), avoidance learning in aged rats (31), open field locomotion in 6-OHDA-lesioned rats (32), and development of epileptic phenomena in 6-OHDA-lesioned (17) or fimbria-fornix lesioned rats (33). The studies summarized in this chapter indicate that locus ceruleus grafts can suppress seizure development in hippocampal kindling both after implantation in the hippocampus and in the amygdala-pyriform cortex. Thus the same functional effect is obtained when the graft is placed close to the stimulating electrode (i.e., in the area of epileptogenesis) and in a more distant region, the amygdala-pyriform cortex, which is of critical importance for the generalization and spread of seizures.

Previous studies (17,34) have clearly demonstrated that intrahippocampal locus ceruleus grafts, prepared as a cell suspension, reinnervate the denervated hippocampal formation, forming a noradrenergic terminal plexus similar to the normal one. By electron microscopy, the grafted noradrenergic neurons (labeled with antibodies against tyrosine hydroxylase) have been observed to form synaptic contacts with host hippocampal neurons and probably also to receive afferent inputs from the host (35). The results presented here strongly suggest that the seizure-suppressant action of intrahippocampal locus ceruleus grafts in hippocampal kindling is mediated via noradrenergic mechanisms. Thus, the retardation of kindling is correlated with the degree of noradrenergic reinnervation of the stimulated hippocampus. The microdialysis study shows that kindled seizures lead to a marked, transient increase of NA release from the grafts, similar to what has been observed from the intrinsic locus ceruleus system. As in the normal brain, NA released from the grafts seems to influence kindling rate via activation of alpha-2-adrenergic receptors. It is of particular interest that a pathological situation, in this case an epileptic seizure, can activate grafted noradrenergic locus ceruleus neurons, which then are able to exert their normal functional effect. Despite their ectopic location in the hippocampus, these neurons thus seem to be anatomically, biochemically, and functionally well-integrated into the host brain.

The experimental evidence obtained so far indicate that the functional effects exerted by the locus ceruleus grafts in the kindling model, at least to a large extent, occur through synaptic release of NA (i.e., via classical "wiring transmission") (36). However, it seems highly warranted to explore the possibility that the high seizure-induced extracellular levels of NA could influence the recipient's brain also in a more humoral way, via so-called volume transmission (36). The long-term action and the high degree of divergence of the signals, which characterize "volume transmission," may in fact be highly important for the seizure-suppressant action of grafted locus ceruleus neurons.

ACKNOWLEDGMENTS

Our own research reviewed here was supported by grants from the Swedish Medical Research Council (14X-8666), the Torsten and Elsa Segerfalk Foundation, and the Bank of Sweden Tricentenary Fund.

REFERENCES

1. Goddard GV, McIntyre DC, Leech CK. A permanent change in brain function resulting from daily electrical stimulation. *Exp Neurol* 1969; 25:295–300.
2. Racine RJ, Burnham WM. The kindling model. In: Schwartzkroin PA, Wheal HV, eds. *Electrophysiology of epilepsy.* London: Academic Press, 1984;153–171.
3. Racine RJ. Modification of seizure activity by electrical stimulation. II. Motor seizure. *Electroencephalogr Clin Neurophysiol* 1972;32:281–294.
4. Wada JA, Sato M, Corcoran ME. Persistent seizure susceptibility and recurrent spontaneous seizures in kindled cats. *Epilepsia* 1974;15:465–478.
5. Girgis M. Kindling as a model for limbic epilepsy. *Neuroscience* 1981;6:1695–1706.
6. McNamara JO. Kindling: an animal model of complex partial epilepsy. *Ann Neurol* 1984;16 (suppl):S72–S76.
7. Gastaut H, Gastaut JL, Goncalves e Silva GE, Fernandez Sanchez GR. Relative frequency of different types of epilepsy: a study employing the classification of the International League Against Epilepsy. *Epilepsia* 1975;16:457–461.
8. McIntyre DC, Saari M, Pappas BA. Potentiation of amygdala kindling in adult or infant rats by injection of 6-hydroxydopamine. *Exp Neurol* 1979;63:527–544.
9. McIntyre DC, Edson N. Facilitation of amygdala kindling after norepinephrine depletion with 6-hydroxydopamine in rats. *Exp Neurol* 1981;74:748–757.
10. Bortolotto ZA, Cavalheiro EA. Effect of DSP4 on hippocampal kindling in rats. *Pharmacol Biochem Behav* 1986;24:777–779.
11. Carre G, Harley C. DSP-4 treatment accelerates kindling. *Exp Neurol* 1986;94:812–816.
12. Corcoran ME, Mason ST. Role of forebrain catecholamines in amygdaloid kindling. *Brain Res* 1980;190:473–484.
13. Jimenez-Rivera C, Voltura A, Weiss GK. Effect of locus coeruleus stimulation on the development of kindled seizures. *Exp Neurol* 1987;95: 13–20.
14. Callaghan DA, Schwark WS. Involvement of catecholamines in kindled amygdaloid convulsions in the rat. *Neuropharmacology* 1979;18: 541–545.
15. Gellman RL, Kallianos JA, McNamara JO. Alpha-2 receptors mediate an endogenous noradrenergic suppression of kindling development. *J Pharmacol Exp Ther* 1987;241:891–898.
16. McIntyre DC, Giugno L. Effect of clonidine on amygdala kindling in normal and 6-hydroxydopamine-pretreated rats. *Exp Neurol* 1988;99: 96–106.
17. Barry DI, Kikvadze I, Brundin P, Bolwig TG, Björklund A, Lindvall O. Grafted noradrenergic neurons suppress seizure development in kindling-induced epilepsy. *Proc Natl Acad Sci USA* 1987;84:8712–8715.
18. Barry DI, Wanscher B, Kragh J, et al. Grafts of fetal locus coeruleus neurons in rat amygdala-pyriform cortex suppress seizure development in hippocampal kindling. *Exp Neurol* 1989;106: 125–132.
19. McIntyre DC, Stuckey GN, Stokes KA. Effects of amygdala lesions on dorsal hippocampus kindling in rats. *Exp Neurol* 1982;75:184–190.
20. Araki H, Aihara H, Watanabe S, Yamamoto T, Ueki S. Role of the amygdala in the hippocampal kindling effects of rats. *Jpn J Pharmacol* 1985; 37:173–179.
21. Le Gal La Salle G, Feldblum S. Role of the amygdala in development of hippocampal kindling in the rat. *Exp Neurol* 1983;82:447–455.
22. Kokaia M, Kalén P, Bengzon J, Lindvall O. Noradrenaline and 5-hydroxytryptamine release in the hippocampus during seizures induced by hippocampal kindling stimulation: an *in vivo* microdialysis study. *Neuroscience* 1989;32:647–656.
23. Ungerstedt U. Measurement of neurotransmitter release by intracranial dialysis. In: Marsden CA, ed. *Measurement of neurotransmitter release* in vivo. *IBRO handbook series: methods in neuroscience,* vol 6. New York: Wiley, 1984;81–107.
24. Kalén P, Kokaia M, Lindvall O, Björklund A. Basic characteristics of noradrenaline release in the hippocampus of intact and 6-hydroxydopamine-lesioned rats as studied by *in vivo* microdialysis. *Brain Res* 1988;474:374–379.
25. Bengzon J, Brundin P, Kalén P, Kokaia M, Lindvall O. Host regulation of noradrenaline release from grafts of seizure-suppressant locus coeruleus (*in press*).
26. Kokaia M, Bengzon J, Kalén P, Lindvall O. Noradrenergic mechanisms in hippocampal kindling with rapidly recurring seizures. *Brain Res* 1989; 491:398–402.
27. Bengzon J, Kokaia M, Brundin P, Lindvall O. Seizure suppression in kindling epilepsy by intrahippocampal locus coeruleus grafts: evidence for an alpha-2-adrenoreceptor mediated mechanism. *Exp Brain Res* 1990;81:433–437.
28. Palacios JM, Wamsley JK. Catecholamine receptors. In: Björklund A, Hökfelt T, Kuhar MJ, eds. *Handbook of chemical neuroanatomy. Classical transmitters and transmitter receptors*

in the CNS, part II, vol 3. Amsterdam: Elsevier, 1984;325–351.

29. Buchanan JT, Nornes HO. Transplants of embryonic brainstem containing the locus coeruleus into spinal cord enhance the hindlimb flexion reflex in adult rats. *Brain Res* 1986;381:225–236.
30. Yakovleff A, Roby-Brami A, Guezard B, Mansour H, Bussel B, Privat A. Locomotion in rats transplanted with noradrenergic neurons. *Brain Res Bull* 1989;22:115–121.
31. Collier TJ, Gash DM, Sladek JR Jr. Transplantation of norepinephrine neurons into aged rats improves performance of a learned task. *Brain Res* 1988;448:77–87.
32. Semenova TP, Gromova EA, Grischenko NI, et al. Behavioural, biochemical and histochemical effects of locus coeruleus transplantation in rats with neurotoxic lesions of the catecholaminergic system. *Neuroscience* 1987;22:993–1002.
33. Buszaki G, Ponomareff G, Bayardo F, Shaw T, Gage FH. Suppression and induction of epileptic activity by neuronal grafts. *Proc Natl Acad Sci USA* 1988;85:9327–9330.
34. Björklund A, Nornes H, Gage FH. Cell suspension grafts of noradrenergic locus coeruleus neurons in rat hippocampus and spinal cord: reinnervation and transmitter turnover. *Neuroscience* 1986;18:685–698.
35. Murata Y, Chiba T, Brundin P, Björklund A, Lindvall O. Formation of synaptic graft-host connections by noradrenergic locus coeruleus neurons transplanted into the adult rat hippocampus. (*Submitted.*)
36. Fuxe K, Agnati LF, Zoli M, Bjelke B, Zini I. Some aspects of the communicational and computational organization of the brain. *Acta Physiol Scand* 1989;135:203–216.

Volume Transmission in the Brain: Novel Mechanisms for Neural Transmission,
edited by K. Fuxe and L.F. Agnati.
Raven Press, Ltd., New York © 1991.

39

Thoughts on the Role of Volume Transmission in Normal and Abnormal Mass Sustained Functions

Paul Bach-y-Rita

University of Wisconsin, Madison, Wisconsin; and Departments of Neurology, (Karolinska-Huddinge) and Physical Medicine and Rehabilitation (Karolinska Institutet), Stockholm, Sweden

Synaptic transmission is the basis of many known neural functions. However, volume transmission may play an important role in other normal and abnormal activity. I will comment on the possible role of volume transmission in one normal mass sustained function, sleep, and in one abnormal function, convulsive activity. A possible role in neurologic rehabilitation will then be considered.

Neural effects that can be produced without direct connections have been known for a long time. For example, in 1955 Ingvar (1) noted such effects in isolated cortical slabs; they appeared to be produced by blood-carried substances, which can be considered to be hormones. More recently, a class of substances called *modulators* have been described: These can include previously identified neurotransmitters as well as peptides and small molecules such as adenosine, which exhibit neuroactivity but do not appear to be neurotransmitters in the classical sense (2). However, here I will discuss actual neurotransmission by neurotransmitters released into the extracellular fluid, with responses produced by the activation of receptors, extrasynaptic as well as those reached by diffusion into synaptic clefts, at some limited distance. The time scales of the responses and the duration of the activation are in this model of volume transmission well beyond the time scales and duration of synaptic transmission. This may be an efficient mechanism for transmission in normal mass sustained functions, but abnormal mass sustained functions (e.g., epilepsy) may also be produced in this way.

In 1964, Bach-y-Rita (3) demonstrated the existence of a class of late responses that, after a series of experiments to exclude other possibilities, were postulated to result from neurohumoral (local diffusion) transmission. This is entirely consistent with volume transmission, discussed by Agnati et al. (4) and by Fuxe in regard to brain plasticity (summarized in ref. 5). It is further discussed elsewhere in this volume. In the 1964 study, unit responses were recorded in various parts of the reticular formation to many types of electrical and natural stimuli, in anesthetized as well as nonanesthetized cats. A large number of neurons responded with both early and late responses. Late-response latencies of up to 4 sec and more were recorded. Virtually all the cells with early and late responses were widely convergent. The latency of the late response was essentially the same to all types of stimuli to which the cell responded, although the latency of the early response was dependent on the conduction distance to the cell (e.g., longer to tail or hind limb stimulation than to anterior limb or whisker stimulation). In several cases, a stimulus producing a late response was repeated one or more times between the early and late responses. In no case was the latency of the late response modified, and irrespective of the number of times the stimulus was repeated between the early and late responses, only one late response was produced.

In 1964, extrasynaptic transmission mechanisms were implied, but it remained for the age of biochemical and molecular neurobiology (which began with the identification in the 1970s of the opiate receptor) to consider the possibility of demonstrating volume transmission. It had long been known that the entire muscle fiber membrane becomes sensitive to acetylcholine after denervation. With the identification of receptors, this is now interpreted as receptor upregulation. In fact, the neuromuscular model has been useful for receptor studies because it is possible to examine it in isolation. Thus, extrasynaptic receptors have been identified in that model, whereas the complex nature of cerebral morphology may have delayed the identification of extrasynaptic receptors in the brain. For example, Klein et al. (6) recently reviewed the evidence for receptor positional plasticity. They noted that in early development, nicotinic acetylcholine receptors in skeletal muscles are scattered over the entire surface of uninnervated myotubes. Innervation causes a major clustering of the receptors at the synaptic site. Surface density of nicotinic receptors increases 50-fold even as 95% of the total number of receptors per cell are eliminated. When muscles are denervated, the pattern of placement reverses and receptors again become spread over the entire surface. However, Klein et al. (6) noted that in the brain it has been particularly difficult to establish where on cell surfaces receptors occur, because axons and dendrites from innumerable cells are intermixed. Thus, if receptor plasticity comparable to neuromuscular receptor plasticity occurs in the brain, its recognition (and thus the recognition of volume transmission) may be delayed.

Elsewhere (7,8), the effect of the conceptual framework [or "conceptual substance" (9)] of a field in delaying the recognition and the incorporation of new data has been discussed. Considered in particular was the delay in the acceptance of the importance of brain plasticity, caused by, in part, the conceptual framework of strict cerebral localization that dominated the neurosciences for 100 years after Broca's (10) identification of the localization of speech in the left temporal cortex. Comparably, the conceptual framework of neural transmission is so dominated by synaptic mechanisms that the merited recognition of volume transmission may be delayed. This volume should help to establish its importance.

SLEEP

The microinjection of choline agonists into the brainstem pontine tegmentum of adult cats elicits immediate and prolonged rapid eye movement (REM) or desynchronized sleep signs (11). The network involved in the generation of sleep is widely distributed, although there are regional differences in regard to neurotransmitter production (12). Adrenergic mechanisms, also, are implicated in the sleep–waking cycle. Deep sleep is enhanced by blockade of beta-adrenergic receptors (13), and Aston-Jones and Bloom (14) found that the spontaneous discharge of norepinephrine containing locus ceruleus neurons in unanesthetized rats covaried consistently with the stages of the sleep–waking cycle, firing fastest during waking, more slowly with slow-wave sleep, and becoming virtually silent during paradoxical sleep.

Robust locus ceruleus discharge may disrupt behavioral states such as sleep, and locus ceruleus discharge is reduced not only for periods of low arousal (e.g., during sleep) but aso during certain specific behaviors (e.g., grooming and drinking) when animals are highly aroused. Locus ceruleus discharge associated with orienting behavior is most intense when automatic, tonic behaviors (e.g., sleep, grooming, or consumption) are suddenly disrupted and the animal orients toward the external environment. These effects may be related primarily to the relation of locus ceruleus unit discharge to vigilance. In general terms, the locus ceruleus may function to influence the overall orientation of behavior or mode of sensorimotor activities, favoring either automatic or vegetative behavioral programs or phasic adaptive responses to salient environmental stimuli (15).

In regard to the possible role of volume transmission in sleep–wakefulness, the studies discussed above implicate the locus ceruleus in the selectivity and responsiveness to peripheral stimuli and also demonstrate a covariance of locus ceruleus discharge with depth of sleep. Aston Jones (15) discussed the possibility that the locus ceruleus is strongly influenced by two general classes of extrinsic afferents: excitatory inputs mediating sensory-evoked activity in locus ceruleus neurons, and a more tonically active set of inhibitory afferents serving to modulate overall locus ceruleus excitability in accordance with the state associated with the current behavior. Strong tonic inhibition could serve to prevent locus ceruleus neurons from responding to environmental stimuli (15). Volume transmission should be considered to mediate this mass sustained activity, especially in view of the evidence discussed by Beaudet and Descarries (16); they presented a strong case for the presence of both synaptic and nonsynaptic varicosities in brain norepinephrine and serotonin (5-HT) cells, and suggested that the biogenic amines released from the nonsynaptic varicosities may act not only on adjacent postsynaptic interfaces, but also more diffusely in tissue on more distant receptor elements. The population of reticular formation cells that reveal early and late unit responses (3) may have the late responses produced by volume transmission mechanisms comparable to those discussed by Beaudet and Descarries (16).

In sleep production and maintenance, the responses to peripheral stimulation must be inhibited to a point that allows sleep, but not totally inhibited, because dangerous and important stimuli (e.g., the cry of a baby) must still be capable of interrupting sleep. The role of volume transmission in this mass sustained function, suggested by the evidence discussed above, remains to be fully determined by future experiments, which should be undertaken and interpreted with a conceptual framework that allows full consideration of volume transmission.

CONVULSIVE ACTIVITY

Among the causes of convulsive activity, several may be related to volume transmission. The foci often occur in areas where the primary functions are highly selective (rather than mass sustained functions) such as in hippocampus and motor cortex. Under certain pathological conditions, convulsive foci are developed, and the mass sustained convulsive activity may be produced and sustained by volume transmission.

Kindling, chemical irritation, and stroke are among the causes of convulsive activity that may be considered as possibly volume transmission–related. In one study, 77 (7.7%) of 1,000 stroke patients had seizures related to their strokes (17). Forty-four had focal seizures and 33 had generalized seizures. Sixty of the 77 patients were studied during extended periods; of these, 33 had the first seizure within the first 2 weeks of the stroke (the seizure preceded motor deficits in 13 patients), whereas the 27 late patients had the first seizure at an average of 6 to 12 months after the stroke, and of these, 22 had recurrent seizures. In another study, 17.3% of 104 patients with severe carotid occlusive disease and 10.4% of 37 patients with middle cerebral artery stenosis had epileptic discharges during the acute stroke (18).

In clinical epilepsy, excessive repetitive discharge of neurons is the basic dysfunction at the cellular level. Epileptic discharge requires synchronization of a population of neurons that have a propensity to discharge. Meyer (19) suggests that this synchronization may be synaptically mediated, but volume transmission should be considered, especially if the presence of an up-regulation of relevant extrasynaptic receptors could be demonstrated. Recently, De Keyser et al. (20) demonstrated an up-regulation of D1 dopamine receptors in the neocortex of persons whose death was due to a recent unilateral infarct of the ventral midbrain, producing a unilateral relative dopamine depletion. In autopsies of patients 9, 19, and 27 days poststroke, there was a 27% to 37% increase of receptors on the lesioned side. It remains for other studies to demonstrate whether the up-regulation occurs extrasynaptically (as occurs in denervated muscle fibers) and whether similar up-regulation of other receptors may occur in such cases. It would also be important to know if extrasynaptic receptor up-regulation is found in other clinical seizure disorders (e.g., the seizures that result from a kindling effect in cocaine addicts, in whom the "high" may be related to a large release of dopamine) (discussed in ref. 21).

The changes in excitatory amino acid receptor ligand binding induced by transient cerebral ischemia were studied in rat hippocampal subfields by Westerberg et al. (22). Ten minutes of ischemia was induced by common carotid artery occlusion combined with hypotension, and the rats were allowed 1 day to 4 weeks of recovery. The binding of 3H-alpha-amino-3-hydroxy-5-methyl-4-isoxazolepropionic acid (AMPA) to quisqualate receptors, 3H-kainic acid (KA) to kainate receptors, and 3H-glutamate to *N*-methyl-D-aspartate (NMDA) receptors was determined by quantitative autoradiography.

One week after ischemia, the CA1 region of the hippocampus displayed a severe (90%) dendrosomatic lesion with preservation of presynaptic terminals. This was associated with a 60% decrease in AMPA binding and a 25% decrease in glutamate binding to NMDA receptors. At 4 weeks (the maximum duration of the observations), both AMPA and NMDA sites were greatly reduced. Although the dentate granule cells are resistant to an ischemic insult of this magnitude, this region showed marked changes in receptor binding. One week after ischemia, the AMPA and NMDA binding decreased by approximately 40% and 20%, respectively. After 2 weeks of recovery, the NMDA binding was not significantly different from control level, whereas the AMPA binding remained depressed through the 4 weeks. The high-density KA binding in the inner molecular layer of the dentate gyrus was unaffected by the ischemic insult, despite an extensive degeneration of cells in the hilus of the dentate gyrus, which projects glutamatergic afferents to this area.

Westerberg et al. (22) commented that their results demonstrate a lack of correlation between receptor changes in the early recovery period after ischemia and the development of neuronal necrosis in different hippocampal regions and also noted long-lasting receptor changes in areas considered resistant to an ischemic insult. Some receptors are down-regulated whereas others appear to be up-regulated on the surviving cells. The loci of the up-regulated receptors remain to be demonstrated. These results, together with those of De Keyser, et al.(20), provide the basis for experiments to determine if receptor loci changes after brain cell loss may lead to volume transmission–related convulsive activity.

NEUROLOGIC REHABILITATION

In addition to the up- and down-regulation of receptors after brain damage that have been demonstrated in animal (e.g., ref 22) and human (20) studies, a number of other changes occur after brain damage that would be expected to affect synaptic as well as extrasynaptic receptors. These include changes in brain "tone," partial denervation of regions of the brain resulting in the unmasking of the input from secondary pathways, and diaschisis, which will be briefly discussed below, and several others, most of which have been discussed elsewhere (23–25). They have a direct bearing on the efficacy of neurologic rehabilitation programs, and volume transmission may play a role in all of them.

Changes in brain "tone"

Specific brain functions appear to require an appropriate background level of sustained mass activity, or "tone." Lashley (26) spoke of "some sort of polarization of activity and . . . rhythmic discharges which pervade the entire brain, influencing the organization of activity everywhere," and a "substratum of postural and tonic activity (that) is constantly present. . . ." Bach-y-Rita (23) and Aston-Jones (15), (see above) also refer to brain tone. It is apparently a mass sustained activity that greatly influences all specific activity.

Unmasking

The increase of secondary pathway activity after damage to the major input to a neural region has been discussed elsewhere by Wall (27). He reviewed a series of studies from his laboratory demonstrating the unmasking of "normally ineffective connections which may become active if the dominant inputs are put out of action." Hyvarinen et al. (28) provided interesting evidence for the unmasking of somatosensory pathways in monkeys that had spent the first year of life with the eye lids sutured. Microelectrode studies of cells in area 19 revealed that 20% of the cells responded to somesthetic stimuli, whereas in the normal monkey no somesthetic responses could be recorded. In another study, approximately half of the cells in normal cat visual cortex that responded to light also responded to pin-prick and/or auditory stimuli but with longer latencies and less secure responses (29). Those nonvisual connections may be unmasked after loss of vision: Wanet-Defalque et al. (30) found, with positron emission tomographic human visual cortex studies, that glucose use is highest in the striate and prestriate areas and was higher in blind subjects than in blindfolded sighted subjects, whether at rest or during an auditory or tactile task. However, in the case of the unmasking of connections that lead to the reestablishment of selective functions (e.g., motor control after a stroke), the receptor upregulation would be expected to be localized primarily in the synapses of the previously "weak" connections. Examples of this may include the spinal cord cells responding to sensory stimulation of a distant dorsal root after the loss of closer dorsal roots (27), the somatosensory input to the visual cortex in the studies of Murata, et al. (29) that are apparently unmasked in blindness, and the unmasked somatosensory pathways to visual associative cortex in vision-deprived monkeys (28).

Diaschisis

Elsewhere, Feeney (summarized in ref. 5) has discussed the concept of diaschisis: that functional disturbances that would spontaneously recover (to varying degrees) were primarily due to depressions in tissue remote from but connected to the primary injury site. As the depression (which was thought to be caused by the loss of excitation) abated in the remote structures, functional recovery occurred in the organism.

Diaschisis may operate by means of changes in neurotransmitter functioning. Boyeson (summarized in ref. 5) has shown that after a unilateral sensorimotor cortex lesion in rats, a depression in norepinephrine functioning occurs in the cerebellum contralateral to the injury. The norepinephrine depression and the hemiplegia can be resolved by infusions of norepinephrine into the contralateral cerebellum. In animals that have recovered from the sensorimotor injury, phenoxybenzamine (an alpha-1-receptor blocker), administered either peripherally or infused into the contralateral cerebellum, reinstated the lesion. Boyeson (summarized in ref. 5) thus suggested that diachisis may occur not only by a loss of excitation to a remote area of the brain (transneuronal effects), but also by effects mediated by bifurcating axonal systems, in which one of the bifurcating axons dies and the other survives.

Striker (summarized in ref. 5) recently discussed neurochemical plasticity of the catecholamine system and in particular its relation to hypothalamic damage. Ventricular administration of the neurotoxin 6-hydroxydopamine (6-OHDA) in desmethylimipramine pretreated animals produces profound behavioral effects (in addition to the more selective effects of hypothalamic lesions on aphagia and adipsia) on arousal if dopamine depletion exceeds 95%. Nevertheless, recovery of function occurs over time, caused by, in large part, changes in residual dopamine activity (e.g., increased release, decreased deactivation, and up-regulation of dopamine receptors).

It remains to be demonstrated whether the up-regulation of receptors in the studies of Striker (5) and the receptor regulation that may occur in the cases discussed above occur in extrasynaptic cell membrane. This would be expected, in view of the extrasynaptic up-regulation demonstrated in denervated muscle fibers and in view of the results reported by Beaudet and Descarries (16; also elsewhere in this volume). If so, volume transmission would have to be considered to be important in brain plasticity, as Fuxe (summarized in ref. 5) suggested. Another intriguing area of emerging interest is the role of glia in volume transmission–related plasticity; this has been discussed by Agnati (summarized in ref. 5) and elaborated further by Hansson elsewhere (*this volume*).

Neurologic rehabilitation, which until now has been poorly grounded in theory (31), would greatly benefit from serious consideration of mechanisms such as those discussed above in the development of future treatment programs. Other possible volume transmission–related considerations include the following:

1. Vigilance. Active patient participation is apparently important for effective neurologic rehabilitation (8). McCormick (32) noted that during drowsiness and sleep, the ascending modulatory neurotransmitters are decreased; he stated that both acetylcholine and norepinephrine can provide a state of excitability consistent with cognition, which is consistent with the observation of Aston-Jones (15) that the locus ceruleus is inhibited during lack of vigilance.

2. Combining neurotransmitter therapy with appropriate rehabilitation. Feeney, et al. (33) demonstrated that neurotransmitter therapy was effective in a rat brain damage model of motor rehabilitation only when combined with appropriate motor experience (rehabilitation). Crisostomo et al. (34) confirmed this finding in human stroke patients. It is relevant to volume transmission, because brain damage produces receptor down- and up-regulation that probably includes extrasynaptic receptors. It may also be related to memory mechanisms (comparable to the translocation of protein kinase C in the hippocampus) that are in-

volved in the motor learning obtained with successful rehabilitation (discussed in ref. 8).

CONCLUSIONS

Volume transmission concepts are not new. They have been discussed for more than 25 years. A considerable body of evidence on volume transmission has accumulated, much of which forms the basis for this volume. To firmly establish volume transmission in the mainstream of neuroscience, further experimental evidence is required, such as the ultrastructural demonstration of extrasynaptic up-regulated dopamine receptors in cases such as those presented by De Keyser, et al. (20) and in cocaine addicts. In addition, a change in the conceptual framework (or conceptual "substance") of this field is necessary and should occur with further clear demonstrations of the existence of volume transmission and its functional roles.

ACKNOWLEDGMENT

Partial support from the National Institute of Disability and Rehabilitation Research, grant number G0087C2008, is appreciated.

REFERENCES

1. Ingvar D. Extraneuronal influences upon the electrical activity of isolated cortex following stimulation of the reticular activating system. *Acta Physiol Scand* 1955;33:169–193.
2. Cooper JR, Bloom FE, Roth RI. *The biochemical basis of neuropharmacology*. Oxford: Oxford, 1986.
3. Bach-y-Rita P. Convergent and long-latency unit responses in the reticular formation of the brain. *Exp Neurol* 1964;9:327–344.
4. Agnati LF, Fuxe K, Merlo Pich E, et al. Aspects on the integrative capabilities of the central nervous system: evidence for "volume transmission" and its possible relevance for receptor–receptor interactions. In: Fuxe K, Agnati LF, eds. *Receptor–receptor interactions* London: Macmillan, 1987;236–249.
5. Boyeson MG, Bach-y-Rita P. Determinants of brain plasticity. *J Neurol Rehab* 1989;3:35–57.
6. Klein WC, Sullivan J, Skorupa A, Aguilar S. Plasticity of neuronal receptors. *FASEB J* 1989; 3:2132–2140.
7. Bach-y-Rita P. Brain Plasticity. In: Goodgold J, ed. *Rehabilitation medicine*. St. Louis: Mosby, 1988;113–118.
8. Bach-y-Rita P, Wicab Bach-y-Rita E. Biological and psychosocial factors in recovery from brain damage in humans. *Can J Psychol* (Special Issue of Neural Plasticity and Recovery of Function) (*in press*).
9. Frank RG. The Columbian exchange: American physiologists and neuroscience techniques. *Fed Proc* 1986;45:2665–2672.
10. Broca P. Remarques sur le siege de la faculte du langage articule; suivies d'une observation d'a phemie (perte de la parole). *Bull Soc Ana Paris* 1861;6:330–357.
11. Hobson JA, Steriade M. Neuronal basis of behavioral state control. In: Mountcastle VM, ed. *Handbook of physiology: the nervous system*, sec. 1, vol 4. Bethesda: American Physiological Society, 1986;701–823.
12. Quattrochi JJ, Mamelak AN, Madison RD, Macklis JD, Hobson JA. Mapping neuronal inputs to REM sleep induction sites with carbachol-fluorescent microspheres. *Science* 1989; 245:984–986.
13. Denlinger S, Patarca R, Hobson JA. Differential enhancement of rapid eye movement sleep signs in the car: a comparison of microinjection of the cholinergic agonist carbachol and the b-adrenergic antagonist propranolol on pontogeniculo-occipital wave clusters. *Brain Res* 1988;473: 116–126.
14. Aston-Jones G, Bloom FE. Activity of norepinephrine-containing locus coeruleus neurons in behaving rats anticipates fluctuations in the sleep-waking cycle. *J Neurosci* 1981;1:876–886.
15. Aston-Jones G. Behavioral functions of locus ceoruleus derived from cellular attributes. *Physiol Psychol* 1985;13:118–126.
16. Beaudet A, Descarries L. The monoamine innervation of rat cerebral cortex: synaptic and nonsynaptic axon terminals. *Neuroscience* 1978; 3:851–860.
17. Louis S, McDowell F. Epileptic seizures in nonembolic cerebral infarction. *Arch Neurol* 1967; 17:414–418.
18. Cocito L, Favale E, Reni L. Epileptic seizures in cerebral arterial occlusive disease. *Stroke* 1982;13:189–195.
19. Meyer FB. Calcium, neuronal hyperexcitability and ischemic injury. *Brain Res Rev* 1989;14: 227–243.
20. De Keyser JD, Ebinger G, Vauquelin G. Evidence for a widespread dopaminergic innervation of the human cerebral cortex. *Neurosci Lett* 1989;104:281–285.
21. Holden C. Flipping the main switch in the central reward system? *Science* 1989;246:1378–1379.
22. Westerberg E, Monaghan DT, Kalimo H, Cotman CW, Wieloch TW. Dynamic changes of ex-

citatory amino acid receptor in the rat hippocampus following transient cerebral ischemia. *J Neurosci* 1989;9:798–805.
23. Bach-y-Rita P. *Brain mechanisms in sensory substitution*. New York: Academic Press, 1972.
24. Bach-y-Rita P. Brain plasticity as a basis for therapeutic procedures. In: Bach-y-Rita P, ed. *Recovery of function: theoretical considerations for brain injury rehabilitation*. Bern, Sweden: Hans Huber, 1980;225–263.
25. Bach-y-Rita P. Brain plasticity as a basis for the development of rehabilitation procedures for hemiplegia. *Scand J Rehab Med* 1981;13:73–80.
26. Lashley KS. In search of the engram. *Symp Soc Exp Biol* 1950;4:454–482.
27. Wall PD. Mechanisms of plasticity of connection following damage in adult mammalian nervous systems. In: Bach-y-Rita P, ed. *Recovery of function: theoroetical considerations for brain injury rehabilitation* Bern, Sweden: Hans Huber, 1980;91–105.
28. Hyvarinen J, Carlson S, Hyvarinen L. Early visual deprivation alters modality of neuronal responses in area 19 of monkey cortex. *Neurosci Lett* 1981;26:239–243.
29. Murata K, Cramer H, Bach-y-Rita P. Neuronal convergence of noxious, acoustic and visual stimuli in the visual cortex of the cat. *J Neurophysiol* 1965;28:1223–1239.
30. Wanet-Defalque MC, Veraart C, De Volder A, et al. High metabolic activity in the visual cortex of early blind human subjects. *Brain Res* 1988; 446:369–373.
31. Bach-y-Rita P. Applications of principles of brain plasticity and training to restore function. In: Young RR, Delwaide PJ, eds. *Principles of restorative neurology*. London: MacMillan, (*in press*).
32. McCormick DA. Cholinergic and noradrenergic modulation of thalamocortical processing. *Trends Neurosci* 1989;12:215–221.
33. Feeney DM, Gonzalez A, Law WA. Amphetamine, haloperidol and experience interact to affect rate of recovery after motor cortex injury. *Science* 1982;217:855–857.
34. Crisostomo EA, Duncan PW, Propst M, Dawson DV, Davis JN. Evidence that amphetamine with physical activity promotes recovery of motor function in stroke patients. *Ann Neurol* 1988; 23:94–97.

Volume Transmission in the Brain: Novel Mechanisms for Neural Transmission,
edited by K. Fuxe and L.F. Agnati.
Raven Press, Ltd., New York © 1991.

40

Neurotransmission and the Control of Pain: Segmental Inhibitory Mechanisms

Jean-Marie Besson and Marie-Christine Lombard

Unité de Recherches de Pharmacologie du Système Nerveux de l'INSERM and Ecole Pratique des Hautes Etudes, 75014 Paris, France

The past 20 years have seen a major step forward in our knowledge of nociception. This subject has greatly benefited from multidisciplinary studies combining electrophysiological, neuroanatomical, neurochemical, immunohistochemical, and behavioral techniques. After a brief survey of peripheral and spinal mechanisms of nociception, this chapter will be devoted to pharmacological aspects of the segmental modulation of nociceptive transmission at the dorsal horn level of the spinal cord. In fact there is much evidence suggesting that volume transmission could be, at least in part, involved in these mechanisms of control.

A SCHEMATIC VIEW OF PAIN PATHWAYS

The transmission of nociceptive message is initiated via activation of free endings of mechano-, thermo-, and chemosensitive peripheral afferent fibers of the skin, muscle, and viscera (see references in refs. 1,2). At the peripheral level, the role of cutaneous afferent fibers is relatively well-demonstrated in animals. In general, the mechano-heat-nociceptor afferents are thinly myelinated A delta fibers; the polymodal nociceptors are unmyelinated C fibers. Experiments in humans have attempted to correlate the sensation of pain with the responses to heat of C polymodal nociceptors. In general, there is a good correlation (a) between the threshold for activation of nociceptors and the threshold of pain sensation and (2) between the discharge level of nociceptors and the sensation of pain. In this regard the intensity of the sensation can be equally well-correlated with the skin temperature as with the activity of these receptors.

The specific roles of articular, muscular, and visceral fine afferent fibers in nociception is less well understood. A partial reason for this is the difficulty in finding adequate stimulation procedures.

The activation of the nociceptors associated with A delta and C fibers, which tend to overlap in sensitivity to mechanical, thermal, and chemical stimuli, probably results from both direct effects of the stimulus and indirect effects mediated by the release of various chemicals. The molecular basis of the action of these algogenic chemicals requires further studies.

An involvement of substance P (SP), contained in fine afferent fibers, in neurogenic inflammation is mainly supported by the effects of its antagonists. In addition the use of capsaicin has revealed the role played by fine afferent fibers not only in nociception and pain, but also in the maintenance of inflammatory states.

The use of various methods confirms that nociceptive fibers terminate mainly at the level of the superficial layers of the dorsal horn: laminae I (marginal zone) and II (substantia gelatinosa). Many substances (e.g., peptides, excitatory amino acids, adenosine 5′-triphosphate) have been proposed as neurotransmitters released at the spinal cord level by nociceptive afferents. SP has long been a candidate, but the multiplicity of peptides revealed by immunohistochemical techniques and their coexistence, on occasions, in the same dorsal root ganglion cells questions a unique role of SP. A promising approach is to combine immunohistochemistry with functional identification to analyze the action of specific SP antagonists on the activity of nociceptive neurons of the dorsal horn. The production of antagonists is of even greater importance when one considers that SP could be coreleased with other peptides or nonpeptides.

Two major classes of neurons activated by nociceptive stimuli exist in the dorsal horn. One of these groups responds only to nociceptive stimuli (nociceptive-specific neurons) and is located mainly in the superficial zones, whereas the other group receives both nociceptive and nonnociceptive cutaneous messages (nociceptive-nonspecific neurons). These cells, located mostly in lamina V are activated by a variety of stimuli and their responses depend on the intensity of the stimulation. Thus, as the intensity of a mechanical stimulus is progressively increased from light touch through moderate pressure, to firm nonnoxious pressure, and then finally to frankly noxious pinch, the neuron responds with an increase in the frequency of its discharge. Increments in thermal stimuli also demonstrate these graduated responses but the threshold of these neurons to a thermal stimulus corresponds, in general, with that of nociceptors. The involvement of nociceptive nonspecific neurons in pain processes is naturally less obvious than that of nociceptive-specific neurons. However, their role is strongly suggested by their capacity for coding the intensity of nociceptive stimulation, by the major visceral, muscular, and cutaneous convergence onto these cells, by their projections to supraspinal areas, and by the variety of modulatory mechanisms acting on them.

We will now consider in detail some pharmacological aspects related to these controls.

INHIBITORY EFFECTS INDUCED BY TRANSCUTANEOUS NERVE STIMULATION

As initially predicted (3), the stimulation of large diameter cutaneous primary afferent fibers produces inhibition of the spinal transmission of nociceptive information. These inhibitory effects have been clearly established by several groups (see references in ref. 1) on both nociceptive-specific and nonspecific dorsal horn neurons and can also be obtained by dorsal column stimulation in spinalized animals. Despite the fact that the inhibitory effects wore off within seconds or minutes after the cessation of stimulation whereas pain relief can outlast stimulation for many minutes or hours, these inhibitions have been put forward as an explanation of the analgesia observed in humans after transcutaneous nerve stimulation using high-frequency and low-intensity stimuli. The pharmacological substrates of these segmental controls, induced by the activation of A fibers, are still poorly understood. The endogenous opioids are probably not involved because, under these conditions in humans, the analgesia, segmentally distributed, is generally not naloxone-reversible (4–6). In general, pharmacological evidence obtained in animals favors a role for gamma-aminobutyric acid (GABA) because the inhibitory effects induced by dorsal column stimulation on dorsal horn nociceptive neurons was reduced by intravenous or electrophoretically applied bicuculline but not by strychnine (7). It has been postulated that both postsynaptic (build-up of hyperpolarization induced by the repetitive stimulation) and presynaptic mechanisms could be involved in these inhibitory effects. The presence of GABAergic cells in the dorsal horn often forming axo-axonic synapses with afferent fibers (8,9) and of GABA A and GABA B receptors on A delta and C fibers (10) favor the latter hypothesis.

However, although the role of GABA in nociception has had renewed interest, one must recognize that we are still far from a complete pharmacological knowledge of the inhibitory mechanisms activated by large-diameter cutaneous fibers. The goal is worthwhile because novel drugs could then be used to potentiate the analgesic effects. In this regard, laminae I and II of the dorsal horn of the spinal cord are extremely rich in GABA-benzodiazepine sites (11). A recent study in our laboratory suggests a near total lack of effect of benzodiazepines on the activity of dorsal horn cells induced by nociceptive stimulation. It would be, however, of considerable interest to study the effects of benzodiazepines on the previously described inhibitory effects because these drugs can clearly potentiate the effects of GABA (12).

In humans, a second form of transcutaneous electrical stimulation is used to produce analgesia or hypoalgesia, consisting of the application of low-frequency (2–4 Hz) currents of high intensity (at the limit of tolerance by the subject), which seem to activate fine fibers. Many of the clinical studies on this subject suggest that the analgesic effects are dependent on endogenous opioid mechanisms because the analgesia is totally or, more generally, partly reversed by the administration of naloxone (4). Overall, in considering the results for the variety of experimental approaches in a number of species, these effects can be explained by, at least in part, the simultaneous activation of two major control systems, the final effect of which is to block or modulate the transmission of nociception at the spinal level:

- One, which we will not consider further here, is via a negative feedback loop involving several supraspinal structures, notably certain regions of the brainstem (see ref. 13).
- The other is produced by purely spinal mechanisms, which are either segmental or propriospinal.

In the animal, the effects of Transcutaneous Electrical Nerve Stimulation (TENS) have been studied in detail by Chung et al. (14–16) on the activity of dorsal horn neurons at the origin of the spinothalamic tract in the monkey. Prolonged repetitive transcutaneous stimulation strongly inhibits the responses of the neurons to the activation of unmyelinated fibers. It was clearly shown that the most important afferent fibers responsible for the antinociceptive effect were the A delta group, although significant additional effects were also produced by activation of both the A and C groups.

The antinociceptive effect, largely dependent on spinal mechanisms, consists of two components, one naloxone-reversible and the other nonnaloxone-reversible. The nonnaloxone-reversible component is powerful and develops quickly but only lasts a short period, whereas the naloxone-reversible component is less powerful, develops slowly, and lasts a long time.

These observations clearly show that the spinal opioid systems can be activated by high-intensity peripheral stimulation; we will discuss later the biochemical studies that have a bearing on this hypothesis. With regard to the nonnaloxone-reversible component, the gauging of the pharmacological manipulation by GABA blockers (picrotoxin, bicuculline), glycine antagonists (strychnine), or serotonin and catecholamine blockers is required. As suggested by Chung (17), it is possible that another characterized chemical agent (e.g., one of the neuropeptides other than enkephalin) or accumulation of K^+ (18) mediates the nonnaloxone-reversible component. Thus it is reasonable to speculate that at this level, volume transmission may play a significant role (see Chap. 25, *this volume*).

DIRECT SPINAL DEPRESSIVE EFFECTS OF OPIATES ON NOCICEPTIVE SPINAL TRANSMISSION

The direct depressive effect of morphine at the spinal level has been clearly established (see references in refs. 1,19,20). By the end of the 1970s, mechanisms of presynaptic inhibition were foremost as an explanation of these effects based on the following observations:

- the presence of opiate receptors on the terminals of primary afferent fibers (21–26);
- the reduction in the levels of opiate receptors in the superficial dorsal horn after dorsal rhizotomy (27–30), peripheral nerve transection (31), and capsaicin treatment of neonatal rats (32,33);
- the blockade by opioids, in a naloxone-reversible manner, of the K^+-evoked release of SP in slices of trigeminal complex (34).

However, the possibility of a direct presynaptic control of the release of SP by enkephalinergic axo-axonic synapses has not been found despite many anatomical studies (35–39). As Sumal et al. (37) have proposed, intermediate neurons may be involved in the interactions between enkephalin-containing neurons and primary afferents. Nevertheless, the rarity of axo-axonic synapses on afferent fibers suggests that if these presynaptic interactions take place in the spinal cord, they could be based on dendro-axonic synapses. In this respect, the dendrites of some enkephalin neurons contain vesicles (36,38,40) and are potential sites of synaptic output. Furthermore, nonsynaptic interactions could be involved; this type of mechanism, already put forward for the release of SP in the dorsal horn, has been proposed (38) to explain how enkephalin neurons could modulate the excitability of fine nociceptive afferent fibers (see also refs. 41,42). In this respect it must be emphasized that electron microscopic autoradiographic examination of the localization of opioid receptors in another CNS area strongly suggests that opioids act primarily at nonjunctional interfaces (42). Consequently, the data related to the presynaptic action of morphine (present section) and some of those concerning the functional significance of the segmental intrinsic opioid systems of the dorsal horn (see below) probably implicate volume transmission.

The postsynaptic effect of opioids have also been well-documented on the basis of the following studies:

- Electrophysiological studies have provided several possible postsynaptic actions of morphine: an inhibitory or blocking action on receptor-activated sodium channels (43–46), an increased potassium conductance (47), and an increased chloride conductance (46).
- Immunohistochemical studies have clearly established that some laminae I and IV neurons at the origin of the spinothalamic tract are directly contacted by enkephalinergic terminals (48,49).
- In freely moving mice it has been shown that intrathecal morphine can supress the behavioral syndrome (biting and scratching) elicited by intrathecal administration of SP (50).
- Iontophoretic morphine depresses the excitatory responses induced by glutamate on primate spinothalamic neurons (51).

However, the ensemble of these studies does not allow any conclusion to be made as to the relative importance of pre- and post-synaptic inhibitions in the direct depressive action of opiates on the transmission of nociceptive messages at the dorsal horn level. Recently we have approached this question in a quantitative manner.

With the use of autoradiographic techniques with extended rhizotomies we have quantitatively measured the relative proportion of pre- and postsynaptic binding sites for different ligands for the opioid receptors (52). To gauge the importance of pre- and postsynaptic mechanisms, we compared, using electrophysiological techniques (53), the effects of morphine on the spontaneous hyperactivity of nociceptive dorsal horn neurons in the spinalized decerebrated deafferented rat (C5-Th1) to those induced by the same dose of morphine [2 mg/kg intravenous (i.v.) in the intact spinalized decerebrated arthritic rats, in which dorsal horn nociceptive neurons display high spontaneous activity (54).

Relative Proportion of Mu, Delta, and Kappa Receptors in the Superficial Dorsal Horn and Their Pre- and Postsynaptic Location

Numerous studies using receptor autoradiography have visualized the three main types of opioid receptors in the superficial layers of the dorsal horn of the rat, primate, and human. As already mentioned, a high proportion of these receptors are located on thin primary afferent fibers. However, with the exception of a few studies (55–57), the previous studies did not include data on their relative proportions, and no quantitative data are available concerning their pre- and postsynaptic location (see, however, ref. 30). Using quantitative autoradiography and highly selective ligands, we determined, at the level of the superficial dorsal horn (laminae I and II), the relative percentage of mu, delta, and kappa binding sites and their relative pre- and postsynaptic locations. For this latter purpose we used a large unilateral dorsal root rhizotomy (C4-Th2) corresponding to the brachial plexus.

(3H) DAGO (Tyr-D-Ala-Gly-(Me)Phe-Gly-ol), (3H)DTLET (Tyr-D-Thr-Gly-Phe-Leu-Thr), and (3H) ethylketocyclazocine (EKC) in presence of DAGO and DTLET were used to label selectively mu-, delta-, and kappa-opioid binding sites.

As shown in Fig. 1, there is a high percentage of mu-opioid binding sites ($69 \pm 5\%$) in the superficial layers of the spinal cord, whereas delta and kappa sites were respectively $23 \pm 2\%$ and $7 \pm 1\%$. The density of kappa receptors was low and contrasted with previous reports (55–59) showing a high density of kappa binding sites in the rat spinal cord. In these other studies, labeling of kappa sites was performed by using nonspecific tritiated ligands (etorphine, bremazocine, diprenorphine) in the presence of high concentration of mu and delta blockers. It has been shown, in the same experimental conditions, that the quantity of kappa sites measured with (3H)etorphine was greater than that with (3H)EKC (60). This could be related to the fact that (3H)etorphine still binds to a number of sites defined as mu and delta sites in the presence of high concentrations of mu and delta ligands, suggesting that previous studies overestimated the number of kappa sites. Our interpretation agrees well with the results obtained by Mack et al. (56), who reported low quantities of kappa sites measured with (3H)EKC, with those of Lahti et al. (61) using (3H)U.69593 and those of Morris and Herz (62), who used (3H)bremazocine in presence of high concentrations (1 μM) of DAGO and DADL (Tyr-D-Ala-Gly-Phe-D-Leu).

From the functional point of view, it must be underlined that the high levels of mu, intermediate delta, and low kappa binding sites are in good agreement with several electrophysiological (63,64) and pharmacological studies (65,66; and see ref. 67) based on intrathecal injection of opioid agonists, showing that the analgesic effects or potency of mu agonists was greater than that produced by kappa agonists, with delta ligands eliciting intermediate effects.

Eight days after the large unilateral dorsal root rhizotomy from C4-Th2, dramatic decreases were seen on the side ipsilateral to the lesion by comparison with the intact side. In the C7 segment, these decreases were, respectively, $76 \pm 1\%$, $61 \pm 1\%$, and $53 \pm 3\%$ for mu, delta, and kappa binding sites (Fig. 1). The loss of kappa sites re-

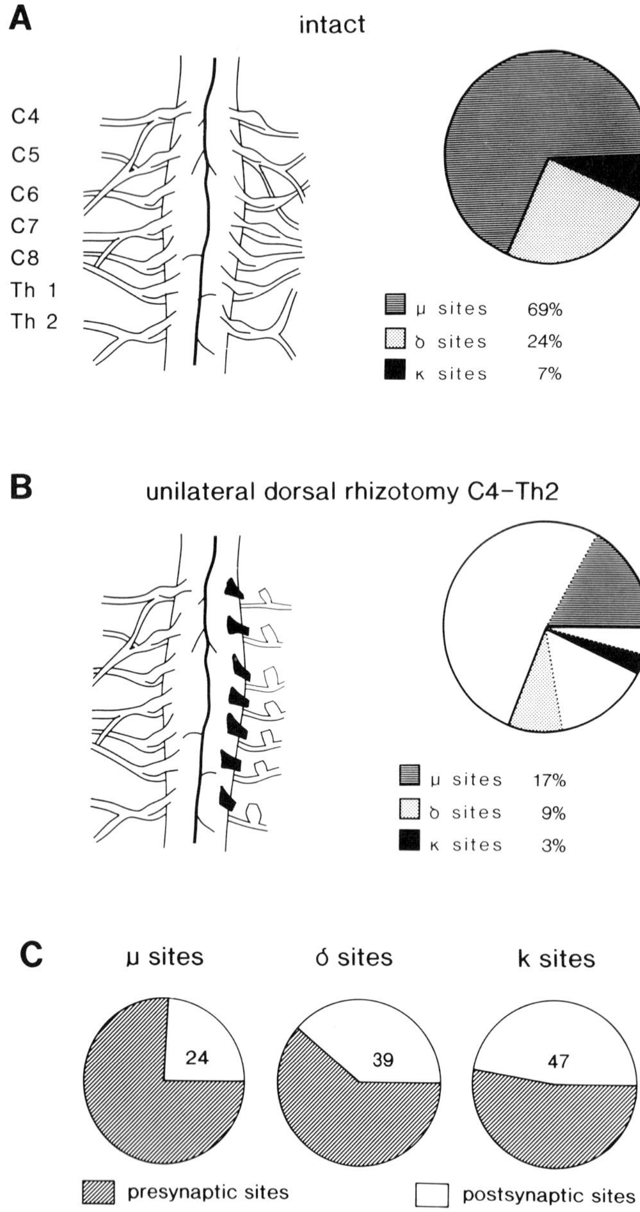

FIG. 1. Proportions of mu-, delta-, and kappa-opioid binding sites in superficial layers of the dorsal horn at the C7 segment level. **A:** Proportions of three types of sites in intact rats. **B:** Proportions of remaining binding sites after a unilateral dorsal rhizotomy C4-Th2 compared with total population. Notice that after rhizotomy, the sum of remaining binding sites represents 29% of the total amount observed in control rats (*white areas* corresponding to the loss of binding sites). **C:** Respective percentages of pre- and postsynaptic distribution of mu (DAGO), delta (DTLET), and kappa (EKC) binding sites.

ported here must be taken with care because the binding measured with (3H)EKC under kappa-selective conditions (60) was low and the accuracy of the measurement therefore is poor.

The loss of mu and delta binding sites is higher than those observed in previous studies (28–31). This is probably due to the larger unilateral rhizotomy performed in this study to include fine afferent fibers over several segments. These fibers do not seem to project over distances greater than two to three segments rostrally and caudally from their entry in the spinal cord (68–71). Thus, it is reasonable to assume that after a large deafferentation, such as that made here, the C7 segment, corresponding to the center of the lesion, will no longer receive any fine afferent fibers. Such an hypothesis is in good agreement with the fact that after section of seven dorsal roots, the percentages of opioid binding sites are not significantly different in the segments C6, C7, and C8. Not surprisingly, the decrease is more progressive on the edge of the deafferented area. Thus, at the level of the C7 segment, it is reasonable to conclude that the decrease in labeling is due to the loss of presynaptic binding sites located on the primary afferent fibers.

The remaining binding sites probably reflect the existence of significant populations of postsynaptic binding sites. This assertion is supported by the fact that in a previous study (30) we reported the same decrease in mu and delta binding sites at both 15 days and 4 months after the rhizotomy, suggesting that the transynaptic degeneration, which generally appears 3 weeks after the rhizotomy (see references in ref. 72), does not eliminate a significant number of opioid binding sites.

The existence of postsynaptic opioid binding sites is directly supported by immunohistochemical demonstrations of the existence of enkephalinergic synapses on dorsal horn neurons at the origin of the spinothalamic tract (48,49). In this respect, it must be noted that after rhizotomy, the decrease in delta sites was less than that observed for mu sites.

The quantitative studies discussed seem to indicate that the effects of morphine on the transmission of nociceptive messages at the spinal level is predominantly due to a presynaptic action because 76% of the mu receptors are located on fine-diameter primary afferent fibers. We have followed up this question with functional studies on the electrophysiological effects of morphine on the spontaneous activity of nociceptive neurons in the dorsal horn of the deafferented rat.

Relative Importance of Pre- and Postsynaptic Effects of Morphine on the Transmission of Noxious Messages in the Dorsal Horn of the Rat Spinal Cord

Results from these investigations performed in spinal rats have been published in detail elsewhere (53). In brief, the first series of experiments was performed in unanesthetized spinal rats (C1), which underwent a large rhizotomy (C5-Th1) at least 10 days before the recording sessions. Because these neurons have no detectable receptive fields, we have therefore attempted to classify the cells by pharmacological means. Many electrophysiological studies have clearly shown that the depressive effect of morphine is selectively exerted on dorsal horn neurons responding to noxious stimuli rather on cells responding exclusively to nonnociceptive stimuli. Using either 2 mg/kg i.v. morphine or 4 mg/kg, we presume that the cells that were depressed by the opiate could be considered as nociceptive neurons. Obviously, this is somewhat arbitrary, but it does not prejudice our conclusions as to a postsynaptic action of morphine, because with the doses of 2 + 4 mg/kg, nearly two-thirds of the cells had their activity depressed by morphine.

In these experiments we found that the depressive effects of morphine on spontaneous hyperactivity of dorsal horn neurons recorded in the deafferented rats were dose-dependent, and the mean firing rate was decreased to 76.48±8.39% ($p<0.01$) and 45.45±6.16% ($p<0.01$) of the control values, respectively, with 2 and 2 + 4 mg/kg of morphine. These effects were clearly reversed by the opiate antagonist naloxone.

To evaluate the relative roles of pre- and postsynaptic mechanisms in the inhibitory effects of morphine, we then compared the above findings with those obtained in nondeafferented animals. We have chosen the arthritic rat because in this model dorsal horn neurons have a marked spontaneous activity (54) compared with the low levels or complete lack found in normal rats.

In these experiments performed on spinal animals all the units were characterized as nociceptive nonspecific neurons with clear C fiber inputs. Their mean spontaneous activity (35.72±6.62 Hz, n = 13) was similar to that found in the deafferented rat (28.10±5.78 Hz, n = 14). In arthritic rats, morphine strongly depressed the spontaneous activity of all neurons tested. With the dose of 2 mg/kg, their firing rate was reduced to 46.03±7.47% ($p<0.001$) of the control values (Fig. 2). These effects were clearly reversed by naloxone.

From these comparative electrophysiological studies, it appears that morphine at the dose of 2 mg/kg i.v. was only half as effective in the deafferented rat compared with the normal rat. Thus the removal of presynaptic opioid receptors diminishes by about half the depressive effects of morphine on the spontaneous activity of nociceptive dorsal horn neurons. These experiments are the first attempts to define physiologically the relative involvement of pre- and postsynaptic mechanisms in the direct spinal action of morphine. Although speculative, our data suggest that both mechanisms are of similar importance.

These results can also explain why morphine is more effective in combating pain originating in nociceptors than for pain arising from deafferentation.

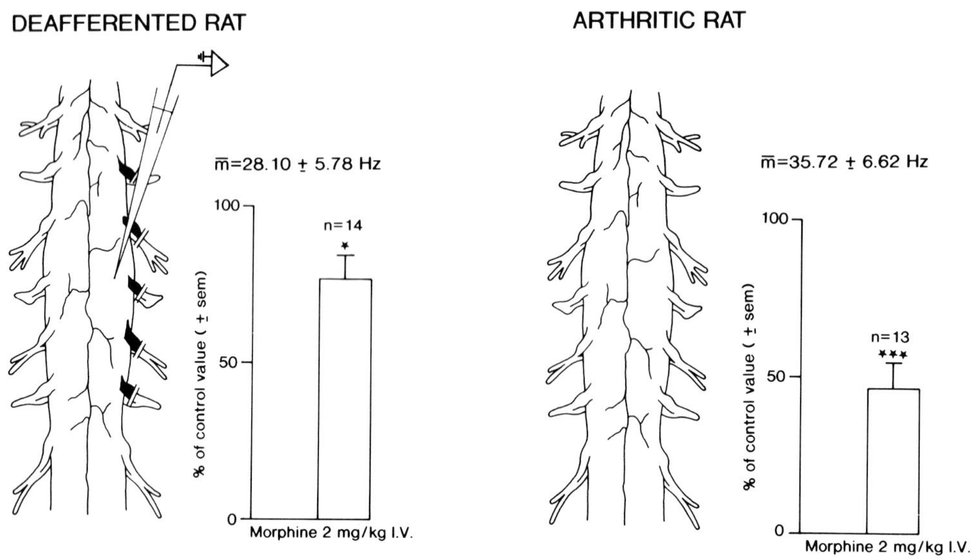

FIG. 2. Comparison of the effect of 2 mg/kg i.v. of morphine on spontaneous activity of dorsal horn neurons in the deafferented and the arthritic rat (see text).

FUNCTIONAL SIGNIFICANCE OF SEGMENTAL INTRINSIC OPIOID SYSTEMS OF THE DORSAL HORN

As discussed, the studies on the direct, spinal depressive action of morphine have arrived at a present good consensus, those on the roles, at the spinal level, of the endogenous opioids are far from being homogeneous and are still the source of contradictions.

A large number of experimental approaches have been applied to this problem, which often gives rise to difficulties in interpretation. We will mention here some of the major points:

- When a drug is administered systematically, its observed effects could result from an action at any level of the nervous system or even at other sites producing indirect effects.
- The use of microinjection techniques requires a consideration of the diffusion of the injected substance and the high local concentration obtained, which could be much greater than that after systemic injection or the concentration in the synaptic cleft under physiological conditions. Similar caution must be applied to iontophoretic studies in which, additionally, the observed effect could result from an indirect action on neurons other than those recorded.
- Another difficulty results from the various tests (some depend on spinal mechanisms, whereas others also involve supraspinal structures) used in evaluating the behavioral effects. In some behavioral experiments only one nociceptive test is used, and the exact method can vary from author to author.

A major problem relating to the functional significance of the intrinsic opioid systems of the dorsal horn is to understand how these systems are triggered. Many questions remain to be answered.

Tonic Activity in the Spinal Cord Intrinsic Opioid System

Because inhibition of firing is the most frequently reported effect of opioids, the opiate antagonist naloxone has been often used to evaluate the activity of this system. Analysis of the action of naloxone on spinal reflexes in spinal animals has frequently resulted in reports of a facilitation of mono- and polysynaptic reflexes. These latter reflexes were in certain cases induced by nociceptive C-fiber stimulation (see references in ref. 19). However, in paraplegic patients (73), low doses of naloxone are without effect on both mono- and polysynaptic reflexes.

Results obtained in intact animals are difficult to interpret because the effects of naloxone could result from actions of the drug at either spinal or, indeed, supraspinal structures. If we restrict discussion to results obtained in the spinal animal even here, the effects of naloxone on the activity of dorsal horn neurons are not uniform in the literature, because Duggan et al. (74) found only two of 13 neurons in spinal cats that were excited by naloxone, whereas others have shown a general facilitatory effect on responses of deep dorsal horn neurons to noxious stimulation (75–77). In addition, topical application of naloxone (e.g., by microiontophoresis or pneumatic application) has also been shown to increase or decrease (lowest doses) the firing rate of dorsal horn neurons (78). Several suggestions have been put forward to explain this discrepancy. However, we believe that the main reason could be because most experimental studies on pain in animals are concerned with nociception, whereas chronic pain, present in many clinical diseases, is known to be associated with long-lasting pathophysiological neural mechanisms, which differ markedly from those observed in the acute situation. We therefore recently evaluated the effect of naloxone on the activities of nociceptive dorsal horn neurons in an experimental model of chronic inflammatory pain, the arthritic rat (see reference in ref. 79). Interestingly, in these animals that experience a diffuse persistent inflammatory state, there are behavioral signs of hyperalgesia (80) and changes in endogenous opioids such that both their levels and biosynthesis are increased in the dorsal horn (81–87).

In these experiments (Fig. 3), performed on 23 nociceptive-nonspecific dorsal horn cells in the spinal rat (88), naloxone (1 mg/kg i.v.) induced a highly significant increase in the spontaneous firing rate of these neurons. In addition, the same dose of naloxone facilitates the transmission of noxious messages at the spinal level as revealed by the large enhancement of the responses of these neurons to C-fiber stimulation. These results are in good agreement with behavioral data showing that such a relatively high dose of naloxone induces well-reproducible hyperalgesia (see reference in ref. 89) and with the previously discussed biochemical observations showing an increase in the levels and biosynthesis of endogenous opioids in the spinal cord of the arthritic rat.

These observations favor the idea of tonic activity of segmental spinal endogenous systems during chronic pain.

Activation of the Segmental Opioid System

In the earlier discussion of the results obtained in animals and in humans with transcutaneous stimulation, we saw that certain

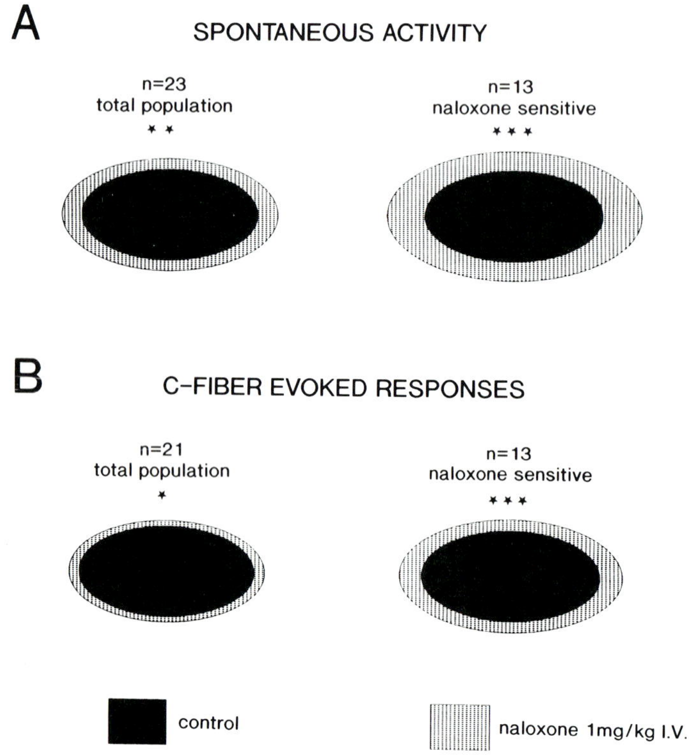

FIG. 3. Effects of naloxone (1 mg/kg i.v.) on activity of nociceptive nonspecific neurons in the arthritic rat. **A:** Effects on spontaneous activity; **B:** effects on C-fiber-evoked discharge.

analgesic or inhibitory effects on neurons at the origin of the spinothalamic tract are partially reversed by naloxone. Chung (17) clearly showed that these effects were principally due to segmental mechanisms, which suggests that the activation of afferent nociceptive fibers can drive enkephalinergic neurons in the dorsal horn, which in turn, by pre- and postsynaptic modulation, can inhibit the transmission of nociceptive information at the spinal level.

In a similar situation to that seen previously, there seems to be an increase in the tonic activity of the segmental intrinsic opioid system in the arthritic rat (88). It can therefore be suggested that nociceptive stimuli activate segmental opioids controls, which have the functional role of damping down excessive nociception; this would agree with certain neurochemical studies on the release of a variety of endogenous opioids at the segmental level.

Modification of the Activity of the Spinal Endogenous Opioid Systems by Nociception and Pain

After the discovery of endogenous opioids, many groups attempted to correlate cerebrospinal fluid (CSF) levels of these substances with pain. The CSF content was presumed to reflect the activity of systems whose terminals lie close to the surface of the CNS. However, it must be underlined that peptides are unstable and that materials found in the CSF samples may represent metabolites.

Various studies have been carried out in animal preparations in which the spinal cord is perfused with artificial CSF. Under these conditions, Yaksh and Elde (90) provided evidence (even in the spinal state) for the spontaneous release of material closely related to Met-enkephalin. The spontaneous release of authentic Met-enkephalin was also reported in spinal superfusates of intact rats (91). In addition, in the cat, spontaneous release was not modified by peripheral electrical stimulation that activates only A fibers; in contrast, increasing the stimulation intensity to excite additionally A delta and C fibers led to an increase in the level of Met-enkephalin-like material (90,92). Tooth pulp stimulation in the cat has also been shown to increase levels of Met-enkephalin-like material in the cisternal CSF (93).

However, these experiments are difficult to interpret because the release of peptides seen in these latter cases could equally result from an activation of segmental mechanisms and/or descending controls, giving rise to negative feedback mechanisms (94).

We will limit discussion here to those studies devoted to the segmental release of endogenous opioids without detailing too much the different substances.

Based on *in vitro* and *in vivo* release studies, it has been shown that SP can cause the release of several endogenous opioids such as Met-enkephalin, ME-Arg-Phe^7, and ME-Arg^6-Gly^7-Leu^8 (95–97). This finding is in good agreement with many other approaches using superfusion of the cord with artificial CSF, in particular the finding that a nociceptive stimulation is required for production of this release. This has clearly been shown after stimulation of the sciatic nerve with intense electrical stimulation (90,98,99) after administration of bradykinin into the femoral artery (90), after nociceptive thermal stimulation (100), or after subcutaneous injection of formalin (101). These results clearly suggest that the segmental intrinsic endogenous opioids system is activated by nociceptive stimulation. As already mentioned, in some experiments it has been demonstrated that the addition of morphine to the spinal superfusate results in a significant reduction in the levels of Met-enkephalin-like immunoreactivity released by sciatic nerve stimulation.

These studies agree well with the fact that increased biosynthesis of opioid peptides in the spinal cord was demonstrated immunologically in rat models of chronic arthritis (81,88). Interestingly, the same disease has been shown to cause increased expression of messenger ribonucleic acid (mRNA) for prodynorphin in several dorsal horn neurons (81,83–86,88). More recently, Noguchi et al. (87) reported an increase in the expression of preproenkephalin-mRNA in dorsal horn neurons after subcutaneous injection of formalin. These results indicate that opioid neurons, which are situated postsynaptically in the spinal cord, are closely related to information processing of the pain sensation.

These biochemical results implicate the spinal cord endogenous opioid systems in nociception and pain. Overall the system is complex, with interplay between various opioid substances. The changes in individual opioid need to be analyzed in relation to nociception and different forms of chronic pain. In addition, the substances released (e.g., Met-enkephalin and dynorphin) have different affinities for the various opioid receptors. Changes in the characteristics of these in relation to their location, either presynaptic on afferent fibers or postsynaptic on dorsal horn neurons, need to be studied.

CONCLUSION

The nonopioid mechanisms implicated in the segmental modulation of nociceptive transmission appear to be difficult to manipulate pharmacologically. A better understanding will undoubtedly give rise to important clinical progress. In contrast, the direct depressive effects of morphine and enkephalins on the transmission of nociceptive messages at the spinal level are clearly established, because their systemic or iontophoretic administration selectively reduces the responses of dorsal horn neurons to nociceptive inputs. The presence of high percentages of opioid receptors on thin primary afferent fibers, the lack of evidence in favor of the existence of enkephalinergic axo-axonic synapses, and the fact that in other CNS areas the opioids seem to act primarily at nonjunctional interfaces suggests that some of these depressive effects are relevant to volume transmission.

The functional implication of the spinal opioid system is well-supported by the powerful and long-lasting analgesic effects seen in animals after intrathecal administration of morphine. Clinical application of this finding was immediate, and today many clinicians use this route of administration for the relief of intractable pain. Clinical effects clearly demonstrate that activation of spinal opioid receptors inhibits the transmission of nociceptive messages. It seems essential to identify precisely the respective roles of the different opioid receptors, which could differ according to different types of pain. Indeed, different opioid agonists applied intrathecally produce analgesic effects that depend on the type of cutaneous nociceptive stimulus (mechanical, thermal, and chemical), visceral or muscular, used to gauge the effects (102).

REFERENCES

1. Besson JM, Chaouch A. Peripheral and spinal mechanisms of nociception. *Physiol Rev* 1987; 67:67–186.
2. Willis DW. *The pain system. The neural basis of nociceptive transmission in the mammalian nervous system.* Basel: S. Karger, 1985.
3. Melzack R, Wall PD. Pain mechanisms: a new theory. *Science* 1965;150:971–979.
4. Sjölund BH, Eriksson MBE. The influence of naloxone on analgesia produced by peripheral conditioning stimulation. *Brain Res* 1979;173: 295–301.
5. Willer JC, Roby A, Boulou P, Albe-Fessard D. Depressive effect of high frequency peripheral conditioning stimulation upon the nociceptive component of the human blink-reflex. Lack of naloxone effect. *Brain Res* 1982;239:322–326.
6. Freeman TB, Campbell JN, Long DM. Naloxone does not affect pain relief induced by electrical stimulation in man. *Pain* 1983;17:189–195.
7. Duggan AW, Foong FW. Bicuculline and spinal inhibition produced by dorsal column stimulation in the cat. *Pain* 1985;22:249–259.
8. Barber RP, Vaughn JE, Roberts E. The cytoarchitecture of GABAergic neurons in rat spinal cord. *Brain Res* 1982;238:305–328.
9. Barber RP, Vaughn JE, Saito K, Mc Laughlin BJ, Roberts E. GABAergic terminals are pre-

synaptic to primary afferent terminals in the substantia gelatinosa of the rat spinal cord. *Brain Res* 1978;141:35–55.

10. Desarmenien M, Feltz P, Occhipinti G, Santangelo F, Schliter R. Coexistence of GABA-A and GABA-B receptors on A and C primary afferents. *Br J Pharmacol* 1984;81:327–333.
11. Richards JG, Schoch P, Haring P, Takacs B, Mohler H. Resolving GABA A/Benzodiazepine receptors: cellular and subcellular localization in the CNS with monoclonal antibodies. *J Neurosci* 1987;7:1866–1886.
12. Haefely W, Polc P. Physiology of GABA enhancement by benzodiazepines and barbiturates. In: Olsen RW, Venter JC, eds. *Benzodiazepines GABA receptors and chloride channels: structural and functional properties*. New York: Alan R. Liss, 1986;97–133.
13. Le Bars D, Chitour D, Kraus E, Dickenson AH, Besson JM. Effects of naloxone upon diffuse noxious inhibitory controls (DNIC) in the rat. *Brain Res* 1981;204:387–402.
14. Chung JM, Fang ZR, Hori Y, Lee KH, Willis WD. Prolonged inhibition of primate spinothalamic tract cells by peripheral nerve stimulation. *Pain* 1984;19:259–275.
15. Chung JM, Lee KH, Hori Y, Endo K, Willis WD. Factors influencing peripheral nerve stimulation produced inhibition of primate spinothalamic tract cells. *Pain* 1984;19:277–293.
16. Lee KH, Chung JM, Willis WD. Inhibition of primate spinothalamic tract cells by TENS. *J Neurosurg* 1985;62:276–287.
17. Chung JM. Neurophysiological mechanisms of acupuncture analgesia in experimental animal models. In: Pomeranz B, Stux G, eds. *Scientific bases of acupuncture*. Berlin: Springer-Verlag, 1988;137–155.
18. Kriz N, Sykova E, Vycklicky L. Extracellular potassium changes in the spinal cord of the cat and their relation to slow potentials, active transport and impulse transmission. *J Physiol (Lond)* 1975;249:167–182.
19. Duggan AW, North RA. Electrophysiology of opioids. *Pharmacol Rev* 1984;35:219–281.
20. Zieglgänsberger W. Central control of nociception. In: Mountcastle VB, Bloom FE, Geiger SR, eds. *Handbook of physiology. The nervous system*, vol IV. Baltimore, Maryland: Williams and Wilkins, 1986.
21. Atweh SF, Kuhar MJ. Autoradiographic localization of opiate receptors in rat brain. I—Spinal cord and lower medulla. *Brain Res* 1977; 124:63–67.
22. Atweh SF, Murrin L, Kuhar MJ. Presynaptic localization of opiate receptor in the vagal and accessory optic-systems: an autoradiographic study. *Neuropharmacology* 1978;17:65–71.
23. Hiller JM, Simon EJ, Crain SM, Peterson ER. Opiate receptors in cultures of fetal mouse dorsal root ganglia (DRG) and spinal cord: predominence in DRG neurites. *Brain Res* 1978; 145:396–400.
24. Fields HL, Emson PC, Leigh BK, Gilbert RFT, Iversen LL. Multiple opiate receptor sites on primary afferent fibres. *Nature* 1980;284:351–353.
25. Young WS III, Wamsley JK, Zarbin MA, Kuhar MJ. Opioid receptors undergo axonal flow. *Science* 1980;210:76–78.
26. Ninkovic M, Hunt SP, Gleave JRW. Localization of opiate and histamine H1-receptors in the primate sensory ganglia and spinal cord. *Brain Res* 1982;241:197–206.
27. Lamotte C, Pert CB, Snyder SH. Opiate receptor binding in primate spinal cord: distribution and changes after dorsal root section. *Brain Res* 1976;112:407–412.
28. Ninkovic M, Hunt SP, Kelly JS. Effect of dorsal rhizotomy on the autoradiographic distribution of opiate and neurotensin receptors and neurotensin-like immunoreactivity within the rat spinal cord. *Brain Res* 1981;230:111–119.
29. Daval G, Verge D, Basbaum AI, Bourgoin S, Hamon M. Autoradiographic evidence of serotoninl binding sites on primary afferent fibres in the dorsal horn of the rat spinal cord. *Neurosci Lett* 1987;83:71–76.
30. Zajac JM, Lombard MC, Peschanski M, Besson JM, Roques BP. Autoradiographic study of μ and opioid binding sites and neutral endopeptidase 24-11 in rat after dorsal root rhizotomy. *Brain Res* 1989;477:400–403.
31. Jessell TM, Tsungo A, Kanazawa I, Otsuka M. Substance P: depletion in the dorsal horn of rat spinal cord after section of the peripheral processes of primary sensory neurons. *Brain Res* 1979;168:247–259.
32. Gamse R, Holzer P, Lembeck F. Indirect evidence for presynaptic location of opiate receptors on chemosensitive primary sensory neurones. *Naunyn Schmiedebergs Arch Pharmacol* 1979; 308:281–285.
33. Nagy JI, Hunt SP, Iversen LL, Emson PC. Biochemical and anatomical observations on the degeneration of peptide-containing primary afferent neurons after neonatal capsaicin. *Neuroscience* 1981;6:1923–1934.
34. Jessell TM, Iverson LL. Opiate analgesics inhibit substance P release from rat trigeminal nucleus. *Nature* 1977;268:549–551.
35. Hunt SP, Kelly JS, Emson PC. The electron microscopic localization of methionine enkephalin within the superficial layers (I and II) of the spinal cord. *Neuroscience* 1980;5:1871–1890.
36. Aronin N, Difiglia M, Liotta AS, Martin JS. Ultrastructural localization and biochemical features of immunoreactive Leu-enkephalin in monkey dorsal horn. *J Neurosci* 1981;1:561–577.
37. Sumal KK, Pickel VM, Miller RJ, Reis DJ. Enkephalin-containing neurons in substantia gelatinosa of spinal trigeminal complex: ultrastructure and synaptic interaction with primary sensory afferents. *Brain Res* 1982;248:223–236.
38. Glazer EJ, Basbaum AI. Opioid neurons and pain modulation: an ultrastructural analysis of enkephalin in cat superficial dorsal horn. *Neuroscience* 1983;10:357–376.
39. Lamotte CC, de Lanerolle NC. Ultrastructure

of chemically defined neuron-systems in the dorsal horn of the monkey. II—Methionine-enkephalin immunoreactivity. *Brain Res* 1983; 274:51–63.

40. Bennett GJ, Ruda MA, Gobel S, Dubner R. Enkephalin immunoreactive stalked cells and lamina IIb islet cells in cat substantia gelatinosa. *Brain Res* 1982;240:162–166.
41. Del Fiacco M, Paxinos G, Cuello C. Neostriatal enkephalin immunoreactive neurons project to the globus pallidus. *Brain Res* 1982; 231:1–17.
42. Hamel E, Beaudet A. Electron microscopic autoradiographic localization of opioid receptors in rat neostriatum. *Nature* 1984;312:155–157.
43. Zieglgänsberger W, Bayerl H. The mechanism of inhibition of neuronal activity by opiates in the spinal cord of the cat. *Brain Res* 1976; 115:111–128.
44. Barker JL, Smith TG, Neale JH. Multiple membrane actions of enkephalin revealed using cultured spinal neurons. *Brain Res* 1978;154: 153–158.
45. Zieglgänsberger W, Tulloch IF. The effects of methionine- and leucine-enkephalin on spinal neurones of the cat. *Brain Res* 1979;167:53–64.
46. Barker JL, Gruol DL, Huang M, MacDonald JF, Smith TG. Peptide receptor functions on cultured spinal neurons. In: Costa E, Trabucchi M, eds. *Neural peptides and neuronal communications*. New York: Raven Press, 1980; 409–423.
47. Yoshimura M, North RA. Substantia gelatinosa neurones hyperpolarized *in vitro* by enkephalin. *Nature* 1983;305:529–530.
48. Ruda MA. Opiates and pain pathways: demonstration of enkephalin synapses on dorsal horn projection neurones. *Science* 1982;215: 1523–1525.
49. Ruda MA, Coffield J, Dubner R. Demonstration of postsynaptic opioid modulation of thalamic projection neurons by the combined techniques of retrograde horseradish peroxidase and enkephalin immunohistochemistry. *J Neurosci* 1984;4:2117–2132.
50. Hylden JLK, Wilcox GL. Pharmacological characterization of substance P–induced nociception in mice: modulation by opioid and noradrenergic agonists at the spinal level. *J Pharmacol Exp Ther* 1983;226:398–404.
51. Wilcockson WS, Kim J, Shin HK, Chung JM, Willis WD. Actions of opioids on primate spinothalamic tract neurones. *J Neurosci* 1986; 6:2509–2520.
52. Besse D, Lombard MC, Zajac JM, Roques BP, Besson JM. Pre- and postsynaptic location of mu, delta and kappa opioid receptors in the superficial layers of the dorsal horn of the rat spinal cord. In: Quirion R, ed. *International narcotic research conference* (INRC). New York: Alan R. Liss, 1989.
53. Lombard MC, Besson JM. Attempts to gauge the relative importance of pre- and postsynaptic effects of morphine on the transmission of noxious messages in the dorsal horn of the rat spinal cord. *Pain* 1989;37:335–345.
54. Menétrey D, Besson JM. Electrophysiological characteristics of dorsal horn cells in rats with cutaneous inflammation resulting from chronic arthritis. *Pain* 1982;13:343–364.
55. Gouarderes C, Cros J. Opioid binding sites in different levels of the rat spinal cord. *Neuropeptides* 1984;5:113–116.
56. Mack KJ, Killian A, Weyhenmeyer JA. Comparison of mu, delta, and kappa opiate binding sites in rat brain and spinal cord. *Life Sci* 1984;34:281–285.
57. Traynor JR, Wood MS. Distribution of opioid binding sites in spinal cord. *Neuropeptides* 1987;10:313–320.
58. Slater P, Patel S. Autoradiographic localization of opiate k-receptors in the rat spinal cord. *Eur J Pharmacol* 1983;92:159–160.
59. Gouarderes C, Cros J, Quirion R. Autoradiographic localization of mu, delta and kappa opioid receptor binding sites in rat and guinea pig spinal cord. *Neuropeptides* 1985;6:331–342.
60. Delay-Goyet P, Roques BP, Zajac JM. Differences of binding characteristics of non-selective opiates towards mu and delta receptors types. *Life Sci* 1987;41:723–731.
61. Lahti RA, Mickelson MM, McCall JM, Von Voigtlander PF. (^{3}H)U.69593 a highly selective ligand for the opioid k receptor. *Eur J Pharmacol* 1985;109:281–284.
62. Morris BJ, Herz A. Distinct distribution of opioid receptor types in rat lumbar spinal cord. *Naunyn Schmiedebergs Arch Pharmacol* 1987;336:240–243.
63. Dickenson AH, Sullivan AF, Knox R, Zajac JM, Roques BP. Opioid receptor subtypes in the rat spinal cord: electrophysiological studies with μ and δ-opioid receptor agonists in the control of nociception. *Brain Res* 1987;413:36–44.
64. Knox RJ, Dickenson AH. Effects of selective and non-selective k-opioid receptor agonists on cutaneous C-fibre-evoked responses of rat dorsal horn neurones. *Brain Res* 1987;415:21–29.
65. Porreca F, Mosberg HI, Hurst R, Hruby VJ, Burks TF. Roles of mu, delta and kappa opioid receptors in spinal and supraspinal mediation of gastrointestinal transit effects and hot-plate analgesia in the mouse. *J Pharmacol Exp Ther* 1984;230;2:341–348.
66. Leighton GE, Rodriguez RE, Hill RG, Hughes J. K-opioid agonists produce antinociception after i.v. and i.c.v. but not intrathecal administration in the rat. *Br J Pharmacol* 1988;93: 553–560.
67. Yaksh TL, Durant P, Onofrio B, Stevens CW. The effect of spinally administered agents on pain transmission in man and animals. In: Besson JM, Lazorthes Y, eds. *Substances opioïdes médullaires et analgésie (Spinal opioids and the relief of pain)*. Paris: INSERM (Colloque INSERM), 1984;267–306.
68. Chung K, Langford LA, Applebaum AE,

Coggeshall RE. Primary afferent fibers in the tract of Lissauer in the rat. *J Comp Neurol* 1979;184:587–598.

69. Sugiura Y, Lee CL, Perl ER. Central projections of identified, unmyelinated (C) afferent fibers innervating mammalian skin. *Science* 1986;234:358–361.
70. Mendell LM, Koerber HR, Traub RJ. The spared root preparation: evidence for selective changes of projections from surviving fibers. In: Pubols CM, Sessle BJ, eds. *Neurology and neurobiology: effects of injury on trigeminal and spinal somatosensory systems*. New York: Alan R. Liss, 1987;30:249–259.
71. Sugiura Y, Terui N, Hosoya Y, Kohno K. Distribution of unmyelinated primary afferent fi bers in the dorsal horn. In: Cervero F, Bennett GJ, Headley PM, eds. *Processing of sensory information in the superficial horn of the spinal cord*. NATO Asi Series, vol 176. New York: Plenum Press, 1989;15–27.
72. Csillik B, Knyihar-Csillik E. *The protean gate*. Budapest: Adademiai Kiado, 1986.
73. Willer JC, Bussel B. Evidence for a direct spinal mechanism in morphine-induced inhibition of nociceptive reflexes in humans. *Brain Res* 1980;187:212–215.
74. Duggan AW, Hall JG, Headley PM, Griersmith BT. The effect of naloxone on the excitation of dorsal horn neurones of the cat by noxious and non-noxious cutaneous stimuli. *Brain Res* 1977;138:185–189.
75. Henry JL. Naloxone excites nociceptive units in the lumbar dorsal horn of the spinal cat. *Neuroscience* 1979;4:1485–1491.
76. Fitzgerald M, Woolf CJ. The stereospecific effect of naloxone on rat dorsal horn neurones: inhibition in superficial laminae and excitation in deeper laminae. *Pain* 1980;9:293–306.
77. Henry JL. Diurnal variation in excitation of dorsal horn units by naloxone in the spinal cat suggests a circulating opioid factor. *Neuroscience* 1981;6:1935–1942.
78. Mayer C, Zieglgänsberger W. Peptidergic neurotransmission in the dorsal horn of the rat spinal cord. In: Schmidt RF, Schaible HG, Vahle-Hinz C, eds. *Fine afferent nerve fibers and pain*. Weinheim: VCH, 1987;251–262.
79. Besson JM, Guilbaud G. *The arthritic rat as a model of clinical pain?* Amsterdam: Experta Medica ICS 837, 1988.
80. Calvino B, Crepon-Bernard MO, Le Bars D. Parallel clinical and behavioural studies of adjuvant-induced arthritis in the rat: possible relationship with "chronic pain." *Behav Brain Res* 1987;24:11–29.
81. Cesselin F, Montastruc JL, Gros C, Bourgoin S, Hamon M. Met-enkephalin levels and opioid receptors in the spinal cord of chronic suffering rats. *Brain Res* 1980;191:25–31.
82. Faccini E, Uzumaki H, Govoni S, et al. Afferent fibers mediate the increase of Met-enkephalin elicited in rat spinal cord by localized pain. *Pain* 1984;18:25–31.
83. Hollt V, Haarmann I, Millan MJ, Herz A. Prodynorphin gene expression is enhanced in the spinal cord of chronic arthritic rats. *Neurosci Lett* 1987;73:90–94.
84. Iadarola MJ, Brady LS, Draisci G, Dubner R. Enhancement of dynorphin gene expression in spinal cord following experimental inflammation: stimulus specificity, behavioral parameters and opioid receptor binding. *Pain* 1988; 35:313–326.
85. Iadarola MJ, Douglass J, Civelli O, Naranjo JR. Differential activation of spinal cord dynorphin and enkephalin neurons during hyperalgesia: evidence using cDNA hybridization. *Brain Res* 1988;455:205–212.
86. Ruda MA, Iadarola MJ, Cohen LV, Young WS. *In situ* hybridization histochemistry and immunocytochemistry reveal an increase in spinal dynorphin biosynthesis in a rat model of peripheral inflammation and hyperalgesia. *PNAS* 1988;85:622–626.
87. Noguchi K, Morita Y, Kiyama H, Sato M, Ono K, Tohyama M. Preproenkephalin gene expression in the rat spinal cord after noxious stimuli. *Mol Brain Res* 1989;5:227–234.
88. Lombard MC, Besson JM. Electrophysiological evidence for a tonic activity of the spinal cord intrinsic opioid systems in a chronic pain model. *Brain Res* 1989;477:48–56.
89. Kayser V, Besson JM, Guilbaud G. Paradoxical effects of low doses of naloxone in experimental models of inflammatory pain. In: Fields HL, Besson JM, eds. *Descending brainstem controls of nociceptive transmission. Progress in Brain Research,* Amsterdam: Elsevier, 1988; 303–314.
90. Yaksh TL, Elde RP. Factors governing release of methionine enkephalin-like immunoreactivity from mesencephalon and spinal cord of the cat *in vivo*. *J Neurophysiol* 1981;46:1056–1075.
91. Cesselin F, Le Bars D, Bourgoin S, et al. Spontaneous and evoked release of Met-enkephalin material from the rat spinal cord *in vivo*. *Brain Res* 1985;339:305–313.
92. Jhamandas K, Yaksh TL, Go VLW. Acute and chronic morphine modifies the *in vivo* release of methionine enkephalin-like immunoreactivity from the cat spinal cord and brain. *Brain Res* 1984;297:91–103.
93. Cesselin F, Oliveras JL, Bourgoin S, et al. Increased levels of Met-enkephalin-like material in the CSF of anaesthetized cats after tooth pulp stimulation. *Brain Res* 1982;237:325–338.
94. Hamon M, Bourgoin S, Le Bars D, Cesselin F. *In vivo* and *in vitro* release of central neurotransmitters in relation to pain and analgesia. In: Fields HL, Besson JM, eds. *Pain modulation. Progress in brain research,* vol 77. Amsterdam: Elsevier, 1988;431–444.
95. Del Rio J, Naranjo JR, Yang HYT, Costa E. Substance P-induced release of Met^5-enkephalin from striatal and periaqueductal gray slices. *Brain Res* 1983;279:121–126.

96. Tang J, Chou J, Yang HYT, Costa E. Substance P stimulates the release of Met^{5}-enkephalin-Arg^{6}-Phe^{7} and Met^{5}-enkephalin from rat spinal cord. *Neuropharmacology* 1983;22:1147–1150.
97. Cesselin F, Bourgoin S, Artaud F, Hamon M. Basic and regulatory mechanisms of *in vitro* release of Met-enkephalin from the dorsal zone of the rat spinal cord. *J Neurochem* 1984;43:763–773.
98. Yaksh TL, Terenius L, Nyberg F, Jhamandas K, Wang JY. Studies on the release by somatic stimulation from rat and cat spinal cord of active materials which displace dihydromorphine in an opiate-binding assay. *Brain Res* 1983; 268:119–128.
99. Iadarola MJ, Tang J, Costa E, Yang HYT. Analgesic activity and release of Met^{5}-enkephalin-Arg^{6}-Gly^{7}-Leu^{8} from rat spinal cord *in vivo*. *Eur J Pharmacol* 1986;121:38–48.
100. Cesselin F, Bourgoin S, Artaud F, et al. The release of Met-enkephalin-like material at the spinal level. *In vivo* and *in vitro* studies. In: Besson JM, Lazorthes Y, eds. *Spinal opioids and the relief of pain*. Paris: INSERM, 1985; 241–267.
101. Bourgoin S, Le Bars D, Clot AM, Hamon M, Cesselin F. Subcutaneous formalin induces a segmental release of Met-enkaphalin-like material from the rat spinal cord. *Pain* 1990;41: 323–329.
102. Yaksh TL, Noveihed R. The physiology and pharmacology of spinal opiates. *Annu Rev Pharmacol Toxicol* 1985;25:433–462.

Volume Transmission in the Brain: Novel Mechanisms for Neural Transmission,
edited by K. Fuxe and L.F. Agnati.
Raven Press, Ltd., New York © 1991.

41

Cerebral Sodium–Angiotensin Interactions Concerning Fluid Balance and ACTH-Cortisol Secretion

Bengt Andersson and Jörgen Junkergård

Department of Physiology, Faculty of Veterinary Medicine, Swedish University of Agricultural Sciences, S-750 07 Uppsala, Sweden

An efficient thirst mechanism and appropriate regulation of antidiuretic hormone (ADH; arginine vasopressin, AVP) secretion are indispensable for optimal water balance, the maintenance of which has high homeostatic priority. The deviations from fluid balance that most effectively elicit thirst and hypersecretion of ADH are a deficit of water without corresponding loss of Na^+ (hypovolemic hypernatremia), and an osmotic shift of water from the cells to the extracellular fluid, induced by excessive salt intake (hypervolemic hypernatremia). Characteristics in common for both conditions are cellular dehydration, hyperosmolality of the body fluids, and elevated extracellular Na^+ concentration. Of these three factors, cellular dehydration has generally been considered to constitute the crucial ADH stimulus, and that kind of regulation has become designated "osmotic" in the light of Verney's (1) fundamental investigations implying that the ADH release is negatively correlated to the volume of hypothalamic osmoreceptors. When microinjections of hypertonic NaCl into the medial hypothalamus were found to induce polydipsia in water-replete goats (2), it was regarded as evidence that osmoreceptors are also involved in the regulation of water intake.

CEREBRAL SODIUM SENSITIVITY

Whether the cerebral sensory mechanism, disclosed by Verney, in fact consists of osmoreceptors or elements more specifically sensitive to Na^+ (3) has subsequently become a controversial question, because it has been shown by use of intracerebroventricular (ICV) infusions that thirst and ADH secretion are positively correlated to the NaCl concentration of the cerebrospinal fluid (CSF), and that these effects of an osmotic stimulus applied from the outside of the blood-brain barrier (BBB) become attenuated by a lowering of the CSF NaCl concentration. Also the renal sodium excretion has been shown to be influenced in the same manner by alterations in the NaCl concentration of the CSF, and ICV infusions of hypertonic NaCl solution have been found to induce moderate rises in the arterial blood pressure (4). Taken together, these observations imply that juxtacerebroventricular sodium sensors play an important role in the homeostatic control of body fluid balance. An indication that such sensors may also be involved in the cerebral regulation of adrenocorticotropic hormone (ACTH)-cortisol secretion has recently been provided in the goat (5). It was shown in this species that infusions of 0.5 M

NaCl into the lateral cerebral ventricle induced a pronounced rise in plasma cortisol concentration.

CENTRALLY MEDIATED ANGIOTENSIN EFFECTS

During the past three decades it has gradually become evident that vasoconstriction with subsequent elevation of the arterial blood pressure is not the only effect of a high blood level of angiotensin II (AII) (6). The discovery that AII, by a direct action on the adrenal cortex, stimulates the synthesis and secretion of aldosterone (7) and the subsequent findings that thirst (8) and release of ADH (9) can be elicited by intravenous (i.v.) administration of the octapeptide have revealed that the renin–angiotensin system (in addition to its role in the regulation of the blood pressure) also forms an important link in the overall control of electrolyte and water balance. Among other effects of circulating AII are stimulation of the ACTH-cortisol secretion (10), heart acceleration, which in some species may mask (11) or even overrun (12) the baroreflex bradycardia associated with an elevation of the arterial blood pressure.

Nearly 30 years ago, Bickerton and Buckley (13) demonstrated that the hypertensive effect of circulating AII to some extent is due to an action of the octapeptide at the medullary level. Conclusive evidence has subsequently been provided that also other effects of systemic AII, in part or entirely, are manifestations of an influence of blood-borne AII on cerebral structures. As regards the dipsogenic effect of systemic AII, it has been shown that drinking in sheep in response to intracarotid infusion of AII can be prevented by the ICV administration of the competitive AII inhibitor saralasin (14). ICV saralasin has recently also been found to inhibit ACTH-cortisol re-

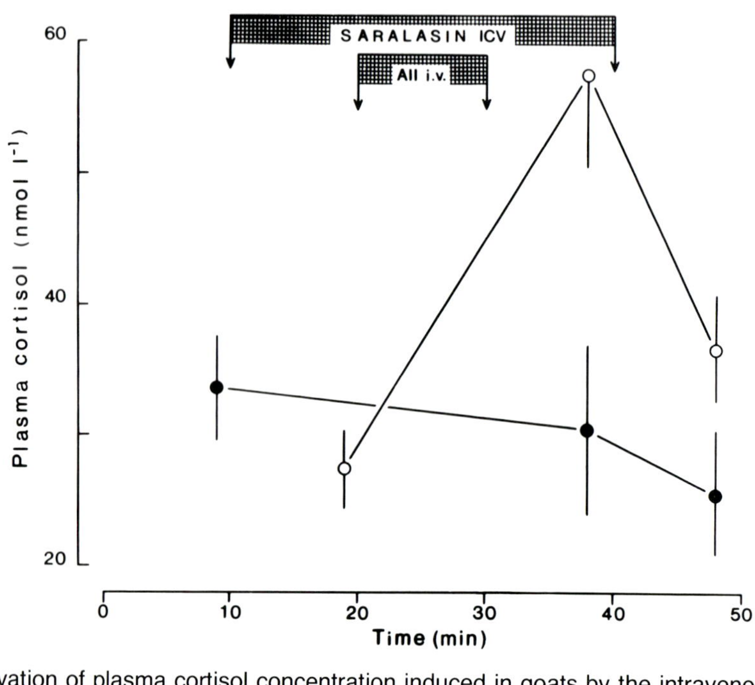

FIG. 1. Elevation of plasma cortisol concentration induced in goats by the intravenous infusion of angiotensin II (40 pmol/kg/min) (○ — ○) compared with the lack of response when a 30-min intracerebroventricular (ICV) infusion of saralasin (13 pmol/kg/min) preceded and outlasted the angiotensin infusion by 10 min (● — ●) (From ref. 16, with permission.)

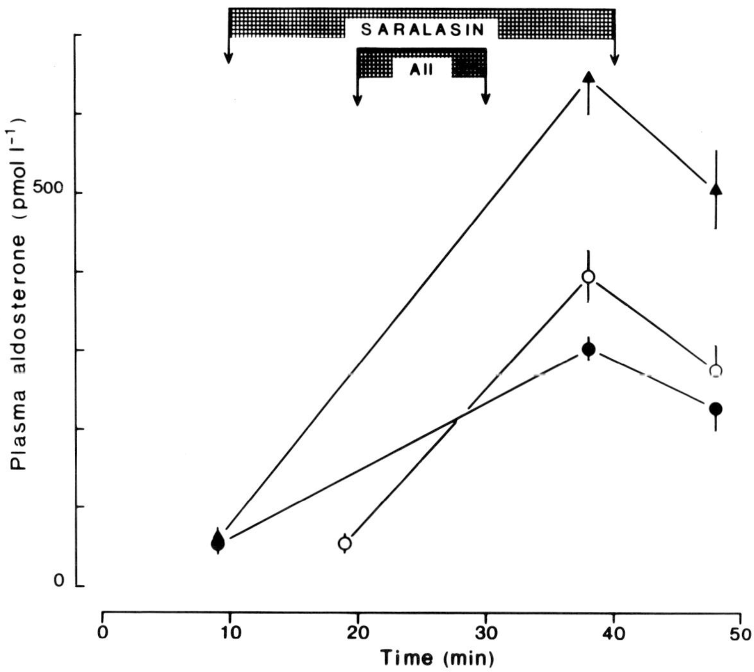

FIG. 2. Increase in plasma aldosterone concentration in goats, elicited by the separate intravenous infusion of angiotensin II (40 pmol/kg/min) (○ — ○). The response became significantly reduced when the infusion of angiotensin was performed during an intracerebroventricular infusion of saralasin (13 pmol/kg/min) (● — ●), suggesting a centrally mediated component in the aldosterone response to systemic AII. When saralasin instead was infused intravenously, the aldosterone response to angiotensin became accentuated (▲ — ▲), indicating that angiotensin and its competitive inhibitor saralasin act synergistically as local stimulators of the adrenal cortex. (From ref. 16, with permission.)

lease as an effect of systemic AII in rats (15) and goats (16). In the latter species, it was demonstrated that infusions of systemically ineffective amounts of saralasin into the lateral cerebral ventricle extinguished the rise in plasma cortisol otherwise obtained in response to the intravenous infusion of AII (Fig. 1). Some reduction of the aldosterone response to intravenous AII was simultaneously observed (Fig. 2), indicating that also this effect of systemic AII to a minor extent is centrally mediated and ACTH-dependent.

The fact that many effects of systemic AII can be reproduced by application of the octapeptide from the inside of the BBB supports that they are centrally mediated but cannot be regarded as any conclusive evidence for this concept. From a physiological point of view, observed responses to the intracerebral application of AII may as well be regarded as replications of an activation of the intrinsic, cerebral renin–angiotensin system (17,18). Nevertheless, it has been shown that intrahypothalamic or ICV administration of AII elicits thirst (19), apparent release of ADH (20), elevation of the arterial blood pressure (21), natriuresis (22), and corticosteroid release (23) (i.e., effects that all can also be obtained in response to the intravenous infusion of AII).

SODIUM–ANGIOTENSIN INTERACTION IN FLUID BALANCE

The dipsogenic, ADH-releasing, hypertensive, and natriuretic effects of ICV administered AII have all been shown to be positively correlated to the NaCl concentration of the CSF, which indicates the presence of a sodium–angiotensin interaction at the cerebral level (22). The evidence for a corresponding sodium dependence of centrally mediated responses to bloodborne AII is less conclusive. However, investigations in the goat have demonstrated a positive correlation between the NaC1 concentration of the CSF and the antidiuretic effect of systemic AII (24). Coherent with that observation is the recent finding in the same species that the plasma ADH and antidiuretic responses to systemic AII are more pronounced in sodium-replete than in mildly sodium-deficient animals (25), and the observation in the dog that intravenous infusions of amounts of AII that are themselves ineffective lower the thirst threshold to elevated plasma Na^+ concentration (26).

SODIUM–ANGIOTENSIN INTERACTION IN ACTH-CORTISOL SECRETION

In view of the apparent influence of CSF Na^+ concentration on AII stimulation of mechanisms involved in the control of salt–water balance, it has become of interest to study whether AII-induced ACTH-cortisol

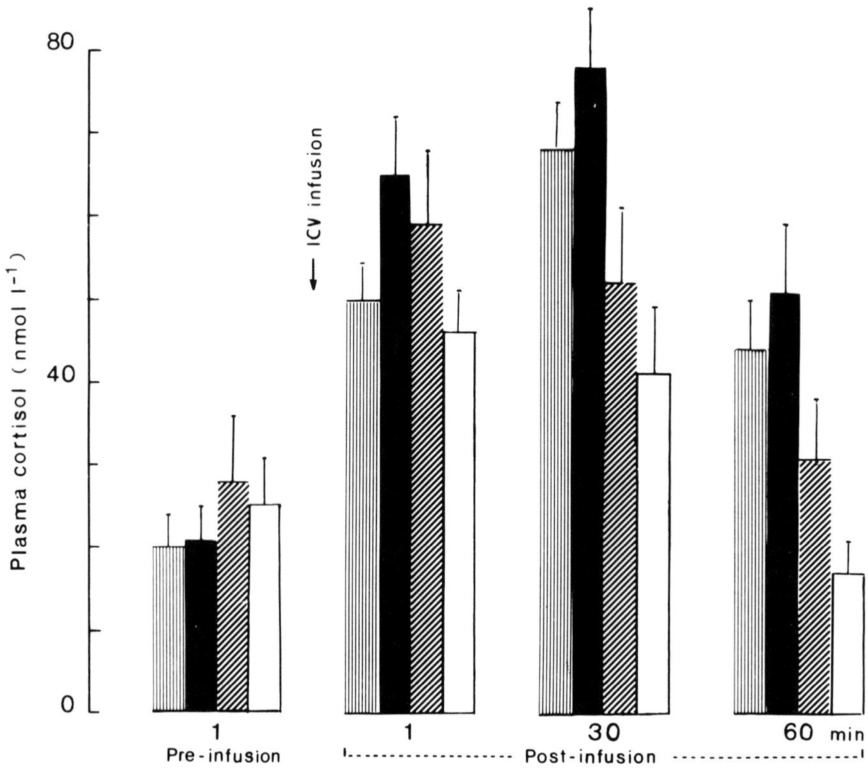

FIG. 3. Stimulation of cortisol secretion in goats by intracerebroventricular (ICV) infusion (20 μl/min for 20 min) of hypertonic NaCl (0.5 M) and angiotensin II (2 pmol/kg/min) dissolved in hypertonic NaCl, isotonic NaCl, respectively, isotonic glucose. ▥, Hypertonic (0.5 M) NaCl; ■, angiotensin II in 0.5 M NaCl; ▨, angiotensin II in 0.15 M NaCl; □, angiotensin II in isotonic *d*-glucose. (From ref. 5, with permission.)

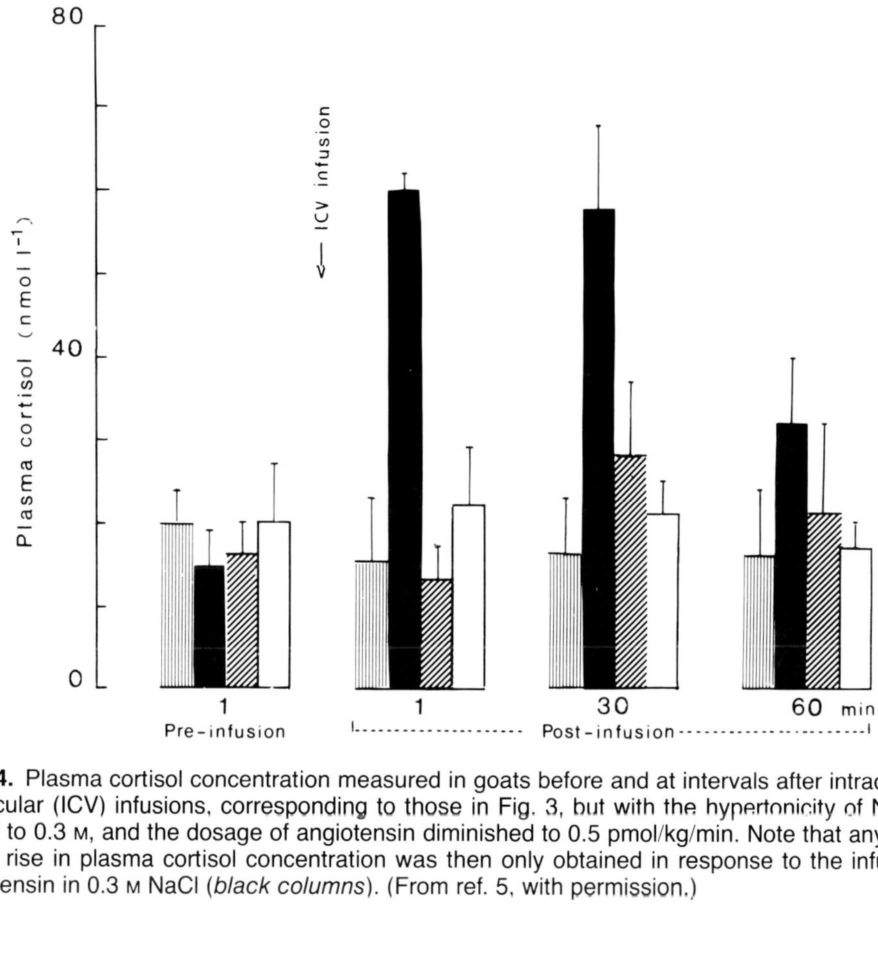

FIG. 4. Plasma cortisol concentration measured in goats before and at intervals after intracerebroventricular (ICV) infusions, corresponding to those in Fig. 3, but with the hypertonicity of NaCl reduced to 0.3 M, and the dosage of angiotensin diminished to 0.5 pmol/kg/min. Note that any appreciable rise in plasma cortisol concentration was then only obtained in response to the infusion of angiotensin in 0.3 M NaCl (*black columns*). (From ref. 5, with permission.)

secretion is affected in the same manner. A comparison between the cortisol-releasing effects of ICV infusions of AII (2 pmol/kg/min) dissolved in hypertonic (0.5 M) NaCl, in isotonic (0.15 M) NaCl, respectively, in isotonic (0.3 M) *d*-glucose was made in the conscious goat (5). As shown in Fig. 3, the most pronounced rise in plasma cortisol concentration was elicited by AII in 0.5 M NaCl. A response of nearly the same order was obtained to AII in isotonic saline, whereas AII in glucose induced a significantly smaller rise. When the dosage of angiotensin was reduced to 0.5 pmol/kg/min, and the hypertonicity of the NaCl to 0.3 M, only the infusion of AII in 0.3 M NaCl elicited any appreciable cortisol increase (Fig. 4). Moreover, when an ICV infusion of isotonic glucose (lowering the CSF Na^+ concentration) preceded and outlasted the intravenous infusion of AII, the rise in plasma cortisol obtained as effect of a separate intravenous infusion of AII became extinguished (Fig. 5). The results suggest that the potency of ICV, as well as of systemic AII to stimulate the pituitary–adrenocortical axis is positively correlated to the Na^+ concentration of the CSF. It might explain the observation that mild sodium depletion significantly reduces the cortisol response to intravenous AII in the goat (25).

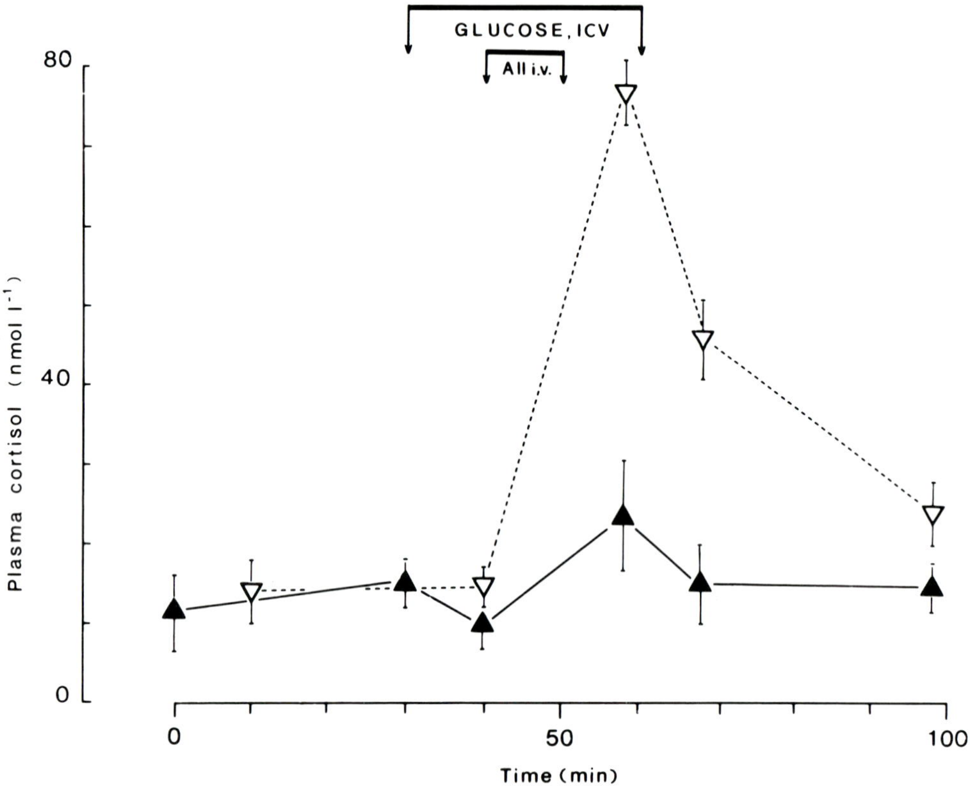

FIG. 5. More or less complete inhibition of the cortisol response to intravenous infusion of angiotensin II (40 pmol/kg/min) in goats by a simultaneous intracerebroventricular (ICV) (20 μl/min) infusion of isotonic *d*-glucose. ▽ – – – ▽, Increase in plasma cortisol concentration elicited by the separate 10-min intravenous infusion of angiotensin II; ▲ — ▲, lack of response to the corresponding infusion of angiotensin when it was preceded and outlasted by the ICV infusion of glucose. (From ref. 5, with permission.)

CEREBRAL SITES OF SYSTEMIC ANGIOTENSIN ACTION

Blood-borne AII does not to any appreciable extent pass the BBB (27). This fact and a bulk of experimental evidence suggest that centrally mediated effects of systemic AII are manifestations of binding of the octapeptide to cerebral receptors located outside the BBB. Therefore, attention has become focused on circumventricular organs, especially the subfornical organ (SFO) and the organum vasculosum of the lamina terminalis (OVLT) (both located in the anterior wall of the third cerebral ventricle; Fig. 6), and the area postrema of the fourth ventricle. These organs are situated outside the BBB and possess a high density of specific binding sites for blood-borne AII; a binding that becomes markedly reduced by intravenous, as well as by ICV administration of the competitive blocker saralasin (27,28). As regards thirst and release of ADH, it has been reported that these effects of systemic AII become severely impaired by a rather selective ablation of the SFO in rats (29,30). In the sheep, however, brain lesions mainly restricted to the SFO do not to any appreciable extent affect drinking to intravenous AII, suggesting that receptors in the OVLT may be of greater importance in this species (31).

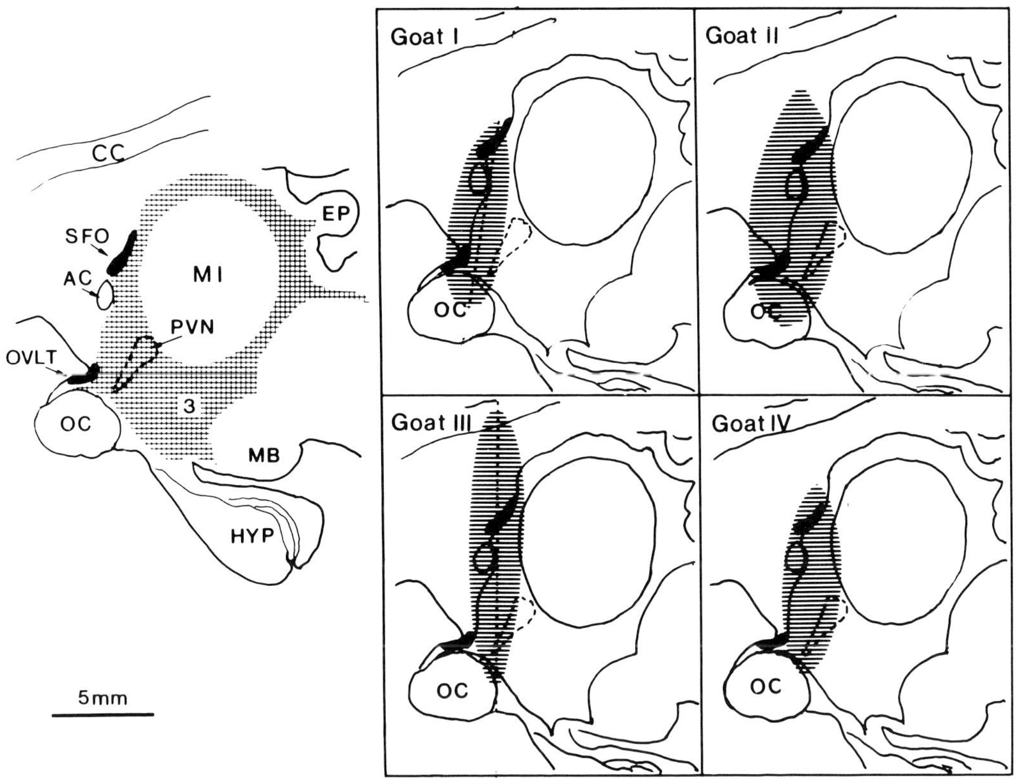

FIG. 6. Drawings of midsagittal sections through the preoptic/hypothalamic region showing the sagittal extent of bilaterally narrow lesions (*horizontally striped*) extinguishing (goats I and II) or reducing by about 50% (goats III and IV) the rise in plasma cortisol concentration obtained prelesioning in response to intravenous infusion of angiotensin II. Note that both circumventracular organs (the SFO and the OVLT) were ablated in the goats in which a total extinction of the cortisol response was seen, whereas the OVLT remained largely intact in the animals showing merely a reduction of the response. AC, Anterior commissure; CC, corpus callosum; EP, epiphysis; HYP, hypophysis; MB, mammillary body; MI, massa intermedia; OC, optic chiasm; OVLT, organum vasculosum of the lamina terminalis; PVN, paraventricular nucleus; SFO, subfornical organ; 3, third ventricle. (From ref. 40, with permission.)

The ultimate cause of ACTH-cortisol release is generally considered to be the liberation of the corticotropin-releasing factor (CRF) into the hypothalamic-pituitary portal vessels. Immunohistochemical studies (32) imply that the principal site for production of CRF is the parvocellular portion of the paraventricular nucleus (PVN), which seems to explain the finding in the rat that a complete bilateral destruction of the PVN reduced the median eminence content of CRF by 90% (33). Therefore, with regard to the possibility that ACTH-cortisol release as an effect of systemic AII is mediated by receptors within the SFO and/or OVLT, it is of interest to consider the connections between these circumventricular organs and the PVN. Neuroanatomical tracing studies have demonstrated direct pathways between the SFO and the PVN (34,35). The connection between the SFO and the PVN has also been elucidated electrophysiologically in the rat. It was found that the intravenous administration of AII activated cells in the parvocellular portion of the nucleus and that this response to systemic AII became markedly reduced after lesions involving the main part of the SFO (36). Furthermore, it has been reported that electrical stimulation of the SFO increases the liberation of CRF into the hypothalamic-pituitary portal vessels of anesthetized rats, where the effect was assumed to be mediated by the PVN (37). As regards the OVLT, this circumventricular organ projects to the PVN via the median preoptic nucleus of the lamina terminalis (38).

It has previously been shown in goats that bilaterally narrow lesions, involving the entire anterior wall of the third ventricle, induce severe distortion of the salt–water balance, including permanent adipsia and absence of ADH release in response to blood-borne AII (39). Similarly placed lesions have recently been made in that species to determine the site of the cerebral target mediating the ACTH-cortisol response to systemic AII (40). It was found that lesions embracing the SFO, the OVLT, and the interjacent lamina terminals (Fig. 6) extinguished the cortisol response to systemic AII, whereas lesions destroying the SFO but leaving the OVLT largely intact only caused about 50% reduction of the response. It suggests that ACTH-cortisol release as an effect of systemic AII is not exclusively mediated by receptors for the octapeptide confined to the SFO. Angiotensin receptors in the OVLT may be equally important in that respect.

CONCLUDING REMARKS

A bulk of experimental evidence favors the view that most hitherto known, centrally mediated effects of blood-borne AII are elicited via binding of the octapeptide to receptors on cells in those circumventricular organs, which are located outside the BBB in the anterior wall of the third ventricle (the SFO and the OVLT). That the very same effects can be induced by the administration of small amounts of AII into the lateral and third cerebral ventricle does not by necessity imply but does suggest that the same cells are also accessible to AII stimulation from the inside of the BBB. In support of that assumption is the fact that the centrally mediated effects of blood-borne AII can be extinguished by the ICV administration of systemically ineffective amounts of the competitive AII inhibitor saralasin. Of particular interest are the facts that all known, centrally mediated effects of systemic AII can be reproduced by raising the NaCl concentration of the CSF and that an artificially induced reduction of its NaCl concentration inhibits the responses to ICV administered AII, as well as the centrally mediated responses to intravenous AII. It speaks in favor of a unitary, cerebral sensory target for Na^+ and AII. Cells of the tanycyte type seem structurally well-adapted for that purpose, because such cells provide a morphological connection between the blood, the CSF, and nerve cells and are abundant in the circumventricular organs (41).

REFERENCES

1. Verney EB. The antidiuretic hormone and factors which determine its release. *Proc R Soc (Lond) (Biol)* 1947;135:25–106.
2. Andersson B. The effect of injections of hypertonic NaCl-solutions into different parts of the hypothalamus of goats. *Acta Physiol Scand* 1953;28:188–201.
3. Andersson B. Thirst—and brain control of water balance. *Am Sci* 1971;59:408–415.
4. Andersson B, Leksell LG, Rundgren M. Regulation of body fluids: intake and output. In: Staub NC, Taylor AE, eds. *Edema*. New York: Raven Press, 1984;299–318.
5. Junkergård J, Augustinsson O. Evidence for an influence of CSF sodium concentration upon ACTH-mediated cortisol response to intravenous and intracerebroventricular angiotensin. *Acta Physiol Scand* 1990;138:301–306.
6. Reid IA. Actions of angiotensin II on the brain: mechanisms and physiologic role. *Am J Physiol* 1984;246:F533–F543.
7. Blair-West JR, Coghlan JP, Denton DA, et al.

Humoral stimulation of adrenal cortical secretion. *J Clin Invest* 1962;41:1606–1627.

8. Fitzsimons JT, Simons BJ. The effect on drinking in the rat of intravenous infusion of angiotensin, given alone or in combination with other stimuli of thirst. *J Physiol* 1969;203:45–57.
9. Bonjour JP, Malvin RL. Stimulation of ADH release by the renin–angiotensin system. *Am J Physiol* 1970;218:1555–1559.
10. Blair-West JR, Reid IA, Ganong WF. Stimulation of angiotensinogen release by raised blood angiotensin concentration in the dog. *Clin Sci Mol Med* 1974;46:665–669.
11. Ismay MJA, Lumbers ER, Stevens AD. The action of angiotensin II on the baroreflex response of the conscious ewe and the conscious fetus. *J Physiol* 1979;288:467–479.
12. Andersson B, Augustinsson O, Bademo E, et al. Systemic and centrally mediated angiotensin II effects in the horse. *Acta Physiol Scand* 1987; 129:143–149.
13. Bickerton RK, Buckley JP. Evidence for a central mechanism in angiotensin induced hypertension. *Proc Soc Exp Biol Med* 1961;106:834–836.
14. Abraham SF, Denton DA, McKinley MJ, et al. Effect of angiotensin antagonist Sar[1]-Ala[8]-angiotensin II on physiological thirst. *Pharmacol Biochem Behav* 1976;4:243–247.
15. Murakami K, Ganong WF. Site at which angiotensin II acts to stimulate ACTH secretion *in vivo*. *Neuroendocrinology* 1987;46:231–235.
16. Junkergård J, Jónasson H. Intracerebroventricular and systemic saralasin used to investigate angiotensin-elicited activation of the pituitary–adrenocortical axis in the goat. *Acta Physiol Scand* 1988;134:271–275.
17. Fischer-Ferraro C, Nehmod VE, Goldstein DJ, et al. Angiotensin and renin in rat and dog brain. *J Exp Med* 1971;133:353–361.
18. Ganten D, Minnich JL, Granger P, et al. Angiotensin-forming enzyme in brain tissue. *Science* 1971;173:64–65.
19. Epstein AN, Fitzsimons JT, Rolls BJ. Drinking induced by injection of angiotensin into the brain of the rat. *J Physiol* 1970;210:457–474.
20. Andersson B, Westbye O. Synergistic action of sodium and angiotensin on brain mechanisms controlling water and salt balance. *Nature* 1970; 288:75.
21. Severs WB, Summy-Long J, Taylor JS, et al. A central effect of angiotensin: release of pituitary pressor material. *J Pharmacol Exp Ther* 1970; 174:27–34.
22. Andersson B, Eriksson L, Fernández O, et al. Centrally mediated effects of sodium and angiotensin II on arterial blood pressure and fluid balance. *Acta Physiol Scand* 1972;85:398–407.
23. Daniels-Severs A, Ogden E, Vernikos-Danellis J. Centrally mediated effects of angiotensin II in the unanaesthetized rat. *Physiol Behav* 1971; 7:785–787.
24. Appelgren B, Eriksson S, Jónasson H, et al. Potentiation of antidiuretic response to systemic angiotensin II by elevation of CSF NaCl concentration. *Acta Physiol Scand* 1982;116:297–302.
25. Junkergård J. Influence of mild sodium depletion upon peripheral and centrally mediated effects of intravenous angiotensin II in the goat. *Acta Physiol Scand* 1990;138:67–73.
26. Kozlowski S, Drzewiecki K, Sobocinska J. Relationship between osmotic reactivity of the thirst mechanism and the angiotensin and aldosterone blood levels in the dog. *Acta Physiol Pol* 1972;23:417–425.
27. Van Houten M, Schiffrin EL, Mann JFE, et al. Radioautographic localization of specific binding sites for blood-borne angiotensin II in the rat brain. *Brain Res* 1980;186:480–485.
28. Van Houten M, Mangiapane ML, Reid IA, et al. [Sar[1] Ala[8]] angiotensin II in cerebrospinal fluid blocks the binding of blood-borne [^{125}I] angiotensin to the circumventricular organs. *Neuroscience* 1983;10:1421–1424.
29. Simpson JB, Routtenberg A. Subfornical lesions reduce intravenous angiotensin-induced drinking. *Brain Res* 1975;88:154–161.
30. Iovino M, Steardo L. Vasopressin release to central and peripheral angiotensin II in rats with lesions of the subfornical organ. *Brain Res* 1984;322:365–368.
31. McKinley MJ, Denton DA, Park RG, et al. Circumventricular organs and water intake: a comparative view. *Proc Int Union Physiol Sci* 1983;15:117.
32. Hökfelt T, Fahrenkrug J, Tatemoto K, et al. The PHI (PHI-27)/corticotropin-releasing factor/enkephalin-immunoreactive hypothalamic neurons: possible morphological basis of integrated control of prolactin, corticotropin, and growth hormone secretion. *Proc Natl Acad Sci USA* 1983;80:985–998.
33. Bruhn TO, Plotsky PM, Vale WW. Effect of paraventricular lesions on corticotropin-releasing factor (CRF)-like immunoreactivity in the stalk-median eminence: studies on the adrenocorticotropin response to either stress and exogenous CRF. *Endocrinology* 1984;114:57–62.
34. Akert K. Das Subfornicalorgan: morphologische Untersuchungen mit besonderer Berucksichtigung der cholinergen Innervation und der neurosekretorischen Aktivität. *Schweiz Arch Neurol Psychiatr* 1967;100:217–231.
35. Miselis RR, Shapiro RE, Hand PJ. Subfornical organ efferents to neural systems for control of body water. *Science* 1979;205:1022–1025.
36. Ferguson AV. Systemic angiotensin acts at the subfornical organ to control the activity of paraventricular neurons with identified projections to the median eminence. *Neuroendocrinology* 1988;47:489–497.
37. Plotsky PM, Sutton SW, Bruhn TO, et al. Analysis of the role of angiotensin II in mediation of adrenocorticotropin secretion. *Endocrinology* 1988;122:538–545.
38. McKinley MJ, Allen AM, Chai SY, et al. The lamina terminalis and its neural connections: neural circuitry involved in angiotensin action

and fluid and electrolyte homeostasis. *Acta Physiol Scand* 1989;suppl 583:113–118.
39. Andersson B, Leksell LG, Lishajko F. Perturbations in fluid balance induced by medially placed forebrain lesions. *Brain Res* 1975;99:261–275.
40. Junkergård J, Andersson B, Jónasson H. Inhibition of the cortisol response to systemic angiotensin II by forebrain lesions involving the anterior wall of the third cerebral ventricle. *Acta Physiol Scand* 1990;138:293–299.
41. Jackson IMD. Significance and function of neuropeptides in cerebrospinal fluid. In: Wood JH, ed. *Neurobiology of cerebrospinal fluid.* New York: Plenum Press, 1980;625–650.

Volume Transmission in the Brain: Novel Mechanisms for Neural Transmission, edited by K. Fuxe and L.F. Agnati. Raven Press, Ltd., New York © 1991.

42

Intrinsic Central Neural Regulation of Cerebral Blood Flow and Metabolism in Relation to Volume Transmission

Donald J. Reis and Costantino Iadecola

Division of Neurobiology, Department of Neurology and Neurosciences, Cornell University Medical College, New York, New York 10021

VOLUME TRANSMISSION AND INTRINSIC CENTRAL NEURAL REGULATION OF CEREBRAL BLOOD FLOW AND METABOLISM

One of the aspects of brain function that can support the concept of volume transmission relates to newly recognized functional systems in rat brain that regulate cerebral blood flow (CBF) and metabolism (1). These systems, which, when activated, can elicit widespread modifications in CBF in association with or independent of metabolic activation of cortex, act through the action of intrinsic networks in brain. Such networks may engage the cerebral vasculature in wide-ranging and imprecise ways, apparently in the absence of localized synaptic contacts.

These systems have not been regarded, until relatively recently, as important in regulating CBF. Rather, the traditional view of the regulation of CBF has been considered to be entirely explicable by a subtle interplay between cerebrovascular autoregulation, local metabolic demands, and a superimposed control of the cerebrovascular bed by neuronal networks arising from ganglia lying outside the CNS (2–5). Autoregulation has been demonstrated by experiments in which arterial pressure (AP) is modified by passively altering the systemic AP. Under these conditions, CBF remains constant over a range of pressures between 60 and 140 mm Hg. In any area of brain, autoregulation may be overridden by local metabolic demand reflecting, in large measure, the local activity of neurons. Regional CBF (rCBF) and regional metabolism [usually assessed by measuring regional cerebral glucose use (rCGU) (6)] are closely matched (coupled) and proportional (see ref. 4 for review). Finally, extrinsic nerves (i.e., neurons), whose cell bodies reside outside of the CNS may modulate the metabolically coupled changes in rCBF. These nerves consist of sympathetic postganglionic neurons with their stores of norepinephrine and neuropeptide Y acting as vasoconstrictors, parasympathetic neurons of intracranial ganglia, which release the vasodilators acetylcholine (ACh) and vasoactive intestinal peptide (VIP), and branches of the trigeminal nerve releasing the vasodilator neuropeptides substance P (SP) and calcitonin gene-related peptide (CGRP) (7,8).

Thus, with the exception of global changes in CBF uncoupled to metabolism and elicited by changes in blood gases (CO_2 and O_2) (2), the cerebral circulation was viewed as a tidy and self-regulated system with flow serving to mirror cerebral metabolic activity. [This premise has led to the development of techniques for measuring CBF in humans to serve as an index of local cerebral activity (9).]

To these views must now be added the concept that the brain can regulate its own flow by the activity of **intrinsic** neuronal networks. This view, suggested by a number of early studies largely based on examining the effects of brain stimulation or lesions on CBF (see ref. 1), has been validated by application of autoradiographic methods for quantitative assessment of rCBF (10,11) and, as an index of cerebral metabolism, rCGU (6,12). Such regulation has been designated an **intrinsic central neurogenic control of the cerebral circulation** (1,13–20) and may serve as a model of volume transmission in brain. In this chapter we would like to review some of our own studies on several systems in rat brain, which indicate the power and variety of effects of these systems on rCBF.

GLOBAL ELEVATIONS IN CEREBRAL BLOOD FLOW UNCOUPLED FROM METABOLISM (PRIMARY VASODILATION) IN THE CEREBELLAR MODEL

Fastigial Pressor Response and CBF

In 1969, Miura and Reis (21) demonstrated that electrical stimulation of the rostral ventromedial pole of the cerebellar fastigial nucleus (FN) elicits a potent elevation of AP and heart rate (HR), the fastigial pressor response (FPR). The FPR is associated with widespread changes in regional organ blood flow simulating the cardiovascular adjustments to upright posture (the orthostatic reflex) (22).

The possibility that the cerebellum could modify the rCBF was first noted by Doba and Reis (22) in the anesthetized cat using electromagnetic flowmetry. This observation was confirmed by McKee et al. (23), who observed in the unanesthetized monkey that FN stimulation increased hemispheric flow without concomitant changes in oxygen use, suggesting the possibility that the increase in CBF might be independent of metabolism. The effect of FN stimulation on cerebral cortical CBF was further demonstrated in rabbit using the H_2 clearance method (24). In this study, it was shown that the stimulus-locked increases in rCBF could not be attributed to the changes in AP and were not dependent on sympathetic nerve activity and that, during stimulation, autoregulation was abolished.

Although these earlier investigations were provocative, they left many questions unanswered. These included such issues as to whether the elevations in rCBF elicited by electrical stimulation of the FN were (a) restricted to the cerebral cortex or generalized; (b) mediated by intrinsic neuronal pathways; (c) independent within regions of brain of rCGU, a much more precise measure of cerebral metabolism; (d) dependent on the release of a particular class of neurotransmitter; and (e) initiated by excitation of fibers or intrinsic neurons within the FN.

During the past few years, we and others (25–35) have sought to answer these questions. In our studies (32,36–38) in anesthetized paralyzed rats, we have measured rCBF by the Kety principle using 14-C-iodoantipyrine (IAP) as a tracer. rCGU was measured by the 2-DG method. Measurements of either rCBF or rCGU were made in brain homogenates or in sections prepared for autoradiography. A technique was developed whereby AP was maintained in the autoregulated range during the period of stimulation, whereas blood gases were carefully controlled.

Cerebrovascular Vasodilation is Global and Intrinsic

Electrical stimulation of FN in chloralose-anesthetized rats results in a marked increase in rCBF across the brain (Fig. 1)

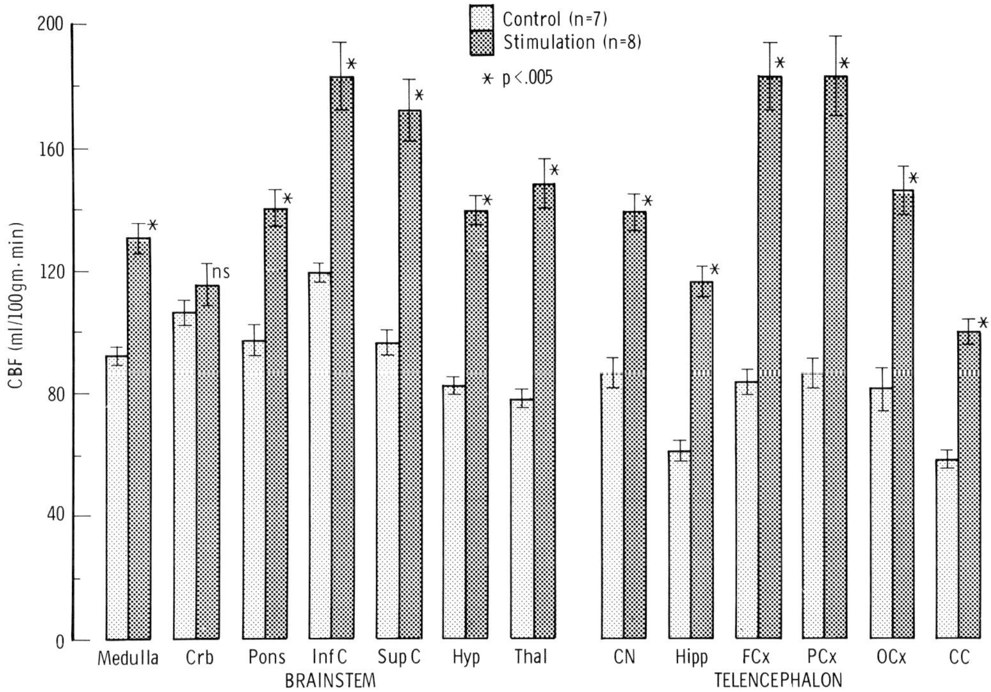

FIG. 1. Regional distribution of changes in CBF (ml/100 g × min) elicited by electrical stimulation of FN in anesthetized, paralyzed rats. Note that FN stimulation increases CBF in all brain regions except cerebellum, with maximal CBF responses occurring in cerebral cortex. Crb, cerebellum; InfC, inferior colliculus; SupC, superior colliculus; Hyp, hypothalamus; Thal, thalamus; CN, caudate putamen; Hipp, hippocampus; FCx, PCx, OCx, frontal, parietal, and occipital cortices; CC, corpus callosum. (From ref. 32, with permission.)

(32,38) and spinal cord (C. Iadecola and D. Reis, *unpublished data*). Despite the powerful elevations in flow, there are no changes in electroencephalogram (EEG) (39). The greatest elevations, up to 250% of control, are to be seen within the cerebral cortex (Fig. 1). However, significant elevations in flow are found in virtually every area examined. Thus, the elevations in rCBF elicited from FN are not anatomically distinct but global.

The elevations in rCBF elicited by FN stimulation persist after bilateral cervical sympathectomy, spinal cord transection, adrenalectomy, or transection of the greater superficial petrosal nerve (24,30,32). However, a unilateral electrolytic lesion of the basal forebrain abolishes the increase in rCBF within the ipsilateral cerebral cortex distal to the lesion without modifying the cerebrovascular vasodilation elsewhere in brain (29). The loss of the vasodilatation distal to the lesion cannot be attributed to diminution of vascular reactivity because cerebrovascular autoregulation as well as the vasodilatation elevated by hypercarbia was preserved (29) in the affected area. These observations prove that the cerebrovascular response to FN stimulation are neural and mediated over intrinsic central pathways, which synapse in or project through the basal forebrain.

FN Stimulation Dissociates CBF and the Metabolism Role of Local Neurons

The relation of FN stimulation to rCGU was examined by comparing the relation

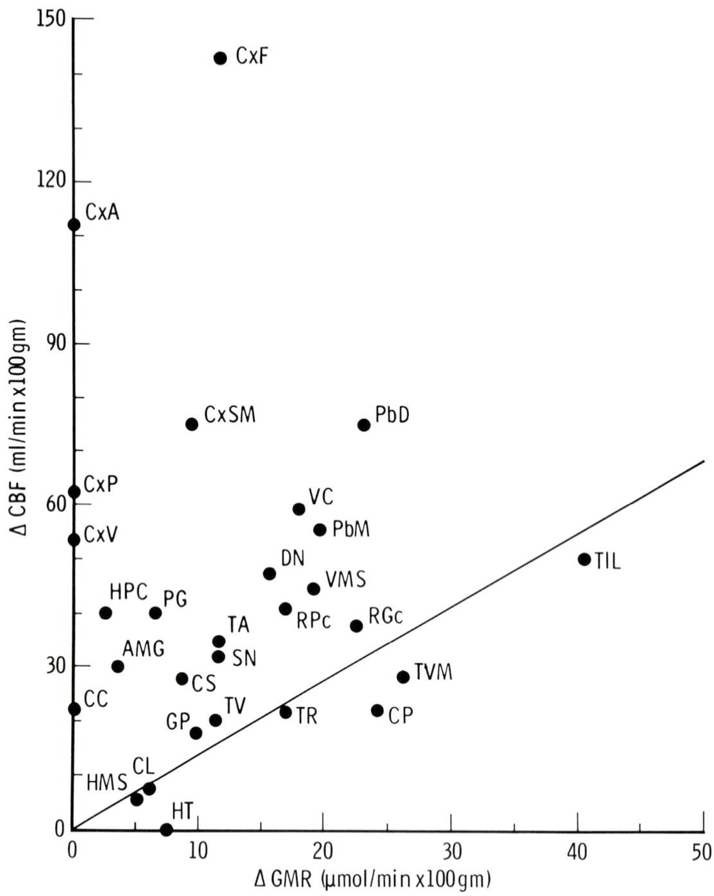

FIG. 2. Relation between changes in rCBF (ΔCBF) and changes in regional cerebral glucose use (ΔGMR) in different brain areas of the rat. *Straight line* represents the relation between rCBF and rGMR observed in unstimulated rats. Note that in most brain regions, changes in CBF are out of proportion with respect to changes in GMR, so that most of the points lie away from the proportionality line. AMG, amygdala; CC, corpus callosum; CL, inferior colliculus; CP, caudate putamen; CS, superior colliculus; CxA, auditory cortex; CxF, frontal cortex; CxP, parietal cortex; CxSM, sensory motor cortex; CxV, visual cortex; DN, cerebellar dentate nucleus; GP, globus pallidus; HMS, cerebellar hemisphere; HPC, hippocampus; HT, hypothalamus; PbD, dorsal parabrachial nucleus; PbM, medial parabrachial nucleus; PG, pontine gray matter; RGc, gigantocellular reticular nucleus; RPc, parvocellular reticular nucleus; SN, substantia nigra; TA, thalamus anterior nucleus; TIL, thalamus intralaminar nucleus (centromedial and paracentral); TR, thalamus reticular nucleus; TV, thalamus ventral nucleus; TVM, thalamus ventromedial nucleus; VC, vestibular complex; VMS, cerebellar vermis. (From ref. 38, with permission.)

between rCGU and rCBF autoradiographically. In contrast to the effects on rCBF, the changes in rCGU elicited by FN stimulation were highly restricted and, across brain, unrelated to flow (Fig. 2). This dissociation was most pronounced in the cerebral cortex wherein the enormous elevations in rCBF were not associated with any changes in metabolism. Although rCGU was increased in other areas, those regions were mono- or disynaptically related to the FN (in contrast to the cerebral cortex, which is not innervated by FN). The findings, therefore, demonstrated that excitation of a circuit involving the FN can initiate a potent increase in rCBF transneuronally. We have termed the response a **primary vasodilation** in recognition of the evident uncoupling of flow and metabolism in several areas by FN stimulation.

Several years ago, it was initially proposed that the primary vasodilation elicited in cerebral cortex by stimulation of FN might represent the release onto local vessels of transmitters by the activity of neurons either innervating the cerebral cortex from brainstem or a small pool of local neurons (15,16). These purported transmitters would result in local vasodilation. A model of the possible innervation patterns is seen in Fig. 3. To ascertain whether local neurons were necessary for the vasodilation elicited from FN, we examined the effects of destruction of neurons but not axons within a restricted region of the somatosensory cortex (28). This was accomplished by local injection of the excitotoxin ibotenic acid. After demonstrating that the dose of toxin indeed destroyed all local neurons but preserved fibers, we examined the effects on local CBF (lCBF) of electrical stimulation of the FN. We observed that such treatment abolished the vasodilation elicited by FN stimulation without affecting either resting rCBF nor the vasodilation evoked by hypercarbia. These findings suggest that the cortical vasomotor response to FN stimulation is not mediated by a direct neurovascular projection originating from, or passing through, basal forebrain (Fig. 3). Rather, the integrity of local neurons is essential. Because the increase in metabolism in the cortex is minimal and FN stimulation does not produce evident motor activity, the possibility is raised that the primary vasodilation results from activation of a few local interneurons, conceivably even dedicated to vascular engagement.

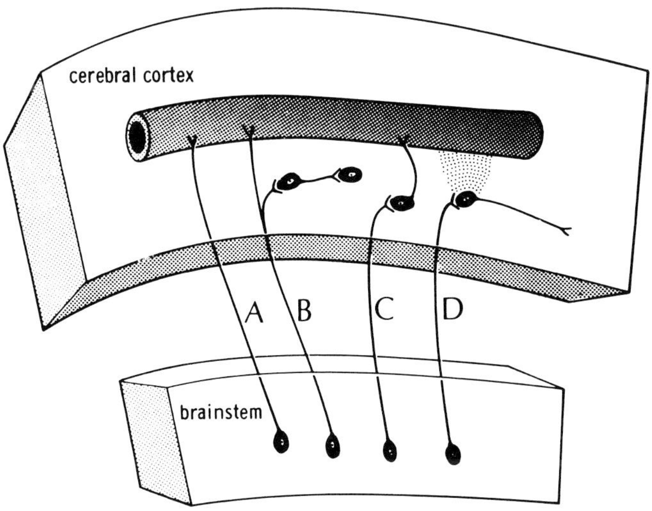

FIG. 3. Four potential mechanisms whereby brainstem neurons may regulate CBF in the cerebral cortex. **A:** Direct neurovascular; **B:** collateral neurovascular; **C:** indirect neurovascular; **D:** coupled metabolic (From ref. 16, with permission.)

Role of Acetylcholine in Primary Vasodilation

There is, at present, some evidence that the primary vasodilation elicited within the cerebral cortex by electrical stimulation of the FN is dependent on the local release of ACh. First, the cerebrovascular vasodilation elicited from the FN can be entirely abolished by the systemic administration of atropine (30). Second, the topical application of atropine to the surface of the cerebral cortex will reduce by more than 60% the evoked increase in rCBF, indicating that a substantial portion of the response depends on the local release of ACh (26). The source of the ACh is not certain. In the somatosensory cortex choline acetyltransferase (ChAT), the biosynthetic enzyme for ACh is found in three sites. The majority is contained within incoming axons arising preponderantly from the basal forebrain. A smaller amount is contained within local interneurons. Finally, the least amount is contained within capillary endothelial cells (see refs. 26,40 for discussion). The specific activity of capillary ChAT (which includes some apposed neuronal processes) is substantially greater than that found in synaptosomes (40). Because the amount of ACh released by FN stimulation in cortex is undetectable (26) in contrast to the substantial release evoked by basal forebrain stimulation (41), it seems probable that the source of ACh involved in the vasodilation may be either from local cholinergic interneurons and/or capillary endothelium (40).

Pathways from FN Evoking Cerebrovascular Vasodilation

Until recently, the identity of the neural elements in FN initiating the cardio- and cerebrovascular responses to electrical stimulation of FN was not known. Was the effect mediated by intrinsic neurons of FN or by local axons? If axons, what is their origin? Recently, studies have been undertaken to illuminate the question.

It is now known that the FPR and the cerebrovascular vasodilation are a consequence of stimulation of local axons in FN (see refs. 27,42–44) because (a) stimulation of local FN neurons with the excitatory amino acids kainic acid, quisqualate, or *dl*-homocysteic acid (but, interestingly, not glutamate or aspartate) will elicit a dose-dependent fall of AP, HR, and rCBF, effects abolished by destruction of intrinsic FN neurons with ibotenic acid; and (b) after destruction of local neurons with ibotenic acid, the FPR and the cerebrovascular vasodilation persists. It has also been demonstrated that the cerebrovascular vasodilation elicited from FN depends on the integrity of the C1 area of the rostral ventrolateral medulla (RVL), a region critical in maintaining resting AP over its reticulospinal pathways directly innervating preganglionic sympathetic neurons in the intermediolateral columns of the spinal cord (45).

The neurons from which these axons originate and the pathways through which information ascends to innervate cortex and the remainder of the neuraxis are unknown. We have proposed (42,43) that the vasodilation and elevations of AP elicited by electrical stimulation of the FN result from excitation of the collateral branch of brainstem neurons innervating the cerebellar cortex and projecting in turn to the C1 area of RVL. Several candidates have been described including projections from parabrachial, solitary, periaqueductal, and hypothalamic nuclei. How information goes from the RVL to cortex is presently under investigation.

GLOBAL CEREBROVASCULAR VASODILATION SECONDARY TO NEURALLY MEDIATED ELEVATED METABOLISM: THE ROLE OF ADRENAL MEDULLA

The question arose in our studies whether all areas of brain that were sympathoexcitatory (e.g., the FN) would elicit a primary vasodilation in the cerebral cortex. As a control, therefore, we compared the effects of FN stimulation with that of stimulating a region of the dorsal medullary reticular formation (DMRF), which, in previous investigations, we defined functionally as the distortion-sensitive area of the brainstem responsible for mediating the Cushing and cerebral ischemic reflexes (46,47). The DMRF lies partially within the confines of the parvocellular reticular nucleus and also within the trajectory of pathways traversed by reticulospinal fibers from the C1 area projecting both to the intermediolateral column and to structures in the upper brainstem (48). It also resides within the central tegmental tract of the brainstem, a major thoroughfare for aminergic fibers. We discovered that although electrical stimulation elevated rCBF profoundly, the elevation was fundamentally different from that produced from FN because (a) it was associated with widespread and proportional changes in cerebral metabolism and (b) it depended for its full expression on the release of adrenomedullary catecholamines.

Stimulation of the DMRF Elicits Global Elevations in CBF Dependent on Metabolism

Electrical stimulation of the DMRF elicits increases of rCBF in anesthetized rabbit (24) and rat (36,37,39,49–51). In rat, the increase in rCBF involves the entire neuraxis (Fig. 4), including the spinal cord (52), is unaffected by sympathectomy, and is not associated with changes in EEG (36,37,39). The largest responses occur in cerebral cortex (as much as 300%).

Unlike the elevation in rCBF elicited from electrical stimulation of the FN, which it simulates with respect to distribution and magnitude, the increase in rCBF elicited from the DMRF is tightly coupled to changes in rCGU (Fig. 5). It therefore represents a secondary cerebral vasodilation.

Secondary Vasodilation Largely Depends on Circulating Catecholamines

The global increase in cerebrovascular vasodilation elicited from the DMRF depends, in large measure, on the remarkable elevation of circulating catecholamines released from the adrenal glands by DMRF stimulation, most probably epinephrine (50,51). Bilateral adrenalectomy, adrenal demedullation, or systemic administration of β-adrenergic blocking agents will significantly reduce the increase in rCBF in some brain areas (e.g., cortex) (Fig. 6) and abolish it in others (e.g., thalamus) (Fig. 6), revealing, thereby, **adrenal-dependent** and **adrenal-independent** components of the response, the former mediated via β-adrenergic receptors.

The mechanism through which adrenal catecholamines increase cerebral metabolism and blood flow is unclear. Normally, adrenal catecholamines do not penetrate the blood-brain barrier and, when elevated in blood, have little effect on rCBF or rCGU (see ref. 50). When introduced directly into brain by the intraventricular route, however, they will markedly increase rCBF (53). We have demonstrated in adrenalectomized rats (50,54) that in thalamus or frontal cortex (Fig. 6), (a) elevating circulating epinephrine to levels obtained in intact animals with stimulation of the DMRF has no effect on rCBF; (b) when epinephrine is comparably elevated and the DMRF stimulated, rCBF is increased to the same extent as seen in intact rats with DMRF stimulation (Fig. 6); and (c) DMRF stimulation does not increase the permeability of the blood-brain barrier, at least to the nonselective indicator amino-isobutyric acid.

This remarkable experiment suggests that in some manner, stimulation of neurons in or passing through the DMRF may "sensitize" the brain to circulating catecholamines. The effect might occur by opening the blood-brain barrier selectively to catecholamines or to other agents released from peripheral organs by an action of catechol-

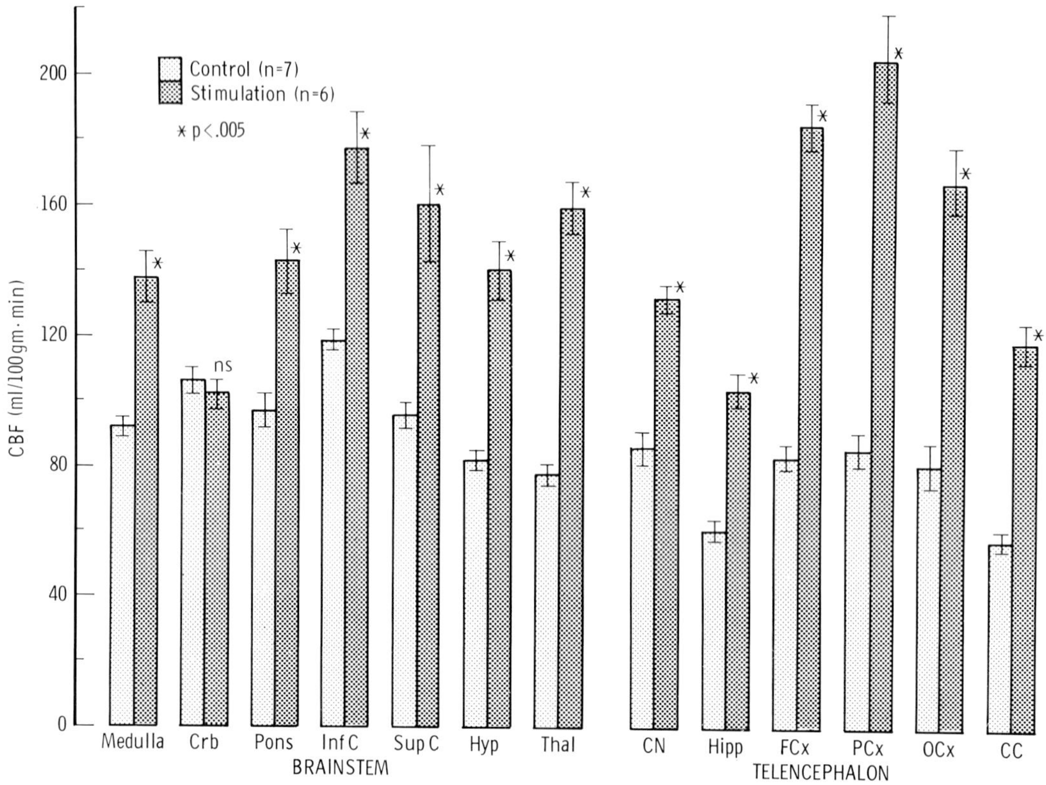

FIG. 4. Regional distribution of change in CBF (ml/100 g × min) elicited by electrical stimulation of DMRF in anesthetized, paralyzed rats. CBF measurements were done in 13 dissected brain regions after ~10 min of electrical stimulation (see text for details). Note that DMRF stimulation increases CBF in all brain regions except cerebellum, with maximal CBF responses occurring in cerebral cortex. Crb, cerebellum; InfC, inferior colliculus; SupC, superior colliculus; Hyp, hypothalamus; Thal, thalamus; CN, caudate putamen; Hipp, hippocampus; FCx, PCx, OCx, frontal, parietal, and occipital cortices; CC, corpus callosum.

amines. Alternatively, it might "gate" the brain so as to permit epinephrine to activate at some trigger site, possibly a circumventricular organ, a circuit with a widespread action on cerebral metabolism.

There is some concordant evidence for a "state-dependent" effect of catecholamine on cerebral metabolism and/or rCBF. Thus, the increase in rCBF elicited in rabbit by immobilization stress (55,56) is also adrenal-dependent.

GLOBAL REDUCTIONS IN CBF ELICITED BY STIMULATION OF CNS

Although the brain can profoundly elevate the CBF through two distinct and independent intrinsic mechanisms, activation of brain can also down-regulate CBF globally from at least two nuclear areas. First, as indicated above, chemical stimulation of the FN, which also elicits a fall of AP and HR (27,42–44), the fastigial depressor response (FDR), also produces a reduction in rCBF (Fig. 7). The reductions in flow involve virtually all regions of the CNS and are maximal in the cerebral cortex where they reduce flow by more than 50%. The reductions in rCBF are paralleled by proportional reductions in rCGU across brain, indicating that the fall in rCBF reflects a **secondary cerebrovascular vasoconstriction** (i.e., is secondary to the reduction in metabolism) (Fig. 7). The reduction in rCBF,

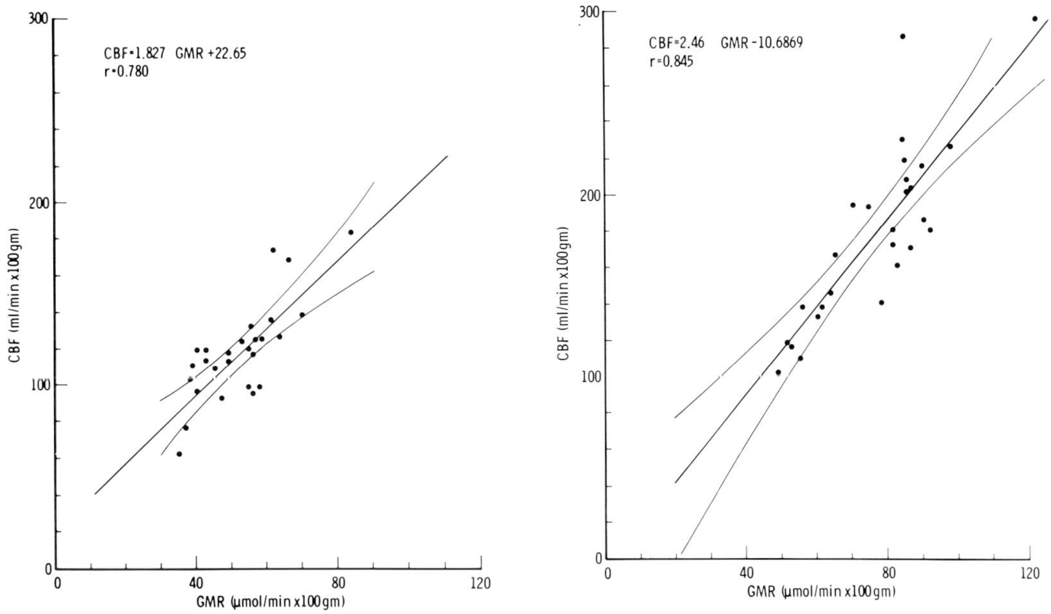

FIG. 5. A: Plotting of rCBF as function of regional cerebral metabolism (rGMR) in 28 brain regions in sham-operated unstimulated rats. Animals were anesthetized and paralyzed. Straight lines represent linear regression lines, and symmetrical curves drawn above and below represent the 95% confidence interval. Note that most of the points lie close to the regression line, with a correlation coefficient, $r = 0.780$ ($p < 0.001$). **B:** Plotting of rCBF as function of rGMR in 28 brain regions during electrical stimulation of DMRF in anesthetized, paralyzed rats. Note that as observed in controls A, most of the points lie close to regression line, within the 95% confidence interval ($r = 0.845$; $p < 0.001$). (From ref. 36, with permission.)

like the fall of AP, is dependent on the integrity of the C1 area of the RVL. Bilateral lesions of that nucleus, although not affecting resting rCBF, will entirely abolish the cerebrovascular vasodilation elicited from the FN. Comparably, electrical stimulation of regions of the parabrachial nucleus, a major autonomic nucleus of pons receiving afferent inputs from NTS and projecting broadly to a variety of cortical and subcortical areas, will also decrease rCBF (57). The effect is topographically restricted within the medial portion of the nucleus and reduces flow across the brain. Maximal decreases, to 35% of control, occurred in regions of the cerebral cortex. The reductions in flow are associated across brain with proportional reductions in rCGU.

These two studies demonstrate, first, that resting rCBF and probably metabolism lie within an intermediate level of their operating ranges. Thus, they can be elevated or depressed with appropriate stimuli. Second, neural mechanisms can primarily reduce metabolism in widespread regions of brain and presumably, secondarily, flow. Third, the proposal implies that there is a "tonic" drive maintaining flow and metabolism in midrange. In this manner, the regulation of the CBF parallels the control of AP through the CNS by activation of preganglionic sympathetic neurons. With respect to AP, preganglionic sympathetic neurons are maintained, at a tonic level of activity in mid-operating range, by reticulospinal neurons in the C1 area of RVL (45). However, neurons of the C1 area do not serve a comparable role with respect to the cerebral circulation because lesions of the area (when AP is maintained) have no effect

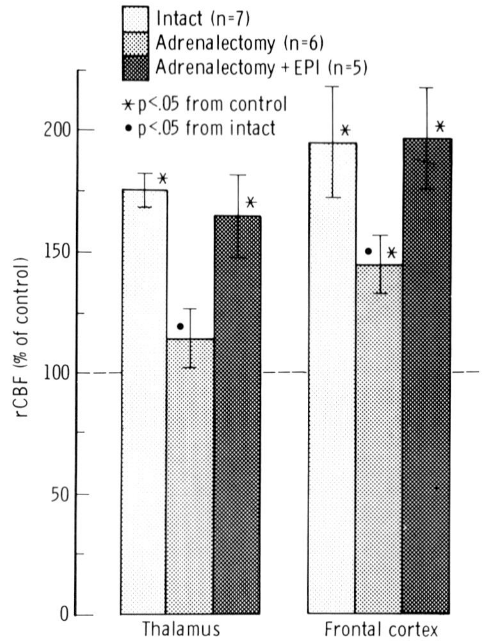

FIG. 6. Effect of total adrenalectomy with and without epinephrine (Epi) infusion on increases in rCBF elicited by DMRF stimulation in the rat.

on either resting rCBF or rCGU (58). The source of the tonic regulation of CBF and metabolism is, therefore, unknown.

EFFECTS OF CNS ON CEREBROVASCULAR AUTOREGULATION AND RESPONSES TO BLOOD GASES

In our studies we have also demonstrated that the CNS may also regulate, in part, autoregulation, a vascular function usually ascribed to local control of the vasculature. We have found that lesions of the nucleus tractus solitarii (NTS) can, in anesthetized rat, selectively impair cerebrovascular autoregulation throughout the whole brain without affecting cerebrovascular reactivity (59). The abolition of autoregulation is not replicated by transection of the IXth and Xth cranial nerves (thereby interrupting all input from arterial baro- and/or chemore-

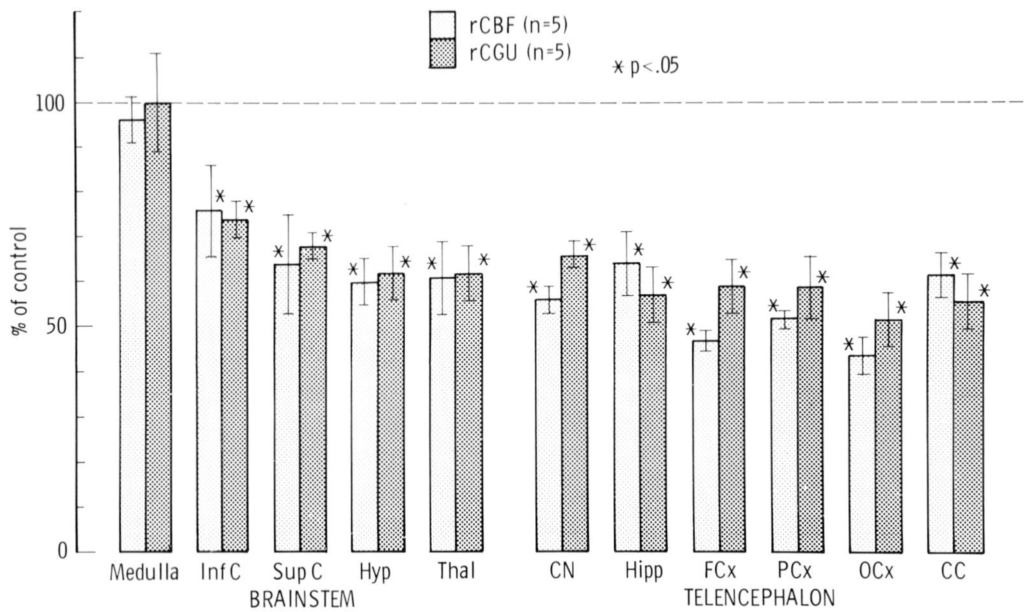

FIG. 7. Effect of microinjection of kainate into FN on rCBF and rCGU (% of vehicle-injected control). Excitation of neurons within FN reduces rCBF and rCGU in all brain regions except the medulla, but reductions in cerebellar cortex are greatest. Reductions of rCBF are proportional to that of rCGU. InfC, inferior colliculus; SupC, superior colliculus; Hyp, hypothalamus; Thal, thalamus; CN, caudate putamen; Hipp, hippocampus; FCx, PCx, OCx, frontal, parietal, and occipital cortices; CC, corpus callosum.

ceptors as well as other cardiopulmonary afferent nerves). Thus, the effect cannot be secondary to deafferentation of a peripheral receptor. This study demonstrates that neurons of the NTS may tonically regulate rCBF throughout brain by some as yet unknown regulatory process influencing cerebrovascular autoregulation. Moreover, it indicates that cerebrovascular autoregulation is not only a process regulated at the local level of the microvasculature but may itself be regulated by neuronal networks represented at remote sites. Further support for the possible role of neurogenic mechanisms in cerebrovascular autoregulation comes from recent findings (60–62) that autoregulation of CBF may be influenced by a number of neuropeptides including β-endorphin, α-melanocyte-stimulating hormone (MSH), and thyrotropin-releasing hormone (TRH).

In addition, the cerebrovascular response to hypoxia, another function believed to reflect regulation within the microvascular bed, also appears to be controlled by the brain. Bilateral lesions of the C1 area will globally reduce, by about 50%, the cerebral vasodilation elicited by hypoxia (58,63), a response that characteristically consists of an increase in rCBF without change in rCGU (2). The attenuation of the response to hypoxia occurs after bilateral transection of the IXth and Xth cranial nerves, indicating that it is not related to interruption by the lesions of signals arising in arterial chemoreceptors. In contrast, bilateral electrolytic lesions of the C1 area have no effect on the cerebrovascular vasodilation elicited by hypercarbia or on resting rCBF, rCGU, and cerebrovascular autoregulation.

This study is of interest in that it suggests that the cerebrovascular response to hypoxia is, in part, reflex and initiated by stimulation of receptors in the brainstem sensitive to hypoxia. That the C1 area may be a sensor for oxygen has also been suggested (45) on the basis of its participation in circulatory adjustments elicited by stimuli related to oxygen need, including pain, exercise, hemorrhagic shock, or cerebral ischemia (see ref. 45) and by the observations that C1 neurons are in intimate connection with capillaries (64) and have a high resting blood flow, metabolism, and capillary density (65).

PRIMARY AND SECONDARY VASODILATION

It is evident that intrinsic central neural networks have the capacity to modify the cerebral circulation profoundly. Moreover, such regulation may be independent of metabolism. Excitation by stimulation, electrically or chemically, of various structures within brain (see ref. 1 for review) may modify the cerebral circulation in complex ways. In summary, the data have shown that neural networks can elicit both vasodilation and vasoconstriction of the cerebrovascular bed. These can be further categorized first, on the basis of metabolic coupling into **primary** (i.e., unassociated with changes in metabolism) or **secondary** (presumably coupled to metabolism) vasodilation, and second, with respect to distribution to **topographically restricted** or **diffuse** (global) activation.

At present it is clear, at least with respect to the cerebral cortex, that primary vasodilation can be elicited that is global and initiated by electrical stimulation of the FN. There is also strong evidence to suggest that a topographically restricted primary vasodilation occurs as the initial response to natural somatosensory stimulation in humans (66) and may be followed, temporally, by a secondary vasodilation (67).

Because the primary elevation of rCBF in cortex is dependent on local neurons (28) whereas activation of the FN does not elicit widespread motor effects (68), it is conceivable that there may be a dedicated interneuron in cortex uniquely involved in mediating vasodilation and itself separately innervated. The presence of such neurons have been proposed on morphological grounds by Hendry et al. (69). In contrast, secondary vasodilation may, at least in

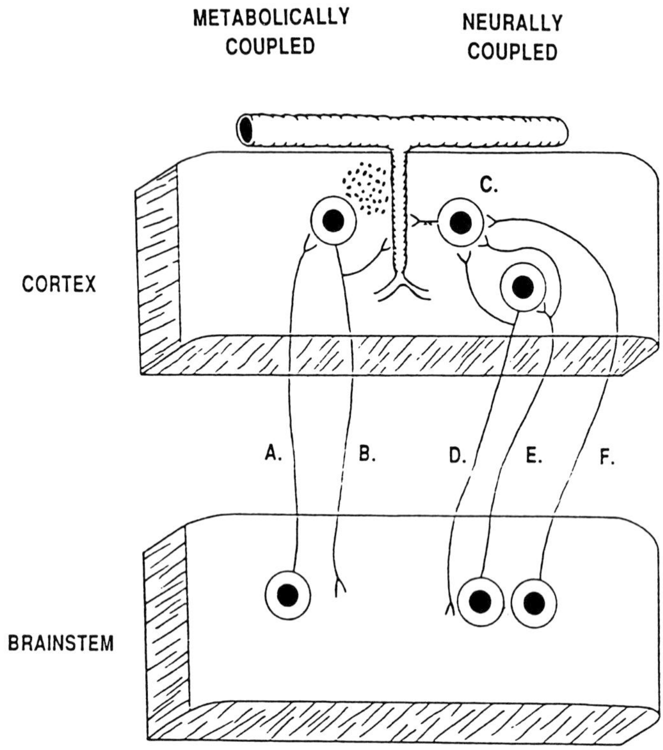

FIG. 8. A proposed model for neural mechanisms in cerebral cortex underlying neuronally (primary) and metabolically (secondary) coupled vasodilation elicited by stimulation of brainstem pathways (From ref. 1, with permission.)

part, be related to local activation of circuit neurons in cortex as elsewhere in brain (see ref. 6).

In Fig. 8, a model is proposed to differentiate between neurally coupled (primary) and metabolically coupled (secondary) vasodilations elicited by activation of the cerebral cortex (as a convenient target) from subcortical sources. The metabolically coupled inputs (Fig. 8A) connect with output neurons (Fig. 8B), which will elicit local vasodilation either by release of local vasoactive agents (possibly with glial interdiction) (Fig. 8B; stipple) and/or through a collateral branch of an output (or conceivably input) axon acting on the local vasculature. In contrast, the neurally coupled system depends on a "dedicated" interneuron (Fig. 8C) driven either by a local output (Fig. 8D), input (Fig. 8E), or dedicated input (Fig. 8F). In this model, metabolic coupling is invariably linked to neuronal activity rather than related to vascular control. However, neuronally coupled responses, such as the primary vasodilation elicited by FN stimulation, may be a response to activation, exclusively, of vascularly coupled neurons.

The CNS can also selectively initiate vasoconstriction, which may be primary or secondary and mostly diffuse. Secondary vasoconstriction has been elicited from brain from the parabrachial complex (31) or the parafascicular thalamic nucleus (70) and by chemical stimulation of FN (42). Primary vasoconstriction may occur in response to reduction of P_{CO_2}, but this is, as yet, not clearly central neurogenic. The neural circuitry and cellular mechanisms that underlie the secondary vasoconstriction are unclear.

CONCLUSIONS AND SOME PROPOSED GENERAL PRINCIPLES OF INTRINSIC REGULATION OF THE CEREBRAL CIRCULATION

1. The CBF may be regulated by intrinsic neural networks of brain acting independently of the extrinsic innervation of cerebral vessels.
2. Intrinsic neuronal systems tonically maintain rCBF in midrange probably through an action on metabolism.
3. The effects of intrinsic neural networks on the cerebral circulation are multiple and probably represent the action of distinct systems. Thus:
   (a) They can increase rCBF globally and independently from metabolism (i.e., by a **primary vasodilation**). Such vasodilation may involve pathways represented in FN, parafascicular thalamus, and basal forebrain and may, in part, be cholinergic. They may function to produce increases in rCBF in anticipation of, rather than reactively to, brain work.
   (b) They can globally increase metabolism and, secondarily, flow (**secondary vasodilation**) by networks represented in DMRF.
   (c) They can increase flow in topographically highly restricted distributions, which may be linked to, or in part independent from, metabolism as exemplified in somatosensory projections to cortex.
   (d) They can reduce flow globally by an action on metabolism (**secondary vasoconstriction**) as exemplified by the responses to electrical stimulation of the medial parabrachial nucleus.
4. The representation of networks governing rCBF encompasses all segments of the neuraxis and are closely allied with brain nuclei involved in central autonomic regulation (the central autonomic core) (see ref. 1 for review).
5. Central neural mechanisms may activate cerebral metabolism and, indirectly, CBF by sensitizing brain through β-adrenergic receptors to concentrations of circulating epinephrine, which themselves would have no cerebral effects when excluded from brain.
6. The central neurogenic elevations in flow unassociated with metabolism depend on the integrity of neurons within the target areas. These neurons may increase flow either by a direct action of neurotransmitter and/or mediators on vascular smooth muscle or through the intervening elements of glia and/or vascular endothelium.
7. Cerebrovascular autoregulation and the cerebrovascular vasodilation in response to hypoxia may, in part, be regulated by intrinsic neural systems represented in the medulla oblongata. Thus, these variables in flow regulation are not, as traditionally believed, entirely dependent on responses confined to the blood vessel wall and/or endothelium.
8. Central neurogenic mechanisms, by modifying autoregulation, may dynamically vary the dependence of the cerebral circulation on systemic AP.
9. The brain, through its intrinsic neural networks, may profoundly modify its own blood flow and metabolism.
10. Systems regulating rCBF and metabolism may reflect activation of systems in brain that influence cerebral function via volume transmission.

REFERENCES

1. Reis DJ, Iadecola C. Central neurogenic regulation of cerebral blood flow. In: Seylaz J, Sercombe R, eds. *Neurotransmission and cerebrovascular function II*. Amsterdam: Elsevier, 1989;369–390.
2. Heistad DD, Kontos HA. Cerebral circulation. In: Abboud F, Shephard J, eds. *Handbook of*

physiology: the cardiovascular system, vol III. Bethesda: American Physiological Society, 1983; 137–182.

3. Kety SS. The cerebral circulation. In: Field J, Magoun HW, eds. *Handbook of physiology: neurophysiology,* vol III. Washington, DC: American Physiological Society, 1960;1751–1760.
4. Lou HC, Edvinsson L, MacKenzie ET. The concept of coupling blood flow to brain function: revision required? *Ann Neurol* 1987;22:289–297.
5. Purves MJ. *The physiology of the cerebral circulation.* Cambridge: University Press, 1972.
6. Sokoloff L. Relationships among local functional activity, energy metabolism, and blood flow in the central nervous system. *Fed Proc* 1981;40:2311–2316.
7. Owman C. Regulatory peptides in cerebrovascular nerves and their effects upon the brain circulation—from a decade of research. In: Edvinsson L, McCulloch J, eds. *Peptidergic mechanisms in the cerebral circulation.* Chichester: Ellis Horward, 1987;191–213.
8. Uddman R. Peptidergic innervation of the cerebrovascular bed. In: Edvinsson L, McCulloch J, eds. *Peptidergic mechanisms in the cerebral circulation.* Chichester: Ellis Horward, 1987;15–33.
9. Raichle ME. Circulatory and metabolic correlates of brain function in normal humans. In: Plum F, ed. *Handbook of physiology: the nervous system,* vol V. Bethesda: American Physiological Society, 1987;643–674.
10. Reivich M, Jehle J, Sokoloff L, Kety SS. Measurement of regional cerebral blood flow with antipyrine ^{14}C in awake cats. *J Appl Physiol* 1969;27:296–300.
11. Sakurada O, Kennedy J, Jehle JD, Brown GL, Carbin GL, Sokoloff L. Measurement of local cerebral blood flow with iodo-[^{14}C]-antipyrine. *Am J Physiol* 1978;234:H59–H66.
12. Sokoloff L, Reivich M, Kennedy C, et al. The [^{14}C]deoxyglucose method for measurement of cerebral glucose utilization: theory, procedure and normal values in the conscious and anesthetized albino rat. *J Neurochem* 1977;28:897–916.
13. Goadsby PJ, Lance JW. Brain stem effects on intra- and extracerebral circulations. Relation to migraine and cluster headache. In: Olesen J, Edvinsson L, eds. *Basic mechanisms of headache.* Amsterdam: Elsevier Science Publishers, 1988;413–427.
14. Pinard E, Seylaz J. Controle nerveux intrinsique du debit sanguin cerebral. In: Bes A, Geraud G, eds. *Cerveau and hypertension arterielle.* Paris: Masson, 1986;90–104.
15. Reis DJ. Nervous control of cerebral blood flow in normal health and in relationship to cerebrovascular disease. In: Castaigne P, Lhermitte F, Gautier JC, Moriniere A, eds. *Maladies vasculaires cérébrales. cerebrovascular diseases.* Deuxième Conférence de la Salpêtrière. Paris: Bailliere, 1979;197–224.
16. Reis DJ. Central neural control of cerebral circulation and metabolism. In: MacKenzie ET, Seylaz J, Bes A, eds. *Neurotransmitters and the cerebral circulation,* vol 2. New York: Raven Press, 1984;91–119.
17. Reis DJ, Iadecola C. Regulation by the brain of its blood flow and metabolism: role of intrinsic neuronal networks and circulating catecholamines. In: Owman C, Hardebo JE, eds. *Neural regulation of brain circulation,* vol 8. Eric K. Fernstrom Symposium. Amsterdam: Elsevier Science Publishers, 1986;129–145.
18. Reis DJ, Iadecola C. Intrinsic neural control of cerebral circulation and metabolism. In: Bes A, Geraud G, eds. *Cerveau et hypertension arterielle.* Paris: Masson, 1986;76–89.
19. Reis DJ, Iadecola C, Nakai M. Control of cerebral blood flow and metabolism by intrinsic neural systems in brain. In: *Cerebrovascular diseases.* 14th Research (Princeton-Williamsburg) Conference, vol 13. Plum F, Pulsinelli WA, eds. New York: Raven Press, 1985;1–22.
20. Seylaz J, Pinard E, Mraovitch S. Influence of specific intracerebral structures on the regulation of the cerebral circulation. In: Owman C, Hardebo JE, eds. *Neural regulation of brain circulation.* Amsterdam: Elsevier, 1986;147–167.
21. Miura M, Reis DJ. Cerebellum: a pressor response elicited from the fastigial nucleus and its efferent pathway in brainstem. *Brain Res* 1969;13:595–599.
22. Doba N, Reis DJ. Changes in regional blood flow and cardiodynamics evoked by electrical stimulation of the fastigial nucleus in the cat and their similarity to orthostatic reflexes. *J Physiol* (Lond)1972;227:729–747.
23. McKee JC, Denn MJ, Stone HL. Neurogenic cerebral vasodilatation from electrical stimulation of the cerebellum in the monkey. *Stroke* 1976;7:179–186.
24. Reis DJ, Iadecola C, MacKenzie E, Mori M, Nakai M, Tucker LW. Primary and metabolically coupled cerebrovascular dilation elicited by stimulation of two intrinsic systems of brain. In: Heistad DD, Marcus ML, eds. *Cerebral blood flow: effects of nerves and neurotransmitters.* New York: Elsevier/North Holland, 1982;475–484.
25. Goadsby PJ, Lambert GA. Electrical stimulation of the fastigial nucleus increases total cerebral blood flow in the monkey. *Neurosci Lett* 1989;107:141–144.
26. Arneric SP, Iadecola C, Underwood MD, Reis DJ. Local cholinergic mechanisms participate in the increase in cortical cerebral blood flow elicited by electrical stimulation of the fastigial nucleus in rat. *Brain Res* 1987;411:212–225.
27. Chida K, Iadecola C, Reis DJ. Lesions of rostral ventrolateral medulla abolish some cardio- and cerebrovascular components of the cerebellar fastigial pressor and depressor responses. *Brain Res* 1990;508:93–104.
28. Iadecola C, Arneric SP, Baker HD, Tucker LW, Reis DJ. Role of local neurons in the cerebrocortical vasodilation elicited from cerebellum. *Am J Physiol* 1987;252:R1082–R1091.
29. Iadecola C, Mraovitch S, Meeley MP, Reis DJ. Lesions of the basal forebrain in rat selectively

impair the cortical vasodilation elicited from cerebellar fastigial nucleus. *Brain Res* 1983; 279:41–52.
30. Iadecola C, Underwood MD, Reis DJ. Muscarinic cholinergic receptors mediate the cerebrovasodilation elicited by stimulation of the cerebellar fastigial nucleus in rat. *Brain Res* 1986;368:375–379.
31. Mraovitch S, Lasbennes F, Calando Y, Seylaz J. Cerebrovascular changes elicited by electrical stimulation of the centromedian-parafascicular complex in rat. *Brain Res* 1986;380:42–53.
32. Nakai M, Iadecola C, Reis DJ. Global cerebral vasodilation by stimulation of rat fastigial cerebellar nucleus. *Am J Physiol (Heart Circ Physiol)*, 1982:243;H226–H235.
33. Iadecola C, Springston ME, Reis DJ. Dissociation by chloralose of the cardiovascular and cerebrovascular responses evoked from the cerebellar fastigial nucleus. *J Cereb Blood Flow Metab* 1990;10:375–382.
34. Iadecola C, Reis DJ. Cholinergic cerebrovasodilation elicited from the fastigial nucleus: temporal profile and susceptibility to general anesthetics. *Fed Proc* 1990 (*in press*).
35. Mraovitch S, Pinard E, Seylaz J. Two neural mechanisms in rat fastigial nucleus regulating systemic and cerebral circulation. *Am J Physiol* 1986;251:H153–H163.
36. Iadecola C, Nakai M, Mraovitch S, Ruggiero DA, Tucker LW, Reis DJ. Global increase in cerebral metabolism and blood flow produced by focal electrical stimulation of dorsal medullary reticular formation in rat. *Brain Res* 1983; 272:101–114.
37. Iadecola C, Nakai M, Arbit E, Reis DJ. Global cerebral vasodilatation elicited by the focal electrical stimulation within the dorsal medullary reticular formation in anesthetized rat. *J Cereb Blood Flow Metab* 1983;3:270–279.
38. Nakai M, Iadecola C, Ruggiero D, Tucker L, Reis DJ. Electrical stimulation of the cerebellar fastigial nucleus increases cerebral cortical blood flow without changes in focal metabolism: evidence for an intrinsic system in brain for primary vasodilation. *Brain Res* 1983;260:35–49.
39. Underwood M, Iadecola C, Reis DJ. Stimulation of brain areas increasing cerebral cortical blood flow and/or metabolism does not activate the electroencephalogram. *J Cereb Blood Flow Metab* 1983;3:S214–S215.
40. Arneric SP, Honig MA, Milner TA, Greco S, Iadecola C, Reis DJ. Neuronal and endothelial sites of acetylcholine synthesis and release associated with microvessels in rat cerebral cortex: ultrastructural and neurochemical studies. *Brain Res* 1988;454:11–30.
41. Inanami O, Meguor K, Ohno K, Sato A. Contribution of cholinergic vasodilators on the increase in cerebral cortical blood flow responses to the intravenous administration of thyrotropin releasing hormone in anesthetized rats. *Neurosci Lett* 1988;88:184–188.
42. Chida K, Iadecola C, Reis DJ. Global reduction in cerebral blood flow and metabolism elicited from intrinsic neurons of fastigial nucleus. *Brain Res* 1989;500:177–192.
43. Chida K, Iadecola C, Reis DJ. Sympathoinhibition elicited by stimulation of glutamate receptors in the fastigial nucleus: the fastigial depressor response. *Brain Res* (*submitted*).
44. Chida K, Iadecola C, Underwood M, Reis DJ. A novel vasodepressor response elicited from the rat cerebellar fastigial nucleus: the fastigial depressor response. *Brain Res* 1986;370:378–382.
45. Reis DJ, Morrison S, Ruggiero DA. The C1 area of the brainstem in tonic and reflex control of blood pressure. *Hypertension* 1988;11(suppl): I8–I13.
46. Doba N, Reis DJ. Localization within the lower brainstem of a receptive area mediating the pressor response to increased intracranial pressure (the Cushing response). *Brain Res* 1972;47:487–491.
47. Kumada M, Dampney RAL, Reis DJ. Profound hypotension and abolition of the vasomotor component of the cerebral ischemic response produced by restricted lesions of medulla oblongata: relationship to the so-called tonic vasomotor center. *Circ Res* 1979;45:63–70.
48. Ruggiero DA, Ross CA, Anwar M, Park DH, Joh TH, Reis DJ. Distribution of neurons containing phenylethanolamine *N*-methyltransferase in medulla and hypothalamus of rat. *J Comp Neurol* 1985;239:127–154.
49. Iadecola C, Arbit E, Nakai M, Mraovitch S, Tucker L, Reis DJ. Increased regional cerebral blood flow and metabolism elicited by stimulation of the dorsal medullary reticular formation in the rat: evidence for an intrinsic neural system in brain regulating cerebral metabolism. In: Heistad DD, Marcus ML, eds. *Cerebral blood flow: effects of nerves and neurotransmitters*. New York: Elsevier/North Holland 1982;485–492.
50. Iadecola C, Lacombe PM, Underwood MD, Ishitsuka T, Reis DJ. Role of adrenal catecholamines in cerebrovasodilation evoked from brainstem. *Am J Physiol* 1987;252:H1183–H1191.
51. Lacombe PM, Iadecola C, Underwood MD, Sved AF, Reid DJ. Plasma epinephrine modulates cerebrovasodilation evoked from brain stem. *Brain Res*. 1990;506:93–100.
52. Iadecola C, Chida K, Lacombe P, Underwood MD, Reis DJ. Global increase in spinal cord blood flow elicited from the dorsal medullary reticular formation in rat. *J Cereb Blood Flow Metab* 1989;9(suppl 1):S480.
53. MacKenzie ET, McCulloch J, O'Keane M, Pickard JD, Harper AM. Cerebral circulation and norepinephrine: relevance of the blood brain barrier. *Am J Physiol* 1976;231:483–488.
54. Lacombe P, Sercombe R, Correze JL, Phillipson V, MacKenzie ET, Seylaz J. Cortical vasodilation induced by electrical stimulation of the substantia innominata in anesthetized rats. *J Cereb Blood Flow Metab* 1989;7:S393.
55. Lacombe P, Seylaz J. Significance of the cerebrovascular effects of immobilization stress in

the rabbit. *J Cereb Blood Flow Metab* 1984;4: 397–406.

56. Pinard E, Lacombe P, Reynier-Rebuffel AM, Seylaz J. Intrinsic and extrinsic mechanisms involved in the cerebrovascular reaction elicited by immobilization stress in rabbits. *Brain Res* 340 1985;340:305–314.
57. Mraovitch S, Iadecola C, Ruggiero DA, Reis DJ. Widespread reduction in cerebral blood flow and metabolism elicited by electrical stimulation of the parabrachial nucleus in rat. *Brain Res* 1985;341:283–296.
58. Underwood MD, Iadecola C, Reis DJ. Neurons in C1 area of rostral ventrolateral medulla mediate global cerebrovascular responses to hypoxia but not hypercarbia. *J Cereb Blood Flow Metab* 1987;7(suppl 1):S226.
59. Ishitsuka T, Iadecola C, Underwood MD, Reis DJ. Lesions of nucleus tractus solitarii globally impair cerebrovascular autoregulation. *Am J Physiol* 1986;251:H269–H281.
60. Sandor P, de Jong W, Cox-van-Put J, de Wied D. Influence of centrally administered alpha- and gamma 2–melanocyte-stimulating hormone on hypothalamic blood flow autoregulation in the rat. *Brain Res* 1987;424:189–192.
61. Sandor P, de Jong W, de Wied D. Endorphinergic mechanisms in cerebral blood flow autoregulation. *Brain Res* 1986;386:122–129.
62. Sandor P, de Jong W, de Wied D. Naloxone-like influence of TRH and ACTH-(4-7) on hypothalamic blood flow autoregulation in the rat. *Peptides* 1988;9:215–219.
63. Underwood MD. *Control of the cerebral circulation and metabolism by the rostral ventrolateral medulla: possible role in the cerebrovascular response to hypoxia.* Ph.D. Dissertation. New York: Cornell University, 1988.
64. Milner TA, Pickel VM, Park DH, Joh TH, Reis DJ. Phenylethanolamine *N*-methyltransferase-containing neurons in the rostral ventrolateral medulla of the rat: I. Normal ultrastructure. *Brain Res* 1987;411:28–45.
65. Glöbel U, Schröck H, Seller H, Kuschinsky W. Glucose utilization, blood flow and density of perfused capillaries in the subretrofacial nucleus of the rat brainstem. *J Cereb Blood Flow Metab* 1989;9:S321.
66. Fox PT, Raichle ME. Focal physiological uncoupling of cerebral blood flow and oxidative metabolism during somatosensory stimulation in human subjects. *Proc Natl Acad Sci USA* 1986;83:1140–1144.
67. Greenberg J, Hand P, Sylvestro A, Reivich M. Localized metabolic-flow couple during functional activity. *Acta Neurol Scand* 1979;72:12–13.
68. Reis DJ, Doba N, Nathan MA. Predatory attack, grooming and consummatory behavior evoked by electrical stimulation of cerebellar nuclei in cat. *Science* 1973;182:845–847.
69. Hendry SH, Jones EG, Beinfeld MC. Cholecystokinin-immunoreactive neurons in rat and monkey cerebral cortex make symmetric synapses and have intimate associations with blood vessels. *Proc Natl Acad Sci USA* 1983;80:2400–2404.
70. Mraovitch S, Seylaz J. Metabolism-independent cerebral vasodilation elicited by electrical stimulation of the centromedian-parafascicular complex in rat. *Neurosci Lett* 1987;83:269–274.

Volume Transmission in the Brain: Novel Mechanisms for Neural Transmission,
edited by K. Fuxe and L.F. Agnati.
Raven Press, Ltd., New York © 1991.

43

Transmitter Release, Ion and Water Fluxes, and Ischemic Brain Damage

Bo K. Siesjö, Anders Ekholm, and Barbro Asplund

Laboratory for Experimental Brain Research and Department of Anesthesiology, University of Lund, Lund, Sweden

It has been known for decades that the transmittal of point-to-point information in the central nervous system (CNS) involves pre- and postsynaptic fluxes of ions and water. For example, when an impulse reaches a nerve ending, the release of transmitter is thought to be mediated by an influx of calcium via voltage-sensitive calcium channels (VSCCs). The transmitter released may then hyperpolarize the postsynaptic membrane by increasing its permeability to K^+ or Cl^-, or it may depolarize it by enhancing influx of Na^+ and/or Ca^{2+}. Calcium may enter either via VSCCs or via agonist-operated calcium channels (AOCCs) (i.e., channels that are opened in response to agonist activation of membrane receptors). If Na^+ enters with Cl^- and osmotically obliged water, the postsynaptic structure will swell. All these changes involve alterations in the composition of the extracellular fluid, which thus serves as a communication channel in the CNS (1,2).

The changes are considered normal events in chemical transmission at synapses. However, it has been recognized that such information transfer involves extrasynaptic structures as well. For example, a whole class of neurotransmitters/neuromodulators may, when released, exert their effects on extrasynaptic neuronal sites or on glial cells. Furthermore, when K^+ accumulates in the extracellular fluid (ECF) during activity, glial cells are believed to aid in its removal and spatial redistribution. Glial cells also participate in the reuptake of presynaptically released transmitter. Indeed, one of their tasks may be to help modulating (or regulating) the composition of the ECF.

Events during chemical transmission become exaggerated and distorted during some pathological states [e.g., spreading depression (SD), anoxia/ischemia, and hypoglycemia] (1–3). SD, which can be elicited by a stab wound, by high K^+ concentrations, or by electrical stimulation, is a propagated disturbance of electrical functions in brain tissues and retina, involving massive efflux of K^+ from cells, influx of Na^+, Cl^-, and Ca^{2+}, and swelling of cellular elements. However, once the stimulus eliciting SD is removed, the ionic changes are spontaneously reversible, and even if repeatedly elicited for several hours SD does not lead to tissue damage (4).

Similar fluxes of ions and water are observed in ischemia and hypoglycemia, but in these conditions, the changes are usually not reversible, and provided that the ischemic and hypoglycemic transients are of a certain minimal duration, cell damage is incurred (3,5,6). In fact, it is believed that the damage is caused by the cellular influx of calcium and/or of Na^+ plus Cl^-, with osmotically obligated water, leading to calcium-related and to osmolytic damage, respectively (5–8). This means that when pathologically enhanced and pathologically prolonged, those ionic fluxes that serve to transmit information between cells via the ECF become a threat to the viability of cells.

Our objectives with this manuscript are to discuss ion and water fluxes in brain ischemia, attempt delineating the mechanisms by which these occur, and speculate about their pathogenetic role. Because the subject has been recently reviewed (5,6,9,10), our discussion will be brief and centered on major concepts and recent findings. For the same reasons, the number of references will be restricted.

ION AND WATER FLUXES DURING ISCHEMIA

Changes during complete ischemia have been described in some detail (for reviews, see refs. 3,5,6). Figure 1 summarizes information on ion and water fluxes, as these have been observed in the rat (see also refs. 11,12). As pointed out by ref. 3, K^+ fluxes occur in three phases, encompassing a slow initial rise of K^+_e (phase 1), a sudden marked efflux (phase 2), and a gradual increase to the final value of 60 to 80 mM (phase 3). During the transition between phase 1 and phase 2 (i.e., when K^+ rapidly leaves cells, and a DC potential difference develops between the ECF and an extracranial site), there is sudden influx of Ca^{2+}, Na^+, and Cl^-. Because the ECF volume is reduced to about half, the influx of Na^+ and Cl^- must be accompanied by osmotically obliged water.

Changes in extracellular pH (pH_e) have their own time course. Thus, pH_e decreases gradually during phase 1. The decrease continues until pH_e reaches its minimal value after 3 to 4 min; however, the record is interrupted by an alkalotic transient at the time of the DC potential shift (see refs. 13–15). While the shape of the pH_e curve and the presence of an alkalotic shift are similar, the magnitude of the pH_e shift varies with the preischemic plasma glucose concentration, the major determinant of the severity of the lactic acidosis (16). However, contrary to expectations, pH_e does not vary continuously with tissue lactate concentration but shows two plateau values (ΔpH_e ~0.5 and ~1.1, respectively), with a transition zone at ischemic lactate concentration of 17 to 19 μmol/g (13,14,17).

As discussed by ref. 3, the initial rise in K^+_e (phase 1) is associated with an increased membrane conductance, suggesting that ion channels are opened, e.g., to K^+. Subsequent results have supported this contention (18,19). The transition between phase 1 and phase 2 seems to enhance greatly membrane conductances (i.e., to involve a "shock" opening of ionic channels).

Ion and water fluxes during ischemia are strikingly similar to those occurring during spreading depression, although the latter are spontaneously reversible before extracellular K^+ concentration reaches the peak values observed during ischemia. On first sight, it appears that in SD, such phase 2 ionic and water fluxes occur without a preceding rise in K^+_e. However, the experiments by ref. 20, conducted on the catfish brain at 25°C, clearly showed that K^+_e reaches about 10 mM before the DC potential swings negative, and the other ions equilibrate across the (depolarized) membranes. Thus, if the comparison is made at 37°C, the crucial differences are the time course of the K^+_e changes and the reversibility of the changes during spreading depression. The latter is probably possible because, in contrast to ischemia, SD is not associated with energy failure (see below).

Whereas complete ischemia leads to a sustained equilibration of Ca^{2+}, Na^+, K^+, H^+, and Cl^- across depolarized membranes, the so-called penumbra zone around the densely ischemic focus of a stroke lesion may show irregularly occurring K^+ transients (21), which are associ-

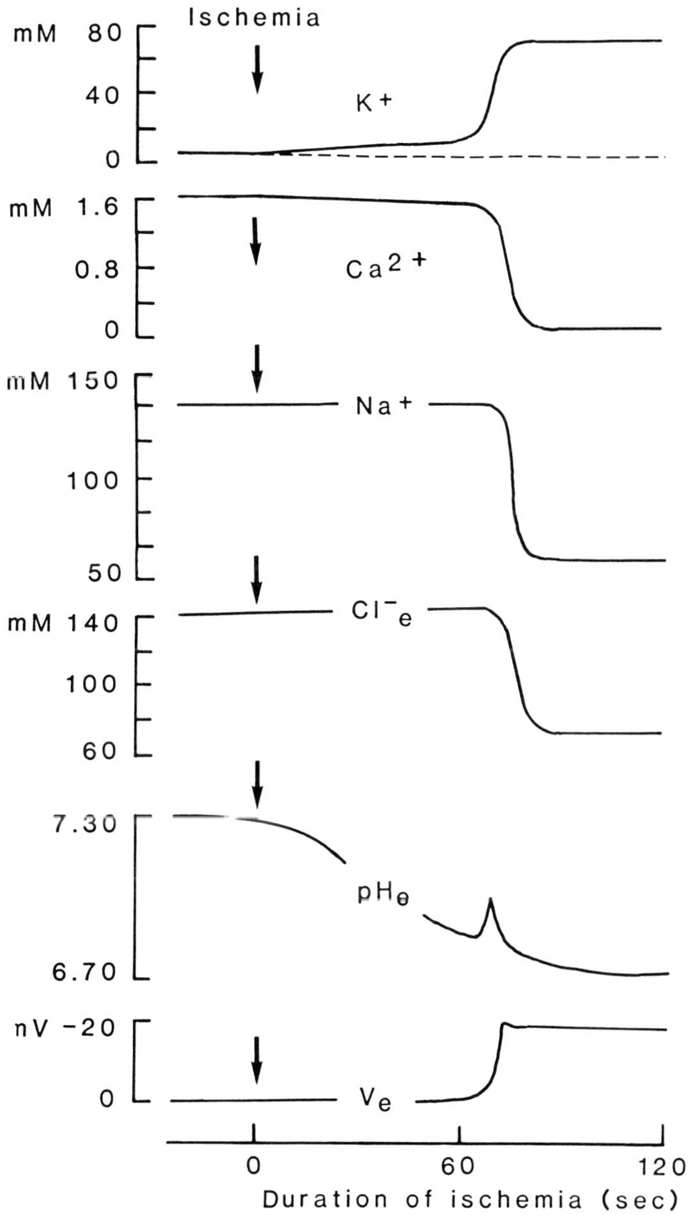

FIG. 1. Schematic diagram illustrating changes in extracellular concentrations of K^+, Ca^{2+}, Na^+, and Cl^-, in pH, and in DC potential during the first 120 sec of anoxia. (Modified from ref. 10, with permission.)

ated with transient reductions of $Ca^{2+}{}_e$, suggesting uptake of calcium by cells (4). It is tempting to assume that these represent SD-like ionic transients affecting a tissue that is energetically strained, containing cells that are viable but "at risk."

MECHANISMS INVOLVED

Before discussing the mechanisms involved it seems justified to discuss the routes whereby the ion fluxes occur. Figure 2 depicts the pre- and postsynaptic ion channels that are probably involved (5,7,9,22,23). The upper part of the figure focuses interest on excitatory events elicited by the opening of an (unspecified) Na^+ channel, with the secondary activation of VSCCs, whereas the lower part shows activation by glutamate of agonist-operated channels. Clearly, the various channels depicted may coexist in the same postsynaptic element.

Presynaptically, we envisage that membrane depolarization by an incoming nerve impulse leads to Ca^{2+} influx via the N type of VSCCs. Although presynaptic depolarization may release a variety of neurotransmitters/neuromodulators, we will concentrate on endings releasing glutamate or a related excitatory amino acid (EAA).

Presynaptic glutamate release is depicted to activate a postsynaptic Na^+ conductance. The ensuing Na^+ influx depolarizes the membrane, thereby triggering calcium influx through the T and/or L type of VSCCs and further depolarization. Nor-

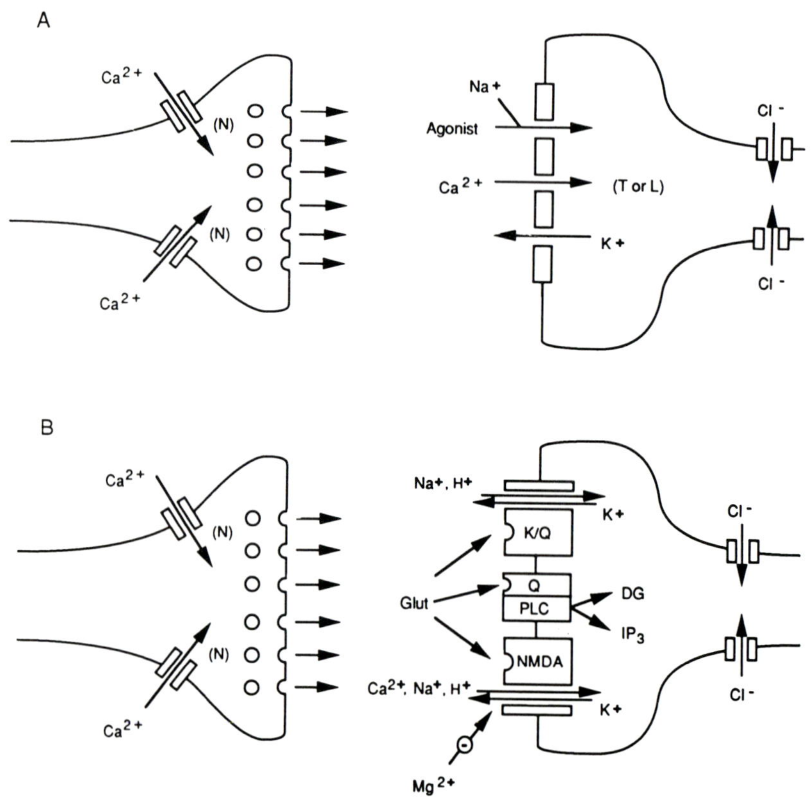

FIG. 2. Schematic diagram illustrating pre- and postsynaptic ion channels involved in excitatory events. The upper panel **(A)** has the focus on voltage-sensitive calcium channels, and the lower panel **(B)** on agonist-operated ones. (Modified from refs. 5,10, with permission.)

mally, this excitatory transient is terminated by activation of a K^+ conductance, which, by triggering K^+ efflux repolarizes the membrane. As will be discussed below, the K^+ conductance can be activated by a rise in Ca^{2+} or by a fall in adenosine triphosphate (ATP) concentration. Activation of a Cl^- conductance, which will tend to clamp the membrane potential at the equilibrium potential for Cl^-, will reinforce inhibition. However, a major factor causing termination of the excitatory transient is reuptake of glutamate. This occurs both into presynaptic terminals and into glial cells, probably driven by an electrogenic $3Na^+$/glutamate symporter (i.e., by an ion translocase, which is energetically fueled by the Na^+ gradient, and by the negative membrane potential, the latter electrophoretically dragging the positively charged ion complex into the cell) (24,25).

Three major subtypes of glutamate receptors have been described, selectively activated by *N*-methyl-D-aspartate (NMDA), by kainate (K), and by quisqualate (Q) (for brief reviews, see refs. 26 28). Because the K/Q receptor is coupled to an ion channel that is indiscriminately permeable to monovalent cations, its activation will lead to Na^+ influx and, thereby, to depolarization. The NMDA-gated channel, which is permeable to Ca^{2+} as well, is normally blocked by Mg^{2+} in physiological concentrations. However, because this block is relieved by depolarization, the simultaneous activation of K/Q and NMDA receptors by EAAs gives rise to downhill fluxes of Na^+, K^+, and Ca^{2+}. In many cells, the NMDA-gated channel is probably the major route of calcium entry during excitation.

Recent results have demonstrated that apart from activating ionotropic receptors, glutamate stimulates a metabolotropic receptor, which is coupled to phospholipase C and which thus triggers release of the two second messengers inositoltrisphosphate (IP_3) and diacylglyceride (DG), the former releasing Ca^{2+} from the ER and the latter serving to activate protein kinase C (for literature, see refs. 7,29). The activation of such a receptor, which seems to be Q-preferring, thus provides another mechanism leading to a rise in $Ca^{2+}{}_i$ and to modulation of membrane channels by protein kinase C.

Before we can attempt synthesizing information contained in Figs. 1 and 2, it seems warranted to recall the coupling of energy failure and ion fluxes in ischemia. We have recently restudied changes in labile organic phosphates during the first 120 sec of ischemia and correlated them to changes in the activity of phosphorylase, a calcium-dependent enzyme (30). In another study, we specifically explored whether the rapid ion fluxes during phase 1 to 2 transition occurred before or after depletion of the tissue ATP stores (A Ekholm, B Asplund, BK Siesjö, *unpublished observations*). The results revealed that phosphorylase b to a conversion occurred after 15 sec (i.e., long before calcium had entered cells from ECF, and at a time when tissue ATP content still remained at ≈95% of the control value). As Fig. 3 shows, the phase 1 to 2 transition occurred when the phosphocreatine stores were almost depleted, and the ATP stores began decreasing at an accelerated rate.

Based on these results and many previous ones, we interpret the events depicted in Fig. 1 as follows. The induction of anoxia interrupts aerobic production of ATP, and since Na^+, K^+, and Ca^{2+} fluxes continue along channels open already at "rest," the consumption of ATP by the various ATPases leads to hydrolysis of ATP to ADP and P_i. However, because a rise in ADP shifts the creatine kinase reaction in the direction of PCr hydrolysis, the ATP stores are maintained >90% of control until the PCr stores have been tapped off. When this occurs, ATP depletion is rapid, probably accelerated by the massive ion fluxes at the phase 1 to 2 transition.

The coupling of energy failure and downhill ion fluxes can be envisaged as follows. Only seconds after induction of ischemia,

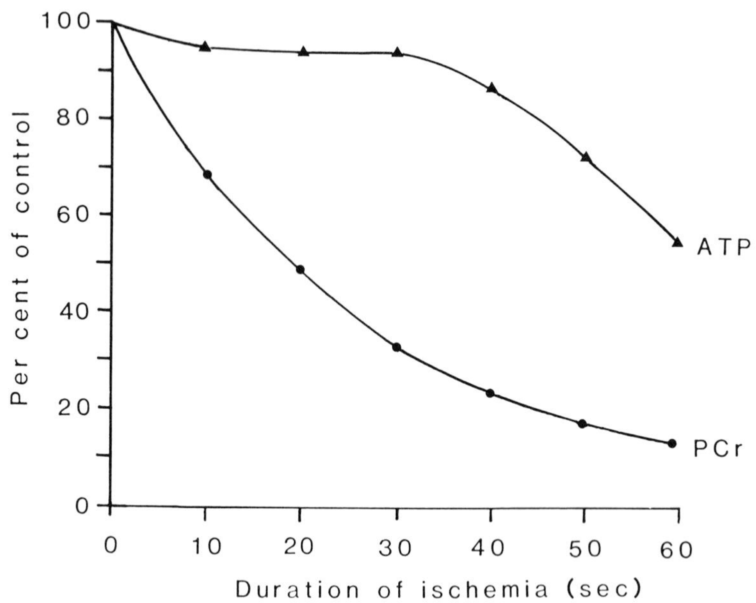

FIG. 3. Relation between changes in extracellular K^+ concentration and in PCr and ATP concentrations of cerebral cortex during the first 60 sec of anoxia.

$Ca^{2+}{}_i$ rises, possibly because surface receptors coupled to phospholipase C (PLC) are stimulated. This is at present a speculative deduction, particularly because agonists achieving this have not been identified. The rise in $Ca^{2+}{}_i$ activates a Ca^{2-}dependent K^+ conductance, leading to enhanced K^+ efflux and hyperpolarization. Because $K^+{}_e$ then rises, one must envisage presynaptic endings to depolarize. Although the early activation of a K^+ conductance causing hyperpolarization has been established, it is still an open question whether activation of an ATP-dependent K^+ conductance is involved (19,31). Conceivably, both Ca^{2+}- and ATP-dependent channels may be opened.

We thus envisage the phase 1 ionic changes to encompass K^+ efflux, postsynaptic hyperpolarization, and presynaptic depolarization. As previously suggested, the phase 2 fluxes of ions and water may be triggered by release of EAAs (32) (i.e., we may assume that presynaptic release of EAAs occurs when depolarization caused by the rise in $K^+{}_e$ reaches a certain critical value (9). This hypothesis receives strong support from the data of ref. 33, which showed release of EAA to the ECF at the point of the direct current (DC) potential shift.

Activation of K/Q and NMDA receptor–gated ion channels may be viewed as causing a massive ionic leak to Na^+, K^+, and Ca^{2+}, and since the channels are unspecific, also H^+. This and the possibility that unspecific anion channels are opened, allowing downhill fluxes of both Cl^- and HCO_3^-, probably explain the transient alkaline shift in the pH_e record (34,35).

There is a lingering suspicion that massive depolarization, or a marked rise in $Ca^{2+}{}_i$, can activate a nonspecific cation channel, perhaps also a corresponding anion channel (for discussion, see ref. 9). If this is so and if that channel is permeable to Ca^{2+}, calcium can enter depolarized cells by AOCCs, VSCCs, and a nonspecific channel activated under adverse conditions. We also recall that release of Ca^{2+} from the endoplasmic reticulum (ER) (by IP_3) may be followed by uptake of calcium and refilling of the ER via a separate channel (36,37) and that calcium may enter cells via a reversal of the electrogenic $3Na^{+}/Ca^{2+}$ antiporter (for literature, see ref. 10). Clearly, multiple pathways of calcium entry exist.

PATHOGENESIS OF CELL DAMAGE

As discussed elsewhere, there is now good evidence that ischemic, hypoglycemic, and epileptic damage is calcium-related, also that calcium "overload" of cells is related to release of EAAs (5–7,9). Probably, the coupling of release of EAAs and calcium-related damage reflects the importance of the NMDA-gated ion channel in translocating Ca^{2+} into the cell; however, we cannot dismiss the importance of stimulation of metabolotropic Q-preferring receptors and intracellular release of Ca^{2+}.

Obviously, we have to explain why hypoglycemia and ischemia but not SD lead to neuronal damage. This becomes explainable if one assumes that the rise in $Ca^{2+}{}_i$ must exceed a certain threshold value and/or be sustained for a certain minimal period to yield cell damage. Probably, the rise in $Ca^{2+}{}_i$ reflects the pump–leak relation [i.e., it is a function of the efficiency of the pump (via ATP-driven transport or the $3Na^{+}/Ca^{2+}$ exchanger) and the magnitude of the leak (via the multiple channels discussed)].

By adhering to the concept of a disturbed pump-leak relation and by considering the importance of a remaining energy source, we can make predictions about the pathogenesis of brain damage and the therapeutic effects of NMDA and calcium antagonists (9,10). In SD, energy production is not compromised. Thus, when calcium enters cells it should be so quickly extruded that $Ca^{2+}{}_i$ never reaches values that trigger cell damage. Furthermore, sequestration of calcium by ER and mitochondria should be operative. Finally, the presence of an adequate energy source allows the EAAs released to be promptly taken up. It follows from this that if SD is repeatedly elicited in a tissue, whose energy production is compromised by hypoxia or ischemia, cell damage may well be incurred. This may be the scenario in *the penumbra zone of a stroke lesion* and also the basis for a successful therapy with NMDA antagonists (which probably act by reducing calcium influx) and with some calcium antagonists (which could improve the pump function by raising blood flow). In hypoglycemic coma, energy production is even more compromised, and in principle, loss of ion homeostasis is permanent unless glucose is given. However, because NMDA antagonists ameliorate hypoglycemic damage, we must assume that the remaining energy source can maintain $Ca^{2+}{}_i$ at sublethal values if calcium influx is curtailed. In *complete ischemia* finally, energy production is nil. This and the presence of multiple entry pathways for calcium explain the severity of the lesions incurred and the absence of marked therapeutic effects by NMDA and calcium antagonist.

VOLUME TRANSMISSION AND BRAIN DAMAGE

Although information transfer between different parts of the brain involve classical chemical transmission at synapses, the interplay between pre- and postsynaptic events, the participation of glial cells, and the transfer of gases and solutes between blood and glia/neurons involve the ECF as a communication channel. This is even more obvious in such pathological states as SD, hypoglycemia, and ischemia. In all these states, a pathological stimulus is conveyed by adverse factors spreading from cell to cell via the ECF.

The stimulus causing such spread of excitation is not known, but for SD, both K^+ and glutamate have been discussed (9,38). Thus, one may envisage that depolarization of a group of neighboring cells causes such release of K^+ or glutamate that the depolarization spreads to other cells in a wave-like manner. If K^+ is the "transmitter," we may simply assume that the rise of $K^+{}_e$

is what triggers release of glutamate, exciting neighboring cells. The participation of EAAs is obvious because NMDA antagonists block SD (39,40).

In the perifocal penumbra zone, which is marginally but not critically underperfused, SD-like depolarization waves could well be elicited in the border zone between penumbra and focus (e.g., by K^+ or glutamate released from energy-deprived cells in the latter). In fact, one may envisage a domino-like mechanism that does not stop until the process reaches an area with normal perfusion or until the K^+ or EAAs released from sick cells have been absorbed by normal ones or proliferating glia cells.

Probably, depolarization during hypoglycemia starts locally at one or several foci, which release K^+ when sufficiently deprived of energy, and then spreads by mechanisms similar to those in true SD. However, although recovery may be initially achieved, at least for seconds or a few minutes (39,40), it does not normally occur, probably because the ionic leaks of depolarized cells overtax the reduced capacity of the ion pumps. In complete anoxia/ischemia, finally, the depolarization may well start locally, but because energy failure develops rapidly, the whole tissue is quickly depolarized. In this situation, therefore, it is hardly appropriate to talk about a domino mechanism causing the initial depolarization. However, it is still possible that volume conduction of depolarizing or toxic agents contributes to the final neuronal (or tissue) damage incurred after transient ischemia. This damage, which is conspicuously delayed, often affects cells in cluster or laminae, suggesting that dying cells "infect" neighboring ones, which are thereby recruited in the disease process. Evidently, the "toxins" involved may be glutamate or related EAAs released by cells whose cell membrane integrity is lost. However, we cannot exclude the possibility that partially reduced oxygen species (e.g., H_2O_2 or $\cdot O_2^-$) are involved, leading to a free radical–induced propagation of the disease process. Whatever the substances involved, they must use the pathways of so-called volume transmission.

REFERENCES

1. Nicholson C. Brain-cell microenvironment as a communication channel. In: Worden FSAF, ed. *The Neurosciences*. Cambridge, Massachusetts: MIT Press, 1979;457–476.
2. Nicholson, C. Dynamics of the brain cell microenvironment. *Neurosci Res Prog Bull* 1980; 18:177–322.
3. Hansen AJ. Effects of anoxia on ion distribution in the brian. *Physiol Rev* 1985;65(1):101–148.
4. Nedergaard M, Hansen AJ. Spreading depression is not associated with neuronal injury in the normal brain. *Brain Res* 1988;449:395–398.
5. Siesjö, BK. Calcium, ischemia, and death of brain cells. *Ann NY Acad Sci* 1988;522:638–661.
6. Siesjö BK. Mechanisms of ischemic brain damage. *Crit Care Med* 1988;16:954–963.
7. Choi DW. Glutamate neurotoxicity and diseases of the nervous system. *Neuron* 1988:1:623–634.
8. Rothman SM, Olney JW. Glutamate and the pathophysiology of hypoxicischemic brain damage. *Ann Neurol* 1986;19:105–111.
9. Siesjö BK, Bengtsson F. Calcium Fluxes, calcium antagonists, and calcium-related pathology in brain ischemia, hypoglycemia, and spreading depression: a unifying hypothesis. *J Cereb Blood Flow Metab* 1989;9:127–140.
10. Siesjö BK, Bengtsson F, Grampp W, Theander S. Calcium, excitotoxins, and neuronal death in the brain. *NY Acad Sci*.
11. Harris RJ, Wieloch T, Symon L, Siesjö BK. Cerebral extracellular calcium activity in severe hypoglycemia: relation to extracellular potassium activity and energy state. *J Cereb Blood Flow Metab* 1984;4:187–193.
12. Harris RJ, Symon L, Branston NM, Bayhan M. Changes in extracellular calcium activity in cerebral ischaemia. *J Cereb Blood Flow Metab* 1981;1:203–209.
13. Kraig RP, Pulsinelli WA, Plum F. Hydrogen ion buffering during complete brain ischemia. *Brain Res* 1985;342:281–290.
14. Kraig RP, Pulsinelli WA, Plum F. Carbonic acid buffer changes during complete brain ischemia. *Am J Physiol* 1986;250:R348–R357.
15. Mutch WA, Hansen AJ. Extracellular pH changes during spreading depression and cerebral ischemia: Mechanisms of brain pH regulation. *J Cereb Blood Flow Metab* 1984:17–27.
16. Ljunggren B, Norberg K, Siesjö BK. Influence of tissue acidosis upon restitution of brain energy metabolism following total ischemia. *Brain Res* 1974;77:173–186.
17. Smith M-L, Hanwehr RV, Siesjö BK. Changes in extra- and intracellular pH in the brain during and following ischemia in hyperglycemic and in

moderately hypoglycemic rats. *J Cereb Blood Flow Metab* 1986;6:574–583.
18. Fujuwara N, Higashi H, Shimoji K, Yoshimura M. Effects of hypoxia on rat hippocampal neurones *in vitro*. *J Physiol* 1987;384:131–151.
19. Leblond J, Krnjevic K. Hypoxic changes in hippocampal neurons. *J Neurophysiol* 1989;62:1–13.
20. Kraig RP, Nicholson C. Extracellular ionic variations during spreading depression. *Neuroscience* 1978;3:1045–1059.
21. Strong AJ, Venables GS, Gibson G. The cortical ischaemic penumbra associated with occlusion of the middle cerebral artery in the cat: 1. Topography of changes in blood flow, potassium ion activity, and EEG. *J Cereb Blood Flow Metab* 1983;3:86–96.
22. Miller RJ. Multiple calcium channels and neuronal function. *Science* 1987;235:46–52.
23. Reynolds IJ, Miller RJ. Multiple sites for the regulation of the *N*-methyl-D-aspartate receptor. *Mol Pharmacol* 1988;33:581–584.
24. Nicholls DG. The release of glutamate, aspartate and GABA from isolated nerve terminals. *J Neurochem* 1989;52:331–341.
25. Schousboe A. Transport and metabolism of glutamate and GABA in neurons and glial cells. *Int Rev Neurobiol* 1981;22:1–45.
26. Asher P, Nowak L. Electrophysiological studies of NMDA receptors. *TINS* 1987;10:284–288.
27. Fagg GE. L-Glutamate, excitatory amino acid receptors and brain functions. *TINS* 1985;8:207–210.
28. Mayer ML, Westbrook GL. The physiology of excitatory amino acids in the vertebrate central nervous system. *Prog Neurobiol* 1987;28:197–276.
29. Sladeczek F, Récasens M, Bockaert J. A new mechanism for glutamate receptor action: phosphoinositide hydrolysis. *TINS* 1988;11:545–549.
30. Folbergrova J, Minamisawa H, Ekholm A, Siesjö BK. Phosphorylase *a* and labile metabolites during anoxia: evidence for a rise in cytosolic free calcium before calcium enters from extracellular fluid. *J Neurochem* (*in press*).
31. Mourre C, Ari YB, Barnardi H, Fosset M, Lazdunski M. Antidiabetic sulfonylureas: localization of binding sites in the brain and effe on the hyperpolarisation induced by anoxia in hippocampal slices. *Brain Res* 1989;486:159–164.
32. van Harreveld A. A mechanism for fluid shifts specific for the central nervous system. *Top Prob Psychiatry Neurol* 1970;10:62–70.
33. Scheller D, Heister U, Peters U, Höller M. Glutamate and aspartate are released concomitantly with the terminal DC-negativation after global cerebral ischemia. *J Cereb Blood Flow Metab* 1989;9:S372.
34. Obrenovitch TP, Garofalo O, Harris RJ, et al. Brain tissue concentration of ATP, phosphocreatine, lactate and tissue pH in relation to reduced cerebral blood flow following experimental acute middle cerebral artery occlusion. *J Cereb Blood Flow Metab* 1988;8:866–874.
35. Siesjö BK. Acidosis and ischemic brain damage. *Neurochem Pathol* 1988;9:31–88.
36. Berridge MJ, Irvine RF. Inositol phosphates and cell signalling. *Nature* 1989;341:197–205.
37. Putney JW Jr. A model for receptor-regulated calcium entry. *Cell Calcium* 1986;7:1.
38. Bures J, Buresova O, Krivanek J. *The Mechanism and Applications of Leao's Spreading Depression of Electroencephalographic Activity. Prague and New York: Academia, 1974.*
39. Goroleva NA, Koroleva VI, Amemori T, Pavlik V, Bures J. Ketamine blockade of cortical spreading depression in rat. *Electroencephalog Clin Neurophysiol* 1987;66:440–447.
40. Pelligrino D, Almqvist L-O, Siesjö BK. Effects of insulin-induced hypoglycemia on intracellular pH and impedance in the cerebral cortex of the rat. *Brain Res* 1981;221:129–147.

Volume Transmission in the Brain: Novel Mechanisms for Neural Transmission,
edited by K. Fuxe and L.F. Agnati.
Raven Press, Ltd., New York © 1991.

44

Neurotensin and Its Possible Significance in the Pathophysiology of Schizophrenia

*Garth Bissette, †Beth Levant, and *†Charles B. Nemeroff

**Departments of Psychiatry and *†Pharmacology, Duke University Medical Center, Durham, North Carolina 27710*

Although several theories have been promulgated concerning the etiology of schizophrenia, at present the cause of this major psychiatric disease remains obscure. Although the beneficial effects of the antipsychotic drugs in ameliorating some of the symptoms of schizophrenia are undisputed, the mechanism by which their known pharmacologic actions produce these therapeutic effects has not yet been elucidated. A major tenet of the dopamine hyperactivity hypothesis of schizophrenia is the recognition that many of the classical antipsychotic drugs exhibit a remarkably close correlation between their potency in blocking D_2 dopamine (DA) receptors and their clinical antipsychotic efficacy. However, several new drugs with demonstrable antipsychotic effects and relatively low affinity at D_2 DA receptors (e.g., rimcazole and remoxapride) have now been discovered. In addition, examination of neurochemical changes in laboratory animals treated with clinically relevant doses of both classical and novel antipsychotic compounds have demonstrated alterations in certain neuropeptide neurotransmitter systems. The concentrations of Met-enkephalin (1) and cholecystokinin (2) were reported to be increased, and the concentrations of substance P (3) and somatostatin (4) were reported to be decreased in several brain regions of rats treated with daily doses of clinically effective antipsychotic drugs for 2 to 3 weeks. However, only one peptide is intimately associated, both anatomically and pharmacologically, with DA neurons, is altered in concentration by clinically effective antipsychotic drugs in laboratory animals, and has been reported to be altered in postmortem brain tissue and cerebrospinal fluid (CSF) from schizophrenic patients. This neuropeptide is neurotensin (NT).

Neurotensin was first isolated from bovine hypothalamus by Carraway and Leeman (5) and has since been found in brain and gut of a variety of mammalian species, including humans (6,7). NT is comprised of 13 amino acids and is localized heterogeneously in neuronal cell bodies and nerve terminals located in several brain regions including the hypothalamus, bed nucleus of the stria terminalis, central amygdaloid nucleus, ventral tegmental area (VTA), substantia nigra, striatum, nucleus accumbens, septum, diagonal band of Broca, and frontal cortex (8). NT is reported to be colocalized within DA neurons of the arcuate nucleus of the hypothalamus (9,10) and with both DA and cholecystokinin (11) in the VTA. The gene for NT consists of four exons, and NT is expressed on the last exon, which also contains the gene for neuromedin N, a structurally related peptide (12). Although the ratio of NT–neuromedin N concentration varies among brain regions (13), the messenger ribonucleic acid (mRNA) for NT and neuromedin N contains both sequences and has recently been mapped by *in situ* hybridization. This NT–neuromedin mRNA was found in the septum, bed nucleus of the stria terminalis, preoptic area, hypothalamus, amygdala, nucleus accumbens, striatum, and cortex (14); a distribution that matches the brain tissue content of these peptides well. However, it is also found in high concentrations in the subiculum of the hippocampus where NT concentrations are relatively low. High-affinity binding sites (putative receptors) for NT have been characterized in several species (15) and have recently been reported to also bind neuromedin N (16). NT is degraded by peptidases (17) and released by physiological stimuli in a calcium-dependent manner (18). Thus, NT fulfills most of the requisite criteria for a brain transmitter and is synthesized, released, and bound to receptors in brain regions that are possibly involved in schizophrenia.

Neurotensin injected intraventricularly (19,20) or applied directly to CNS DA neurons (21,22) increases the concentrations of DA metabolites. Recently, Cador et al. (23) injected NT, substance P, or enkephalin onto VTA neurons and observed increased 3,4-dihydroxyphenylacetic acid (DOPAC)–DA concentrations in the nucleus accumbens and septum after all three peptides, but only in the amygdala after NT. NT increases the binding affinity of DA agonists (e.g., *N*-propylnorapomorphine) (24,25), and recently NT was proposed to bind directly to the DA molecule (26), although these data have not been confirmed by other laboratories at present. The colocalization of NT with DA in neurons of the ventral tegmental area have been reported by several groups (9,11,27). As much as 90% of the NT staining cells of the VTA also contain cholecystokinin as well as the DA neuronal marker, tyrosine hydroxylase, and these neurons predominantly project to the nucleus accumbens and prefrontal cortex (11). In addition, the close communication of NT nerve terminals with dendrites and cell bodies of VTA DA neurons has been demonstrated using immunocytochemistry (28).

A dense population of NT immunoreactive neurons has also recently been described running through the ventromedial striatum, the bed nucleus of the stria terminalis–diagonal band region, the nucleus accumbens, and the olfactory tubercles (29–31). These are known terminal regions of the mesencephalic DA neuron projections and correspond to regions where the binding of classical antipsychotics to D_2 DA receptors is seen.

Lesions of brain dopamine systems have been reported to alter NT concentrations in several areas. Although the density of immunoreactive staining for NT is initially decreased in brain areas receiving mesencephalic projections after 6-hydroxydopamine (OHDA) lesions of the ventral tegmental area, within 3 days after the lesion the number of NT neurons is increased in the striatum compared with before the lesion (32). However, the total content of NT in the striatum and nucleus accumbens is not altered when measured 2 weeks after 90% to 95% depletion of regional brain DA by intracisternal 6-OHDA (33). This is probably due to the disproportionately large amount of NT contained in nerve terminals originating in NT cell bodies that do not contain DA compared with the total amount of NT contained within DA neurons. Interestingly, in Parkinson's disease, no regional changes in NT concentrations in DA perikarya or terminal regions were noted (34), although NT receptors, which are found on the DA neurons of the substantia nigra, were reported to be decreased (35). NT receptors in the VTA, caudate nucleus, and lateral nucleus accumbens were reduced by lesions of the VTA, whereas NT receptors in the medial nucleus accumbens and adjacent limbic areas were unchanged (36). Similarly, 6-OHDA lesions of the VTA have been reported to decrease the concentration of NT in the frontal cortex (37) and increase the number of NT receptors in this region (38). 1-methyl-4-phenyl-1,2,3,6-tetrahydropyridine (MPTP) has also been used to lesion DA cells in mouse brain with concomitant decreases in NT concentration in several DA terminal regions (39). Other evidence that the DA–NT system in the prefrontal cortex is differentially regulated compared with striatal or nucleus accumbens NT–DA interactions was reported by Bean et al. (40). Reserpine administrations produced a dose- and time-dependent decrease in both NT and DA concentrations in the frontal cortex, whereas the same treatment increased NT and decreased DA in the striatum and nucleus accumbens. Similarly, electrical stimulation of the median forebrain bundle increased release of both DA and NT in the prefrontal cortex as measured with microdialysis sampling (41). This data and that of others (42) suggest that the prefrontal region receive DA–NT input from the same neurons, whereas in the nucleus accumbens and striatum, a large component of the NT present is not found within DA neurons or terminals.

The investigation of the similarities between NT and antipsychotic drugs began when NT was found to be active in several pharmacological and behavioral paradigms used initially to screen compounds for potential neuroleptic activity (see refs. 43,44 for review). This data presaged the discovery that the concentration of NT in the nucleus accumbens and striatum could be significantly increased by administration of clinically effective antipsychotic drugs (45). Within 16 hours of a single acute dose of haloperidol, NT content was increased in the nucleus accumbens; clinically ineffective but structurally related compounds did not have this effect on NT concentrations. Further investigation revealed that clozapine, an antipsychotic devoid of extrapyramidal effect liability and presumably also without risk for tardive dyskinesia, increased NT content like haloperidol in the nucleus accumbens but was without effect in the caudate nucleus (46). The concentration of NT in the prefrontal cortex was also found to be decreased by both clozapine and haloperidol in this micropunch dissection study. We have recently shown a similar effect of S(+)-*N*-*n*-propyl-norapomorphine (NPA) in increasing NT concentrations of the nucleus accumbens but not the caudate nucleus (47). An increase in the number of NT staining neurons in the nucleus accumbens, olfactory tubercles, and ventromedial caudate-putamen after neuroleptic treatment has also been reported (48). The long-acting neuroleptic pipotiazine palmitic ester has been reported to increase the density of NT receptor fields in the prefrontal cortex, nucleus accumbens, and striatum (38).

Thus, the prefrontal cortex exhibits several peculiarities in the interactions between DA and NT that are not shared by other brain regions receiving DA projections from the mesencephalon. That this region is one of the brain regions implicated in schizophrenia may not be coincidental. It is well known that many antipsychotic drugs exert effects at receptors other than the D_2 DA type. Our laboratory has examined the effect of stereoisomers of butaclamol on NT concentrations. The (+) isomer of butaclamol blocks the D_2 DA receptor and is a clinically effective antipsychotic, whereas the (−) isomer does not have either of these properties. Both isomers, however, share similar affinities for sigma, cholinergic, serotonergic, and alpha-adrenergic receptors, which is a characteristic of many neuropeptic drugs (49). NT concentrations in the nucleus accumbens and caudate-putamen were increased by administration of (+) butaclamol, whereas (−) butaclamol was without effect on NT concentrations in these regions (50). Effects of haloperidol on cholinergic receptors were also ruled out by the failure of atropine to alter NT concentrations (51).

In another collaborative study, the effects of chronic (6-month) administration of haloperidol on NT concentrations were examined (52). The ability of haloperidol to elevate NT concentrations was not diminished by this chronic treatment. Two months after withdrawal of haloperidol, NT concentrations showed a significant decrease to below pretreatment levels. Thus the increase in NT concentration does not show tachyphylaxis and therefore remains a possible component of the antipsychotic mechanism.

Recently, we destroyed rat brain DA systems with 6-OHDA and then treated them with the neuroleptic drug haloperidol. There was no change in the ability of haloperidol to increase NT concentrations in the nucleus accumbens or caudate nucleus after destruction of DA neurons, and there was no correlation between the magnitude of DA depletion and the increase in NT concentrations in individual rats (53). Thus, the site of the D_2 DA receptor blockade and subsequent induction of NT concentration increases is apparently postsynaptic to the DA neuron. Further investigations have now examined the possible involvement of the "sigma" receptor in NT increases induced by antipsychotic agents. This receptor is thought to mediate the psychotomimetic effects of certain benzmorphans and may be involved in the mechanism of antipsychotic drugs (54).

As many of the "novel" antipsychotic compounds have effects at sigma receptors (54), as well as such classical ones as haloperidol, chloropromazine, and fluphenazine (55), we examined the effect of haloperidol and the purportedly specific sigma receptor antagonist BMY 14802 for ability to increase NT concentrations. We found that BMY 14802, which has no activity at D_2 receptors, is as effective as haloperidol in increasing NT concentrations (56). The relative density of sigma binding sites, however, is equal to D_2 DA receptors in the nucleus accumbens, much higher than D_2 DA receptors in cortex, hippocampus, and amygdala and lower than D_2 DA receptors in the caudate nucleus (57). Thus these receptors are anatomically well-placed to mediate NT neuronal systems and possibly also participate in higher cognitive functions. However, (−) butaclamol is a potent sigma antagonist and is without effect on NT concentrations in the nucleus accumbens and striatum (50), and this suggests that more than blockade of sigma receptors is required for this response.

Several studies have examined the concentration of NT in CSF from living schizophrenic patients and in postmortem brain tissue from patients dying with a diagnosis of schizophrenia. These data are often confounded by previous neuroleptic drug treatment, although patients are usually withdrawn from drug treatment for 1 to 2 weeks before sampling in the spinal fluid studies and neuroleptics are often discontinued in the terminal medical crises that precede the postmortem tissue studies. Mean NT concentrations in CSF were reported to be decreased in a population of patients in our original study (58). A subgroup of patients with markedly low levels of NT during a 2-week interval in antipsychotic drug therapy responded to renewed treatment with antipsychotics with NT concentrations that increased toward the normal range as measured in a second CSF sample. In a following study, the mean decrease in NT concentrations in the CSF of schizophrenics on a 2-week drug holiday was confirmed, but in this study no change in NT levels was seen after reinstatement of antipsychotic treatment (59). We have reviewed these data and several other studies on NT in CSF of other diagnostic groups besides schizophrenia (60). More recently, a population of schizophrenic and schizo-affective patients were found to show resistance to antipsychotic drug therapy when NT concentration in CSF was low, although there were no mean group differences between the schizophrenics and controls (61). All three patients treated with haloperidol that agreed to a second lumbar puncture showed elevated NT concentrations in CSF compared with their initial CSF sample. Thus, neuroleptic treatment can indeed alter human, as well as animal, concentrations of NT. It has been previously demonstrated that haloperidol treatment of both rats and humans increases the number of NT receptors in the substantia nigra (62). As NT in the CSF may reflect the extracellular fluid concentrations of NT and thus indicate changes in the ability of NT to communicate through "volume" transmission, CSF concentrations of NT may be potentially useful in predicting treatment response to neuroleptic drugs in schizophrenia. In agreement with the hypothesis that "volume" transmission is of increased importance during development, we have recently reported markedly increased CSF concentrations of NT in infants that decline to adult levels at about 2 years of age (63).

There have now been four studies of postmortem alterations in regional brain concentrations of NT in schizophrenic patients. As in CSF studies, the presence of chronic neuroleptic drug treatment must be taken into consideration in interpreting these data. Using brain tissue provided by the McLean Hospital Brain Bank in Boston, our group examined NT concentrations in several subcortical and cortical brain regions of schizophrenics compared with psychiatrically normal controls (64). Although there was no difference in caudate, amygdala, hypothalamus, or nucleus accumbens NT concentrations between these groups, the schizophrenic group exhibited a statistically significant increase in NT concentrations of Brodmann's area 32 of the frontal cortex but not in the other cortical areas provided.

Crow's group in London examined postmortem brain tissue from both type I and type II schizophrenic patients but found no alterations in NT concentration in several cortical (frontal, temporal, cingulate, parietal) and subcortical (hippocampus, amygdala, globus pallidus, putamen, thalamus) brain region (65). Kleinman at the National Institute of Mental Health (NIMH) has examined the nucleus accumbens, hypothalamus, and globus pallidus from schizophrenic and control patients but found no changes in NT concentration (66). Finally, another British group has reported that there is no difference in the concentration of NT in the amygdala of schizophrenic patients compared with normal patients (67). Thus the available postmortem data have not reproduced the decreases in NT concentrations that are seen in CSF studies. The true test of the hypothesis that lowered NT concentrations contribute to the symptoms of schizophrenia awaits an NT receptor agonist that can cross the blood-brain barrier. It is hoped that such compounds will become available in the near future.

ACKNOWLEDGMENTS

The author's research is supported by NIMH MH-39415 and a grant from the Scottish Rite Schizophrenia Research Foundation. The excellent secretarial assistance of Sheila Walker and Ward Virts is greatly appreciated.

REFERENCES

1. Hong JS, Yang H-YT, Fratta W, Costa E. Rat striatal methionine-enkephalin content after chronic treatment with cataleptogenic and noncataleptogenic antischizophrenic drugs. *J Pharmacol Exp Ther* 1978;205:141–147.
2. Frey P. Cholecystokinin octapeptide levels in rat brain are changed after sub-chronic neuroleptic treatment. *Eur J Pharmacol* 1983;95:87–92.
3. Hong JS, Yang H-YT, Costa E. Substance P content of substantia nigra after chronic treatment with antischizophrenic drugs. *Neuropharmacology* 1978;17:83–85.
4. Beal MF, Martin JB. Effects of neuroleptic drugs on brain somatostatin-like immunoreactivity. *Neurosci Lett* 1984;47:125–130.
5. Carraway RE, Leeman SE. The isolation of a new hypotensive peptide, neurotensin, from bovine hypothalami. *J Biol Chem* 1973;248:6854–6861.
6. Manberg PJ, Youngblood WW, Nemeroff CB, et al. Regional distribution of neurotensin in human brain. *J Neurochem* 1982;38:1777–1780.
7. Mai JK, Triepel J, Metz J. Neurotensin in the human brain. *Neuroscience* 1987;22:499–524.
8. Emson PC, Goedert M, Mantyh PW. Neurotensin-containing neurons. In: Bjorklund A, Hokfelt T, eds. *Handbook of chemical neuroanatomy,* vol 4. Amsterdam: Elsevier, 1985;355–405.

9. Hokfelt T, Everitt BJ, Theodorsson-Norheim E, Goldstein M. Occurence of neurotensin-like immunoreactivity in sub-populations of hypothalamic, mesencephalic and medullary catecholamine neurons. *J Comp Neurol* 1984; 222:543–549.
10. Ibata Y, Okamura FH, Kawakami T, et al. Coexistence of dopamine and neurotensin in hypothalamic arcuate and periventricular neurons. *Brain Res* 1983;269:177–179.
11. Seroogy K, Ceccatelli S, Schalling M, et al. A sub-population of dopaminergic neurons in rat ventral mesencephalon contains both neurotensin and cholecystokinin. *Brain Res* 1988;455: 88–98.
12. Kislauskis E, Bullock B, McNeil S, Dobner PR. The rat gene encoding neurotensin and neuromedin N: structure, tissue-specific expression, and evolution of exon sequences. *J Biol Chem* 1988;263:4963–4968.
13. Carraway RE, Mitra SP. The use of radioimmunoassay to compare the tissue and subcellular distributions of neurotensin and neuromedin N in the cat. *Endocrinology* 1987;120:2092–2100.
14. Alexander MJ, Miller MA, Dorsa DM, et al. Distribution of neurotensin/neuromedin N mRNA in rat forebrain: unexpected abundance in hippocampus and subiculum. *Proc Natl Acad Sci* 1989;86:5202–5206.
15. Kitabgi P, Carraway R, Van Rietschoten J, et al. Neurotensin: specific binding to synaptic membranes from rat brain. *Proc Natl Acad Sci USA* 1977;74:1846–1850.
16. Checler F, Vincent JP, Kitabgi P. Neuromedin N: high affinity interaction with brain neurotensin receptors and rapid inactivation by brain synaptic peptidases. *Eur J Pharmacol* 1986; 126:239–244.
17. Dupont A, Merand Y. Enzymic inactivation of neurotensin by hypothalamic and brain extracts of the rat. *Life Sci* 1978;22:1623–1630.
18. Iversen LL, Iversen SD, Bloom F, Douglas C, Brown M, Vale W. Calcium-dependent release of somatostatin and neurotensin from rat brain *in vitro*. *Nature* 1978;273:161.
19. Widerlov E, Kilts CD, Mailman RB, et al. Increase in dopamine metabolites in rat brain by neurotensin. *J Pharmacol Exp Ther* 1982;222: 1–6.
20. Nemeroff CB, Luttinger D, Hernandez DE, et al. Interactions of neurotensin with brain dopamine systems: biochemical and behavioral studies. *J Pharmacol Exp Ther* 1983;225:337–345.
21. Hetier E, Boireau A, Dubedat P, Blanchard C. Neurotensin effects on evoked release of dopamine in slices from striatum, nucleus accumbens and pre-frontal cortex in rat. *Arch Pharmacol* 1988;337:13–17.
22. Gudelsky GA, Berry SA, Meltzer HY. Neurotensin activates tuberoinfindibular dopamine neurons and increases serum corticosterone concentrations in the rat. *Neuroendocrinology* 1989;49:604–609.
23. Cador M, Rivet JM, Kelley AE, LeMoal M, Stinas L. Substance P, neurotensin and enkephalin injections into the ventral tegmental area: comparative study on dopamine turnover in several forebrain structures. *Brain Res* 1989;486:357–363.
24. Agnati LF, Fuxe K, Benefenati F, Battistini N. Neurotensin *in vitro* markedly reduces the affinity in subcortical limbic [^{3}H] *N*-propyl-norapomorphine binding sites. *Acta Physiol Scand* 1983;119:459–461.
25. von Euler G, Fuxe K, Benefenati F, Hansson T, Agnati LF, Gustafsson JA. Neurotensin modulates the binding characteristics of dopamine D_2 receptors in rat striatal membranes also following treatment with toluene. *Acta Physiol Scand* 1989;135:443–448.
26. Adachi DK, Kalivas PW, Schenk JO. Neurotensin binds to dopamine. *J Neurochem* 1990;54: 1321–1328.
27. Kalivas PW, Miller JS. Neurotensin neurons in the ventral tegmental area project to the medial nucleus accumbens. *Brain Res* 1984;300:157–160.
28. Woulfe J, Beaudet A. Immunocytochemical evidence for direct connections between neurotensin-containing axons and dopaminergic neurons in the rat ventral midbrain tegmentum. *Brain Res* 1989;479:402–406.
29. Zahm DS, Heimer L. Ventral striatopallidal parts of the basal ganglia in the rat: I. Neurochemical compartmentation as reflected by the distributions of neurotensin and substance P immunoreactivity. *J Comp Neurol* 1988;272:516–535.
30. Zahm DS. Neurotensin-immunoreactive neurons in the ventral striatum of the adult rat: ventromedial caudate-putamen, nucleus accumbens and olfactory tubercle. *Neurosci Lett* 1987;81: 41–47.
31. Jennes L, Stumpf WE, Kalivas PW. Neurotensin: topographical distribution in rat brain by immunohistochemistry. *J Comp Neurol* 1982;210: 211–224.
32. Zahm DS, Johnson SN. Asymmetrical distribution of neurotensin immunoreactivity following unilateral injection of 6-hydroxydopamine in rat ventral tegmental area (VTA). *Brain Res* 1989; 483:301–311.
33. Bissette G, Jennes L, Prange AJ Jr, Breese GR, Nemeroff CB. Neurotensin and dopamine are not co-localized in rat brain. (*Abstr.*) *Soc Neurosci Abstr* 1983;9:290.
34. Bissette G, Nemeroff CB, Decker MW, Kizer JS, Agid Y, Javoy-Agid F. Alterations in regional brain concentrations of neurotensin and bombesin in Parkinson's disease. *Ann Neurol* 1985; 17:324–329.
35. Uhl GR, Whitehouse PJ, Price DL, Tourtelotte WW, Kuhar MJ. Parkinson's disease: depletion of substantia nigra neurotensin receptors. *Brain Res* 1984;308:186–190.
36. Dilts RP, Kalivas PW. Autoradiographic localization of μ-opioid and neurotensin receptors within the mesolimbic dopamine system. *Brain Res* 1989;488:311–327.
37. Studler JM, Kitabgi P, Tramu G, Herve D,

Glowinski J, Tassin JP. Extensive co-localization of neurotensin with dopamine in rat mesocortical-frontal dopaminergic neurons. *Neuropeptides* 1988;11:95–100.
38. Herve D, Tassin JP, Studler JM, et al. Dopaminergic control of ^{125}I-labeled neurotensin binding site density in corticolimbic structures of the rat brain. *Proc Natl Acad Sci* 1986;83:6203–6207.
39. Levant B, Bissette G, Parker YM, Nemeroff CB. Alterations in neurotensin- and bombesin-like immunoreactivity in MPTP-treated mice. *Ann NY Acad Sci* 1988;537:515–517.
40. Bean AJ, Adrian TE, Modlin IM, Roth RH. Dopamine and neurotensin storage in colocalized and noncolocalized neuronal populations. *J Pharmacol Exp Ther* 1989;249:681–687.
41. Bean AJ, During MJ, Roth RH. Stimulation-induced release of coexistent transmitters in the prefrontal cortex: an *in vivo* microdialysis study of dopamine and neurotensin release. *J Neurochem* 1989;53:655–657.
42. Tassin JP, Kitabgi P, Tramu G, et al. Rat mesocortical dopaminergic neurons are mixed neurotensin/dopamine neurons: immunohistochemical and biochemical evidence. *Ann NY Acad Sci* 1988;537:531–533.
43. Levant B, Bissette G, Nemeroff CB. Further evidence for neurotensin–dopamine interactions in the central nervous system. In: Beart P, Woodruff GN, Jackson DM, eds. *Pharmacology and functional regulation of dopaminergic neurons*. London: Macmillan Press, 1988;257–265.
44. Levant B, Nemeroff CB. The psychobiology of neurotensin. In: Ganten D, Pfaff D, eds. *The neuroendocrinology of mood,* vol 8. Berlin: Springer-Verlag, 1988;231–262.
45. Govoni S, Hong JS, Yang H-YT, Costa E. Increase of neurotensin content elicited by neuroleptics in nucleus accumbens. *J Pharmacol Exp Ther* 1980;215:413–417.
46. Kilts CD, Anderson CM, Bissette G, Ely TD, Nemeroff CB. Differential effects of antipsychotic drugs on the neurotensin content of discrete rat brain nuclei. *Biochem Pharmacol* 1988;37:1547–1554.
47. Nemeroff CB, Kilts CD, Levant B, et al. Effects of aporphines on regional brain neurotensin concentrations: implications for development of novel antipsychotic drugs (*Abstr.*). *Soc Neurosci Absts* 1989;15:1122.
48. Eggerman KW, Zahm DS. Numbers of neurotensin-immunoreactive neurons selectively increased in rat ventral striatum following acute haloperidol administration. *Neuropeptides* 1988; 11:125–132.
49. Richelson E. Pharmacology and clinical considerations of the neuroleptics. In: *Neuropharmacology of central nervous system and behavioral disorders*. Orlando: Academic Press, 1981;123–147.
50. Bissette G, Dauer WT, Kilts CD, O'Connor L, Nemeroff CB. The effect of the stereoisomers of butaclamol on neurotensin content in discrete regions of the rat brain. *Neuropsychopharmacology* 1988;1:329–335.
51. Levant B, Bissette G, Nemeroff CB. Effects of anticholinergic drugs on regional brain neurotensin concentrations. *Eur J Pharmacol* 1989; 165:327–330.
52. Radke JM, MacLennan AJ, Beinfeld MC, et al. Effects of short- and long-term haloperidol administration and withdrawal on regional brain cholecystokinin and neurotensin concentrations in the rat. *Brain Res* 1989;480:178–183.
53. Bissette G, Dole K, Johnson M, et al. Antipsychotic drugs increase neurotensin concentrations after destruction of dopamine neurons by 6-hydroxydopamine (*Abstr.*). *Soc Neurosci Absts* 1988;14:1211.
54. Snyder S, Largent BL. Receptor mechanisms in antipsychotic drug action: focus on sigma receptors. *J Neuropsychiat Clin Neurosci* 1989;1: 7–15.
55. Largent BL, Wikstrom H, Gundlach AL, Snyder SH. Structural determinants of σ receptor affinity. *Mol Pharmacol* 1987;32:772–784.
56. Levant B, Bissette G, Nemeroff CB. Sigma antagonist MJ 14802 increases neurotensin concentration in the rat nucleus accumbens and caudate nucleus (*Abstr.*). *Soc Neurosci Absts* 1989;15: 811.
57. Contreras PC, Quirion R, Gehlert DR, Contreras ML, O'Donohue TL. Autoradiographic distribution of non-dopaminergic binding sites labelled by [H^3]haloperidol in rat brain. *Neurosci Lett* 1987;75:133–140.
58. Widerlov E, Lindstrom LH, Besev G, et al. Subnormal CSF levels of neurotensin in a subgroup of schizophrenic patients: normalization after neuroleptic treatment. *Am J Psychiatry* 1982; 139:1122–1126.
59. Lindstrom LH, Widerlov E, Bissette G, Nemeroff CB. Reduced CSF neurotensin concentration in drug-free schizophrenic patients. *Schizophr Res* 1988;1:55–59.
60. Nemeroff CB, Bissette G, Widerlov E, et al. Neurotensin-like immunoreactivity in cerebrospinal fluid of patients with schizophrenia, depression, anorexia nervosa-bulimia and premenstrual syndrome. *J Neuropsychiat Clin Neurosci* 1989;1:16–20.
61. Bissette G, Garver D, Kelly K, Nemeroff CB. CSF neurotensin concentrations in psychoses (*Abstr.*). *Soc Neurosci Absts* 1989;15:1122.
62. Uhl GR, Kuhar MJ. Chronic neuroleptic treatment enhances neurotensin receptor binding in human and rat substantia nigra. *Nature* 1984; 309:350–352.
63. Hedner J, Hedner T, Lundell KH, Bissette G, O'Connor L, Nemeroff CB. Cerebrospinal fluid concentrations of neurotensin and corticotropin-releasing factor in pediatric patients. *Biol Neonate* 1989;55:260–267.
64. Nemeroff CB, Youngblood W, Manberg PJ, Prange AJ Jr, Kizer JS. Regional brain concentrations of neuropeptides in Huntington's chorea and schizophrenia. *Science* 1983;221:972–975.
65. Ferrier IN, Crow TJ, Roberts GW, et al. Alterations in neuropeptides in the limbic lobe in schizophrenia. In: Trimble MR, Zaritan E, eds.

Psychopharmacology of the limbic system. London: Oxford University Press, 1984;244–254.
66. Kleinman JE, Iadorola M, Govoni S, Hong J, Gillin JC, Wyatt RJ. Post-mortem measurements of neuropeptides in human brain. *Psychopharmacol Bull* 1983;19:375–377.
67. Biggins J, Perry EK, McDermott JR, Smith IA, Perry RH, Edwardson JA. Post-mortem levels of thyrotropin-releasing hormone and neurotensin in the amygdala in Alzheimer's disease, schizophrenia and depression. *J Neurol Sci* 1983;58:117–122.

Volume Transmission in the Brain: Novel Mechanisms for Neural Transmission,
edited by K. Fuxe and L.F. Agnati.
Raven Press, Ltd., New York © 1991.

45

Geometry of Massively Parallel Neural Interconnectedness in Wired Networks and Wireless Volume Transmission

A. J. Pellionisz

Department of Physiology and Biophysics, New York University Medical Center, New York, New York 10016

INTERCONNECTEDNESS AS A THEORETICAL PROBLEM

We are in the midst of a major conceptual revolution of neuroscience. Several orthodox axioms have already undergone or presently experience a crisis. Some dogma have already been rendered untenable. Worst of all, professional brain theory is not easily available as yet to cope with the required conceptual reform.

For instance, it has long been taken for granted that the exclusive means of communication of neurons is the synaptic "wired" transmission. This book provides ample fresh evidence for nonsynaptic "wireless' volume conduction. Although this phenomenon has been "in the books" for some time (1), "volume transmission" is now becoming an acceptable concept. Nonetheless, it is probably fair to say that arguments–counterarguments over actual details and precise mechanisms of this new type of communication will go on for decades.

In the meantime, this chapter focuses on some theoretical implications of the downfall of the axiom of exclusively "wired" communication and the meaning of this conceptual shift on theories of computational organization of the brain. Indeed, to discern the fundamental (mathematical) principles of information processing of the central nervous system (CNS) is ultimately one of the most important tasks in neuroscience in general and in this field in particular (2). Therefore, in view of the strong evidence for "volume transmission," the first question that one may ask is whether discoveries of "wireless" on top of "wired" transmission change in any way our basic theoretical understanding of the mathematical principle of brain function? This question is difficult to answer as long as there is no general agreement about such principle. Indeed, as some biologists imply, brain theory may not turn out to be mathematical at all (3).

THEORY AND DOGMA ABOUT COMMUNICATIONAL AND COMPUTATIONAL ORGANIZATION OF THE BRAIN

For those standing on the basis of a reductionalist (mathematical) neurophilosophy (4), probably the most common dogma is that the CNS is a kind of computer whose functioning is also based on the mathematical language of present-day computational machines: Boolean *algebra* of mathematical logics (5). This rudimentary notion goes back to an (over)simplified model of neurons as "flip-flops" (6), connected by a wiring discerned in the style of classical neuroanatomists (7). "Loops" of neurons, thought to be representative of networks, were thus interpreted in terms of operations of mathematical logics. The computer = brain "axiom," in fact, was discredited a generation ago by no else than the inventor of traditional computers, of "von-Neumann machines" (8). Still, primarily because of its simplicity, secondarily for the scarcity of alternative concepts, the dogma of equating the computer with the brain still lingers. We can point out at the outset that volume transmission itself is probably the most devastating phenomenon that one could think of for the idea that the intrinsic language of CNS is mathematical logics (Boolean algebra). A "wireless" general influence on a whole circuit would effectively destroy the logical structure of any meticulously laid-out network that implements logical algebra, just like a rainstorm would short-circuit and thus incapacitate a computer or a telephone network. Fortunately, the false assumption of (logical) algebra as the mathematical basis of brain function has long been rejected by professional brain theorists. Otherwise, the whole new field of neurocomputing, seeking precisely the *unknown* mathematical language of CNS, would not have emerged at all (9).

Some diametrically opposite fundamental principles between the computer and the brain are evident. Computers are serial information-processors, executing one operation at a time albeit up to a billion of them one after the other in each second. However, the most generally accepted axiom in neurocomputing is that the CNS is a massively parallel processor (10). Nobody would deny that in the CNS large arrays of active elements (up to a billion neurons) simultaneously perform their mathematical operations (whatever those may be), but the operation of each neuronal element is orders of magnitude slower than those of manmade electronic devices. Volume transmission is obviously a mechanism that can effectively ensure, or even impose, such parallelism, which appears to be a key principle of CNS function and thus attracts intense theoretical scrutiny.

NEURAL NETS FOR GAZE CONTROL: MODEL OF MASSIVELY PARALLEL NEUROCOMPUTING

The drastically different theoretical principles, parallel rather than serial processing, can be best demonstrated and studied in greatest detail in the example of gaze control. Following the trend set by pioneers (7), the anatomical basis of gaze control was a loop of a handful of individual neurons connected to one another. The "loop" concept is fully consistent with the Sherringtonian–Pavlovian view of the CNS as consisting of "reflexes." Of course, those with a new concept are the first to caution against its misuse. Sherrington himself expressed early that "an isolated reflex is a purely abstract conception, because all parts of the nervous system are fully connected" (11). Still, in the overreductionalist post-Sherringtonian trend the seemingly intractable CNS networks were simplified into separable loops, or reflexes. For instance, the epitome of such chains, the classical "three-neuron reflex arc" was featured as the neuronal mechanism that drives the extraocular muscles in a compensatory manner to head movements (measured by the vestibular semicircular canals). The vestibulo-ocular reflex (VOR) was singled out half a century ago as a serial loop (12); "according to these diagrams the nervous system is to be considered as a system of neuron chains involving a varying number of cells superimposed as the elementary two or three neuron reflexes"

(12). In all fairness, we should emphasize that Lorente de No immediately went on saying that "the elementary reflex arc, however, is still only an anatomic idealization of the structure of the nervous system," and his paper literally ended by the warning that "in the vestibular system there is no 'localization' of reflexes in anatomic nuclei; the whole system is a functional unit; reflex reaction may be produced as long as the reflex arc is closed through one of the multiple pathways; on the other hand, lesions in any part of the vestibular system modify all the reflexes" (12). With all these caveats, why is it that in case of gaze control, the three-neuron reflex arc was taken literally, leading to the long-prevailing axiom that VOR can be deduced to separable horizontal-horizontal, vertical-vertical, and torsional-torsional semicircular canal and extraocular muscle reflexes (13) as seen in Fig. 1?

First of all, the "loop-axiom" enjoyed the tremendous initial boost generated by cybernetics (14) when everything was seen in terms of "feedback control systems." Control system theory almost totally permeated quantitative neuroscience by the end of the 1960s (15,16) and still lingered in the early 1980s (17,18). This success was based, not to a small extent, on the well-established theoretical-mathematical foundation and readily available instrumentation of measuring of the gain and phase of feedback amplifiers, borrowed from the potent field of electrical engineering. The input–output of "biological amplifiers" can be easily established and thus the system characterized by its gain and phase (displayed, e.g., in a Bode diagram of gain as a function of frequency). This is applicable when a single controlled biological variable (such as gaze direction, or even a hormone concentration) can be both stimulated and measured in a quantitative, and often quite precise, manner. Thus, for such cases, the concept, elaborate methodology, and technical jargon could be simply ported into biology from engineering. In this manner, for instance, a maddeningly complex hierarchial gaze (eye–head coordination) neural net could be collapsed onto the simple "model" of horizontal vestibular canal to horizontal extraocular muscle "feedback control loop." As a generalization of this organizing principle, from the CNS to the entire biological organism, the at least as complex interactions among a high number of hormonal agents and receptors could be simplified into an agent–receptor feedback-control system of a single biological variable, such as a hormone concentration only (see, e.g., ref. 19).

Approaching the 1990s, after all the progress made in gaze research on the basis of system theory of serial loops, its pioneer (D. A. Robinson) is strongly in favor of a drastic change to "distributed parallel processing in the vestibulo-oculomotor system" (20). He proposes the phasing-out of the approach of system theory of serial connections and replacing it by other mathematical paradigms, which are based on parallel computation. His strongest rationale is that earlier system theoretical models "have generally taken the form of black-box diagrams [for example, ref. 18] representing the flow of hypothetical signals between idealized signal-processing blocks. They approximate overall oculomotor behavior but indicate little about how real eye-movement signals would be carried and processed by real neural networks" (20). In this chapter we might wish to emphasize that system theory of isolated serial loops does not appear to be any more useful for theoretically interpreting volume transmission either. Thus, a conceptual reform now appears inevitable, not just in gaze research but in neuroscience at large.

Indeed, an alternative concept, inherently based on distributed parallel processing, was proposed a decade ago. Tensor network theory was developed for explaining the function of massively parallel neuronal networks in terms of transformation of multidimensional generalized vectors (tensors) (21). Implications of this principle

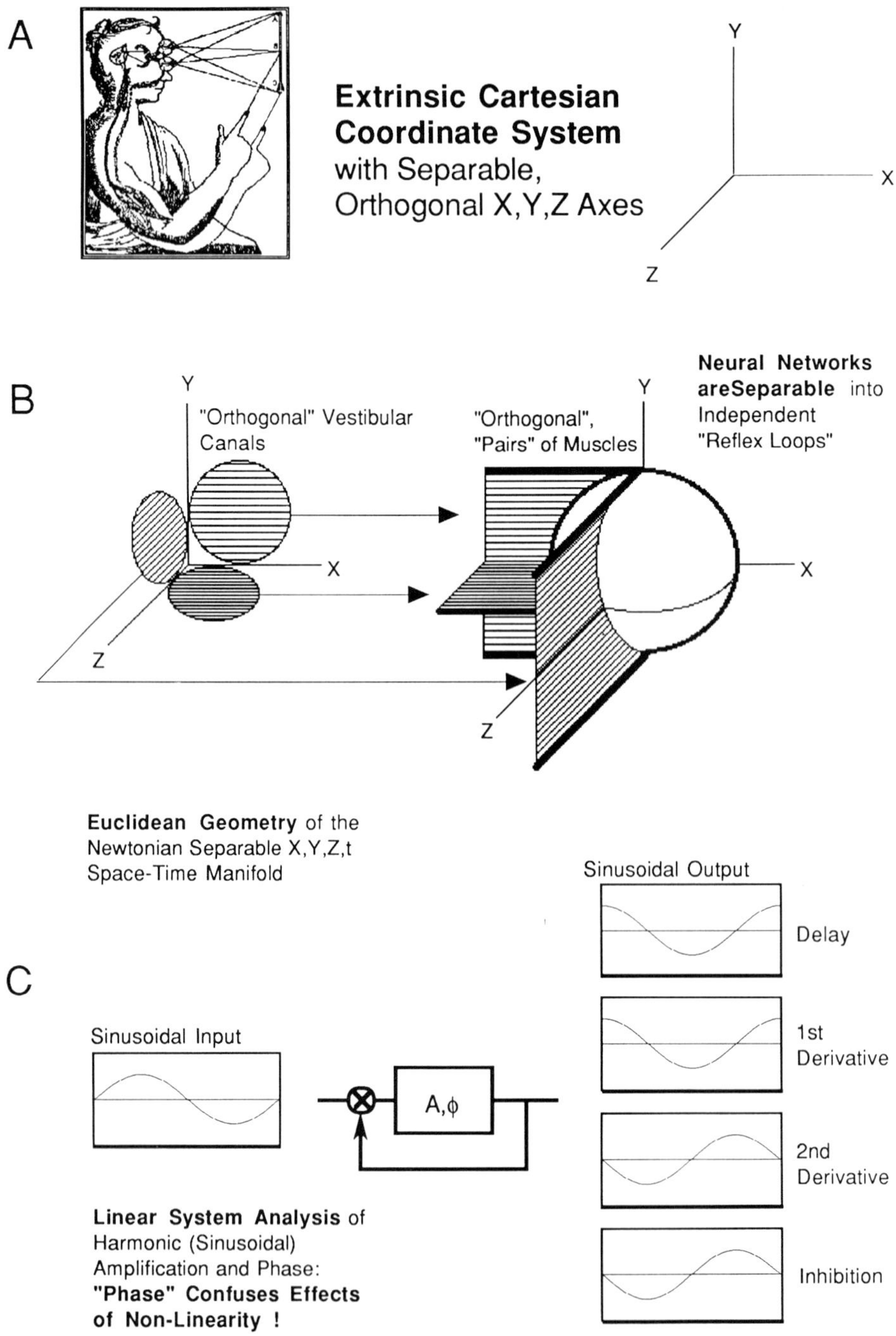

FIG. 1. *The classical concept of "separable reflexes" applied to analysis of gaze control.* The traditional engineering view assumes extrinsic Cartesian coordinate systems and underlying Euclidean geometry. **A:** Cartesian reference frame. **B:** Cartesian paradigm of sensorimotor transformation: orthogonal, separable sensors (vestibular canals) driving orthogonal, separable pairs of muscles. These assumptions simplify sensorimotor transformations into one-dimensional reflex loops, such as horizontal vestibular canals driving the "pair" of lateral and medial rectus muscles (B). Use of Cartesian x,y,z,t frame also simplifies the space–time manifold into a Newtonian separable space and time domain in which Euclidean geometry governs. With the added assumption of linearity (excluding delays and temporal differentiation), harmonic system analysis can be applied **(C)**. This method, although appropriate for amplitude and phase measurements, confuses nonlinear effects.

on theory and modeling of gaze reflexes, such as VOR, were immediately pursued, as evidenced by the chapter "The Vestibulo-Ocular 'Reflex' as a Tensorial Response" in 1980 (22). However, the novel generalized vectorial approach of intrinsic coordinates, similar to other new theories based on a shift of scientific axiom, encountered fierce opposition. This is a natural reaction, even if not fair, from those heavily committed to the concept of control system analysis that may earlier have been revolutionary but now is admittedly obsolete (ref. 18 versus ref. 20) or from those who were (and more than a decade later still are) unable to compete with any viable, biologically applicable alternative sensorimotor theory (23).

In the face of such opposition, through a decade, there was much progress in modeling gaze, with a clear trend toward multidimensional approaches; both using extrinsic vectors (20,24,25) or intrinsic generalized vectors (tensors); (22,26–39). As for tensor network theory, it concluded in experimentally verifiable quantitative theoretical hypotheses that were tested and confirmed both by independent workers (40) as well as in collaborative experimental research (41,42). Some more amibitious generalizations of tensor network theory have also been attempted (43,44). In a general sense, the fundamental principle of the CNS as a distributed parallel processor that uses a vectorial language was also experimentally confirmed (45).

NON-EUCLIDEAN METRIC TENSOR OF NEURAL GEOMETRY ESTABLISHED BY "WIRED" INTERCONNECTEDNESS: A CONCEPTUAL PRECURSOR TO MODELS OF MASSIVELY PARALLEL "WIRELESS" INTERCONNECTION

Vestibulo-oculomotor research came to a full circle in a single decade regarding the prevailing mathematical concepts of distributed parallel processing. It is one of the interesting features of this successful conceptual reform that it is justified now, in retrospect, by the purely empirical findings of massively parallel connections among the real neural networks of this three-neuron reflex arc (20). Indeed, the underlying fundamental theoretical problem, the question of the intrinsic mathematical language of brain function, may not be important, or even seen, within the narrow perspective of gaze analysis only. The most profound issue of intrinsic mathematical language of neuronal interconnectedness is, however, crucial for neural net (neurocomputing) theory. Therefore, a case is made below to illuminate that there is much more in gaze reflexes to teach us than the need to switch from one- to three-dimensional analysis of VOR (46), or the inevitability of scrapping system theory of loops (18) in favor of massively interconnected circuits (20).

Indeed, the most important issue is to conceive CNS function in terms of geometry (as opposed to algebra) as its intrinsic mathematical language. It is in this sense that research of sensorimotor (gaze and cerebellar) neural nets could become the stronghold of a geometrical school of theoretical-experimental neuroscience. There is a rapidly increasing number of workers adhering to the principle that the mathematical language of brain function is not (logical) algebra, as formally thought, but multidimensional abstract geometry (32,36, 37,40,47–49). It can be realistically expected that workers encountering volume transmission, with its inherently massively parallel nature, could use some lessons learned from the early struggle and successful emergence of this geometrical conceptual school.

The concept that a system of parallel interconnections incorporate an abstract functional geometry is illustrated in Figs. 1 and 2. For centuries it has been implicit (50) that even the assumption of a separable set of horizontal, vertical, and torsional reflex arcs implies a geometry. Separability would, of course, eliminate interactions

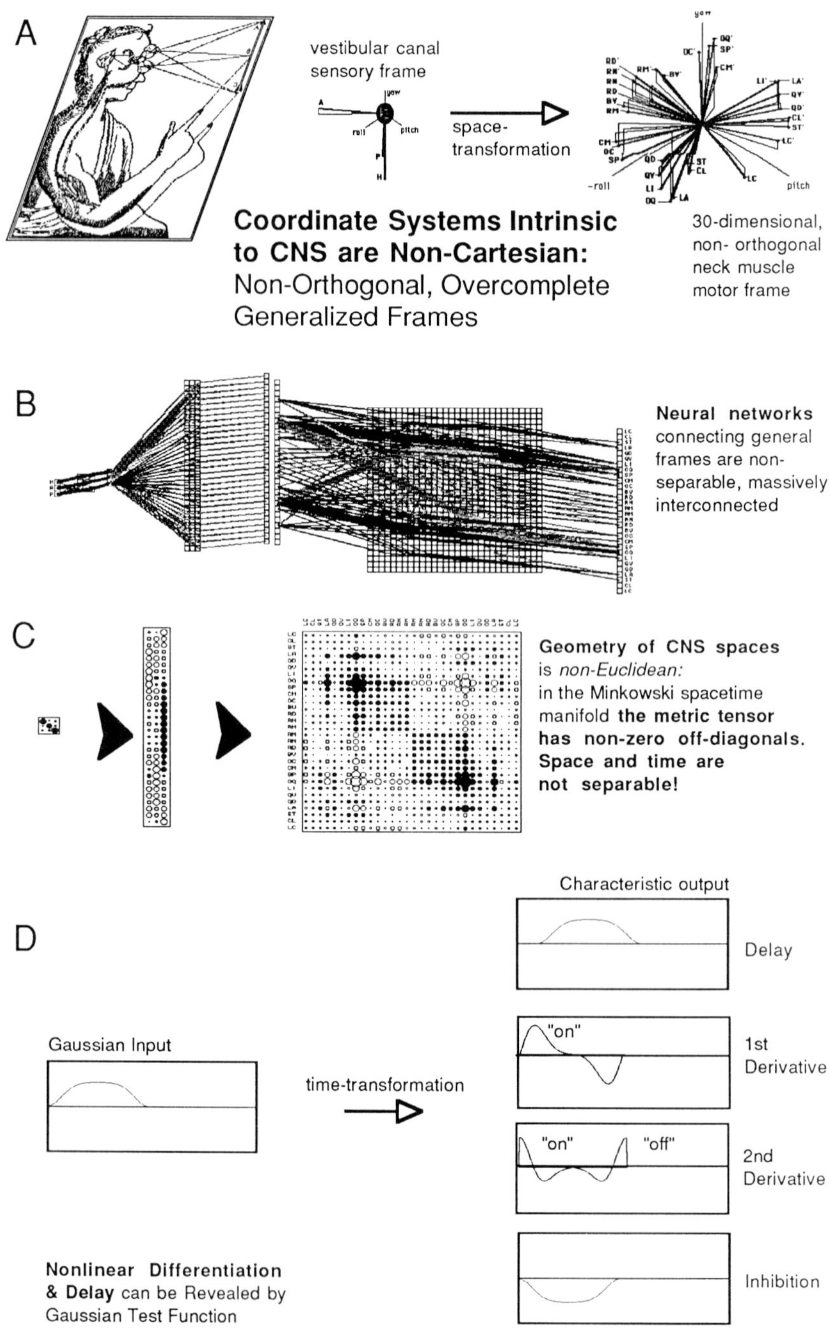

FIG. 2. *The approach of distributed processing (based on generalized non-Cartesian, overcomplete coordinates) implies non-Euclidean geometry established by wired interconnections, a conceptual precursor to massively parallel wireless communication.* **A:** Axiomatic difference from the classical approach (Fig. 1 versus Fig. 2) is based on nature's use of nonorthogonal and overcomplete frames of reference. Generalization of Cartesian frames to nonorthogonal coordinate systems leads to a tensor model of vestibulo-collic reflex. Generalized frames lead to a network model with massively parallel interconnections and to a functional geometry in the neck motor space that is demonstrably non-Euclidean (see the 30 × 30 patch matrix of the neck motor metric tensor, in **C**, in which the off diagonals are nonzeros, and space and time in the Minkowski manifold are inseparable). Because sensorimotor transformations include nonlinearities (differentiations and delays), use of Gaussian testing (in addition to sinusoidal harmonic analysis) is warranted. (From ref. 42.)

between independent x,y,z reflex arcs and thus effectively ignore any massively parallel phenomenon such as (chemical) volume transmission, or even (the anatomically well-documentable) cross-connections among horizontal, vertical, and torsional vestibular canals and corresponding extraocular muscles. Mathematically, the orthodox view of separable loops would mean that the orthogonal (Cartesian) axes define an utterly simple Euclidean geometry.

As shown in Fig. 2, the departure from Cartesian orthogonal coordinate systems leads directly to a replacement of the concept of separability by the alternative idea of massive interconnections. Morphologically, interrelation is expressed by a set of intricate primary, secondary, and tertiary cross-connections among vestibular canals and extraocular muscles. The existence of such interconnections has long been verified by both physiological and morphological methods (51–53). The mathematics involved in such distributed networks could not be made explicit until relatively recently, when the concept and formalism of nonorthogonal intrinsic coordinates (generalized vectors; tensors) was introduced (22,31). Tensor geometry of massively parallel transformations (see Fig. 2B) can quantitatively address the issue of almost total interconnectedness among a set of inputs (e.g., vestibular canals) and set of output elements (oculomotor or neck motor muscles).

The new alternative to a serial three-neuron reflex arc is therefore a set of three parallel transformations, as elaborated in detail for the VOR in (31,54), and for the vestibulo-collicular reflex (VCR) in ref. 41. As shown in Fig. 2, the morphological fact (B) of "everything-to-everything-type" interconnections among input and output element is mathematically expressed in the motor metric tensor in which the off-diagonals are nonzeros. As seen in Fig. 2C, hardly any component of the metric tensor is zero, as practically all input elements are connected to all output elements. In terms of mathematics, these nonzero off-diagonals establish the fact that the functional motor geometry is non-Euclidean, as the metric tensor in Euclidean geometry would be a matrix of unitary diagonal elements and all-zero off-diagonals.

Looking at the network expression of this metric (Fig. 2B) one may realize that volume transmission can be conceptually seen as a wireless equivalent of establishing the kind of massively parallel interconnectedness that is incorporated, e.g., in the wired VCR in Fig. 2B. In this sense, the tensorial model of VCR, with its massively parallel set of wired interconnections, could serve as a conceptual precursor to a future mathematical (abstract geometrical) model of wireless massively parallel connections by means of volume transmission among neurons.

Because the quantitative set of interrelations in such systems will be extremely difficult to experimentally ascertain, it is noteworthy that a multidimensional geometrical model of the endocrine system could be helpful as an interim project that is less "point-to-point" than the "hard-wired" VOR or VCR but perhaps more specific than the highly unconstrained wireless volume transmission. At least, some cross-talks among formerly separable hormonal control mechanisms have already been geometrically formulated, as pointed out in a germinal concept outlined recently (43).

NEURAL GEOMETRIES CENTURIES AFTER EUCLID AND DESCARTES

When, in search of "neural geometry," we start looking beyond the simplest Euclidean structures we are cautioned that progress is unlikely to stop at the boundaries of Euclidean, Riemannian, or even metrical spaces. On one hand, the simplest kind of sensorimotor embedding of the external (Euclidean) physical space into an internal (non-Euclidean, Riemannian) gaze space in which the external space is represented is already documented in the above-quoted studies conducted at system and neural levels. Indeed, a gradual relaxation of overly constrained reflex arcs toward multidimensional geometrical models is realistically expected to start with the simplest sensorimotor neural networks such as gaze control. Progress is likely to be followed with similar analyses of cerebellar coordination, in terms of coordinates (55), vectorial transformations in the accessory optic system, where the existence of intrinsic vectors of the CNS is documented (56), the optic tectum and motor cortex, where the vectorial operation is evident at a neural level, although presently interpreted in extrinsic x,y,z,t-type coordinate systems (45,57), and a representation of the structure of external space by a metric incorporated by the hippocampus, where the function is presently interpreted as generating an inverse matrix for construction of a Cartesian representation (58,59).

On the other hand, even today some evidence is already in that certain neural geometries, also relying on a massively parallel (volume transmission type) communication, are distant from Euclidean and Cartesian roots and in fact are close to the other extreme of the spectrum of geometries.

FRACTAL NEURAL GEOMETRY, AFFECTED BY INTERCONNECTEDNESS VIA WIRELESS VOLUME CONDUCTION

A grossly non-metrical (characteristically non-Euclidean) fractal neural geome-

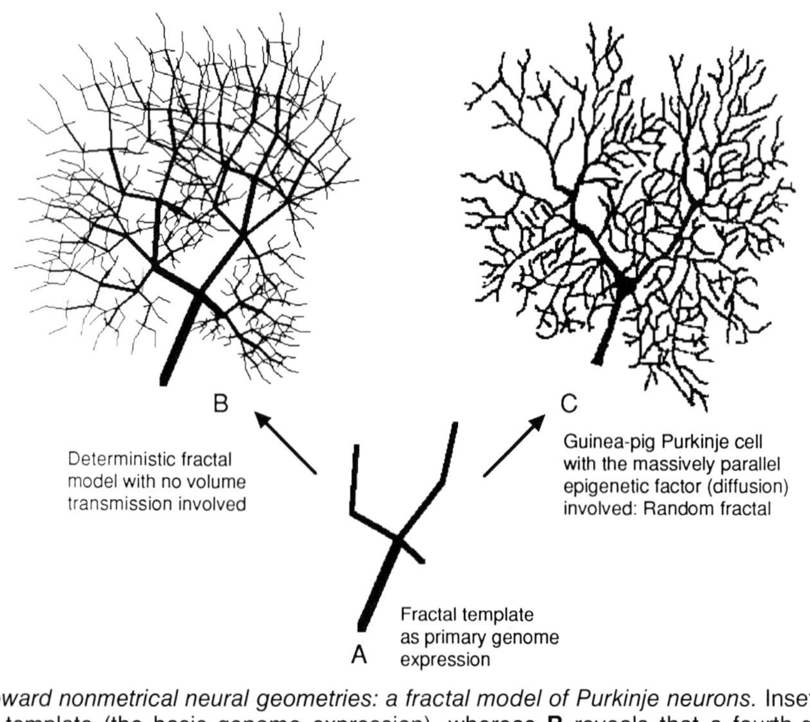

FIG. 3. *Toward nonmetrical neural geometries: a fractal model of Purkinje neurons.* Inset **A** shows the fractal template (the basic genome expression), whereas **B** reveals that a fourth-generation deterministic fractal dendritic tree is an encouraging match of the real Purkinje cell in the guinea pig. Some of the variance of hitherto deterministic fractal model from the much more stochastic biological Purkinje cells are undoubtedly caused by massively parallel volume conduction (by means of diffusion) between developing dendritic branchlets. (From refs. 60,61,64.)

try is manifested by the morphology and morphogenesis of dendritic trees of, e.g., Purkinje cells (60,61). Early results in modeling of perhaps the most obvious expression of neural geometry, the arbor of a single neuron, are presented in Fig. 3. At least two potentially important theoretical aspects of fractal models are deemed relevant to the topic of this volume. First, the substantiation of the conjecture (62) that neural form is fractal (61) stakes out the spectrum of neural geometries from the rigid and simple (quasi)Euclidean and (quasi)linear geometries, spun by Cartesian orthogonal frames, to the other extreme of nonlinear dynamics implying nonmetrical geometries. This broad scope should embolden approaches that presently only gingerly inch away from Cartesian (extrinsic) coordinates and trivial Euclidean geometries, as if the CNS were not permitted to use anything more sophisticated and more natural.

Second, the neuron model presented in Fig. 3 is hitherto based on fully deterministic fractal geometry, in which the morphogenesis is determined exclusively by the code incorporated in the fractal template (which, in turn, is evidently encoded in the double helix). Although the predominant information of morphogenesis resides undoubtedly in the genetic code, it is well known that epigenetic factors are capable of substantially modulating genome expression. This likelihood relates to the fractal model of neurons shown in Fig. 3 in the sense that the morphogenesis of cells (although primarily determined by an internal code) is obviously affected by the microenvironment in which the cellular structure emerges. This factor will certainly have to be considered in further generations of fractal neural models. Ultimately, such models have to become less than fully deterministic representations, accounting for stochastic properties, as modulated, e.g., by diffusion in the extracellular milieu (63). By means of volume transmission, neighboring dendritic branches (and neighboring neurons) would surely interact. The potential importance of these interactions for both the development of fractal neuronal models and studies of volume transmission is that fractal geometry establishes a mathematical linkage between these two seemingly different studies. Volume transmission research could thus become instrumental for a mathematical evolution of modeling as exemplified in Fig. 3, progressing from deterministic to random fractals (62). Such progress could be kept in constant relation with experimentation by monitoring the (required) convergence of a preliminarily acceptable model of dendritic arbor to the obviously more complex and less rigid existing neuronal dendritic trees (see Fig. 3C).

CONCLUSION: VOLUME TRANSMISSION IS EVIDENCE AND A GREAT CHALLENGE FOR GEOMETRICAL PRINCIPLES OF MASSIVELY PARALLEL INTERCONNECTEDNESS, LEADING TO A MATHEMATICAL LANGUAGE ESSENTIAL TO NEUROCOMPUTING

The summary figure (Fig. 4) points to neurocomputing as a major force behind the thrust toward establishing the mathematical principles of CNS function. Some (theoretical) tasks involved in such an enormous challenge are clearly beyond the traditional means and preparedness of experimental neuroscience. A role of volume transmission in the required conceptual evolution is that wireless parallel communication, which presently largely is an enigma, is likely to demand progress in developing its theoretical foundations. At the least, questions arising about volume transmission will be provocative in triggering the emergence of novel computational paradigms, as well as it will be forceful to reject those that ignore or deny massive interactions in CNS function.

In conclusion, it is already established that volume transmission is one of the strongest phenomena to support the prin-

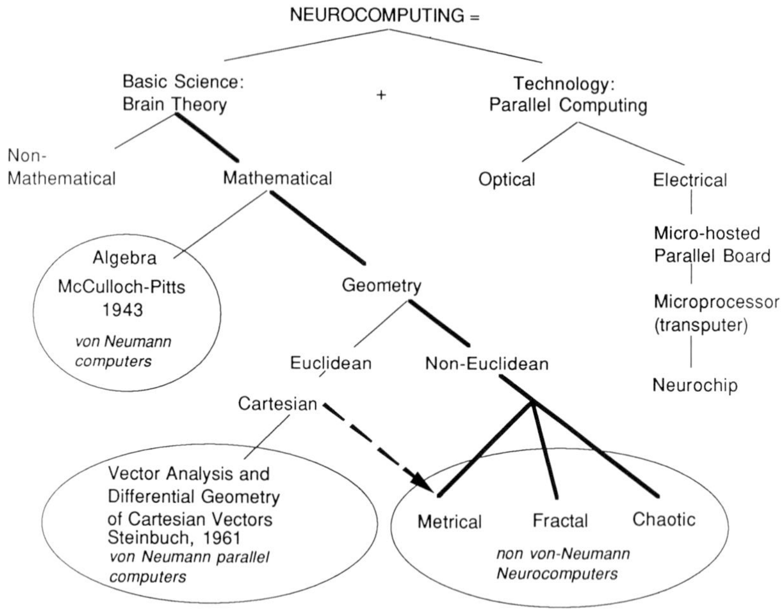

FIG. 4. *Mathematical language of brain function as a formal scientific basis of neurocomputing.* Traditional computers are based on algebra of serial processors, whereas modern neural net theories are geometrical representations of massively parallel interconnections. Transcending such Euclidean-Cartesian philosophy appears as a subtle axiomatic shift (see *dashed arrow*), yet it opens up non-Euclidean geometries to neural net theory. Metrical (flat and curved) spaces are but the simplest neural geometries, demonstrable in gaze systems. Nonmetrical and nonlinear (fractal and chaotic) neural geometries are also researched, with *non-Euclidean geometry rapidly becoming a novel formal basis of neurocomputing.*

ciple that CNS function is based on massively parallel interconnections. Its primary role is supported by evolution, because such (chemical) interactions among cells were probably the earliest and thus most fundamental means of communication. It is also plausible that the general effect of volume transmission, permeating the entire organism, be described as imposing an overall functional geometry on CNS action. However, to measure experimentally the quantitative implications of parallel interconnections that volume transmission imposes and to reveal basic mathematical features of the functional geometries that they implement will undoubtedly take a long time. Regrettably, until such quantitation begins to emerge, based on conceptually sound theory, volume transmission will probably mean to most workers hardly more than an intriguing exception to the rule of regular wired communication among brain cells.

REFERENCES

1. Vizi ES. *Non-synaptic interaction between neurons: modulation of neurochemical transmission.* New York: John Wiley, 1984.
2. Fuxe K, Agnati LF, Zoli M, Bjelke B, Zini I. Some aspects of the communicational and computational organization of the brain. *Acta Physiol Scand* 1980;135:203–216.
3. Edelman G. *Neural Darwinism.* New York: Academic Press, 1987.
4. Churchland P. *Neurophilosophy: towards a unified understanding of the mind-brain.* Boston: Bradford Books/MIT Press, 1985.
5. Shannon C. A mathematical theory of communication. *Bell System Technology J* 1948;27:3–4.
6. McCulloch WS, Pitts W. A logical calculus of the ideas immanent in nervous activity. *Bull Math Biophys* 1943;5:115–133.
7. Cajal S. *Histologie du systeme nerveux de l'homme et des vertebres.* Paris: Maloine, 1911.

8. Neumann J von. *The computer and the brain.* New Haven: Yale University Press, 1957.
9. Anderson J, Rosenfeld E. *Neurocomputing: I. Foundations of research.* Cambridge, Massachusetts: MIT Press, 1987.
10. Hinton GE, Anderson JA. *Parallel models of associative memory.* Hillsdale, New Jersey: Lawrence Erlbaum, 1981.
11. Sherrington C. *The integrative action of the nervous system.* New York: Scribner, 1906.
12. Lorente de No R. Vestibulo-ocular reflex arc. *Arch Neurol Psychiatry* 1933;30:245–291.
13. Robinson D. The oculomotor control system: a review. *Proc IEEE* 1968;56:1032–1049.
14. Wiener N. *Cybernetics, or control and communication in the animal and the machine.* Cambridge, Massachusetts: MIT Press, 1948.
15. Stark L. *Neurological control systems.* New York: Plenum Press, 1968.
16. Ito M. Neurophysiological aspects of the cerebellar motor control system. *Int J Neurol* 1970;7:162–176.
17. Houk J. Principles of system theory as applied to physiology: systems and models. In: Mountcastle V, ed. *Medical physiology* St. Louis: CV Mosby, 1980.
18. Robinson D. The use of control systems analysis in the neurophysiology of eye movements. *Annu Rev Neurosci* 1981;4:463–503.
19. Stolwijk JAJ, Hardy JD. Regulation and control in physiology. In: Mountcastle VB, ed. *Medical physiology,* vol II. St. Louis: CV Mosby, 1974;1343–1358.
20. Anastasio TJ, Robinson DA. Distributed parallel processing in the vestibulo-oculomotor system. *Neural Comp* 1989;1:230–241.
21. Pellionisz A, Llinás R. Brain modeling by tensor network theory and computer simulation. The cerebellum: distributed processor for predictive coordination. *Neuroscience* 1979;4:323–348.
22. Pellionisz A, Llinás R. Tensorial approach to the geometry of brain function: cerebellar coordination via metric tensor. *Neuroscience* 1980;5: 1125–1136.
23. Arbib M, Amari SI. Sensorimotor transformations in the brain with a critique of tensor network theory of the cerebellum. *J Theor Biol* 1985;112:123–155.
24. Robinson DA. The use of matrices in analyzing the three-dimensional behavior of the vestibulo-ocular reflex. *Biol Cybern* 1982;46:53–66.
25. Robinson D. The coordinates of neurons in the vestibulo-ocular reflex. *In: Berthoz A, Melvill-Jones G, eds. Adaptive mechanisms in gaze control facts and theories.* Amsterdam: Elsevier, 1985;297–311.
26. Pellionisz A, Llinás R. Space-time representation in the brain. The cerebellum as a predictive space-time metric tensor. *Neuroscience* 1982; 2949–2970.
27. Pellionisz A. Coordination: a vector-matrix description of transformations of overcomplete CNS coordinates and a tensorial solution using the Moore-Penrose generalized inverse. *J Theor Biol* 1984;110:353–375.
28. Simpson J, Pellionisz A. The vestibulo-ocular reflex in rabbit as interpreted using the Moore-Penrose generalized inverse transformation of intrinsic coordinates (*Abstr.*). *Soc Neurosci Absts* 1984;10:909.
29. Graf W, Pellionisz A. Morphology and a tensor network model of the three-neuron vestibulo-ocular reflex arc in cat (*Abstr.*). *Soc Neurosci Absts* 1985;11:1041.
30. Pellionisz A, Llinás R. Tensor network theory of the metaorganization of functional geometries in the CNS. *Neuroscience* 1985;16:245–274.
31. Pellionisz A. Tensorial aspects of the multidimensional approach to the vestibulo-oculomotor reflex and gaze. In: Berthoz A, Melvill-Jones G, eds. *Adaptive mechanisms in gaze control,* vol I. Amsterdam: Elsevier, 1985;281–296.
32. Pellionisz A, Soechting J, Gielen C, Simpson J, Peterson B, Georgopoulos A. Workshop: multidimensional analyses of sensorimotor systems (*Abstr.*). *Soc Neurosci Absts* 1986;12:1.
33. Daunicht W, Pellionisz A. Spatial arrangement of the vestibular and the oculomotor system in the rat. *Brain Res* 1987;435:48–56.
34. Liverneaux P, Pellionisz AJ, Lestienne FG. Morpho-anatomy and muscular synergy of sub-occipital muscles in Macaca Mulatta: study of head-trunk coordination. *Neuroscience* 1987;22 (suppl):S847.
35. Lestienne A, Liverneaux P, Laczkó J, Pellionisz A. Tensor model of the musculo-skeletal head-neck system of the monkey. *Neuroscience* 1987; 22(suppl):S658.
36. Bloedel JR, Tillery S, Pellionisz AJ. Experimental-theoretical analysis of the intrinsic geometry of limb movements (*Abstr.*). *Soc Neurosci Absts* 1988;14/2:953.
37. Berthoz A, Benhamou M, Pellionisz AJ. Postural positions yielding aligned Eigenvectors of coordinate frames intrinsic to vestibular and head-neck muscle systems in human. (*Abstr.*). *Soc Neurosci Absts* 1988;14/2:1235.
38. Laczkó J, Pellionisz A, Jongen H, Gielen CCAM. Computer modeling of human forelimb muscle activation in multidimensional intrinsic coordinate frames (*Abstr.*). *Soc Neurosci Absts* 1988;14/2:955.
39. Laczkó J, Pellionisz AJ, Peterson BW, Buchanan TS. Multidimensional sensorimotor "patterns" arising from a graphics-based tensorial model of the neck-motor system (*Abstr.*). *Soc Neurosci Absts* 1987;13:372.
40. Gielen CCAM, van Zuylen EJ. Coordination of arm muscles during flexion and supination: application of the tensor analysis approach. *Neuroscience* 1986;17:527–539.
41. Peterson BW, Pellionisz AJ, Baker JA, Keshner EA. Functional morphology and neural control of neck muscles in mammals. *Am Zool* 1989;29: 139–149.
42. Pellionisz A, Peterson BW. A tensorial model of neck motor activation. In: Peterson B, Richmond F, eds. *Control of head movement.* Oxford: Oxford University Press, 1988;178–186.

43. Gaál G. Non-orthogonal hormonal coordinate axes calculated from correlation matrices according to tensor network theory. *Neurosci Suppl* 1987;22:S146.
44. Collins TJ. Cerebral cortical parallel processing using a metric tensor. *Proc Int Neur Network Soc, 2nd Annual Meeting*, Boston, 1988;85.
45. Georgopoulos A, Schwartz A, Kettner R. Neuronal population coding of movement direction. *Science* 1986;233:1416–1419.
46. Henn W, Cohen B. *Proc. of the Symp: Representation of 3-Dimensional Space in the Vestibular, Oculomotor and Visual Systems*. Italy: Bologna, Bárány Society, 1988.
47. Koenderink JJ. Geometrical structures determined by the functional order in nervous nets. *Bio Cybern* 1984;50:43–50.
48. Droulez J, Berthoz A. Spatial and temporal transformations in visuomotor coordination. In: Eckmiller R, Malsburg C, eds. *Neural computers*, Heidelberg: Springer, 1987;345–359.
49. Eckmiller R. Neural networks for motor program generation. In: Eckmiller R, Malsburg CV, eds. *Neural computers*. Düsseldorf, Berlin: Springer, 1988;359–370.
50. Descartes R. *Treatise of man* (original in 1629 in Latin and French). Cambridge, Massachusetts: Harvard University Press, 1972.
51. Högyes E. Az associált szemmozgások idegmechanismusaról. Ertekezések a természet-tudományok köréböl. German translation: Högyes, A. (1912) über den Nervenmechanismus der assoziierten Augen-bewegungen. *Monatsschr Ohrenheilk u Laryngo-Rhinol* 1880;46:685–740, 1027–1083, 1554–1571.
52. Bárány R. Untersuchungen über den vom Vestibularapparat des Ohres reflektorisch ausgelösten rythmischen Nystagmus und seine Begleiterscheinungen. *Monatsschr Ohrenheikd* 1906;43: 191.
53. Szentágothai J. Die zentrale Innervation des Augenbewegungen. *Arch Psychol Nervkrankh* 1943;721–760.
54. Pellionisz A, Graf W. Tensor network model of the "three-neuron vestibulo-ocular reflex-arc" in cat. *J Theor Neurobiol* 1987;5:127–151.
55. Bloedel J, Dichgans J, Precht W. *Cerebellar functions*. Berlin: Springer, 1985.
56. Simpson JI. The accessory optic system. *Annu Rev Neurosci* 1984;7:13–41.
57. Sparks DL, Pollack JG. The neural control of saccadic eye movements: the role of the superior colliculus. In: Brooks BA, Bajandas FJ, eds. *Eye movements*. New York: Plenum, 1977;179–219.
58. Keefe JO. Computations the hippocampus might perform. In: Nadel L, Cooper LA, Culicover P, Harnish M. eds. *Neural connections, mental computation*, Cambridge, Massachusetts: MIT Press, 1989;225–284.
59. McNaughton BL. Neural mechanisms for spatial computation and information storage. In: Nadel L, Cooper LA, Culicover P, Harnish M, eds. *Neural connections, mental computation*, Cambridge, Massachusetts: MIT Press, 1989;283–350.
60. Pellionisz A. Fractal geometry of Purkinje neurons: relationships among metrical and nonmetrical neural geometries (*Abstr.*). *Soc Neurosci Absts* 1989;15:180.
61. Pellionisz A. Neural geometry: towards a fractal model of neurons. In: Cotterill RM, ed. *Models in brain function*. Cambridge: Cambridge University Press, 1990.
62. Mandelbrot B. *The fractal geometry of nature*. New York: WH Freeman, 1977.
63. Nicholson C. Diffusion from an injected volume of a substance in brain tissue with arbitrary volume fraction and tortuosity. *Brain Res* 1985; 333:325–329.
64. Tank DW, Sugimori M, Connor JA, Llinás RR. Spatially resolved calcium dynamics of mammalian Purkinje cells in cerebellar slice. *Science* 1988;242:773–777.

Volume Transmission in the Brain: Novel Mechanisms for Neural Transmission, edited by K. Fuxe and L.F. Agnati. Raven Press, Ltd., New York © 1991.

46
Concluding Remarks

Jean-Pierre Changeux

Institut Pasteur, Neurobiologie Moléculaire, Département des Biotechnologies, UA CNRS D1284, 75724 Paris, France

In physics, theory possesses a well-defined status. It is recognized as a specific discipline distinct from the experimental work and even sometimes far detached from it. Yet, apart from some recent developments of the cognitive sciences, this is not the case in biology. An ambiguity persists about the importance of theory: Either the theoretical speculations are considered too far ahead of the empirical work and disregarded as irrelevant, or they lie too close to the experiments and found obvious or unnecessary. Even experimental techniques as such are sometimes viewed as more crucial than theory. This does not mean that theory in biology is not necessary or even is useless. It signifies that to be accepted as an intrinsic element of scientific research, a biological theory must keep a finite distance from ongoing experimental work. Of course, it should anticipate the experiments actually being done. But, also, it should "immediately" lead to otherwise unexpected applications, thereby opening new lines of research. Its recognition (and the credit given to it) thus hinges on stringent constraints: It should be formulated, if possible, in a rigorous and hopefully understandable formalism but should also appear at an adequate moment in the history of science.

At this particular time, observations dispersed in distinct fields reach a threshold such that they start to "crystallize" into a unifying concept or theory. In the following days, of course, the idea will be tested, attacked, rejected, adopted, sometimes turned into pieces, and often forgotten. Hopefully, in such a struggle-for-ideas-to-survive, some of these theoretical expectations will be selectively stabilized according to their objective fitness with the real world. Sometimes, they will even be recognized by the scientific community. . . .

Throughout this volume organized by Kjell Fuxe and Luigi Agnati, we have been living a moment in which a complementary, if not alternative, mechanism to classical chemical synaptic transmission was presented and further discussed within the general framework of intercellular communications in the brain. This process was named "volume transmission," to contrast with the more classical "wiring transmission"(1–3). In these concluding remarks, I shall try to specify these two mechanisms, to evaluate the experimental basis of their distinction and discuss their physiological significance.

Before doing so, I shall recall that communication, according to its latin root (*communicatio*), means sharing or distributing. Shannon and Weaver in their classical theory of 1949 view communication as taking place between a source (or emitter) and a destination (or receiver) through a "communication channel." It is achieved by encoding a "message," which cannot travel, into a "signal," which can. At the other end, the signal is decoded by the receiver. In the central nervous system, the complexity of the message varies with the level of organization considered (4). It ranges, for instance, from discrete "plus" signals up to elaborate "mental representations," or "mental objects," plausibly encoded into patterns of correlated neuronal activities (see refs. 5–8). At the meeting, the level of organization, primarily considered, was the cellular level. Yet, one of the most positive aspects of volume transmission is perhaps that, as discussed in the final section of these remarks, it may create a link with higher levels, leading to more "global" regulations in the brain.

THE CLASSICAL SCHEME OF CHEMICAL SYNAPTIC TRANSMISSION

Most of our knowledge about synaptic transmission derives from studies with peripheral synapses (9,10), which, as the central ones, are characterized by (a) a short length of the communication channel and (b) the use of chemical substances (i.e., neurotransmitters and/or coexisting messengers) (11) for intercellular signaling. Direct electrical communication between nerve cells also exists in the brain. It is mediated by gap junctions and can be controlled in many ways (12). Yet, it does not seem directly relevant to the present theme and will not be further discussed here.

At the vertebrate motor endplate (see ref. 13), the nerve ending covers a few square micrometers of surface area of the muscle fiber, and the intercellular distance between cytoplasmic membranes ranges from 20 to 40 nm, thus delimiting a confined space of a few cubic micrometers referred to as the synaptic cleft. Moreover, the junction displays a straightforward morphological polarity. On the nerve ending side, small clear vesicles, 50 nm in diameter, accumulate in front of differentiated release sites, with a few interspersed dense-core vesicles (ref. 14; and Chap. 13, *this volume*). On the muscle side, the sarcolemmal membrane makes multiple folds with thickenings in front of the release sites, and a complex set of filaments densely decorates these folds (see refs. 13,15). The junction thus displays elaborate pre- and postsynaptic specializations.

In the motor nerve ending, the small vesicles contain approximately 10,000 molecules of the neurotransmitter acetylcholine (ACh), and their content is released in the cleft in a discrete manner, as packets, or quanta, which elicit "miniature" endplate potentials. On invasion of the motor nerve ending by an electrical impulse, Ca^{2+} ions enter the terminal (16,17). About 200 to 300 quanta are liberated (about $2–3 \times 10^6$ molecules), which, because of the small volume of the cleft, causes a rise in the local concentration of ACh of as much as 3×10^{-4} M in a fraction of a millisecond (18,19). In a few tenths of a millisecond, a high concentration pulse of ACh is generated, crosses the synaptic cleft, and reaches the postsynaptic membrane. Thus, chemical communication at the motor endplate takes place across the synaptic cleft through brief "chemical impulsions."

After crossing the synaptic cleft, ACh binds to closely packed receptor molecules (about 20,000 molecules/μm^2) located at the tip of the postsynaptic folds (20), and there causes the "collective," fast, opening of the acetylcholine receptor (AChR) ionic channels in the milliseconds timescale (16). It is thought that these channels open only once and that the excitatory postsynaptic potential recorded in the muscle cell corresponds to the sum of their all-or-none individual openings. In the postsynaptic membrane, the receptor molecules "decode" the chemical signal into an electrical one (review ref. 21).

The resulting communication process is fast, local, discrete, robust, phasic, and of high fidelity, thus the metaphor of wiring transmission (Chap. 1, *this volume*) to designate this mode of signal transfer.

In addition,

1. An almost perfect match exists between the motor nerve ending and the receptor molecules present in the sarcolemmal membrane. The match holds both for the pharmacological specificity of the receptor toward the dominant excitatory neurotransmitter (ACh) and for the topological distribution of the postsynaptic receptor molecules (see ref. 14; and Chap. 6, *this volume*). Yet, the motor neurons contain several neuropeptides that coexist with ACh such as calcitonin gene-related peptide (CGRP), vasoactive intestinal peptide (VIP), and somatostatin (see ref. 22). Receptor sites for CGRP have been identified in the sarcolemmal membrane (23,24) and their density appears higher at the endplate level (25). Thus, in this system, the match plausibly holds for both the classical neurotransmitter and at least one coexisting neuropeptide.

2. There are several modes of ACh release at the motor endplate. In addition to the Ca^{2+}-dependent release of ACh evoked by the nerve impulse, ACh steadily leaks from the nerve ending under resting conditions. Katz and Miledi (26) have estimated that 98% of the spontaneous release of ACh is nonquantal. Thus, most of the neurotransmitter produced by the motor nerve ending contributes to a steady-state "tonic" level of about 10^{-8} M ACh in the synaptic cleft (in the presence of cholinesterase inhibitor). The functional significance of this leak, which also seems to exist in the brain with GABA (Chap. 34, *this volume*), of course differs from that of the neurally evoked release of neurotransmitter in the cleft (see Chap. 5, *this volume*).

In other peripheral systems such as the sympathetic innervation of smooth muscle (10) and in the central nervous system, the physiology of neurotransmitter liberation has been analyzed in detail by statistical methods. For instance, in the case of fish Mauthner cell, inhibitory innervation by defined interneurons (27) yields an highly branched axon terminal, which contacts the neuronal soma by multiple "boutons." Interestingly, each bouton may release one vesicle but with a probability smaller than one. Moreover, the time course of the electrical response for a single bouton strikingly resembles that of a miniature endplate potential (27; see also ref. 10). Thus, in both peripheral and central systems are involved similar, if not identical, elementary mechanisms of transmitter release.

In the central nervous system, microphysiological methods further reveal that adjacent synapses may interact possibly because of "a diffuse action of the transmitter over domains larger than those restricted under a given release site" (27; see also ref. 28). Thus, the neurotransmitter serving as communication signal in wiring transmission may also take part in a more delocalized process.

Finally, coexisting neuropeptides (11) are also released in the course of chemical synaptic transmission (Chap. 33, *this volume*). They are stored in vesicles at least partially distinct from those involved in classical neurotransmitter storage (see Chap. 13, *this volume*), and in the autonomic control of peripheral organs, they can be differentially released under different patterns of electrical stimulation (Chap. 33, *this volume*).

VOLUME TRANSMISSION: A PROPOSAL

According to classical wiring transmission, communication between nerve cells is confined to the synaptic cleft. Fuxe and Agnati proposal stresses the possibility that in the nervous system, signaling may also efficiently take place via extracellular spaces. An enlargement and divergence of the communication channel occurs. They have called this mode of communication "volume transmission," by reference with the theory of "volume conduction" for electrocorticogram recordings, which "describes the flow of current through the extracellular fluid under various conditions of brain activity" (Chap. 1, *this volume*). According to the authors, volume transmission includes altogether diffusion of chemical messengers and of electrical signals via electronic currents ("ephaptic" transmission) (see refs. 29,30).

Fuxe and Agnati give in their introductory remarks (see Table 1 in Chap. 1, *this volume*) an historical outlook of the progressive development of this general concept. It bears first on Claude Bernard's notion of "milieu intérieur" and of its constancy via homeostatic regulation, which, of course, is a prerequisite for efficient signaling to take place. Neurohumoral transmission by "local diffusion" was mentioned by Bach-y-Rita (31; and Chap. 39, *this volume*), and "action upon some distant central site" after injection of carbachol in brain ventricles by Routtenberg (32; and Chap. 22, *this volume*). In fact, decades ago, Feldberg and collaborators already showed that injection of neuroactive substances in the cerebral ventricles brings about a wide spectrum of behavioral changes (see references Chap. 21, *this volume*). Then, a clearcut distinction emerged, depending on the nature of the medium used for signal transfer: paracrine signaling bypasses the bloodstream and uses the extracellular fluids (see, e.g., refs. 33–35; see Chap. 7, *this volume*), whereas endocrine communication operates via the circulation.

In their initial synthesis, Fuxe and Agnati (refs. 1–3) have grouped together these different means of communication within the definition of volume transmission (see also Chaps. 9 and 10, *this volume*). In my opinion, the distinction between such mode of communication and classical wiring transmission possesses an important heuristic value and hopefully may stimulate altogether theoretical and experimental work.

Yet, the experimental demonstration of such a mechanism is not straightforward. In the case of chemical signaling, which will be primarily discussed here, it has to include

1. demonstration of the release of neurotransmitter or neuropeptide at a defined locus (i.e., the identification of the source);
2. follow-up of the diffusion "à distance" of this particular chemical signal through brain tissue (i.e., the mapping of the communication channel);
3. recording of the response to this particular chemical by the postulated target cells [i.e., the identification of the receiver(s), the delineation of the "sphere of influence" (Chap. 5, *this volume*) of the released signal at the neuronal population level, and the definition of the time scale of the whole process].
4. demonstration that the chemical circuit thus described is physiologically relevant and can be reconstructed by locally injecting the signaling substance and assaying the response.

Data pertinent to such demonstration have been presented repeatedly with several different systems throughout the meeting and will, now, be briefly discussed. Yet, the number and diversity of the contributions was such that an exhaustive and fair discussion of all the papers appeared difficult to achieve in a few pages. I hope the reader will forgive my personal bias and prejudices, which will necessarily become apparent as a consequence of the required drastic condensation of the available information.

DIFFUSION THROUGH BRAIN TISSUE

Endocrinology is based on the fact that "informational substances" may exert their action on targets far distant from their release site by traveling through the bloodstream. This is a well-established notion. It is less commonly accepted that signaling molecules may passively diffuse through a nervous tissue like brain parenchyma. Yet, this is a *sine qua non* condition for volume transmission to take place in the brain. This is why it will be considered first and shown, indeed, to be grounded on empirical observations, quantitative measurements, and relevant theoretical models.

Nicholson and Rice (Chap. 21, *this volume*) have used radiotracers, microdialysis, ion-selective microelectrodes, or carbon fiber voltammetry to follow the diffusion of tetramethylammonium or endogenous ions, Ca^{2+}, K^+, serotonin or dopamine within cerebral cortex, cerebellum, vertical lobe, from several species (rat, turtle, cuttle fish). In all systems, the movement of the substances was accounted for by diffusion through intercellular interstices according to Fick's law modified to incorporate two new parameters: the tortuosity λ related to the shape of the intercellular volumes and to the fact that the diffusing particles execute random walks to avoid membrane obstructions, and the volume fraction, α, which gives the size of the extracellular space compared with the total volume of brain tissue. In all the systems studied, the volume fraction was found to be about 20% of the total volume and the tortuosity about 1.6.

The time scale of the experiments ranged from 20 to 80 sec for distances separating ejection micropipette and recording electrodes close to 200 μm (i.e., several orders of magnitude slower than the milliseconds time scale of synaptic wiring transmission). Yet, the "porosity" of neural tissues appears sufficient to allow diffusion of neuroactive substances across distances of several hundreds of micrometers, up to the entire brain, with, of course, the expected decrease of concentration. Thus, the texture of brain tissue appears adequate to mediate volume transmission.

However, several molecules of biological importance were found to move much more slowly than predicted. For instance, with K^+ the volume fraction routinely exceeds 100% as a consequence of the ability of K^+ to cross membranes. With dopamine, migration is strongly curtailed by a specific uptake mechanism that obeys Michaelis-Menten kinetics. Such uptake mechanisms, known to occur in nerve as in glial cells for a variety of neurotransmitters and ions, may create significant obstacles to volume transmission.

Another important obstacle to diffusion is the inactivation of the neuroactive substances by degradative enzymes. The case of ACh hydrolysis by "its specific" and nonspecific esterases is well-documented at the neuromuscular junction (19,36).

Neuropeptides (Chap. 27, *this volume*) may as well diffuse through extracellular spaces and serve as signals in volume transmission (see Chaps. 9,10,12,33, *this volume*). They are not subject to reuptake but to active degradation by peptidases. An extensively studied case is that of the enkephalins, which are substrates of two-membrane-bound, integral, ectopeptidases with their active site oriented toward the extracellular space (Chap. 29, *this volume*). One of them, enkephalinase, is a zinc enzyme, primarily but not exclusively located on neuronal cell bodies, axons, and synapses. Its localization largely coincides with the presence of enkephalinergic and/or substance P-ergic fibers, and its inhibition elicits numerous opioid-like naloxone reversible responses. Amino peptidase M is another membrane-bound aminopeptidase, primarily localized in brain microvessels, which breaks down enkephalins but also a wide variety of peptides. Its inhibition potentiates the opioid-like responses observed with enkephalinase. Both enzymes thus should prevent the diffusion of released neuropeptides at large distance, even though as stated by Schwartz et al. (Chap. 29, *this volume*), this hypothesis has never been tested directly in the mammalian brain.

Other important obstacles to diffusion of neuroactive compounds through the brain are epithelial barriers such as between blood and brain or the choroid plexus. The origins of the cerebrospinal fluid (CSF), from periventricular brain (acting somewhat as a "nephron") and from choroid plexus secretion, or the regulation of its composition under normal and pathological conditions has been dealt with extensively in several communications in this volume (Chaps. 23–25,32). In particular, the CSF contains at least four endopeptidases active on neuropeptides: angiotensin-converting enzyme, dynorphin-converting enzyme, substance P endopeptidase, and the already mentioned enkephalinase. As suggested by Terenius and Nyberg (Chap. 32, *this volume*), these enzymes might be secreted from neurons during synaptic transmission and "invade" the CSF by some kind of global volume transmission. Their role in the CSF is still hypothetical, but they may plausibly act as "scavengers" of neuroactive peptides. An additional possibility is that these peptidases in brain tissues and/or fluids contribute to the processing of neuropeptides by the formation of active molecule(s) from inactive precursor and/or by the elaboration of new classes of neuroactive substances (Chaps. 27,31,32, *this volume*).

The blood itself through the intrinsic network of brain capillaries may, of course, transport circulating substances (e.g., hormones) from distant sites of release to their target cells in shorter time scales than in the CSF because of differences in their volume flow (see Chaps. 14–16, *this volume*). Yet, the major obstacle to diffusion remains, as mentioned, the blood-brain barrier.

In conclusion, despite a number of defined obstacles to the diffusion of neuroactive substances through the brain, efficient pathways for the distant transport of these chemical signals, including neurotransmitters, neuropeptides, and steroid hormones, exist within the brain tissue itself and in and out to it.

PHYSIOLOGICAL RELEASE OF NEUROACTIVE SUBSTANCES IN THE BRAIN: VOLUME OR WIRING TRANSMISSION?

Volume transmission requires a source from which the "neuroactive" substance is released in the brain. This source might be identified, in principle, by anatomical and cytochemical techniques and by assay of the substances in the intercellular fluids.

Morphology of the Synaptic Contacts

Inference of function from morphology becomes particularly challenging when it deals with the release of neurotransmitters or peptides. The structural criteria used with the aim to distinguish volume and wiring transmission originate from the early electron microscopic description of aminergic axonal varicosities in the neocortex from adult rats (Chaps. 2,4, *this volume*). Many nerve endings were reported to lack junctional membrane specializations and, in particular, postsynaptic interfaces (Chap. 5, *this volume*). On this basis, it was hypothesized that the amines released from the "nonsynaptic" varicosities may diffuse to reach relatively distant targets. Yet, the actual morphology of the aminergic synapses in the brain is still debated. The view that aminergic neurons may establish morphologically differentiated "conventional" junctions with cortical neurons and thus are involved in wiring transmission was defended (37; and Chap. 2, *this volume*). In fact, quantitative analysis of serial ultrathin sections immunostained for dopamine, norepinephrine, or serotonin lead to the conclusion that the two categories of terminals do occur but to different extents, depending on the neurotransmitter and the region of the brain considered (Chap. 5, *this volume*). For instance, dopamine innervation appears predominantly junctional (see also Chap. 26, *this volume*) but with significant interregional differences. However, cortical norepinephrine innervation is mostly nonjunctional. In the case of serotonin, the incidence of conventional synapses extrapolated for whole varicosities range from 28% to 46%, and the junctions are found systematically asymmetrical and on dendritic spines or shafts. Thus, the suggestion is that, in the brain, norepinephrine and serotonin inputs are characterized by transmission at a distance, whereas dopamine acts on its target neurons by conventional synaptic transmission (Chap. 5, *this volume*).

The lamprey spinal cord is a particularly suitable system (38) to analyze, in vertebrates, the relation between a defined behavior, locomotion, and simple circuits made up of identifiable neurons (Chap. 11, *this volume*). The initiation of locomotion relies on the activation by excitatory amino acids but is modulated, in particular, by serotonin. Serotonin immunoreactivity is detected at the level of varicosities that, after examination on ultrathin serial sections, does not display any sign of synaptic specialization (e.g., synaptic clefts or postsynaptic membrane thickenings) (Chap. 11, *this volume*). The ultrastructure of the serotonin terminals in lamprey spinal cord thus looks appropriate for volume transmission (see Chap. 5, *this volume*).

Yet, examination of synapse ultrastructure does not suffice to decide in favor of volume transmission. There are indeed documented examples (e.g., the developing synapses) (see refs. 39–41) in which morphologically undifferentiated synaptic contacts mediate conventional chemical synaptic transmission. Moreover, morphologically differentiated synaptic contacts may let neurotransmitters leak from the cleft. Additional evidence (e.g., measurement of neurotransmitter levels in extracellular space and/or assay of the physiological response) is required.

Measurements of Extracellular Neurotransmitter Release

A more direct and complementary approach to the identification of the source is to measure *in vivo* the levels of neuroactive substances liberated in the extracellular space, for instance, by push-pull cannulae (42), microdialysis (U. Ungerstedt, *personal communications*), or electrochemistry (Chap. 26, this volume).

Quantitative data about extracellular release of dopamine have been reported in this volume by Gonon et al. (Chap. 26, *this volume*). In rat striatum, the extracellular concentration of dopamine varies, depending on impulse flow, from 10 to 100 nM and results from bursts of three to 20 spikes. Dopamine is presumably released at the level of strings of varicosities, which make symmetric synaptic contacts mostly on the neck of spines. Thus, according to Gonon et al. (Chap. 26, *this volume*), the level of dopamine in the synaptic cleft would, as in the case of the neuromuscular junction (see Chap. 1, *this volume*), reach levels as high as 10 to 100 μM. Yet, dopamine can be found in extracellular spaces; the lower recorded levels there would result from diffusion and dilution from the clefts. Interestingly, the high estimates of the cleft concentrations are, according to Gonon et al. (Chap. 26, *this volume*), consistent with the low affinities of D1 dopaminergic receptors. Also, local injections of exogenous dopamine elicit electrophysiological responses, which differ from those observed when the release of endogenous dopamine is triggered by physiological stimulation of the dopaminergic pathways, thus suggesting that diffusion of the neurotransmitters through intercellular spaces does not contribute significantly to the physiological response. Thus, altogether, the electrochemistry and anatomical data lead Gonon et al. (Chap. 26, *this volume*) to conclude that dopamine mediates classical wiring transmission in the brain, a view consistent with the anatomical data of Descarries et al. (Chap. 5, *this volume*) presented in this volume.

The importance of dopamine deficit in Parkinson disease consecutive to loss of nigral mesotelencephalic neurons (43) is well-recognized (review refs. 44,45). Dopamine agonists or L-DOPA given to the patients yields, at least, transient remission of some of the symptoms, supporting the view that these pharmacological agents reach dopamine receptors by diffusion. Accordingly, it was mentioned that under normal conditions, "dopamine formed in the terminal networks is continuously leaking out from the terminals to reach adjacent dopamine receptors by *diffusion*" (46), and the possibility may be considered that such "volume effects" become dominant when, in the patients, the dopaminergic neurons start to die (see ref. 43).

Using the same electrochemical methods, Gonon et al. (Chap. 26, *this volume*) have attempted to extend their conclusions to norepinephrine in the paraventricular nucleus. They note that, in addition to standard chemical synaptic communication, a signal transmission mediated by low norepinephrine extracellular concentration might exist. Moreover, in rat tail artery the concentration of norepinephrine reached at the arterial surface after electrical stimulation is 120 nM, in the range of the norepinephrine concentration, which elicits the same response. Thus, in this peripheral tissue, sympathetic terminals exert their contractile effect through diffusion of norepinephrine (see ref. 10) and/or neuropeptides (Chap. 33, *this volume*) from a source distant from their target site.

In conclusion, measurements of transmitter release favor the notion that in the case of dopamine, classical wiring transmission still holds but that volume effects may take place in addition and/or alternatively with other biogenic amines and neurotransmitters.

Evidence from Grafts

A particularly convenient method to demonstrate "à distance" transfer of chemical signals is to graft neurons either at previously lesioned targets or at ectopic positions and to monitor release of neurotransmitters from these grafts (Chap. 35, *this volume*). Measurements by microdialysis of ACh, norepinephrine, serotonin, and dopamine show that the levels of transmitters released by these grafts are near-normal both under resting conditions and after K^+-induced depolarization, blockade of transmitter uptake, and neuronal electrical activity by tetrodotoxin. Despite their ectopic localization, the grafted neurons display electrical activity-dependent release of neurotransmitter in a manner virtually identical to that in intact animals (Chap. 35, *this volume*).

Another elegant experiment bears on the effect of locus ceruleus transplants on kindling (Chap. 38, *this volume*). In the rat, repeated electrical stimulation with subthreshold stimulus strength leads to induced seizures (kindling). These seizures are markedly inhibited by locus ceruleus and by noradrenergic agonists acting via α2- (but not α1- or β-) adrenergic receptors. Interestingly, bilateral grafts of locus ceruleus in norepinephrine-depleted animals either close to the stimulating electrode (hippocampus) or in a more distant region (amygdala-pyriform cortex) cause a marked retardation of kindled seizure as compared with sham-grafted animals. The reinstatement of norepinephrine release seems sufficient to normalize the susceptibility to kindled seizure. According to Lindvall et al. (Chap. 38, *this volume*), the functional effects exerted by the graft occur via classical wiring transmission. Yet, the "possibility that the high seizure-induced extracellular levels of noradrenaline could influence the recipient brain via volume transmission" (Chap. 38, *this volume*) seems worthwhile to explore.

Strong evidence in favor of this last mode of communication is offered by adenohypophysical transplants in rat neostriatum (Chap. 37, *this volume*). The gland cells secreting luteinizing hormone (LH), adrenocorticotropic hormone (ACTH), and prolactin survive in their ectopic positions, and prolactin was found to be released into the neuropil of the neostriatum, most likely through extracellular fluid pathways. No catecholamine terminals could be demonstrated within the transplant itself. Yet, 6-hydroxydopamine denervation of the dopamine afferents to the neostriatum caused increased release of prolactin. Thus, endogenous dopamine modulates prolactin secretion into striatal neuropil from the transplanted adenopituitaries by extracellular diffusion of dopamine from striatal terminals into the transplanted adenopituitary.

Finally, clearcut demonstration of "à distance" diffusion of chemical effectors in the brain has been achieved by following the effect of a peptide that, up to now, has not been shown to act as a "classical neurotransmitter": the nerve growth factor (NGF) (Chap. 36, *this volume*). First, administration of purified NGF by simple infusion into the lateral ventricle promotes recovery of cholinergic biochemistry and function after lesions of nucleus basalis, magnocellularis, or after axotomy of septohippocampal cholinergic neurons. Also, implantation of NGF-producing cells, such as mouse submandibular gland implants or genetically modified cell lines, stimulates cholinergic functions (e.g., induces sprouting of host cholinergic terminals, supports grafts of fetal cholinergic neurons, and rescues lesioned adult septal cholinergic neurons). This exogenous NGF thus acts as a "trophic signal" by diffusion through brain parenchyma, and it may further be speculated that during development endogenous NGF "not only acts to attract and maintain a cholinergic innervation, but also by local diffusion as a paracrine diffusion signal" (Chap. 36, *this volume*).

In conclusion, grafting experiments show that ectopic implants may release neuroactive substances in a manner similar to that found with the same neurons at their original position. If, in some instances, this may result from a nearly normal outgrowth of nerve terminals and target innervation by the implant, strong evidence exists that defined effects, in particular those of NGF, take place via diffusion through extracellular spaces (i.e., via volume transmission). Thus, volume transmission may concern, in addition to "classical" neurotransmitters and neuropeptides, "trophic substances" (Chaps. 17,18,20), which may have to cross important distances by diffusion to act on their physiological target.

RECEPTOR MECHANISMS INVOLVED IN VOLUME TRANSMISSION

In the standard theory of communications, the definition and properties of the "receiver" are as crucial as those of the "source." It must "decode" the signal it receives, here the variation in time of the extracellular level of a neuroactive agent. The "receptive substance" postulated by Langley at the turn of the century (1905–1906) is no longer a "working hypothesis." Several receptors have been identified in the course of the past decades, thanks to a concerted approach grouping together pharmacology, biochemistry, physiology, and recently, molecular biology. As a consequence, several receptors linked with ion channels (review in refs. 21,47), enzymes, or G proteins (48,49) and several voltage-sensitive ion channels (review refs. 50,51) have been sequenced and their transmembrane organization investigated.

Volume transmission involves, of course, receptor mechanisms. Yet, the identification of the physiological target of an extracellularly released neuroactive signal appears more difficult to establish than that engaged in wiring transmission, at classical synapses where an almost perfect match exists between the release and receptor sites across the synaptic cleft (see Chap. 2, *this volume*).

Physiological Response to Extracellular Signals

Electrophysiological recordings, despite their intrinsic limitations to changes in electrical parameters, have been used to monitor cell responses to extracellular signals within time scales that range from microseconds to hours. In the lamprey spinal cord, it was mentioned by Christenson et al. (Chap. 11 *this volume*) that serotonin is present in varicosities that do not form authentic synaptic specializations and thus appear appropriate for volume transmission. Additional experimental evidence supports this view. Blocking serotonin uptake (which increases extracellular levels of serotonin) slows down *in vitro* the rate of bursting (by increasing both intensity and duration of burst discharges) of the spinal cord neurons associated with locomotion. Moreover, bath application of serotonin has a similar effect as the endogeneously released neurotransmitter. Serotonin does not affect the resting membrane properties or the mechanisms underlying the action potential but acts directly by depressing calcium-dependent potassium channels responsible for the regulation of firing frequency in several types of interneurons and motoneurons. In other words, there is a "collective" effect of serotonin over a defined population of spinal cord nerve cells, and volume transmission is a plausible mechanism that may allow the overall "gating" of a group of neurons that carry its receptors (Chap. 11, *this volume*).

In addition, in this volume, several chapters dealt with responses of a variety of systems to neuroactive substances plausibly liberated in brain extracellular space. The techniques used were those of pharmacology (Chap. 33, *this volume*), electrophysiology (Chaps. 28,40, *this volume*), biochemistry (Chaps. 10,22,30, *this volume*), and others (Chaps. 41–43, *this volume*). The cellular targets in most instances were neurons. But special emphasis was placed on an eventual participation of glial cells to the response to diffusing extracellular signals (Chaps. 17–20, *this volume*). Conventional synaptic contacts with glial cells are not commonplace in the vertebrate nervous system. These target cells, thus, can only be reached via diffusion through extracellular spaces. Along these lines, we can mention the case of the glutamate-kainate receptor, which, according to Gregor et al. (52), is almost exclusively found on glial cells.

Receptor Mismatch

Receptor mismatch (53) has become a rather popular notion despite the fact that from an experimental point of view, it is essentially based on the contrast between "negative" results and positively obvious ones and that its relation with volume transmission is not easy to establish in a causal manner. A match between nerve endings and receptor clusters, which resembles that found at the motor endplate, has been demonstrated in central glycinergic synapses (54), and this may not be a unique disposition in the brain. Yet, Herkenham (Chap. 6, *this volume*) has listed an impressive number of cases in which anatomical techniques disclose either "high terminal density and sparse receptor" or "low terminal density and dense receptors" for more than 15 neuropeptides and seven classical neurotransmitters. According to Herkenham (Chap. 6, *this volume*), the number of situations in which an "informational substance" in terminals coincides with the complementary receptor appears smaller, in particular for neuropeptides.

Herkenham carefully examines four main categories of explanations for receptor mismatches:

1. The first and most difficult one to eliminate is that of technical failures and/or incomplete knowledge of the transmitters and receptors considered. Indeed mismatch means that using histochemical techniques, which are often different and nonlinear, we do not find qualitatively the same staining intensity for informational substances and their receptor. In any case, absolute ratios of both partners are rarely established in quantitative terms.

2. A second of explanations deals with biochemical difficulties in evaluating functional receptors strictly by binding techniques. For instance, the receptor sites could (a) be occupied by endogenous ligands, (b) possess low affinity (see Chap. 1, *this volume*), or (c) not be coupled with any effector protein or, even, are "drug binding sites" without any evident function. None of these possibilities is easy to reject.

3. Also, at the cellular level, the neurotransmitters may have a distribution that is not directly relevant to actual intercellular communication but to their biosynthesis, metabolism, transport, and targeting to the relevant synapse.

4. The last but not least possibility is that the nonsynaptic localization indeed signifies transmitter action at a distance. Yet, it is misleading to say that volume transmission implies receptor mismatch. Rather, we should say that as in any communication process, it requires a "match" between the source and the receiver. But this match is expected to take place "à distance," whereas it occurs in a more confined space in the case of conventional transmission. There is thus a requirement for "à distance match," which appears more difficult to establish than the so-called receptor mismatch.

Recent developments on the molecular biology of receptors bring additional complexity to Herkenham points 2 and 3. In the case of brain nicotinic receptor in chick and rat, at least five genes coding for the ACh binding α-subunits have been identified (55). These α2 to α5 subunits are expressed in different (sometimes overlapping) brain areas and may combine together with non-α subunits into functional ligand-gated ion channels (55). A variety of oligomers (possibly pentamers) result, which possess a variety of channel and ligand binding properties (review refs. 56,57). A similar situation has been reported with gamma-aminobutyric acid (GABA) receptors (58) from brain (review ref. 47). Several different complementary deoxyribonucleic acids, (cDNAs) which belong to four main categories of subunits (α,β,γ,δ), have been cloned and sequenced. The different polypeptide chains may combine into a wide variety of oligomers, which differ, in addition to their channel properties, by their intrinsic affinity for the neurotransmitter, their ability to desensitize, and their regulation by allosteric effectors (such as the benzodiazepines). As a consequence, some of them may escape to detection even though they are functional, and others may be easily detected but are nonfunctional.

Such diversity of receptors' structure and function does not simplify the mismatch issue nor its relation with volume transmission. Is the receptor mismatch notion valid for any of the subspecies of receptor? Which one of these subspecies is involved in a given physiologically significant distant communication? These are questions still to answer.

Topological Distribution of Receptors at the Cell Level

In the adult motor endplate (20,21) as in conventional central synapses (54,59), the distribution of the receptor almost exactly coincides with that of the motor nerve terminal. Yet, in the adult skeletal muscle, such match can be severely perturbed by sectioning the motor nerve. After a few days (chick, rat) or weeks (frog), the surface of the muscle fiber starts to respond to

ACh. A denervation "hypersensitivity" develops (60,61). ACh receptor molecules become incorporated into extrajunctional areas (62) as a consequence of the transcriptional activation (63,64) of ACh receptor genes in myo-myonuclei (65,66) located in these areas. A striking "mismatch" develops. In the denervated muscle, the ACh receptor is produced in large amounts in the absence of the cholinergic nerve endings (review ref. 67). Yet, nobody will argue that such disposition is designed for volume transmission. The most plausible explanation is that the denervated muscle fiber becomes ready to form synaptic contacts at any point of its surface with reinnervating axons, whereas the adult innervated muscle appears refractory to innervation (68).

Along these lines, it is worth noting that the development of ACh receptor distribution during embryogenesis follows an evolution in a direction opposite to that of denervation (review ref. 67). In the embryonic myotube, the ACh receptor is "isotropically" distributed all over the surface of the cell. As soon as a contact forms with an exploratory motor axon, the receptor accumulates under the motor nerve ending as a consequence of a recruitment from surrounding noninnervated areas (69) and the maintenance of ACh receptor gene expression in junctional nuclei, whereas the extrajunctional ones are turned off (see ref. 65). In extrasynaptic areas, the neurally evoked electrical activity of the muscle fiber represses the transcription of the ACh receptor genes (70), thereby creating a "symmetry breaking" (71) in the distribution of the receptor protein. The "receptor match" of the adult junction is thus progressively achieved via the compartmentalized expression of ACh receptor genes in topographically distinct nuclei (66).

An important component of the system, then, becomes the "anterograde" signal, which is released by the motor nerve ending and positively regulates the ACh receptor gene in the subjunctional nuclei. Among several plausible candidates [acetylcholine receptor inducing activity (ARIA) (72), ascorbic acid (73)], some of the neuropeptides that coexist (11) with ACh in the motor nerve ending such as CGRP (67,74–76). Receptor match requires a transsynaptic exchange of signals. Thus, CGRP could be one of these "matching factors."

In nerve cells, the situation looks *a priori* simpler than in the motor endplate, because there exists only one nucleus per cell. The differential regulations of gene expression leading to the known anisotropic distribution of neurotransmitter receptors in the dendrites, soma, and axon will thus essentially rely on posttranscriptional processes. These mechanisms are still largely unexplored but may concern the transit of the receptor proteins through the endoplasmic reticulum, the Golgi apparatus, their transport to the cytoplasmic membrane, their targeting to defined subcellular compartments of the nerve cell, and their redistribution within the cytoplasmic membrane (for review, see ref. 67).

In any case, the achievement of receptor match involves sophisticated mechanisms of gene expression, which require a "cooperative" intercellular communication between the two juxtaposed partners during synapse formation. It is thus not unrealistic to assume that receptor mismatch may frequently exist even without apparent physiological significance. During development, a "gratuitous" expression of genes may well take place simply for morphogenetic reasons. The evolution of the cell adhesion molecule (N-CAM) in the course of embryogenesis (77; Chap. 3, *this volume*) which includes stages of transient gene expression, down-regulation, and stabilization, illustrates how complex the topological evolution of a membrane-bound protein involved in intercellular interaction may be.

Volume Transmission and "Receptor Efficacy"

In the case of distant communication, the specificity of the receptor for the diffusing signal creates the specificity of the communication line. Steady low concentrations of the neuroactive substance serve as signals instead of brief, high concentration pulses of neurotransmitter. The resolution in time and space thus appears less stringent than in the case of wiring transmission. High-affinity receptors, which do not efficiently desensitize, are thus expected to be primarily involved in such a process. This is often the case for neuropeptides and hormone receptors, which are, in general, coupled with enzymes and second-messenger systems (G proteins).

As discussed (see, for instance, refs. 78–81), in addition to a direct role in signal recognition and transduction, receptors for neurotransmitters may be the targets of neuroactive substances, which regulate their response efficacy. Indeed, an important physiological role of volume transmission could be the modulation of wiring transmission through allosteric properties of receptors present in conventional synapses. The diffusing signal may bind to allosteric sites, such as the glycine binding site of the glutamate N-methyl d-aspartate (NMDA) receptor (82) or the benzodiazepine binding site of the GABA receptor (47). But it may also interact with a neighboring receptor via receptor–receptor interactions (see ref. 83) or through a second-messenger cascade. Conformational equilibria between "activatable" and "desensitized" (or refractory) states of the receptor are known in some cases to exist before ligand binding (see ref. 21). The ratio of these two states might then be a measure of the "efficacy" of the receptor response. The neuroactive substances diffusing through volume transmission may regulate the ratio of these two states or the rates of their interconversion (84). These mechanisms may be extended to the integration of multiple signals by a single receptor molecule within a given time-window and offer models of the Hebb synapse under conditions of chemical signaling (78,80,81). In this respect, the nonquantal leak of neurotransmitter (see Chaps. 1,34, *this volume*) may play a particular role in "pre-setting" the ratio of activatable and desensitized states of receptors via a steady-state volume transmission.

MODELS OF HIGHER BRAIN FUNCTIONS AND VOLUME TRANSMISSION

The message coded by chemical signals propagated by diffusion through intercellular space is not expected to resemble that carried by standard chemical synaptic transmission. The "specificity" of the connectivity, which characterizes wiring transmission (see ref. 7), and the speed and repetitivity of signal transfer are lost to the benefit of a slower collective regulation of large ensembles of neurons. From an informational point of view, volume transmission makes possible a regulation at a higher level of organization. It provides an efficient process to create a coordination between cells. It yields a highly "divergent" gating function, which is distinct from what is usually referred to as "parallel" computation (Chap. 45, *this volume*).

As mentioned, a rather well-established case is the modulation of Lamprey swimming behavior by serotonin (Chap. 11, *this volume*). In mammalian brain, the biogenic amines released by the highly divergent midbrain neurons have been assumed to play a role in regulating states of wakefulness (see ref. 85), mood (Chap. 2, *this volume*), and/or focusing of attention. In the case of the visual cortex, such "global" aminergic regulation has been shown to affect the signal-to-noise ratio of recorded single units (85).

This last observation has served to build a computational model for the effect of catecholamines on the brain (86). According to this model, catecholamine tone was assumed to affect the gain of individual "formal neuron" units composing the network. An interesting property of such a simple network is that a change in individual unit response does not lead to a change in the signal detection performance of these units. Yet, the performances of the system as a whole in signal detection appears strikingly improved by catecholamines.

Another formal application of a collective regulation of neuron properties by a chemical signal is included in a recent model of prefrontal cortex function in delayed-response tasks (87). Learning is assumed to take place by selection among a repertoire of possible rules encoded in defined clusters of neurons. A reinforcement signal monitors the success (or failure) of a given action on the outside world through a feedback "reward" loop. It governs both the occurrence of random modification of rule clusters activity (generator of diversity) and the global modifications of synaptic efficacies (selection). Such effects of the reward loop can be modeled in terms of a global action of a volume-transmitted chemical signal (e.g., a catecholamine) on the allosteric transitions of neurotransmitter receptors themselves involved in wiring transmission.

In conclusion, from both an experimental and theoretical point of view, volume transmission of a neuroactive substance may efficiently serve to regulate, in a coordinate fashion, the state of activity of populations of neurons in the central nervous system of vertebrates.

PERSPECTIVES

The theoretical proposal of Fuxe and Agnati (Chap. 1, *this volume*) of a mode of intercellular communication by diffusion through extracellular spaces, mainly using chemical signals, appears altogether plausible and useful.

Its plausibility is supported by a wealth of experimental data, which include:

1. The ability of classical neurotransmitters and other substances including peptides to diffuse (in the seconds to minutes time scale) through brain tissue;
2. The occurrence of sources from which the neuroactive compounds are released and diffuse into the extracellular spaces. These sources may be free varicosities without clearcut morphological synaptic differentiations; yet, morphologically differentiated synapses may also let chemical signals diffuse outside the cleft under conditions of both evoked release and passive leakage;
3. The recording of the response of potential target cells to neuroactive compounds (including neuropeptides and hormones) released at distant sites far from the target. The recorded responses might either be macroscopic changes in peripheral organ properties, variations in biochemical and/or electrophysiological parameters, but also the biosynthesis of defined proteins, the survival of cells, and the outgrowth of neurites.

Compared with classical wiring transmission, volume transmission is much slower, uses lower concentrations of effector, and spreads over larger populations of cells. But, the information it carries, in many instances, appears different. Conventional wiring transmission contributes to discrete communication between individual nerve cells, which exploit the "singularities" of their connectivity. It plays a basic role in standard reflex activities at low levels of organization. But it is expected to play an even more critical function in the elaboration of the complex neuronal maps, which underlie high level "mental" representations. However, volume transmission submits populations of cells to a common global control. It is expected to gate ensembles of neurons possessing, in common, a receptor(s) for the diffusing signal.

The concept of volume transmission thus appears useful in stimulating research in still rather unexplored areas. Future work should, in particular, further specify the cellular and molecular bases of these global communication processes and their relations with classical wiring transmission. It should also lead to an analysis of their eventual perturbation in pathological conditions. In any case, already at this stage, the aim to contrast volume transmission and classical wiring transmission appears a stimulating and thought-provoking attitude.

REFERENCES

1. Agnati LF, Fuxe K, Merlo Pich E, et al. Aspects on the integrative capabilities of the central nervous system: evidence for "volume transmission" and its possible relevance for receptor–receptor interactions. In: Fuxe K, Agnati LF, eds. *Receptor–receptor interactions.* Wenner-Gren International Symposium Series, vol 48. 1985; 236–249.
2. Agnati LF, Fuxe K, Zoli M, Zini I, Toffano G, Ferraguti F. A correlation analysis of the regional distribution of central enkephalin and b-endorphin immunoreactive terminals and of opiate receptors in adult and old male rats. Evidence for the existence of two main types of communication in the central nervous system: the volume transmission and the wiring transmission. *Acta Physiol Scand* 1986;128:201–207.
3. Fuxe K, Agnati L, Härfstrand A, et al. Morphofunctional studies on the neuropeptide Y/adrenaline co-storing nerve terminal systems in the dorsal cardiovascular region of the medulla oblongata. Focus on receptor–receptor interactions in co-transmission.. *Prog Brain Res* 1986; 68:303–320.
4. Changeux JP, Dehaene S. Neuronal models of cognitive functions. *Cognition* 1989;33:63–109.
5. Georgopoulos AP. Neural integration of movements: role of motor cortex in reaching. *FASEB J* 1988;2:2849–2857.
6. Von der Malsburg C, Singer W. Principles of cortical network organisation. In: Rakic P, Singer W, eds. *Neurobiology of neocortex.* J. Wiley and Sons, 1988;69–100.
7. Changeux JP. Concluding remarks on the "singularity" of nerve cells and its ontogenesis. *Prog Brain Res* 1983;58:465–478.
8. Edelman G. *Neural Darwinism.* New York: Basic Books, 1987.
9. Katz B. Nerve muscle and synapse. New York: McGraw Hill, 1966.
10. Stjärne L. Basic mechanisms and local modulation of nerve impulse-induced secretion of neurotransmitters from individual sympathetic nerve varicosities. *Rev Physiol Biochem Pharmacol* 1989;112:1–137.
11. Hökfelt T, Holets VR, Staines W, et al. Coexistence of neuronal messengers—an overview. *Prog Brain Res* 1986;68:33–70.
12. Bennett M, Spray DC. Intercellular communication mediated by gap junctions can be controlled in many ways: In: Edelman G, Gall E, Cowan W, eds. *Synaptic function.* New York: J. Wiley, 1987;109–135.
13. Couteaux R. Structure and cytochemical characteristics of the neuromuscular junction. In: Cheymol J, ed. *Neuromuscular blocking and stimulating agents.* (I.E.P.T. Section 14), vol 1. Oxford, New York: Pergamon Press, 1972;7–56.
14. Matteoli M, Haimann C, Torri-Tarelli F, Polak JM, Ceccarelli B, De Camilli P. Differential effect of alpha-latrotoxin on exocytosis from small synaptic vesicles and from large dense-core vesicles containing calcitonin gene-related peptide at the frog neuromuscular junction. *Proc Natl Acad Sci USA* 1988;85:7366–7370.
15. Jasmin BJ, Changeux JP, Cartaud J. Compartmentalization of cold-stable and acetylated microtubules in the subsynaptic domain of chick skeletal muscle fibre. *Nature* 1990;334:673–675.
16. Katz B. The release of neural transmitter substances. Liverpool: *Liverpool Univ. Press,* 1969.
17. Llinas R. Calcium regulation of neuronal function. In: Schmitt FO, Worden FG, eds. *The Neurosciences—fourth study program.* Cambridge, Massachusetts: MIT Press, 1979;555–571.
18. Katz B, Miledi R. The binding of acetylcholine to receptors and its removal from the synaptic cleft. *J Physiol* 1973;231:549–574.
19. Kuffler SW, Yoshikami D. The number of transmitter molecules in a quantum: an estimate from iontophoretic application of acetylcholine at the neuromuscular synapse. *J Physiol (Lond)* 1975; 251:465–482.
20. Salpeter M, Loring RH. Nicotinic acetylcholine receptors in vertebrate muscle: properties, distribution and neural control. *Prog Neurobiol* 1985;25:297–325.
21. Changeux JP. Functional architecture and dynamics of the nicotinic acetylcholine receptor: an allosteric ligand-gated ion channel. Fidia Research Foundation Neuroscience Award Lectures, vol 4. New York: Raven Press, 1989;21–168.
22. Villar MJ, Roa M, Huchet M, et al. Immunoreactive calcitonin gene-related peptide vasoactive intestinal polypeptide, and somatostatin in developing chicken spinal cord motoneurons. Distribution and role in regulation of cAMP in cultured muscle cells. *Eur J Neurosci* 1989;1: 269–287.
23. Jennings CGB, Mudge AW. Chick myotubes in culture express high-affinity receptors for calcitonin gene-related peptide. *Brain Res* 1989;504: 199–205.
24. Roa M, Changeux JP. Characterization and developmental evolution of a high-affinity binding site for calcitonin gene-related peptide (CGRP) on chick skeletal muscle membrane *Neuroscience (in press).*
25. Popper P, Micevych PE. Localization of calcitonin gene-related peptide and its receptors in a striated muscle. *Brain Res* 1989;496:180–186.
26. Katz B, Miledi R. Transmitter leakage from motor nerve endings. *Proc R Soc Lond (B)* 1977;196:59–72.

27. Korn H, Faber DS. Regulation and significance of probabilistic release mechanisms at central synapses. In: Edelman GM, Gall WE, Cowan WM, eds. *Synaptic function*. Neurosciences Research Foundation/John Wiley and Sons, 1987; 57–108.
28. McNaughton B, Douglas RM, Goddard GV. Synaptic enhancement in fascia dentata: cooperativity among coactive afferents. *Brain Res* 1978;157:277–293.
29. Shepherd GM. *Neurobiology*. New York: Oxford University Press, 1988.
30. Faber DS, Korn H. Synergism at central synapses due to lateral diffusion of transmitter. *Proc Natl Acad Sci* 1988;85:8708–8712.
31. Bach-y-Rita P. Convergent and long-latency unit responses in the reticular formation of the brain. *Exp Neurol* 1964;9:327–344.
32. Routtenberg A. Drinking induced by carbachol: thirst circuit or ventricular modification? *Science* 1967;157:838–839.
33. Guillemin R. Peptides in the brain—the new endocrinology of the neuron. New York: *Science* 1978;202:390–402.
34. Vizi ES. Nonsynaptic modulation of transmitter release: pharmacological implication. *TIPS* 1980;172–175.
35. Schmitt FO. Molecular regulators of brain function: a new view. *Neuroscience* 1984;13:991–1001.
36. Magleby KL, Pallotta BS. A study of desensitization of acetylcholine receptors using nerve-released transmitter in the frog. *J Physiol (Lond)* 1981;316:225–250.
37. Parnavelas JG, Papadopoulos GC. The monoaminergic innervation of the cerebral cortex is not diffuse and nonspecific. *Trends Neurosci* 1989;12:315–320.
38. Grillner S, Wallen P, Dale N, Brodin L, Buchanan J, Hill R. Transmitters, membrane properties and network circuitry in the control of locomotion in lamprey. *Trends Neurosci* 1987;10: 34–41.
39. Filogamo G, Gabella G. The development of neuromuscular correlations in vertebrates. *Arch Biol (Liège)* 1967;78:9–60.
40. Giacobini G, Filogamo G, Weber M, Boquet P, Changeux JP. Effect of a snake-neurotoxin on the development of innervated motor muscles in chick embryo. *Proc. Natl Acad Sci USA* 1973;70:1708–1712.
41. Fishbach G, Hume RI, O'Brien RJ, Role LW, Usdin TD. *Synapse formation: a view from the neuromuscular junction*. Liviana Press: Fidia Research Foundation Neuroscience Award Lectures, 1986;147–174.
42. Niéoullon A, Cheramy A, Glowinski J. Nigral and striatal dopamine release under sensory stimuli. *Nature* 1977;269:340–342.
43. Fuxe K, Hökfelt T, Ungerstedt U. Morphological and functional aspects of central monoamine neurons. *Int Rev Neurobiol* 1970;13:93–126.
44. Hornykiewicz O. Neurotransmitter substitution in neuropharmacology and psychiatry. In: Stjärne L, Hedqvist P, Lagercrantz H, Wennmalm A, eds. *Chemical neurotransmission 75 years*. London: Academic Press, 1981;513–525.
45. Carlsson A. Chemical neurotransmission and mental function: pharmacological aspects. In: Stjärne L, Hedqvist P, Lagercrantz H, Wennmalm A, eds. *Chemical neurotransmission 75 years*. London: Academic Press, 1981;526–540.
46. Fuxe K, Sedvall G. Pharmacological and clinical implications of chemical neurotransmission: chairmen's overview. In: Stjärne L, Hedqvist P, Lagercrantz H, Wennmalm A, eds. *Chemical neurotransmission 75 years*. London: Academic Press, 1981;541–548.
47. Olsen RW, Tobin AJ. Molecular biology of $GABA_A$ receptors. *FASEB J* 1990;4:1469–1480.
48. Dixon RAF, Kobilka BK, Strader DJ, et al. Cloning of the gene and cDNA for mammalian β-adrenergic receptor and homology with rhodopsin. *Nature* 1986;321:75–79.
49. Emorine L, Marullo S, Sutren MM, et al. Common properties of receptors coupled to GRP binding regulatory proteins. In: Maelicke, ed. *Molecular biology of neuroreceptors and ion channels*, NATO ASI Series, vol H32. Berlin, Heidelberg: Springer-Verlag, 1989;245–257.
50. Numa S. *A molecular view of neurotransmitter receptors and ionic channels*. Harvey Lecture Series 83. 1989;121–165.
51. Dani JA. Site directed mutagenesis and single-channel currents define the ionic channel of the nicotinic acetylcholine receptor. *Trends Neurosci* 1989;12:125–128.
52. Gregor P, Mano I, Maoz I, McKeown M, Teichberg VI. Molecular structure of the chick cerebellar kainate-binding subunit of a putative glutamate receptor. *Nature* 1989;342:689–692.
53. Cuello AC. Nonclassical neuronal communications. *Fed Proc* 1983;42:2912–2922.
54. Triller A, Cluzeaud F, Pfeiffer F, Betz H, Korn H. Distribution of glycine receptors at central synapses: an immunoelectron microscopy study. *J Cell Biol* 1985;101:683–688.
55. Heinemann S, Boulter J, Deneris E, et al. Brain and muscle nicotinic acetylocholine receptor: a gene family. In: Maelicke A, ed. *Molecular biology of neuroreceptors and ion channels*. NATO ASI Series vol. H32. Berlin, Heidelberg: Springer-Verlag, 1989;13–30.
56. Vidal C, Changeux JP. Pharmacological profile of nicotinic acetylcholine receptors in the rat prefrontal cortex: an electrophysiological study in a slice preparation. *Neuroscience* 1989;29: 261–270.
57. Mulle C, Changeux JP. A novel type of nicotinic receptor in the rat central nervous system characterized by patch-clamp techniques. *J Neurosci* 1990;10:169–175.
58. Schofield PR, Darlison MG, Fujita N, et al. Sequence and functional expression of the $GABA_A$ receptor show a ligand-gated receptor superfamily. *Nature* 1987;328:221–227.
59. Triller A, Cluzeaud F, Pfeiffer F, Korn H. Distribution and transmembrane organization of glycine receptors at central synapses: an immu-

nocytochemical touch. In: Montalcini RL, et al., eds. *Molecular aspects of neurobiology*. Berlin, Heidelberg: Springer-Verlag, 1986;101–105.

60. Axelsson J, Thesleff F. A study of supersensitivity in denervated mammalian skeletal muscle. *J Physiol (Lond)* 1959;147:178–193.
61. Miledi R. The ACh sensitivity of frog muscle fibers after complete or partial denervation. *J Physiol (Lond)* 1960;151:1–23.
62. Miledi R, Potter LT. Acetylcholine receptors in muscle fibres. *Nature* 1971;233:599–603.
63. Buonanno A, Merlie JP. Transcriptional regulation of nicotinic acetylcholine receptor genes during muscle development. *J Biol Chem* 1986; 261:11452–11455.
64. Tsay H, Schmidt J. Skeletal muscle denervation activates acetylcholine receptor genes. *Cell Biol* 1989;108:1523–1526.
65. Fontaine B, Sassoon D, Buckingham M, Changeux JP. Detection of the nicotinic acetylcholine receptor α-subunit mRNA by *in situ* hybridization at neuromuscular junctions of 15-day old chick striated muscles. *EMBO J* 1988;7:603–609.
66. Fontaine B, Changeux JP. Localization of nicotinic acetylcholine receptor α-subunit transcripts during myogenesis and motor endplate development in the chick. *J Cell Biol* 1989;108: 1025–1037.
67. Laufer R, Changeux JP. Activity-dependent regulation of gene expression in muscle and neuronal cells. *Mol Neurobiol* 1989;;3:1–53.
68. Lømo T. Formation of ectopic neuromuscular junctions: role of activity and other factors. In: Changeux JP, Konishi M, eds. *Dahlem Konferenzen, The cellular and molecular bases of learning*. New York: John Wiley and Sons, 1987;359–374.
69. Anderson MJ, Cohen MW, Zorychta E. Effects of innervation on the distribution of acetylcholine receptors on cultured cells. *J Physiol (Lond)* 1977;268:731–756.
70. Klarsfeld A, Laufer R, Fontaine B, Devillers-Thiéry A, Dubreuil C, Changeux JP. Regulation of muscle AChR α-subunit gene expression by electrical activity: involvement of protein kinase C and Ca^{2+}. *Neuron* 1989;2:1229–1236.
71. Turing AM. The chemical basis of morphogenesis. *Philos Trans Soc Lond (B)* 1952;237:37–72.
72. Usdin TB, Fischbach GD. Purification and characterization of a polypeptide from chick brain that promotes the accumulation of acetylcholine receptors in chick myotubes. *J Cell Biol* 1986; 103:493–507.
73. Knaack D, Podleski T. Ascorbic acid mediates acetylcholine receptor increase induced by brain extract on myogenic cells. *Proc Natl Acad Sci USA* 1985;82:575–579.
74. Fontaine B, Klarsfeld A, Changeux JP. Calcitonin-gene related peptide and muscle activity regulate acetylcholine receptor α-subunit mRNA levels by distinct intracellular pathways. *J Cell Biol* 1987;105:1337–1342.
75. New HV, Mudge AW. Calcitonin gene-related peptide regulates muscle acetylcholine receptor synthesis. *Nature* 1986;323:809–811.
76. Duclert A, Piette J, Changeux J. Induction of acetylcholine receptor α-subunit gene expression in chicken myotubes by electrical activity blockade requires ongoing protein synthesis. *Proc Natl Acad Sci USA* 1990;87:1391–1395.
77. Edelman G. *Topobiology*. New York: Basic Books, 1988.
78. Heidmann T, Changeux JP. Un modèle moléculaire de régulation d'efficacité d'un synapse chimique au niveau postsynaptique. *CR Acad Sci* 1982;295:665–670.
79. Changeux JP. Coexistence of neuronal messengers and molecular selection. *Prog Brain Res* 1986;68:373–403.
80. Changeux JP, Heidmann T. Allosteric receptors and molecular models of learning. In: Edelman G, Gall WE, Cowan WM, eds. *Synaptic function*. New York: John Wiley, 1987;549–601.
81. Finkel LH, Edelman GM. Regulation rules for synapses in networks. In: Edelman G, Gall WE, Cowan WM, eds. *Synaptic function*. New York: John Wiley, 1987;711–757.
82. Johnson JW, Ascher P. Glycine potentiates the NMDA response in cultured mouse brain neurons. *Nature* 1987;325:529–531.
83. Fuxe K, Agnati LF. Receptor–receptor interactions in the central nervous system. A new integrative mechanism in synapses. *Med Res Rev* 1985;5(4):441–482.
84. Mulle C, Benoit P, Pinset C, Roa M, Changeux JP. Calcitonin gene-related peptide enhances the rate of desensitization of the nicotinic acetylcholine receptor in cultured mouse muscle cells. *Proc Natl Acad Sci USA* 1988;85:5728–5732.
85. Livingstone MS, Hubel DH. Effects of sleep and arousal on the processing of visual information in the cat. *Nature* 1981;291:554–561.
86. Servan-Schreiber D, Printz H, Cohen J. A neural network model of catecholamine effects: enhancement of signal detection performance is an emergent property of changes in individual behavior (*submitted*).
87. Dehaene S, Changeux JP. A simple model of prefrontal cortex function in delayed-response tasks. *J Cogn Sci* 1989;1:244–261.

Index

Note: References to figures are in *italics*.